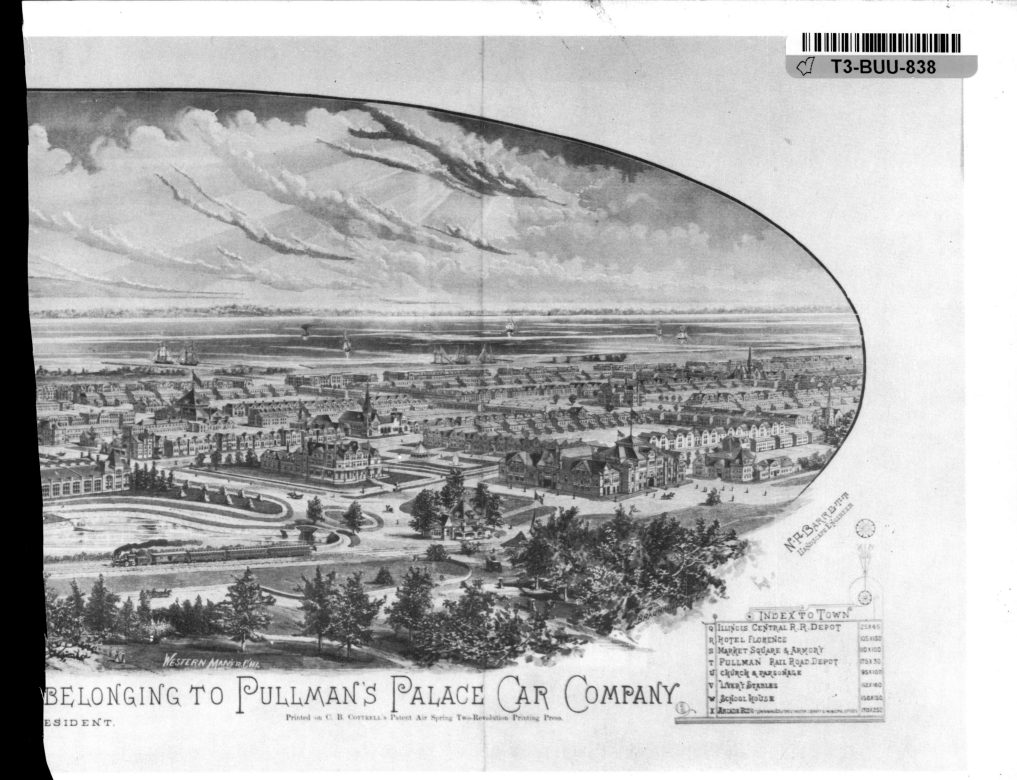

BELONGING TO PULLMAN'S PALACE CAR COMPANY.

ESIDENT.

Printed on C. B. Cottrell's Patent Air Spring Two-Revolution Printing Press.

WESTERN MANFR.CHI.

N.F.BARRETT
LANDSCAPE ENGINEER

T3-BUU-838

INDEX TO TOWN

Q	ILLINOIS CENTRAL R.R. DEPOT	25X45
R	HOTEL FLORENCE	105X150
S	MARKET SQUARE & ARMORY	110X100
T	PULLMAN RAIL ROAD DEPOT	175X150
U	CHURCH & PARSONAGE	95X100
V	LIVERY STABLES	112X160
W	SCHOOL HOUSE	150X150
X	ARCADE BLDG (COMMERCIAL, HOTEL, HEALTH, LIBRARY & MUNICIPAL OFFICES)	170X250

SCHAUMBURG TOWNSHIP DISTRICT LIBRARY

3 1257 00720 1386

Schaumburg Township District Library

32 West Library Lane

Schaumburg, Illinois   60194

ADDITIONS and CORRECTIONS
A Century Of Pullman Cars, Volume One, Alphabetical List #10-6715 (Second Printing)

PAGE	CAR NAME	CORRECTION
23	ABSECON	Add new line CAR NAME ABSECON, CAR TYPE 24ChBDr, CAR BUILDER Pullman, DATE BUILT 6/6/05, PLAN 1988A, and LOT 3170.
23	ADAIR (PLAN 17)	Delete entire line.
23	ADELINE #128	Delete PLAN number.
24	ALBANIA	Change CAR TYPE to OS.
25	ALLEGHENY E&A-24	Delete PRR from CAR BUILDER and Alt2 from LOT.
25	ALLENDE	Change LOT C4 to C3.
25	ALMA #77	Add DATE BUILT 7/79 and PLAN 755?
26	ALSTON (PLAN 1581F)	Delete entire line.
28	ANTHONY (PLAN 14B)	Change ST to N.
28	ANTHONY (PLAN 19)	Delete entire line.
28	ARAMINTA	Change DATE BUILT to 8/31/84 and add LOT 199.
29	ARKANSAS #108	Change PLAN number to 698.
29	ARVERNE #118	Delete PLAN number.
30	ATHENS (PLAN 155)	Change CAR TYPE to 8SecSrB (NG).
30	ATLANTA (PLAN 2061)	Change LOT to 3218.
30	ATLANTIC #1	Change PLAN number to 7.
30	ATLANTIC #41	Add PLAN number 197 and LOT Day.
30	ATLANTIC #106	Change CAR BUILDER to B&O RR (WS&PCC).
30	ATLANTIC AVE. #128	Delete PLAN number.
30	AURELIA #77	Change ST to RN and PLAN to 755.
33	BENGAL (H&H)	Add PLAN number 114.
34	BEVETON #142	Add new line CAR NAME BEVETON #142, CAR TYPE DDR, CAR BUILDER PCW Det, DATE BUILT 1/20/70, PLAN 17, LOT D4.
36	BRAZIL #32	Change ST to RN and add LOT Han1.
36	BROCKVILLE (Grand Trunk)	Change PLAN number to 18.
36	BROOKFIELD	Change ST to RN.
36	BROOKHAVEN #108	Change PLAN number to 698.
36	BROOKLYN #156	Delete PLAN number.
38	CAMDEN #105	Change CAR BUILDER to B&O RR (WS&PCC).
40	CATALINA (Plan 149)	Change ST to RN and PLAN to 149R.
40	CATALINA (PLAN 149R)	Delete entire line.
40	CATARACT E&A-32	Change CAR TYPE to 18Ch7StsB, CAR BUILDER to (E&ASCCo) and PLAN to 147.
40	CEDAR BROOK #77	Change CAR TYPE to 28 Ch and PLAN number to 755.
41	CHABALLE	Delete entire line.
41	CHARITON (PLAN 17)	Change ST to RN, CAR TYPE to 14SecO, CAR BUILDER to CB&Q, PLAN to 17E, LOT to Aur1, and add DATE BUILT 1/71.
41	CHELSEA #170	Change ST to RN.
41	CHEMUNG E&A-13	Change CAR TYPE to Slpr.
42	CHICKASAW (PLAN 6)	Change PLAN number 6 to 26.
43	CITY of NEW YORK #5	Add CAR TYPE 10SecB and PLAN 759.
43	CITY of ST. LOUIS #4	Add CAR TYPE 10SecB and PLAN 759.
45	COLORADO #153	Change CAR TYPE to 10SecDrB and PLAN number to 696, add DATE BUILT 7/86.
45	CONEMAUGH #169	Change PLAN number to 17.
45	CONESTOGA #182	Change PLAN number to 21.
46	CONVOY	Change PLAN 99A to 90?
47	CRESTLINE #129	Change PLAN number to 14.
50	DIADEM (PLAN 4C)	Change DATE BUILT to 1/71 and PLAN to 27.
50	DIONE #157	Delete PLAN number.
50	DOLPHIN (PLAN 14C)	Change ST from N to RN.
50	DOPHIN	Delete entire line.
51	DUBUQUE #80	Change CAR BUILDER to H&H (CM&StP) and add PLAN 449?
51	DUBUQUE #104	Add PLAN number 231.
51	DULUTH (14SecO)	Change BUILT DATE to 7/71 and add PLAN 20.
52	DUNROBIN (PLAN 39)	Change PLAN number to 30 and LOT to D158.
52	DUNNVILLE #178	Add ST RN.
52	E. KIRBY SMITH	Change LOT to 3218.
52	EDELSTEIN	Change ST N to RN, CAR TYPE from Slpr to 10SecDr and add PLAN 142A.
53	ELDEN	Delete entire line.
53	ELDON	Change CAR NAME to ELDEN.
54	ELINOR	Change PLAN number to 772D.
54	ELIZABETH #39	Change PLAN number to 753.
54	ELLA San#2	Add BUILT DATE 7/71 and PLAN 20.

PAGE	CAR NAME	CORRECTION
55	ELVIRA #127	Delete PLAN number.
55	ERLON #76	Change ST to P, add DATE BUILT 7/79 and PLAN 755.
56	ETELKA GERSTER #110	Delete PLAN number.
57	F Sanderson	Change CAR NAME to F SANDERSON.
59	FLATBUSH #129	Delete PLAN number.
60	GENEVA (PLAN 14)	Change PLAN 14 to 19.
64	HALSEY (PLAN 2028R)	Delete entire line.
64	HAMILTON (PLAN 27)	Change ST from P to RN and CAR TYPE from OS to 12SecO.
64	HAMPSHIRE #85	Add PLAN number 197.
64	HARBELL (#181)	Change ST to RN and CAR TYPE to 12Ch36Sts SUV.
64	HASELMERE	Change CAR NAME to HASLEMERE.
65	HAZELWOOD #540	Change CAR TYPE to 10SecO, CAR BUILDER to B&S(B&O RR), and PLAN number to 656.
66	HIGHLAND #118	Delete PLAN number.
66	HOLLIS (PLAN 729A)	Delete entire line.
66	HOLLIS #101	Change CAR TYPE to 12SecDr and PLAN number to 729.
67	HURON #5	Add CAR TYPE 10SecB and PLAN 759.
67	IDA #26	Change PLAN number to 833.
67	IDLEWILD (LOT D7)	Change PLAN number 22? to 29P.
68	ILLINOIS #84	Add PLAN number 197.
68	ILLINOIS #106	Add PLAN number 231.
69	IONE #46	Delete PLAN number.
69	ITASKA #164	Change PLAN number to 663.
70	ITURBIDE	Change ST to RN.
70	JACKSON (PLAN 90?)	Change CAR TYPE to 10SecO.
70	JAMESTOWN	Add new line CAR NAME JAMESTOWN (Steel), CAR TYPE 12SecDr, CAR BUILDER Pullman, DATE BUILT 3/07 PLAN 1963D, LOT 3402.
70	JANISCH	Delete PLAN number.
70	JESSICA #130	Delete PLAN number.
70	JOCKEY CLUB #130	Delete PLAN number.
70	JOHN A. LOGAN	Change LOT to 3218.
72	KEYSVILLE #171	Change ST to RN.
72	KILBOURNE #73	Add PLAN number 197.
73	LA PALOMA (PLAN 1001)	Change CAR TYPE to 8SecSrB (NG).
74	LaRABIDA	Change CAR NAME to LA RABIDA.
76	LILLIE San#3	Add BUILT DATE 7/71 and PLAN 20.
76	LINDELL (PFtW&C)	Change PLAN number 17 to 14.
77	LONG ISLAND #155	Delete PLAN number.
78	MADISON #76	Add PLAN number 197.
80	MANITOBA #62	Add PLAN number 197.
80	MANSFIELD #184	Change PLAN number to 26.
82	MAYFLOWER (OS)	Add PLAN number 27 and change LOT to MC.
83	MEXICANO	Change CAR TYPE to 10SecO(NG).
83	MIDVILLE #176	Change ST to RN.
84	MINNEAPOLIS #61	Add PLAN number 197.
84	MINNESOTA #28	Add CAR TYPE 12SecDr and PLAN number 197.
85	MISSISSIPPI #44	Add PLAN number 197.
85	MISSISSIPPI #107	Add PLAN number 231.
87	myles STANDISH	Change CAR NAME to MYLES STANDISH.
87	NANKING (PLAN 37F)	Delete entire line.
87	NASSAU #157	Delete PLAN number.
87	NATHALENA #38	Add CAR TYPE Sleeper? and change PLAN number to 753.
88	NEW YORK #83	Add PLAN number 197.
88	NEWBURGH #90	Add PLAN number 197.
89	NIAGARA #150	Change ST to P and PLAN number to 720.
89	3008B	Change line to read CAR NAME NYLES, ST P, CAR TYPE 12SecDr, CAR BUILDER (Wagner), DATE BUILT /75, PLAN 3008B.
89	NINA #45	Delete PLAN number.
89	NINITA (Slpr)	Change ST to RN, CAR TYPE to 10SecDrB, CAR BUILDER to (E&ASCCo) and add DATE BUILT /72 and PLAN 147B?
89	NIOBRARA (H&H)	Add PLAN number 197.
89	NORMAN (PLAN 48)	Change PLAN number to 46.
90	NYDIA #45	Delete PLAN number.
90	OCONOMOWOC #69	Add PLAN number 197.
92	OPAL (12SecDr)	Change PLAN number 1581A to 3581E.
92	ORIENTAL #154	Delete PLAN number.
93	OSCOLA	Change CAR NAME to OSCODA.
93	PACIFIC #43	Add PLAN number 197.
95	PAVILION	Change CAR BUILDER to PCW Det, add PLAN number 45K and LOT D508.
95	PEGASUS #165	Change PLAN number to 663.
95	PEMBINA #66	Add PLAN number 197.
96	PENOBSCOT #165	Change PLAN number to 663.
98	PLANET (G&B/GMP)	Change CAR TYPE to 10SecO and add PLAN 47.
98	PLANET #256	Change CAR TYPE to 12SecO.
98	POCAHONTAS (OP Lines)	Add PLAN 47?
98	PONTIAC (PLAN 2061)	Change LOT to 3218.
99	PRESIDENT (PLAN 13)	Change PLAN number to 7.
99	PRINCESS (PLAN 48)	Change PLAN number to 46.

PAGE	CAR NAME	CORRECTION
99	PROSPECT PARK #127	Delete PLAN number.
100	QUANTZINTECOMATZIN (LN Diner)	Change CAR BUILDER to PCW (NP RR), add DATE BUILT 4/83, PLAN 165, and LOT 54.
100	QUOGUE #102	Add CAR TYPE 12SecDr and change PLAN to 729.
100	RACINE (G&B/GMP)	Change CAR TYPE to 10SecO and add PLAN 47.
101	RAPIDAN #152	Change PLAN number to 696.
101	RAPPAHNNACK #151	Change PLAN number to 696.
102	RED WING #68	Add PLAN number 197.
102	REVERE (CB&Q RR)	Change PLAN number 17 to 14.
103	ROCKAWAY #101	Change CAR TYPE to 12SecDr and PLAN number to 729.
103	ROCKLAND (PCW Det)	Change PLAN number 19A to 26.
105	SABINA #219	Change CAR NAME to SABRINA #219.
105	SABULA #78	Change CAR BUILDER to H&H (CM&StP) and add PLAN 499.
106	SAN JACINTO	Change PLAN number to 27 and add LOT Chi2.
106	SAN JUAN SP-51	Add new line CAR NAME SAN JUAN SP-51, CAR TYPE 10SecDrB, CAR BUILDER PCW, DATE BUILT 4/6/88, PLAN 445F, and LOT 1377.
106	SANGAMON	Add new line CAR NAME SANGAMON, ST P, CAR TYPE 12SecDr, CAR BUILDER Wagner, DATE BUILT 4/88, PLAN 3061B, and LOT Wg10.
107	SARAH BERNHARDT	Add new line CAR NAME SARAH BERNHARDT, ST N, CAR TYPE Private, CAR BUILDER PRR, DATE BUILT 8/72, PLAN 29, and LOT AltP1.
107	SAVANNAH #79	Change CAR BUILDER to H&H (CM&StP).
107	SAXON (PLAN 48)	Change PLAN number to 46.
108	SEMINOLE (Pull/Sou)	Add PLAN 47?
109	SHAMROCK #5	Add CAR TYPE 10SecB and PLAN 759.
111	ST. GEORGE (PLAN 4C)	Change DATE BUILT to 1/71 and PLAN to 27.
112	ST. PAUL (I&StLSCCo)	Change DATE BUILT to 7/71 and add PLAN 20.
112	ST. PAUL #55	Add PLAN number 197.
113	SUMATRA (PLAN 993A)	Change ST to RN, CAR TYPE to 16Sec and PLAN number to 1877.
113	SWANNANOA #336	Add CAR TYPE 10SecDrB and PLAN 565B.
114	SYBIL #46	Delete PLAN number.
114	T W PIERCE GH&SA	Change CAR NAME to T. W. PIERCE GH&SA.
114	Figure I-38	Change Smithsonian Photograph number to P638.
116	TONGA	Change CAR BUILDER to PCW BUF and add DATE BUILT 3/30/03, PLAN 1581C, and LOTB-15
117	TRANSIT (PLAN 48)	Change PLAN number to 46.

PAGE	CAR NAME	CORRECTION
118	TROSACHS	Change CAR TYPE to 10SecDrB, CAR BUILDER to (E&ASCCo) and add DATE BUILT /72.
120	VALERIA #40	Add CAR TYPE Sleeper and change PLAN number to 753.
120	VELMAR	Add PLAN number 147B.
122	VICEROY (PLAN 13)	Change PLAN number to 7.
122	VICTORIA (Diner blt/76)	Change CAR BUILDER from PCW to PCW Det, add PLAN 45 and LOT D31.
122	VICTORIA (Hotel)	Add PLAN number 7?
122	VICTORIA #76	Change ST to RN, CAR TYPE to 28Ch and PLAN number to 755.
122	VOLUSIA (PLAN 28A)	Add ST RN and change CAR TYPE to 10SecDrB, PLAN to 149S and LOT to Ren2.
123	WALTHAM #168	Change ST to RN.
123	WANATAH (PLAN 2089A)	Change LOT to 3549.
123	WAPITIA #323	Change CAR NAME to WAPITI, ST to N and add DATE BUILT /89.
124	WESTFIELD #141	Change PLAN number to 17.
125	WILTON	Change ST to RN and PLAN to 114.
126	WISCONSIN #27	Add CAR TYPE 12SecDr? and PLAN 197?
178	K-100	Add ST RN.
132	ARISTOTLE	Change CAR TYPE to 12SecDrCpt.
185	LAS VEGAS	Remove stray "T" between CAR NAME and CAR TYPE.
253	BAD AXEL RIVER GN-1264	Change CAR NAME both lines to BAD AXE RIVER.
256	BRIGHAM YOUNG D&RG-1272	Change CAR NAME to BRIGHAM YOUNG D&RGW-1272.
266	DAVID MOFFAT D&RG-1271	Change CAR NAME to DAVID MOFFAT D&RGW-1271.
268	Figure I-102	Change PLAN 4253A to 4153A.
270	GEORGE M. PULLMAN	Add new line CAR NAME GEORGE M. PULLMAN, CAR TYPE 3DbrCptDrBLngObsRE, DATE BUILT /33, CAR BLDR PS, PLAN 4028 and LOT 6411.
273	HEBER C. KIMBALL D&RG-1273	Change CAR NAME to HEBER C. KIMBALL D&RGW-1273.
278	JOHN EVANS D&RG-1270	Change CAR NAME to JOHN EVANS D&GRW -1270.
312	1865	Change second line to read: into service. Eight-wheel trucks . . .
315	LN	Add SYMBOL LN for Leased Name.

PAGE	CAR NAME	CORRECTION
315	Tourist	Change ABBREVIATION to TC for Tourist Car.
315	CTR	Add ABBREVIATION CTR for Center.
315	CP	Change ABBREVIATION to CPR for Canadian Pacific Railroad.
316	H&StJ RR	Change spelling of Hanibal to Hannibal.
316	I&StLSCCo	Change NAME to Indianapolis & St. Louis Sleeping <u>Coach</u> Company.
317	SouTransCo	Change ABBREVIATION to SouTrans.
317	WS&PCC	Change NAME to Woodruff Sleeping & <u>Parlor</u> Coach Company.
318	COUNTESS	Change PLAN number to 1031B.
318	EAST NEWARK	Add "S" for Steel after name.
318	Medical Dept:	Add "W" PLAN 3061B.

Listed below are NYNH&H Parlor cars that were leased back to Pullman and had not been included in Volume One. They have been catagorized according to car type. These are all Lightweight Cars (Pages 244-311).

26 Chair 14 Seat Lounge, built 10/48 by PS, PLAN 4502, LOT W6788.

NEW BEDFORD NH-403 (2/50 from BLACKSTONE RIVER).
NEW BRITIAN NH-400 (2/50 from CONNECTICUT RIVER).
NEW HAVEN NH-404 (4/50 from HUDSON RIVER).
NEW MILFORD NH-402 (4/50 from HOUSATONIC RIVER).
NEW ROCELLE NH-401 (4/50 from THAMES RIVER).

36 Chair Parlors, built 7-8/48 by PS, PLAN 4500, LOT W6788

*ATTLEBORO NH-312
BRIDGEPORT NH-301
BROCKTON NH-303
CRANSTON NH-306
*DANBURY NH-317
*GROTON NH-318
*GUILFORD NH-314
HARTFORD NH-304
*HOLYOKE NH-319
*LYME NH-324
MANSFIELD NH-303
*MERIDEN NH-320
*MILFORD NH-321

*MYSTIC NH-316
*NEWPORT NH-322
PITTSFIELD NH-300
*PAWTUCKET NH-313
PROVIDENCE NH-311
ROWAYTON NH-305 (6/49 renamed WOONSOCKET NH-305)
STAMFORD NH-308
STONINGTON NH-309
*TAUNTON NH-323
TORRINGTON NH-310
*WESTERLY NH-315
WORCESTER NH-307

Baggage 28 Seat Lounge, 2 Roomette, 3 Drawing Room, Buffet built 10/48 by PS PLAN 4503B, LOT W6788

BERKSHIRE COUNTY NH-208
BRISTOL COUNTY NH-202
DUKES COUNTY NH-203
ESSEX COUNTY NH-204
KINGS COUNTY NH-209
MIDDLESEX COUNTY NH-200
NEW LONDON COUNTY NH-207
PLYMOUTH COUNTY NH-205
PUTNAM COUNTY NH-206
SUFFOLK COUNTY NH-201

Baggage, 14 Chair Parlor, 11 Seat Lounge, Drawing Room, 2 Roomette, Buffet Built 10/48 by PS PLAN 4501, LOT W6788

BRONX COUNTY NH-219
FAIRFIELD COUNTY NH-214
HAMPSHIRE COUNTY NH-218
KENT COUNTY NH-210
LITCHFIELD COUNTY NH-216
QUEENS COUNTY NH-215
TOLLAND COUNTY NH-213
ULSTER COUNTY NH-212
WASHINGTON COUNTY NH-211
WINDHAM COUNTY NH-217

* Built as 52 Chair Parlor PLAN 4500, rebuilt to 36 Chair during 1949.

A Century of
PULLMAN CARS

Volume One
Alphabetical List

Ralph L. Barger

Cover Photographs Top: PARISIAN, Pullman Photograph 1777. Courtesy of Robert Wayner. Middle: HENRY W. GRADY, Pullman Photograph 33900. Courtesy of Robert Wayner. Bottom: NATIONAL EMBASSY, Railroad Avenue Enterprises Photograph PN-2834.

End Papers: (Inside front and rear cover.) Courtesy of the Chicago Historical Society, negative number ICHi-01918.

Greenberg Publishing Company, Inc.
Sykesville, Maryland

SCHAUMBURG TOWNSHIP DISTRICT LIBRARY
32 WEST LIBRARY LANE
SCHAUMBURG, ILLINOIS 60194

625.23
BAR

3 1257 00720 1386

Copyright © 1988

By Ralph L. Barger
Columbia, MD 21045

Greenberg Publishing Company, Inc.
7566 Main Street
Sykesville, MD 21784
(301) 795-7447

First Edition   *Second Printing*

Manufactured in the United States of America

All rights reserved. No part of this book may be reproduced in any form or by any means including electronic, photocopying, or recording or by any information storage system without written permission of the author, except in the case of brief quotations embodied in critical articles and reviews. Published by the Greenberg Publishing Company, Inc., which offers the world's largest selection of Lionel, American Flyer, LGB, Ives, and other toy train publications as well as a selection of books on model and prototype railroading, dollhouse miniatures, toys, and other collectibles. For a copy of our current catalogue, please send a large self-addressed stamped envelope to Greenberg Publishing Company, Inc. at the address above.

ISBN 0-89778-061-2

Library Of Congress Cataloging-in-Publication Data

Barger, Ralph L., 1929-
    A century of Pullman cars.

    Bibliography: p.
    Contents: v. 1. Alphabetical list.
    1. Railroads--Pullman cars. I. Title.
TF459.B37 1987   625.2'3   87-23611
ISBN 0-89778-061-2

DEDICATION

To my late mother, Ruble E. Barger, for her
limitless patience with my enthusiasm for trains
and her encouragement of my endeavors.

ACKNOWLEDGMENTS

I have spent twenty-five years researching and gathering the material that will be presented in this publication. Hopefully, a great deal more material will be assembled before the sixth and last volume goes to press, but the fact that this immense amount of information is ready to present to you is due largely to the support and assistance of a number of knowledgeable and interested parties. It gives me great pleasure to record my appreciation to these people who have been so generous in giving aid from their own writings, collections, time and memories.

Robert Wayner has contributed a vast amount of information, collected over a number of years and lovingly organized, so that when our two collections came together the story of the Pullman wooden era was ready to tell. He has been for me a "source supreme" and I am most grateful for his interest, energy and contributions. His time, given to editing, has been invaluable.

My sincere gratitude must be extended to **Arthur D. Dubin** for his forethought in preserving records and information which made possible the solution of questions that might otherwise have gone unanswered. His generosity in sharing photographs and other materials from his collection places all of us in his debt.

The leading historian of railroading in this country is **John H. White, Jr.**, Senior Historian for the Division of Transportation at The Smithsonian Institution. Jack's encouragement has been a major inspiration to me, and his vast knowledge of source material has been invaluable. His book, **The American Railroad Passenger Car** was the first major scholarly work on the subject, and its critical and popular success helped convince me that the field was ready for a comprehensive catalog of Pullman cars. I am also very grateful for the assistance of Smithsonian staff members **John Stein** and **Susan Tolbert**. **Robert Harding**, at the Smithsonian Archives, has been overly helpful and I extend to him and his staff my thanks for their time and unlimited patience.

I have drawn on the records from many organizations, archives, and libraries, and I gratefully acknowledge them along with their staff members who were ever present to assist with questions and requests. They are: **J.S. Aubrey**, the Newberry Library, Chicago, Illinois; **Marian Smith**, the B&O Railroad Museum, Baltimore, Maryland; **Stephen E. Drew**, the California State Railroad Museum Library, Sacramento, California; **Linda Ziemer**, the Chicago Historical Society; the Delaware State Archives, Dover, Delaware; the Enoch Pratt Library, Baltimore, Maryland; **Mary Lou Neighbour**, Hagley Museum & Library, Wilmington, Delaware; the Howard County Library, Columbia, Maryland; Interstate Commerce Commission Library, and the National Archives, Washington, D.C.

Those in organizations and societies who have lent a great deal of assistance are: **John Berry**, N.P. Railroad Historical Society; **Leroy Hutchinson**, B&M Railroad Historical Society; **Randolph Kean**, C&O Historical Society; **Norman C. Keyes, Jr.**, Great Northern Railway Historical Society; **Alvin Lawrence**, New Haven Railroad & Technical Society; **Gary Schlerf** and **Walter C. Figiel** of The B&O Railroad Historical Society, and **Fred A. Stindt** of the Railway & Locomotive Historical Society, Inc.

Photographs, essential additions to this publication, came from many willing sources: special thanks to **Wilber C. Whittaker**, who has photographed railroad equipment for over 40 years and has made his extensive collection available to all of us; **Howard Ameling**; **Arthur D. Dubin**; **Bob Pennisi**; **Robert Wayner**; The B&O Railroad Museum and The B&O Railroad Historical Society; and The Smithsonian collection.

The collaboration with new and old friends who share the same labor of love is one of the most enjoyable aspects of a project of this magnitude. To those faithful to the cause, please take some credit for the success of this venture. My tribute is paid to: **Jim Bennett**, a long-time correspondent who has exchanged mountains of reference material with me over a number of years; **David Briggs**, on whom I depend for the identification of the cars that are now privately owned, on exhibit in museums, serving as restaurants, or rusting in the weeds throughout America; **George Cockle**, a superb photo journalist, Union Pacific enthusiast and fellow 1951 graduate of the U.S. Army Ground School; **Thomas H. Donnelly**, a long-time Pullman enthusiast and my "Chicago Connection;" **Major General (USA Ret) Hugh F. Foster, Jr.**, my authority on U.S. Government Medical Department railway equipment; **John P. Hankey**, my "Redactor Extraordinaire;" **Bruce Heiner**, who has found and brought to my attention vital railroad documentation; **William F. Howes**, one of the last directors of The Pullman Company who provided information on the company's final days; **Robert A. LeMassena**, an indispensable fact finder; **Robert MacDougall**, with whom I have shared an occasional search for railroad cars that turned out to be more fun than stars, spangles and sawdust; **Patrick O. McLaughlin**, an insatiable pursuer of private car history and the leading authority on the subject; **William Mitchell**, my photographer pal who maintains I'm no fun any more since I began work full time on the publication of this material; **Jackson Thode**, a devotee of three-rail railroading; my fellow members of the B&O Railroad Historical Society whose dedication, persistence and fellowship have meant a great deal to me as we unearthed the history of B&O rolling stock, and finally, my oldest friend and most faithful fellow enthusiast, **J.P. Barger**.

During my endeavors over the last twenty-odd years, many people have contributed to my files. It is possible that some have been omitted from these acknowledgments. If so, please be assured it was unintentional. To those who may find their name missing — thank you.

Last, my heartfelt thanks to the members of my family, for without their assistance and support the publication of this work would be years away: my wife, **Lois**, who has written, edited, researched, and spent months at the computer; my daughter, **Keven Sargent**, who was instrumental in introducing the AT&T 6300 XT computer system to my study; my son, **Ralph III**, who assisted me in programming and setting up the files; and my daughter, **Carol Amante**, who is responsible for the title lettering and is presently engaged in designing the jacket covers for future volumes.

At **Greenberg Publishing Company, Inc.** the project was managed by **Cindy Lee Floyd**. She organized the work flow for the readers, editors, typesetting and art department. **Maureen Crum**, senior staff artist, designed the book with the assistance of **Angel Dunchack** and **Betty Meade**, the staff artist. The book's design involved the organization of text, placement of photographs and drawings. **Marsha Davis** attended to the necessary editorial corrections. **Donna Price** has read the text and made comments regarding consistency and clarity.

Ralph L. Barger
Columbia, Maryland
January, 1988

TABLE OF CONTENTS

FOREWORD

It is difficult to believe that The Pullman Company as an operating entity has been gone for nearly twenty years. With each passing year its memory grows more dim because fewer people are of an age to remember sleeping and parlor car service as it was when deluxe traveling hotels were operated under the watchful management of The Pullman Company. Today younger people, and surely generations to come, cannot possibly imagine the amenities once available on first class trains that included barber shops, valets and maids. Luxury, style and service were not empty words. They were performed consistently on a daily basis. Travelers today must go to Western Europe, Japan, Australia or South Africa to experience what once was so common in the United States and Canada.

And so it is the duty of present day historians to record the story of Pullman. It is a large and fascinating story for it encompasses management, labor, finance, technology, promotion, civil rights and public relations. The business history of the firm could alone fill a fat volume. The biography of George Mortimer Pullman awaits an author. The operations and technology of The Pullman Company and portions of its rolling stock have been outlined in several books already published. The author of this volume has catalogued a century of Pullman cars from the original records.

In this series, of which this is Volume One, Ralph L. Barger has taken on the formidable task of listing and illustrating the Pullman fleet in considerable detail. This is not a job for the timid, casual or indifferent. Patience, energy and dedication are required, for every experienced historian understands that research involves considerable time, leg work and eyestrain. It is not an armchair occupation. Ten years ago the prospect for one man to accomplish the assembly of an all-time Pullman roster would have seemed impossible or at least unlikely. However, the fortuitous recovery of the Pullman car record books since the firm's dissolution offered much needed documentation on the wooden cars. And yet, for all the names and data available the history of some cars remains unknown.

Undoubtedly you will ask if the information presented in this volume is absolutely correct. Please understand that some errors may be found in such a large corpus of data and omissions are inevitable principally because the available documents are not complete. Yet the record presented here is surely more extensive and accurate than anything available until the present time. Serious students of railroad cars and car building — and there are legions — will appreciate Ralph Barger's efforts to record what was once termed "The world's greatest hotel."

Those readers with additional documentation are urged to contact the author so that those additions and/or corrections can be made in the descriptive volumes and subsequent editions. A history of this kind is based on a collection of records that must be consolidated and compiled before they are lost forever.

Arthur D. Dubin   John H. White

INTRODUCTION

During the 1950s, my lifelong fascination with railroading coalesced around the great standard fleet of The Pullman Company. As with any serious study, it took years simply to survey the available sources and begin a systematic collection of photos, diagrams, and documentation. In those years, the serious study of railroad history was also in its infancy, and I was informed that early Pullman records no longer existed. Good source material of any kind was difficult to obtain.

The cessation of Pullman Company operations in 1968 arbitrarily defined the era I was interested in. What began with my first Car Builder's Cyclopedia in 1936 had by 1961 been systematized into a catalogue of United States passenger cars filed on 3" x 5" notecards. Through the collection of diagrams, a network of associates and first hand observation and research, that catalogue expanded to over 40,000 cards and countless three ring binders, all carefully organized and indexed. Yet there were many gaps, and in the 1960s various works on passenger equipment found a ready audience among railfans, hobbyists, scholars, car owners, and researchers such as myself.

The invaluable works by Arthur Dubin and the anecdotal writings and illustrations of Lucius Beebe brought passenger train history to life and hinted at the availability of source material. The 1968 compilations of the so-called heavyweight Pullman fleet by Robert Wayner and William Kratville quickly became classics, along with consist books, cyclopedia, railroad passenger diagram reprints, and pictorial railroad histories. As companies ceased operating or disposed of records, more and more data found its way into the collections of private individuals, libraries and museums.

While my first real inspiration dates from Arthur Dubin's 1964 **Some Classic Trains**, it was the publication in 1978 of John H. White's work, **The American Railroad Passenger Car** that convinced me to assemble a comprehensive Pullman catalogue. Excellent source books on passenger equipment were beginning to appear, but no comprehensive survey, much less a complete illustrated list, had been attempted.

Beginning in the 1970s, the availability of original Pullman Company car records at The Newberry Library provided primary documentation for much of the fleet. The accession of the early Pullman Cost of Car Records at The Smithsonian greatly furthered the work, as did the discovery and dispersal of the last Pullman Company records in 1980. With these primary sources, and the unqualified support of the major scholars and collectors in the field, I felt confident in announcing the project and beginning the list in earnest.

The great number of recent publications on standard and lightweight cars obscure the fact that little has been brought together into a "union list," and still less published on Pullman Palace Cars. No less than sixty percent of the 1920 fleet were wooden cars; even as the last two standard sleeping cars were being built in 1931 (Lot 6396), eighteen percent of the active cars were still of wood origin. As the fleet itself evolved chronologically in several major stages of car body construction, yet as a continuum, the catalogue will likewise be arranged.

This first of six proposed volumes constitutes the Alphabetical List of the over 21,600 cars in the fleet, divided into Wood, Standard and Lightweight sections. It provides basic information on every named car known to have operated in Pullman service and will serve as a directory to later volumes. The Alphabetical List is both the most general and comprehensive book of the set, and it is the foundation for expanded treatment, corrections and additional research to appear in subsequent volumes. The data was derived from The Pullman Palace Car Company's handwritten records of the production, modification, purchase and disposal of all Company-owned cars from 1870 to December, 1899. From 1900 to 1969, similar Pullman Company logs, lists, rosters, Additions & Betterments, construction records, descriptive lists and memoranda form the basis of The Alphabetical List, supplemented by original railroad documentation for cross-checking and illustration. Additional secondary sources provided further illustration, verification and context, but only original records from Pullman have been used for primary data. The illustrations chosen for Volume I are but a

fraction of those available. They were selected to be representative of major car types, innovations, practices and modifications. As specific types are treated in further detail in later volumes, photos and drawings will be used extensively to amplify lists and text.

The tentative organization for future volumes is as follows:

Volumes Two and Three are planned at this time to contain the detailed rosters and illustrations for the wooden cars that were built and named by Pullman. They will begin with the first 37 cars which were the property of George Pullman and follow the Pullman production of Palace Cars (including those built in Detroit and those early cars purchased from other car builders that were designed by Pullman) through 1910, along with the history of each car until it was retired. These volumes will give the assignments when new, their former names and dispositions, and provide as many photos and/or illustrations as possible. Because extensive photo coverage for this period is a major challenge, your help is solicited to review the many collections of photographs throughout the country for possible inclusion in this series. It is impossible for me to have direct access to and identify Pullman-related photographs made by the railroads and those in private collections. If in fact a major portion of the available photographs and/or illustrations are to be included in these books, your assistance is absolutely vital. Credit will be given to the photographer, if known, the finder, and the collection from which it comes.

Volume Four should cover the "Standard" Named Cars (so-called Heavyweights.) It is my objective to present extensive photo coverage and/or floor plans for these cars along with their initial assignments and many details, including Government Storage and extended dispositions.

Volume Five will list all cars purchased by Pullman from sleeping car companies, the railroads and the Pullman/Railroad Associations. This volume will list all Pullman Private cars, Wood and Steel numbered cars including Tourist, Pullman Supply and Porter (S&P), and Hospital cars with representative consists from the era of "Black Jack" Pershing's Mexican Border train to World War II. Also, there will be a roster of the railroad-owned baggage cars converted to World War II troop diners. Detailed lists of the cars sold to circus-related companies will be included, along with those of the Western Union Telegraph Company, Interstate Commerce Commission, Bureau of Mines, State Boards of Health, American Red Cross, U.S. Army and other Departments of the U.S. Government. The major buyers of used Pullman cars (i.e., F.M. Hicks & Company, John Ringling, and Hotchkiss Blue & Company), Pullman-owned coaches, other non-sleeping cars and more will be treated.

Volume Six will contain the all-time Pullman Lot List. Though the Lots retained for Pullman ownership and service will be listed, emphasis will be placed on the cars sold to the U.S. Railroads and Transit Companies including delivery dates, car numbers when known and many odd or unusual lots. By the time this data is ready for publication, more than one volume may be required to adequately accommodate this vast amount of information and extensive illustration.

As with any technological history stretching back over a century, we owe gratitude to those who preserved records and placed them in suitable research facilities. Also, we owe thanks to hundreds of amateur historians, clerks and photographers, many of whom never realized they were contributing to a historical work. The volume of data involved in this project is immense, and its compilation was possible only with the use of an AT&T 6300 XT computer system and several years of diligent data entry and processing.

Though this is truly a collective undertaking, I assume responsibility for the errors, omissions and gremlins that must appear in work wrought by humans. Because research is never finished and new sources continually turn up, I solicit any additions, corrections and suggestions the reader may wish to contribute. This work can never be one hundred percent complete, but it can stand as a tribute to the fleet and to the men and women who, together, constituted the world's grandest hotel — to one hundred years of Pullman cars and service.

EDITOR'S NOTE

There is need for explanation of the format used in this book. Taking into consideration the mass of information assembled for this and later volumes, you will understand that a computer data base was required to meet the demands of accuracy, legibility, space and ease of reading. You might say it also made the project possible before another century passed.

HEADINGS

The following are the headings used in this book:

CAR NAME   ST   CAR TYPE   CAR BUILDER   DATE   PLAN   LOT
                           (OWNER)   BUILT

CAR NAME: The car names in this reference appear as they did on the cars themselves and are printed in UPPER CASE.

ST: represents the status of the car when new or at the last change. Following are the designations for status:

"**L**" indicates that Pullman leased the car from or to a railroad or leased it to a private party.

"**N**" indicates that the car had been renamed.

"**P**" indicates the car was purchased despite the fact that it may have been built by Pullman or another builder.

"**R**" indicates that the car had been rebuilt and is usually accompanied with an "N" designating it as rebuilt and renamed.

"**S**" indicates that Pullman sold the car, normally when new, as was the procedure for most of the Lightweight cars. Sometimes you will find the "S" accompanied by an "N" which normally means a used car was sold and the name was changed. Most of the cars that were sold to the Mexican Railroads had their names changed and many were leased back to Pullman for operation.

CAR TYPE: The abbreviations for all car types are listed in the GLOSSARY.

CAR BUILDER (OWNER): The car builder is always given when known; however, the OWNER (railroad or sleeping car company) is essential for reference to further details on the car in later volumes. Owner abbreviations may be found in the GLOSSARY.

DATE BUILT: The date of construction was derived from the "Record of Lots" books which list the date the cars were ready for shipment from the Pullman Car Works. In the case of other manufacturers, it is the date Pullman received the car. "Dates of construction" were also given in other references which vary considerably

with the above data; I have chosen to use the "Record of Lots" as the standard reference. The construction dates of cars that were built by non-Pullman facilities are more difficult to ascertain. These dates were arrived at through other Pullman references but they vary considerably and where some doubt exists, I have included only the year or left them blank. These Pullman records do not differentiate between the date cars were purchased, rebuilt, placed in service, or originally constructed.

PLAN: As far as can be determined, Pullman Plan numbers were assigned to all cars in Pullman service. Despite this, plan numbers for several hundred cars have not been found. Most of these cars were purchased from other sleeping car companies and were scrapped or rebuilt with no note of their plan number. Some names simply appeared in the records and disappeared without leaving a trace of their origin, service, or disposition.

LOT: All Pullman-built cars were assigned a Lot Number. This number was, in effect, the order or job number. I have carried the original lot number with each car throughout its life. Lot numbers for the other manufacturers vary but are listed when known. Many of these lots were simply the location of the facility and a group number.

ABBREVIATIONS

A list of abbreviations may be found in GLOSSARY of this volume. This list is extensive as a result of the nature and volume of information compiled. The space required by unabbreviated words and phrases would almost double the size of this work.

QUICK REFERENCE

Alphabetical notations have been added to the upper corners of the pages much the same as you would find in a telephone book, for ease in finding the listings within each of the sections. Hopefully, you will be able to find any named car in the Pullman fleet in seconds.

ILLUSTRATIONS

A listing of all illustrations in this volume may be found in INDEX. This list will be carried forward and added to the lists in Volumes Two through Five for ease of reference throughout the set. Volume Six, The Lot List, will chronicle the cars that were sold and will have its own listing of illustrations.

THE SLEEPING CAR

From Concept to Monopoly

For the past century, rail travelers have referred to overnight accommodations aboard a sleeping car as "Pullman", making an entire mode of travel synonymous with its leading promoter. George M. Pullman traditionally has been credited as the inventor of the sleeping car, a reputation he enjoyed but one that is not accurate. Just as Westinghouse did not invent the air brake nor Janney the automatic coupler, Pullman did not "invent" the sleeping car. What he did was perfect a workable version and market it to a willing public at a time of explosive growth in railway travel. Pullman was an entrepreneur who borrowed ideas freely from colleagues, competitors, common sense, and tradition. He was a genius at organization and promotion, and within two decades of entering the railway arena he controlled virtually the entire sleeping car business. His work will inevitably dominate any history of sleeping car service; nevertheless, Pullman did not single-handedly create this more sophisticated mode of travel. How, then, did it occur?

Since the late middle ages, Hollanders and other Europeans have lived in homes with limited space so that "upper beds" were built into the walls over others at floor level. Passengers aboard ships which crossed the Atlantic in the early 1800s enjoyed onboard accommodations that offered sleeping and dining facilities much like those later adopted for railroad service. Double-tier and even triple-tier bedding arrangements in wagons were utilized by the gypsies during their nomadic wanderings through Europe. These facilities may well have been the origin of the sleeping car concept. The use of this space-saving device cannot be credited to "the inventor of the sleeping car" for, as such, he never existed. It seems obvious that the concept of a sleeping car evolved from a requirement, was adopted from facilities that already existed, and then was redesigned and refined for use on railroad cars.

As this evolution occurred, numerous patents were issued to protect designs, many of which never progressed further than the paper they were on or the model that was used to apply for the patent. By the early 1850s development of the concept was actively pursued by both existing and newly-founded manufacturing companies. As is the case with most successful technologies, several factors converged to make sleeping cars a practical necessity rather than a novelty. After the Civil War, the economy in the United States grew explosively as the country entered the prime of its industrial revolution. The great waves of railroad building, steelmaking, oil refining, and fortune building all created a tremendous demand for comfortable overnight rail travel. People now had a need to go somewhere and the means to pay for it. George Pullman capitalized on this, but he built on the work of earlier designers.

Most of the early car builders were experienced cabinet and carriage makers who made a fairly easy transition to the construction of wooden passenger and later, sleeping cars. A few of the companies formed in the mid-1800s which survived to become the country's leading manufacturers were Allison & Murphy in Philadelphia, Barney & Smith in Dayton, Jackson & Sharp and Harlan & Hollingsworth, both engaged in ship building in Wilmington, Delaware, and the Gilbert Car Works in Troy, New York. One of the earliest and most talented pioneers was Richard Imlay, a former carriage craftsman who started with the B&O Railroad in Baltimore and later settled in Philadelphia. The railroads hired master car builders such as Joseph Jones at the New York Central who later helped to found the Master Car Builders Association in 1867. Thomas Bissell is well known for perfecting wooden car framing and for patenting a design for the end platform and open vestibule. He served many years with the Wagner Palace Car Company in Buffalo, New York. James Leighton was an early sleeping car concessionaire in New Haven, Connecticut and founded the New Haven Car Company. An early patent in 1854 by H. B. Meyer was a hinged coach chair that could be folded into a bed. By 1838 a number of manufacturing firms were already established to build passenger and freight cars along with other firms which designed and constructed the necessary sub-assemblies and hardware. Out of this existing industrial activity, the daytime/nighttime car developed and was instrumental in the creation of sleeping car concessionaires.

It is accepted that the Cumberland Valley Railroad placed the first successful sleeping car into service on its line between Harrisburg and Chambersburg, Pennsylvania. The exact date is uncertain, but historians have established 1838 as the year this innovative service was introduced on the CV Railroad.[1] Previously, travelers bound for Philadelphia from Pittsburgh arrived in Chambersburg exhausted by the arduous stagecoach journey over the Allegheny Mountains. On the railroad, they faced further discomfort from connections made in the middle of the night and hard upright coach seats. The sleeping car **CHAMBERSBURG**, though rough and unsophisticated in design, represented a welcome relief and gradually other railroads recognized the potential for developing a much-needed service that might expand their revenues. Within ten years, seven additional railroads were involved in operating sleeping-car service. Richard Imlay, who was responsible for the design and construction of many of the early sleeping cars, provided the car for the Cumberland Valley. There is no clear record of whether it was built to order or rebuilt from an existing coach.

In the ensuing years, railroad-owned shops converted coaches to sleeping cars and, inspired by the public's demand for either wayside hotels or better railway equipment, companies were established to design cars for day and night travel and sell or lease them as concessionaires to the railroads. Dozens of railroad sleeping car

---

1. **History of the Cumberland Valley Railroad** by Paul J. Westhaeffer, Washington D.C. Chapter, National Railway Historical Society, pp. 31, 32.

lines and private sleeping car companies were active before Pullman began consolidating them in the 1870s. They will be treated in greater detail in subsequent volumes of this work. Yet as they all ultimately became part of Pullman's empire, it seems appropriate to begin with the great capitalist himself.

George Mortimer Pullman was born March 3, 1831 in Brockton, New York. He mastered woodworking under the tutelage of his brother, Albert, and in his early twenties he accepted a contract to move buildings during the widening of the Erie Canal. He moved to Chicago in late 1855 to set up a business raising buildings above the level of Lake Michigan, a successful operation which made him well known in Chicago long before his sleeping cars rolled into its station. Two years later he returned to New York and became reacquainted with his old friend, Benjamin C. Field, a former member of the New York Senate who retired from the legislature in 1856. Field had acquired some sleeping car patents from T. T. Woodruff and was negotiating with the Chicago & Alton Railroad to remodel two coaches for sleeping car service. Pullman has been quoted by numerous sources as stating that a ride in a sleeper in 1858 from Buffalo to Westfield aboard one of the Woodruff cars convinced him that the accommodations had to be made more comfortable.

Pullman and Field became partners and set about rebuilding two coaches, Nos. 9 and 19, for the Chicago & Alton. The work commenced at the railroad's repair shops in Bloomington, Illinois, and the cars went into service in late 1859. The following is an excerpt from an affidavit of George M. Pullman in a suit against the Baltimore & Ohio Railroad: "In the spring of 1859, I made a contract with the Chicago & Alton Railroad Company to supply its line with sleeping cars of the most approved pattern of that day. The managers of the road selected as the best the Meyers & Furniss type, and I purchased a right to make use of the Meyers & Furniss patent for that road, and commenced, in connection with Mr. Field, to alter the cars belonging to the company into sleepers. I was convinced from the first that a practical sleeping car must be made of two tiers of berths, and not three, and that the interior of the

car by day should be free from all obstructions in the shape of standing rods or fixed stanchions or permanent partitions, and I began to apply these ideas to the Meyers & Furniss plan." [2]

No pictures or drawings exist of No. 9 as originally rebuilt. It was destroyed by fire, and in 1897 Pullman remodeled Tourist Car 402 as a replica. This car had been purchased from AT&SF in January 1889 and was originally built by Barney & Smith. It had twelve double berth sections and was lit by overhead candle lamps. The No. 9 replica was exhibited at Nashville in 1897 and the Omaha Exposition in 1898. It was then transferred from the Tourist Car Account to Miscellaneous Equipment in December 1937. Retirement was authorized on July 6, 1953, and the replica of No. 9 was destroyed by fire on August 12, 1953. [3]

In early 1860, Pullman left Chicago for Colorado. He did not relinquish his interest in the sleeping car business. While in Colorado, he bought land in the vicinity of Pike's Peak, and he was involved in silver mining and the sale and delivery of provisions and supplies to prospectors from his own line of wagons. Pullman returned to Chicago in 1863 with a considerable amount of money from these and other business enterprises. In the meantime, Field had purchased a new car for the Chicago & Alton Railroad in 1861, No. 30 (later **JACKSONVILLE**) from Barney, Parker & Company of Dayton. By the time Pullman rejoined his partner, there were three Field/Pullman cars in service on the C&A. Field had also opened negotiations with the Galena and Chicago Union Railroad.

2.   Excerpt from Affidavit of George M. Pullman in suit against The Baltimore & Ohio Railroad, 1880, reprinted in Statement of Documentary Evidence, United States of America vs. The Pullman Company, Civil Action No. 994, p. 18. The Smithsonian Institution.

3.   Authority for Expenditure or Retirement C-3181, The Pullman Company, approved September 3, 1953. The Ralph L. Barger Collection.

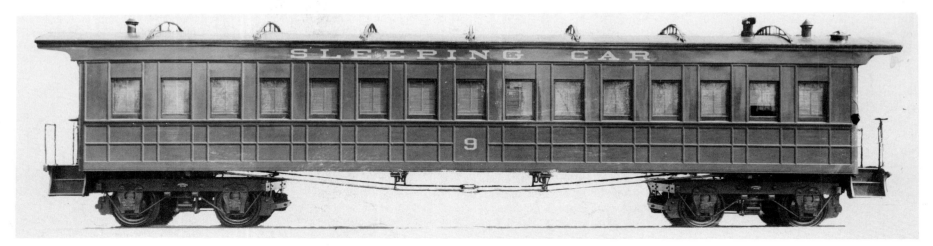

Figure I-1   No photographs or drawings exist of the original No. 9 as rebuilt from the Chicago & Alton coach of the same number. This is a publicity photograph from the Pullman files of the No. 9 Replica. The available facts do not verify that it was a faithful reproduction of the original sleeping car.   Smithsonian Institution Photo No. P-3659.

Anticipating the end of the Civil War and the country's vast growth potential, Pullman wagered on the expansion of railroads and their need for sleeping-car service. He and Field contracted to design a car for the Chicago & Alton in space rented to them by that railroad.

The first car was designated "A," later named **PIONEER**. It was completed in 1865 but, unfortunately, no records survive that describe the car in detail as it went into service. There are conflicting reports on its original dimensions and descriptions of the interior. It was remodeled several times and described, at one time, to have been 54 feet in length with a large clerestory roof, a design Pullman claimed would provide better ventilation and allow gentlemen to pass through the aisle without removing their top hats. Inside, there were twelve two-tier berths instead of the three-tier arrangement standard in existing sleeping cars. The heavily upholstered sofas were converted into lower berths. Overhead, and folded into the top of the car, were upper berths which pulled down to form sleeping spaces. Like others in its day, the car was heated by a wood stove. It was wider and higher than existing passenger coaches. Because of its width and height, it could not operate over the Chicago & Alton due to close clearances. The railroad finally rebuilt many wayside structures to accommodate the **PIONEER** when it carried Illinois dignitaries as part of the consist of the funeral train for President Lincoln. This established an important precedent based on a risky venture, as indicated by court testimony submitted by Pullman himself: "It became apparent, then, that before a sleeping car, such as the 'Pioneer' could be used, a radical change had to be made in the structure of cars; and we were confronted with the question whether railroads would consent to the increased size of cars, and submit to the alterations which they would have to make in their permanent structures to allow such cars to pass." [4]

The interior of **PIONEER** incorporated few innovations in design, yet the cost of construction was an astronomical $20,000 due to postwar inflation. There are no known photographs of **PIONEER** as originally built and no Pullman documentation on its service after inclusion in the Lincoln funeral train in 1865. There is evidence of a sleeping car named **PIONEER #1** in Southern Transportation Company service between 1875 and 1884. [5] It is assumed this means that the **PIONEER** was sold (passed to) Southern Transportation Company about 1875, and that the car was sold back to Pullman's Palace Car Company sometime later. It was scrapped in 1903.

Pullman's early cars were test beds, experiments that brought together as many amenities as he could incorporate in a single car body. After the reluctant acceptance on the part of the railroads of the larger and heavier design represented by **PIONEER**, Pullman began a systematic program of railroad negotiation, car building, marketing, and takeovers that created a near monopoly in the space of two decades. By offering the public superior accommodations, he built a reputation as a first class hotelier. By controlling the entire car-building process, he ensured the

---

4.   Excerpt from Affidavit of George M. Pullman in suit against The Baltimore & Ohio Railroad, 1880. The Smithsonian Institution.

5.   **PIONEER** is listed in the inventory for Southern Transportation Company in **COST OF CARS**, Book I, July 31, 1875, p. 28 through **COST OF CARS**, Book II, July 31, 1884, p. 128. The Smithsonian Institution. **NOTE:** Concurrently, **PIONEER** is missing from the inventory for Pullman's Palace Car Company during this period of time.

Figure I-2: The third car in the Pullman inventory was PIONEER, the first new car designed and built by Field and Pullman six years after they rebuilt Nos. 9 and 19 for the Chicago & Alton Railroad. This car represents new dimensions and design which Pullman felt would provide more comfortable accommodations. It was the first shot fired in the revolution of sleeping-car design that would follow. Smithsonian Institution Photo No. 40000A.

price, quality, and availability of his fleet. By skillfully manipulating the individual railroads, Pullman expanded his route miles and began the process of standardization that set The Pullman Company apart from the rest of 19th Century railroading. Lastly, he used tactics that might be judged illegal or unethical today to absorb every competitor in the sleeping car business.

His competitors were also men of vision, ability, and resources, yet all eventually succumbed to the ever-growing operations of Pullman. A survey of these early companies is essentially a survey of the roots of The Pullman Company. Thus, the following summary histories are not complete, but offer a sense of the business as it coalesced in the late 19th Century.

THE FOUNDING OF SLEEPING CAR COMPANIES

The **T. T. WOODRUFF & COMPANY** was founded by Theodore T. Woodruff (1811-1892), one of the first to operate daytime-nighttime cars on a concession basis. Mr. Woodruff was a master car builder for the Terre Haute, Alton & St. Louis Railroad and his designs for improved sleeping cars were exceptional for the times. He secured two patents in 1856 and the following year formed his own company. His first car was built by Wason & Company of Springfield, Massachusetts and later in 1858, he landed a prized contract with the Pennsylvania Railroad to design four cars (Nos. 22-25). These were built by Murphy & Allison of Philadelphia.

Silver Palace Cars were introduced in 1866 by Theodore's brother, Jonah Woodruff. Three of these cars, **NEW YORK**, **CHICAGO**, and **PITTSBURGH**, were built by the New Jersey Central Railroad. Central Pacific purchased Silver Palace cars from Harlan & Hollingsworth and Jackson & Sharp. By 1870, Woodruff had 21 cars in operation and eight additional cars were built for him by Barney & Smith in 1875.

THE CENTRAL TRANSPORTATION COMPANY was formed in 1862 as a consolidation between Woodruff and Edward C. Knight, an enormously wealthy sugar merchant from Philadelphia who was introducing cars in growing competition with Woodruff. It is believed that Eli Wheeler's patents, along with six other inventors, may have been a part of this consolidation. Central Transportation controlled the Eastern routes and most sleeping-car patents. They rapidly became the largest sleeping-car company in the country.

SOUTHERN TRANSPORTATION COMPANY, a subsidiary of Central Transportation, was formed in 1865 to fill the demands for new markets in sleeping cars during the reconstruction of the railroads in the Southeast at the end of the Civil War.

The **PULLMAN, KIMBALL & RAMSEY SLEEPING CAR COMPANY** was incorporated on May 10, 1866 in the State of Tennessee. The formation of this company with Robert H. Ramsey and Hannibal I. Kimball, both of Atlanta, was an effort by Pullman to counter the control of sleeping-car service in the South by the Southern Transportation Company.

In 1866, it was apparent that if the Pullman business was to expand and acquire additional capital it would be necessary to incorporate. It was during this year that Field chose to retire. **PULLMAN'S PALACE CAR COMPANY** was incorporated in Illinois on February 22, 1867. George Pullman sold his assets which included 37 railroad cars to the new company. These 37 cars consisted of 26 belonging to George Pullman, eleven of which he owned part interest (seven with Michigan Central and four with Chicago, Burlington & Quincy Railroad), one from Pullman, Kimball & Ramsey Sleeping Car Company, and two under construction.

By 1868, the race for transcontinental service was under way between Pullman and the Central Transportation Company. Andrew Carnegie withdrew his support from Central Transportation in favor of George Pullman, and **THE PULLMAN PACIFIC CAR COMPANY** was incorporated in the State of New York on April 24, 1868 to provide sleeping-car service on the Union Pacific Railroad. Carnegie joined the newly-formed company as a major stockholder in return for his assistance in acquiring this coveted contract. Later, stock control was obtained by Pullman's Palace Car Company on November 15, 1870. The gradual demise of the Central Transportation Company originated in the struggle with Pullman to gain the contract with this western railroad. Pullman Pacific itself experienced a short career due to disappointing consumer response to transcontinental rail travel. Central Pacific continued to use its own Silver Palace cars, while Union Pacific operated its extensive emigrant service concurrently with its Pullman Pacific contract. In October 1871, an Association Agreement was signed with the UP, Pullman Pacific ceased operating, and was officially liquidated in 1884. Central Pacific signed with Pullman in 1883, thus consolidating transcontinental sleeping-car service for the first time.

Pullman's Pacific Car Company provided equipment for the first chartered Transcontinental excursion from Atlantic to Pacific sponsored by the Boston Board of Trade. The eight-car train traveled round trip from Boston to San Francisco between May and July, 1870. It was supplemented along the way by open observation cars provided by the railroads. George Pullman was aboard the train from Boston to Sterling, Illinois. His brother, Albert, General Superintendent of Pullman's Palace Car Company at that time, remained aboard for the entire excursion to San Franciso as the company's representative.[6] It is interesting to note that the train left Chicago by the Galena division of the Chicago & North-Western Railway, the former Galena & Chicago Union Railroad with which Benjamin Field and George Pullman had negotiated contracts in the early 1860s.

By 1870, Pullman had acquired the Detroit Car & Manufacturing Company which became Pullman's Detroit Shops where the first Pullman-built cars were constructed. Pullman signed a 99 year lease for control of the Central Transportation Company on January 1, 1870. The agreement included all rolling stock, patents, and contracts for 4,400 miles on 16 railroads. At the same time, Southern Transportation was leased by Pullman, an agreement which included 17 cars and five contracts for providing service on 1,500 miles of line at $20,000 yearly. Pullman purchased all the stock of Southern Transportation in 1878 for $90,000. The records indicate that there were ten cars remaining in service in 1884, of which one was **PIONEER**. Pullman then turned his attention to the full acquisition of Central Transportation Company. Following many years of bitter legal battles, a final settlement occurred in 1899 in which Central Transportation's few remaining assets were liquidated.

---

6. **Trans-Continental,** Vol I, No. 1, May 24, 1870, and Vol I, No. 6, May 31, 1870, a newspaper published aboard the Boston Board of Trade Pullman Hotel Express train. Reprinted by Timothy J. Hughes, Williamsport, PA, 1981.

On March 11, 1870, the **CRESCENT CITY SLEEPING CAR COMPANY** was incorporated in the State of Kentucky. An amendment dated February 13, 1871 changed the name of this company to **PULLMAN SOUTHERN CAR COMPANY.**[7] The date of organization was March 11, 1871. On October 10, 1872, Pullman, Kimball & Ramsey Sleeping Car Company and the so-called "Paine Lines" were acquired by Pullman Southern in exchange for capital stock, along with eleven sleeping cars from Louisville & Nashville Railroad (seven of which were ex-McComb cars) and part interest in two Memphis & Charleston Railroad sleeping cars. The net result of this consolidation created a company with 46 sleeping cars which with the Southern Transportation Company lease meant that Pullman now controlled the Southeast.

The **INDIANAPOLIS & ST. LOUIS SLEEPING COACH COMPANY,** founded in 1862, was a joint stock association which operated sleeping cars between Terre Haute and Columbus under contracts with the Columbus & Indianapolis Central Railway Company and the Terre Haute & Indianapolis Railroad Company. The company filed suit to prevent operation of Pullman cars over these railroads. In order to settle this litigation Pullman purchased the assets of the sleeping car company in 1871, the so-called Sanderson cars.

THE ERIE & ATLANTIC SLEEPING COACH COMPANY was a joint stock association organized in 1865 in New York State by officers and others interested in the New York, Lake Erie & Western Railroad Company, one of the predecessors of the present Erie Railroad. The company acquired and serviced sleeping cars for the railroad. In 1872 Pullman purchased two-thirds of the capital stock of The Erie & Atlantic Sleeping Coach Company. At the time the Erie was broad gauge (six foot), so that these cars were restricted to service on the Erie system. This transaction enabled Pullman to gain control of the sleeping-car service on a through line between Chicago and New York. In 1883, Pullman sold to the railroad enough stock in the Sleeping Coach Company to give it a half interest, and at this time, it was renamed Pullman-New York Lake Erie & Western Association. A dispute arose over the amounts due the Pullman Company, and in 1888 the Association was dissolved when Pullman acquired all of its stock and assets.

The "Association" was used principally by Pullman to get a "foot in the door." It was an arrangement whereby Pullman would initially purchase a percentage of the value of railroad-owned sleepers and parlor cars and would operate them. The cost of operation and maintenance of the cars and the profits were shared based on the percentage of ownership. This arrangement led to the exclusive purchase of Pullman-built cars, eventually resulting in Pullman ownership and operation in all cases except the Chicago Milwaukee & St. Paul Railroad. There, all interests and cars were sold to the railroad in November 1890. Pullman's Palace Car Company was fast becoming the major supplier and operator of sleeping-car equipment in the United States.

In 1873 Pullman entered the foreign market under an operating contract with Midland Railway Company of England. The agreement was for six cars built at Detroit and shipped disassembled in 1874. However, Britons favored the first-class accommodations on their own lines as well as the overnight hotel facilities which were prevalent along the railway routes. There was no stampede by the British railway lines to make use of the Pullman sleeping cars and the venture was unsuccessful. An attempt to organize a company to use Pullman-built cars in Europe failed also, principally because the open-section arrangement was not acceptable to Europeans who were accustomed to the privacy of closed compartments as provided by the Mann Boudoir Car. Italy as the owner and operator of its own railroad lines was the only country to sign with Pullman.

Georges Nagelmackers, a formidable Pullman competitor, was operating the Companie Internationale des Wagon Lits et Grand Expresses (CIWL) and was very much in control of the European sleeping-car market. His equipment was better suited to European standards and by virtue of his family's royal court connections, he had already negotiated the complex arrangements for moving foreign travelers across national borders. Pullman suggested a merger, Mr. Nagelmackers declined, and Pullman abandoned the European market. He sold his assets to The Pullman Company, Ltd. about 1906.

Pullman turned his attention to Mexico in 1884 and provided service between El Paso and Mexico City. This operation was successful and, except for a discontinuance of service during the Mexican Revolution between 1914-1920, it continued until The Pullman Company was dissolved.

WOODRUFF SLEEPING & PARLOR COACH COMPANY was incorporated May 19, 1871 in the State of Pennsylvania by Jonah Woodruff under patents issued in that year. It was an effort to prevent a Pullman monopoly. Two cars, **LADY WASHINGTON** and **LADY FRANKLIN,** were completed by Harlan & Hollingsworth in 1871. Jonah died in 1876. The company continued, however, and purchased the Lucas Sleeping Car Company of Atlanta in 1878. By 1880 Woodruff had 64 cars in service, but Pullman had 800.

During the organization of the Woodruff Company another contender for sleeping-car contracts surfaced. William D'Alton Mann (1839-1920) received his patent in 1872. His designs adopted closed compartments rather than the open sections of Pullman-built cars. However, they were not popular with the American people. He left Alabama for Europe where he became involved in an alliance with Georges Nagelmackers. The Mann cars eventually ran on 12 lines transporting roughly 3,000 passengers per day between four European cities. In 1878, Mann received another U.S. patent, sold his interests in CIWL, and returned to America.

THE MANN BOUDOIR CAR COMPANY was chartered on March 23, 1883 in New York State. It seems apparent that the use of the word "boudoir" was an unfortunate selection. American society of the times, Victorian as it was, found the word too suggestive and stories circulated which clouded the respectability of travel aboard these luxurious cars. Actually it was a compartment, the standard accommodation on European trains, and in the lightweight era became the most popular traveling space for two people. The Mann Company attempted to establish a business of leasing private, parlor, and luxury sleeping cars to wealthy individuals and railroads for first-class service. It operated at a loss during the entire five years of its existence, but during this period the Company acquired 38 buffet boudoir, five

---

7. Amendment to An Act to Incorporate the "Crescent City Sleeping Car Company," which dissolved that company and created Pullman Southern Car Company. The original document is in the Newberry Library.

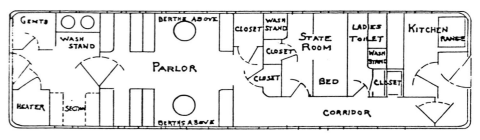

Figure I-3: Jackson & Sharp records show these hand-drawn floor plans for the Woodruff rotunda car MARY C. as it arrived December 1, 1886 to be rebuilt into private car YELLOWSTONE. MARY C. was originally built by Harlan & Hollingsworth in July 1878. Above is the floor plan for YELLOWSTONE as delivered September 7, 1887. Smithsonian Institution.

Figure I-4: Velvet, silk tassels and fringe, mirrors and gilt, heavily upholstered sofas, and even a piano (not visible in this view) were used by Mann in the interior of ADELINA PATTI, the private car used by the diva during her American tour with the Italian Opera Company. It was purported to be "the richest in decoration, finish and furniture of any car ever built in the world." Note Madame Patti's initials inlaid over the window and the musical theme of the mural painted by Nicholas Rossignoli. The Arthur D. Dubin Collection.

sleeping, three private, and four parlor cars, which was not a bad record in such a fiercely competitive field. The private cars built and named for the opera singers, Etelka Gerster and Adelina Patti and used by them during their American tours in the late 1880s will be recalled by historians as the most lavish in the annals of car building. They eventually fell to Pullman ownership when the Union Palace Car Company was purchased. **ADELINA PATTI** had by then been renamed to **ADELAIDE MOORE** by Mr. Mann and Pullman renamed it **CORONET**. It was sold to Buffalo Bill through Fitzhugh & Company and was named **CODY**. **ETELKA GERSTER** was sold to The Bangor & Aroostook Railroad and became that line's business car No. 97.

The **UNION PALACE CAR COMPANY** was incorporated on September 24, 1888 in New Jersey to secure control of the Mann and Woodruff companies through an exchange of common stock. It was a plan instigated by the two car builders, Jackson & Sharp and Harlan & Hollingsworth of Wilmington, Delaware, holding substantial notes from both Mann and Woodruff. The principal promoters of the stock exchange controlled the Richmond & Danville Railroad Company which entered into an operating agreement with the Union Company. The new agreement was not satisfactory to them, and before the Union Company began operation, Pullman purchased it at the solicitation of the promoters in January 1889 for $2.5 million. The Mann, Woodruff, and Union cars maintained separate identities in the Pullman Descriptive List until 1909.

After the acquisition of the Union Palace Car Company, the strongest Pullman competitor, The Wagner Palace Car Company, was the only sleeping car concessionaire left in this country to contest Pullman's Palace Car Company monopoly.

Webster Wagner (1817-1882) in partnership with T. N. Parmelee, Morgan Gardner, and George B. Gates, operated sleeping cars between Albany and Buffalo on the New York Central Railroad in 1858. He made use of Woodruff designs and there appears to have never been a patent issued in Mr. Wagner's name. He contracted with Commodore Cornelieus Vanderbilt to provide sleeping-car service on the roads owned and operated by this venerable millionaire. With Vanderbilt's backing the success of the company became assured by the end of the Civil War. In an effort to expand his business, Wagner founded **THE NEW YORK CENTRAL SLEEPING CAR COMPANY** in 1866. George Gates (**GATES SLEEPING CAR COMPANY**) joined the company and brought with him the contract for Lake Shore & Michigan Southern. New York Central was a closed corporation and enjoyed additional success as the Vanderbilt interests expanded. A bitter feud ensued between Pullman and Vanderbilt when the latter gained control of the Michigan Central Railroad in 1875. The railroad was required to use Wagner's cars and Pullman's were removed from service on the major portion of the lines.

Ironically, Webster Wagner was killed in a rear-end collision on one of his own cars at Spuyten Duyvil in January 1882. Commodore Vanderbilt had passed away in 1877, shortly after a disastrous wreck with an enormous loss of lives on one of his affiliates, the Lake Shore & Michigan Southern. The New York Central Sleeping Car Company would see several presidents appointed by the Vanderbilt family including its last, a son-in-law, William Seward Webb.

Webb felt a change of name might spirit some railroads away from Pullman and into the fold of the New York Company. Consequently, the name was changed to

THE WAGNER PALACE CAR COMPANY in 1886. During Webb's fifteen-year tenure he expanded the company's Buffalo shops and secured an experienced car builder, Thomas Bissell, superintendent of Barney & Smith and at one time head of Pullman's shops at Detroit. As many as 55 cars a year were built prior to 1894. The company invested heavily in equipment from virtually unlimited Vanderbilt funds. Yet with no substantial increase in new customers, profits fell off and little return was made on its investments. The Vanderbilts eventually concluded that it was time to give up the conflict with Pullman.

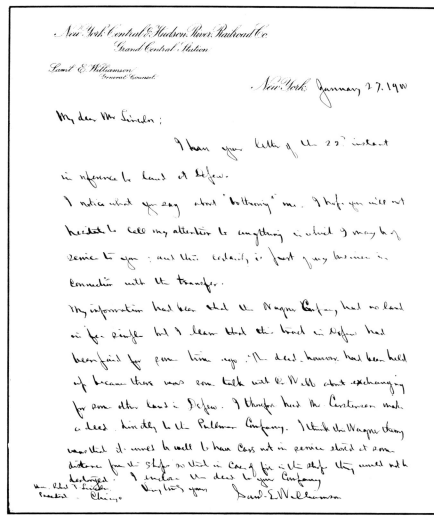

Figure I-5: Letter from Sam'l E. Williamson, General Counsel for New York Central & Hudson River Railroad Company to Robert Todd Lincoln, President, The Pullman Company, transferring the deed of Wagner property at Depew to Pullman. The Ralph L. Barger Collection.

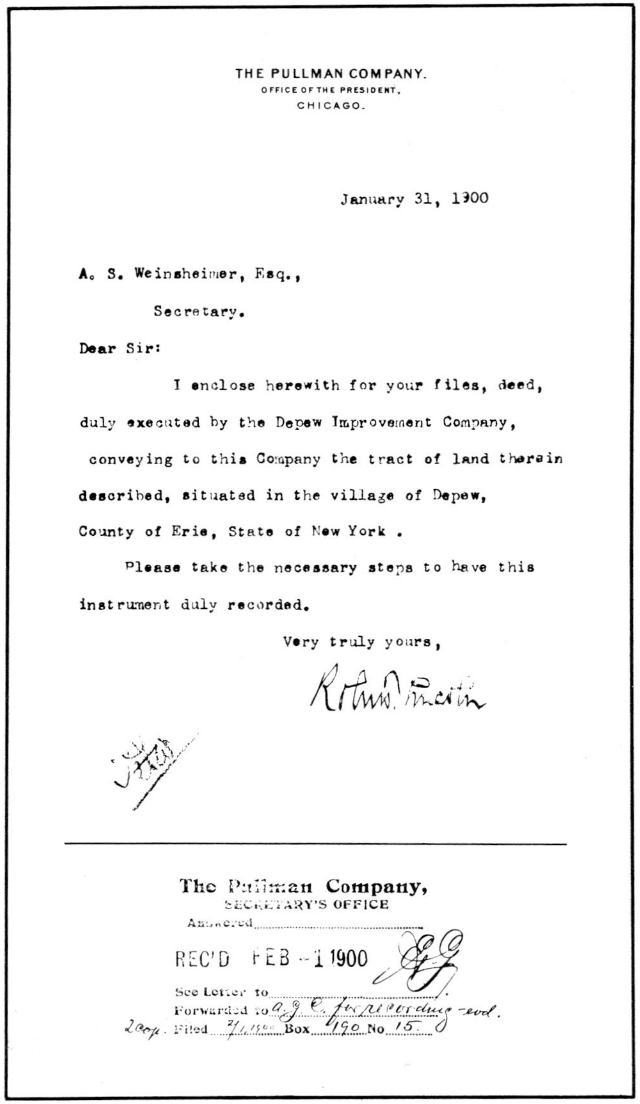

THE PULLMAN COMPANY.
OFFICE OF THE PRESIDENT,
CHICAGO.

January 31, 1900

A. S. Weinsheimer, Esq.,

Secretary.

Dear Sir:

I enclose herewith for your files, deed,
duly executed by the Depew Improvement Company,
conveying to this Company the tract of land therein
described, situated in the village of Depew,
County of Erie, State of New York .

Please take the necessary steps to have this
instrument duly recorded.

Very truly yours,

The Pullman Company,
SECRETARY'S OFFICE
Answered
REC'D FEB - 1 1900
See Letter to
Forwarded to a. g. for recording ans.
2cop. Filed 2/1900 Box 190 No 15

Figure I-6: The covering letter from Mr. Lincoln forwarding the deed for the former Wagner property to be recorded. The Ralph L. Barger Collection.

After a prolonged negotiation, agreement was made for a one-for-one exchange of Wagner shares with Pullman and the 36 million dollar sale was consummated. William K. Vanderbilt and William Seward Webb were given seats on the Pullman Board of Directors, and the Wagner Palace Car Company was absorbed into Pullman's Palace Car Company on January 1, 1900.

In 1881 the Pullman factory had been constructed south of Chicago between the banks of the Calumet River and the Illinois Central Railroad tracks. The model town of Pullman was erected on the prairie adjacent to the car works to house the workers and their families who rented their homes from the company. Pullman's concept for the town was based on the premise that better living conditions in a clean urban environment would attract a better class of workingmen, provide a happier family atmosphere for them, and meld labor and industry into a community that would enjoy congenial relations. He believed also that it would relieve labor problems because his employees would be away from Chicago, which was troubled by conflicts between rich and poor and constant labor disputes. Here there would be no radical class system. His concept was well meaning, but the enterprise failed. The town's fatal weakness was the total power of the company and the absence of self government. When the depression occurred in 1893 the company took measures to cut back on costs. Bids on new contracts were lowered to the extent that, in some cases, the company took a loss. Its reasoning was to keep the men working but the company also cut their wages and held their rent at the same rate, an act the workers felt was inequitable. Progressively, the men fell heavily into debt and although the company did not foreclose or evict anyone in arrears, fuel had been poured onto already smoldering labor relations and a major confrontation between labor and management flared.

In his book, "**PULLMAN**, An Experiment in Industrial Order and Community Planning," Stanley Buder refers to the Pullman strike in May 1894 as "synonymous with the most serious American labor conflict of the century." The prolonged walkout drew attention from the newly-organized American Railway Union, headed by Eugene Debs. The union, sympathetic to the worker's plight, offered membership to the Pullman employees on the broad technicality that since the Illinois Central Railroad had trackage through the town, the workers were essentially railroad men. In point of fact, they had nothing to do with the IC except to use its services as passengers to and from Chicago. The strike eventually became a national crisis as the union boycott against Illinois Central spread to other railroads serving Pullman cars. The railroads refused to arbitrate, and so did Pullman. Railroad men from Ohio to California walked out when required to service a Pullman car in any capacity. Rail traffic was disrupted to the extent that Chicago became a city besieged by railroad riots and President Cleveland ordered Federal troops (a first in American labor disputes) to ensure the delivery of mail into the city. The rioters gradually dispersed when confronted with this superior force. The strike failed, and the Pullman plant reopened in August. The laborers were by then desperate for work and were rehired on the proviso that they relinquish their union membership.

The effects of adverse press and public condemnation of policies which Pullman believed were in the best interests of the company and its working force may have contributed to the founder's failing health. He died in 1897 without seeing his company become the victor as the sole remaining sleeping-car concessionaire. On January 1, 1900, Pullman's Palace Car Company became what would be recognized by yet another generation of travelers as **THE PULLMAN COMPANY.**

Robert Todd Lincoln, son of the former U.S. President and General Counsel of Pullman's Palace Car Company, was named acting president of the company after Pullman's death. At this time, the corporation controlled nearly 90 percent of the

CHART OF ACQUISITIONS

	NAME	INCORPORATION	CONTROL ACQUIRED BY PULLMAN COMPANY	OTHER INFORMATION
1	The Pullman Company	See #2		
2	Pullman's Palace Car Company	By Act of Illinois, Feb 22, 1867		Name change to #1, Dec 30, 1899
3	Pullman Southern Car Company	See #4	Oct 10, 1872, operations began	Assets sold to #2, Oct 1, 1894
4	Crescent City Sleeping Car Company	Authorized by Special Act of Kentucky, Mar 16, 1870, Chartered as #2, Mar 1891		Name changed to #3, Feb 13, 1871
5	Pullman, Kimball and Ramsey Sleeping Car Company	May 10, 1866 in Tennessee	May 10, 1866	Consolidated with #3, Oct 10, 1872
6	Paine Lines (so-called)	Group of 4 partnerships	Oct 10, 1872	Consolidated with #3, Oct 10, 1872
7	The Pullman Pacific Car Company	By Special Act, New York, Apr 24, 1868	Apr 24, 1868 through common officers and substantial amount of stock—stock control obtained Nov 15, 1870	
8	Erie and Atlantic Sleeping Coach Company	Joint Stock Association, New York, 1871	Jun 7, 1872	All stock with assets acquired Feb 29, 1888
9	Woodruff Sleeping and Parlor Coach Company	By act of Pennsylvania, May 19, 1871	Date of control of this company is in dispute and no stipulation is made.	Assets sold to #1, Oct 31, 1909
10	Lucas Sleeping Car Company	Fulton Co., Georgia, Oct 4, 1875	Never acquired by Pullman Company	Sold to #9, Sep 10, 1878
11	Mann's Boudoir Car Company	New York, Mar 23, 1883, General Laws	Date of control of this company is in dispute and no stipulation is made.	Assets sold to #1, Oct 31, 1909
12	The Union Palace Car Company	New Jersey, Sep 24, 1888, General Laws	Date of control of this company is in dispute and no stipulation is made.	Assets sold to #1, Oct 31, 1909
13	Wagner Palace Car Company	See 14	Dec 31, 1899	
14	New York Central Sleeping Car Company	Joint Stock Association, New York, Jan 1, 1866		Name changed to #13, Dec 22, 1886
15	Gates Sleeping Car Company	No information	Never acquired by Pullman Company	Sold to #14 in 1869

Figure I-7: Chart of aquisitions.

sleeping-car service in the United States and it owned the largest car manufacturing plant in the world. Lincoln served as president for fourteen years, during which the company enjoyed soaring profits with mounting passenger traffic and no competition. He was succeeded in 1911 by John S. Runnells.

A public outcry based on concern for the safety of passengers aboard the deteriorating Pullman wooden cars, along with other complaints, caused a decision in 1906 by the Interstate Commerce Commission to place The Pullman Company under its regulation. Pullman management fought back in the courts, and after a four-year debate on its liability to Federal regulation, the company agreed to report its financial condition to the ICC.

CONVERSION TO STEEL

In March 1907, The Pullman Company produced its first steel sleeping car, **JAMESTOWN,** for the Jamestown Exposition in Virginia. The rush to provide the public with steel cars was an enormously expensive undertaking. The wooden cars were depreciated drastically, new shops had to be built, and repair shops remodeled. Nevertheless, the result was a fleet of stronger and safer equipment which added both prestige to the operating company and profit to its manufacturing division because practically every railroad in the country ordered new passenger equipment.

The last major acquisition of outside rolling stock by The Pullman Company took place on December 31, 1912 when it purchased 252 wooden cars from the New York, New Haven & Hartford Railroad.[8] This purchase came about principally because of the restriction on the use of wooden cars in the tunnels of the new Grand Central Station in New York City. The **NYNH&H** could not afford to replace its fleet and The

8.   Agreement between The New York, New Haven & Hartford Railroad Company and The Pullman Company, in force January 1, 1913, p. 14. The Alvin A. Lawrence Collection.

Pullman Company viewed the purchase as the elimination of another competitor. The Pullman Company purchase consisted of 60 sleeping, 8 sleeping/observation, 148 parlor, 13 parlor/observation, and 23 composite cars at a cost of 3.3 million dollars.[9] Many of the parlors were converted to World War I Hospital Cars and sold to the U.S. Government. In 1922 they were sold to Ringling Bros., Barnum & Bailey Circus where they served until 1947.

During the 1920s, production of new cars reached an all-time peak and gradually Pullman absorbed almost all the sleeping-car operations from the railroads. A major monopoly in American business enterprises had developed. In 1922, the presidency was passed to Edward F. Cary who negotiated a merger with the freight car manufacturer, Haskell and Barker in Michigan City. **THE PULLMAN CAR AND MANUFACTURING COMPANY** was organized in 1924 to allow an expansion into the manufacture of freight cars, and a holding company, **PULLMAN, INC.,** was chartered in Delaware. At approximately the same time, The Pullman Company purchased the Standard Steel Car Company, operating plants in Hammond, Indiana; Butler, Pennsylvania and Sagamore, Massachusetts, in addition to the Osgood Bradley Car Works. Standard was controlled by the Mellons, who owned the Pittsburgh-based banking empire. At the time, these were sound business decisions but they proved disastrous with the sudden and devastating stock market crash.

The Great Depression held a grip on the financial world and, as railroads faced bankruptcy, orders for new equipment declined. The Pullman Company built its last new standard sleeping car in February 1931. The company's surplus profits kept it alive, but other measures were necessary for it to survive. No stock dividends were paid for two years, plans for the new freight car manufacturing facilities were abandoned, and the number of employees in its plants was cut drastically. Many of the older sleepers were sold for scrap and attention was turned toward rebuilding and modernizing the fleet in order to salvage the sleeping-car trade. With these cutbacks, along with the reduction of fares, the company managed to weather the depression.

THE LIGHTWEIGHT ERA

As the 1940s dawned, the plants were again humming with the manufacture of the lightweight cars, but The Pullman Company monopoly finally fell under the shadow of the Sherman & Clayton Antitrust Laws.

The Edward G. Budd Manufacturing Company initiated a complaint by the U.S. Justice Department in the Federal District Court in Philadelphia on July 12, 1940. It alleged The Pullman Company controlled sleeping-car orders and claimed unfair competition by monopoly. The complex litigation ended in 1944 when Pullman was ordered to sell either its operating or its manufacturing companies. It chose to sell the operations division and offered its fleet of cars to the railroads. They were not especially interested until Robert R. Young, a financier who had voiced sharp criticism of coast-to-coast rail service, filed a petition in the courts for permission to purchase Pullman. The railroads saw Young as a less desirable sleeping-car manager than Pullman was and agreed to block this maneuver. A stock transfer was completed with a consortium of 57 railroads. They took possession on June 30, 1947 and established a new Pullman Company under their control.[10] In December 1948, 2,367 cars from the Pullman fleet were sold to the railroads.[11]

A CHANGE OF FORTUNE

After the tremendous demand for sleeping car service during World War II, the following years saw a continual decline in Pullman revenues. The American people, released from wartime gas rationing, were driving their own cars, traveling on buses which offered less expensive fares, and discovering the conveniences offered on the greatest enemy of all — the airplane. One could fly, in many cases, for less than it would cost for an overnight journey by rail and without the added expense for Pullman fares, meals, and tipping. Businessmen in particular appreciated the savings in time, and a major defection occurred from rail to air travel.

The last Pullman Standard lightweight sleeping cars (Lot 6959) were delivered to the Union Pacific Railroad in April 1956. They were five Double Bedroom, Buffet, Lounge cars. After a steady decline, culminating in heavy debts, the 101-year-old Pullman Company ceased operations on January 1, 1969.

It might be said that The Budd Company had the last word. They were responsible for building the very last lightweight sleeping cars, leased to the Northern Pacific Railroad and delivered in December 1959. They were built as twenty-four Single, eight Double-Room Slumber coaches.

The major facilities of The Pullman Company were located as follows:

Pullman Car Works, Pullman, Illinois, built in 1881.

General Office Building, Chicago, Illinois, built between 1882 and 1884.

Repair Shops, Calumet, Illinois, built July 1901.

Canal Street Storehouse, Chicago, Illinois, built in February 1903.

Richmond, California Repair Shops, built in December 1909.

St. Louis, Missouri Repair Shops, built in August 1880.

Wilmington, Delaware Repair Shops, built on the former property of Bowers & Dure, August 1886.[12]

LOOKING BACK

The historical accounts of the railroad empires sprawled across this nation credit many persons with contributions to the success of that major undertaking. Some of those contributions were self-serving while others served the welfare of the traveling public. If there is a chart of credits, George Pullman will have to be listed as one of the more gifted in understanding the public's needs and improving the railroad car to provide safer and more comfortable accommodations for extended travel. Although not a trained engineer, he recognized inventions and new technologies which would improve rail travel. His triumph in a highly competitive field says a great deal for his vision and acute business judgments. For instance, by testing the

9. Memorandum, The Pullman Company, Chicago, dated February 28, 1913, Newberry Library.

10. Pullman News, July 1947, p. 3. The Smithsonian Institution.

11. Authority for Expenditures or Retirement, C-3088, The Pullman Company, approved March 12, 1949. The Ralph L. Barger Collection.

12. Valuation Docket 1079, Interstate Commerce Commission, submitted November 19, 1928, p. 893. The ICC Library.

much-criticized Allen Paper Wheels on his sleeping cars, he was instrumental in their promotion to standard equipment. He is rightfully credited with the development of the closed vestibule which, with its accordian-style diaphragm, created a safer passage between cars through a flexible connection and was designed for added strength so that the platform could not ride on top of the other. This eliminated the so-called "telescoping" of cars in the event of an accident. The first dining car, **DELMONICO**, was placed into operation by Pullman on the Chicago & Alton Railroad between Chicago and St. Louis in 1868.[13] He also provided the first "hotel car," **WESTERN WORLD**, a sleeping car with dining facilities and eighteen enclosed private rooms. Pullman was responsible for the first dining car in service on British railways in 1879. It was the parlor car **OHIO**, rebuilt and renamed **PRINCE OF WALES**. Britons also traveled for the first time in 1870 on a Pullman-built parlor car, the **VICTORIA**. Unlike other businessmen of his era, he hired Negroes, assigned them to the sleeping cars as porters and, consequently, was responsible for at least a chapter in the history of black Americans.

Pullman believed the public would accept readily a railroad car that would provide the comfort of their own homes away from home. He also believed they would pay extra for that comfort and some added luxury as well. For one hundred years, he was absolutely right. Pullman service was excellent and the conductors

---

13.   Chicago Tribune, February 17, 1882.

and porters courteous, which speaks highly of a system that trained and managed what amounted at one time to over 14,000 railroad car employees. The Pullman Company was responsible for providing all onboard staff, equipment, and supplies. In 1947 its inventory included 2,270,000 towels, 2,803,000 sheets, 2,237,000 pillow slips, 389,000 blankets, 325,700 pillows, 256,000 head rest covers, 175,000 mattresses, and 165,000 porter's jackets. In this same year, purchases of incidental supplies included: 345,078,700 drinking cups, 16,398,000 cakes of soap, 4,070,000 boxes of safety matches, and 103,000 gallons of liquid soap.[14] The Company was also responsible for all cleaning services. It was an enormous system of management and housekeeping which could be compared to a mammoth hotel chain with reservations for approximately 175,000 rooms each night. In an average year, Pullman's overnight guests numbered almost one-fifth of the country's population.

Scores of successful businesses and corporations listing assets in the millions have been built by the energy and hard work of people from humble origins. Among these one has to include The Pullman Company and its founder. The railroad cars that were built during the century of Pullman represent a period in American history which will never be duplicated and will survive only because of those who care to keep the records intact and the memory alive.

---

14.   Excerpts from Press Release by Carl Byoir & Associates, Chicago, June 30, 1947. The Arthur D. Dubin Collection, Smithsonian Institution.

NAMING THE CARS
Rhyme or Reason?

Pullman cars have borne distinctive and intriguing names since George Pullman's first went into service in 1865 as PIONEER. From the beginning, names have been a symbol of an individualized service unexcelled in transportation. Seasoned travelers recognized Pullman cars by their names and greeted them as they would an old friend.

The naming of cars owned and operated by Pullman started in a rather random manner. The first sleeping cars designed by George Pullman and Benjamin Field were No. 9, followed by No. 19. Both retained the original numbers of the Chicago & Alton coaches from which they were rebuilt and continued in service under these designations. The next five cars were assigned letters of the alphabet but since there were only twenty-six letters from which to select, the system of numbering cars returned to use. However, this resulted in a conflict with the system in use by the railroads. Naming cars seemed the best solution, and ample precedent existed on the railroads. Since the 1830s, locomotives had been named for railroad officials, places, animals, U.S. Presidents, State Governors, and prominent men of the time. Thus the first car built as a sleeping car was called "A" but soon became PIONEER, an appropriate name for the first Pullman car and the beginning of a colorful tradition.

The cars that followed carried the names of states, cities, and designations such as WESTERN WORLD, SPRINGFIELD, CITY OF BOSTON, OMAHA, MISSOURI, ILLINOIS, etc., and were normally accompanied by a number. The names of countries became popular by 1870; however, many cars carried the names of towns and cities along the lines they served to promote the territory through which the trains traveled. A few Greek and Roman names appeared during the 1870s and were very popular. The first parlor cars were given feminine names, a tradition which lasted to the final days of The Pullman Company. Other parlors carried the names of flowers, cities, and towns, but a wide assortment of other names were used as well. In general, naming the wood cars had no real rhyme or reason.

The practice of naming cars illustrates George Pullman's entrepreneurial genius. From its earliest days, the sleeping- car business involved superimposing a separate, usually quite different operation on existing passenger train services. While part and parcel of a given railroad's service, sleeping car lines nonetheless strove to differentiate their product from coach travel. It was often to their advantage to distance themselves from the contracting carrier in the public's eye, as in the case of operating difficulties, labor disputes, or contractual difficulties. These concerns suggested naming sleeping cars to create a new marketing identity for sleeping-car service.

Secondly, there were good reasons from an operational standpoint. The earliest coaches had been named but the practice fell out of favor in the 1840s. Likewise, naming locomotives ceased on most large railroads in the 1870s. Each of the hundreds of railroads in the U.S. used numbers for motive power, passenger, freight, and work equipment, and there never was nor will there ever likely be a central arbiter of numbering schemes. As the sleeping-car business was extremely fluid, with car assignments, routes, contracts, and carriers changing constantly, Pullman early decided that uniquely named cars would rarely or never conflict with railroad equipment.

This was vital as, even in the 1860s, railroad accounting and record keeping was complex. The business was unique, with centralized administration and decentralized operations. Small armies of clerks kept track of millions of individual transactions by hand, relying usually on the car number or name for accuracy. The foundation of the American railroad industry was the ability to interchange equipment between railroads, account for mileages, repairs, revenues, costs, and assignments of equipment, and equitably settle those accounts on a timely basis. Sleeping-car companies were all part of that process and the opportunity for fraud, theft, or error was drastically lessened when this class of equipment could not be mistaken for anything else.

Sleeping cars have been named for rivers, lakes, cities, towns, countries, universities, famous people in history, soldiers, generals, battlefields, camps, forts, poets, operas, the arts, Greek mythology, flowers, indian tribes, royalty, authors, and so forth. It is apparant that in the mid-twenties Pullman officials on the committee of nomenclature developed a system of names that would indicate the different types of accommodations. For example, in 1925 a series of 12-section drawing room cars was given prefix names of the colors RED, ORANGE, and GREEN, such as RED LION, ORANGE RIVER, and GREEN SPRING. Other cars of this description used prefixes of ST., Mc, and EAST. In 1923 some 10-section lounge observation cars received the prefix MOUNTAIN. In later years others followed with MT., MOUNT, MONTE, and GENERAL, like MT. HOOD, MOUNT ROYAL, MONTE GRANDE, and GENERAL EWELL. Some 14-section cars carried ALPINE, PARK, OVER, and STAR. In 1920 a series of 10-section drawing room and 2-compartment cars were given the prefix FORT and these were followed by CAMP, CAPE, LAKE, and LOCH.

America was ready for creative car-naming in the late 19th Century. With the post Civil War economic boom came a leap in public education and general literacy and a new sense of the country's history. The emerging American middle class fancied itself more refined and enlightened than ever before. It still looked to Europe for cultural guidance, but the 1876 Centennial Exposition in Philadelphia marked the emergence of an identifiable American culture.

Pullman, Wagner, and Woodruff all played on this developing sense of identity. They cannily named cars for powerful public figures, revered patriots, American ideals, and well known places. The name of a famous person on the side of a sleeper was like an endorsement to the average 19th Century citizen. Likewise, place names were important in this era of sectional differences and native pride.

The more whimsical names ranged from pompous to expedient to lofty. Greek and Roman names gave the obligatory nod to ancient culture when America was undergoing a classical revival in architecture, learning, and the arts. Some cars were named for the wives, daughters, and girl friends of railroad men; others for equally personal reasons. Most cars after 1900 carried euphonious names of no particular import, or rather mundane ones reflecting the maturing of the business and increasing sophistication of the nation.

The traveling public enjoyed named cars. They were evocative of different places, different cultures, or of home. Names had powerful associational value, like all good advertising working on a subliminal level. Exotic names made this exotic form of travel even more so and added to the cachet of luxury and service. Public interest in Pullman car names was emphasized in 1939, when The Pullman Company staged a nationwide contest to select a name for a sleeping car on exhibition at the New York World's Fair. A total of 780,000 names were submitted in a 60-day period. The name selected was AMERICAN MILEMASTER, a car which later became second SP-400 while in SP Lark service, then SP-9500, and eventually was sold and converted by General Motors to Engine Test #800.

Some cars received names similar or related to the name of the train to which they were assigned. Examples of these are NATIONAL for the "National Limited;" ALEXANDER HAMILTON and THOMAS JEFFERSON for "The Congressional;" CENTURY for "The 20th Century Limited," and GOLDEN for the "The Golden State." The Santa Fe "Chief" had cars with the names of Indian towns of the Southwest, including BILTABITO, CHUSKA, and DENEHOTSO. The "City of San Francisco" carried CHINATOWN, UNION SQUARE, and SEAL ROCKS.

Many names were changed over the years due to various circumstances, such as the acquisition of a car with a duplicate name to one that already existed in Pullman service. This happened frequently during the days when Pullman was eliminating his competition and acquiring their rolling stock. Rebuilding or changing the accommodations of cars as well as their assignment to a specific train or railroad usually resulted in a name change. Seven cars received appropriate temporary names in June 1926 for assignment to The Cardinals Train. The Rexall Train is another good example. All the cars that were sold to the Mexican Railroads were given new names. The classic example of name-changing is best illustrated by BEAVERDALE, a wood car which became DIXIE, SALUTA, LACKAWAXEN, and finally CORNING in its thirty-year history. The steel cars built in Pullman Plan 3584A Lot 4899 for the "Crescent Limited" in September and October of 1925 were given gentlemens' names, a custom for cars assigned to this train. When new replacement equipment was built in 1929, the original cars received new names. Then between 1937 and 1941 most of the cars were rebuilt resulting in a third name. One car, THOMAS RUFFIN, was changed to JAMES W. WILSON in October 1929, in December 1933 it became POINT WASHINGTON, and in December 1950 it was rebuilt and renamed FIR SPRINGS.

In the records column, credit for the longest name goes to The National Railroad of Mexico for its sleeper GENERAL PEDRO ANTONIO DE LEON, the former C&O 12 section drawing room, MARY WASHINGTON. The shortest name was shared by several cars in the 1860s which were given letters of the alphabet. The longest single name was QUANTZINTECOMATZIN while it was assigned to dining car leased to the American Tourist Association for Mexican tours. The Aztec Indian translation is said to be "The Noble Eater of the Royal Dish". The name ILLINOIS was used eleven times over a span of sixty-four years. Running second at ten each were CHICAGO and MISSISSIPPI. To give you an idea of the number of name changes, 1,950 changes were made from a total of 9,045 wood cars, and 1,506 standard cars had their names changed.

Two hundred and forty-three names were changed within months, some virtually overnight, when Pullman acquired 725 cars from the Wagner Sleeping Car Company on December 31, 1899. The last major changes of car names took place between 1948 and 1951 when Pullman modernized seventy-three standard cars by including roomette accommodations. Their names were preceded by ELM, OAK, ASH, FIR, and LOCUST, plus two cars for the C&O Railroad named CASCADES and OLD WHITE, KEYSTONE BANKS for the Pennsylvania Railroad and two cars that continued with their original names for the Erie Railroad. Four of the above cars were again rebuilt in May 1948 to names beginning with BEECH.

Naming cars had its costs. In the 19th Century, data transmission was by mail or telegraph. In either case, names were far more tedious than numbers. In the early 1950s, with the advent of accounting by computer, the new railroad-owned Pullman Company developed a simplified numerical system whereby owning railroads were assigned code numbers. For example, the Baltimore & Ohio Railroad became "003." At the same time, a Pullman Car Code was developed whereby each car was initially assigned a number based upon its listing in alphabetical order. An example of the system is "003 2616," computer language for the B&O-owned GARDEN BROOK. This car code number, once assigned, remained the same even after the car was sold and its name changed.

Imagine also a newly-arrived immigrant carman's helper, circa 1915, sent out into a large coach yard in Chicago to find the MASSACHUSETTS, or a clerk mixing up the RONKONKOMA with the RONCEVERTE, or porters, ticket sellers, passengers with limited language skills dealing with names like WISSINOMING, YPSILANTI, and CANAJOHARIE. It also stands to reason that some Americans, traveling on L'ETOILE DU NORD, never realized they had boarded THE NORTH STAR.

As the Pullman enterprise grew, named cars humanized and personalized what had become a large, frightfully efficient corporation. There was always a dichotomy present in Pullman service. On one hand, its product was individualized service that demanded the training of skilled public ambassadors in hundreds of widespread locations. It demanded that thousands of patrons "out there somewhere" be treated consistently well by a largely unsupervised staff. At the same time, it was a superb quartermaster operation, manipulating millions of inventory items, thousands of cars and employees and shareholders' money. It was a steely-eyed, cold-hearted economic enterprise at the same time it was naming cars PETUNIA, BUTTERCUP, and KICKAPOO.

In the final count, over 21,600 names are listed in this first volume of **A CENTURY OF PULLMAN CARS**. Towards the end of Pullman service, names had outlived their usefulness and existed only as a grand tradition. Yet for eighty years they were a tool, a useful device perfected like so many others by George himself. It matters not that there was never real rhyme or reason, or what individual names were. That they were named at all is characteristic of the whole Pullman experience.

THE PALACE CARS
The Wooden Cars 1859-1910

Figure I-8 COURTIER, an example of Pullman's finest of 1890 was acquired by John Ringling in October 1914, a car that was originally built for "The Richmond & Danville Limited." The intricate design used for the observation platform railing has never been surpassed. It is not commonly known that this world-renowned circus manager also owned and operated a used railroad car business. No records have been found of the dispositions of cars that were sold to John Ringling. Smithsonian Institution Photograph P-1281.

CAR NAME	ST	CAR TYPE	CAR BUILDER (Owner)	DATE BUILT	PLAN	LOT
A		Bagg	PCW Det	5/19/70	1	D4
A		12SecO	G.M.Pullman	/64	5	Chi
A	N	10SecLibObs	Pullman	10/29/02	1827	2927
ABASIA	RN	16SecB	Pull Cal	8/8/02	1794	C3
ABBEVILLE		12SecDr	Pull Buf	5/23/04	1963	B37
ABBOTSFORD	P	12SecDr	G&B (Wagner)	11/80	3023	
ABBOTTSFORD	N	12SecDr	Pullman	5/2/08	1963C	3616
ABBOTTSFORD #567	P	12SecB	PCW	3/4/85	227	1034
ABBOTTSFORD CTC-23	R		PFtW&C	rblt/70		
ABELARD	N	12SecDr	B&S (Wagner)	5/81	3022A	
ABERCORN	N	14SecDr	Pullman	3/13/03	1895	2982
ABERDEEN #557	P	14SecSr	H&H (B&O)	8/87	654	
ABERLIN		Diner	PCW	2/11/92	964A	1874
ABIGAIL		18Ch4StsDrB	PCW	7/20/93	1039A	1995
ABILENE		12SecDr	Pullman	12/15/06	1963C	3448
ABINGTON	N	10SecDrB	PCW	2/11/88	445F	1369
ABINGTON	P	36Ch SUV	Pull P-2494	3/11	3275	3821
ABOUKIR		12SecDr	Pullman	4/8/03	1581D	2984
ABOUKIR	N	18Ch3Sts	(Wagner)	6/84	3129	
ABRONIA		12SecDr	PCW	2/25/91	784B	1798
ABRONIA		16Sec	Pullman	10/25/01	1721	2766
ABSARAKA	RN	12SecDr	PCW Det	3/21/83	1467A	D184
ABSECON #38	RN		H&H (WS&PCC)	5/1/78	753	
ABSOL		14SecO	PCW	2/25/93	1021	1969
ABYLO		10SecDrB	PCW	12/31/91	920B	1880
ABYO		12SecDr	PCW	2/7/91	784B	1797
ABYSSINIA #404	P	12SecDr	Wagner	11/4/89	3067A	Wg28
ACACIA		12SecDr	Pullman	7/25/08	1963C	3631
ACACIA #38	N		H&H (WS&PCC)	5/1/78	753A	
ACADIAN	N	12SecDr	PCW	11/26/92	990	1941
ACADIAN		Private	Pullman	12/14/01	1671	2746
ACAMBARO		8SecDrB (NG)	PCW	2/18/96	1175	2161
ACAMBARO	N	12SecDrB	PCW Det	12/1/83	1468C	D212
ACAMPO		10Sec2Dr	Pullman	8/27/02	1746A	2925
ACANTHA		22Ch6StsB	PCW	8/13/89	645	GT9
ACANTHUS		10SecLibObs	Pullman	4/9/09	2351H	3688
ACESTES		10Sec2Dr	PCW	3/20/96	1173A	2153
ACHAIA	P	BaggBLibSmk	Wagner	5/93	3092	
ACHATES		10SecDrB	PCW	10/13/84	149D	1007
ACHATES	R	16Sec	PCW	10/13/84	1877A	1007
ACHILLES		BaggDiner	PCW	7/3/91	783B	1845
ACHILLES #220	P	Parlor	J&S	7/31/73	35	Wil1
ACHMET		14SecO	PCW Det	4/26/93	1021	D518
ACHRAY	N	DDR	PRR	9/70	17M	Alt1
ACHRAY	N	12SecDr	B&S (Wagner)	5/81	3022B	
ACILIUS		BaggDiner	PCW	9/30/90	783A	1705
ACILIUS	R	BaggBLib	PCW	10/1/90	920E	1705
ACILIUS	N	12SecDr	Pullman	5/23/08	1963C	3616
ACME		10SecDr	PCW Det	7/26/80	59	D89
ACOMA		10SecDr3Cpt	Pullman	4/19/06	2078C	3342
ACRA	P	12SecDrB	PCW (Wagner)	5/29/86	3050	
ACTIUM	N	8SecDrHotel	G&B (Wagner)	6/75	3091	
ACTON		12SecDr	Pull Buf	7/21/03	1581E	B20
ACTON	R	30Ch SUV	Pull Buf	7/21/03	3581E	B20
ADA	N	12SecO		114A?		
ADAIR	N	DDR Hotel	CB&Q RR	2/71	14	Aur1
ADAIR	R	12SecO	CB&Q RR	2/71	17	Aur1
ADAIR	N	12SecDr	PCW	/98	1318Q	2370
ADAIR	P	16SecO	Wagner	10/93	3072	Wg101
ADALBERT		12SecDrB	Pullman	4/27/08	2326	3605
ADALIA	N	12SecDrB	Wagner	12/92	3073	Wg65
ADALINA	N	26Ch3StsDr	PCW	4/28/98	1278C	2258
ADAMSDALE		12SecDr	Pullman	10/15/07	2163A	3562
ADAMSDALE	P	24Ch2Sts	B&S	11/93	3264	
ADANA		10SecDrB	PCW	10/25/87	445E	1334
ADAPTA		16Sec	Pullman	6/25/09	2184D	3700
ADDISON		12SecDr	Pull Buf	2/2/04	1581F	B32
ADELA		14Ch13StsDrB	PCW	7/23/87	505	1376
ADELA		24ChDr	Pull Buf	4/12/07	2156	B58
ADELAIDE		22Ch6StsB	PCW	8/13/89	645	GT9
ADELAIDE MOORE #101	N	Private	G&B (MannBCC)	/83	691	Troy
ADELANTE	N	10SecO (NG)	PCW Det	8/6/80	73A	D76
ADELANTE		12SecDrB	Pullman	4/23/08	2326	3605
ADELANTE	P	14SecDr SUV	Pull P-1855	1/03	3207	2900
ADELE	P	16SecO	Wagner	10/93	3072	Wg101
ADELINA PATTI #101	P	Private	G&B (MannBCC)	/83	691	Troy
ADELINE		20Ch4StsDrB	Pullman	6/11/01	1518E	2703
ADELINE #128	N	Parlor (3'gauge)	J&S (WS&PCC)	5/17 84	528?	
ADELPHI		Parlor	DetC&MCo		22	D2
ADELPHI	N	12SecDr	B&O (Wagner)	5/81	3022A	
ADELPHIA	P	10SecDr	H&StJ RR	7/31/73	19	Han2
ADEN		10SecDrB	PCW	12/10/89	725B	1629
ADEONA		24ChDr	Pullman	7/8/05	1980A	3171
ADERNO		12SecDr	PCW	1/17/91	784	1785
ADIRONDACK #218		Parlor	PCW Det	8/31/71	22	D7
ADIRONDACK #406	P	12SecDr	Wagner	1/90	3067A	Wg28
ADLER	N	10Sec2DrB	Wagner	10/99	3079	Wg124
ADMETUS		10SecLibObs	Pullman	4/9/09	2351H	3688
ADMIRAL	N	DDR	DetC&MCo	/70	17K	D3
ADMIRAL		12SecDrSr	PCW	5/23/99	1318C	2423
ADONIA		12SecDr	Pullman	7/22/08	1963C	3631
ADONIS		27StsParlor	PCW Det	/76	45	D31
ADONIS	P	Parlor	Bowers, Dure	6/30/75	35	Wil1
ADONIS		12SecDr	Pullman	11/16/08	1963E	3659
ADONIS	N	16SecO	Wagner	10/93	3072	Wg101
ADOUR #452		10SecDrB	PCW	12/30/85	240A	1144
ADRA	P	10Sec2DrB	Wagner	10/99	3079	Wg124

CAR NAME	ST	CAR TYPE	CAR BUILDER (Owner)	DATE BUILT	PLAN	LOT	CAR NAME	ST	CAR TYPE	CAR BUILDER (Owner)	DATE BUILT	PLAN	LOT
ADRASTUS		10SecLibObs	Pullman	4/9/09	2351H	3688	ALAMO	N	12SecDrSr	Pullman	1/25/01	1581A	2614
ADRIA	N	10SecDr	PCW Det	4/5/81	112	D149	ALAMOSA		10SecO (NG)	PCW Det	6/28/81	73A	D108
ADRIAN	P	10Sec2DrB	Pull Buf	2/15/00	3079A	Wg134	ALAMOSA	P	8CptLngObs	Wagner	11/97	3078A	Wg114
ADRIANA		20Ch6StsB	PCW	1/22/85	167B	200	ALANDALE		12SecDrB	PCW	8/22/96	1204C	2187
ADRIANA	N	8CptLngObs	Wagner	11/97	3078	Wg114	ALANNA	P	18Ch6StsB	Wason	7/88	3256	
ADRIANOPLE		12SecDr	Pullman	4/14/04	1963	3082	ALARIC	N	10SecDrB	J&S (Wagner)	11/82	3046B	
ADRIATIC		10SecDr	PCW Det	3/25/72	26A	D10	ALASKA	N	10SecB	(IC RR)		231	
ADRIATIC		12SecDrSr	PCW	7/12/99	1318C	2424	ALASKA	P	12SecDr	G&B (Wagner)	11/80	3022	
ADRIATIC #405	P	12SecDr	Wagner	1/90	3067A	Wg28	ALASKAN		12SecDrB	Pullman	4/25/08	2326	3605
ADRIENNE		20Ch4StsDrB	Pullman	6/17/01	1518E	2703	ALBA	P	10Sec2DrB	Wagner	10/99	3079	Wg124
ADVANCE		10SecO (NG)	PCW Det	8/6/80	73A	D76	ALBACORE		16Sec	Pullman	6/2/08	2184C	3623
ADVANCE		12SecDrObs	Pullman	8/28/06	2194	3395	ALBANIA		12SecDr	PCW Det	4/4/83	93B	D184
ADVENTURE		12SecDr	Pullman	10/6/08	1963E	3649	ALBANIA	P	10SecDrB	J&S (Wagner)	11/82	3046B	
ADVOCATE	N	5CptLibObs	Pullman	8/15/05	2089	3222	ALBANIA	P	Slpr	(PK&RSCC)			
AEOLUS		12SecDr	Pullman	1/22/02	1581A	2792	ALBANY	N	5CptDrLibObs	Pullman	10/29/02	1845F	2926
AFGHAN		12SecDrB	PCW	10/13/88	375F	1481	ALBANY		14SecDr	Pullman	3/13/03	1895	2982
AFRICA		16SecO	PCW	11/10/90	780	W5	ALBANY	P	28Ch5StsDr	Wagner	11/92	3125B	Wg74
AFTON	P	16SecO	Wagner	10/93	3072	Wg101	ALBANY O&C-8000	P	10SecDrB	PCW	5/22/85	240	1048
AFTON #370		10SecDrB	PCW	12/21/82	149	21	ALBATROSS		24ChDrB	Pullman	6/28/00	1518C	2553
AGATHA		18Ch4StsDrB	PCW	7/8/93	1039A	1995	ALBEMARLE		10SecDrHotel	PCW Det	4/15/72	33	D11
AGINCOURT		12SecDrSr	Pullman	2/13/01	1581B	2638	ALBEMARLE		8Sec4Cpt SUV	Pullman	2/6/06	2144	3300
AGNES		20Ch6StsDr	PCW	7/3/91	928A	1826	ALBERT						
AGRA		12SecDr	PCW	12/18/90	784	1780	EDWARD		27StsParlor	PCW Det	9/77	45	D31
AGRICOLA	N	12SecDr	Jones C&MC	8/82	3023B		LEA		12SecDr	PCW Det	7/26/82	93B	D162
AGUAS CALIENTES		10SecDrB	PCW Det	4/19/84	149D	D242	SYDNEY JOHNSTON	N	20Ch4StsObs	Pullman	7/15/05	2057B	3217
AIDA	P	10Sec2DrB	Wagner	10/99	3079	Wg124	VICTOR		40StsParlor	PCW	12/88		
AIDA #109	P	8BoudB	J&S (MannBCC)	/84	373B		ALBERTA		22ChDr	Pullman	6/17/02	1647F	2863
AIKEN	RN	16SecB	Pull Cal	9/10/02	1794	C3	ALBERTA	N	16Ch2StsDrB	(Wagner)	6/85	3115D	
AILEEN	P	28Ch3StsB	J&S	12/95	2367		ALBERTA #103		16Ch15Sts	PCW	8/88	638	
AINSLIE		12SecDr	Pull Cal	7/10/03	1581D	C5	ALBERTUS		7CptLngObs	Pullman	6/16/04	1977B	3091
AIRDRIE		12SecDr	Pull Cal	6/10/03	1581D	C5	ALBIA		12SecDr	PCW	11/22/90	784	1767
AIRLIE	N	10SecDrB	PCW	5/24/84	149D	192	ALBINA		12SecDrB	PCW	3/17/90	375J	1662
AJAX		12SecDr	PCW	10/15/90	784	1766	ALBION		SDR	DetC&MCo	/68	12D	Det
AJAX	P	10Sec2DrB	Wagner	10/99	3079	Wg124	ALBION	N	10SecDrB	PCW Det	2/11/84	149	D227
AKISTA	P	14Sec	B&S	12/82	3201		ALBION		27StsParlor	PCW Det	10/9/76	45	D31
AKRON		12SecDr	PCW Det	9/15/82	93B	D162	ALBION #71	P	13Ch9StsDr	(Wagner)	/82	3110	
ALABAMA		10Cpt	Pullman	1/4/02	1704	2758	ALBUQUERQUE		10SecDr	PCW Det	9/23/80	59	D78
ALABAMA #23	P	10SecDr	H&H	4/72	26A	7	ALBUQUERQUE		12SecDr	Pullman	4/20/07	1963C	3533
ALABAMA #98	P	Slpr	(IC RR)		2/127		ALCALDE		12SecDrSr	PCW	1/6/99	1318C	2379
ALABAMA #327	P	8BoudB?	J&S (UnionPCC)	/89	373?		ALCALDE		12SecDr	Pull Buf	8/31/07	1963C	B66
ALADDIN		8SecBObs	PCW	1/12/88	535	1407	ALCAMO		12SecDr	PCW	1/10/91	784	1785
ALADDIN	P	7CptObs	Wagner	7/92	3062C	Wg78	ALCAMO		14SecO	PCW	3/1/93	1021	1969
ALAMEDA		12SecDrSr	PCW	1/9/99	1318C	2379	ALCATRAZ		12SecDrSr	PCW	1/17/99	1318C	2379
ALAMEDA	P	8CptLngObs	Wagner	11/97	3078	Wg114	ALCATRAZ CP-30	P	10SecDr	(CP Co)	6/23/84	37K	
ALAMEDA GH&SA-4	P	Slpr	(SP Co)	7/1/83	37?		ALCAZAR		Diner	PCW	1/12/88	538C	1408
ALAMITOS		12SecDrB	PCW	9/6/88	375F	1480	ALCAZAR		12SecDrSr	PCW	1/5/99	1318C	2379
ALAMITOS		10SecDr2Cpt	Pullman	3/22/07	2078D	3528	ALCAZAR	P	8CptLngObs	Wagner	11/97	3078	Wg114
ALAMITOS	N	20Ch4Sts	(Wagner)	8/86	3102		ALCESTER		12SecDr	Pull Buf	6/2/04	1963	B37
ALAMO	N	10SecB	Bowers, Dure	10/17/73	37	Wil11	ALCETAS	N	12SecDr	Wagner	1/90	3067A	Wg28
ALAMO	N	6Sec4Dr	(WS&PCC)	4/8/86	720G		ALCIDES	N	10SecDrB	J&S (B&O)	4/88	663	

CAR NAME	ST	CAR TYPE	CAR BUILDER (Owner)	DATE BUILT	PLAN	LOT
ALCIDES	N	10SecDrB	J&S (Wagner)	3/82	3046B	
ALCIRA		12SecDr	Pull Buf	8/26/07	1963C	B66
ALCONA		12SecDrB	Pullman	4/23/08	2326	3605
ALCONA	P	12SecDr	B&S (Wagner)	/75	3008	
ALCYONE		12SecDr	Pullman	7/23/08	1963C	3631
ALDEBARAN #100	P	20Ch8StsDrB	Wagner	7/90	3126	Wg35
ALDEN		12SecDr	Pullman	10/8/03	1581F	3022
ALDEN	R	30Ch SUV	Pullman	10/8/03	3581E	3022
ALDERNEY		16SecO	PCW	6/10/93	1040	1990
ALDERSON		12SecDr	Pullman	1/9/07	2163	3449
ALDHAM		12SecDr	Pullman	10/10/07	2163A	3562
ALDINE	N	10SecB	(IC RR)		231	
ALDINE		12SecDr	Pullman	3/5/08	1963C	3600
ALECTO		12SecDr	Pullman	7/23/08	1963C	3631
ALENCON		12SecDr	Pullman	9/6/07	1963C	3550
ALENTO		12SecDr	Pullman	7/22/08	1963C	3631
ALEPPA		12SecDr	PCW	12/7/92	990	1941
ALERION		12SecDr	Pullman	10/6/08	1963E	3649
ALERT	N	10SecDrB	PCW Det	4/19/84	149D	D242
ALERT		12SecDr	Pullman	9/2/08	1963C	3646
ALESIA		10SecDrB	PCW	10/25/87	445E	1334
ALESIA	N	16SecO	Wagner	10/93	3072	Wg101
ALEXANDER		12SecDr	Pullman	5/3/01	1581A	2726
ALEXANDER	R	30Ch SUV	Pullman	5/3/01	3581E	2726
ALEXANDRA (1st)		27StsParlor	PCW Det	/77	45	D31
ALEXANDRA (2nd)	RN	34StsParlor	PCW Det	9/77	45	D31
ALEXANDRA (3rd)	N	27Sts Parlor	PCW Det	1/25/74	39	D20
ALEXANDRA		22Ch7StsDrObs	Pullman	6/13/02	1634B	2862
ALEXANDRA		12SecO	PCW Det	5/30/76	58	D42
ALEXANDRIA		12SecDr	Pullman	4/15/03	1581D	2984
ALEXANDRIA	P	12SecDr	G&B (Wagner)	11/80	3023A	
ALEXIA	N	26Ch3StsDr	PCW	4/25/98	1278C	2258
ALEXIS	N	DinerParlor	H&H	7/31/73	40C	Wil2
ALEXIS		12SecDrSr	PCW	1/14/99	1318C	2379
ALEXIS	P	10SecDrB	J&S (Wagner)	3/82	3046B	
ALFARATA		BaggLib	PCW	6/7/87	477	1325
ALFORD		12SecDr	Pullman	7/24/08	1963C	3631
ALFREDA	P	20Ch6StsDrB	J&S	1/02	2369	
ALGARDI		12SecDr	Pullman	7/27/01	1581A	2727
ALGERIA	N	OS	GW Ry/GMP	/65	13	Ham
ALGERIA		12SecDrSr	PCW'	1/16/99	1318C	2379
ALGERIA #14	P	20Ch12StsDr	G&B (Wagner)	5/74	3104	
ALGIERS		12SecDr	Pullman	7/7/03	1581F	3019
ALGINET	P	22ChDrB	(Big-4)	/98	2377	
ALGOMA	P	Hotel	B&S	/70		Day3
ALGOMA	P	12SecO	B&O (UnionPCC)	/80	665F	
ALGOMAH		12SecDr	Pullman	2/8/06	1963C	3316
ALGONQUIN		12SecDrSr	PCW	1/7/99	1318C	2379
ALHAMBRA		Diner	PCW	12/31/87	538C	1408
ALHAMBRA		12SecDrSr	PCW	12/19/98	1318C	2370
ALHAMBRA	P	8CptLngObs	Wagner	11/97	3078	Wg114
ALIANZA	N	10SecDrB	PCW Det	12/22/83	149	D213
ALIANZA		12SecDr	Pull Buf	8/29/07	1963C	B66
ALICANTE	P	10SecDr	Bradley	2/91	3200	
ALICE		28ChDr	Pullman	6/23/04	1986A	3087
ALICE	P	12SecO	LS&MS (Wagner)	1/73	3031	
ALICIA		22ChDr	Pull Buf	5/12/02	1647C	B8
ALINDA #161	N	18Ch7StsB	J&S (WS&PCC)	6/27/87	711A	
ALISA		20Ch6StsDr	PCW	7/24/91	928A	1826
ALIXE		14SecO	PCW	3/10/93	1021	1969
ALLAIRE		26ChDr	Pullman	5/25/07	1987B	3524
ALLEGHANEY			CNJ (SilverPC)	/66		
ALLEGHENY	P	10SecDr	PRR	2/22/72	31	Alt2
ALLEGHENY	N	7CptLngObs	PCW	10/25/95	1159A	2126
ALLEGHENY #84	N		(WS&PCC)	7/1/80		
ALLEGHENY #183	P	10SecB	PRR	9/74	37U	Alt2
ALLEGHENY #407	P	12SecDr	Wagner	1/90	3067A	Wg28
ALLEGHENY #565	P	11SecDr	PCW	2/23/85	222	1033
ALLEGHENY E&A-24	P	10SecDr	PRR (E&A Assn)	2/22/72	21	Alt2
ALLEGRIPPUS		12SecDr	Pullman	10/20/05	2049B	3232
ALLENDE		10SecDrB	PCW	5/7/87	445E	1292
ALLENDE	R	12SecDr	Pullman		1581C	C4
ALLENHURST		12SecDrSr	Pullman	4/8/05	2049A	3169
ALLENWOOD		26ChDr	Pullman	3/28/06	1987A	3335
ALLER		12SecDr	PCW Det	8/8/91	784D	D470
ALLERTON	N	12SecDr	Pullman	10/30/07	1963C	3570
ALLIANCE #131	P	DDR	PFtW&C	7/71	17	FtW2
ALLIANCE PRR-16		16Ch6StsDrCafe	Pullman	1/00	1536	
ALLOA	N	10SecDr	PCW Det	6/16/81	112	D118
ALLOWAY		28ChDr	Pullman	4/5/06	1986A	3336
ALMA		12SecDr	PCW Det	9/15/82	93B	D162
ALMA	P	12SecDrB	Wagner	10/92	3070	Wg76
ALMA #77	RN	28Ch	(WS&PCC)			
ALMAGRA	N	10SecDrB	J&S (Wagner)	3/82	3046B	
ALMEIDA		20Ch4StsDrB	Pullman	6/14/01	1518E	2703
ALMERIA		12SecDr	PCW Det	4/19/83	93B	D184
ALMERIA		12SecDr	Pull Buf	9/4/07	1963C	B66
ALMERIA	P	10SecDrB	J&S (Wagner)	3/82	3046B	
ALMONDE		Diner	PCW	11/1/90	662D	1776
ALMONDE		12SecDrSr	PCW	1/13/99	1318C	2379
ALOHA	N	22Ch6Sts	PCW	8-9/92	976B	1910
ALONZO		12Sec2Dr	Pullman	3/31/01	1580A	2646
ALOUETTE		12SecDr	Pullman	10/9/08	1963E	3649
ALPENA		12SecDr	Pullman	9/9/03	1581F	3022
ALPENA	P	12SecDr	B&S (Wagner)	11/85	3008	
ALPHA	RN	6SecBCh	PRR	11/70	17S?	Alt1
ALPHEUS		12SecDr	Pullman	10/15/08	1963E	3649
ALPINE		12SecDrSr	PCW	1/12/99	1318C	2379
ALPINE #15	P	OS	(SouTranCo)		49B	
ALPINE SPA-8242	P	10Sec2Dr	PCW	3/6/99	1158	2376

SCHAUMBURG TWP. DISTRICT LIBRARY

Figure I-9 ALPINE was a car on the roster of the Southern Transportation Company. In this circa 1896 photo, it was near the end of its service. Note the berths are numbered and are opened with handles instead of a berth key. Smithsonian Institution Photograph P2301.

CAR NAME	ST	CAR TYPE	CAR BUILDER (Owner)	DATE BUILT	PLAN	LOT
ALROY		8SecBObs	PCW	12/31/87	535	1407
ALROY	N	12SecDrSr	PCW	12/10/99	1318C	2370
ALROY	N	7CptObs	Wagner	7/92	3062C	Wg78
ALSACE		10SecDrB	PCW Det	8/31/83	149	D206
ALSACE	R	16Sec	PCW Det	8/31/83	1877C	D206
ALSATIA		12SecDr	PCW Det	5/17/82	93B	D154
ALSATIA	P	10SecDrB	J&S (Wagner)	3/82	3046B	
ALSTON	R	12SecDr	Pullman		1581F	
ALSTON		12SecDr	Pull Buf	7/30/03	1581E	B20
ALTADENA		12SecDr	Pull Buf	7/22/07	1963C	B64
ALTAIRE		14SecO	PCW Det	3/25/93	1021	D517
ALTAMAHA		10SecO	PCW Det	12/15/74	47	D27
ALTAMAHA		12SecDr	Pull Buf	7/25/07	1963C	B64
ALTAMONT		12SecDrSr	PCW	1/10/99	1318C	2379
ALTAMONT		12SecDr	Pull Buf	5/31/04	1963	B37
ALTAMONT #99	P	20Ch8StsDrB	Wagner	7/90	3126	Wg35
ALTAMONT #199		OS	B&O RR	4/30/73	27 1/2	B&O2
ALTAMONT #722	P	20Ch12StsB	B&S		678	
ALTAMONT SP-40	P	10SecDrB	PCW	3/12/87	445B	1299
ALTANO		12SecDr	Pull Buf	8/5/07	1963C	B64
ALTATA	N	10SecObs	PCW	5/5/88	445W	1414
ALTATA SP-58	P	10SecDrB	PCW	5/5/88	445F	1414
ALTENWALD		12SecDr	Pull Buf	7/29/07	1963C	B64
ALTHEA	P	16Ch6StsDrObs	(B&MRR)		1202	
ALTON	P	10SecDr	B&S	10/71	26	Day6
ALTON		16SecO	Pullman	4/25/05	2068	3177
ALTOONA		12SecDrSr	Pullman	1/17/01	1581A	2614
ALTOONA			CNJ (SilverPC)	/66		
ALTOONA #133	P	DDR	PRR	9/70	17	Alt1
ALTRURIA		12SecDr	Pull Buf	8/1/07	1963C	B64
ALTURAS		12SecDr	Pull Buf	9/7/07	1963C	B66
ALTURAS SP-39	P	10SecDrB	PCW	3/8/87	445B	1299
ALTUS		10SecDrSr	PCW	6/16/90	784A	1715
ALUMINA		12SecDr	Pullman	7/20/08	1963C	3631
ALVA		10SecDrB	PCW	5/27/91	920B	1808
ALVA	P	12SecDrB	Wagner	10/92	3070	Wg76
ALVARADO	N	12SecDrSr	PCW	1/10/99	1318C	2379
ALVARADO	R	30Ch SfV	PCW	1/10/99	3581M	2379
ALVEDO		10SecDrB	PCW	9/17/87	445E	1362
ALVEDO	R	10SecObs	PCW	9/17/87	445W	1362
ALVERTON		12SecDrSr	Pullman	3/30/05	2049A	3169
ALVIN	N	12SecDr	Pullman	10/15/07	2163A	3562
ALVISO	R	12SecDrB	PCW	2/18/87	1566	1298
ALVISO SP-38	P	10SecDrB	PCW	2/18/87	445A	1298
ALYSSUM		24ChDrSr	Pullman	7/14/04	1980A	3088
AMADEO		12SecDr	Pullman	8/9/01	1581A	2727
AMADEUS	N	16SecO	Wagner	10/93	3072	Wg101
AMADOR SPA-8243	P	10Sec2Dr	PCW	3/8/99	1158	2376
AMAGANSETT #94	P		J&S (WS&PCC)	7/23/81		
AMANITA		12SecDr	Pullman	10/7/08	1963E	3649

CAR NAME	ST	CAR TYPE	CAR BUILDER (Owner)	DATE BUILT	PLAN	LOT
AMANTIA		12SecDr	Pullman	6/8/03	1581E	3010
AMANUS	N	12SecDrSr	PCW	6/29/99	1318C	2424
AMAPALA		12SecDr	Pullman	9/19/06	1963C	3397
AMARANTH		24ChDr	Pull Buf	3/26/06	2156	B50
AMARILLO		12SecDr	Pull Buf	5/25/04	1963	B37
AMARYLLIS		20Ch5StsB	PCW	8/8/84	167A	199
AMARYLLIS	N	12SecDrB	Wagner	10/92	3070	Wg76
AMAZON #390		12SecB	PCW Det	6/20/83	168	D154
AMAZONIA		12SecDr	Pullman	9/1/00	1572	2585
AMBASSADOR	N	12SecDr	PCW	6/7/87	473A	1323
AMBIA	N	12SecDr	PCW	10/1/98	1318B	2339
AMBOY	P	Composite	(IC RR)			
AMBOY	N	28Ch5StsDr	Wagner	11/92	3125B	Wg74
AMBRIDGE		BaggClub	Pullman	1/21/07	2136A	3461
AMBROSIA		20Ch3StsB	PCW Det	9/22/83	167A	D209
AMEER	RN	6SecCh	PRR	7/28/73	29D	Alt1
AMELIA		18Ch4StsDrB	PCW	6/20/93	1039A	1994
AMELIAN	N	12SecDr	Pullman	/01	1581A	2726
AMERICA	P	12SecO	C&NW RR	12/70	5	Fdl
AMERICA	P	DDR	Ohio Falls	1/12/70	17	Jef
AMERICA		12SecDr	PCW	4/12/87	473A	1323
AMERICA		10SecDrSr	PCW	4/23/93	1011A	1968
AMERICA	N	Diner	PCW	4/23/93	1030A	1976
AMERICAN		10SecDrB	Grand Trunk	11/10/87	445E	GT8
AMERICANO		10SecO (NG)	PCW Det	7/12/80	73A	D76
AMETHYST		12SecDrSr	PCW	4/8/99	1318C	2390
AMHERST	N	10SecB	Grand Trunk	7/31/74	37	Mon5
AMHERST		12SecDr	PCW Det	12/18/90	784	D455
AMHERST		12SecDr	Pullman	11/24/05	1963C	3252
AMIENS		12SecDr	PCW	12/20/92	990	1941
AMIRANTE		12SecDrB	PCW	9/12/88	375F	1480
AMITE #38	P	Slpr	(IC RR)			2/237
AMOMO		12SecDr	Pullman	10/8/08	1963E	3649
AMON	N	12SecDrB	Wagner	8/91	3065	Wg54
AMORET		26ChDrSr	Pullman	6/3/04	1987A	3085
AMORITA		12SecDr	Pullman	10/8/08	1963E	3649
AMPARO	N	10SecDrB	PCW	12/20/88	445E	1519
AMPERSAND #523	P	12SecDrB	Wagner	8/92	3070	Wg72
AMPHION		12Sec2Dr	PCW Det	1/2/88	537A	D334
AMPHION	N	12SecDrSr	Pullman	1/26/01	1581A	2614
AMPHORA		12SecDr	Pullman	12/11/06	1963C	3448
AMSBRY		26ChDr	Pullman	5/25/07	1987B	3524
AMSTERDAM		12SecDr	Pullman	8/26/01	1581A	2756
AMURATH	N	12SecDr	Pullman	5/19/08	1963C	3616
AMURATH	N	12SecDr	G&B (Wagner)	11/80	3022	
ANACONDA		10SecDrCpt	PCW	5/8/90	668A	1709
ANACORTES		10SecDrSr	PCW	8/12/91	937A	1838
ANAHUAC		8SecSrB (NG)	PCW	2/18/96	1175	2161
ANAKRON	N	12SecDrSr	Pullman	1/14/01	1581A	2614
ANAN #451		10SecDrB	PCW	12/26/85	240A	1144
ANASTASIA	N	27ChDrObs	Pullman	6/8/08	2346	3624
ANATOLI		14SecO	PCW	5/16/93	1021	1972
ANCHISES		10Sec2Dr	PCW	3/3/96	1173A	2153
ANCHORIA		12SecDr	PCW Det	12/21/81	93B	D135
ANCILLA	RN	10SecO	PCW Det (rblt)	12/70/75	47	D29
ANCONA	N	10SecDr	Grand Trunk	6/72	31C	Mon3
ANCORA		24ChDrB	Pullman	6/5/05	1988A	3170
ANDALUSIA		12SecDr	PCW Det	5/31/82	93B	D154
ANDAMAN		12SecDr	Pull Buf	3/9/03	1581C	B14
ANDARA	N	12SecDr	Pullman	3/14/06	2163A	3333
ANDES #389		12SecB	PCW Det	6/19/83	168	D193
ANDORNO		12SecDrB	Pullman	4/27/08	2326	3605
ANDOVER	N	10SecDrB	PCW Det	1/9/83	149	D176
ANDOVER	R	16Sec	PCW Det	1/9/83	1877C	D176
ANDOVER #83	N		(WS&PCC)	7/1/80		
ANDROMEDA		16Ch6StsB	PCW Det	8/3/76	54	D40
ANDROMEDA		22Ch7StsDrObs	Pullman	6/13/02	1634B	2862
ANELO		14SecO	PCW	2/4/93	1017C	1962
ANEMONE		12SecDr	PCW	2/24/91	784B	1798
ANEMONE		12SecDrB	Pullman	4/25/08	2326	3605
ANGELICA		26Ch4StsObsB	Pullman	8/6/04	1993B	3111
ANGELINE	P	18Ch3StsDrObs	(B&M RR		1203	
ANGELO		7Cpt2Dr	PCW	5/26/97	1065A	2226
ANGELUS	N	14Sec	(Wagner)	6/86	3049C	
ANGLESEA		12SecDr	Pullman	8/20/01	1581A	2756
ANGOLA		12SecDr	PCW	1/8/91	784	1785
ANGORA		10SecDrB	PCW Det	12/29/83	149	D223
ANGORIA	N	12SecDr	Wagner	1/90	3067A	Wg28
ANICIUS		10SecLibObs	Pullman	4/9/09	2351H	3688
ANITA	P	16SecO	Wagner	10/93	3072	Wg101
ANITA #101	N	12SecDr	(MannBCC)	/89	878	
ANITRA	N	24ChDrB	Pullman	6/3/05	1988A	3170
ANJOU		12SecDr	PCW Det	9/18/91	784D	D481
ANKONA	N	12SecDr	Pullman	10/21/05	2049B	3232
ANNABEL		24ChDr	Pull Buf	2/23/07	2156	B57
ANNAN	N	10SecDrB	PCW	12/26/85	240A	1144
ANNANDALE	P	12SecDr	G&B (Wagner)	11/80	3023	
ANNAPOLIS	N	10SecDrB	PCW	9/27/84	149D	1006
ANNAPOLIS		12SecDr	Pull Buf	5/16/04	1963	B37
ANNAPOLIS #312	P	10SecB	B&S (B&O)	7/31/75	37	Day11
ANNE		18Ch4StsDrB	PCW	7/15/93	1039A	1995
ANNETTE		14Ch13StsDrB	PCW	7/23/87	505	1376
ANNETTE		24ChDr	Pull Buf	2/22/07	2156	B57
ANNIE #78	P		(WS&PCC)	2/17/79		
ANNISQUAM		12SecDr	Pullman	10/8/08	1963E	3649
ANNISTON	N	10SecO	PCW Det (rblt)	11/20/74	47A	D27
ANNISTON		12SecDr	PCW	11/1/98	1318B	2340
ANNISTON #180	RN	12Ch36Sts SUV	Pull Buf	10/3/03	3581B	B23
ANOKA		12SecDr	Pullman	6/12/07	2271A	3458
ANSELM	N	12SecDrSr	Pullman	1/14/01	1581A	2614

CAR NAME	ST	CAR TYPE	CAR BUILDER (Owner)	DATE BUILT	PLAN	LOT
ANSHAN	N	12SecDrSr	Pullman	3/1/01	1581B	2654
ANSLEY		12SecDr	Pullman	7/20/08	1963C	3631
ANSONIA		12SecDr	Pullman	7/20/03	1581F	3015
ANSONIA #84	N		(WS&PCC)	7/1/80		
ANTARES		14SecO	PCW	5/23/93	1021	1972
ANTELOPE	R	OS	G,B&Co/GMP	/68	27A	Troy
ANTELOPE #29		12SecO	G,B&Co/GMP	/68	5	Troy
ANTHARIS	N	12SecDr	G&B (Wagner)	11/80	3022	
ANTHEA	P	14SecDr SUV	B&S	6/11	3207C	
ANTHONY	P	10SecB	H&StJ RR	8/5/71	14B	Han2
ANTHONY	R	10SecDr	H&StJ RR	8/71	19	Han2
ANTHONY		BaggBLibSmk	PCW	5/10/92	970	1905
ANTHRACITE	N	20Ch6StsDrObs	Pullman	6/16/04	1979A	3084
ANTICOSTI	P	10SecDr	Grand Trunk	10/73	31	Mon4
ANTILLES		10SecDrB	PCW Det	5/2/84	149D	D242
ANTILLES	R	12SecDrB	PCW Det	5/2/84	1611	D242
ANTIOCH		12SecDr	Pullman	7/21/02	1581B	2852
ANTIOCH #132	RN	7BoudB	(MannBCC)	/84	776A	
ANTIOPE #116	RN	7BoudB	(MannBCC)	/84	697A	
ANTONIO		12SecDrCpt	Pullman	5/23/04	1982	3090
ANTONITO		10SecO (NG)	PCW Det	10/3/81	73A	D108
ANTRIM		14SecO	PCW Det	4/26/93	1021	D518
ANTWERP		10SecDrHotel	PCW Det	11/22/80	63 1/2	D77
APACHE		12SecDr	Pullman	12/10/06	1963C	3448
APELLES		10Sec2Dr	PCW	7/7/96	1173A	2153
APHRODITE		26Ch3StsDr	PCW	4/27/98	1278C	2258
APHRODITE		12SecDr	Pullman	9/27/09	1963E	3744
APOLLO		27Sts Parlor	PCW Det	/76	45	D31
APOLLO		7Cpt2Dr	PCW	5/24/97	1065A	2226
APOLLO #213	P	DDR	CB&Q RR	12/9/71	34A	
APPALACHIA	N	BaggBLibSmk	Pullman	4/24/07	2087	3534
APPIAN		14SecDr	Pullman	6/11/00	1523	2559
APPIUS		BaggDiner	PCW	10/2/90	783A	1705
APPIUS	R	BaggBLib	PCW	10/4/90	920E	1705
APPLETON		12SecDr	Pullman	3/1/02	1581B	2818
APPONAUG	P	34Ch SUV	Pull P-1484A	6/00	3265	2518
APULIA		16Sec	Pullman	10/18/01	1720A	2766
ARABELLA	N	18Ch4StsB	PCW	6/20/93	1039A	1994
ARABIA	N	10SecDr	PRR	2/72	26	Alt2
ARABIA		12SecDrSr	PCW	6/27/99	1318C	2424
ARABIA #22	P	15Ch8StsDrB	(Wagner)	4/85	3115	
ARABIAN		10SecDr	H&H	2/71	19	Wil2
ARABIAN	P	10Sec2DrB	PCW (Wagner)	2/14/00	3079A	Wg134
ARADUS	N	12SecDr	PCW	8/1/98	1318A	2313
ARAGON		10SecDrB	PCW Det	4/21/84	149D	D242
ARAGON		12SecDr	Pullman	11/16/08	1963E	3659
ARAGON	P	12SecDr	B&S (Wagner)	5/81	3022A	
ARAMINTA	N	22Ch3StsB	PCW	/84	167Q	
ARAMIS	P	16Ch4StsB	B&S	8/82	3252B	
ARANCO		12SecDr	Pullman	7/24/08	1963C	3631

CAR NAME	ST	CAR TYPE	CAR BUILDER (Owner)	DATE BUILT	PLAN	LOT
ARANDA	N	14SecO	Elmira Car W	6/30/73	37A	Elm1
ARANDA		12SecDrB SUV	Pullman	4/25/08	2326	3605
ARANSAS	RN	12SecDrB	PCW	3/12/87	1566	1299
ARAPAHOE		12SecDr	PCW Det	9/8/82	93	D169
ARAPAHOE		12SecDr	PCW Denver	9/23/03	1581F	Den1
ARBACES		10Sec2Dr	PCW	5/20/96	1173A	2153
ARBELA	N	28Ch5StsDr	Wagner	9/89	3125B	Wg18
ARBELLA	N	16Ch2StsDrB	(Wagner)	6/85	3115C	
ARBUTUS		24ChDrSr	Pullman	7/13/04	1980A	3088
ARCADIA		BaggBSmkBarber	PCW	10/3/95	1155A	2122
ARCADIA	P	12SecDr	B&O (Wagner)	5/81	3022A	
ARCADIA #233		Parlor	PCW Det	7/14/73	35 1/4	D17
ARCHDALE		12SecDrB	Pullman	4/23/08	2326	3605
ARCOLA	N	12SecDr	Pull Buf	7/29/07	1963C	B64
ARCOLA	N	12SecDr	Wagner	7/88	3061B	Wg12
ARCOLA #268	P	10SecDr	J&S	12/72	26	Wil8
ARCTIC	RN	14SecO	PRR	9/70	17N	Alt1
ARCTURUS	RN	10SecLibObs	PCW Det	8/2/92	1014C	D509
ARDARA		26ChDr	Pullman	3/26/06	1987A	3335
ARDEN		12SecDr	PCW	8/3/91	784C	1828
ARDEN		12SecDr	Pullman	3/25/03	1581D	2983
ARDEN	N	14SecO	(Wagner)	/86	3032A	
ARDENHEIM		12SecDr	Pullman	10/16/07	2163A	3562
ARDENNES		12SecDrSr	Pullman	2/6/01	1581B	2638
ARDENT		12SecDr	Pullman	8/7/08	1963C	3636
ARDFERN		12SecDr	Pullman	7/25/08	1963C	3631
ARDMORE		12SecDrSr	Pullman	1/14/01	1581A	2614
ARDRION	N	26StsDrLngObs	Pullman	6/18/04	1417K	3083
ARDSLEY		12SecDr	Pullman	11/17/03	1581F	3043
ARECIBO		12SecDr	Pullman	8/15/04	1963	3098
ARENTIS		12SecDrB	Pullman	4/27/08	2326	3605
ARETHUSA		22ChDr	Pull Buf	5/12/02	1647C	B8
ARETINO		12SecDr	Pullman	8/2/01	1581A	2727
ARETINO	N	12SecDr	Pullman	5/16/08	1963C	3616
ARGENTA		10SecDrB	PCW	12/6/86	445C	1261
ARGENTA	N	12SecDrSr	PCW	1/12/99	1318C	2379
ARGENTINE		12SecDr	Pullman	7/22/02	1581B	2852
ARGO		10SecO (NG)	PCW Det	8/6/80	73A	D76
ARGO	P	16SecO	Wagner	10/93	3072	Wg101
ARGO #189	N	10SecB	PRR	9/74	37M	2
ARGONAUT	N	10SecDrB	PCW	4/3/88	445F	1377
ARGONAUT		10SecB	PCW	10/1/84	209	202
ARGONAUT		12SecDr	Pullman	5/19/08	1963C	3615
ARGONIA		12SecDrSrB	PCW	2/24/99	1319A	2400
ARGOSY		20Ch5StsB	PCW	11/24/83	167A	89
ARGUS		14SecO	PCW Det	8/30/81	115	D114
ARGYLE		10SecDrB	PCW	8/11/87	445E	1333
ARGYLE #84	P	28Ch4Sts	PCW (Wagner)	7/29/86	3124	1200
ARGYLE #111	P		(WS&PCC)	7/83		
ARIADNE		16Ch6StsB	PCW Det	7/12/76	54	D40

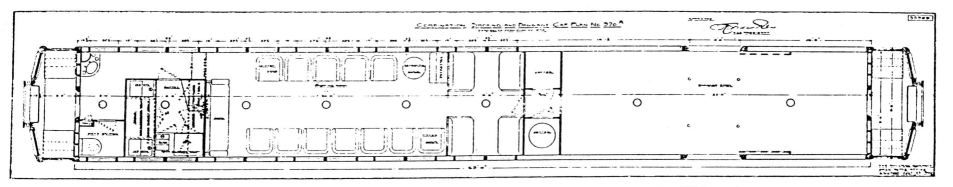

Figure I-10 ANTHONY was 1 of 3 cars built under Plan 970, Lot 1905 at The Pullman Car Works in May 1892 for service on the "Colonial Express." All three were sold to the Chicago & Alton in July 1920. The Newberry Library Collection.

CAR NAME	ST	CAR TYPE	CAR BUILDER (Owner)	DATE BUILT	PLAN	LOT	CAR NAME	ST	CAR TYPE	CAR BUILDER (Owner)	DATE BUILT	PLAN	LOT
ARIADNE		22Ch6StsDrObs	Pull Buf	5/27/03	1634D	B17	ARONA	N	16Sec	Pullman	7/15/03	1721	3014
ARIADNE	N	16Sec	Pullman	7/10/03	1721A	3014	AROOSTOOK	N	12SecDr	Pullman	10/5/01	1581A	2765
ARIEL		27StsParlor	PCW Det	8/76	45	D31	AROOSTOOK #85	P	28Ch3Sts	PCW (Wagner)	6/14/84	3124	D256
ARIEL #238	P	Parlor	Bowers,Dure	11/72	35	Wil1	ARPINO		12SecDr	PCW	1/31/91	784	1787
ARIEL #470	P	12SecDrB	Wagner	8/91	3065	Wg54	ARRAN		14SecO	PCW	3/2/93	1021	1969
ARIES		14SecO	PCW Det	3/25/93	1021	D517	ARSARTA		10secDr2Cpt SUV	Pullman	3/18/10	2471	3784
ARIMO		12SecDr	PCW Det	9/23/91	784D	D481	ARSENE	N	26ChDrB	Pullman	6/26/07	1988B	3526
ARION	N	12SecDr	PCW (Wagner)	6/5/87	3049A		ARTEMIS		26StsDrLngObs	Pullman	6/18/04	1417I	3083
ARISTIDES	·	12SecDrB	PCW	9/7/85	191	1074	ARTHURET	N	10SecB	PCW Det	10/7/72	26Y	D13
ARISTIDES	R	12SecDrB	PCW	9/7/85	1611	1074	ARTIGAN	P	20Ch6StsDr	B&S	7/89	3258	
ARISTON	N	12SecDr	G&B (Wagner)	11/80	3023A		ARTLAND	N	12SecDr	Pullman	12/18/09	1963E	3766
ARIUS #471	P	12SecDrB	Wagner	8/91	3065	Wg54	ARTOIS	P	36Ch SUV	Pull P-2494	3/11	3275	3821
ARIZONA	N	12SecO	Nash&ChatRR	/67	5		ARUNDEL		14SecO	PCW Det	5/27/93	1021	D525
ARIZONA		12SecDr	Pull Buf	12/12/05	1963C	B49	ARVA		10SecDr	PCW	12/9/89	717B	1630
ARIZONA	P	10SecO	MC RR (Wagner)	12/67	3000		ARVA	P	12SecDr	PCW (Wagner)	6/5/87	3049A	
ARKANSAS	N	12SecDr	Pullman	7/22/08	1963C	3631	ARVERNE #118	N		(WS&PCC)	7/83	395?	
ARKANSAS		10Cpt	Pullman	1/8/02	1704	2758	ASBURY		12SecDr	Pullman	10/14/01	1581F	3025
ARKANSAS #108	N	20Ch11StsDr	(WS&PCC)	/73	570?		ASBURY PARK		24ChDr	Pull Buf	5/3/06	2156	B52
ARLEE	N	10SecDrCpt	PCW	5/8/90	668A	1709	ASCALON	N	10SecDrB	J&S (B&O)	4/88	663	
ARLETTA	N	24Ch6StsDr	(Wagner)	7/86	3100		ASCOLI		12SecDr	PCW	1/7/91	784	1785
ARLINE		30Ch	Pullman	5/16/01	1646D1	2700	ASCOT		12SecDr	PCW	5/22/91	784D	1806
ARLINGHAM		12SecDr	Pullman	5/2/08	1963C	3616	ASCUTNEY		12SecDr	Pullman	9/10/03	1581F	3022
ARLINGTON	P	DDRHotel	CB&QRR	/70	17	Aur1	ASHBY		12SecDr	PCW	6/12/91	784D	1806
ARLINGTON		12SecDr	PCW	10/21/98	1318B	2340	ASHCROFT		16SecO	Pullman	5/1/05	2068	3177
ARLINGTON	P	12SecDr	B&S (Wagner)	5/81	3022A		ASHEFOO	RN	10SecO	PCW Det (rblt)	12/20/75	47	D29
ARMADA		20Ch3StsB	PCW Det	10/4/83	167A	D209	ASHEVILLE		12SecDr	Pullman	10/15/03	1581F	3025
ARMAGH #412		10SecDrB	PCW (UnionPCC)	11/3/83	149	D207	ASHFORD		12SecDrB	Pullman	4/30/08	2326	3605
ARMENIA		12SecDr	PCW Det	12/30/81	93B	D135	ASHLAND WC- 45	P	12SecO	CentCarCo	2/82	532	
ARNO	R	12SecDrB	PCW	/83	1468	21	ASHLAND WC- 45	R	10SecDrCafe	CentCarCo	2/82	532B	
ARNO	P	10Sec2DrB	Wagner	10/99	3079	Wg124	ASHLAND O&C-8001	P	10SecDrB	PCW	6/2/85	240	1048
ARNO #369		10SecDrB	PCW	1/17/83	149	21	ASHLEY	N	12SecDr	PCW	5/6/91	784D	1805
ARONA		12SecDr	PCW	11/21/90	784	1767	ASHLEY FALLS	P	BaggSmk	B&S	8/92	3213A	

CAR NAME	ST	CAR TYPE	CAR BUILDER (Owner)	DATE BUILT	PLAN	LOT
ASHMORE		12SecDr	Pullman	12/11/00	1963C	3448
ASHTABULA		12SecDr	Pullman	4/20/07	1963C	3533
ASHTABULA	P	10SecO Tourist	(Wagner)		3037	
ASHTON		12SecDr	PCW	8/1/91	784C	1828
ASHUMET		12SecDr	Pullman	10/15/08	1963E	3649
ASIA		14SecO	PCW Det	12/9/70	20	D4
ASIA		12SecDrCpt	PCW	10/4/89	622A	1529
ASIATIC		14SecDr	PCW	8/115/99	1318D	2447
ASPEN		10SecDr	PCW	11/1/87	508B	1368
ASPERN	N	16SecO	Wagner	10/93	3072	Wg101
ASPINWALL		12SecDr	Pullman	1/11/04	1581H	3053
ASPIRANT		16Sec	Pullman	6/5/06	2184	3361
ASSYRIA	N	16SecO	J&S (Wagner)	3/82	3072A	Wg108
ASSYRIA #208	P	Parlor	PRR	7/6/72	29	Alt1
ASTER		24ChDrSr	Pullman	7/15/04	1980A	3088
ASTEROID		12SecO	PCW Det	7/30/81	114	D113
ASTOLFO		12SecDrCpt SUV	Pullman	5/23/04	1982	3090
ASTOLFO #106	N	29Ch	J&S (WS& PCC)	7/1/82	750F	
ASTOR	N	Diner	PCW	11/12/89	662D	1641
ASTORIA		10Sec2Dr	PCW	2/18/93	995A	1942
ASTRILD		12SecDr	Pullman	10/9/08	1963E	3649
ATALANTA	N	10SecDrHotel	PCW Det	6/12/77	63	D45
ATCHISON	N	OS	B&S	/67	27	
ATCHISON		12SecDrB	Pullman	3/20/02	1611B	2820
ATGLEN		12SecDrSr	Pullman	3/30/05	2049A	3169
ATHELSTAN		12SecDr	Pullman	5/3/01	1581A	2726
ATHEMON		12SecDr	Pullman	8/1/01	1581A	2727
ATHENE		26Ch3StsDr	PCW	4/29/98	1278C	2258
ATHENIA #468	P	12SecDrB	Wagner	8/91	3065	Wg54
ATHENIAN	P	10Sec2DrB	PCW (Wagner)	3/16/00	3079A	Wg135
ATHENS		8Sec5rB (NG)	PCW Det	11/4/82	155	D173
ATHENS		12SecDr	Pullman	7/23/02	1581B	2852
ATHERTON		10SecDrB	PCW Det	4/3/72	33	D11
ATHERTON	R	12SecDr	Pullman		1581F	
ATHERTON		12SecDr SUV	Pull Buf	6/26/03	1581C	B19
ATHLONE		12SecDr	PCW Det	1/28/92	784E	D488
ATHOL	N	10SecB	Grand Trunk	7/31/74	37	Mon5
ATHOL		10SecDrB	PCW	11/16/86	445	1262
ATHOS		12SecDr	PCW	2/10/92	784E	1881
ATKA	P	10Sec2DrB	Wagner	10/99	3079	Wg124
ATLANTA	P	OS	(PK&RSCC)	/73	74	
ATLANTA		29ChDr	Pullman	7/16/05	2061	3217
ATLANTA	P	12SecDr	G&B (Wagner)	11/80	3022	
ATLANTA #341	P	10SecDrB	(UnionPCC)		565B	
ATLANTIC	N	BaggBSmkBarber	PCW	4/23/93	1029A	1975
ATLANTIC #41	P	12SecDr	B&S	/76-/78		
ATLANTIC #106	P	26Ch	J&S (WS&PCC)	7/1/82	750	
ATLANTIC E&A-35	P	Parlor	(E&A Assn)			
ATLANTIC #1	P	OS	CB&Q/GMP	2/66	13C	Aur
ATLANTIC #2		DDR	PCW Det	11/8/70	17	D4
ATLANTIC AVE. #128	P	Parlor (3' gauge)	J&S (WS&PCC)	5/17/84	528?	
ATLANTIS		26StsDrLngObs	Pullman	6/18/04	1417I	3083
ATLAS	P	10SecDr	B&S	12/71	26	Day7
ATOKA	N	Slpr	(IC RR)		X2/81	
ATOSSA		24ChDr	Pullman	6/26/05	1980A	3171
ATOYAC		10SecDrB	PCW	5/17/88	445F	1413
ATREUS	N	12SecDr	Pullman	5/20/08	1963C	3616
ATREUS	R	BaggSmkBLib	PCW?		1706	
ATRICILLA	P	12SecSr SUV	Pull P-2488	10/10	3209	3809
ATSINA		14SecDr	Pullman	8/20/09	2438	3716
ATTALIA	N	10SecDrB	PCW	6/26/85	240	1057
ATTICA	N	DDR	PFtW&C	7/71	17	FtW2
ATTICA #469	P	12SecDrB	Wagner	8/91	3065	Wg54
ATTICUS	P	10SecB	H&StJ RR	8/5/71	14B	Han2
ATTICUS		6CptDrLibObs	Pullman	3/2/03	1751B	2986
ATTLEBORO	P	34Ch SUV	Pull P-1484A	6/00	3265	2518
ATWATER		BaggClub	Pullman	1/21/07	2136A	3461
AUBREY		10SecB	PCW	6/5/84	206	198
AUBURN	P	DDR	DetC&MCo	C:/70	17	D3
AUBURN	P	12SecDr	G&B (Wagner)	11/80	3023	
AUCILLA	RN	10SecO	PCW Det (rblt)	12/20/75	47	D29
AUCILLA	N	16Sec	Pullman	7/7/03	1721	3014
AUCKLAND		12SecDrB	PCW	1/16/87	446A	1260
AUDREY	N	26ChB	PCW (Wagner)	5/29/86	3124A	1194
AUDUBON		10SecDrCpt	PCW	6/5/89	668	1587
AUGUSTA	N	10SecDr	B&S	4/71	19	Day5
AUGUSTA	RN	16SecB	Pull Cal	9/16/02	1794	C3
AUGUSTANA		12SecDr	Pullman	11/24/05	1963C	3252
AUGUSTIN		12SecDr	Pullman	8/7/01	1581A	2727
AURANIA #20	P	15Ch8StsDrB	(Wagner)	4/85	3115A	
AURANIA #407		12SecDrB	PCW	2/23/84	180	100
AURELIA		20Ch6StsDr	PCW	7/23/91	928A	1827
AURELIA #77	P	Slpr	H&H (WS&PCC)	7/1/79	524?	
AURELIAN		12SecDr	Pullman	5/3/01	1581A	2726
AURIGA		12SecDr	PCW Det	2/27/91	784B	D460
AURIGA	N	24StsDrSrB	Pullman	6/16/04	1988A	3086
AURORA	R	12SecDr	CB&Q/GMP	2/66	13B	Aur
AURORA	P	OS	CB&Q/GMP	2/66	7	Aur
AURORA		27StsParlor	PCW Det	4/29/76	45	D31
AURORA	P	18Ch3Sts	(Wagner)	6/84	3129	
AUSTERLITZ #387		10SecDr	PCW	4/83	112A	37
AUSTIN	RN	OS	I&StLSCCo	6/69	27 3/4	
AUSTIN		12SecDr	Pullman	2/23/04	1963	3070
AUSTIN	P	18Ch3Sts	(Wagner)	6/84	3129	
AUSTRALIA		14SecO	PCW Det	12/9/70	20	D4
AUSTRALIA		22Berths	PCW Det	6/20/76	48	D32
AUSTRALIA		12SecDrCpt	PCW	3/15/89	622	1529
AUSTRALIA	N	16Sec	Pullman	10/22/01	1720A	2766
AUSTRIA		12SecDr	PCW	6/7/87	473A	1323
AUSTRIA		12SecDr	Pullman	10/5/01	1581A	2765
AUTHION	RN	14Sec	PCW	12/26/89	717F	1630

CAR NAME	ST	CAR TYPE	CAR BUILDER (Owner)	DATE BUILT	PLAN	LOT
AUTOCRAT	P	10SecDr	J&S		37	Wil10
AUVERGNE		12SecDrB	PCW	11/17/88	375G	1513
AUVERGNE	P	14SecDr SUV	Pull P-2069A	6/05	3207B	3187
AVALON	N	10SecB	J&S (Pull/Sou)	/73	37U	
AVALON		12SecDr	Pull Wilm	1/27/03	1581C	W8
AVALON #576	P	10SecDrB	J&S (B&O)	4/88	663	
AVENDON	P	16Sec SUV	B&S	7/11	3212	
AVENEL	N	10SecDrB	(Wagner)		3046	
AVERIL	P	19Ch3Sts	B&S	7/82	3252	
AVIGNON #138	RN	7BoudB	(MannBCC)	/84	776A	
AVILLA	P	12SecO	B&O (UnionPCC)		665E	
AVIS		12SecDr	PCW	2/6/91	784B	1797
AVOCA		10SecDrB	PCW Det	11/17/83	149	D208
AVON		12SecDr	PCW	8/8/90	784	1707
AVON	P	16SecO	Wagner	10/93	3072	Wg101
AVON E&A-6	P	Slpr	Erie (E&AAssn)	C:/74		
AVONDALE		12SecDrB	PCW	8/25/96	1204C	2187
AVONLEA	N	26StsDrLngObs	Pullman	6/18/04	1417K	3083
AVONMORE		12SecDr	Pullman	10/21/05	2049B	3232
AXMINSTER		12SecDr	Pullman	8/20/01	1581A	2756
AYER #169	R	30Ch SUV	Pullman	2/2/03	3581E	2962
AZALEA		12SecDr	PCW	2/26/91	784B	1798
AZORES		12SecDr	Pull Buf	3/14/03	1581C	B14
AZOV		14SecO	PCW	1/27/93	1017C	1962
AZTEC		10SecO (NG)	PCW Det	7/12/80	73A	D76
AZTEC	RN	8SecBObs	PCW	1/30/93	1021S	1962
AZTEC	N	12SecDrCpt	Pullman	5/25/04	1985	3095
AZTLAN		BaggLib	PCW	6/7/87	477	1325
AZUL		12SecDr	PCW	4/25/91	784D	1805
AZUSA		10SecDrB	PCW	12/14/87	445E	1360
B		Bagg	PCW Det	8/30/70	1	D4
B		OS	G,B&Co	/66		Troy
B		12SecO	C&NW/GMP	/66	5	Chi
B (Cuban Special)	N	40Ch6DinDrObs	Pull P-2266A	4/07	3276A	3493
BABYLON		10SecDrB	PCW	6/26/85	240	1057
BABYLON #88	P		B&D (WS&PCC)	6/7/81		
BACA		10SecDr	PCW Det	9/23/80	59	D78
BACCHANTE	N	30Ch	PCW	4/28/98	1278A	2293
BACK BAY	P	27Ch8Sts2DrBObs	Pull P-1997	7/04	3270	3116
BADEN		10SecDrB	PCW Det	7/5/84	149D	D248
BAGDAD		14SecO	PCW Det	8/2/92	1014A	D509
BAHAMA	P	10SecDr	H&H	10/30/72	26 1/4	W6
BAHAMA	P	10SecDr	G&B (Wagner)	10/75	3011	
BAIREUTH #114	P	20Ch8StsDrB	Wagner	12/90	3126	Wg44
BAJARDO #171	N	BaggBLib	J&S (WS&PCC)	6/23/88	747B	
BAJAZET	N	16Ch2StsDrB	(Wagner)	4/85	3115	
BALA		12SecDr	PCW	5/13/91	784D	1805
BALAYAN		12SecDr	Pullman	8/15/04	1963	3098
BALKAN		10SecDrB	PCW Det	4/24/84	149D	D242
BALKIN	N	12SecDr	Wagner	4/92	3065C	Wg67
BALLANTRAE		12SecDr	Pullman	1/10/06	1963C	3276
BALLSTON #69	P	25Ch2StsB	G&B (Wagner)	10/81	3100B	
BALMORAL	N	12SecO	Grand Trunk	4/71	23M	Mon1
BALMORAL	S	8SecCpt	PCW Det	1/8/83	130	D158
BALMORAL		12SecDrSr	PCW	7/1/99	1318C	2424
BALMORAL #98	P	20Ch8StsDrB	Wagner	1/90	3126	Wg35
BALTIC		10SecDr	H&H	2/28/71	19	Wil2
BALTIC		14SecDr	PCW	8/3/99	1318D	2447
BALTIC #261	P	12SecDr	Wagner	7/88	3061B	Wg12
BALTIMORE	N	26Ch3StsDr	PCW	4/28/98	1278C	2258
BALTIMORE #51	P	6Sec12Ch	(Wagner)	5/85	3094	
BALTIMORE #152	P	10SecDr	J&S (B&O)	4/30/73	19	Wil1
BALTUSROL		12SecDr	Pullman	11/17/03	1581F	3043
BANBURY		12SecDr	Pullman	9/11/03	1581F	3022
BANCROFT	N	26ChBarB SUV	Pull P-2265	8/07	3274	3497
BANDA		12SecDr	PCW	11/25/90	784	1767
BANDIT	P	12SecDrTourist	(Wagner)		3008A	
BANDURA		12SecDrB	PCW Det	4/27/88	375G	D346
BANGOR	N	10SecDr	B&S	4/71	19	Day5
BANGOR		12SecDr	Pull Buf	3/11/04	1581G	B35
BANGOR #17	P	10SecO	G&B (Wagner)	9/70	3017	
BANNOCK		12SecDrSr	Pullman	3/15/01	1581A	2654
BANQUO		12Sec2Dr	Pullman	7/16/01	1580C	2742
BARABOO		10SecDr	PCW Det	11/13/72	26	D14
BARABOO		14SecCptDr	Pullman	11/1/05	2099	3237
BARACOA		12SecDr	Pullman	8/15/04	1963	3098
BARBARA		22Ch6StsDr	PCW	8/30/92	976A	1910
BARBARA		32ChDr	Pullman	3/9/07	2274	3509
BARCELONA	N	12SecDr	Pull Buf	4/22/04	1581F	B36
BARCELONA #404		12SecDrB	PCW	2/18/84	180	100
BARCELONA	R	30Ch SUV	PullBuf	4/22/04	3581C	B36
BARCENA	N	12SecDrCpt	Pullman	5/25/04	1985	3095
BARCENA 2510		14SecO	PCW Det	8/2/92	1014A	D509
BARFLEUR		16Sec	Pullman	6/1/08	2184C	3623
BARHAM		16Sec	Pullman	6/23/08	2184C	3627
BARMEN		12SecDr	Pullman	4/14/04	1963	3082
BARMOUTH		26ChDr	Pullman	5/28/07	1987B	3524
BARNEGAT #260	P	12SecDr	Wagner	5/24/88	3061B	Wg12
BARNESBORO		23ChDrLibObs	Pullman	8/28/06	2206	3388
BARNSTABLE		32ChDr SUV	Pullman	12/12	2671	4045
BARONESS		20Ch5StsB	PCW	6/17/84	167A	104
BARRANCA		12SecDr	Pullman	7/27/06	1963C	3385
BARREE		26ChDr	Pullman	5/25/07	1987B	3524
BARRINGTON		14SecCptDr	Pullman	11/2/05	2099	3237
BARSTOW	N	12SecDr	Pullman	11/10/01	1581A	2784
BARSTOW		10Sec2Dr	Pullman	11/8/02	1746A	2928
BARTHOLDI		14SecO	PCW StL	8/31/88	611A	StL4
BARTHOLDI		12SecDr SUV	Pull Buf	9/11/07	1963C	B67
BARTOLOMEO BONO		12Sec	PCW Det	7/1/76	50	D35
BASALT	R	10SecDr	PCW Det	8/30/81	115H	D114

CAR NAME	ST	CAR TYPE	CAR BUILDER (Owner)	DATE BUILT	PLAN	LOT	CAR NAME	ST	CAR TYPE	CAR BUILDER (Owner)	DATE BUILT	PLAN	LOT
BASEL		12SecDr	PCW	5/10/92	784E	1899	BAYONNE	N	12SecDrB	Wagner	2/90	3065	Wg32
BASILISK		16Sec	Pullman	6/24/08	2184C	3627	BAYPORT		12SecDr	Pullman	12/13/06	1963C	3448
BASSANIO		7Cpt2Dr	PCW	1/6/96	1065A	2140	BAYRIDGE #121	P		(WS& PCC)	7/83		
BATANGAS		12SecDr	Pullman	8/17/04	1963	3098	BAYSIDE	P	12SecDrB SUV	B&S	7/11	3211	
BATAVIA		12SecDrSrB	PCW	2/27/99	1319A	2400	BAYVIEW	N	10SecDrB	PCW	11/1/87	445E	1334
BATAVIA #210	P	Parlor	PRR	9/72	29	Alt1	BAYWOOD	N	12SecDr	PCW Det	1/9/82	93B	D135
BATH			(DL& W RR)				BEACHMONT		12SecDr	Pullman	1/5/06	1963C	3276
BATH		12SecDr	Pullman	10/20/03	1581F	3025	BEACON		10SecDr	PCW Det	11/24/81	59	D93
BATHURST		12SecDr	Pullman	10/16/06	1963C	3408	BEACON	N	12SecDrCpt	Pullman	5/26/04	1985	3095
BAVARIA		12SecDr	PCW Det	4/14/83	93B	D184	BEACONSFIELD	RN	14Sec	PCW	12/9/86	445C	1261
BAVARIA	R	12SecDrB	PCW Det	4/14/83	1468	D184	BEATRICE	RN	20StsDr Parlor	PCW Det	/77	45	D31
BAVARIA		12SecDr	Pullman Buf	4/23/03	1581C	B16	BEATRICE		20Ch5StsB	PCW	8/19/84	167A	199
BAVARIA #37	P	16Ch3StsB	(Wagner)	3/85	3114		BEATRICE #731	P	17Ch3StsB	J&S (B&O)	6/29/87	686	2518
BAY SHORE #113	P		(WS& PCC)	7/83			BEAUFORT		16SecO	PCW	6/16/93	1040	1990
BAYAMON		12SecDr	Pullman	8/15/04	1963	3098	BEAUMONT		12SecDrB	PCW	9/17/98	1319A	2314
BAYARD		12SecDr	Pullman	5/24/06	1963C	3360	BEAUMONT M&LT-3	N	Slpr	(M&LT RR)		37?	
BAYCHESTER		32ChDr SUV	Pullman	12/12	2671	4045	BEAUPRE		12SecDr	Pullman	1/10/06	1963C	3276
BAYFIELD	N	12SecDr	PCW Det	2/14/82	93B	D144	BEAUREGARD #262	P	12SecDr	Wagner	7/88	3061B	Wg12
BAYFIELD		12SecDr	Pull Buf	9/21/07	1963C	B67	BEAUVAIS	N	12SecDrB	Wagner	10/92	3070	Wg76
BAYONNE		10SecDrB	PCW	5/24/84	149D	192	BEAUVOIR		16SecO	Pullman	5/1/05	2068	3177

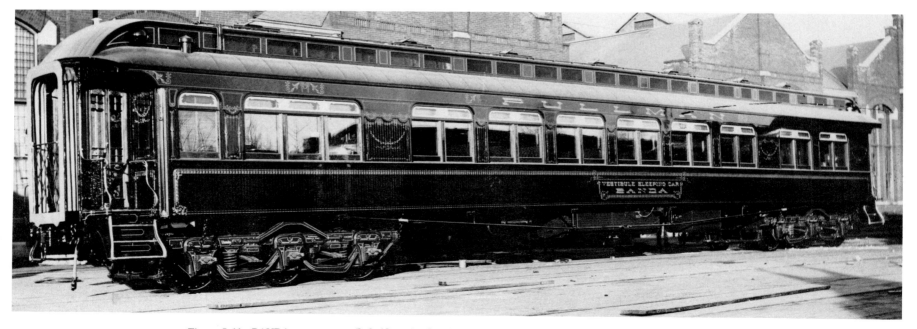

Figure I-11 BANDA, a narrow vestibule 12-section drawing room sleeper with paper wheels, was built in the same lot as ARONA (See page 29) for service on the Chicago, Burlington & Quincy RR. Note the decorative painting on the truck which includes the car name, the builder, and wheel numbers. Smithsonian Institution Photograph P1230.

CAR NAME	ST	CAR TYPE	CAR BUILDER (Owner)	DATE BUILT	PLAN	LOT
BEAVERDALE		23ChDrLibObs	Pullman	8/28/06	2206	3388
BEAVERTON		12SecDr SUV	Pullman	9/16/07	1963C	3550
BEDFORD		10SecDrB	PCW	6/18/87	445C	1322
BEDFORD #259	P	12SecDr	Wagner	7/88	3061B	Wg12
BEDOUIN		12SecDrB	Pullman	2/19/06	2142	3318
BEECHFIELD		28ChDr	Pullman	4/5/06	1986A	3336
BEESTON		12SecDr	Pullman	3/1/02	1581B	2818
BEETHOVEN		14SecDr	PCW	9/1/88	600	W1
BEGONIA		12SecDrB	PCW Det	10/31/88	375G	D358
BEGONIA		24ChDrObs	Pull Buf	4/2/06	2156A	B50
BELFAST		12SecDr	PCW	7/4/92	784E	1915
BELGIC		10SecDrHotel	PCW Det	11/22/80	63 1/2	D77
BELGIUM		12Sec2Dr	PCW Det	12/88	537A	D334
BELGRADE	N	OS	(PK & RSCC)	/73	74	
BELGRADE #458	P	12SecDrB	Wagner	4/3/91	3065	Wg50
BELGRAVIA		20Ch3StsB	PCW Det	9/22/83	167A	D209
BELGRAVIA		12SecDr	PCW	8/5/98	1318A	2313
BELGRAVIA #97	P	20Ch8StsDrB	Wagner	7/90	3126	Wg35
BELIZE		10SecDrB	PCW	8/26/87	445E	1333
BELLAIRE		12SecDr	Pull Buf	6/23/05	1963	B42
BELLAIRE #191	P	10SecDr	B&O RR	4/30/73	19	B&O1
BELLE #19	N		(WS& PCC)	7/1/75		
BELLE ROSE #6	P	Slpr	(SouTranCo)			
BELLECLAIRE		16Sec	Pullman	6/5/06	2184	3361
BELLEFONTE		12SecDrSr	Pullman	3/24/05	2049A	3169
BELLEFONTE #21	P	OS	PRR	12/71	27	Alt2
BELLEMERE	P	28ChDr SUV	Pull P-1857	1/03	3268	2902
BELLEVILLE		12SecO	Grand Trunk	4/71	18	Mon1
BELLEVILLE		12SecDr	Pullman	10/16/03	1581F	3025
BELLEVUE		Diner	PCW	8/2/93	538D	2010
BELLEVUE #5	P	Slpr	(SouTranCo)			
BELLEVUE #456	P	12SecDrB	Wagner	4/91	3065	Wg50
BELLFIELD		12SecDr SUV	Pullman	2/15/06	1963C	3316
BELLINGHAM		32ChDr SUV	Pullman	12/12	2671	4045
BELLNAP		BaggClub	Pullman	8/20/06	2136A	3386
BELLONA	RN	Slpr	(SouTranCo)			
BELLONA	N	12SecDrB	Wagner	8/91	3065	Wg53
BELLWOOD		12SecDrSr	Pullman	1/21/01	1581A	2614
BELMAR		24ChDrB	Pullman	6/10/05	1988A	3170
BELMONT	P	Slpr	(SouTranCo)			
BELMONT	N	10SecDrB	PCW	6/2/85	240	1048
BELMONT		12SecDrB	PCW	9/10/98	1319A	2314
BELMONT		12SecDr	Pullman	10/16/06	1963C	3408
BELMONT #451	P	12SecDrB	Wagner	4/91	3065	Wg50
BELMORE	N	10SecDrB	PCW	5/22/85	240A	1193
BELOIT		12SecDr	PCW Det	2/14/82	93B	D144
BELOIT		12SecDr	Pullman	10/17/03	1581F	3025
BELVIDERE		Diner	PCW	8/7/93	538D	2010
BELVIDERE		12SecDrSr	Pullman	1/26/01	1581A	2614
BELVOIR		12SecDr	Pullman	12/6/09	1963E	3766

CAR NAME	ST	CAR TYPE	CAR BUILDER (Owner)	DATE BUILT	PLAN	LOT
BEMIS	N	17Ch12StsB	J&S (B&O)	6/29/87	686E	
BEN AHIN		10SecLibObs	Pullman	10/29/02	1827	2927
BEN ALDER		10SecLibObs	PCW	12/5/99	1434B	2483
BEN AMMI		10SecLibObs	Pullman	10/29/02	1827	2927
BEN ARTHUR		10SecLibObs	PCW	12/4/99	1434B	2483
BEN ATTUW		10SecLibObs	Pullman	10/29/02	1827	2927
BEN AVON		10SecLibObs	PCW	11/27/99	1434B	2483
BEN CLIBRICK		10SecLibObs	Pullman	10/21/02	1827	2927
BEN CRUACHAN		10SecLibObs	Pullman	10/29/02	1827	2927
BEN DEARG		10SecLibObs	Pullman	10/29/02	1827	2927
BEN DORAN		10SecLibObs	PCW	12/9/99	1434A	2483
BEN HOPE		10SecLibObs	PCW	12/9/99	1434A	2483
BEN LAWERS		10SecLibObs	Pullman	10/29/02	1827	2927
BEN LEDI		10SecLibObs	Pullman	10/29/02	1827	2927
BEN LOMOND		10SecLibObs	PCW	11/14/99	1434B	2483
BEN MACDHUI		10SecLibObs	Pullman	10/29/02	1827	2927
BEN MORE		10SecLibObs	Pullman	2/5/00	1434A	2483
BEN NEVIS		10SecLibObs	PCW	11/21/99	1434B	2483
BEN VENUE		10SecLibObs	Pullman	10/29/02	1827	2927
BEN VORLICH		10SecLibObs	Pullman	2/5/00	1434A	2483
BEN WYVIS		10SecLibObs	Pullman	2/5/00	1434A	2483
BENA		12SecDr	Pullman	6/10/07	2271A	3458
BENARES		12SecDr	PCW	12/20/92	990	1941
BENARES	N	12SecDr	Pull Buf	4/23/03	1581C	B16
BENBOW		16Sec	Pullman	6/23/08	2184C	3627
BENGAL	N	SDR	H&H	11/71		W6
BENGAL	N	12SecDrB	Wagner	11/92	3073	Wg65
BENICIA	N	12SecDrB	Wagner	2/90	3065H	Wg43
BENICIA CP-31	P	10SecDr	(CP Co)	11/1/84	37K	
BENITO	N	12SecDr	CB&Q RR	7/15/69	12B	Aur
BENNINGTON		12SecDr	Pullman	11/1/00	1581A	2589
BENTON	N	10SecB	Bowers, Dure	9/73	37	Wil11
BENTON	N	12SecDr	PCW	10/21/98	1318B	2340
BENVOLIO		12Sec2Dr	Pullman	7/19/01	1580C	2742
BENWOOD		10SecB	B&O	7/31/75	37	B&O
BENWOOD	N	10SecDr	PCW	2/23/85	222A	1033
BEREA		12SecDr	Pullman	11/25/05	1963C	3252
BERENDA		12SecDr	Pullman	1/6/06	1963C	3276
BERENGER	N	20Ch8StsDrB	Wagner	7/90	3126	Wg35
BERESFORD		12SecDr	Pullman	8/26/01	1581A	2756
BERGEN		12SecDr	Pullman	4/14/04	1963	3082
BERGEN		12SecDr	Pullman	12/13/06	1963C	3448
BERGEN 2516		14SecO	PCW Det	1/27/93	1014A	D511
BERHERA		12SecDrB	PCW	10/6/88	375F	1481
BERHERA	N	12SecDr	PCW	8/11/98	1318A	2313
BERKELEY		12SecDr	Pullman	4/17/03	1581B	3003
BERKELEY	P	10SecDr	(Wagner)	/86	3000A	
BERKELEY #512	P	12SecO	B&O RR		665	
BERKSHIRE		12SecDr	Pullman	7/11/03	1581F	3019
BERKSHIRE	N	12SecDr	Pullman	4/7/04	1963	3080

CAR NAME	ST	CAR TYPE	CAR BUILDER (Owner)	DATE BUILT	PLAN	LOT
BERKSHIRE #16	P	10SecO	G&B (Wagner)	/70	3017	
BERKSHIRE #143	N	6Sec4Dr	(WS&PCC)	4/8/86	720?	
BERLAMONT	N	BaggClub	Pullman	3/1/06	2136	3319
BERLIN		12SecDr	Pullman	7/25/02	1581B	2852
BERLIN #188	P	10SecDr	B&S	2/23/71	19	Day5
BERMUDA	P	10SecDr	H&H	10/30/72	26 1/4	W6
BERMUDA	N	12SecDrB	Wagner	2/91	3065	Wg46
BERNARDINO		10SecDr2Cpt	Pullman	9/2/05	2078	3203
BERNARDO		7Cpt2Dr	PCW	1/7/96	1065A	2140
BERNE		12SecDr	PCW	2/10/92	784E	1881
BERNICE		22ChDr	Pull Buf	5/16/02	1647C	B8
BERTHA		22ChDr	Pull Buf	5/20/02	1647C	B8
BERTRAND	N	12SecDr	B&S (Wagner)	8/82	3023A	
BERWICK		16SecO	PCW	6/3/93	1040	1990
BERWYN		12SecDr	PCW	8/5/91	784C	1828
BERYL	N	30Ch	PCW	4/27/98	1278A	2293
BESSEMER		12SecDr	Pullman	1/6/06	1963C	3276
BETHANY		12SecDr	Pullman	9/12/03	1581F	3022
BETHEL	P	36Ch SUV	Pull P-2263	8/07	3275	3492
BETHESDA		12SecDr	Pullman	1/8/06	1963C	3276
BETHLEHEM		12SecDr	Pullman	2/9/04	1963	3070
BETHULIA	P	12SecDrB SUV	Pull P-2497	12/10	3211	3824
BETZWOOD		26ChDr	Pullman	5/25/07	1987B	3524
BEULAH	N	26ChDr	Pullman	3/26/06	1987A	3335
BEULAH #110	P		(WS&PCC)	/75	698	
BEVARD		12SecDr	Pullman	1/8/06	1963C	3276
BEVERLEY		10SecB	B&S	7/73	37	Day10
BEVERLEY	N	12SecDr	PCW	11/28/90	784	1767
BEVERLY	P	10SecDr	(Wagner)	1/85	3055	
BEXLEY		16SecO	PCW	5/3/93	1040	1990
BIANCA		14Ch13StsDrB	PCW	10/27/87	505	1376
BIANCA		26ChDrSr	Pullman	6/4/04	1987A	3085
BICESTER		12SecDr	Pullman	12/13/09	1963E	3766
BIENVILLE		12SecDr	Pullman	2/7/08	1963C	3598
BIG SIX		OS	MC RR/GMP	8/1/67	27	Det
BILLINGS		12SecDr	PCW	10/10/82	93B	22
BILLINGS		16Sec	Pullman	3/28/07	2184	3523
BILOXI	N	10SecDr	H&H	10/30/72	26 1/4	W6
BILTMORE		12SecDr	Pullman	1/10/06	1963C	3276
BINGEN		10SecDrB	PCW Det	8/31/83	149	D206
BINGEN	R	12SecDr	PCW Det	8/31/83	1602B	D206
BINGHAMTON	P	14Ch5StsB	(DL&W RR)	4/75	211	
BINGHAMTON	N	23ChDrLibObs	Pullman	8/28/06	2206	3388
BINGHAMTON #56	P	20Ch4Sts	(Wagner)	8/86	3102	
BINSTEAD		12SecDr	Pullman	10/15/07	2163A	3562
BION	N	28Ch3Sts	PCW (Wagner)	7/29/86	3124	1200
BIONDELLO		12SecDrCpt	Pullman	5/23/04	1982	3090
BIRCHTON		12SecDr	Pullman	10/18/06	1963C	3408
BIRMINGHAM	N	12SecO	(L&N RR)	/72	114	
BIRMINGHAM		12SecDr	Pull Cal	6/2/04	1581G	C8
BIRMINGHAM #6	P	22Ch4StsDr	(Wagner)	6/85	3108D	
BISCAY		10SecDrB	PCW Det	7/11/84	149D	D248
BISCAY	N	10Sec2DrB	Wagner	8/97	3074B	Wg115
BISERTA		14SecO	PCW	5/13/93	1021	1972
BISMARCK		14SecO	PCW Det	6/28/82	115	D153
BISMARCK		12SecDr	Pull Buf	9/18/07	1963C	B67
BISMARCK	P	14SecO	(Wagner)	/76	3030	
BISSAO		12SecDrCpt SUV	Pullman	8/17/05	2103	3227
BISUKA		10SecDrB	PCW	3/13/88	445E	1361
BITHYNIA		12SecDr	Pull Buf	9/24/03	1581F	B23
BITOLIA		12SecDr SUV	Pull Buf	9/26/03	1581F	B23
BITTERN	N	26ChDrB	Pullman	6/26/07	1988B	3526
BLACK HALL	P	BaggSmk	B&S	8/93	3213B	
BLACKHAWK #103	P	Slpr	(IC RR)			X2/81
BLACKMORE		12SecDr	Pullman	12/10/09	1963E	3766
BLACKSTONE		12SecDr2Sr	Pullman	11/4/09	2447	3745
BLAIRSDEN	N	5CptLngObs	PCW	1/10/98	1299C	2283
BLAIRSVILLE		12SecDrSr	Pullman	3/23/05	2049A	3169
BLANCHE		18Ch4StsDrB	PCW	7/8/93	1039A	1995
BLANCO		12SecDr	PCW Det	1/28/92	784E	D488
BLANCO	N	12SecDr	Pullman	9/11/03	1581F	3022
BLENHEIM		10Cpt	PCW	1/5/93	993A	1951
BLENHEIM #459	P	12SecDrB	Wagner	4/91	3065	Wg50
BLITHEDALE #428	P	8Sec4Dr	Wagner	8/90	3068	Wg37
BLOCK ISLAND	P	16Ch23StsBObs	New Haven RR	7/02	3266	
BLODGETT	N	12SecDr	Pull Buf	9/18/07	1963C	B67
BLOOMFIELD		14SecDr	PCW	12/14/99	1318J	2450
BLOOMINGTON	P	10SecDr	B&S	10/71	26	Day6
BLOOMINGTON		16SecO	Pullman	4/25/05	2068	3177
BLOOMINGTON	P	12SecO	(WS&PCC)	6/27/74	833?	
BLUFFTON		12SecDr	Pullman	3/3/02	1581B	2818
BLUFFTON	R	30Ch SUV	Pullman	3/3/02	3581E	2818
BOBOLINK	N	26ChDrB	Pullman	6/27/07	1988B	3526
BOCACCIO		12SecDrCpt SUV	Pullman	8/18/05	2103A	3227
BOGOTA		10SecDrB	PCW	2/20/84	149D	101
BOHEMIA		12SecDr	PCW	6/83	93B	38
BOHEMIA		12SecDrB	PCW	2/13/89	375G	1557
BOHEMIA	P	16Ch3StsB	(Wagner)	5/85	3114	
BOHEMIAN GIRL #127	P	7BoudB	(MannBCC)	/84	776	
BOHIO	N	12SecDr	Pullman	10/16/05	2049B	3232
BOISE	RN	10SecDrSr	PCW	6/16/90	784R	1715
BOISE CITY		12SecB	PCW Det	8/15/76	58	D42
BOKHARA		12SecDr	Pullman	7/12/03	1581F	3019
BOLERO	N	14SecO	J&S	9/73	37A	Wil12
BOLERO	N	16Sec	Pullman	11/28/06	2184	3409
BOLINAO		12SecDr	Pullman	8/15/04	1963	3098
BOLIVAR	P	10SecDr	J&S (McComb)	/71	19A	Wil
BOLIVAR		12SecDr	Pullman	9/14/03	1581F	3022
BOLIVAR	N	12SecDr	Pullman	5/24/02	1581B	2853

Figure I-12 The interior of BEXLEY has a minimum of marquetry but a fine example of "French Polish" on the upper berths which was a process that incorporated multiple coats of varnish and was hand-rubbed with pumice and oil, resulting in the impression of a mirrored surface. The car reflects the European-inspired classical revival motifs which became popular in the United States beginning with the Columbian Exposition. Note the Italian-inspired clerestory leaded windows, Fleur-de-lis motif on the window shades, and brocade upholstery. Smithsonian Institution Photograph P2224.

CAR NAME	ST	CAR TYPE	CAR BUILDER (Owner)	DATE BUILT	PLAN	LOT
BOLIVIA		12SecDrSr	PCW	6/17/99	1318C	2424
BOLIVIA #31	RN	12SecO	(H&StJ RR)	2/71	23	
BOLTON	N	10SecDrB	PCW	5/20/85	240A	1193
BOLTON	R	16Sec	PCW	5/20/85	1881	1193
BOMBAY	N	12Sec	H&StJ RR	1/73	114	3
BOMBAY		12SecDr	Pullman	4/6/03	1581D	2984
BOMBAY	P	16SecO	Wagner	6/96	3072A	Wg108
BONAIR	N	10SecDrB	PCW Det	9/29/83	149	D206
BONANZA		10SecO (NG)	PCW Det	10/25/79	73	D61
BONAVENTURE		16Sec	Pullman	6/24/08	2184C	3627
BONHAM	N	12SecDrSr	Pullman	3/2/01	1581B	2654
BONHEUR		32ChDr	Pullman	3/4/07	2274	3509
BONITA		10SecDrCpt	PCW	6/12/89	668	1587
BONITA	N	30Ch	PCW	4/28/98	1278A	2293
BONNYBROOK #434	P	12SecDr	Wagner	9/90	3065B	Wg38
BOONTON		16Ch6StsDrB	PCW Det	6/29/87	167F	D283
BOONVILLE		12SecDr SUV	Pull Buf	5/9/04	1963	B41
BOOTHWYN #575	P	10SecDrB	J&S (B&O)	4/88	663	
BORACHIO		7Cpt2Dr	PCW	2/1/96	1065A	2143
BORDEAUX		12SecDr	Pull Buf	5/18/03	1581C	B18
BORDEAUX #127	RN	7BoudB	(MannBCC)	/84	776A	
BORDENTOWN #294		10SecB	J&S	9/73	37	Wil12
BORNA		14SecO	PCW	1/19/93	1017C	1962
BORNEO		10SecDrB	PCW	3/29/84	149B	101
BORNEO	P	16SecO	J&S (Wagner)	3/82	3072A	Wg108
BORODINO	RN	10SecB?	PCW Det	11/13/71	27M	D8
BORODINO	N	10SecDrB	J&S (Wagner)	3/82	3046B	
BOSCOBEL		10SecDrB	PCW	/83	149	21
BOSNIA		16Sec	Pullman	10/23/01	1721	2766
BOSPORUS		10SecDrB	PCW Det	7/21/84	149D	D248
BOSTON	N	12SecDr	CB&Q/GMP	/67	13B	Aur
BOSTON	P	Parlor	Vermont Cent	/71	8	StA
BOSTON		16SecO	Pullman	3/13/03	1894	2981
BOSWELL	N	12SecDr	PCW	9/27/98	1318B	2339
BOTHNIA		10SecDr	PCW Det	2/29/76	52	D33
BOTHNIA		12SecDr	PCW	8/8/98	1318A	2313
BOTHNIA #45	P	6Sec12Ch	(Wagner)	9/70	3095	
BOULOGNE		12SecDr	Pullman	10/19/03	1581F	3025
BOURBON		12SecDr	Pullman	6/25/01	1581A	2726
BOURGES		16SecO	PCW	8/29/93	1040	2008
BOUVARDIA		12SecDrCpt	PCW	8/16/92	983	1919
BOUVINES	N	15Ch8StsDrB	(Wagner)	4/85	3115A	
BOWDOIN 2559		14SecO	PCW Det	2/28/93	1021	D516
BOYLSTON		12SecDr2Sr SUV	Pullman	11/6/09	2447	3745
BOYNE		12SecDr	PCW Det	9/21/91	784D	D470
BOZEMAN		12SecDr	PCW Det	5/26/84	93D	D241
BRABANTIO		12SecDrCpt	Pullman	5/23/04	1982	3090
BRADDOCK		12SecDrSr	Pullman	3/27/05	2049A	3169
BRADFORD	R	12SecDrB	PCW?		1611	
BRADFORD #252	P	10SecDr	PC&StL RR	9/72	26H	Col2

CAR NAME	ST	CAR TYPE	CAR BUILDER (Owner)	DATE BUILT	PLAN	LOT
BRAEBURN		12SecDr	Pullman	11/17/03	1581F	3043
BRAESIDE		12SecDr	Pullman	10/18/06	1963C	3408
BRAGANZA		12SecDr	Pullman	6/21/01	1581A	2726
BRAINERD		12SecDr	PCW Det	5/27/84	93D	D241
BRAINTREE		32ChDr SUV	Pullman	12/12	2671	4045
BRAMCOTE		28ChDr	Pullman	4/27/07	1986B	3527
BRAMPTON		12SecDr	Pullman	10/18/06	1963C	3408
BRANCHVILLE	P	36Ch SUV	Pull P-2263	8/07	3275	3492
BRANDON #420	P	12SecDrB	Wagner	4/90	3065	Wg33
BRANDYWINE		12SecDr	Pullman	3/1/07	1963C	3525
BRANDYWINE #147	P	6Sec4Dr?	J&S (WS&PCC)	5/17/86	720?	
BRANDYWINE #226	P	10Ch15StsDr	J&S	11/72	35	Wil2
BRANDYWINE #724	P	31StsB	B&S (B&O)	/85	653	
BRANHAM		12SecDr	Pullman	12/11/09	1963E	3766
BRANKSMERE #433	P	12SecDr	Wagner	9/90	3065B	Wg38
BRANTFORD		10SecDr	Grand Trunk	9/71	26	Mon2
BRANTFORD		12SecDr	Pull Buf	10/28/01	1581A	B6
BRAQUEMONDE #432	P	12SecDr	Wagner	9/90	3065B	Wg38
BRAYTON		12SecDr	Pullman	2/5/08	1963C	3598
BRAZIL		16SecO	PCW	11/10/90	780	W5
BRAZIL #32	N	12SecO	(H&StJRR)	2/28/71	27	
BRAZIL H&StJ #3	P	OS	(H&StJRR			
BRAZITO		7Cpt2Dr	Pullman	9/25/02	1845	2926
BRAZORIA	N	10SecDr	H&H	2/71	19K	Wil2
BRAZORIA		12SecDr	Pullman	10/2/01	1581A	2765
BRAZOS	RN	12SecDrB	PCW	6/12/88	1566	1414
BREDA		14SecO	PCW	2/24/93	1021	1969
BREMEN		10SecDrHotel	PCW Det	10/22/80	63 1/2	D77
BREMEN		12SecDr	Pull Buf	4/12/04	1581F	B36
BRENNAN	N	12SecDr	Pull Buf	4/12/04	1581F	B36
BRENO 2521		14SecO	PCW Det	1/30/93	1014A	D511
BRENTFORD	RN	12SecDr	Pull Buf	10/28/01	1581C	B6
BRENTWOOD #99	N		(WS&PCC)	7/29/81	360?	
BRENTWOOD #430	P	8Sec4Dr	Wagner	8/90	3068	Wg37
BRESLAU		12SecDr	Pull Buf	4/22/04	1581F	B36
BRESLAU #113	P	20Ch8StsDrB	Wagner	12/90	3126	Wg44
BRESLAU #129	RN	7BoudB	(MannBCC)	/84	776A	
BRESLIN		12SecDr	Pullman	12/10/07	1963C	3585
BRETAGNE		12SecDr	PCW	9/20/93	1040A	2008
BRETON	N	10SecDr	PCW Det	5/13/73	26 3/4	D15
BRETON	N	12SecDr	PCW	10/21/98	1318B	2340
BREVOORT		Diner	CB&Q RR	6/30/71		Aur
BREVOORT		10SecDrB Hotel	PCW Det	7/78	58 1/2	D48
BREVOORT		12SecDr	Pull Buf	9/14/07	1963C	B67
BREWSTER	N	14Ch5StsB	(DL&W RR)		211	
BREWSTER	P	28ChDr SUV	Pull P-1857	1/03	3268	2902
BRIARCLIFF		8CptLibObs	Pullman	12/22/06	2260	3502
BRIAREUS	N	8CptLngObs	Wagner	11/97	3078	Wg114
BRIDGETON		12SecDr	Pull Buf	8/8/03	1581E	B20
BRIDGEWATER	P	26ChBarB SUV	Pull P-2493	1/11	3274	3820
BRIENZE		14SecO	PCW Det	5/5/93	1021	D518
BRIERLEY		12SecDr	Pullman	1/5/06	1963C	3276
BRIGHTON		10SecB Hotel	PCW Det	1/6/79	37C	D50
BRIGHTON		16SecO SUV	Pullman	5/1/05	2068	3177
BRIGHTON	P	10SecDr	(Wagner)	/85	3033	
BRIGHTWOOD		14SecDr	Pull Buf	9/17/00	1523D	B1
BRILHART		26ChDrB	Pullman	6/28/07	1988B	3526
BRILLIANT		8Sec4SrB	PCW	5/1/91	894B	1807
BRILLIANT	N	12SecDr	Pull Buf	9/14/07	1963C	B67
BRINDISI		12SecDrB	PCW Det	10/10/88	375G	D358
BRINSMEAD		16Sec	Pullman	5/27/09	2184D	3700
BRINSMERE	N	26ChDr	Pullman	3/28/06	1987A	3335
BRINXWORTH #556	P	14SecSr	H&H (B&O)	8/87	654	
BRIONE		14SecO	PCW	2/26/93	1017C	1962
BRISBANE		12SecDr	Pullman	7/8/03	1581F	3019
BRISTOL		12SecDr	Pullman	4/23/03	1581B	3003
BRISTOL	R	30Ch SUV	Pullman	4/23/03	3581E	3003
BRISTOL #181	P	10SecDr	H&H	1/72	26	Wil6
BRITAIN		12SecDr	Pullman	8/26/01	1581A	2756
BRITANNIA		27Sts Parlor	PCW Det	3/15/74	39	D20
BRITANNIA		12SecDrSr	PCW	12/9/98	1318C	2370
BRITTANIC		10SecDrHotel	PCW Det	10/22/80	63 1/2	D77
BRITTANY	N	10SecDr	Grand Trunk	10/31/72	26	Mon2
BRITTANY #275	P	9CptB	Wagner	4/89	3062B	Wg14
BRIXTON		16SecO	PCW	6/6/93	1040	1990
BROAD BROOK	P	BaggSmkB SUV	New Haven RR	11/04		3218
BROADVIEW		12SecDr SUV	Pullman	10/18/06	1963C	3408
BROADWAY		12SecDr SUV	Pullman	2/6/08	1963C	3598
BROCKPORT		12SecDr SUV	Pull Buf	5/16/04	1963	B41
BROCKTON		12SecDr	Pull Buf	8/11/03	1581E	B20
BROCKVILLE		12SecO	Grand Trunk	4/71	23M	Mon1
BROCKVILLE		16SecO	Pullman	5/2/05	2068	3177
BROCTON		10SecDr	PCW Det	10/20/70	26A	D4
BROMLEY		10SecLibObs	Pullman	11/16/09	2351A	3748
BRONTE		10SecLibObs	Pullman	11/17/09	2351A	3748
BROOKFIELD	P	OS	H&StJ RR	2/28/71	27	
BROOKFIELD		12SecDr2Sr	Pullman	11/5/09	2447	3745
BROOKHAVEN	N	12SecDr	Pullman	7/11/03	1581F	3019
BROOKHAVEN	N	12SecDrCpt	Pullman	5/23/04	1985	3095
BROOKHAVEN #108	P	20Ch11StsDr	(WS&PCC)	/73	570?	
BROOKHAVEN #435	P	12SecDr	Wagner	9/90	3065B	Wg38
BROOKLAND		16Sec	Pullman	6/1/09	2184D	3700
BROOKLINE		12SecDr2Sr	Pullman	11/3/09	2447	3745
BROOKLYN		16Sec	Pullman	7/7/03	1721	3014
BROOKLYN #110	P	Parlor	(WS&PCC)	/75	698?	
BROOKLYN #156	P	22ChSr	J&S (WS&PCC)	5/15/87	516?	
BROOKVILLE		16Sec	Pullman	12/3/06	2184	3409
BROOKWOOD	P	24Ch6StsDr	PCW	3/91	3259B	
BROWNFIELD	N	12SecDr	B&O RR	11/80	657B	
BROWNING		14SecDr	Pullman	12/16/01	1523F	2798

36

CAR NAME	ST	CAR TYPE	CAR BUILDER (Owner)	DATE BUILT	PLAN	LOT
BROZTELL		12SecDr	Pullman	3/28/08	1963C	3600
BRUCIA	N	28ChDr	Pullman	4/27/07	1986B	3527
BRUGES		12SecDr	Pullman	4/11/03	1581D	2984
BRUNEHILDE	N	Private	PCW	1/15/98	1273A	2251
BRUNHILDA		24StsDrSrB	Pullman	6/16/04	1988A	3086
BRUNSWICK	N	12SecDr	PCW	5/4/92	784E	1899
BRUNSWICK No.1 #259	N	OS	CB&Q RR	5/72	36B	Hinc1
BRUNSWICK No.2	N	10SecDr	H&StJ RR	8/71	19	Han2
BRUNSWICK No.2	P	DDR	PFtW&C	7/71	17	FtW2
BRUSHWOOD	N	12SecDr	Wagner	8/91	3065J	Wg45
BRUSSELS		10SecDrB	PCW Det	4/30/84	149D	D242
BRUSSELS	N	16Ch3StsDrB	(Wagner)	4/85	3108B	
BRUTUS		BaggBLibSmk	PCW	5/11/92	970	1905
BRYMERE		10SecDr2Cpt	Pullman	11/10/09	2451	3747
BRYN MAWR		12SecDrSr	Pullman	1/25/01	1581A	2614
BRYN MAWR #275	P	10SecDr	Elmira Car W	7/73	37	Elm1
BUCHAREST		12SecDr	Pullman	7/7/03	1581F	3019
BUCKINGHAM		10SecDrB Hotel	PCW Det	5/20/76	56	D39
BUCKINGHAM		12SecDr	Pull Buf	9/25/07	1963C	B67
BUCYRUS		14SecO	PCW Det	5/17/93	1021	D525
BUDA		10SecDrB	PCW	6/5/91	920B	1808
BUDAPEST		12SecDr	Pullman	7/11/03	1581F	3019
BUENA VISTA		29ChBObs SUV	Pullman	2/16/09	2397	3684
BUENA VISTA SP-31	P	10SecDrB	PCW	1/7/87	445A	1297
BUFFALO	P	14Ch5StsB	(DL&W RR)	8/83	211	
BUFFALO	R	12SecDrB	PCW?		1611	
BUFFALO		27ChDrObs	Pullman	6/8/08	2346	3624
BUFFALO #171	P	BaggBLib	J&S (WS&PCC)	6/23/88	747?	
BULGARIA		10Cpt	PCW	2/2/93	993A	1951
BULWER		12SecDr	PCW	10/14/92	990	1931
BULWER		10SecLibObs	Pullman	11/18/09	2351A	3748
BUMALDA		16Sec SUV	Pullman	3/31/10	2184G	3808
BURBANK		32ChDr	Pullman	3/2/07	2274	3509
BURGUNDY		12SecDr	Pullman	6/26/01	1581A	2726
BURHAM		10SecDrB	PCW	3/8/84	149	102
BURLINGTON	P	SDR	CB&Q RR	7/15/69	12	Aur
BURLINGTON	N	12SecDr	CB&Q/GMP	/66	13	Aur
BURLINGTON	P	12SecDr	(Wagner)	/86	3026	
BURLINGTON WC-101	P	12SecO	CentCarCo	2/86	532	
BURMAH	N	12SecDr	PCW	12/25/98	1318B	2340
BURMONT		12SecDr	Pullman	1/10/06	1963C	3276
BURNABY	P	12SecDrB SUV	Pull P-2497	12/10	3211	3824
BURNET		12SecDr	Pullman	2/6/08	1963C	3598
BURNHAM	N	12SecDrB	PCW	12/3/88	375G	1513
BURNLEY		12SecDr	PCW	12/20/92	990	1941
BURNS		12SecDr	PCW	10/26/92	990	1931
BURNSIDE #421	P	12SecDrB	Wagner	4/90	3065	Wg33
BURRISTON		10SecDr2Cpt	Pullman	3/23/07	2078D	3528
BURTON #2560		14SecO	PCW Det	2/25/93	1021	D516
BUSHTON		12SecDr	Pullman	9/6/07	1963C	3550
BUTTE		12SecDr	Pullman	3/2/04	1963	3080
BUTTERFIELD		12SecDr	Pullman	2/5/08	1963C	3598
BUXTON		12SecDr	Pull Buf	8/6/03	1581E	B20
BUZZARDS BAY	R	Bagg4DrBClub	New Haven	7/02	3267	
BUZZARDS BAY	P	Bagg2CptB	New Haven	/02	3216A	
BYRON		10SecDrB	PCW	10/30/86	445	1262
BYRON		10SecLibObs	Pullman	11/19/09	2351A	3748
BYRON	R	16Sec	PCW	10/30/86	1880	1262
BYRONTON		12SecDr	PCW	10/26/98	1318B	2340
C	P	12SecO	CB&Q/GMP	8/1/67		Aur
C	P	Baggage	CB&Q RR	10/70		
CABALLERO	P	12SecHotel	J&S	1/99	2371A	
CABANAS		12SecDr	Pullman	8/17/04	1963	3098
CABRA		14SecO	PCW	3/9/93	1021	1969
CABUL		12SecDr	PCW Det	11/15/90	784	D449
CACIN	N	6Sec12Ch	(Wagner)	5/85	3094	
CACOUNA	P	10SecDr	Grand Trunk	10/31/72	31	Mon3
CACTUS		12SecDr	Pullman	10/29/08	1963E	3658
CADENUS		12SecDr	Pull Buf	8/31/05	1963B	B45
CADI		12SecDr	PCW	2/11/91	784B	1797
CADILLAC		16SecSr	Pullman	8/29/04	2033A	3135
CADIZ		10SecDrB	PCW	6/18/84	149D	192
CADIZ	P	10Sec2Dr	Wagner	11/98	3073D	Wg110
CADMUS	N	20Ch8StsDrB	Wagner	8/93	3126B	Wg94
CADMUS #444		10SecDrB	PCW	8/22/85	240	1058
CAESAR		BaggBLibSmk	PCW	5/9/92	970	1905
CAHAWBA		10SecDr	PCW Det	3/22/82	112A	D141
CAHAWBA		12SecDr	Pullman	4/25/03	1581B	3003
CAIRO		8SecSrB (NG)	PCW Det	10/30/82	155	D173
CAIRO		12SecDr	Pullman	2/9/04	1963	3070
CALABRIA		10SecB	PCW Det	9/5/73	37	D16
CALABRIA		14SecDr	Pullman	6/15/00	1523	2559
CALADA		10SecDr3Cpt	Pullman	4/16/06	2078B	3342
CALAIS		10SecDrB	PCW	6/28/84	149D	192
CALALONIA		12SecDr	Pullman	10/9/01	1581A	2765
CALANTHE		12SecDr SUV	Pullman	10/29/08	1963E	3658
CALAPAN		12SecDr	Pullman	8/22/04	1963	3098
CALCUTTA #332	P	12SecDr	Wagner	5/87	3061D	Wg3
CALCUTTA #391		12SecB	PCW Det	7/2/83	168	D193
CALDENO	P	14Sec	New Haven	5/82	3201A	
CALDER		12SecDrB	PCW Det	12/28/88	375H	D378
CALDERON		14SecDr	Pullman	12/30/01	1523F	2798
CALEDONIA		10SecDr	PCW Det	4/30/71	26	D6
CALEDONIA		12SecDr	PCW	8/9/98	1318A	2313
CALEDONIA #331	P	12SecDr	Wagner	5/87	3061C	Wg3
CALERA	P	10SecB	Bowers, Dure	5/31/73	37	Wil11
CALERA	N	12SecDr	Wagner	5/87	3061	Wg3
CALGARY #416	P	12SecDrB	Wagner	5/90	3065	Wg31
CALIENTE		10SecDr3Cpt	Pullman	4/18/06	2078C	3342
CALIFORNIA	R	12SecDrB	PCW?		1611	

Figure I-13 "The Overland Limited" acquired ten new 12-section drawing room smoking room sleepers, Plan 2079, in October 1905. These cars were given names that were appropriate for their service. Pullman Photograph P8095. Courtesy Robert Wayner.

CAR NAME	ST	CAR TYPE	CAR BUILDER (Owner)	DATE BUILT	PLAN	LOT
CALIFORNIA	N	7Cpt2Dr	Pullman	10/29/02	1845	2926
CALIGULA	N	26ChB	Wagner	11/92	3125	Wg74
CALIPH		12SecDrSr	PCW	5/25/99	1318C	2423
CALISTA #170	N	BaggBLib	(WS&PCC)	6/23/88	747B	
CALIURO		7Cpt2Dr	Pullman	10/29/02	1845	2926
CALLAO	N	12SecO	Grand Trunk	11/71	27C	Mon2
CALLAO		12SecDrB	Pullman	5/4/08	2326	3605
CALLIOPE		26ChDrSr	Pullman	6/6/04	1987A	3085
CALLISTO	N	7Cpt2Dr	PCW	9/26/94	1065A	2025
CALLISTO		6CptDrLibObs	Pullman	3/13/02	1751B	2799
CALMAR		12SecDrB	PCW Det	10/27/88	375G	D358
CALPE		10SecDrB	PCW	1/11/92	920B	1880
CALUMET		12SecO	PCW Det	1/15/81	91	D106
CALUMET		12SecDr	Pullman	9/24/09	1963E	3744
CALUMPIT		12SecDr	Pullman	8/18/04	1963	3098
CALVE	N	Private	PCW	12/14/01	1671	2746
CALVERT	N	10SecDrB	PCW Det	12/28/82	149	D174
CALVERT	R	12SecDrB	PCW Det	12/28/82	1611	D174

CAR NAME	ST	CAR TYPE	CAR BUILDER (Owner)	DATE BUILT	PLAN	LOT
CALYPSO		12SecDr	PCW Det	1/20/81	93	D88
CALYPSO		26ChDr	Pullman	7/27/04	1987A	3097
CALYPSO #339		12SecDr	PCW	9/6/84	93D	201
CALYPSO	N	20Ch6Sts	G&B (Wagner)	5/74	3014	
CAMANCHE		DDR	DetC&MCo	7/31/73	17	2
CAMANCHE	R	12SecDr	DetC&MCO	7/31/73	93B	2
CAMAS		12SecDr	Pullman	6/8/07	2271A	3458
CAMBODIA		12SecDrSr	Pullman	2/15/01	1581B	2638
CAMBRIA		12SecDr	PCW	12/15/98	1318C	2370
CAMBRIA No.1 #139	P	6SecBCh	PRR	7/31/73	17S?	Alt1
CAMBRIA No.2	P	10SecDr	B&S	1/71	26	Day6
CAMBRIDGE	N	Parlor	Vermont Cent	/71	8	StA
CAMBRIDGE	N	14SecO	PCW Det	11/2/75	53E	D34
CAMBRIDGE #500	P	12SecO	B&O RR	10/80	657	
CAMBYSES	N	6Sec12Ch	(Wagner)	9/70	3095	
CAMDEN	N	12SecO	B&O RR	11/80	657	
CAMDEN	N	30Ch	PCW	4/27/98	1278A	2293
CAMDEN #105	P	26Ch	(WS&PCC)	7/1/82	750	

38

CAR NAME	ST	CAR TYPE	CAR BUILDER (Owner)	DATE BUILT	PLAN	LOT	CAR NAME	ST	CAR TYPE	CAR BUILDER (Owner)	DATE BUILT	PLAN	LOT
CAMDEN #285		10SecDr	PCW Det	4/18/73	26 3/4	D15	CAPE MAY #240	P	SummerParlor	H&H	7/73	40	Wil2
CAMELON		12SecDrB	Pullman	2/19/06	2142	3318	CAPET	N	14SecO	(Wagner)	/76	3030	
CAMELOT		10SecDr	B&S	6/73	37H	Day9	CAPITANO	N	ClubLib	H&H	8/72	35	1
CAMELOT	N	12SecDrB	Wagner	11/92	3073A	Wg80	CAPITANO		BaggSmk	H&H	8/72	534	
CAMEO		12SecDrSr	PCW	4/5/99	1318C	2390	CAPITOL	P	Diner	PCW	/90	321A	?1235
CAMERON	P	14SecO	H&StJ RR	9/10/70	20	Han1	CAPITOLA CP-8070	P	10Sec2Dr	PCW	2/25/93	995A	1942
CAMERON	RN	10SecDrB	PCW	2/23/85	222E	1033	CAPPOQUIN	N	12SecDr	Pull Cal	6/15/04	1581G	C8
CAMERON		10SecDr3Cpt	Pullman	4/30/06	2078C	3342	CAPRI	N	6SecBCh	PRR	1/71	17S?	Alt1
CAMILLA	P	16Ch15Sts	PCW	/88	638		CARACAS		12SecDr	Pull Buf	9/5/05	1963B	B45
CAMILLO		12Sec2Dr	Pullman	4/4/01	1580A	2646	CARACAS		12SecDr	Pullman	4/15/04	1963	3082
CAMPANIA		12SecDr	PCW Det	6/14/82	93B	D154	CARAVEL		12SecDr	Pull Buf	8/29/05	1963B	B45
CAMPANIA		Private	PCW	1/27/98	1273A	2251	CARAVEL	N	16Sec	Pullman	7/14/03	1721	3014
CAMPANIA		12SecDr	Pull Buf	7/12/07	1963C	B63	CARAX		16Sec SUV	Pullman	4/1/10	2184G	3808
CAMPANIA	P	12SecDr	PCW Det	6/14/82	93B	D154	CARBONDALE	P	12SecO	(OPLines)	/72	4	
CAMPELLO	P	24ChB SUV	Pull P-1858	1/03	3269	2903	CARBONDALE		12SecDr	Pull Buf	7/18/07	1963C	B63
CAMPO SECO SP-46	P	10SecDrB	PCW	1/28/88	445E	1361	CARDENAS	N	10SecDr	PCW Det	3/15/82	112A	D141
CAMPVILLE	P	34Ch SUV	Pull P-1996C	7/04	3265	3114	CARDIFF		12SecDr	PCW	5/10/92	784E	1899
CANAAN	N	10SecDr	(Wagner)	1/85	3055		CARDIFF		12SecDr	Pull Wilm	7/19/04	1581C	W16
CANADA		16SecO	PCW	11/12/90	780	W5	CARDIGAN		12SecDr	Pullman	9/10/01	1581A	2756
CANADIAN		10SecDrB	PCW	11/10/87	445E	GT8	CARDINAL		12SecDrSr	PCW	5/24/99	1318C	2423
CANADIAN	N	6CptDrLibObs	Pullman	10/29/02	1845F	2926	CARDINAL		12SecDr	Pullman	4/24/03	1581B	3003
CANAJOHARIE	P	10SecDrB	J&S (Wagner)	4/81	3046		CARILLO		10Sec2Dr	Pullman	8/21/02	1746A	2925
CANANDAIGUA		12SecDrSr	Pullman	4/6/05	2049A	3169	CARLETON		Diner	PCW	11/1/90	662D	1776
CANARSIE		12SecDr	Pullman	11/2/08	1963E	3658	CARLETON		12SecDr	PCW	10/7/98	1318B	2339
CANASTOTA		12SecDr	Pullman	2/26/07	1963C	3525	CARLETON	P	14SecO	(Wagner)	/86	3032	
CANAWACTA E&A-25	P	Slpr	(E&A Assn)				CARLISLE	N	12SecDrB	PCW Det	9/29/88	375G	D358
CANDELA		12SecDr	Pull Buf	10/4/02	1581A	B10	CARLISLE	P	14SecO	(Wagner)	/86	3032	
CANDELA	R	30Ch SUV	Pull Buf	10/4/02	3581E	B10	CARLOMAN		10SecDr2Sr	Pullman	8/27/09	2078K	3722
CANDIDA		12SecDrB	Pullman	5/1/08	2326	3605	CARLOTA		22ChDr	Pullman	6/18/02	1647F	2863
CANELLA	P	16Ch4StsDr	B&S	7/82	3251		CARLSBAD		16SecO	PCW	6/15/93	1040	1990
CANELLE #318	N	8BoudB	(UnionPCC)		373		CARMANIA		12SecDr	Pullman	6/22/03	1581F	3010
CANETA		10SecDrB	PCW	12/9/86	445C	1261	CARMEL		12SecDr	Pullman	9/16/03	1581F	3022
CANISIUS		12SecDr	Pullman	12/16/05	1963C	3252	CARMEN		10SecDrB	PCW	11/3/87	445E	1334
CANMORE #417	P	12SecDrB	Wagner	5/90	3065	Wg31	CARMEN	P	10Sec2Dr	Wagner	11/97	3073D	Wg110
CANNES		12SecDr	PCW	2/10/92	784E	1881	CARMEN #118	P	7BoudB	(MannBCC)	/84	776	
CANNING		12SecDr	Pullman	9/15/03	1581F	3022	CARMENCITA	N	30Ch	PCW	4/28/98	1278A	2293
CANON CITY	N	12SecB	PCW Det	8/25/76	58	D42	CARMI	RN	12SecDr	PCW	6/22/88	1603	1415
CANONICUS	P	10SecDrB	J&S (Wagner)	4/81	3046		CARMINA	N	12SecDrB	PCW	9/16/98	1319A	2314
CANOPUS	N	12SecO	B&O RR	11/80	657		CARMITA		12SecDr	Pullman	10/30/08	1963E	3658
CANOPUS		16Sec	Pullman	6/25/08	2184C	3627	CARMONA		12SecDrB	Pullman	5/1/08	2326	3605
CANOSA	N	12SecO	B&O RR	11/80	657		CARNARVON		16Sec	Pullman	6/25/08	2184C	3627
CANOSA		12SecDr	Pull Buf	10/11/02	1581A	B10	CARNATION		24ChDrSr	Pullman	7/15/04	1980A	3088
CANTABRIA		16Sec	Pullman	10/26/01	1721	2766	CARNEROS	N	12SecDrB	PCW	2/18/87	1566	1298
CANTON	R	12SecDr	PCW?		1603		CAROLINA		10Cpt	Pullman	1/6/02	1704	2758
CANTON #158	P	12SecO	J&S (McComb)	12/70	18	Wil3	CAROLINA #325	P	8BoudB?	J&S (UnionPCC)	/89	373?	
CANUTE	N	20Ch8StsDrB	Wagner	12/90	3126	Wg44	CAROLYN		24ChDr	Pull Buf	3/25/07	2156	B58
CANYON		12SecDrCpt	Pullman	10/3/05	2079	3204	CARONDELET		12SecDr	Pull Buf	8/8/07	1963C	B65
CANYON CITY		12SecB	PCW Det	8/25/76	58	D42	CARPATHIA		12SecDrSr	PCW	10/5/99	1318E	2449
CAPE COD	P	34Ch SUV	Pull P-1484	1/03	3265	2904	CARRARA		12SecDr	Pullman	7/15/03	1581F	3019
CAPE MAY		24ChDrB	Pullman	6/6/05	1988A	3170	CARRIZO		7Cpt2Dr	Pullman	10/29/02	1845	2926

CAR NAME	ST	CAR TYPE	CAR BUILDER (Owner)	DATE BUILT	PLAN	LOT
CARROLLTON		12SecDr	PCW	10/17/98	1318B	2340
CARSON		10SecB	PCW Det	7/16/73	37	D16
CARSON		12SecDr	Pullman	9/13/01	1581A	2756
CARTAGENA		12SecDr	Pull Buf	4/15/04	1581F	B36
CARTHAGE	N	10SecB	Bowers,Dure	10/73	37	Wil11
CARTHAGE		12SecDrB	Pullman	3/21/02	1611B	2820
CARVUS	N	14SecO	PCW Det	3/9/93	1021	D516
CARYSTUS	R	BaggSmkBLib	PCW?		1706	
CARYVILLE	P	36Ch SUV	Pull P-2263	8/07	3275	3492
CASA GRANDE	R	12SecDr	PCW	2/19/87	1603	1298
CASA GRANDE SP-35	P	10SecDrB	PCW	2/19/87	445A	1298
CASA MONICA		Diner	PCW	12/7/88	538D	1411
CASANOVA		12SecDr	Pull Buf	9/28/07	1963C	B68
CASCADE		10SecO (NG)	PCW Det	12/7/81	73A	D120
CASCADE		12SecDrCpt	Pullman	10/2/05	2079	3204
CASCADE #415	P	12SecDrB	Wagner	2/90	3065	Wg32
CASCO	N	6Sec2DrB	Grand Trunk	11/70	17J	Mon1
CASCO	P	12SecDr	PCW (Wagner)	4/87	3049A	
CASHION		10SecDr2Cpt	Pullman	11/20/07	2078D	3583
CASIMIR		12SecDr	Pullman	5/10/01	1581A	2726
CASIMIR	N	12SecDr	(Wagner)	/86	3026	
CASINO	N	BaggClub	Pullman	8/18/06	2136A	3386
CASPAR	N	16SecB	J&S	1/99	2371B	
CASPIAN		10SecDr	PCW Det	5/20/72	26A	D10
CASPIAN		12SecDr	PCW	11/25/98	1318C	2359
CASSANDRA		20Ch6StsB	PCW	1/22/85	167B	200
CASSANDRA		16Sec SUV	Pullman	4/8/10	2184G	3808
CASSELTON	RN	14Sec	PCW	5/28/89	668F	1587
CASSIO		12SecDrCpt	Pullman	5/24/04	1982	3090
CASSIUS		BaggDiner	PCW	10/1/90	783A	1705
CASSIUS		BaggClubBarber	PCW	12/17/98	1372B	2365
CASSIUS		BaggClubBarber	Pullman	8/24/04	1372B	3130
CASSOPOLIS		16SecSr	Pullman	8/30/04	2033A	3135
CASTAH		14SecDr	Pullman	8/24/09	2438	3716
CASTALIA		14SecDr	PCW Det	11/2/75	53	D34
CASTALIA		22Berths	PCW Det	4/28/76	48	D32
CASTALIA		14SecDr	PCW	8/7/99	1318D	2447
CASTANO	N	12SecDrSr	Pullman	1/17/01	1581A	2614
CASTANO M&LT-21	N	10SecDrB	PCW	1/28/88	445F	1361
CASTARA	N	10SecDrB	PCW	9/21/87	445E	1365
CASTILE	R	12SecDr	PCW	6/2/88	1603	1414
CASTILE SP-59	P	10SecDrB	PCW	6/2/88	445F	1414
CASTILLO	N	Parlor	PRR	9/72	29	Alt1
CASTLE GATE		10SecB (NG)	PCW Det	10/16/83	178	D211
CASTLE ROCK		12SecDr	PCW Det	4/12/82	93B	D142
CASTLEFORD #424	P	12Sec2Dr	Wagner	6/90	3065A	Wg34
CASTLEMAINE #425	P	12Sec2Dr	Wagner	6/90	3065A	Wg34
CASTLETON		12SecDr	PCW	9/29/98	1318B	2339
CASTLEWOOD		14SecDr	Pull Buf	9/30/00	1523D	B1
CATAKA		10secDr2Cpt SUV	Pullman	3/10/10	2471	3784

CAR NAME	ST	CAR TYPE	CAR BUILDER (Owner)	DATE BUILT	PLAN	LOT
CATALDO	N	10SecDr2Cpt	Pullman	3/18/07	2078D	3528
CATALINA	N	10SecDrB	PCW	3/8/84	149	102
CATALINA	?	10SecDrB	PCW		149R	
CATALINA		10SecDr2Cpt	Pullman	3/18/07	2078D	3528
CATALONIA	N	12SecDr	PCW?		1581A	
CATALPA		24ChDr	Pull Buf	4/7/06	2156	B51
CATALPA #16	N	12SecO	(WS&PCC)	6/30/73	833	
CATANIA		12SecDr	Pull Buf	1/17/02	1581A	B7
CATANIA	R	30Ch SUV	Pull Buf	1/17/02	3581E	B7
CATARACT		10SecDrHotel	PCW Det	5/28/72	33	D11
CATARACT E&A-32	P	Parlor	(E&A Assn)		147A	
CATHAY		10SecDrB	PCW	11/5/83	149	80
CATHAY	R	12SecDrB	PCW	11/5/83	1468	80
CATO		10SecDrB	PCW	6/3/91	920B	1808
CATOHA		10SecDr2Sr	Pullman	8/20/09	2078J	3721
CATORCE		8SecSrB (NG)	PCW Det	11/92	1001	D514
CATSKILL		10SecDrB	PCW Det	4/25/83	149	D194
CATSKILL		12SecDr	Pullman	3/28/03	1581D	2983
CATSKILL #9	P	24Ch6Sts	(Wagner)	6/85	3119	
CAUCASUS		10SecDrB	PCW Det	7/28/84	149D	D248
CAUCASUS	R	12SecDrB	PCW Det	7/28/84	1468	D248
CAUGHNAWAUGA		10SecB	Grand Trunk	7/31/74	37	Mon5
CAVALIER		BaggSmk	PCW	8/3/88	534C	1410
CAVALIER		21Ch6StsDrObs	Pullman	3/22/06	2091B	3334
CAVARISTA		12SecDrB SUV	Pullman	4/28/08	2326	3605
CAVATINA		12SecDrB	Pullman	5/1/08	2326	3605
CAVITE		12SecDr	Pullman	8/18/04	1963	3098
CAWTHON		12SecDr	Pullman	2/10/08	1963C	3598
CAXTON	N	10SecDrB	PCW Det	5/5/83	149N	D194
CAXTON	N	12SecDr	Pullman	7/25/02	1581B	2852
CAYENNE		12SecDr	Pullman	12/28/03	1581H	3053
CAYUGA	P	DDR	DetC&MCo	/71	17	D2
CAYUGA		12SecDr	Pullman	11/4/08	1963E	3658
CAZENOVIA		12SecDr	Pull Buf	1/12/04	1581F	B27
CECELIA		16Ch12StsDrB	PCW	3/12/89	614B	1508
CECIL		12SecDr	PCW	5/27/91	784D	1806
CECROPIA		12SecDr	PCW Det	6/7/82	93B	D154
CEDAR BROOK #77	N	Sleeper	(WS&PCC)	7/1/79	524?	
CEDAR FALLS #301	P	20Ch12StsB	(MannBCC)	/89	708	
CEDAR HILL	P	BaggSmkB SUV	H&H	/88	3219	
CEDAR RAPIDS		12SecDr	Pull Buf	7/5/07	1963C	B63
CEDAR RAPIDS #325	P	12SecO	PCW (Wagner)	6/68	3013A	
CEDARHURST #94	N		J&S (WS&PCC)	7/23/81		
CEDARVALE		12SecDr	Pull Buf	8/15/07	1963C	B65
CEDRO		12SecDr	PCW	4/21/91	784D	1801
CELANDINE		12SecDrB	Pullman	5/1/08	2326	3605
CELAYA	RN	8SecSrB (NG)	PCW Det	9/26/82	1001	D120
CELAYA	N	12SecDr	Pullman	5/16/08	1963C	3616
CELESTE		20Ch6StsDr	PCW	7/3/91	928A	1826
CELESTINE		21Ch7StsDrObs	Pullman	10/6/05	2091	3228

CELIA
Wood

CAR NAME	ST	CAR TYPE	CAR BUILDER (Owner)	DATE BUILT	PLAN	LOT
CELIA		18Ch6StsDrB	PCW Det	6/13/90	772D	D424
CELILO		10SecDrB	PCW	3/26/87	445C	1281
CELTIC		10SecDrHotel	PCW Det	8/12/80	63 1/2	D77
CELTIC		14SecDr	PCW	8/5/99	1318D	2447
CENTAUR		12SecDr	Pullman	9/27/09	1963E	3744
CENTAUR #153	RN	10SecDrB	J&S (WS&PCC)		696	
CENTENNIAL		CafeSmk	Pullman	4/6/04	2007	3033
CENTENNIAL CTC-76	R	Slpr	PRR rblt/73			
CENTERDALE	P	26ChBarB SUV	Pull P-2493	1/11	3274	3820
CENTRAL #202	P	Parlor	CB&Q RR	7/31/71	29	Aur1
CENTRAL CITY		12SecO	B&S/GMP	1/67	5	Day
CENTRALIA	N	14SecO	PRR	7/73	69A	
CENTRALIA		12SecDr	Pullman	7/21/03	1581F	3015
CENTURION		12SecDr	Pull Buf	9/7/05	1963B	B45
CEPHALONIA		12SecDr	Pullman	9/7/00	1572	2585
CEPHEUS	N	12SecDrSr	Pullman	1/21/01	1581A	2614
CEPHEUS #131	N	12SecB	J&S (WS&PCC)	9/22/84	789A	
CERALVO		12SecDrB	Pullman	5/4/08	2326	3605
CEREDO	N	OS	C&NW Ry	2/71	27	
CERES		27StsParlor	PCW Det	/77	45	D31
CERES		26Ch3StsDr	PCW	4/25/98	1278C	2258
CERES		26ChDr	Pullman	7/27/04	1987A	3097
CERES #17	P	18Ch4StsDr	(Wagner)	6/85	3109	
CERES #228	P	Parlor	H&H	6/72	35	1
CERIGO		14SecO	PCW	2/4/93	1017C	1962
CERIGO	R	10SecDrB	PCW	2/4/93	1021W	1962
CERIMON		12SecDr	Pullman	3/26/10	1963E	3807
CERRO GORDO SP-41	P	10SecDrB	PCW	3/8/87	445B	1299
CERVANTES		12SecDr	PCW	10/13/92	990	1931
CERVERA		12SecDr	Pull Buf	9/25/02	1581A	B10
CESSNA		28ChDr	Pullman	4/9/06	1986A	3336
CETUS		12SecDr	PCW	1/26/91	784	1786
CEYLON		10SecDrB	PCW	3/14/84	149	102
CEYLON	P	10Sec2Dr	Wagner	11/98	3073D	Wg110
CEYLON	R	16Sec	PCW	3/14/84	1877A	102
Chair #1903		72Chair	Pullman	4/6/04	2006	3032
CHALDEA		10SecDrB	PCW	7/1/85	240	1057
CHALDEA	R	16Sec	PCW	7/1/85	1881	1057
CHALFONTE		16Sec	Pullman	6/6/06	2184	3361
CHALLENGER		10SecLibObs	Pullman	9/1/08	2351C	3634
CHALMETTE		12SecDr	Pull Buf	5/4/04	1581F	B36
CHALONS		16SecO	PCW	8/29/93	1040	2008
CHAMBERLAIN		10SecDr2Cpt	Pullman	1/2/08	2078D	3594
CHAMBLY	N	12SecDr	Pull Buf	7/15/07	1963C	B63
CHAMITA		8SecDrB	PCW	7/27/88	556A	1463
CHAMOUNI		16SecO	PCW	8/23/93	1040	2008
CHAMPAIGN #96	P	Slpr	(IC RR)			2/173
CHAMPION		12SecDr	Pullman	1/23/02	1581A	2792
CHAMPLAIN	P	10SecDr	Vermont Cent	4/72	31	StA2
CHAMPLAIN		12SecDr	Pullman	11/5/08	1963E	3658
CHAMPLIN		10SecDr3Cpt	Pullman	4/14/06	2078C	3342
CHANCELLOR		10SecDr2Cpt	Pullman	1/16/08	2078D	3594
CHAPALA		12SecDr	Pullman	7/14/06	1963C	3385
CHAPERONE		12SecDr	Pullman	8/7/08	1963C	3636
CHAPULTEPEC	P	12SecHotel	B&S	1/99	2370A	
CHARITON	N	DDR	C&NW Ry		17	Fdl
CHARITON		12SecDr	Pullman	11/2/00	1581A	2589
CHARLEMAGNE		12SecDr	Pullman	6/26/01	1581A	2726
CHARLEMONT		16Sec (SUV)	Pullman	10/29/09	2184G	3746
CHARLEROI		12SecDr	Pullman	4/21/04	1963	3082
CHARLESGATE		16Sec (SUV)	Pullman	10/30/09	2184G	3746
CHARLESTON		12SecDr	PCW	10/18/98	1318B	2340
CHARLEVOIX		16SecSr	Pullman	8/29/04	2033A	3135
CHARLIETTA	P	20Ch6StsDrB	J&S	1/02	2369	
CHARLOTTE	P	34Ch	Wason & Co	8/87	1900	
CHARLTON	P	12SecDr	CB&Q	1/71	17E	
CHARMARY		12SecDr	Pullman	10/30/08	1963E	3658
CHARMION		12Sec2Dr	PCW Det	1/2/88	537A	D334
CHARO		12SecDr	PCW	2/10/92	784E	1881
CHARTER OAK	P	Bagg16ChB	New Haven RR	11/04	3272	
CHARTLEY	P	36Ch SUV	Pull P-2263	8/07	3275	3492
CHARTRES		12SecDr	Pull Buf	8/19/05	1963B	B45
CHASKA	N	12SecDr	Pullman	12/19/02	1581B	2852
CHATAUQUA		DDR	PCW Det	/73	17	D2
CHATAUQUA		12SecDr	Pull Cal	6/15/04	1581A	C8
CHATAWA		10SecDr	PCW Det	9/23/82	112A	D171
CHATFIELD	N	12SecDr	PCW	10/27/98	1318B	2340
CHATHAM		16SecO	PCW	6/7/93	1040	1990
CHATTANOOGA	P	OS	(PK&RSCC)			
CHATTOOGA #338	P	10SecDrB	(UnionPCC)		565B	
CHAUCER		14SecDr	Pullman	12/12/01	1523F	2798
CHAUGA #342	P	10SecDrB	(UnionPCC)		565B	
CHAUTAUQUA	P	DDR	DetC&MCo	/71	17	D2
CHAVEZ		10SecDr	PCW Det	8/16/80	59	D78
CHEAT RIVER #502	P	12SecO	B&O RR	/80	657	
CHEBALLE	N?	12SecO	(NP Assn)			18D
CHECOTAH		12SecDrB	PCW Det	4/28/88	375E	D328
CHEHALIS	N	12SecO	(NP Assn)			18D
CHEHALIS		14SecDr	Pullman	5/16/06	2165	3352
CHELMSFORD		12SecDr	Pull Buf	10/16/01	1581A	B6
CHELSEA	N	12SecDr	PCW Det	9/8/82	93	D169
CHELSEA	P	Parlor	J&S	6/72	35	Wil1
CHELSEA	P	36Ch SUV	Pull P-2263	8/07	3275	3492
CHELSEA #170	R	30Ch SUV	Pull Buf	5/5/03	3581E	B16
CHELTENHAM	N	12SecDrB	PCW	8/4/88	375E	1395
CHEMNITZ		14SecO	PCW	4/25/93	1021	1972
CHEMUNG E&A-13	P	Slpr	(E&A Assn)			
CHENOA		10SecDr	PCW Det	1/31/79	37 1/4	D51
CHERBOURG		16SecO	PCW	8/5/90	780	1706
CHEROKEE		12SecDrB	Pullman	3/21/02	1611B	2820

41

CAR NAME	ST	CAR TYPE	CAR BUILDER (Owner)	DATE BUILT	PLAN	LOT
CHEROKEE #102	P	10SecB	(IC RR)		231	1/228
CHEROKEE CTC-57	R		Det rblt/74			
CHERRY BROOK	P	Bagg2CptB	New Haven RR	/02	3216A	
CHERRY BROOK	R	Bagg4DrBClub	New Haven RR	7/02	3267	
CHESAPEAKE #365		10SecDr	PCW Det	7/12/82	112A	D152
CHESHIRE		16SecO	PCW	8/8/93	1040	2008
CHESILHURST		24ChDrB	Pullman	6/9/05	1988A	3170
CHESTER	N	30 Ch	PCW	4/27/98	1278A	2293
CHESTER #27	P	19Ch4StsDr	(Wagner)	7/85	3117	
CHESTER #159		12SecO	J&S	12/70	18	Wil3
CHESTERTON		12SecDr	Pullman	11/26/00	1581A	2589
CHESWICK		12SecDr	Pullman	10/23/05	2049B	3232
CHESWOLD		12SecDr	Pullman	10/19/05	2049B	3232
CHETOLAH		12SecDrB	PCW Det	4/7/88	375G	D346
CHETOPA		14SecO	PCW	/88	603	
CHETOPAH	N	10SecDr	H&StJ RR	8/5/71	19	Han2
CHETWOOD		16Sec	Pullman	6/6/06	2184	3361
CHETWYND		12SecDr	Pullman	9/17/03	1581F	3022
CHEVALIER		10SecB14ChObs	PCW	12/23/90	868	1779
CHEVIOT	N	10SecB	B&S	9/5/73	37M	Day10
CHEVIOT	N	16Sec	Pullman	7/9/03	1721	3014
CHEWAH		14SecDr	Pullman	8/26/09	2438	3716
CHEYENNE		10Sec2Dr	PCW	2/21/93	995A	1942
CHEYENNE #28	P	OS	DetC&MCo/GMP	/69	5	Det
CHICAGO	RN	16SecB	PCW	6/7/93	1040I	1990
CHICAGO	P	8SecDrHotel	G&B (Wagner)	6/75	3091	
CHICAGO #23	P	OS	CNJ (WS&PCC)	/66		
CHICAGO #53	P	12SecDr	(CRI&P Assn)	3/18/83	527	
CHICAGO #58	P	12SecDr	B&S	/76-/78	197	
CHICAGO #68	P	25Ch2StsB	(Wagner)	10/81	3100A	
CHICAGO #101	P	12SecDr?	(MannBCC)	/89	878	
CHICAGO #506	P	12SecO	B&O RR		657	
CHICAGO #131	P	12SecB	J&S (WS&PCC)	9/22/84	789	
CHICHESTER		12SecDrSr	PCW	9/15/99	1318E	2449
CHICKAHOMINY		12SecDr	Pullman	11/29/01	1581A	2784
CHICKASAW	P	10SecDr	H&StJ RR	11/20/72	6	Han3
CHICKASAW	N	12SecDrCpt	Pullman	5/25/04	1985	3095
CHICO		10SecDrB	PCW	10/23/86	445	1262
CHICO	P	12SecDrB	PCW (Wagner)	4/81	3050A	
CHICOPEE		12SecDr	Pullman	7/17/03	1581F	3019
CHICORA		12SecDr	Pull Buf	9/29/02	1581A	B10
CHICOSA		8SecDrB	PCW	7/24/88	556A	1463
CHICOTA		12SecDr	Pullman	11/4/08	1963E	3658
CHIEFTAIN		12SecO	PCW Det	11/22/81	114	D122
CHIEFTAIN		12SecDr	PCW	12/1/98	1318C	2359
CHIHUAHUA		10SecDr	PCW Det	4/5/81	112	D149
CHILI		7Cpt2Dr	PCW	4/11/94	1065A	2025
CHILI #27	N	10SecDr	H&StJ RR	/71	19	
CHILLICOTHE #2		OS	H&StJ RR			
CHILTON		12SecDr	Pullman	10/3/00	1581A	2588
CHIMNAPUM		10secDr2Cpt SUV	Pullman	3/18/10	2471	3784
CHINA		12SecDrCpt	PCW	6/25/89	622	1529
CHINA #143	P	10SecDr	C&NW RR	10/70	19	Fdl3
CHINOOK	R	12SecDr	PCW	12/17/88	1603	1519
CHINOOK SP-52	P	10SecDrB	PCW	12/17/88	445H	1519
CHIPPETA		10SecO (NG)	PCW Det	2/23/83	73A	D120

Figure I-14 This rare photograph of the Detroit-built 10-section drawing room CHESAPEAKE #365 was found in the Archives at the B&O Railroad Museum, Baltimore, Maryland. Pullman Photograph 2797.

CAR NAME	ST	CAR TYPE	CAR BUILDER (Owner)	DATE BUILT	PLAN	LOT
CHIPPEWA	P	OS	CB&Q/GMP	/66		Aur
CHIPPEWA		OS	PCW Det	12/12/70	27	D4
CHIPPEWA		12SecDr	Pullman	11/12/01	1581A	2784
CHIPPEWA FALLS #11	P	14SecO	PCW (Wagner)	3/19/86	3018	1191
CHIQUITA		12SecDrB	Pullman	4/28/08	2326	3605
CHISPA SP-34	P	10SecDrB	PCW	2/18/87	445A	1298
CHITTENDEN		10SecDr2Cpt	Pullman	1/16/08	2078D	3594
CHIVINGTON		12SecDrB	PCW Det	5/31/88	375E	D328
CHIVOTA		7Cpt2Dr	Pullman	9/24/02	1845	2926
CHLORIS		20Ch5StsB	PCW	8/30/84	167A	199
CHOCTAW		12SecDr	Pull Buf	10/3/07	1963C	B68
CHOLULA		12SecDr	Pullman	7/17/06	1963C	3385
CHOPIN		12SecDr	PCW	4/29/92	784E	1899
CHOPIN	N	16Sec	Pullman	12/4/06	2184	3409
CHOTEAU		10SecDr2Cpt	Pullman	11/20/07	2078D	3583
CHRISTABEL		26ChDrSr	Pullman	6/7/04	1987A	3085
CHRISTIANA #149	P	6Sec4Dr?	J&S (WS&PCC)	5/17/86	720?	
CHRISTINE		22ChDr	Pullman	7/10/01	1647C	2702
CHRYSES		10Sec2Dr	PCW	5/15/96	1173A	2153
CHUSCA		7Cpt2Dr	Pullman	9/25/02	1845	2926
CICERO		10SecDrB	PCW	8/14/85	240	1069
CICOLA		14SecO	PCW	1/25/93	1017C	1962
CILICIA		12SecDr	Pullman	6/13/03	1581E	3010
CIMARRON		10SecB (NG)	PCW Det	10/23/83	178	D211
CIMBRIA	N	10SecDrB	Grand Trunk	11/87	445E	Mont8
CIMBRIA	N	12SecDr	Pullman	5/20/08	1963C	3616
CINAOLA		10SecDrB	PCW	5/18/88	445F	1413
CINAOLA	R	16Sec	PCW	5/18/88	1880	1413
CINCINNATI	?	?	?	8/1/67		
CINCINNATI		12SecDr	Pullman	4/26/07	1963C	3533
CINCINNATI #195	P	10SecDr	B&O RR	4/30/73	19	B&O1
CINCINNATI #417		14SecO	PCW	3/25/84	115	114
CINCINNATI #508	P	12SecO	B&O RR		657	
CINDERELLA	P	20Ch6StsDrB	J&S	1/02	2369	
CINEAS	N	BaggBLibSmk	Pullman	4/24/07	2087	3534
CINNABAR		10SecDrCpt	PCW	6/25/89	668	1587
CIPANGO		10SecDrB	PCW	5/28/89	623B	1567
CIPANGO	N	12Sec2Dr	Pullman	4/4/01	1580A	2646
CIRCASSIA		12SecDr	PCW Det	1/9/82	93B	D135
CIRCASSIA		14SecDr	Pullman	6/16/00	1523	2559
CISCO		12SecDr	PCW	6/12/91	784D	1806
CITRA		12SecDr	PCW	4/24/91	784D	1801
CITY of ALBANY E&A-9	P	Slpr	(E&A Assn)			
BALTIMORE #82	R		PW&B	/73		
BINGHAMTON E&A-7	P	Slpr	(E&A Assn)			
BOSTON		12SecDr Hotel	CB&Q/GMP	/67	13B	Aur
BUFFALO E&A-40	P	Parlor	(E&A Assn)			
CITY of CHICAGO	P	12SecO	C&NW RR	12/70	5	Fdl
CHICAGO	P	OS	CB&Q/GMP	2/66		Aur
CHICAGO	P	DDR	Ohio Falls		17	Jef
DETROIT	P	OS	MC RR/GMP	8/1/67	7C	Det
DUBUQUE		OS				
FOND DU LAC	P	OS	C&NW RR			Fdl
GALVESTON	P	OS	Ohio Falls	-/75		Jef
HUNTSVILLE	P	SDR	(Mem&CharRR)			
LONDON	N	12SecO	B&S/GMP	/68	5	Day
MADISON	P	OS	C&NW RR			Fdl
MEMPHIS	P	SDR	(Mem&CharRR)			
MOBILE #99	P	Slpr	(IC RR)			
NASHVILLE	P	12SecO	Nash&ChatRR	/67	5	
NEW ORLEANS	P	OS	(OPLines)	/72	75	
NEW YORK		12SecO	CB&Q/GMP	/66	13	Aur
NEW YORK #5	P	12SecO	J&S (WS&PCC)	6/73	833?	
NEW YORK E&A-39	P	18Ch7StsB	(E&A Assn)	/72	147	
OSHKOSH	P	OS	C&NW RR	/65	27	Fdl
QUINCY	P	SDR	CB&Q RR	C:12/68	12	Aur
ROCHESTER		SDR	DetC&MCo	/68	12	Det
ST. LOUIS #4	P	12SecO	(WS&PCC)	6/73	833?	
CIUDAD BLANCA	RN	12SecDrB	Pullman	5/23/06	2165C	3352
CIVRY		12SecDr	PCW	2/10/92	784E	1881
CLAGHORN		BaggBLibSmk	Pullman	4/24/07	2087	3534
CLARA		Slpr	J&S (WS&PCC)	/73		
CLARE	N	24Ch6StsB	PCW Det	8/3/76	54B	D40
CLAREMONT		12SecDrB	PCW	9/16/98	1319A	2314
CLAREMONT	P	12SecDr	B&S (Wagner)	5/81	3022B	
CLAREMONT #287		10SecB	B&S	6/30/75	37	Day9
CLAREMORE		12SecDr	Pull Buf	8/12/07	1963C	B65
CLARENDON	P	Commissary	CB&Q RR			Aur
CLARENDON	P	10SecDrB Hotel	PullPac C.C.	7/31/75	27	
CLARENDON		12SecDr	PCW	9/28/98	1318B	2339
CLARENDON	P	12SecDr	(Wagner)	6/81	3022A	
CLARENDON No.2	N	DDR	PFtW&C	8/70	17C	FtW2
CLARIBEL		18Ch4StsDrB	PCW	7/19/93	1039A	1995
CLARICE		24ChDr	Pull Buf	4/1/07	2156	B58
CLARIDGE		26ChDr	Pullman	5/29/07	1987B	3524
CLARINE		30Ch	Pullman	5/17/01	1646D1	2700
CLARION		12SecDr	PCW Det	1/7/82	93B	D135
CLARION		12SecDr	Pullman	1/22/02	1581A	2792
CLARISSA		12SecDrB	PCW StL	4/14/88	375E	StL1
CLARKSVILLE	P	12SecO?	(L&N RR)	/69	114?	
CLARNIE		10SecDrB	PCW	10/7/87	445E	1334
CLATSOP		12SecDr	Pullman	6/11/07	2271A	3458
CLAUDIA		18Ch4StsDrB	PCW	6/30/93	1039A	1994
CLAUDINE		30Ch	Pullman	5/21/01	1646D1	2700
CLAUDIO		12Sec2Dr	Pullman	4/4/01	1580A	2646
CLAUDIUS	N	Slpr	(IC RR)			2/173

CLAUDIUS
Wood

CAR NAME	ST	CAR TYPE	CAR BUILDER (Owner)	DATE BUILT	PLAN	LOT
CLAUDIUS		BaggClubBarber	Pullman	5/17/04	1372B	3094
CLAUDIUS		BaggClubBarber	PCW	12/3/98	1372B	2365
CLAVERACK	P	12SecDr	B&S (Wagner)	6/81	3022A	
CLAYMONT		12SecDrCpt	Pullman	8/8/05	2088	3221
CLAYTON		12SecDr	Pullman	10/6/00	1581A	2588
CLEANTHUS		12SecDrB	Pullman	4/28/08	2326	3605
CLEARFIELD	N	16Sec	Pullman	7/16/03	1721B	3014
CLEARFIELD		12SecDrSr	Pullman	3/30/05	2049A	3169
CLEARFIELD #303	P	10SecDr	Elmira Car W	6/74	37	Elm2
CLEARVIEW		24ChDrB	Pullman	6/3/05	1988A	3170
CLEARWATER		12SecDr	Pull Buf	10/10/07	1963C	B68
CLEBOURNE		10SecDrB	PCW	1/5/88	445E	1360
CLEMATIS		12SecDrCpt	PCW	8/5/92	983	1919
CLEMENT	N	12SecDrB	Wagner	4/91	3065	Wg50
CLEMENTINE		22ChDr	Pullman	7/6/01	1647C	2702
CLEMENTON (#167)	RN	37Ch SUV	Pullman	4/15/03	3581P	2984
CLEMSON		12SecDr	Pullman	12/19/05	1963C	3252
CLEOMENES	N	16Ch3Sts	(Wagner)	5/85	3114	
CLEOPATRA	RN	Private	PCW Det	6/12/77	63I	D45
CLEOPATRA		24ChDr	Pullman	6/27/05	1980A	3171
CLEOPATRA		Slpr	(WS&PCC)	/73		
CLEORA	N	12SecDrSr	PCW	12/20/98	1318C	2370
CLEORA #26	N	12SecO	(WS&PCC)	6/27/74	833	
CLERMONT	N	10SecDrB	PCW	11/3/87	445E	1334
CLERMONT		Diner	PCW	11/3/90	662D	1776
CLETHRA		10SecLibObs	Pullman	5/6/09	2351H	3688
CLEVELAND	N	12SecO	I&StLSCCo	/71		
CLEVELAND	RN	14Sec	PCW	6/8/89	668F	1587
CLEVELAND		12SecDr	PCW	8/1/98	1318A	2313
CLIFFWOOD		24ChDrB	Pullman	6/9/05	1988A	3170
CLIFTON	P	12SecO	B&S/GMP	/68	5	Day
CLIFTON		12SecDr	Pull Buf	8/14/03	1581F	B21
CLIFTON E&A-14	P	Slpr	(E&A Assn)			
CLIFTON FORGE		10SecDr	PCW Det	7/12/82	112A	D152
CLIMAX		12SecDr	Pullman	12/15/06	1963C	3448
CLINTON	N	10SecDrB	PCW Det	5/5/83	149N	D194
CLINTON	N	12SecDrB	(Wagner)	11/82	3025	
CLIO	N	10SecB	B&S	9/15/73	17R	Day10
CLIO		26ChDr	Pullman	7/27/04	1987A	3097
CLIO	R	10SecB	B&S	9/15/73	37M	
CLIOLA		10SecDr	PCW Det	10/5/81	112A	D124
CLIOLA	N	6Sec4Dr	Wagner	6/93	3074A	Wg93
CLIVE		12SecDr	PCW Det	2/5/92	784E	D488
CLIVEDEN	N	14SecO	PCW Det	12/29/75	53E	D34
CLOCHETTE		12SecDr	Pullman	10/31/08	1963E	3658
CLODION	N	16SecO	Wagner	6/96	3072A	Wg108
CLOQUET	RN	12SecDr	PCW	8/12/91	937E	1838
CLORINDA	P	28Ch3StsB	PCW	1/98	1167A	2268
CLORINDA	N	18Ch6Sts	PCW	9/95	1153	2116
CLORITA	P	28Ch3StsB	PCW	1/98	1167A	2268
CLOTAIRE	N	12SecDrB	Wagner	4/91	3065	Wg50
CLOTHO	N	10SecB	PCW Det	10/7/72	26Y	D13
CLOTHO	N	6Sec4Dr	Wagner	6/93	3074A	Wg93
CLOTILDA	P	20ChB	Wason & Co	8/87	1902	
CLOVER		12SecDr	PCW	2/28/91	784B	1798
CLOVERCROFT	N	12SecDr	Pull Buf	7/6/07	1963C	B63
CLOVERDALE		10Sec3Cpt	Pullman	2/6/06	2098	3300
CLOVERDALE #429	P	12Sec2Dr	Wagner	8/90	3065A	Wg37
CLOVIS	N	16Ch3Sts	(Wagner)	3/85	3114	
CLURO		10SecDr2Cpt	Pullman	12/30/05	2078C	3287
CLYDE		10SecDrB	PCW	8/15/87	445E	1333
CLYDE	R	12SecDrB	PCW	8/15/87	1566	1333
CLYDE	P	12SecDrB	(Wagner)	3/81	3050	
CLYMENE		6CptDrLibObs	Pullman	2/25/02	1751B	2799
CLYTIE		22ChDr	Pull Buf	5/28/02	1647C	B8
Coach #1803		72StsCoach	Pullman	4/6/04	2005	3031
Coach #1893		Coach	PCW	4/25/93	1035	1985
Coach #20		Coach	PCW	5/18/92	726C	1909
COANZA		10SecDrB	PCW Det	8/4/84	149D	D248
COANZA	N	12SecDr	Pullman	5/18/08	1963C	3616
COBDEN		12SecDr	Pullman	9/19/03	1581F	3022
COBHAM		12SecDr	Pullman	9/21/03	1581F	3022
COBLENTZ		14SecO	PCW Det	4/15/93	1021	D518
COBLENTZ	N	12SecDr	Pullman	5/23/08	1963C	3616
COBOURG		12SecO	Grand Trunk	7/71	18	Mon1
COBURG	N	12SecDr	Wagner	2/92	3065C	Wg64
COCHISE	N	12SecDrCpt	Pullman	5/26/04	1985	3095
COCONINO		10SecDrB	PCW	12/7/87	445E	1360
COCONINO	R	16Sec	PCW	12/7/87	1880A	1360
COCOPAH GH&SA-9	P	10SecB	(CP Co)	6/11/84	37	
CODRUS		12SecDr	Pull Buf	8/25/05	1963B	B45
COHASSET		12SecDr	Pullman	11/17/03	1581F	3043
COLA		12SecDr	PCW	10/16/90	784	1766
COLANO		10Sec2Dr	Pullman	8/22/02	1746A	2925
COLCHESTER		12SecDrSr	PCW	9/14/99	1318E	2449
COLCHESTER	P	10SecDr	(Wagner)	/85	3056	
COLD SPRINGS	P	12SecO	(Wagner)	/71	3037D	
COLERAINE		16Sec (SUV)	Pullman	10/29/09	2184G	3746
COLERIDGE		14SecDr	Pullman	12/14/01	1523F	2798
COL JAMES FISK JR.	P	Slpr	(E&A Assn)			
COLEUS		10SecDr2Cpt SUV	Pullman	11/29/09	2451	3747
COLFAX		12SecDrB	PCW	3/13/90	375J	1662
COLIMA		12SecDr	Pullman	7/16/06	1963C	3385
COLLINGDALE #579	P	10SecDrB	J&S (B&O)	C:/88	663	
COLLINSVILLE	P	24Ch3StsDr	B&S	9/92	3261	
COLOGNE		12SecDr	Pullman	10/21/03	1581F	3025
COLOGNE #118	RN	7BoudB	(MannBCC)	/84	776A	
COLOMA		12SecDr	Pull Buf	10/17/07	1963C	B68
COLOMA	N	10SecO	(Wagner)	/72	3001	
COLOMA SPofA-1	P	Slpr	(SP Co)	7/1/83	37	

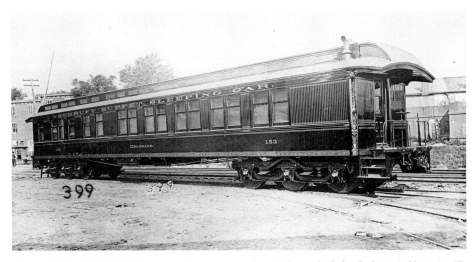

Figure I-15 COLORADO, brown with an ochre-yellow roof, was built by Jackson & Sharp for The Woodruff Sleeping & Parlor Coach Company as a sleeper/parlor car for assignment to the "Wabash Cannon Ball." It was later converted to a 10-section drawing room buffet and renamed CENTAUR. The Robert Wayner Collection.

CAR NAME	ST	CAR TYPE	CAR BUILDER (Owner)	DATE BUILT	PLAN	LOT
COLOMBIA		7Cpt2Dr	PCW	9/26/94	1065A	2025
COLOMBO		12SecDr	Pullman	4/20/04	1963	3082
COLON		12SecDr	PCW	4/20/91	784D	1801
COLONEL E.W. COLE	RN	18ChLngCafe SUV	Pull	6/4/08	3968A	3623
COLONIA		12SecDr	Pull Buf	10/20/03	1581F	B24
COLONIAL		Private	Pullman	12/11/01	1672	2747
COLONNADE	RN	Diner	CB&Q RR	12/9/71	34C	
COLORADO		12SecO	DetC&MCo	C:12/70	5	Det
COLORADO	P	Bagg Smk	PCW	/99	1400	2410
COLORADO #56	P		(CRI&P Assn)	10/6/80		
COLORADO #153	P	SlprParlor	J&S (WS&PCC)		450?	
COLORADO #329	P	12SecDr	Wagner	5/88	3061A	Wg3
COLTON		10SecDr3Cpt	Pullman	4/30/06	2078C	3342
COLUMBA		6SecCptSr	PCW Det	1/16/80	68	D54
COLUMBIA	P	12SecO	B&S/GMP	/68	5	Day
COLUMBIA	N	Private	PCW	10/30/99	1418C	2451
COLUMBIA	N	22Ch6StsDr	PCW	4/25/93	1031	1978
COLUMBIA #41	P	20Ch4Sts	(Wagner)	9/86	3105	
COLUMBINE		12SecDrCpt	PCW	8/1/92	983	1919
COLUMBUS	N	12SecO	PCW Det	11/13/71	27	D8
COLUMBUS		12SecDr	Pullman	11/13/01	1581A	2784
COLUMBUS #16	P	12SecO	(WS&PCC)	6/30/73	833	
COLUMBUS #525	P	12SecO	B&O RR	/81	665	
COLUMBUS CTC-91	R		PRR rblt/73			
COLUSA SPC-8155	P	10Sec2Dr	PCW	3/9/99	1158	2376

CAR NAME	ST	CAR TYPE	CAR BUILDER (Owner)	DATE BUILT	PLAN	LOT
COLVILLE		12SecDrB	Pullman	4/30/08	2326	3605
COMAL		12SecDrB	PCW Det	4/28/88	375E	D328
COMANCHE	P	DDR	DetC&MCo	/71	17	D2
COMANCHE	R	12SecDr	DetC&MCo	/71	93B	D2
COMANCHE	N	14SecO	B&S	7/92	1593	
COMET	S	27StsParlor	PCW Det	/76	45	D31
COMET #260		10SecDr	G,B&Co/GMP	7/10/68	10	Troy
COMETA		10Sec2Dr	Pullman	9/3/02	1746A	2925
COMINIUS		12SecDr	Pull Buf	4/28/10	1963E	B83
COMITAN		12SecDr	Pull Buf	10/24/07	1963C	B68
COMMODORE	N	12SecDr	PCW	6/7/87	473A	1323
COMMODORE VANDERBILT		Parlor	(Wagner)			
COMMONWEALTH		Private	Pullman	1/31/06	2110	3262
COMMUNIPAW #168	P	19ChDrB	J&S (WS&PCC)	12/9/87	731	
COMMUNIPAW #562	P	11SecDr	PCW	2/23/85	222	1033
COMO		12SecDr	PCW Det	9/1/82	93	D169
COMO	N	16SecO	Wagner	3/93	3072	Wg89
COMO #170	P	BaggBLib	(WS&PCC)	6/23/88	747?	
COMPEER		12SecDr	Pullman	11/5/08	1963E	3658
COMPTON	N	10SecDrB	PCW Det	5/3/83	149N	D194
COMPTON		Diner	PCW	11/5/90	662D	1776
COMPTON		10SecDr2Cpt	Pullman	3/22/07	2078D	3528
COMUS	P	23ChDr	Bowers, Dure	7/72	35	Wil1
COMUS		12SecDr	Pullman	7/15/08	1963C	3630
COMYN #77	N	Slpr	(WS&PCC)	7/1/79	755A?	
CONCHO		10SecDrB	PCW	8/11/87	445E	1333
CONCHO	N	12SecDrCpt	Pullman	5/24/04	1985	3095
CONCORD	P	12SecDr	VC RR	8/71	21B	StA1
CONCORD		12SecDr	Pull Buf	3/15/04	1581G	B35
CONCORD	P	10SecDr	(Wagner)	/85	3056	
CONCORDIA		12SecDrSrB	PCW	2/25/99	1319A	2400
CONDOR		24ChDrB	Pullman	7/9/00	1518C	2553
CONDRON		12SecDr	Pullman	10/12/07	2163A	3562
CONEMAUGH		16Sec	Pullman	7/7/03	1721	3014
CONEMAUGH #80	N	SlprParlor	(WS&PCC)	6/1/80		
CONEMAUGH #169	P	10SecDr	PRR	2/71	26 3/4	Alt1
CONESTOGA		26ChDrB	Pullman	6/25/07	1988B	3526
CONESTOGA #182	P	10SecDr	PRR	2/72	26	Alt2
CONEWAGO		16Sec	Pullman	7/7/03	1721	3014
CONEWAGO #288		10SecB	B&S	6/30/75	37	Day9
CONFEDERATION	N	5CptDrLibObs SUV	Pullman	10/29/02	1845F	2926
CONFIANZA	N	10SecDrB	PCW Det	12/22/83	149	D213
CONFUCIUS		7CptLngObs	Pullman	6/16/04	1977B	3091
CONGAREE #3	R	10SecO	PCW Det (rblt)	3/31/75	47	D27
CONGAREE #31	N	8BoudB?	J&S (Union PCC)		373?	
CONGO	N	6Sec2DrB	Grand Trunk	10/70	17J	Mon1
CONGO		12SecDr	Pullman	2/19/03	1581D	2975
CONISTON		12SecDr	Pullman	3/1/07	1963C	3525
CONNECTICUT		12SecB	PCW Det	7/6/76	57	D41

CAR NAME	ST	CAR TYPE	CAR BUILDER (Owner)	DATE BUILT	PLAN	LOT
CONNECTICUT		12SecDr SUV	Pullman	4/21/07	1963C	3533
CONNELLS-VILLE #204	P	10SecDr	H&H (B&O)	C:/71	26	Wil6
CONQUEROR		10SecLibObs	Pullman	9/1/08	2351C	3634
CONSORT		10SecB14ChObs	PCW	12/23/90	868	1779
CONSTANCE		22Ch6StsDr	PCW	8/29/92	976A	1910
CONSTANCE		32ChDr	Pullman	3/13/07	2274	3509
CONSTITUTION		Private	Pullman	1/29/06	2110	3262
CONSUELO		26ChDrSr	Pullman	6/8/04	1987A	3085
CONSUL	N	12SeCDr	PCW	6/11/87	473A	1323
CONTENTO	R	10SecObs	PCW	3/19/88	445V	1377
CONTENTO SP-50	P	10SecDrB	PCW	3/19/88	445F	1377
CONTINENTAL		10SecDrHotel	PCW Det	4/3/72	33	D11
CONTINENTAL		Diner	PCW	5/31/88	538D	1411
CONTRA COSTA	R	10SecObs	PCW	2/18/87	445W	1298
CONTRA COSTA SP-32	P	10SecDrB	PCW	2/18/87	445A	1298
CONVOY	N	10SecO	J&S (McComb)	/71	99A	
CONVOY #61	P	Private	Wagner	11/92	3085	Wg86
CONWAY	P	Parlor	H&H	8/72	35	1
CONWAY		26Ch6StsDr	PCW		1061	
CONYAMOND	P	20Ch6StsDr	Pull	2/91	3259A	
COOPER		12SecDr	PCW	10/31/92	990	1931
COOPERSTOWN #23	P	20Ch4Sts	(Wagner)	8/86	3102	
COOZA	R	10SecO	PCW Det (rblt)	11/20/74	47	D27
COPELAND		12SecDr	Pullman	1/28/07	1963C	3448
COPENHAGEN		12SecDr	Pull Buf	4/27/04	1581F	B36
COPIA		Diner	PCW	11/7/90	662D	1776
COPLEY		12SecDrB SUV	Pullman	12/17/07	2326	3587
COPPEI	P	10SecDr?	(NP RR)			
COQUETTE	N	24ChDrB	Pullman	6/9/05	1988A	3170
CORACLE	N	16Sec	Pullman	1/15/08	2184	3588
CORALINE		10SecDr2Sr	Pullman	8/27/09	2078K	3722
CORANUS	N	12SecDrSr	Pullman	1/22/01	1581A	2614
CORANUS	N	22Ch4StsDr	(Wagner)	6/85	3108D	
CORCORAN		10SecDr2Cpt	Pullman	11/15/07	2078D	3583
CORDELE	P	12SecDrB	(GS&F Ry)		2490	
CORDELIA		18Ch4StsDrB	PCW	7/17/93	1039A	1995
CORDERO		10Sec2Dr	Pullman	10/31/02	1746A	2928
CORDERO O&C-8002	P	10SecDrB	PCW	5/27/88	445F	1377
CORDOVA		12SecDr	PCW Det	6/14/82	93B	D154
CORDOVA #330		12SecDr	Wagner	5/87	3061	Wg3
COREA	N	6SecBCh	PRR	5/71	17S?	Alt1
COREA		12SecDr	Pull Buf	10/2/02	1581A	B10
CORIANDER		24ChDr	Pull Buf	4/26/06	2156	B52
CORINNE	N	12SecO	B&O RR	/81	657	
CORINNE		24ChDr	Pullman	6/27/05	1980A	3171
CORINTH		10SecDr	B&S	4/71	19	Day5
CORINTH	N	12SecDrB	(Wagner)	6/86	3050A	
CORINTHIA		12SecDr	PCW Det	6/7/82	93B	D154
CORINTHIA		12SecDr	Pull Buf	7/6/07	1963C	B63
CORINTHIAN		12SecDrSr	PCW	12/8/98	1318C	2370
CORISCO		12SecDr	Pullman	3/18/03	1581D	2975
CORMORANT	N	26ChDrB	Pullman	6/28/07	1988B	3526
CORNELIA	P	18Ch6Sts	PCW	9/95	1153	2116
CORNELL		12SecDr	Pullman	9/18/03	1581F	3022
CORNING		12SecDr	Pull Buf	7/15/07	1963C	B63
CORNING	N	23ChDrLibObs	Pullman	8/28/06	2206	3388
CORNING E&A-5	P	10SecDr	(E&A Assn)	/73	26-O	
CORNWALL	P	10SecB	(E&A Assn)	/76	37	
CORNWALL		12SecDr	Pullman	7/14/03	1581F	3019
CORNWALL	N	10SecDr2Cpt	Pullman	11/15/07	2078D	3583
CORNWALL	P	12SecO Tourist	(Wagner)	1/71	3037C	
CORONA	N	10SecDr	PCW Det	10/1/81	112A	D124
CORONA		12SecO	PCW Det	5/5/80	58	D74
CORONA		16Sec	Pullman	3/28/07	2184	3523
CORONA BLANCA	RN	12SecDrB	Pullman	5/18/06	2165C	3352
CORONADO		Diner	PCW	1/11/89	538D	1411
CORONADO #1		Diner	PCW	4/26/88	538D	1396
CORONET		16Sec (SUV)	Pullman	11/1/09	2184G	3746
CORONET #101	N	Private	G&B (MannBCC)	/83	691	
COROZAL	N	12SecDr	Pullman	10/12/07	2163A	3562
CORRY		12SecDrCpt	Pullman	5/23/04	1985	3095
CORSAIR		13SecO	J&S	6/82	187	
CORSAIR (2nd)	N	10Cpt	Wagner	6/96	3075A	Wg111
CORSAIR (1st)	RN	Private	Wagner	11/88	3084	Wg16
CORSICAN		10SecDr	PCW Det	5/20/72	26A	D10
CORSICAN		14SecDr	Pullman	6/11/00	1523	2559
CORTEZ		10SecDrB	PCW	12/9/86	445C	1261
CORTEZ	R	14Sec	PCW	12/9/86	445X	1261
CORTLAND		10SecDrB	PCW Det	3/23/87	445C	D282
CORUNNA		12SecDr	Pull Buf	10/8/02	1581A	B10
CORUNNA #114	RN	7BoudB	(MannBCC)	/84	697A	
CORVALLIS		10SecDr3Cpt	Pullman	4/26/06	2078C	3342
CORVETTE	RN	Private	Wagner	6/96	3075A	Wg104
CORVINUS	N	BaggBSmk	PCW	/99	1400A	2410
COS COB	P	Bagg16ChB	Pull P-1995	7/04	3272	3115
COSETTE	N	26ChDr	Pullman	3/26/06	1987A	3335
COSHOCTON	N	10SecDr	Elmira Car W	6/30/73	37	Elm1
COSHOCTON		12SecDr	PCW	10/27/98	1318B	2340
COSMOPOLITAN		Diner	CB&Q	6/30/71		Aur
COSSACK		12SecDr	Pull Buf	8/24/05	1963B	B45
COSTILLA		7Cpt2Dr	Pullman	10/29/02	1845	2926
COTSWOLD	N	12SecDr	Pullman	3/12/06	2163A	3333
COUNCIL BLUFFS	P	OS	B&S/GMP	8/1/67	17 1/2	
COUNTESS	P	14Ch6StsDrB	PCW	11/27/80	154	34
COUNTESS	N	22Ch6StsDr	PCW	4/25/93	1031B	1978
COURIER		10SecDr	PCW Det	11/24/81	59	D93
COURIER	P	Private	Wagner	11/92	3086	Wg118
COURTIER		10SecB14ChObs	PCW	12/23/90	868	1779

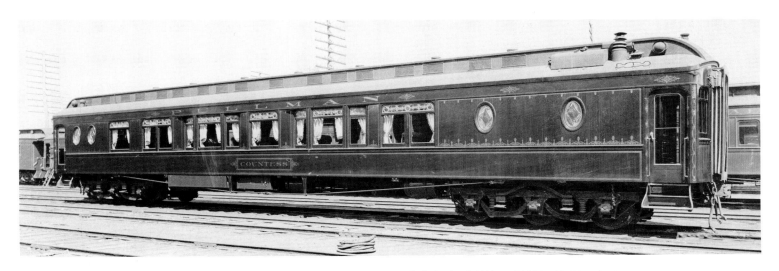

Figure I-16 COUNTESS has an extraordinary life story. It was built in April 1893 as SANTA MARIA at a cost of $38,365.46 for display at the Columbian Exposition. In March 1898 it was renamed COLUMBIA for service on the Pennsylvania Railroad. In January 1899, again it was renamed to COUNTESS and assigned to the Baltimore & Ohio's "The Royal Limited." If you look closely, you will note that this car has "bay windows," the first use of this unique innovation. Outdated, but certainly not outstyled, it spent the last eleven years of its life as a Supply and Porter Car at Buffalo, N.Y. Smithsonian Institution Photograph P4578.

CAR NAME	ST	CAR TYPE	CAR BUILDER (Owner)	DATE BUILT	PLAN	LOT	CAR NAME	ST	CAR TYPE	CAR BUILDER (Owner)	DATE BUILT	PLAN	LOT
COURTNEY		12SecDr	Pull Buf	6/28/05	1963	B42	CRESCENT	P	DDR	DetC&MCo	/70	17	D3
COVENTRY		14SecO	PCW Det	5/6/93	1021	D518	CRESCENT		12SecDr	Pullman	4/18/03	1581B	3003
COVINGTON		12SecDr	PCW	10/1/98	1318B	2339	CRESCENT #211	P	12SecO	(Wagner)	/83		3014
COWLITZ		14SecDr	Pullman	5/17/06	2165	3352	CRESCENT						
COWPER		14SecDr	Pullman	12/30/01	1523F	2798	BEACH	P	16Ch23StsBObs	Pull P-1486A	6/00	3266	2519
COXSACKIE	P	10SecDrB	J&S (Wagner)	4/81		3046	CITY	P	14SecO	(OPLines)	/78	214	
COYETA		12SecDrB	PCW Det	4/14/88	375E	D328	CITY #133	P	10SecDrB	J&S (WS&PCC)	1/15/85	886	
CRACOW		12SecDr	Pullman	7/13/03	1581F	3019	CITY	RN	8SecBLibObs	PCW	6/29/88	1707	1415
CRAFTON		BaggBLibSmk	Pullman	4/24/07	2087	3534	CRESCIUS		16Sec	Pullman	6/8/06	2184	3361
CRAGANOUR	N	10SecDr2Cpt	Pullman	10/1/06	2078D	3398	CRESCIUS	N	BaggClub SUV	Pullman	8/20/06	2136A	3386
CRAIGEL-							CRESCO	P	14Ch5StsB	(DL&W RR)	9/83	211	
LACHIE #423	P	12SecDrB	Wagner	4/90	3065	Wg33	CRESCO	N	12SecDrCpt	Pullman	5/23/04	1985	3095
CRANBROOK	N	12SecDr	Pullman	4/26/07	1963C	3533	CRESHEIM		12SecDrCpt	Pullman	8/9/05	2088	3221
CRANSTON	P	36Ch SUV	Pull P-2494	3/11	3275	3821	CRESHEIM		12SecDrSr	Pullman	5/24/07	2088	3553
CRATHORNE #422	P	12SecDrB	Wagner	4/90	3065	Wg33	CRESPINO		12SecDr	Pull Buf	1/20/03	1581C	B12
CRAYFORD		12SecDr	Pull Buf	10/22/01	1581A	B6	CRESSIDA		26ChDrSr	Pullman	6/6/04	1987A	3085
CREEDMOOR		12SecDr	Pullman	1/24/03	1581D	2962	CRESSON		12SecDrSr	Pullman	1/22/01	1581A	2614
CREIGHTON		12SecDr SUV	Pullman	12/19/05	1963C	3252	CRESSON #132	P	14SecO	PRR	8/70	17	Alt1
CREMONA	N	10SecDr	Grand Trunk	7/73	31E	Mon4	CRESSY #107	N	Parlor?	(WS&PCC)	/73	698A	
CREMONA		12SecDr	Pull Buf	10/18/02	1581A	B10	CRESTLINE		12SecDr	Pullman	2/7/08	2163A	3599
CREOLE	N	10SecB	(IC RR)		231	1/128	CRESTLINE #126	P	28Ch12Sts	J&S (WS&PCC)	5/3/84	749	
CREOLE	N	10Sec2Dr	Wagner	6/93	3074	Wg93	CRESTLINE #129	P	DDR	PFtW&C	8/15/70	4B	FtW2
CREON	N	12SecO	PW&B RR	12/70	18	Wil3	CRESTON		10Sec2Dr	PCW	5/12/93	995A	1942

CAR NAME	ST	CAR TYPE	CAR BUILDER (Owner)	DATE BUILT	PLAN	LOT
CRESWELL		28ChDr	Pullman	4/6/06	1986A	3336
CRESWICK		12SecDr	Pullman	8/10/01	1581A	2727
CRETE		12SecDr	Pullman	2/25/03	1581D	2975
CRIMORA	R	10SecDrB	PCW	3/21/85	240	1037
CRIMORA #438		10SecDrB	PCW	3/21/85	149E	1037
CRISHOLM	P	14SecDr SUV	Pull P-2069A	5/06	3207B	3320
CRISPIAN	P	10Sec2DrB	PCW (Wagner)	3/13/00	3079A	Wg135
CRISTOBAL	N	12SecDrCpt	Pullman	5/26/04	1985	3095
CRITERION		12SecDr	PCW Det	3/18/81	93	D88
CROATIA		12SecDr	Pull Buf	1/21/02	1581A	B7
CROCUS		12SecDr	PCW	2/28/91	784B	1798
CRONSTADT		12SecDr	Pullman	7/18/03	1581F	3019
CROOKSTON	RN	12SecDr	PCW	8/21/91	937E	1838
CROTONA		Slpr	PCW Det	5/10/84	170	D222
CROTONA		10SecDr2Cpt	Pullman	9/29/06	2078D	3398
CROYDON		16SecO	PCW	6/20/93	1040	1990
CRUZ BLANCA	RN	12SecDrB	Pullman	5/17/06	2165C	3352
CRYSANTHA		10SecLibObs	Pullman	5/6/09	2351H	3688
CRYSTAL		10SecDrB	PCW	11/9/86	445	1262
CRYSTAL		12SecDr	Pullman	4/20/03	1581B	3003
CRYSTAL SPRING	P	BaggSmk	NH RR	5/90	3214	
CUBA	P	12SecDrB	Wagner	12/98	3073E	Wg117
CUBA #47		10SecDr	B&S	12/70	19	Day5
CUCHILIO		7Cpt2Dr	Pullman	9/26/02	1845	2926
CUERNAVACA	P	12SecHotel	J&S	1/99	2371A	
CUERO		16SecO	PCW	6/7/93	1040	1990
CULEBRA		7Cpt2Dr	Pullman	10/4/02	1845	2926
CULLODEN	N	12SecDr	Wagner	7/88	3061B	Wg12
CULPEPER		12SecDr	Pullman	1/10/07	2163	3449
CULROSS		8SecCpt Slpr	PCW Det	1/8/83	130	D158
CULVERTON		12SecDr	Pullman	9/10/07	1963C	3550
CUMBERLAND		OS	B&O RR	4/30/73	27 1/2	B&O2
CUMBERLAND		12SecDr	Pull Cal	6/18/04	1581G	C8
CUMBERLAND	P	14SecO	(Wagner)	/86	3032	
CUMBERLAND #142	P	6Sec12ChSr	(WS&PCC)	12/5/85	741B	
CUPID	N	Parlor	PCW Det	11/10/71	34C	D4
CUPID	N	26Ch3StsDr	PCW	4/25/98	1278C	2258
CUPOLA		12SecDr	Pullman	3/12/06	2163A	3333
CURFEW	N	10SecB	PFtW&C	1/71	14D	FtW2
CURLEW	'	24ChDrB	Pullman	6/2/00	1518C	2552
CURRAHEE	N	12SecDrB	PCW	8/4/88	375E	1395
CURTIUS	N	BaggClub	Pullman	8/14/05	2087	3220
CUSTER		10SecDrCpt	PCW	5/28/89	668	1587
CUYAHOGA		16Sec	Pullman	6/15/07	2247	3459
CYBELE	N	20Ch8StsDrB	Wagner	7/90	3126	Wg35
CYCLAMEN	.	24ChDr	Pull Buf	4/27/06	2156	B52
CYDNUS		10SecDrB	PCW Det	12/1/83	149	D212
CYDNUS	R	12SecDrB	PCW Det	12/1/83	1468	D212
CYGNET #119	N		(WS&PCC)	7/83		
CYMBELINE		22ChDr	Pullman	7/6/01	1647C	2702
CYMIA	N	22Ch6Sts	PCW	8-9/92	976B	1910
CYMON	P	16Sec SUV	B&S	7/11	3212	
CYNTHIA	P	18Ch3StsDrObs	(B&M RR)		1203	
CYNWYD		12SecDrSr	Pullman	4/7/05	2049A	3169
CYPRUS		10SecDrB	PCW	5/10/89	623B	1567
CYPRUS	P	10Sec2Dr	Wagner	11/97	3073D	Wg110
CYRENE		22ChDr	Pull Buf	6/2/02	1647C	B8
CYRENE #105	N	29Ch	(WS&PCC)	7/1/82	750F	
CYRIL	N	10SecDrB	PCW	12/6/86	445C	1261
CYRUS	R	BaggSmkBLib	PCW?		1706	
CZARINA	N	26Ch3StsDr	PCW	4/25/98	1278C	2258
D	P	OS	CB&Q RR	?/66		Aur?
D	N	8CptLibObs	Pullman	12/26/06	2260	3502
DACIA	P	20Ch8StsDrB	Wagner	8/93	3126C	Wg95
DACIA #358		12SecDr	PCW Det	5/31/82	93B	D154
DACOTA		12SecDr	Pullman	5/17/02	1581B	2853
DACOTAH	P	SDR	C&NW RR	1/30/69	12	FdL1
DAFFODIL		24ChDr	Pull Buf	3/30/06	2156	B50
DAHINDA	N	16Sec	Pullman	7/7/03	1721	3014
DAHLIA		18Ch6StsDrB	PCW Det	7/11/90	772D	D425
DAHOMEY		12SecDr	Pullman	12/28/03	1581H	3053
DAISY		24ChDrSr	Pullman	7/18/04	1980A	3088
DAKOTA		SDR	?			
DALE WC-100	N	12SecO	CentCarCo	2/86	532	
DALHOUSIE	RN	10SecB	B&S		37	Day11
DALLAS	N	10SecB	PCW Det	9/5/73	37	D16
DALLAS		12SecDr	Pull Buf	3/2/04	1581F	B34
DALLAS	P	12SecB	Wagner	4/92	3065C	Wg67
DALLES CITY		12SecB	PCW Det	8/25/76	58	D42
DALMATIA		12SecDr	Pullman	/03	1581F	3015
DALMATIA #357		12SecDr	PCW Det	5/24/82	93B	D154
DALMORES		10SecLibObs	Pullman	11/29/09	2351A	3748
DALTON	N	10SecB	PCW	10/1/84	209	202
DAMASCUS		10SecDrB	PCW Det	12/29/83	149	D223
DAMASCUS	R	12SecDrB	PCW Det	12/29/83	1468	D223
DAMON	P	12SecDr	Wagner	4/92	3065C	Wg67
DAMORA		12SecDrB	PCW	9/8/88	375F	1480
DAMORA	N	12SecDr	Wagner	4/92	3065C	Wg67
DANA		12SecDr	PCW	5/9/91	784D	1805
DANA	P	18Ch5StsDr	(Wagner)	4/86	3117B	
DANBURY	P	12SecDr	H&H	7/88	1899	
DANSVILLE			(DL&W RR)			
DANSVILLE		10SecDrB	PCW Det	3/23/87	445C	D282
DANTE		12SecDr	PCW	10/13/92	990	1931
DANTZIC		12SecDrB	PCW	10/19/88	375F	1481
DANUBE		12SecDrB	Pull Wilm	2/16/03	1611A	W9
DANUBE #371		10SecDrB	PCW	/83	149	21
DANUBE #492	P	12SecDr	Wagner	3/92	3065C	Wg57
DANVILLE #163	P	10SecDr	TW&W Ry		19	Tol1
DANVILLE #329	N	8BoudB	J&S (UnionPCC)	/89	373A	

CAR NAME	ST	CAR TYPE	CAR BUILDER (Owner)	DATE BUILT	PLAN	LOT
DAPHNE		16Ch6StsB	PCW Det	8/3/76	54	D40
DAPHNE	N	24ChDrSr	Pullman	7/18/04	1980A	3088
DAPHNE	N	28Ch3Sts	PCW (Wagner)	7/14/86	3124	1200
DARDANIA		12SecDr	Pullman	10/3/01	1581A	2765
DARENT		12SecDr	Pullman	10/15/07	2163A	3562
DARIEN		10SecDrB	PCW	6/27/84	149D	192
DARIEN	R	12SecDr	PCW	6/6/88	1603	1414
DARIUS		10SecDrB	PCW	4/29/85	240	1051
DARIUS	N	12SecDr	(Wagner)	6/81	3022A	
DARLINGTON		12SecDr	Pullman	3/2/07	1963C	3525
DARMSTADT		14SecO	PCW Det	5/11/93	1021	D525
DAROCA		14SecO	PCW	2/25/93	1021	1969
DARTFORD		12SecDr	Pull Buf	10/11/01	1581A	B6
DARTMOUTH	N	10Sec2Dr	PCW	3/15/99	1158	2376
DARUNA #133	RN	7BoudB	(MannBCC)	/84	776A	
DARWIN	P	12SecDr	Wagner	4/92	3065C	Wg67
DAUDET		16Sec SUV	Pullman	4/7/10	2184G	3808
DAUNTLESS	P	12SecO	PCW (Wagner)	4/71	3010	1070
DAUPHIN		12SecDrSr	Pullman	3/28/05	2049A	3169
DAVENPORT		10SecDr	PCW Det	9/9/81	112A	D112
DAVENPORT		12SecDr	Pullman	7/9/08	1963C	3630
DAVOUST	N	12SecDr	Wagner	6/93	3065C	Wg85
DAVY CROCKETT	P	Private Hunting	(CH&D RR)	/73	71F	
DAWSON		12SecDr	Pull Buf	8/10/05	1963B	B44
DAYTON		10SecDr	B&S	4/71	19	Day5
DAYTON	N	12SecDr	Wagner	5/87	3061C	Wg3
DAYTON #255	P	10SecDr	H&StJ RR	8/72	26	Han3
DAYTONA	N	6CptDrLibObs	Pullman	2/25/02	1751B	2799
DE KALB		12SecDr	PCW Det	4/23/81	93	D92
DE KALB	P	12SecDr	(Wagner)	4/81	3049	
DE PERE		10SecO	PCW	8/2/88	617	StL5
DE SOTO	N	12SecO	B&O RR	/81	665	
DE WITT		12SecDr	PCW Det	4/29/81	93	D92
DEANWOOD		26ChDrB	Pullman	6/26/07	1988B	3526
DEARBORN	RN	12SecO	B&S/GMP	/69	119C	Day
DEBORAH		18Ch4StsDrB	PCW	7/12/93	1039A	1995
DEBUTANTE		10SecLibObs	Pullman	12/20/09	2351A	3748
DECATUR	P	OS	(PK&RSCC)			
DECATUR	P	Composite	(IC RR)			
DECATUR		12SecDr	Pullman	7/22/03	1581F	3015
DECATUR	P	10SecDr	(Wagner)	/85	3057	
DECATUR #167		10SecDr	TW&W Ry		19	Tol1
DECIUS	R	BaggSmkBLib			1706	
DECURION		12SecDr	Pullman	9/7/07	1963C	3550
DEEPHAVEN	N	12SecDr	Wagner	12/87	3061A	Wg6
DEEPHAVEN	P	14SecDr SUV	Pull P-2069A	6/05	3207B	3187
DEER LODGE		12SecDr	PCW Det	3/6/83	93B	D184
DEER LODGE	RN	12SecDr	PCW	8/17/91	937E	1838
DEER PARK	N	12SecDrSr	Pullman	4/6/05	2049A	3169
DEER PARK #510	P	14SecO	B&O RR	11/80	657C	
DEERCOURT		10SecLibObs	Pullman	4/28/06	2176	3355
DEERFIELD		10Sec3Cpt	Pullman	11/4/05	2098	3236
DEERFIELD #149	RN	BaggBLib?	J&S (WS&PCC)	5/17/86	720D	
DEERMONT		12SecDrB	PCW	9/14/98	1319A	2314
DEERTON	N	10SecDr	H&H	1/72	26A	Wil6
DEERTON		12SecDr	PCW	10/6/98	1318B	2339
DEFENDER		12SecDr	Pullman	8/12/08	1963C	3636
DEFIANCE		12SecDr SUV	Pullman	12/19/05	1963C	3252
DEIOCES		10Sec2Dr	PCW	8/4/96	1173A	2153
DEL MONTE	RN	Diner	H&H	7/73	40C?	Wil2
DEL NORTE SP-60	P	10SecDrB	PCW	6/6/88	445F	1414
DELANCO		26ChDr	Pullman	3/31/06	1987A	3335
DELAND		12SecDrB	Pullman	12/19/06	2142A	3405
DELAVAN	N	12SecO	CB&Q RR	/68	5A	Aur
DELAVAN		12SecDr SUV	Pull Buf	8/11/05	1963B	B44
DELAWARE		12SecO	CB&Q RR	/68	5	Aur
DELAWARE	P	14Ch5StsB	(DL&W RR)	4/75	211	
DELAWARE	P	10SecDr	G&B (Wagner)	10/75	3011	
DELAWARE #81	P	12SecDr	H&H	/80-/82	197	
DELAWARE #223	P	17Ch12StsDr	J&S	7/31/73	35	Wil2
DELAWARE E&A-20	P	Slpr	(E&A Assn)			
DELECTO		8Sec4SrB	PCW	9/29/91	894B	1857
DELEVAN	N	10SecO	(Wagner)	8/85	3038A	
DELGADA	N	10SecDr	H&StJ RR	8/72	26	Han3
DELHI		10SecDrB	PCW	4/30/84	149D	192
DELHI	P	14Sec	B&S	8/93	3204	
DELILAH	P	16Ch6StsDrObs	(B&M RR)			1202
DELLSIDE	P	24Ch6StsDr	PCW	3/91	3259B	
DELMAR	N	SDR	DetC&MCo	/68	12D	Det
DELMAR	N	12SecDrB	Wagner	1/93	3073B	Wg88
DELMONICO	P	Diner	CB&Q RR	1/2/71	16	Aur
DELMONICO	P	DinerHotel	CB&Q RR	/68		Aur16
DELMONICO	RN	Diner	PCW Det	1/25/74	39	D20
DELMONICO		10SecLibObs	Pullman	11/30/09	2351A	3748
DELOS	R	12SecDr	PCW	10/6/85	1603	1080
DELOS #450	P	10SecDrB	PCW	10/6/85	240	1080
DELPHI	P	12Ch24StsObs SU	(CI&L RR)	C:/11		2747
DELPHI	P	12SecDr	Wagner	4/92	3065C	Wg67
DELPHINE		20Ch4StsDrB	Pullman	6/12/01	1518E	2703
DELPHOS		12SecDr	PCW	8/14/91	784C	1828
DELRAY	N	16Sec	Pullman	12/7/06	2184	3409
DELROSA		10Sec2Dr	Pullman	9/4/02	1746A	2925
DELTA	P	10SecDr	J&S (McComb)	/71	19P	
DELTA	P	20Ch8StsDrB	Wagner	8/93	3126C	Wg95
DEMARARA #405		12SecDrB	PCW	2/18/84	180	100
DEMETER		26Ch3StsDr	PCW	4/25/98	1278C	2258
DEMING		12SecDr	Pull Buf	8/3/05	1963B	B44
DENBIGH		16SecO	PCW	8/2/93	1040	2008
DENHAM	N	12SecDrB	PCW Det	8/18/88	375G	D346
DENHOLM		26ChDr	Pullman	3/28/06	1987A	3335

DENISON
Wood

CAR NAME	ST	CAR TYPE	CAR BUILDER (Owner)	DATE BUILT	PLAN	LOT
DENISON		12SecDr	Pullman	2/10/04	1963	3070
DENLEY		12SecDr	Pull Buf	10/31/07	1963C	B69
DENMARK	N	10SecDr	J&S (McComb)	/72	26H	Wil
DENMARK		12SecDr	Pullman	12/29/03	1581H	3053
DENNISON	P	10SecDr	B&S	11/14/72	26	Day8
DENSMORE		12SecDr	Pull Buf	11/11/07	1963C	B69
DENTON		12SecDr	Pull Buf	11/6/07	1963C	B69
DENVER		12SecO	B&S/GMP	/67	5	Day
DENVER	R	12SecDrB	PCW?		1611	
DENVER	P	OS	(MP RR)		27	
DEPTFORD		12SecDr	Pull Buf	11/6/01	1581A	B6
DERBY	N	10SecDr	B&S	11/14/72	26	Day8
DERBY		12SecDr	Pullman	7/18/03	1581F	3019
DERBY	P	20Ch8StsDrB	Wagner	8/93	3126B	Wg95
DERONDA		12SecDr	Pull Buf	11/18/07	1963C	B69
DERRY		26ChDrB	Pullman	6/26/07	1988B	3526
DERWENT		12SecDrB	Pull Wilm	3/30/03	1611A	W9
DERWENT #432		10SecDrB	PCW	11/14/84	149D	1017
DES MOINES	S	14SecDr	Pullman	/00	1515B	2544
DES MOINES		12SecDr	Pull Buf	2/12/04	1581F	B34
DES MOINES	P	12SecDr	(Wagner)	/84	3023A	
DES MOINES #313	P	8BoudB	J&S (MannBCC)	/89	373A	
DESDEMONA		20Ch5StsB	PCW	11/24/83	167A	89
DESHLER	N	BaggClub	Pullman	8/15/05	2087	3220
DESNA 2511		14SecO	PCW Det	8/2/92	1014A	D509
DeSOTO		12SecDr	Pullman	10/22/03	1581F	3025
DETROIT	N	DDR	MC RR/GMP	8/1/67	7C	Det
DETROIT		12SecDr	Pull Buf	3/19/04	1581G	B35
DETROIT	P	12SecDr	(Wagner)	12/75	3027	
DEVASEGO	N	20Ch6StsSr	(Wagner)	6/96	3118	
DEVON	RN	14SecO	PCW Det	7/16/73	37A	D16
DEVON		12SecDrSr	Pullman	3/24/05	2049A	3169
DEVONIA		12SecDr	PCW Det	12/21/81	93B	D135
DEVONIA		14SecDr	Pullman	6/15/00	1523	2559
DEVONSHIRE	S	32StsB Parlor	Pullman	8/00	1376D	2551
DEWITT		12SecDr	PCW Det	4/29/81	93	D92
DEXTER	P	DDR	B&S	/71	17	Day4
DEXTER	P	12SecDr	Wagner	4/92	3065C	Wg67
DIADEM	N	12SecO	CB&Q RR	4/74	4C	Aur
DIADEM	N	12SecDrB	Wagner	12/98	3073E	Wg117
DIAMOND	N	10SecDr	Ohio&Miss RR	6/72	26A	Coch
DIAMOND		12SecDrSr	PCW	4/17/99	1318C	2390
DIAMOND		12SecDr	Pullman	7/10/08	1963C	3630
DIANA	N	34StsParlor	PCW Det	9/77	45	D31
DIANA		20Ch5StsB	PCW Det	6/22/83	167	D203
DIANA #490	P	12SecDr	Wagner	3/92	3065C	Wg57
DIANTHUS		12SecDr	Pullman	5/25/06	1963C	3360
DICKENS		12SecDr	PCW	10/28/92	990	1931
DICKINSON		12SecDr	PCW	9/19/82	93B	22
DIEPPE		14SecO	PCW Det	1/27/93	1014A	D511
DIEPPE		12SecDr	Pullman	7/20/03	1581F	3019
DIGHTON	P	26ChBarB SUV	Pull P-2493	1/11	3274	3820
DINORAH	N	12SecDrSr	Pullman	3/24/05	2049A	3169
DINORAH #117	P	7BoudB	(MannBCC)	/84	776	
DINSMORE	N	10SecDrB	PCW	5/24/84	149D	192
DINWIDDIE #495	P	12SecDr	Wagner	3/92	3065C	Wg57
DIOCLES		10Sec2Dr	PCW	6/15/96	1173A	2153
DIOGENES		10Sec2Dr	PCW	3/7/96	1173A	2153
DIOMEDES		12SecO	PCW	9/83	114	48
DIOMEDES		10Sec2Dr	PCW	3/19/96	1173A	2153
DIONE	P	20Ch8StsDrB	Wagner	8/93	3126B	Wg95
DIONE #157	N	22ChSr	J&S (WS&PCC)	5/15/87	516?	
DIOSMA		12SecDr	PCW	4/8/91	784C	1798
DIPLOMATE	N	12SecDr	PCW	6/7/87	473A	1323
DIVA		18Ch6StsDrB	PCW Det	7/17/90	772D	D424
DIXIE	N	23Ch2StsDrLibObs	Pullman	8/28/06	2206	3388
DIXMONT		12SecDr	Pullman	2/8/08	2163A	3599
DIXON		12SecDr	PCW Det	4/25/81	93	D92
DNEIPER 2519		14SecO	PCW Det	2/4/93	1014A	D511
DNIEPER		12SecDr	Pull Buf	7/20/04	1963	B40
DODGEVILLE	P	36Ch SUV	Pull P-2263	8/07	3275	3492
DOLORES		24ChDr	Pullman	6/28/05	1980A	3171
DOLORES	N	10SecB (NG)	PCW Det	10/16/83	178	D211
DOLPHIN	N	10SecB	PFtW&C	8/15/70	14C	FtW2
DOLPHIN	P	12SecO	PCW (Wagner)	2/70	3010	1070
DOM PEDRO		12SecDrB	PCW	11/24/88	375G	1513
DOMINICA #406		12SecDrB	PCW	2/18/84	180	100
DOMINION		12SecDr	Pullman	9/29/09	1963E	3744
DOMINION #386		10SecDr	PCW	4/83	112A	37
DOMINOE	N	26Ch3StsDr	PCW	4/29/98	1278C	2258
DOMITIUS		BaggDiner	PCW	12/31/90	783A	1705
DON CARLOS #131	P	7BoudB	(MannBCC)	/84	776	
DON PASQUALE #133	P	7BoudB	(MannBCC)	/84	776	
DONALD		12SecDr	Pull Buf	10/4/05	1963C	B47
DONEGAL		16SecO	PCW	6/5/93	1040	1990
DONGOLA	N	12SecB	PCW	3/4/85	227	1034
DONIPHAN		12SecDr	Pull Buf	11/25/07	1963C	B69
DONORA	N	12SecDrSr	PCW	1/5/99	1318C	2379
DOPHIN		OS	PFtW&C/GMP		7	FtW2
DORADO		12SecDr	PCW Det	3/19/91	784B	D460
DORANTE	N	12SecO	B&O RR		657	
DORANTE	N	12SecDr SUV	Pullman	10/14/05	2049B	3232
DORANTE	N	10SecDr	(Wagner)		3042	
DORCAS		20Ch6StsDr	PCW	7/7/91	928A	1826
DORCHESTER		12SecDrSr	PCW	9/28/99	1318E	2449
DOREAS #491	P	12SecDr	Wagner	3/92	3065C	Wg57
DORIAN	N	12SecDrSr	Pullman	3/27/05	2049A	3169
DORIAN 2561		14SecO	PCW Det	3/11/93	1021	D516
DORINDA		12SecDr	Pullman	9/29/09	1963E	3744
DORIS	N	20Ch3StsB	H&H (B&O)		653B	

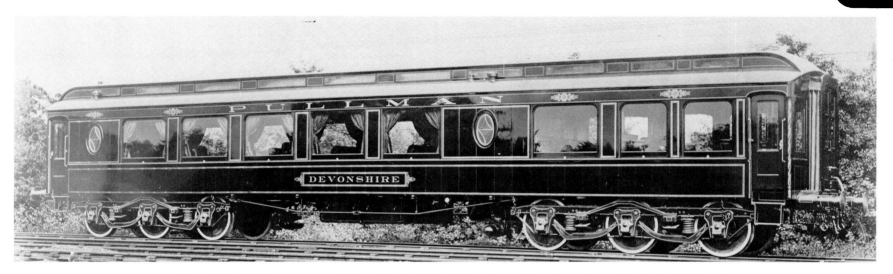

Figure I-17 This beauty never rode the rails in the United States. It spent its entire service on the London-Brighton & South Coast Railroad in England and was condemned in 1932. Smithsonian Institution Photograph P7900. Courtesy Robert Wayner.

CAR NAME	ST	CAR TYPE	CAR BUILDER (Owner)	DATE BUILT	PLAN	LOT
DOROTHEA		20Ch12StsB	PCW	5/29/89	618A	1514
DOROTHY		22ChDr	Pull Buf	6/4/02	1647C	B8
DORSET		12SecDr	PCW	2/10/92	784E	1881
DORSET	P	20Ch8StsDrB	Wagner	8/93	3126	Wg95
DOUGLAS	P	10SecB	(IC RR)		231	1/228
DOUGLAS		12SecDr	Pull Buf	8/13/05	1963B	B44
DOUGLAS	P	10SecDr	(Wagner)	/85	3011	
DOULTON		12SecDr	Pullman	3/3/02	1581B	2818
DOURO #433		10SecDrB	PCW	11/14/84	149D	1017
DOVER		10SecDrB	PCW	6/18/84	149D	192
DOVER	P	20Ch8StsDrB	Wagner	8/93	3126B	Wg95
DRAVE		16SecO	PCW	8/7/93	1040	2008
DRESDEN		10SecDrB	PCW	11/5/83	149	80
DRESDEN	R	12SecDrB	PCW	11/5/83	1611	80
DREUX	N	10SecDr	(Wagner)	/85	3056	1194
DREXEL	N	10SecDr	PCW Det	9/9/81	112A	D112
DREXEL		12SecDr	Pullman	7/9/08	1963C	3630
DRIFTWOOD		26ChDrB	Pullman	6/27/07	1988B	3526
DROMIO		7Cpt2Dr	PCW	1/4/96	1065A	2140
DRONGO	N	20Ch4StsDrB	Pullman	6/13/01	1518E	2703
DRURY		10SecDrB	PCW	1/14/92	920B	1880
DRURY	N	12SecDrB	Wagner	6/93	3073E	Wg82
DRUSILLA		16Sec	Pullman	6/8/09	2184D	3700
DRYAD		20Ch6StsDrObs	Pullman	6/16/04	1979	3084
DRYDEN	N	12SecDrCpt	Pullman	8/9/05	2088	3221

CAR NAME	ST	CAR TYPE	CAR BUILDER (Owner)	DATE BUILT	PLAN	LOT
DRYDEN	P	12SecDr	Wagner	4/92	3065C	Wg67
DUANE #77	P	18Ch4StsDr	(Wagner)	/82	3111	
DUARTE		10SecDrB	PCW	5/21/88	445F	1377
DUBLIN	N	12SecB	PCW	3/4/85	227	1034
DUBUQUE	N	12SecO	C&NW/GMP	/56	5	Chi
DUBUQUE		12SecDr	Pull Buf	3/3/04	1581F	B34
DUBUQUE	R	30Ch SUV	Pull Buf	3/3/04	3581E	B34
DUBUQUE #80	P	10SecDr	H&H	/81-/82		
DUBUQUE #104	P	Slpr	(IC RR)			1/229
DUBUQUE #106	P	12SecDr	(MannBCC)	/89	878	
DUCAS	N	20Ch8StsDrB	Wagner	7/90	3126	Wg35
DUCHESS	P	14Ch6StsDrB	PCW	3/25/83	154	34
DUCHESS	N	20Ch5StsB	PCW	6/17/84	167A	104
DUCHESS of						
ALBANY		32StsParlor	PCW	3/90		
CONNAUGHT		32StsParlor	PCW Det	10/93		D513
FIFE		32StsParlor	PCW	3/90		
NORFOLK		32StsB Parlor	Pullman	1/06	1376C	3223
YORK		26StsBParlor	PCW	6/17/95	1148	2105
DUFFERIN	N	10SecDr	Grand Trunk	2/71	21D	Mon1
DULCINEA		20Ch5StsB	PCW	11/24/83	167A	89
DULUTH	RN	14SecO	I&StLSCCo	/71		
DULUTH		10SecDrCpt	PCW	2/5/90	668	1649
DULUTH #493	P	12SecDr	Wagner	5/92	3065C	Wg57
DUMAS		12SecDr	PCW	4/17/91	784D	1801

CAR NAME	ST	CAR TYPE	CAR BUILDER (Owner)	DATE BUILT	PLAN	LOT
DUNBARTON		12SecDrB SUV	Pullman	2/21/06	2142	3318
DUNBLANE	P	14SecDr SUV	B&S	6/11	3207C	
DUNCANNON		12SecDrSr	Pullman	4/7/05	2049A	3169
DUNDALE		12SecDr	Pullman	1/5/07	2163	3460
DUNDAS 2562		14SecO	PCW Det	3/9/93	1021	D516
DUNDEE		10SecDrB	PCW	11/17/86	445	1262
DUNDEE	N	12SecDrB	Pull Wilm	2/16/03	1611A	W9
DUNDEE #494	P	12SecDr	Wagner	3/92	3065C	Wg57
DUNEDIN		12SecDr	Pullman	7/23/03	1581F	3015
DUNELLEN #175	P	18Ch6StsDr	(WS&PCC)	6/23/88	730A	
DUNIRA	P	14SecDr SUV	B&S	6/11	3207C	
DUNKIRK		12SecDr	Pullman	9/22/03	1581F	3022
DUNLAP		12SecDr	PCW Det	5/7/81	93	D92
DUNMORE		12SecDr	Pullman	2/15/06	1963C	3316
DUNNVILLE #178		16Ch28Sts SUV	Pullman	7/7/03	3581D	3019
DUNROBIN	N	8SecCpt Slpr	PCW Det	1/8/83	39	D20
DUNROBIN		12SecDr	Pullman	1/29/03	1581D	2962
DUNSMUIR		12SecDr	Pullman	3/3/04	1963	3080
DUQUESNE		12SecDr	Pullman	3/2/07	1963C	3525
DURAND		12SecDr	Pull Buf	8/15/05	1963B	B44
DURANGO		10SecDrB	PCW Det	3/26/84	149B	D234
DURANGO	N	10SecB (NG)	PCW Det	10/23/83	178	D211
DURANGO		12SecDr	Pullman	4/21/03	1581B	3003
DURBAN		12SecDr	Pullman	9/23/03	1581F	3022
DURGA		14SecO	PCW	2/3/93	1017C	1962
DURHAM	N	12SecDrB	PCW	12/3/88	375G	1513
DUXBURY		12SecDr	Pull Buf	7/3/05	1963	B42
DWINA #106	RN	12SecDr	(MannBCC)	/89	878A	
DYNAMENE		6CptDrLibObs	Pullman	2/25/02	1751B	2799
DYSART		26ChDrB	Pullman	6/27/07	1988B	3526
E. KIRBY SMITH	N	29ChDr	Pullman	7/15/05	2061	3217
EAGLE		12SecDr	PCW	3/10/91	784B	1801
EAGLE	P	10SecDr	(Wagner)	6/85	3048	
EARLHAM		12SecDr	Pullman	9/14/01	1581A	2756
EARLINGTON	P	25Ch3StsDrB	Wagner	11/93	3126D	Wg102
EARLTON	N	Slpr	(NYLE&W Assn)			
EARLTON		10SecDr2Cpt	Pullman	11/30/07	2078D	3583
EARLVILLE		16SecO	Pullman	5/4/05	2068	3177
EARNSCLIFFE		12SecDr SUV	Pullman	10/19/06	1963C	3408
EASLEY		10SecDr2Cpt SUV	Pullman	6/11/08	2078D	3622
EAST BERLIN	P	BaggSmkB SUV	H&H	/88	3219	
EAST FARMS	P	BaggSmk	Wagner	8/89	3213	Wg20
EAST FARMS	N	BaggSmkB SUV	H&H	/88	3219	
EAST HAVEN	P	BaggSmk SUV	Pull P-1991	4/04	3217	3093
EAST LAKE	N	12SecDr	Pullman	3/15/06	2163A	3333
EAST RIVER	N	40Ch6StsDrObs	Pull P-2266A	4/07	3276	3493
EAST WINDSOR	P	BaggSmk	New Haven RR	5/90	3214	
EASTHAM		12SecDr	Pullman	11/20/01	1581A	2784
EASTHAMPTON	P	24Ch3StsDr	B&S	9/92	3261	
EASTLAKE		12SecDr	Pullman	7/13/08	1963C	3630
EASTLAND		12SecDr	Pullman	12/15/06	1963C	3448
EASTMAN		12SecDrB	Pullman	12/17/07	2326	3587
EASTON	N	12SecDrB	PCW	8/16/88	375G	1480
EASTON	P	34Ch	J&S	/93	3263	
EASTPORT		12SecDr	Pull Buf	3/31/04	1581G	B35
EASTWOOD #168	N	15Ch4StsB	J&S (WS&PCC)	12/9/87	731A	
EATON		16Sec	Pullman	12/5/06	2184	3409
EAU CLAIRE	N	12SecDr	PCW Det	5/19/81	93A	D107
EAU CLAIRE	N	14SecDr	Pullman	12/30/01	1523I	2798
EAU CLAIRE #3	P	14SecO	PCW (Wagner)	3/19/86	3018A	1191
EBRO	R	12SecDrB	PCW	/83	1468	21
EBRO #372		10SecDrB	PCW	1/16/83	149	21
ECHO		12SecDr	PCW Det	9/1/82	93	D169
ECHO		10Sec2Dr	PCW	2/24/93	995A	1942
ECLIPSE		27StsParlor	PCW Det	/77	45	D31
ECLIPSE		10SecDr	PCW Det	6/5/73	37 1/2	D18
ECLIPSE	P	12SecO	PCW (Wagner)	5/85	3010	1070
ECLIPSE CTC-89	R		PRR rblt/72			
ECOLA		10secDr2Cpt SUV	Pullman	3/11/10	2471	3784
ECUADOR		12SecDr	Pullman	7/28/02	1581B	2852
EDDYSTONE		12SecDr	Pullman	8/26/01	1581A	2756
EDELSTEIN	N	Slpr	(NYLE&W Assn)	/72		
EDELSTEIN		12SecDrB	PCW	8/4/88	375E	1395
EDELWEISS		24ChDrSr	Pullman	7/18/04	1980A	3088
EDEN	P	12SecDr	Wagner	5/92	3065C	Wg68
EDENHALL	N	12SecDr	Pullman	10/30/01	1581A	2784
EDENTON		12SecDr	Pullman	10/8/00	1581A	2588
EDGAR	P	26ChBarB SUV	Pull P-2265	8/07	3274	3497
EDGEBROOK		12SecDr	Pull Buf	12/28/03	1581F	B26
EDGEFIELD	RN	12SecB	(L&N RR)	2/72	114A	Han3
EDGEFIELD		12SecDr	Pull Buf	12/13/07	1963C	B71
EDGELEY		12SecDr	Pullman	12/15/06	1963C	3448
EDGEMERE	P	Private	PCW Det	1/25/83	152D	D178
EDGEMONT	N	10SecDrB	PCW	2/17/88	445F	1369
EDGEMONT		12SecDr	Pull Buf	12/13/07	1963C	B71
EDGERTON		12SecDr	PCW	10/20/98	1318B	2340
EDGEWATER		12SecDr	Pullman	11/23/03	1581F	3043
EDGEWOOD		14SecDr	Pull Buf	9/27/00	1523D	B1
EDGEWOOD #41	N	Private	H&H (WS&PCC)	7/25/78	689	
EDGEWORTH		12SecDrCpt	Pullman	8/8/05	2088	3221
EDINA	N	12SecDr	CB&Q RR	C:12/68	12B	Aur
EDINBORO #580	P	10SecDrB	J&S (B&O)	C:/88	663	
EDINBURG		12SecDr	Pullman	4/7/04	1963	3080
EDISTO	R	10SecO	PCW Det (rblt)	3/31/75	47	D27
EDISTO	N	12SecDrB	(Wagner)	9/87	3050A	
EDITH		14SecO	PCW	1/23/93	1017C	1962
EDITHA		20Ch6StsDr	PCW	7/17/91	928A	1827
EDMONTON		12SecDr	Pull Buf	7/15/05	1963B	B43
EDNA		10SecDr	PCW	1/9/90	717B	1630
EDNA	N	32ChDr	Pullman	3/2/07	2274C	3509

CAR NAME	ST	CAR TYPE	CAR BUILDER (Owner)	DATE BUILT	PLAN	LOT
EDNA	P	12SecDr	Wagner	5/92	3065C	Wg68
EDNADA	N	16Sec	Pullman	7/13/03	1721	3014
EDRED	N	14SecO	(Wagner)	/86	3032	
EDSONIA		12SecDr	Pullman	5/26/06	1963C	3360
EDWARD HOPKINS		BaggSmkBLib	Pullman	7/30/03	1945A	3020
EDWARD WINSLOW		BaggSmkBLib	Pullman	7/30/03	1945A	3020
EDWINA		20Ch6StsDr	PCW	7/10/91	928A	1826
EFFIE ELLSLER #11	P	12SecO	(WS&PCC)	6/5/73	694	
EFFINGHAM		12SecDr	Pullman	3/5/04	1963	3080
EGBERT	N	10Sec2Dr	Wagner	11/98	3073D	Wg110
EGERIA		12SecDr	PCW Det	5/17/82	93B	D154
EGG HARBOR #76	N		(WS&PCC)	7/1/79	755?	
EGGLESTON		12SecDr	Pullman	11/21/00	1581A	2589
EGLANTINE		24ChDr	Pull Buf	4/30/06	2156	B52
EGLINGTON	P	25Ch3StsDrB	Wagner	11/93	3126D	Wg102
EGMONT #105	P	20Ch8StsDrB	Wagner	9/90	3126E	Wg41
EGNATIA		30ChObs	PCW	4/28/98	1278B	2299
EGYPT	RN	12SecDr	PRR	11/70	17E	Alt1
EGYPT		12SecDr	Pullman	2/24/03	1581D	2975
EHREN		12SecDr	PCW Det	8/28/91	784D	D481
EIDER		12SecDr	PCW Det	11/29/90	784	D449
EILEEN		22Ch7StsDrObs	Pullman	6/17/04	2023	3126
EKWANOK		12SecDr	Pullman	11/18/03	1581F	3043
EL ALARCON	N	6CptDrLibObs	Pullman	6/16/04	1977D	3091
EL ASTRO ML&T 23	P	10Sec2Dr	PCW	10/25/95	1158	2125
EL CABRILLO	N	6CptDrLibObs	Pullman	6/16/04	1977D	3091
EL CAMPO	N	6CptDrLibObs	Pullman	3/13/02	1751B	2799
EL CAMPO SP-36	P	10SecDrB	PCW	2/23/87	445B	1298
EL CAPITAN	N	BaggLib	PCW	6/7/87	477	1325
EL CAPITAN	P	8Sec8ChClub	(ATSF)		1756	
EL CAPITAN CARRANZA	RN	12SecDrB	Pullman	5/19/06	2165C	3352
EL CARDENAS		10SecLibObs	Pullman	11/28/08	2351A	3663
EL CASTANEDA		10SecLibObs	Pullman	11/28/08	2351A	3663
EL DIA T&NO 33	SN	BaggParlor	PCW	8/88	534C	1410
EL DORADO	P	10SecB	PCW Det	7/16/73	37	D16
EL DORADO #451		12SecDrB	Wagner	2/91	3065	Wg46
EL DORADO SP NM-8301	P	10Sec2Dr	PCW	3/18/99	1158	2376
EL ESCALANTE	N	10SecLibObs	Pullman	2/5/00	1434A	2483
EL GARCES		10SecLibObs	Pullman	12/4/08	2351A	3663
EL GLOBO ML&T 24	P	10Sec2Dr	PCW	10/25/95	1158	2125
EL INDIO T&NO 32	SN	Bagg Parlor	PCW	8/88	534C	1410
EL KAIBAB	N	10SecLibObs	Pullman	11/29/09	2351A	3748
EL LAGO M&LT-15		10SecDrB	PCW	10/13/88	445H	1510
EL MAHDI #5	P	17Ch3StsDrB	(Wagner)	4/64	3108A	
EL MAR M&LT-16	P	10SecDrB	PCW	11/13/88	445H	1510
EL MARCOS		10SecLibObs	Pullman	11/28/08	2351A	3663
EL MENDOZA		10SecLibObs	Pullman	12/5/08	2351A	3663
EL MONTE CP-47	P	10SecDrB	PCW Det	12/22/83	149	D213

CAR NAME	ST	CAR TYPE	CAR BUILDER (Owner)	DATE BUILT	PLAN	LOT
EL MORO		10SecO (NG)	PCW Det	8/22/81	73A	D108
EL MORO	R	8SecSrB (NG)	PCW Det	8/22/81	1662A	D108
EL MUNDO ML&T 25	P	10Sec2Dr	PCW	10/30/95	1158	2125
EL ORBE LW 27	P	10Sec2Dr	PCW	10/30/95	1158	2125
EL ORIENTE LW 28	P	10Sec2Dr	PCW	10/30/95	1158	2125
EL ORO	P	12SecDrB	Wagner	12/90	3065	Wg43
EL ORO ML&	P	10Sec2Dr	PCW	10/25/95	1158	2125
EL PADILLA	N	6CptDrLibObs	Pullman	6/16/04	1977D	3091
EL PARQUE	RN	6CptLibObs	Pullman	8/15/05	2089B	3222
EL PASO		10SecDr	PCW Det	6/16/81	112	D118
EL PASO	N	4CptLngObs	Pullman	6/16/04	1977E	3091
EL PASO #442	P	12SecDrB	Wagner	12/90	3065	Wg43
EL PASTOR SPC-8151	P	7CptLngObs	PCW	10/30/95	1159A	2126
EL PILOTO LW 30	P	7CptLngObs	PCW	10/30/95	1159A	2126
EL PINO LW 29	P	7CptLngObs	PCW	10/25/95	1159A	2126
EL POETA SPC-8150	P	7CptLngObs	PCW	10/30/95	1159A	2126
EL PUEBLO	R	8SecBLibObs	PCW	11/12/88	1707	1510
EL PUEBLO M&LT-14	P	10SecDrB	PCW	11/12/88	445H	1510
EL REPOSA	RN	6CptLibObs	Pullman	8/14/05	2089B	3222
EL RIO CP-42	P	10SecDrB	PCW Det	12/8/83	149	D213
EL SALADO	N	8SecSrB (NG)	PCW Det	11/4/82	155	D173
EL SIGLO T&NO	SN	Composite	PCW	8/88	534C	1410
EL SOL M&LT-18	P	10SecDrB	PCW	12/19/88	445H	1519
EL TOVAR		10SecLibObs	Pullman	11/28/08	2351A	3663
EL TUSAYAN		10SecLibObs	Pullman	12/8/08	2351A	3663
EL VALLE	RN	6CptDrLibObs	Pullman	5/31/07	2089B	3533
EL VALLE M&LT-17	P	10SecDrB	PCW	11/15/88	445H	1510
EL VARGAS	N	10SecLibObs	PCW	12/4/99	1434B	2483
EL VISTA	RN	6CptLibObs	Pullman	8/16/05	2089B	3222
EL VIZCAINO	N	10SecLibObs	Pullman	12/1/09	2351A	3748
EL ZALDIVAR		10SecLibObs	Pullman	12/7/08	2351A	3663
ELAINE	N	10SecB	(IC RR)		231	1/226
ELAINE		26ChDrSr	Pullman	6/8/04	1987A	3085
ELAINE	P	10Sec2Dr	Wagner	6/93	3074	Wg92
ELBA		10SecDr	PCW	11/19/89	717B	1630
ELBA #83	P	26ChB	PCW (Wagner)	7/20/86	3124A	1200
ELBERON	P	10SecDrB	(Wagner)	5/82	3046	
ELBERON #158	P	20Ch4StsB	J&S (WS&PCC)	6/27/87	711	
ELBERTUS		16Sec SUV	Pullman	11/4/09	2184G	3746
ELCAYA	P	14SecDr SUV	Pull P-2069A	5/06	3207B	3320
ELDEN		10SecDr	Grand Trunk		37K	
ELDON	N	10SecDr	Grand Trunk	6/72	31C	Mon3
ELECTRA		12SecDrB	PCW Det	9/1/88	375G	D346
ELECTRA		30ChObs	PCW	4/27/98	1278B	2299
ELECTRA	N	12SecDr	Wagner	1/98	3061B	WG13
ELFIN		20Ch6StsDrObs	Pullman	6/16/04	1979	3084
ELGIN	N	10SecDr	B&S	1/73	26G	Day7
ELGIN		12SecDr	Pullman	2/11/04	1963	3070
ELGIVA		20Ch6StsDr	PCW	7/8/91	928A	1826
ELIDA		12SecDr	Pullman	2/11/08	2163A	3599

CAR NAME	ST	CAR TYPE	CAR BUILDER (Owner)	DATE BUILT	PLAN	LOT
ELINOR		18Ch6StsDrB	PCW Det	7/23/90	1772D	D424
ELIZABETH	N	30 Ch	PCW	4/28/98	1278A	2293
ELIZABETH #39	P		H&H (WS&PCC)	5/1/78	570?	
ELKHART	N	10SecDrB	PCW	3/31/88	445F	1377
ELKHART #84	P		(WS&PCC)	7/1/80		
ELKHORN	P	OS	CB&Q/GMP	/66		Aur
ELKHORN		10SecDrSr	PCW	6/16/90	784A	1715
ELKHORN #450	P	12SecDrB	Wagner	2/91	3065	Wg46
ELKHORN CLUB		Commissary?	C&NW RR	C:7/70		Com1
ELKHURST		10SecDr2Cpt	Pullman	12/28/05	2078C	3287
ELKMONT	N	12SecO	PRR	10/71	27C	Alt2
ELKMONT		12SecDrB	PCW	9/15/98	1319A	2314
ELKO		10SecDr2Cpt	Pullman	9/6/05	2078	3203
ELKO CP-32	P	10Sec	(CP Co)	7/18/84	37L	
ELKTON		12SecDr	PCW	9/2/91	784C	1828
ELKVIEW		12SecDr	Pullman	10/24/05	2049B	3232
ELLA San#2	P	14SecO	DetC&MCo	/71		
ELLENCOURT	N	12SecDrB	PCW	11/5/83	1611	80
ELLENDALE		12SecDr	PCW Det	12/8/88	635	D377
ELLENSBURG	RN	12SecDr	PCW	7/2/92	937E	1925
ELLERSLIE #106	P	20Ch8StsDrB	Wagner	9/90	3126E	Wg41
ELLICOTT #472	P	12SecDrB	Wagner	11/91	3065	Wg54
ELLINGTON	P	24Ch6StsB	B&S	1/94	3262B	
ELLIS	N	12SecDr	Pull Buf	6/13/08	1963C	B78
ELLISTON		12SecDr	Pullman	11/24/00	1581A	2589
ELLITHORPE #453	P	12SecDrB	Wagner	2/91	3065	Wg46
ELLSMERE (1st) #289?	P	Private	Wagner	11/28/88	3088	Wg16
ELLSWORTH		12SecDr	Pull Buf	7/18/05	1963B	B43
ELMDALE		12SecDr	Pullman	9/16/01	1581A	2756
ELMER		12SecDr	PCW	8/19/91	784C	1828

CAR NAME	ST	CAR TYPE	CAR BUILDER (Owner)	DATE BUILT	PLAN	LOT
ELMHURST	N	10SecB	PCW	10/1/84	209	202
ELMHURST		12SecDr	Pullman	11/21/03	1581F	3043
ELMINA		10SecLibObs	Pullman	12/12/08	2351A	3663
ELMIRA		12SecO	(DL&W RR)	5/82		2353
ELMIRA		12SecDr	Pull Buf	7/20/05	1963B	B43
ELMIRA #11	P	20Ch4Sts	(Wagner)	9/86		3102
ELMIRA #272	P	10SecDr	Elmira Car W	1/73	26	Elm1
ELMO		10SecDrB	PCW	5/29/91	920B	1808
ELMORE	N	14SecO	L&N RR (OPL)	/78		214
ELMORE	P	12SecDr	(Wagner)	10/82		3023C
ELMSFORD	N	12SecDrB	Wagner	8/91	3065	Wg53
ELMWOOD		10SecDr	PCW Det	1/5/76	52	D33
ELMWOOD		12SecDr	PCW Buf	12/13/07	1963C	B71
ELNORA		24ChDr	Pull Buf	2/28/07	2156	B57
ELOISE		10SecDrB	PCW	11/9/87	445E	1334
ELOISE	P	10Sec2Dr	Wagner	6/93	3074	Wg92
ELOISE	R	14Sec	PCW	11/9/87	445X	1334
ELORA	N	10SecDr	Grand Trunk	7/73	31K	Mon4
ELOTA	RN	10SecDr	PCW Det	4/5/81	112F	D149
ELREBA		10SecLibObs	Pullman	12/12/08	2351A	3663
ELROSE		10SecDrB	PCW	10/26/86	445	1262
ELROSE		16Sec SUV	Pullman	11/2/09	2184G	3746
ELROY #12	P	14SecO	PCW (Wagner)	3/19/86	3018A	1191
ELSA		12SecDr	PCW Det	2/5/92	784E	D488
ELSA	N	32ChDr	Pullman	3/6/07	2274C	3509
ELSINORE		20Ch5StsB	PCW	8/19/84	167A	199
ELSINORE #452	P	12SecDrB	Wagner	2/91	3065	Wg46
ELTHAM		12SecDr	Pullman	8/26/01	1581A	2756
ELTON		10SecDrCpt	PCW	5/8/90	668A	1709
ELTOPIA		10SecDrCpt	PCW	2/5/90	668	1649

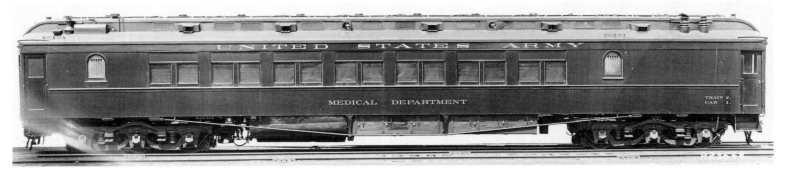

Figure I-18 This United States Army Medical Department car was built by Wagner in 1898 as ELECTRA. It later was converted to Tourist Car 1346 and in February 1918 rebuilt as Hospital Car No. 1 for Train 2. The name BENJAMIN F. POPE was applied to this car while in medical service, and after the war the government sold it to the Western Union Telegraph Company where it was transformed into Bunk Car 1397. Smithsonian Institution Archives Photograph P-23568.

CAR NAME	ST	CAR TYPE	CAR BUILDER (Owner)	DATE BUILT	PLAN	LOT
ELVAS		14SecO	PCW Det	4/22/93	1021	D518
ELVIN	P	24Ch6StsDr	PCW	3/91	3259	
ELVIRA	N	24ChDrB	Pullman	6/8/05	1988A	3170
ELVIRA #127	N	Parlor (3'gauge)	J&S (WS&PCC)	5/17 84	552?	
ELWOOD		16Sec	Pullman	11/27/06	2184	3409
ELWYN		BaggClub	Pullman	8/17/05	2087	3220
ELWYN #39	RN		H&H (WS&PCC)	5/1/78	753A	
ELYRIA		12SecDr	Pull Buf	12/26/07	1963C	B71
ELYRIA	N	20Ch4Sts	(Wagner)	8/70	3101	
ELYSIA		12SecDr	PCW Det	2/4/82	93B	D144
ELYSIA	R	12SecDr	PCW Det	2/4/82	1602B	D144
ELYSIAN		Private	Pullman	12/14/01	1671	2746
ELZEVIR #473	P	12SecDrB	Wagner	11/91	3065	Wg54
EMBLANCHE	N	28ChDr	Pullman	4/5/06	1986A	3336
EMELIA		20Ch6StsDr	PCW	7/20/91	928A	1827
EMELYN		24ChDr	Pull Buf	2/25/07	2156	B57
EMERALD	N	10SecDr	O&M RR	6/30/72	26A	Coch
EMERALD		12SecDrSr	PCW	4/17/99	1318C	2390
EMERSON		12SecDr	Pull Buf	2/10/04	1581F	B32
EMERSON	R	30Ch SUV	Pull Buf	2/10/04	3581E	B32
EMILIA		26ChDrSr	Pullman	6/8/04	1987A	3085
EMILISSA		24ChDrObs	Pull Buf	4/15/07	2091C	B70
EMIR		12SecDr	PCW	10/20/90	784	1766
EMIR	P	12SecDr	Wagner	5/92	3065C	Wg68
EMISON		12SecDr	Pull Buf	12/19/07	1963C	B71
EMISON #164	RN	10SecDrB	(WS&PCC)	9/24/87	663E	
EMISSARY		12SecDr	Pullman	2/13/06	1963C	3316
EMMA		Slpr	H&H (WS&PCC)	/79		
EMPEROR		16Sec	Pullman	5/9/08	2184C	3623
EMPIRE		12Sec	PCW Det	6/11/79	65	D53
EMPIRE		12SecDr	Pullman	3/14/08	1963C	3600
EMPIRE	P	10Sec2Dr	Wagner	6/93	3074	Wg 92
EMPIRE STATE		27ChDrObs	Pullman	6/8/08	2346	3624
EMPORIA		10SecDr	PCW Det	9/11/76	59	D43
EMPORIA		12SecDr	Pull Buf	7/27/05	1963B	B43
EMPORIUM	R	12SecDrB	PCW?		1611	
EMPRESS	S	29Sts Parlor	PCW	5/90		
EMPRESS		20Ch5StsB	PCW	6/17/84	167A	104
EMPRESS	N	26Ch3StsDr	PCW	4/25/98	1278C	2258
ENARIA	N	12SecDrSr	Pullman	1/24/01	1581A	2614
ENARIA		12SecDr	Pull Buf	9/30/03	1581F	B23
ENCHANTRESS	N	30ChObs	PCW	4/28/98	1278B	2299
ENCINO		10Sec2Dr	Pullman	9/5/02	1746A	2925
ENDERLIN		12SecDr	Pullman	10/24/06	1963C	3408
ENDOR		10SecDrB	PCW	1/8/92	920B	1880
ENDOR	R	12SecDr	PCW	1/8/92	1497	1880
ENDYMION		12SecDr	Pullman	10/29/01	1581A	2784
ENECHUR		10SecDr2Sr	Pullman	8/24/09	2078J	3721
ENFIELD		12SecDr	Pullman	7/11/08	1963C	3630
ENGADINE #82	P	26ChB	PCW(Wagner)	7/22/86	3124A	1200
ENGLAND		12SecDr	PCW	4/12/87	473A	1323
ENGLEWOOD		12SecDr	Pull Buf	7/24/05	1963B	B43
ENGLEWOOD #264	P	10SecDr	J&S	11/72	26	Wil8
ENID		10SecDr	PCW	1/4/90	717B	1630
ENOREE #334	P	10SecDrB	(UnionPCC)		565B	
ENSIGN	N	Parlor	Bowers, Dure	11/72	35H	1
ENSIGN	P	10Sec2Dr	Wagner	6/93	3074	Wg92
ENTERPRISE	P	10SecDr	B&S	4/30/73	26	
ENTERPRISE		22Berths Slpr	PCW Det	1/25/74	38	D19
ENTERPRISE		16Sec	Pullman	2/14/07	2184	3508
ENTERPRISE	P	12SecDr	(Wagner)		3001C	
ENUMCLAW		16Sec	Pullman	6/19/07	2247	3459
ENVOY	N	10SecDr	H&H	2/71	19	Wil2
ENVOY		12SecDr	Pullman	9/8/08	1963C	3646
ENVOY	N	10SecDr	(Wagner)	/85	3001C	
ENZILLA	N	20Ch4StsDrB	Pullman	6/12/01	1518E	2703
EOLA		10SecDr	PCW	11/19/89	717B	1630
EPERNAY		14SecO	PCW Det	5/5/93	1021	D518
EPHESUS		10SecDrB	PCW	9/29/84	149D	1006
EPHIALTES		10Sec2Dr	PCW	7/28/96	1173A	2153
EPICUREO	N	BaggClubB	Pullman	7/30/03	1945D	3020
EPICURUS	N	10SecDr	PCW Det	11/24/81	59	D93
EPIRUS	R	BaggSmkBLib	PCW?		1706	
EPIRUS #443		10SecDrB	PCW	8/19/85	240	1058
EPPING	N	Parlor	PCW Det	2/15/72	22	D9
EPPING		12SecDr	Pullman	2/2/03	1581D	2962
EPSILON		12SecDrB	Pullman	2/23/06	2142	3318
EPSOM		16SecO	PCW	8/29/93	1040	2008
EPWORTH		14SecO	PCW Det	5/24/93	1021	D525
EPWORTH		12SecDr	Pullman	11/10/01	1581A	2784
EQUADOR	N	12SecO	B&O RR	/80	665	
EQUINOX		12SecDr	Pullman	3/28/03	1581D	2983
EQUINOX #132	N	12SecB	(WS&PCC)	9/22/84	789A	
ERATO		18Ch6StsDrB	PCW Det	7/23/90	772D	D424
ERICSSON		12SecDr	Pull Buf	2/4/04	1581F	B32
ERIE	P	12SecO	(Wagner)	/85	3010A	
ERIE #189	P	OS	Phil&Erie RR	5/71	27	Erie1
ERIN	N	12SecO	B&O RR	/80	665	
ERINA		20Ch4StsDrB	Pullman	6/14/01	1518E	2703
ERIVAN	N	12SecO	B&O RR		657	
ERIVAN	N	12SecDrSr	Pullman	3/27/05	2049A	3169
ERLAN	N	10SecB	PCW Det	1/6/79	37Q	D50
ERLON #76	RN	28Ch	(WS&PCC)			
ERMINIE		14Ch13StsB	PCW Det	9/7/87	505	D309
ERMITA	N	14SecDr	(Wagner)	5/85	3050D	
ERNANI		12SecDr	Pullman	3/27/08	1963C	3604
ERNANI #119	P	7BoudB	(MannBCC)	/84	776?	
ERNESTINE	N	24ChDrB	Pullman	6/10/05	1988A	3170
ERVIN	N	12SecO Tourist	(Wagner)	6/69	3013	
ESCALUS	?	12SecO	B&O?		665F	

CAR NAME	ST	CAR TYPE	CAR BUILDER (Owner)	DATE BUILT	PLAN	LOT
ESCAMBIA		10SecDr	PCW Det	3/15/82	112A	D141
ESCAMBIA		12SecDr	Pullman	9/8/00	1572	2585
ESCANES #140	N	6Sec12ChSr	J&S (WS&PCC)	12/5/85	741B	
ESCORT	P	10SecDr	J&S	/73	37	Wil10
ESCORT	P	10Sec2Dr	Wagner	6/93	3074	Wg92
ESDRAS		10SecDrB	PCW	10/30/91	920B	1866
ESDRAS	R	12SecDrB	PCW	10/30/91	1940	1866
ESMERALDA		12SecDr	Pullman	12/4/07	1963C	3585
ESMOND		Diner	PCW	1/23/90	662D	1659
ESMOND	P	12SecDrB	Wagner	6/96	3073C	Wg109
ESPANOLA		10SecO (NG)	PCW Det	6/28/81	73A	D108
ESPARTO		10Sec2Dr	Pullman	9/6/02	1746A	2925
ESPEJO	N	10SecLibObs	Pullman	12/12/08	2351A	3663
ESPERANZA	N	10SecDrB	PCW	9/27/84	149D	1006
ESPERANZA		BaggLib	PCW	4/12/87	477	1325
ESPERANZA		12SecDr	Pullman	9/22/06	1963C	3397
ESPINOSA		12SecDr	Pullman	8/14/01	1581A	2727
ESPIRA	R	10SecObs	PCW	6/4/88	445V	1414
ESPIRA SP-61	P	10SecDrB	PCW	6/4/88	445F	1414
ESPIRITO	P	12SecHotel	B&S	1/99	2370A	
ESSEN		12SecDr	PCW	4/26/92	784E	1899
ESSEX		DDR	PCW Det	11/1/70	17	D4
ESSEX		12SecDr	Pullman	1/26/03	1581D	2962
ESSEX	P	12SecO	(Wagner)	/85	3005	
ESTELLA PETOSKEY #15	P		(WS&PCC)	9/28/72		
ESTELLE		24ChDr	Pull Buf	3/5/07	2156	B57
ESTELLE #24	P	14Ch4StsDr	(Wagner)	6/85	3106D	
ESTELLE #97	P		(WS&PCC)	7/23/81		
ESTHER		30ChObs	PCW	4/28/98	1278B	2299
ESTHER		24ChDr	Pullman	7/10/05	1980A	3171
ESTHONIA		12SecDr	Pull Buf	1/16/02	1581A	B7
ESTRELLA	R	12SecDr	PCW	6/6/88	1603	1414
ESTRELLA SP-62	P	10SecDrB	PCW	6/6/88	445F	1414
ESTRELLA BLANCA	RN	12SecDrB	Pullman	5/23/06	2165C	3352
ESWICK		12SecDr	Pullman	3/24/03	1581D	2983
ETELKA GERSTER #110	P	Private	J&S (MannBCC)	/84	6	
ETHEL		20Ch6StsDr	PCW	7/11/91	928A	1827
ETHELBERT		12SecDr	Pullman	5/5/01	1581A	2726
ETHELRIDA		26ChDrSr	Pullman	6/9/04	1987A	3085
ETHERTON		12SecDr	Pullman	9/10/07	1963C	3550
ETHIOPIA		12SecDr	Pull Buf	4/30/03	1581C	B16
ETIDORPHA		12SecDr	Pullman	12/4/07	1963C	3585
ETIWANDA		10SecDrB	PCW	12/20/87	445E	1360
ETIWANDA	N	12SecDr	Pullman	10/17/05	2049B	3232
ETLAH	RN	14SecO	PCW Det	7/16/73	37A	D16
ETNA	P	10SecDr	H&StJ RR	3/71	19	Han2
ETNA	P	12SecDr	Wagner	5/92	3065C	Wg68
ETON		12SecDr	PCW Det	1/7/91	784	D455
ETOWAH	R	10SecO	PCW Det (rblt)	11/20/74	47	D27
ETRURIA		12SecDr	PCW Det	5/24/82	93B	D154
ETRURIA		12SecDr	Pullman	7/10/08	1963C	3630
ETRURIA #26	P	16Ch8StsDrB	(Wagner)	6/85	3115B	
EUCLID	N	Slpr	(IC RR)			2/176
EUCLID	P	12SecDrB	Wagner	6/96	3073C	Wg109
EUCLID #6	P		(WS&PCC)	7/16/72		
EUDEMUS		7CptLngObs	Pullman	6/16/04	1977B	3091
EUDOCIA		20Ch4StsDrB	Pullman	6/17/01	1518E	2703
EUDORA		14Ch13StsDrB	PCW	7/29/87	505	1376
EUDORA	N	12SecDrB	Wagner	11/92	3073H	Wg80
EUDOXUS		6CptDrLibObs	Pullman	3/3/03	1751B	2986
EUFAULA	N	10SecDr	B&S	3/72	26A	Day7
EUFAULA		12SecDr	Pull Buf	2/25/03	1581C	B13
EUGENE	N	10SecDr3Cpt	Pullman	4/26/06	2078C	3342
EUGENIA #735	P	17Ch3StsB	J&S (B&O)	7/2/87	686	2522
EUGENIE	RN	22Ch3Sts	PCW	5/18/92	973E	1909
EULALIA		18Ch6StsDrB	PCW Det	6/13/90	772D	D424
EUMENES		10Sec2Dr	PCW	6/15/96	1173A	2153
EUNICE		18Ch4StsDrB	PCW	6/20/93	1039A	1994
EUNOMIA		20Ch5StsB	PCW Det	6/22/83	167	D203
EUPATORIA		12SecDr	Pull Buf	10/22/03	1581F	B24
EUPHRATES		12SecDr SUV	Pull Buf	7/25/04	1963	B40
EURASIA	RN	10SecDr	PCW	2/23/85	222A	1033
EUREKA		10SecDr	PCW Det	2/28/80	59	D71
EUREKA		12SecDrB	Pullman	3/22/02	1611B	2820
EUREKA #7	P		(WS&PCC)	7/16/72		
EURIPIDES		10Sec2Dr	PCW	3/28/96	1173A	2153
EUROPA		10SecB	PCW Det	9/2/73	37	D16
EUROPA		26Ch3StsDr	PCW	4/25/98	1278C	2258
EUROPA		24StsDrSrB	Pullman	6/16/04	1988A	3086
EUROPA	N	19Ch4StsDr	(Wagner)	7/85	3117	
EUROPEAN #234		Parlor	PCW Det	11/10/71	34	D4
EURYBIA	N	28ChDr	Pullman	4/27/07	1986B	3527
EURYDICE	N	17Ch14StsB	J&S (B&O)	6/29/87	686B	2519
EURYDICE		24StsDrSrB	Pullman	6/16/04	1988A	3086
EURYNOME		6CptDrLibObs	Pullman	3/13/02	1751B	2799
EURYNOME	N	24ChDrObs	Pull Buf	4/15/07	2091C	B70
EUSEBIA		26ChDrSr	Pullman	6/9/04	1987A	3085
EUTERPE		20Ch5StsB	PCW	11/24/83	167A	89
EUTERPE	P	10Sec2Dr	Wagner	6/93	3074	Wg92
EUXINE	N	20Ch8StsDrB	Wagner	6/91	3126A	Wg51
EVA #132	P	12SecB	(WS&PCC)	2/22/84	789	
EVADNE		22Ch6StsDrObs	Pull Buf	5/29/03	1634D	B17
EVANGELINE		12Ch5StsB	PCW (UnionPCC)	5/8/84	167A	D233
EVANGELINE #734	P	17Ch3StsB	J&S (B&O)	7/2/87	686	2521
EVANSTON		12SecDr	PCW	8/29/82	93B	23
EVANSTON	N	12SecDrSr	Pullman	3/24/05	2049A	3169
EVANSVILLE		10SecDr	PCW Det	3/27/72	26A	D10
EVANSVILLE		12SecDr	Pull Buf	3/8/04	1581F	B34

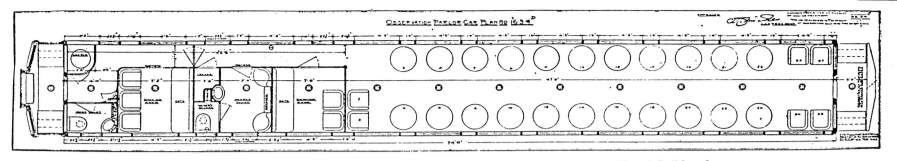

Figure I-19 Two 22-chair, 6-seat, drawing room observation cars were built at the former Wagner Shops in Buffalo under Plan 1634D, Lot B17 in May 1903. One of these cars was EVADNE. Both cars were renamed in October 1905, with EVADNE becoming GLEN SUMMIT, and remained in active service until January 1928 when it was scrapped. The Newberry Library Collection.

CAR NAME	ST	CAR TYPE	CAR BUILDER (Owner)	DATE BUILT	PLAN	LOT
EVANSVILLE	P	12Sec Round End	(WS&PCC)			
EVANSVILLE (2nd)	N	12SecB	(WS&PCC)	9/22/84	789	
EVELINE		30Ch	Pullman	5/23/01	1646D1	2700
EVELYN	N	20Ch6StsB	PCW	1/22/85	167B	200
EVENING						
STAR	N	12SecO	CB&Q RR	/68	5	Aur
STAR E&A-2	P	Slpr	(E&A Assn)			
EVENTIDE		12SecDr	Pullman	8/26/01	1581A	2756
EVERETT	P	12SecDr	J&S (Wagner)	11/82	3022C	
EVERGLADE		12SecDr	Pull Buf	12/1/03	1581F	B26
EVERTON		12SecDr	Pullman	10/30/01	1581A	2784
EVESHAM	N	10Sec2Dr	Wagner	11/97	3073D	Wg110
EVRAN		14SecO	PCW	3/11/93	1021	1969
EXCELSIOR	P	10SecDr	B&S	4/30/73	26	
EXCELSIOR		22Berths Slpr	PCW Det	2/15/74	38	D19
EXCELSIOR		12SecDr	Pullman	8/11/08	1963C	3636
EXCELSIOR CTC-86	R		PRR rblt/74			
EXETER	N	10SecDrB	PCW Det	5/20/76	56	D39
EXETER	P	12SecDrB	Wagner	6/96	3073C	Wg109
EXMOOR		12SecDr	Pullman	11/23/03	1581F	3043
EXMOUTH		12SecDr	Pullman	10/23/03	1581F	3025
EXTON		12SecDr	PCW	3/11/91	784B	1801
EYLAU	N	10SecObs	PCW	11/17/86	445W	1262
EYRIA		16SecO	PCW	8/29/93	1040	2008
F Sanderson	P	OS	I&StLSCCo	6/69		
FABIUS		14SecDr	Pullman	6/12/00	1523	2559
FABRIANO		12SecDr	Pull Buf	11/14/02	1581C	B11
FABYAN		Diner	PCW	11/15/89	662D	1641
FABYAN		12SecDr	Pull Buf	5/21/07	1963C	B60
FAENZA		12SecDr	PCW	1/7/91	784	1785
FAIR OAKS	N	12SecDrSr	Pullman	3/30/05	2049A	3169
FAIRBROOK		12SecDr	Pullman	1/4/07	2163	3460
FAIRBURN		10SecDr2Cpt	Pullman	6/10/08	2078D	3622

CAR NAME	ST	CAR TYPE	CAR BUILDER (Owner)	DATE BUILT	PLAN	LOT
FAIRCHANCE		12SecDrSr	Pullman	4/6/05	2049A	3169
FAIRFAX		14SecO	PCW Det	2/3/93	1014A	D511
FAIRFIELD		10SecDrSr	PCW	6/17/90	784A	1715
FAIRFIELD		12SecDr	Pullman	11/24/03	1581F	3043
FAIRHAVEN #445	P	12SecDrB	Wagner	1/91	3065	Wg45
FAIRLIE		12SecDr	Pullman	3/31/03	1581D	2983
FAIRMONT		12SecDr	Pullman	4/27/03	1581B	3003
FAIRMOUNT #230	P	Parlor	H&H	12/72	35 3/4	Wil1
FAIRPORT		12SecDr SUV	Pull Buf	6/7/04	1963	B37
FAIRVIEW	R	12SecO?	PCW Det	3/7/71	27B	D5
FAIRVIEW	N	10SecDrB	PCW	10/7/87	445E	1334
FAIRVIEW		12SecO	PCW Det	3/7/71	5	D5
FAIRWOOD		14SecDr	Pull Buf	9/23/00	1523D	B1
FALCON	N	10SecDr	C&NW Ry	10/70	19	FdL3
FALCON		24ChDrB	Pullman	6/15/00	1518C	2552
FALCON	N	12SecDr	(Wagner)	6/86	3049A	
FALCONET		12SecDr	Pullman	7/24/01	1581A	2727
FALKA	N	18Ch21Sts	CB&Q RR	4/71	29R	1
FALKA	R	32Ch	Pullman	4/6/04	2063	3032
FALKIRK	P	10SecO	(Wagner)	3/85	3039	
FALKLAND		12SecDr	Pull Buf	5/27/07	1963C	B60
FALKLAND #137	N	12SecB	(WS&PCC)	1/15/85	742B	
FALLS CITY	P	SDR	(OPLines)	/72	114	
FALLS CITY	RN	8SecBLibObs	PCW	5/27/88	1707	1377
FALLSINGTON		26ChDr	Pullman	5/28/07	1987B	3524
FALMOUTH	N	10SecDr	H.Y. Hull	7/71	19	3
FALMOUTH #444	P	12SecDrB	Wagner	6/91	3065	Wg45
FAMA		12SecDr	PCW	1/5/91	784	1781
FAMBRIDGE		16Sec	Pullman	6/1/09	2184D	3700
FAMOSA	N	BaggSmCafe SUV	Pull P-2267	4/07	3278B	3494
FAMOSO	N	BaggSmCafe SUV	Pull P-2267	4/07	3278B	3494
FANCHON		20Ch4StsDrB	Pullman	6/12/01	1518E	2703
FANNIE	P	12SecO	J&S (WS&PCC)	6/30/73	833	

CAR NAME	ST	CAR TYPE	CAR BUILDER (Owner)	DATE BUILT	PLAN	LOT
FANNY	P	OS	(OPLines)			
FANNY FERN		12SecO	J&S (WS&PCC)	6/30/73	833?	
FANO		12SecDr	PCW	1/12/91	784	1781
FANSHAWE		12SecDr	Pullman	12/6/07	1963C	3585
FANTASY		16Sec	Pullman	1/8/09	2184D	3662
FANTOME		16Sec	Pullman	5/23/08	2184C	3623
FANWOOD	N	12SecO	B&O RR	/80	665	
FANWOOD		12SecDr	Pull Buf	1/8/08	1963C	B72
FAREEDA	N	28ChDr	Pullman	4/7/06	1986A	3336
FARENZO		12SecDr	Pull Buf	1/2/08	1963C	B72
FARENZO 2515		14SecO	PCW Det	1/14/93	1014A	D510
FARGO	?	12SecO	(NP Assn)	6/81	18D	
FARGO		12SecDr	Pull Buf	5/23/07	1963C	B60
FARIBAULT		12SecDrSr	PCW	8/7/86	375	1215
FARIBAULT		12SecDr	Pullman	3/7/04	1963	3080
FARMINGTON	N	19Ch3StsB	H&H (B&O)		653A	
FARMINGTON	P	36Ch SUV	Pull P-2494	3/11	3275	3821
FARNHAM	N	12SecDr	Pullman	3/7/04	1963	3080
FARNHURST		BaggBLibSmk	Pullman	4/24/07	2087	3534
FARNLEY		16Sec	Pullman	5/14/09	2184D	3700
FARRAGUT #446	P	12SecDrB	Wagner	6/91	3065	Wg45
FASANO		16SecO	PCW	8/3/93	1040	2008
FASHODA		12SecDr	Pullman	12/29/03	1581H	3053
FATANITZA		10SecLibObs	Pullman	7/21/09	2351A	3715
FATIMA	P	18Ch6Sts	PCW	9/95	1153	2116
FAUST		12SecDr	Pullman	3/17/08	1963C	3604
FAUST #111	P	7BoudB	(MannBCC)	/84	697	
FAUSTINA		16Ch12StsDrB	PCW	5/29/89	614B	1508
FAVERWILM	N	12SecDrSr	Pullman	3/30/05	2049A	3169
FAVORITA		10SecLibObs	Pullman	7/21/09	2351A	3715
FAVORITE		10SecDr	PCW Det	7/9/73	37 1/2	D18
FAYAL		10SecDrB	PCW	1/14/92	920B	1880
FAYAL		12SecDr SUV	Pull Buf	6/3/07	1963C	B60
FAYETTE	N	12SecDr	Pullman	4/27/03	1581D	3003
FAYETTE #447	P	12SecDrB	Wagner	6/91	3065	Wg45
FAZENDA		16Sec	Pullman	5/18/09	2184D	3700
FEATHERSTON		12SecDr SUV	Pullman	12/5/07	1963C	3585
FEDALMA	P	20Ch6StsDrB	J&S	1/02	2369	
FEDORA		Dr	(Wagner)	C:/75		
FEDORA	RN	22Ch3Sts	PCW Det	4/19/83	1928A	D184
FELICIA		16Ch12StsDrB	PCW	10/15/88	614C	1508
FELICIAN		12SecDr SUV	Pull Buf	5/29/07	1963C	B60
FELICITO		8Sec4SrB	PCW	10/6/91	894B	1857
FENELLA #139	N	6Sec12ChSr	J&S (Woodruff)	12/1/85	741B	
FENELON	N	10SecDr	B&S	12/70	19	Day5
FENELON		12SecDrB	Pullman	2/24/06	2142	3318
FENWICK		14SecO	PCW Det	6/13/93	1041	D519
FENWICK	P	10SecO	(Wagner)	3/85		3039
FENWICK (#179)	RN	16Ch28Sts SUV	Pull Buf	1/24/03	3581D	B12
FENWOOD		14SecDr	Pull Buf	9/21/00	1523D	B1
FEODOSIA		12SecDr	Pullman	9/11/00	1572	2585
FERDINAND		10Cpt	PCW	4/22/93	993B	1965
FERDINAND	P	10Sec2DrB	Wagner	8/97	3074B	Wg115
FERMO		12SecDr	PCW	11/26/90	784	1767
FERNANDO		10SecDrB	PCW	11/10/87	445E	1334
FERNCLIFF	N	8CptLibObs	Pullman	12/22/06	2260	3502
FERNCLIFFE #723	P	16Ch12StsB	B&S (B&O)		664	
FERNCROFT	N	28ChDr	Pullman	4/6/06	1986A	3336
FERNDALE		12SecDrB	PCW	8/25/96	1204C	2187
FERNWOOD	RN	8Sec3Dr	PCW	5/13/84	191J	132
FERNWOOD		12SecDrSr	Pullman	4/8/05	2049A	3169
FERONIA		12SecDrB	PCW	8/21/88	375F	1480
FERRIDAY		12SecDr	Pullman	10/29/07	1963C	3570
FIDELIA		16Ch12StsDrB	PCW	2/23/89	614C	1508
FIDELIO #122	P	7BoudB	(MannBCC)	/84	776	
FIFE		12SecDr	PCW	5/11/91	784D	1805

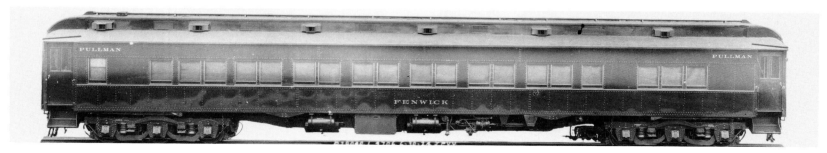

Figure I-20   In the early 1920s, The Pullman Company found a market for some of its obsolete parlor cars. FENWICK, a steel-sheathed steel underframe coach / parlor, was rebuilt in June 1924 from Parlor SALERNO. These coaches were leased and eventually sold to the railroads, including the National Railroad of Mexico. Smithsonian Institution Photograph P28040.   Courtesy Robert Wayner.

CAR NAME	ST	CAR TYPE	CAR BUILDER (Owner)	DATE BUILT	PLAN	LOT
FIGARO	N	12SecB	PCW	3/4/85	227	1034
FILOMENA		24StsDrSrB	Pullman	6/16/04	1988A	3086
FINDLAY		12SecDr	Pullman	9/24/03	1581F	3022
FINESSE	N	26ChDr	Pullman	5/25/07	1987B	3524
FINGAL		14SecO	PCW Det	8/1/92	1014A	D509
FINLAND	N	12SecDrSr	H&H (B&O)	? 8/87	654I	
FINLAND		12SecDr	Pullman	3/18/03	1581D	2975
FIORENZA		20Ch5StsB	PCW	8/23/84	167A	199
FISHKILL	N	16SecO	PCW	5/88	543K	1427
FISK E&A-4	N	Slpr	(E&A Assn)			
FITCHBURG	N	12SecDr	Wagner	1/89	3061E	Wg13
FITCHBURG	P	26Ch SUV	Pull P-2071	6/05	3273	3189
FITCHBURG #246		Parlor	PCW Det	2/5/72	22	D9
FLAMENCOE		16Sec (SUV)	Pullman	4/5/10	2184G	3808
FLAMINGO		24ChDrB	Pullman	6/30/00	1518C	2553
FLATBUSH #129	P	Parlor (3'gauge)	(WS&PCC)	5/17 84	528?	
FLAVIUS	R	BaggSmkBLub	PCW?		1706	
FLEETWING		16Sec	Pullman	1/8/09	2184D	3662
FLEETWOOD		12SecDr	Pullman	5/28/06	1963C	3360
FLEURY	N	10Sec2Dr	Wagner	11/98	3073D	Wg110
FLODDEN	N	25Ch2StsB	(Wagner)	10/81	3100A	
FLORA		22ChDr	Pull Buf	6/7/02	1647C	B8
FLORA #20	P		(WS&PCC)	6/29/72		
FLORA #76	N		(WS&PCC)	7/1/79	755A	
FLORADEL	P	28Ch3StsB	J&S	12/95	2367	
FLORAYME		16Sec	Pullman	6/2/09	2184D	3700
FLORELLA	R	32Ch	Pullman	4/6/04	2063	3031
FLORENA #65	N		(WS&PCC)	5/1/79		
FLORENCE		12SecO	DetC&MCo	/68	5	Det
FLORENCE		16Ch12StsDrB	PCW	10/11/88	614B	1508
FLORENCE RICHMOND #14	P		(WS&PCC)	9/28/72		
FLORENTINE		12SecDr	Pull Buf	1/29/08	1963C	B72
FLORES SP-63	P	10SecDrB	PCW	6/12/88	445F	1414
FLORETTE	N	28ChDr	Pullman	4/5/06	1986A	3336
FLORIAN		12SecDr	Pullman	3/22/10	1963E	3807
FLORIBEL	N	16Ch12StsDrB	PCW	10-11/88	614C	1508
FLORIDA	P	SDR	(PK&RSCC)			
FLORIDA		10Cpt	Pullman	1/10/02	1704	2758
FLORINE		30Ch	Pullman	5/18/01	1646D1	2700
FLORISSANT		10SecDr	PCW	11/3/87	508B	1368
FLOSSMOOR		10SecDr2Cpt	Pullman	10/1/06	2078D	3398
FLUSHING #99	P	Round end	(WS&PCC)	7/29/81	360?	
FOLCROFT		26ChDr	Pullman	5/28/07	1987B	3524
FOLKSTON		12SecDrB	Pullman	12/10/06	2142A	3405
FOLSOM	N	12SecDr	PCW Det	7/26/82	93B	D162
FOND DU LAC		10SecDrSr	PCW	1/2/93	937B	1950
FOND DU LAC WC-41	P	10SecCafe	CentCarCo	2/82	532	
FONDA		10SecDrB	PCW	10/30/86	445	1262
FONDA	R	16Sec	PCW	10/30/86	1880	1262
FONDA	N	BaggClubB	Pullman	7/30/03	1945D	3020
FONDA	N	32Ch	Pullman	4/6/04	2063	3031
FONESCA		20Ch4StsDrB	Pullman	6/15/01	1518E	2703
FONSECA	N	20Ch4StsDrB	Pullman	6/18/01	1518E	2703
FONTANA		12SecDr	Pull Buf	1/15/08	1963C	B72
FONTENOY		14SecO	PCW Det	4/22/93	1021	D518
FORDHAM		12SecDr	Pullman	3/4/07	1963C	3525
FOREST #267	P	10SecDr	J&S	11/72	26	Wil8
FOREST HILLS	P	34Ch SUV	Pull P-1484	1/03	3265	2904
FORMOSA		12SecDr	PCW Det	2/14/82	93	D144
FORMOSA #449	P	12SecDrB	Wagner	1/91	3065	Wg45
FORREST	RN	Private	PCW Det	5/5/83	1929B	D194
FORRESTON		12SecDr	Pullman	11/5/00	1581A	2589
FORSYTHE	RN	14Sec	PCW	5/17/90	668F	1709
FORT BENTON		12SecDr	PCW Det	3/16/83	93B	D184
FORT KEOGH		12SecDr	PCW	9/30/82	93B	22
FORT PLAIN		16Sec	Pullman	4/4/07	2184	3523
FORT SCOTT		SDR B	PCW Det	11/13/71	27	D8
FORT SCOTT		10SecDr	PCW Det	3/25/72	26A	D10
FORT SNELLING		12SecDrSr	PCW	8/7/86	375	1215
FORT WAYNE		12SecDr	Pullman	2/7/08	2163A	3599
FORT WAYNE #128	P	DDR	PFtW&C	8/70	17	FtW2
FORTUNA		6CptLngObs	PCW	1/10/98	1299A	2283
FOSTORIA		12SecDrSrB	PCW	2/28/99	1319A	2400
FOXBORO	P	34Ch SUV	Pull P-1484A	6/00	3265	2518
FOXVALE	P	24Ch2Sts	B&S	11/93	2264	
FRA DIAVOLO #114	P	7BoudB	(MannBCC)	/84	697	
FRANCE		12SecDr	PCW	4/12/87	473A	1323
FRANCES		18Ch4StsDrB	PCW	6/30/93	1039A	1994
FRANCISCA	P	18Ch3StsDrObs	(B&M RR)			1203
FRANCISCO		12Sec2Dr	Pullman	3/31/01	1580A	2646
FRANCITAS		12SecDr	Pullman	9/21/06	1963C	3397
FRANCOIS	P	20Ch6Sts	B&S	7/89	3258A	
FRANCONIA		Parlor	PCW Det	8/7/72	35 1/2	D12
FRANCONIA		12SecDr	Pullman	6/20/01	1581A	2726
FRANCONIA	P	12SecDr	(Wagner)	/75	3008B	
FRANKFORT		16SecO	PCW	8/2/90	780	1706
FRANKLIN	P	OS	MP RR?		27	
FRANKLIN		12SecDrB	PCW Det	7/26/88	375G	D345
FRAULEIN	N	28ChDr	Pullman	4/27/07	1986B	3527
FRAZER		12SecDrSr	Pullman	3/25/05	2049A	3169
FREDA	P	22ChDrB	(Big-4 RR)	/98	2377	
FREDERIC #155	P	10SecDr	J&S (B&O)	4/30/73	19	Wil1
FREDERICKA	N	14Ch6StsDrB	PCW	11/27/80	154	34
FREDERICKS-BURG CTC-39	R		Det rblt/74			
FREDONIA	N	10SecDrB	PCW Det	11/22/80	63S	D77
FREDRICKA		24ChDr	Pull Buf	2/20/07	2156	B57
FREEHOLD		BaggClub	Pullman	8/15/06	2136A	3386
FREELAND		12SecDr	Pullman	1/21/07	1963C	3448
FREEPORT		12SecDr	Pull Buf	4/2/03	1581C	B15
FREEPORT		12SecDr	Pullman	4/21/07	1963C	3533
FREEPORT #104	P	12SecDr?	(MannBCC)	/89	878	
FREMONT		10SecDrSr	PCW	6/19/90	784A	1715

CAR NAME	ST	CAR TYPE	CAR BUILDER (Owner)	DATE BUILT	PLAN	LOT
FREMONT	P	12SecDr	Jones C&MC	8/82	3023B	
FRENCH LICK	P	12Ch24StsObs SU	(CI&L ER)	C:/11	2747	
FRESNO SPofA-6	P	Slpr	B&S	10/3/83	37?	Day
FRESNO SPC-8156	P	10Sec2Dr	PCW	3/10/99	1158	2376
FRISCO		10SecDr2Cpt	Pullman	9/5/05	2078	3203
FRONTENAC		12SecDrSr	PCW Det	5/12/87	375	D292
FRONTENAC #448	P	12SecDrB	Wagner	1/91	3065	Wg45
FRONTERA		12SecDr	Pullman	9/19/06	1963C	3397
FRONTINO	N	12SecO	B&O RR	/81	665	
FRONTINO		12SecDr	Pull Buf	1/22/08	1963C	B72
FRYBURG	N	12SecDr	B&O RR	11/80	657B	
FT. WAYNE #124	P	12SecDr	(WS&PCC)	9/18/83	914	
FUCHSIA		26CH4StsDr	Pull Buf	7/9/00	1519C	B2
FUCINO		12SecDr	PCW Det	3/5/91	784B	D460
FUCINO	N	10SecDrSr	PCW Det	2/ /91	784I	D460
FUEDOS		12SecDr	Pullman	8/22/04	1963	3098
FULDA		6SecDr2Sr	PCW	5/24/88	542A	1404
FULLERTON		12SecDr	Pullman	3/4/07	1963C	3525
FULTON		12SecDr	PCW	9/4/91	784C	1828
FUNDY		14SecO	PCW Det	1/20/93	1014A	D510
FURNESSIA		12SecDr	Pullman	5/26/06	1963C	3360
G	P	OS	G,B&Co/GMP	/56		
GABRIELLE	N	14Ch18StsDrObs	PCW	7/23/87	505	1376
GAETINA	P	16Sec SUV	Pull P-1894	3/03	3212	2981
GAILLARDIA	N	24ChDr	Pull Buf	2/20/07	2156	B57
GAINESVILLE		10SecLngObs	Pullman	9/1/08	2176	3635
GAINSBORO		16Sec	Pullman	8/21/05	2101	3233
GALAHAD		12SecDr	Pullman	12/16/09	1963E	3766
GALANA	#105	P12SecDr	(MannBCC)	/88	878A.#	
GALATEA		16SecO	PCW	5/24/88	543B	1427
GALATEA #307	P	10SecO	PCW (Wagner)	3/15/87	3040	1264
GALATIA		12SecDr	Pullman	8/31/00	1572	2585
GALATZ		12SecDr	PCW	12/20/92	990	1941
GALEN	N	12SecDrB	(Wagner)	3/81	3050	
GALENA	P	DDR	DetC&MCo	/70	17	D3
GALENA #105	P	12SecDr	(MannBCC)	/88	878	
GALESBURG	RN	12SecDr	CB&Q/GMP	/67	13B	Aur
GALETTA #303	N	20Ch12StsB	(MannBCC)	/89	708	
GALETTA		12SecDr	Pull Buf	2/28/08	1963C	B73
GALICIA #104	RN	7BoudB	(MannBCC)	/84	697A.#	
GALILEO		12SecDrB	PCW	11/24/88	375G	1513
GALION		12SecDr	Pull Buf	2/17/08	1963C	B73
GALLATIN	P	12SecO	(L&N RR)	/72	114	
GALLATIN		12SecDr	Pullman	3/9/04	1963	3080
GALLATZIN		16Sec	Pullman	7/9/03	1721	3014
GALLATZIN #279	P	10SecDr	J&S (PRR)	2/73	26	Wil8
GALLIA		10SecDrHotel	PCW Det	8/12/80	63 1/2	D77
GALLIA #18	P	16Ch8StsDrB	(Wagner)	4/85	3108B	
GALLINAS		10SecDrB	PCW	12/12/86	445C	1261
GALLINAS	R	14Sec	PCW	12/12/86	445X	1261
GALVA		14SecO	PCW	1/21/93	1017C	1962
GALVESTON	N	OS	Ohio Falls			
GALVESTON	N	10SecB	Bowers, Dure	9/73	37	Wil11
GALVESTON		12SecDr	Pullman	7/25/03	1581F	3015
GALWAY		12SecDr	Pullman	7/29/02	1581B	2852
GALWAY	N	16Sec	Pullman	8/17/05	2101	3233
GAMBIA	N	12SecDr	H&H (B&O)	? 8/87	654I	
GAMBIER		BaggClub	Pullman	8/14/05	2087	3220
GANGES #373		10SecDrB	PCW	1/18/83	149	21
GANOGA		16Ch7StsDrObs	PCW	5/2/96	1193	2179
GARDA		12SecDr	PCW	11/28/90	784	1767
GARDA	N	10SecDrB	(Wagner)	5/81	3046	
GARDEN CITY	N	16SecO	PCW	5/88	543K	1427
GARDEN CITY #57	P	Slpr	(IC RR)			
GARDEN CITY #100	P	Round end	(WS&PCC)	7/29/81		
GARDEN CITY #271		10SecDr	PCW Det	10/7/72	26 1/2	D13
GARDENIA		24ChDr	Pull Buf	4/5/06	2156	B51
GARDETTA	N	26ChDr	Pullman	3/28/06	1987A	3335
GARDINER	RN	14Sec	PCW	6/20/89	668F	1587
GARFIELD		10Sec2Dr	PCW	5/9/93	995A	1942
GARITA		10SecDrB	PCW	5/17/88	445F	1413
GARLAND		14SecDr	Pullman	8/9/00	1523E	2567
GARNET		12SecDrSr	PCW	4/3/99	1318C	2390
GARNET	N	12SecDr	Wagner	6/93	3065C	Wg85
GARONNE	R	12SecDrB	PCW	11/19/84	1611	1017
GARONNE #434		10SecDrB	PCW	11/19/84	149D	1017
GARRICK		16Sec	Pullman	1/9/08	2184	3588
GARRISON		12SecDr	PCW Det	6/2/84	93D	D241
GARRISON		12SecDr	Pullman	9/5/07	1963C	3550
GARVANZA		10SecDr2Cpt	Pullman	3/23/07	2078D	3528
GARWAY		12SecDr	Pullman	10/11/07	2163A	3562
GASCOGNE		14SecO	PCW Det	1/14/93	1014A	D510
GASCONY	N	12SecDrSr	Pullman	3/30/05	2049A	3163
GASCONY #278	P	9CptB	Wagner	4/89	3062B	Wg14
GASPERAN		12SecDrB	Pullman	2/21/06	2142	3318
GASTON	P	10SecDr	(Wagner)	/85	3011	
GATUN	N	12SecDr	Pullman	3/7/06	2163A	3333
GAVIOTA	N	32Ch	PCW	10/5/87	505	1376
GAVIOTA		24StsDrSrB	Pullman	6/16/04	1988A	3086
GAWAINE		12SecDr	Pullman	3/24/10	1963E	3807
GAYLE	N	22ChDr	Pull Buf	5/20/02	1647C	B8
GEISHA	N	28ChDr	Pullman	4/26/07	1986B	3527
GEM	P	Commissary	C&NW Ry	2/71	27	Chi3
GEMINI		14SecO	PCW	5/18/93	1021	1972
GEMONA		12SecDr	Pull Buf	2/22/08	1963C	B73
GENESCO		12SecDr	Pullman	4/23/07	1963C	3533
GENESEE	P	20Ch8StsDrB	Wagner	8/93	3126B	Wg94
GENESEE E&A-23	P	Slpr	(E&A Assn)			
GENESEO		12SecDrB	Pullman	3/22/02	1611B	2820
GENESTA		16Sec	Pullman	1/15/08	2184	3588
GENESTA #337	P	10SecDrB	(UnionPCC)		565B	
GENEVA	P	10SecDr	H&StJ RR	8/71	14	Han2
GENEVA		10SecDrHotel	PCW Det	11/9/80	63 1/2	D77
GENEVA		12SecDr	PCW	8/3/98	1318A	2313

Figure I-21 GARRISON, one of ten 12-section drawing room cars built in Lot 3550 in September 1907, was used in General Service on the Union Pacific Railroad. This handsome window design was carried forward to the first steel cars. Smithsonian Institution Photograph P10012.

CAR NAME	ST	CAR TYPE	CAR BUILDER (Owner)	DATE BUILT	PLAN	LOT
GENEVA #303	P	28Ch8StsB	(MannBCC)	/89	708	
GENEVA #320	P	12SecO	PCW (Wagner)	3/29/87	3026A	1264
GENEVIEVE		12Ch5StsB	PCW (UnionPCC)	5/10/84	167A	D233
GENEVRA		20Ch6StsDr	PCW	7/21/91	928A	1827
GENOA	P	10SecDr	H&StJ RR	8/5/71	19	Han2
GENOA	P	20Ch8StsDrB	Wagner	8/93	3126B	Wg94
GEORGETOWN		12SecDr	Pullman	3/9/04	1963	3080
GEORGIA		10SecDr	H&H	9/72	26	Wil7
GEORGIA	P	SDR	(PK&RSCC)			
GEORGIA #302	P	10SecO	PCW (Wagner)	3/15/87	3041	1264
GEORGIA #324	P	8BoudB	J&S (UnionPCC)	/89	373	
GEORGIAN	P	10Sec2DrB	PCW (Wagner)	3/16/00	3079A	Wg135
GERAINT		12SecDr	Pullman	12/16/09	1963E	3766
GERALDINE		16Ch12StsDrB	PCW	10/22/88	614C	1508
GERALDINE #733	P	17Ch3StsB	J&S (B&O)	7/2/87	686	2520
GERANIUM		24ChDrSr	Pullman	7/21/04	1980A	3088
GERBER	N	10SecDr2Cpt	Pullman	10/1/06	2078D	3398
GERHILDE		12SecDr SUV	Pullman	3/23/10	1963E	3807
GERMANIA		22Berths	PCW Det	8/30/76	48	D32
GERMANIA		12SecDr	PCW	8/10/98	1318A	2313
GERMANIA #333	P	12SecDr	Wagner	5/87	3061D	Wg3
GERMANIC		10SecDr	PCW Det	1/5/76	52	D33
GERMANTOWN		12SecDr	Pullman	10/15/07	2163A	3562
GERMANY		12SecDr	PCW	6/7/87	473A	1323
GERONIMO		10Sec2Dr	Pullman	11/1/02	1746A	2928
GERTRUDE		18Ch4StsDrB	PCW	6/27/93	1039A	1994
GERTRUDE #10	P	12SecDr	(WS&PCC)	6/5/73		
GERVAIS		14SecO	PCW Det	4/29/93	1021	D518
GHENT		12SecDr	PCW	6/1/92	784E	1915
GHENT	N	12SecDr	(Wagner)	6/86	3049A	
GHIZEH		14SecO	PCW Det	4/4/93	1021	D517
GIBRALTAR		12SecDr	Pullman	8/26/01	1581A	2756

CAR NAME	ST	CAR TYPE	CAR BUILDER (Owner)	DATE BUILT	PLAN	LOT
GIBSON		12SecDr	Pull Buf	2/4/08	1963C	B73
GILA		12SecDr	PCW	4/29/91	784D	1805
GILA	N	12SecDrCpt	Pullman	5/23/04	1985	3095
GILBOA	N	20Ch8StsDrB	Wagner	8/93	3126B	Wg95
GILEAD CTC-36	R		O&M rblt/68			
GILEAD	N	14SecO	(Wagner)	/86	3032A	
GILROSE	RN	Sleeper	(SouTranCo)			
GILSEY		Diner	PCW	7/27/93	538D	2010
GILSEY #3		Diner	PCW	4/25/88	538D	1396
GIPSY		16SecO	PCW	6/23/93	1040	1990
GIPSY	R	12SecDr	PCW	6/23/93	1040F	1990
GIPSY	P	12SecDr	(Wagner)	/75	3008B	
GIRALDA		12SecDr	Pullman	10/1/09	1963E	3744
GIRALDO	P	10SecDrB	(UnionPCC)		565B	
GIRARD	N	10SecDrB	PCW	10/26/86	445	1262
GIRARD		12SecDr	Pullman	3/10/04	1963	3080
GIRONDE		12SecDr	Pull Buf	2/10/08	1963C	B73
GISMONDA		12SecDr	Pullman	10/1/09	1963E	3744
GITA		14SecO	PCW Det	3/9/93	1021	D516
GLACIER #419	P	12SecDrB	Wagner	5/90	3065	Wg31
GLADIOLUS		12SecDrCpt	PCW	8/3/92	983	1919
GLADSTONE #408	P	12SecDrB	Wagner	2/90	3065	Wg30
GLADYS		24ChDr	Pull Buf	4/8/07	2156	B58
GLAMORGAN #409	P	12SecDrB	Wagner	2/90	3065	Wg30
GLASGOW		10SecDrHotel	PCW Det	11/9/80	63 1/2	D77
GLASGOW		12SecDr	Pull Buf	5/7/04	1581F	B36
GLAUCUS		14SecDr	Pullman	6/12/00	1523	2559
GLEN COVE	P	14SecO Tourist	(Wagner)	5/76	3051	
GLEN EASTON	N	12SecDrSr	Pullman	3/25/05	2049A	3169
GLEN ECHO		12SecDr	Pullman	12/10/03	1581F	3043
GLEN ERIE #410	P	12SecDrB	Wagner	2/90	3065	Wg30
GLEN EYRE		Private	PCW	8/10/85	247	1105

CAR NAME	ST	CAR TYPE	CAR BUILDER (Owner)	DATE BUILT	PLAN	LOT	CAR NAME	ST	CAR TYPE	CAR BUILDER (Owner)	DATE BUILT	PLAN	LOT
GLEN EYRE		Private	PCW	3/22/99	1379B	2382	GLENROCK #296		10SecB	J&S	7/73	37	Wil12
GLEN FOREST		16Sec	Pullman	1/14/08	2184	3588	GLENVIEW	RN	10SecDrB	PCW	4/1/85	240	1037
GLEN MILLER		16Sec	Pullman	1/14/08	2184	3588	GLENVILLE	N	12SecDrB	PCW Det	10/31/88	375G	D358
GLEN ONOKO	N	22Ch6StsDrObs	Pull Buf	5/27/03	1634D	B17	GLENWOOD	P	12SecO	(Wagner)		3059	
GLEN SUMMIT	N	22Ch6StsDrObs	Pull Buf	5/29/03	1634D	B17	GLENWOOD #142	N	6SecDrLibObs	J&S (WS&PCC)	12/5/85	741J	
GLENARM #100	N	R-end	(WS&PCC)	7/29/81			GLOBE	P	10SecDr	B&S	12/71	26	Day 7
GLENBROOK		12SecDr	Pull Buf	12/7/03	1581F	B26	GLOBE		27Sts Parlor	PCW Det	/78	45	D31
GLENCAIRN		12SecDr	Pullman	10/17/05	2049B	3232	GLORIANA		24StsDrSrB	Pullman	6/17/04	1988A	3086
GLENCOE		10SecDr	PCW Det	4/30/71	26	D6	GLORIETA		7Cpt2Dr	Pullman	10/29/02	1845	2926
GLENCOE	N	12SecO	(Wagner)	/67	3044		GLOUCESTER		12SecDrSr	PCW	9/27/99	1318E	2449
GLENCOE	N	12SecDrB	Wagner	2/90	3065H	Wg30	GLYNDON		10SecDrCpt	PCW	6/25/89	668	1587
GLENCOVE #98	P		(WS&PCC)	7/23/81			GODERICH		10SecDr	Grand Trunk	9/71	26	Mon2
GLENDALE		12SecDrB	PCW	8/17/96	1204C	2187	GODFREY	N	12SecDr	Wagner	3/92	3065C	Wg57
GLENDALE #254	P	10SecDr	(L&N RR)	2/72	26	Han3	GODIVA #395		20Ch5StsB	PCW	1/2/84	167A	90
GLENDALE #336	P	12SecDr	Wagner	5/87	3061	Wg3	GODWIN	P	20Ch8StsDrB	Wagner	8/93	3126B	Wg95
GLENDIVE		12SecDr	PCW	10/10/82	93B	22	GOETHE		12SecDr	PCW	10/13/92	990	1931
GLENDIVE		12SecDr	Pullman	3/22/04	1963	3080	GOGEBIC		12SecDr	Pullman	10/24/03	1581F	3025
GLENGARRY #335	P	12SecDr	Wagner	3/88	3061A	Wg3	GOLCONDA		12SecDr	Pullman	1/24/02	1581A	2792
GLENHAM	N	12SecDrB	PCW	8-9/88	375G	1480	GOLDEN GATE	N	BaggLibB	PCW	6/7/87	477F	1325
GLENHAM		Diner	PCW	11/21/90	662D	1777	GOLDEN GATE		Diner	PCW	12/1 90	662D	1778
GLENITA		16Sec	Pullman	1/15/08	2184	3588	GOLDENDALE	P	6Sec3CptObs	Pullman	10/08	2354	3637
GLENLYON		26ChDr	Pullman	3/29/06	1987A	3335	GOLDENROD		24ChDr	Pull Buf	3/29/06	2156	B50
GLENMERE	N	26ChDr	Pullman	5/25/07	1987B	3524	GOLDFIELD	N	12SecDrB	(Wagner)	5/81	3050	
GLENMONT		12SecDrB	PCW	9/13/98	1319A	2314	GOLDFINCH		24ChDrB	Pullman	7/10/00	1518C	2553
GLENMOORE		24ChDrB	Pull Buf	5/16/07	2278	B59	GOLDSBORO		12SecDr	Pullman	3/15/04	1963	3080
GLENMORE	P	10SecDrB	J&S (WS&PCC)	C:/88	663F		GOLONDRINA	P	12SecHotel	B&S	1/99	2370A	
GLENMORE		10SecDr2Cpt	Pullman	10/1/06	2078D	3398	GONDOLIER		21Ch6StsDrObs	Pullman	3/22/06	2091B	3334
GLENMORE #411	P	12SecDrB	Wagner	2/90	3065	Wg30	GONZALO		12Sec2Dr	Pullman	3/31/01	1580A	2646
GLENOLDEN		26ChDr	Pullman	5/25/07	1987B	3524	GORDON		16Sec	Pullman	8/19/05	2101	3233
GLENORA	N	12SecDrB	PCW Det	6/14/82	93B	D154	GORDONSVILLE #26		10SecB	B&S	6/73	37	Day9
GLENROCK		12SecDr	Pullman	3/7/06	2163A	3333	GORGONA	N	12SecDr	Pullman	3/10/06	2163A	3333

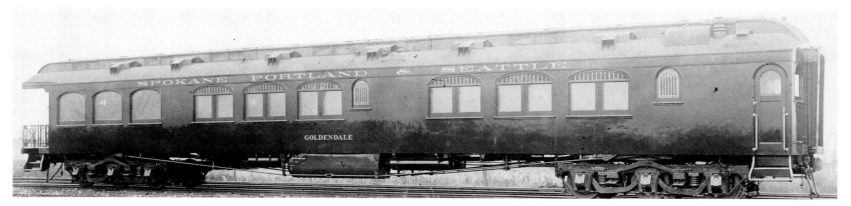

Figure I-22 GOLDENDALE was built in October 1908, along with its sister **WILLIAMETTE**, for service in the last Association to be absorbed by The Pullman Company. The Spokane, Portland & Seattle Association lasted until August 1922, and this car was dismantled in April 1924. Smithsonian Institution Photograph P11053.

CAR NAME	ST	CAR TYPE	CAR BUILDER (Owner)	DATE BUILT	PLAN	LOT	CAR NAME	ST	CAR TYPE	CAR BUILDER (Owner)	DATE BUILT	PLAN	LOT
GORHAM		10SecDr	Grand Trunk	3/71	19	Mon1	GREENBRIER		12Sec2Dr	PCW	5/18/99	1318C	2423
GORHAM	N	12SecDrB	PCW	2/8/89	375G	1557	GREENBUSH	P	34Ch SUV	Pull P-1996C	7/04	3265	3114
GORMA	N	12SecDr	Wagner	5/92	3065C	Wg57	GREENCASTLE	RN	8SecBLibObs	PCW		1707	
GOSHEN		12SecDr	Pullman	1/30/03	1581D	2962	GREENFIELD		14SecDr	PCW	12/15/99	1318J	2450
GOSNOLD		16SecO	Pull Cal	/01	1637A	C2	GREENLAND		14SecDr	Pullman	8/6/00	1523E	2567
GOTHA		14SecO	PCW Det	5/6/93	1021	D518	GREENPORT #96	P		(WS&PCC)	7/16/81		
GOTHAM		BaggClub	Pullman	3/1/06	2136	3319	GREENSBORO		10SecLibObs	Pullman	4/28/06	2176	3355
GOTHAM		16Sec	Pullman	8/17/05	2101	3233	GREENSBURG		12SecDrSr	Pullman	3/28/05	2049A	3169
GOULA #2564		14SecO	PCW Det	3/11/93	1021	D516	GREENUP		12SecDr	Pullman	2/11/08	2163A	3599
GOWAN		12SecDr	Pullman	12/17/09	1963E	3766	GREENWICH	N	12SecDr	PCW Det	7/26/82	93B	D162
GOYA #2513		14SecO	PCW Det	8/1/92	1014A	D509	GREENWICH	N	12SecDr	PCW	9/30/98	1318B	2339
GRACCHUS		12SecDrB	PCW	12/3/88	375G	1513	GREENWOOD		OS	B&O RR	4/30/73	27 1/2	B&O2
GRACCHUS	R	BaggSmkBLib	PCW?		1706		GREENWOOD	N	12SecDr	B&O RR		657B	
GRACE		18Ch4StsDrB	PCW	7/13/93	1039A	1995	GREGORIAN		16Sec	Pullman	1/10/08	2184	3588
GRACE #25	P	12SecO	(WS&PCC)	6/27/74	833		GREGORIAN	R	6SecCafeLgn SUV	Pullman	1/10/08	3986	3588
GRACELAND		10SecDr2Cpt	Pullman	10/1/06	2078D	3398	GREMIO		12SecDrCpt	Pullman	5/24/04	1982	3090
GRACETON		12SecDr	Pullman	10/9/00	1581A	2588	GRENADA	P	10SecDr	J&S	/73	37	Wil10
GRAFTON		12SecDr	Pullman	3/4/02	1581B	2818	GRENADA	N	10SecDrB	PCW	3/8/88	445F	1380
GRAFTON #153	P	10SecDr	J&S (B&O)	4/30/73	19	Wil1	GRENADA	N	12SecDrSr	Pullman	1/19/01	1581A	2614
GRAMATAN		16Sec	Pullman	1/9/08	2184	3588	GRENADIER		21Ch6StsDrObs	Pullman	3/22/06	2091B	3334
GRAMPIAN		16Sec	Pullman	8/18/05	2101	3233	GRENOBLE #313	P	12SecO	PCW (Wagner)	3/21/87	3028	1264
GRANADA		10SecDr	PCW Det	9/13/76	59	D43	GRENVILLE		16Sec	Pullman	1/11/08	2184	3588
GRANADA	P	14SecO	(Wagner)	/72	3020		GRETA		18Ch4StsDrB	PCW	6/28/93	1039A	1994
GRANBY	N	12SecO	C&NW Ry	9/70	19A	FdL3	GRETCHEN		24ChDrObs	Pull Buf	4/15/07	2091C	B70
GRANBY	N	26ChB	(Wagner)	9/96	3124A		GREYCOURT		16Sec	Pullman	1/10/08	2184	3588
GRAND GORGE	P	24Ch6StsDr	(Wagner)	7/86	3100		GREYLOCK	N	12SecDr	PCW Det	9/1/82	93L	D169
GRAND RAPIDS		OS	G&B/D&MSCC	2/68		Troy	GRIDLEY		12SecDrB	Pullman	3/20/02	1611B	2820
GRAND RAPIDS		12SecDr SUV	Pullman	10/31/07	1963C	3570	GRISELDA		24StsDrSrB	Pullman	6/18/04	1988A	3086
GRAND RAPIDS #22	N	12SecO	H&H (WS&PCC)	6/14/71	736		GRISWOLD		16Sec	Pullman	1/9/08	2184	3588
GRANDIFLORA		16Sec SUV	Pullman	4/9/10	2184G	3808	GROSVENOR	P	12SecO	PCW (Wagner)	3/87	3026B	1264
GRANDIN	N	16Sec	Pullman	8/17/05	2101	3233	GROTON	P	36Ch SUV	Pull P-2494	3/11	3275	3821
GRANDIOSO	N	BaggSmkB SUV	Pull P-2496	1/11	3278A	3828	GROVE BEACH	P	BaggSmk	Wagner	8/89	3213	Wg20
GRANGER		10SecDrSr	PCW	6/20/90	784A	1715	GROVEDALE		26ChDr	Pullman	3/29/06	1987A	3335
GRANGER		10SecDr2Cpt	Pullman	9/6/05	2078	3203	GROVELAND	RN	10SecDrB	B&S	3/17/70	149S	Day2
GRANITE		8SecDrB	PCW	7/25/88	556A	1463	GROVETON		12SecDr	PCW	10/4/98	1318B	2339
GRANTS PASS	N	10SecDr2Cpt	Pullman	11/11/07	2078D	3583	GROVETON	R	30Ch SUV	PCW	10/4/98	3581M	2339
GRANVILLE		12SecDrSr	Pullman	2/11/01	1581B	2638	GRUNEWALD		12SecDr	Pullman	3/24/08	1963C	3600
GRAPHIC		12SecDr	Pullman	1/21/07	1963C	3448	GRYPHON	P	14Sec	B&S	11/82	3201	
GRASSE		14SecO	PCW Det	6/16/93	1041	D519	GUADALUPE	N	10SecDr	PCW Det	3/15/82	112A	D141
GRASSMERE #246	P	Private	Wagner	11/87	3089	Wg5	GUADALUPE		12SecDr SUV	Pullman	3/17/04	1963	3080
GRATIANO		12SecDrCpt	Pullman	5/24/04	1982	3090	GUADIANA		12SecDr	Pull Buf	7/28/04	1963	B40
GRATIANO	N	18Ch12StsB	J&S (B&O)	6/29/87	686C	2518	GUADIOLA		Diner	PCW	1/21/90	662D	1659
GRAYDON		12SecDr	Pullman	10/4/09	1963E	3744	GUANAJUATO		10SecDrB	PCW Det	4/10/84	149D	D242
GRAYFORD		16Sec	Pullman	12/3/06	2184	3409	GUANICA		12SecDr	Pullman	8/22/04	1963	3098
GRAYLOCK	N	12SecDrSr	Pullman	4/8/05	2049A	3169	GUARDIAN		12SecDrB	Pullman	2/21/06	2142	3318
GRAYMONT		12SecDrB	PCW	2/14/89	375G	1557	GUARDIOLA	P	12SecHotel	B&S	1/99	2370A	
GREAT NORTHERN	P	OS	J&S (McComb)				GUATEMALA		7Cpt2Dr	PCW	8/31/94	1065A	2025
GRECIAN #303	N	24Ch8StsB	(MannBCC)	/89	708		GUAYMAS		10SecDr	PCW Det	7/23/81	112	D118
GRECIAN		12SecDr	PCW	11/18/98	1318C	2359	GUAYMAS		12SecDr	Pullman	4/22/04	1963	3082
GREECE		12SecDr	Pullman	2/27/03	1581D	2975	GUELPH		12SecO	Grand Trunk	6/71	18	Mon1
GREELEY		12SecDrCpt	Pullman	10/4/05	2079	3204	GUERNSEY		14SecO	PCW Det	7/1/93	1041	D519
GREEN HAVEN	P	BaggSmkB SUV	H&H	/88	3219		GUIANA		7Cpt2Dr	PCW	4/12/94	1065A	2025

CAR NAME	ST	CAR TYPE	CAR BUILDER (Owner)	DATE BUILT	PLAN	LOT
GUILFORD		12SecDr	Pullman	12/21/05	1963C	3252
GUINEA		12SecDr	Pullman	3/14/03	1581D	2975
GUINEVERE		24StsDrSrB	Pullman	6/18/04	1988A	3086
GULF STATES	RN	6SecCafeLng SUV	Pull	11/1/09	3986	3746
GULFCREST	P	16Sec	B&S	12/82	3202	
GULFPORT		12SecDr	Pullman	5/28/06	1963C	3360
GUNILDA	N	20Ch8StsDrB	Wagner	8/93	3126C	Wg95
GUNNISON		10SecO (NG)	PCW Det	12/7/81	73A	D120
GUNNISON		10SecDr2Cpt	Pullman	3/25/07	2078D	3528
GUTHRIE	P	12SecO	(OPLines)	/72	5	
GUTHRIE		12SecDr	Pullman	4/22/03	1581B	3003
GUYANDOT		12SecDr	Pullman	11/29/01	1581A	2784
GWENDOLIN		24StsDrSrB	Pullman	6/17/04	1988A	3086
H		12SecO	GMP/CB&Q	/66	5	Aur
H.Q. SANDERSON	N	12SecO	I&StLSCC		27	
HACIENDA		12SecDr	Pullman	5/29/06	1963C	3360
HACKENSACK E&A-30	P	Slpr	(E&A Assn)			
HADASSAH	P	24Ch3StsB	J&S	2/93	2366	
HADDAM	P	14SecDr SUV	Pull P-2069A	6/05	3207B	3187
HADDON		14SecO	PCW Det	5/5/93	1021	D518
HAGERSTOWN	P	10SecB	B&S(B&O)	7/31/75	37	Day
HAGUE		12SecDr	Pullman	12/31/03	1581H	3053
HAIDEE		14Ch13StsDrB	PCW	7/29/87	505	1376
HAINAULT	N	12SecDr	Wagner	5/92	3065C	Wg67
HALBERT		12SecDr	Pull Buf	10/13/05	1963C	B47
HALCYON		12SecO	PCW Det	11/28/81	114	D123
HALCYON		16Sec	Pullman	5/18/08	2184C	3623
HALESIA		12SecDr	Pullman	10/8/01	1581A	2765
HALIFAX	P	10SecDr	Grand Trunk	5/73	31	Mon4
HALIFAX #485	P	12SecDr	Wagner	1/91	3065C	Wg56
HALLEY	N	12SecDrSr	Pullman	3/1/01	1581B	2654
HALSEY		10SecDr2Cpt	Pullman	9/21/05	2078	3203
HALSEY	R	10SecDr2Cpt	Pullman		2028R	
HAMBURG	N	11SecDr	PCW	2/23/85	222	1033
HAMBURG	N	16Ch2StsDrB	(Wagner)	6/85	3115C	
HAMILTON	P	OS	H&StJ RR		27	Han1
HAMILTON		12SecDr	PCW	10/25/98	1318B	2340
HAMILTON H&StJ #1		OS	H&StJ RR		27	Han1
HAMLET		14SecO	PCW Det	8/9/82	115	D155
HAMLET		12SecDr	Pull Buf	3/12/08	1963C	B74
HAMLET	P	6Sec4Dr	Wagner	6/93	3074A	Wg93
HAMMOND		12SecDr	Pullman	3/18/04	1963	3080
HAMPDEN		12SecDr	Pullman	12/21/05	1963C	3252
HAMPDEN		12SecDr	Pullman	3/21/04	1963	3080
HAMPSHIRE		12SecDr	Pull Buf	6/14/04	1963	B37
HAMPSHIRE #85	P	12SecDr	H&H	/80-/82		
HAMPTON	P	Parlor	J&S	/73	35	Wil1
HAMPTON		16SecSr	Pullman	8/31/04	2033A	3135
HAMPTON #302	P	20Ch12StsB	(MannBCC)	/89	708	
HANDSEL		16Sec	Pullman	1/4/09	2184D	3662
HANNAH	N	16Ch2StsDrB	(Wagner)	6/85	3115E	
HANNIBAL	P	Saloon	H&StJ RR	9/10/70	20	Han1
HANNIBAL	R	14SecO	H&StJ RR	9/10/70	43	Han1
HANNIBAL		12SecDr	Pull Buf	9/12/05	1963B	B46
HANOVER	R	16Sec	PCW Det	11/24/83	1877A	D212
HANOVER #283	P	10SecDr	J&S (PRR)	2/73	26	Wil9
HANOVER #486	P	12SecDr	Wagner	1/91	3065C	Wg56
HANSON	P	20ChBowers, Durer SUVPull P-1856A		1/03	3269A	2901
HAPSBURG		12SecDr	Pullman	6/19/01	1581A	2726
HAPULA	N	10SecDr	C&NW Ry	10/70	19	FdL3
HARBELL (#181)	N	12Ch36Coach SUV	Pull Buf	12/15/03	3581B	B26
HARCOURT		12SecDr	Pull Buf	6/20/04	1963	B38
HARELDA		16Sec	Pullman	12/31/08	2184D	3662
HARGRAVE		12SecDr	Pullman	2/8/08	1963C	3598
HARLAN	P	20ChBowers, Durer SUVPull P-1856A		1/03	3269A	2901
HARLECH	N	12SecB	PCW	3/4/85	227	1034
HARLEM		12SecDr	Pull Buf	3/6/08	1963C	B74
HARLEM #25	N	12SecO	(WS&PCC)	6/27/74	833	
HARLEQUIN		16Sec	Pullman	5/18/08	2184C	3623
HARMONIA		18Ch6StsDrB	PCW Det	8/8/90	772D	D425
HARMONIA	RN	BaggSmkCafe SUV	Pull P-2267	4/07	3278B	3494
HAROLD	N	26ChDrB	Pullman	6/29/07	1988B	3526
HAROLD	P	12SecO	(Wagner)	1/73	3031	
HARPERS FERRY #519	P	12SecO	B&O RR	/87	665	
HARRIET	N	26Ch3Sts	PCW Det	7/89	365	D391
HARRISBURG	RN	11SR	Wagner	9/89	1769C	Wg24
HARRISBURG #170	P	DDR	PRR	5/71	17	Alt1
HARRISVILLE	P	32ChDr SUV	Pull P-2494	3/11	3277	3821
HARTFORD		14SecO	PCW Det	5/27/93	1021	D525
HARTLAND		12SecDrSr	PCW Det	5/7/87	375	D292
HARTLAND		10Sec3Cpt	Pullman	2/6/06	2098	3300
HARTNEY	N	10SecDrCpt	PCW	6/21/89	668D	1587
HARTSDALE		12SecDr	Pullman	1/22/07	1963C	3448
HARTWELL	RN	6CptLngObs	Pullman	12/18/03	1299E	3065
HARTWELL	N	10Sec3Cpt	Pullman	2/6/06	2098B	3300
HARVARD	P	10SecDr	B&S	12/71	26	Day7
HARVARD	N	ClubLibObs	J&S	6/72	35M	1
HARVARD	N	10SecDr	B&S	4/30/73	26V	Day7
HARVARD		12SecDrB	Pullman	2/25/01	1611A	2614
HARVARD	P	10SecO	(Wagner)	/72	3001	
HARVEST MOON	N	18ChLngCafe SUV	Pull	6/4/08	3968A	3623
HARVESTER	N	12SecDr	Pull Buf	9/18/05	1963B	B46
HARWOOD	N	12SecDr	Pullman	10/23/05	2049B	3232
HASELMERE		Private	PCW	1/8/89	564	1494
HASTINGS		12SecDr	PCW	8/10/82	93B	23
HASTINGS	R	12SecDrB	PCW	8/10/82	1468	23
HASTINGS	P	12SecO Tourist	(Wagner)	8/85	3060	
HATCHETA		7Cpt2Dr	Pullman	10/21/02	1845	2926
HATHAWAY	N	12SecDr	Pullman	3/24/08	1963C	3600
HATHOR	N	12SecDrSr	Pullman	3/2/01	1581B	2654
HATTERAS		12SecDrB	PCW Det	10/20/88	375G	D358

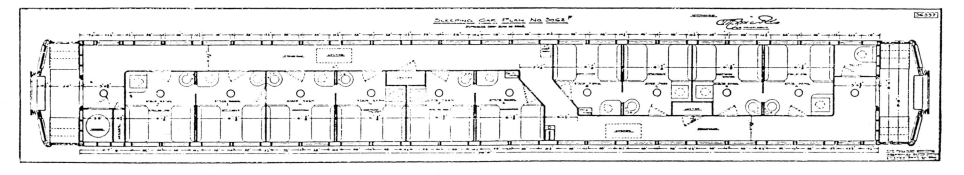

Figure I-23 Floor Plan 3062F was assigned to the former Wagner-built 10-compartment car HEBRON, shortly after it was acquired by Pullman in December 1899. The used railroad equipment dealer Hotchkiss Blue & Company purchased it in October 1913. The Newberry Library Collection.

CAR NAME	ST	CAR TYPE	CAR BUILDER (Owner)	DATE BUILT	PLAN	LOT	CAR NAME	ST	CAR TYPE	CAR BUILDER (Owner)	DATE BUILT	PLAN	LOT
HAVANA	P	12SecDr	(Wagner)	4/87	3049A		HECTOR	P	12SecDrB	(Wagner)	5/81	3050	
HAVANA #3		OS	J&S	3/71	27	Wil4	HEIDELBERG		12SecDr	Pull Buf	9/18/05	1963B	B46
HAVERFORD		12SecDrSr	Pullman	3/1/01	1581B	2654	HELEN	P	24ChDrB	(BIG-4 RR)	/01	2378	
HAVERHILL	N	10SecDr	PCW Det	10/7/72	26 1/2	D13	HELENA		8SecSrB (NG)	PCW Det	11/4/82	155	D173
HAVERHILL	N	16Sec	Pullman	7/1/08	2184C	3627	HELENA		10SecDrCpt	PCW	2/5/90	668	1649
HAVERHILL #488	P	12SecDr	Wagner	1/91	3065C	Wg56	HELENA	P	6Sec4Dr	Wagner	6/93	3074A	Wg93
HAVILAND		12SecDr SUV	Pull Buf	9/15/05	1963B	B46	HELENIUM		10SecDr2Cpt	Pullman	12/2/09	2451	3747
HAVRE		6SecDr2Sr	PCW	5/23/88	542A	1404	HELENUS		12SecDr	Pull Buf	5/12/10	1963E	B83
HAVRE #484	P	12SecDr	Wagner	1/91	3065C	Wg56	HELIADES		18Ch6StsDrB	PCW Det	8/2/90	772D	D425
HAWAII	N	SDR	DetC&MCo	/72	12D	Det	HELIANTHUS	N	12SecDr	PCW	6/7/87	473A	1323
HAWAII	P	6Sec4Dr	Wagner	6/93	3074A	Wg93	HELICON	N	10SecDrB	J&S (B&O)	C:/88	663B	
HAWARDEN #136	RN	7BoudB	(MannBCC)	/89	776A		HELIOTROPE		26Ch4StsDr	Pull Buf	7/11/00	1519C	B2
HAWFINCH	N	26ChDrB	Pullman	6/29/07	1988B	3526	HELMETTA		26ChDr	Pullman	3/26/06	1987A	3335
HAWTHORN #73	P	25Ch2StsB	(Wagner)	6/82	3100A		HELOISE		20Ch4StsDrB	Pullman	6/18/01	1518E	2703
HAWTHORNE	N	Parlor	PCW Det	8/2/71	22	D7	HELVETIA	N	OS	GWRy/GMP	C:7/67	7	Ham
HAYDEN		10SecDr3Cpt	Pullman	4/16/06	2078C	3342	HELVETIA		12SecDrSr	PCW	12/16/98	1318C	2370
HAYTI		12SecDrB	PCW Det	9/29/88	375G	D358	HEMPSTEAD	N	BaggClub	Pullman	1/21/07	2136A	3461
HAYTI	P	6Sec4Dr	Wagner	6/93	3074A	Wg93	HENDERSON		10SecDr	PCW Det	10/7/72	26 1/2	D13
HAZEL		18Ch6StsDrB	PCW Det	6/30/90	772D	D430	HENDERSON		12SecDr	Pullman	9/9/07	1963C	3550
HAZELWOOD		12SecDr	Pull Buf	12/15/03	1581F	B26	HENGIST		12SecDr	Pullman	5/8/01	1581A	2726
HAZELWOOD #540	P	10SecB	B&O RR?		649		HENNEPIN	P	14SecO	B&S	4/88	1593	
HAZLETON	N	10SecDrB	PCW Det	10/22/80	63 O	D77	HENNEPIN		12SecDr	Pullman	1/22/07	1963C	3448
HEBE		16Ch6StsB	PCW Det	7/12/76	54	D40	HENRICO #388		10SecDr	PCW	4/83	112A	37
HEBE	P	12SecDrB	(Wagner)	11/80	3025		HENRIETTA	N	14Ch6StsDrB	PCW	3/25/83	154	34
HEBRIDES		16SecO	PCW	8/28/93	1040	2008	HER MAJESTY	S	26Sts Parlor	PCW	6/17/95	1147	2104
HEBRON		10SecDrB	PCW Det	7/17/84	149D	D248	HERA	P	12SecDrB	(Wagner)	11/80	3025	
HEBRON #455	P	10Cpt	Wagner	4/91	3062	Wg48	HERACLEA		Slpr	PCW Det	5/10/84	170	D222
HECLA	P	10SecDr	H&StJ RR	3/71	19	Han2	HERCULES		BaggDiner	PCW	7/3/91	783B	1845
HECLA		12SecDr	Pullman	5/15/03	1581D	3004	HERCULES	P	12SecDr	(Wagner)	/84	3023A	
HECLA	N	12SecO	(Wagner)	9/72	3027C		HEREDIA	N	10SecDrCpt	PCW	2/11/90	668E	1649
HECTOR		14SecO	PCW Det	8/30/81	115	D114	HEREFORD		16SecO	PCW	6/13/93	1040	1990

CAR NAME	ST	CAR TYPE	CAR BUILDER (Owner)	DATE BUILT	PLAN	LOT
HERKIMER		10SecDrB	PCW Det	5/10/83	149	D194
HERKIMER	N	10Sec2DrB	Wagner	10/99	3079	Wg124
HERMAN		12SecO	B&S	/67	5	Day
HERMES		12SecO	PCW Det	2/16/82	114	D145
HERMES	N	26ChDr	Pullman	3/31/06	1987A	3335
HERMES	P	10Sec2Dr	Wagner	6/93	3074	Wg93
HERMES #727	P	31StsB	H&H (B&O)		653	
HERMIDA		10SecLibObs	Pullman	7/23/09	2351A	3715
HERMIONE	N	14SecO	PCW Det	8/9/82	115	D155
HERMIONE	P	20Ch8StsDrB	Wagner	8/93	3126A	Wg94
HERMIT		12SecDr	Pull Buf	3/18/08	1963C	B74
HERMITAGE		12SecDr	Pull Buf	9/21/05	1963B	B46
HERMOSA		10SecDrB	PCW	5/10/89	623B	1566
HERON		10SecDrCpt	PCW	5/11/90	668A	1709
HERON	P	12SecDrB	(Wagner)	10/80	3025	
HEROT	N	12SecDr	Wagner	3/92	3065C	Wg57
HERSILIA		16Sec	Pullman	10/22/01	1721	2766
HERTHA		18Ch6StsDrB	PCW Det	8/1/90	772D	D426
HERVISE		12SecDr	Pullman	3/21/10	1963E	3807
HESHBON		12SecDr	Pullman	12/27/06	2163	3460
HESIONE		18Ch6StsDrB	PCW Det	8/8/90	772D	D426
HESPERIA		12SecDr	PCW Det	12/30/80	93	D88
HESPERIA		10SecLibObs	Pullman	7/23/09	2351A	3715
HESPERUS	P	10SecDrB	(Wagner)	5/82	3046	
HESTER		18Ch4StsDrB	PCW	7/5/93	1039A	1994
HESTIA		24StsDrSrB	Pullman	6/20/04	1988A	3086
HEYWOOD	N	22Ch4StsDr	Pullman	4/6/04	2003	3029
HIAWASSEE #322	N	8BoudB	J&S (UnionPCC)	/89	373?	
HIAWATHA		10SecDr	PCW Det	2/28/80	59	D71
HIAWATHA	R	10SecDrB	PCW Det	2/28/80	240	D71
HIAWATHA	N	22Ch4StsDr	Pullman	4/6/04	2003	3029
HIAYA	N	12SecDr	Wagner	3/92	3065C	Wg57
HIBERNIA		16SecO	PCW	5/25/88	543B	1427
HIBERNIA		12SecDr	Pullman	10/10/01	1581A	2765
HIBISCUS		12SecDrCpt	PCW	8/4/92	983	1919
HIDALGO	N	8SecSrB (NG)	PCW Det	11/4/82	155	D173
HIDALGO		12SecDr	Pull Buf	9/27/05	1963B	B46
HIGHLAND		12SecDr	Pullman	4/28/03	1581B	3003
HIGHLAND #118	P		(WS&PCC)	7/83	395?	
HIGHLAND #138	P	DDR	PRR	12/70	17	Alt1
HIGHLANDER #1	P	24Ch6Sts	(Wagner)	6/85	3119	
HILDA	P	24Ch3StsB	J&S	2/93	2366	
HILDEGARDE		16Sec	Pullman	7/1/08	2184C	3627
HILLCOTE	N	10Sec2Dr	PCW	3/16/99	1158	2376
HILLSBORO		12SecDr	Pull Buf	9/25/05	1963B	B46
HILLSDALE		10Sec2Dr2Cpt	Pullman	12/29/05	2078C	3287
HILLSIDE		12SecDr	Pullman	12/27/06	2163	3460
HIMALAYA		12SecDrB	PCW	5/10/84	191	132
HIMALAYA	R	12SecDrB			1611	
HIMALAYA?	N	10SecDr	PCW Det	2-3/82	112A	D141
HINAN		14SecO	PCW	2/24/93	1021	1969
HINDOO		10SecDrB	PCW	4/8/84	149	113
HINDOO	R	16Sec	PCW	4/8/84	1877	113
HINDOO	P	12SecDrB	(Wagner)	6/86	3050A	
HINDORA		10SecLibObs	Pullman	7/24/09	2351A	3715
HINDOSTAN		12SecDrB	PCW	1/20/87	446A	1260
HINESBURG #32	P	6Sec8Ch	(Wagner)	6/85	3096	
HINGHAM	P	24Ch6StsDr	PCW	3/91	3259	
HINSDALE		12SecDr	Pullman	12/10/03	1581F	3043
HINSDALE	R	30Ch SUV	Pullman	12/10/03	3581E	3043
HINTON	N	10SecDrB	J&S (B&O)	C:/88	663F	
HIRONDELLE	N	26ChDr	Pullman	3/29/06	1987A	3335
HISPANIA	N	14SecO	PCW Det	8/2/82	115	D155
HISPANIA		12SecDr	Pullman	12/22/09	1963E	3766
HISPIDA	N	12SecDrSr	PCW	5/23/99	1318C	2423
HIWASSEE #322	P	8BoudB	J&S (UnionPCC)	/89	373?	
HOBOKEN	P	14Ch5StsB	(DL&W RR)	5/76	211	
HOCKESSIN #573	P	10SecDrB	J&S (B&O)	C:/88	663	
HOFFMAN		12SecDr	Pullman	3/24/08	1963C	3600
HOISINGTON		12SecDr	Pullman	3/25/02	1581B	2821
HOLBEIN	N	12SecDr	Pullman	7/23/01	1581A	2727
HOLBORN	N	10SecDrB	PCW	4/8/84	149	113
HOLBORN	R	16Sec	PCW	4/8/84	1877B	113
HOLBROOK	P	36Ch SUV	Pull P-2263	8/07	3275	3492
HOLCROFT	N	12SecDr	Pullman	7/15/08	1963C	3630
HOLDEN	N	12SecDr	PCW Det	9/1/82	93	D169
HOLDEN		12SecDr	Pull Buf	3/30/08	1963C	B74
HOLDREGE	N	20Ch6StsDrObs	Pullman	6/16/04	1979A	3084
HOLLAND		14SecDr	Pullman	8/4/00	1523E	2567
HOLLENDEN #600	N	Diner	B&S (Wagner)	6/92	3090	
HOLLINGDALE	N	12SecDr	Pullman	7/23/01	1581A	2727
HOLLIS	R		(WS&PCC)	8/5/81	729A	
HOLLIS #101	N		(WS&PCC)	8/5/81	465?	
HOLLY		24ChDrSr	Pullman	7/19/04	1980A	3088
HOLLY OAK #14	P	Slpr	(SouTranCo)			
HOLLYHOCK		24ChDrObs	Pull Buf	4/3/06	2156A	B50
HOLLYWOOD		12SecDr	Pullman	12/12/03	1581F	3043
HOLLYWOOD	R	30Ch SUV	Pullman	12/12/03	3581E	3043
HOLLYWOOD #159	P	20Ch4StsB	J&S (WS&PCC)	6/27/87	711	
HOLSTEIN		10Cpt	PCW	2/6/93	993A	1951
HOLYOKE #489	P	12SecDr	Wagner	1/92	3065C	Wg56
HOLYROOD		12SecDrSr	PCW	7/10/99	1318C	2424
HOMER		10SecDrB	PCW	10/30/86	445	1262
HOMER		BaggLib	PCW	4-6/87	477	1325
HOMER	N	18Ch4Sts	(Wagner)	4/86	3113	
HOMESTEAD		12SecDr	Pull Buf	6/10/04	1963	B37
HOMEWOOD	N	12SecDr	PCW Det	12/21/81	93B	D135
HOMILDON	P	16Sec SUV	B&S	7/11	3212	
HONDO M&LT-19	N	10SecDrB	PCW	1/28/88	445F	1361
HONDURAS		10SecDrB	PCW	7/10/84	149D	193
HONG KONG		12SecDrB	PCW	1/27/87	446A	1260
HONOLULU		12SecDr	Pull Buf	4/19/04	1581F	B36

CAR NAME	ST	CAR TYPE	CAR BUILDER (Owner)	DATE BUILT	PLAN	LOT
HONOLULU	N	10SecDr	PCW Det	3/15/73	26 3/4	D15
HONORA		22Ch6StsB	PCW	9/26/89	645	GT9
HONORIUS	N	BaggBSmk	PCW	/99	1400A	2410
HOOSAC		12SecDr	Pull Buf	7/7/05	1963	B42
HOPATCONG #169	P	14Ch6StsB	J&S (WS&PCC)	12/9/87	731B	
HOPE		10SecDrCpt	PCW	5/15/90	668A	1709
HOPETON	RN	10SecDrB	J&S (B&O)	4/88	663F	
HOPETON	P	14SecDr SUV	Pull P-1895	3/03	3207A	2982
HOPKINTON	P	32ChDr SUV	Pull P-2264	8/07	3277	3496
HOQUIAM		10SecDrSr	PCW	8/17/91	937A	1838
HORACE		12SecDr	Pull Buf	3/23/08	1963C	B74
HORATIO		7Cpt2Dr	PCW	1/10/96	1065A	2140
HORICON	RN	14SecO	PCW Det	7/73	37A	D18
HORICON	P	20Ch8StsDrB	Wagner	8/93	3126B	Wg94
HORSA		12SecDr	Pullman	5/9/01	1581A	2726
HORTENSE		10SecO (NG)	PCW Det	9/26/82	73A	D120
HORTENSIA		16Ch12StsDrB	PCW	10/22/88	614C	1508
HOT SPRINGS	N	12SecDr	Pullman	7/15/08	1963C	3630
HOTSPUR		12SecDr	Pullman	9/2/08	1963C	3646
HOUGHTON		12SecDr	Pull Buf	5/22/04	1963	B41
HOURI		24ChDr	Pullman	7/12/05	1980A	3171
HOUSATONIC	P	34Ch SUV	Pull P-1996C	7/04	3265	3114
HOUSTON	P	OS	H&H	2/72	27	Wil7
HOUSTON		12SecDr	Pullman	2/13/04	1963	3070
HOUSTON M&LT-4	P	Slpr	(M&LT RR)		37?	
HOYDEN		16Sec	Pullman	1/4/09	2184D	3662
HUDSON	P	10SecDr	B&S	3/72	26	Day7
HUDSON	P	12SecDrB	(Wagner)	6/87	3050A	
HUDSON #141	P	8Sec12ChSr	J&S (WS&PCC)	12/5/85	741	
HUDSON	N	12SecDr	PCW Det	5-6/82	93B	D154
HUELMA		10SecLibObs	Pullman	7/26/09	2351A	3715
HUGO	P	12SecDrB	(Wagner)	11/80	3024	
HUGUENOT		10SecB	PCW	10/1/84	209	202
HUGUENOT		16Sec	Pullman	5/22/08	2184C	3623
HUGUENOT #551	P	14SecSr	H&H (B&O)	8/87	654	
HUMBER		10SecDrB	PCW	11/19/85	149F	1135
HUMBER	R	12SecDrB	PCW	11/19/85	1611	1135
HUMBER #374		10SecDrB	PCW	1/83	149	21
HUMBOLDT	P	12SecO	B&S	12/70	5	Day
HUMBOLDT	R	12SecO	B&S	12/70	43	Day
HUMBOLDT	N?	20Ch5StsB	PCW Det	6/83	167	D203
HUMBOLT #487	P	12SecDr	Wagner	1/92	3065C	Wg56
HUNGARY		7Cpt2Dr	PCW	9/10/94	1065A	2025
HUNTERS MOON	N	18ChLngCafe SUV	Pull	7/1/08	3968A	3623
HUNTINGTON		12SecDrSr	Pullman	3/1/01	1581B	2654
HUNTINGTON #35	P	10SecB	Bowers, Dure	10/73	37	Wil11
HUNTINGTON #109	P	Parlor	(WS&PCC)	/75		
HUNTLEY	N	20Ch6StsDrObs	Pullman	6/16/04	1979A	3084
HUNTRESS	N	30ChObs	PCW	4/27/98	1278B	2299
HURLEY	N	12SecDr	PCW	6/1/92	784E	1915
HURLOCK		26ChDrB	Pullman	6/28/07	1988B	3526
HURON		12SecDr	Pull Buf	12/20/05	1963C	B49
HURON	P	10SecO	(Wagner)	/72	3001A	
HURON #5	N		(WS&PCC)	6/73		
HURON #17		DDR Hotel	B&S	/70	14	Day4
HUSSAR		12SecDr	Pull Buf	9/30/05	1963B	B46
HUTCHINSON		10SecDr2Cpt	Pullman	11/20/07	2078D	3583
HYACINTH		12SecDrCpt	PCW	8/6/92	983	1919
HYANNIS		14SecDrO SUV	Pull	12/12	2673	4047
HYDRANGEA		24ChDr	Pull Buf	3/28/06	2156	B50
HYDRIAD		16Sec	Pullman	12/31/08	2184D	3662
HYGEIA		10SecDr	PCW Det	7/19/82	112A	D152
HYGEIA	N	20Ch6StsDr	PCW	6/26/91	928A	1826
HYGELA	N	12SecDr	B&S (Wagner)	8/82	3023A	
HYPATIA	N	20Ch6StsDr	PCW	7/29/91	928A	1827
HYPERIDES	N	12SecDrB	Wagner	2/91	3065	Wg46
HYPERION		14SecO	PCW Det	10/15/81	115	D125
HYRCANIA	RN	16SecB	Pull Cal	8/11/02	1794	C3
HYTHE		10SecDrB	PCW	1/15/92	920B	1880
I		12SecO	GMP/CB&Q	/66	5	Aur
IACMEE		14SecO	PCW	2/1/93	1017C	1962
IAMBE		24StsDrSrB	Pullman	6/20/04	1988A	3086
IANTHA #37	N		H&H (WS&PCC)	5/1/78	753	
IANTHE		24StsDrSrB	Pullman	6/21/04	1988A	3086
IBER		10SecDr	PCW	12/26/89	717B	1630
IBERIA	RN	16SecB	Pull Buf	12/18/02	1794	B□
IBERIA M&LT-2	N	Slpr	(M&LT RR)	5/18/83	37?	
ICELAND		12SecDr	Pullman	3/18/03	1581D	2975
IDA #26	P	12SecO?	H&H (WS&PCC)	6/27/74	688	
IDAHO	N	12SecO	C&NW/GMP	/66	5	Chi
IDAHO		12SecDrB	PCW	2/12/89	375G	1557
IDAHO	P	12SecO Tourist	(Wagner)	6/69	3013	
IDALIA	N	12SecB	PCW	3/4/85	227	1034
IDANHA	P	14SecDr SUV	B&S	6/11	3207C	
IDEAL		6Dr	PCW	2/8/90	685B	1598
IDELETTE	P	34Ch	Wason & Co	8/87	1900	
IDLEHOUR	P	7DrB Hotel SUV	Pull P-2491	10/10	3210	3811
IDLER	N	Parlor	PCW Det	7/14/73	35 1/4	D17
IDLER	N	26ChDr	Pullman	3/31/06	1987A	3335
IDLER #300	P	Private	Wagner	3/88	3083	Wg8
IDLEWILD	RN	Private	PCW Det	8/31/71	22?	D7
IDLEWILD	RN	Private	PCW Det	11/25/76	561	D39
IDLEWOOD		12SecDrSr	Pullman	3/30/05	2049A	3169
IDOMENEO #139	P	7BoudB	(MannBCC)	/84	776	
IDRIS		16Sec	Pullman	6/5/09	2184D	3700
IDUMEA	N	12SecDrB	Wagner	2/91	3065	Wg46
IDYLLE	N	12SecO	PCW Det	11/22/81	114	D122
IGERIA	N	18Ch12StsB	J&S (B&O)	7/2/87	686C	2521
IGNACE	P	24ChDrB	(Big-4 RR)	/01	2378	
IGNACE SP-2540		10SecO	PCW	11/92	948A	W6
IGRAINE	P	22Ch2Sts	B&S	7/84	3252A	
IL BARBIERE #129	P	7BoudB	(MannBCC)	/84	776	
IL PROFETTA #134	P	7BoudB	(MannBCC)	/84	776	
IL PURITANI #123	P	7BoudB	(MannBCC)	/84	776	

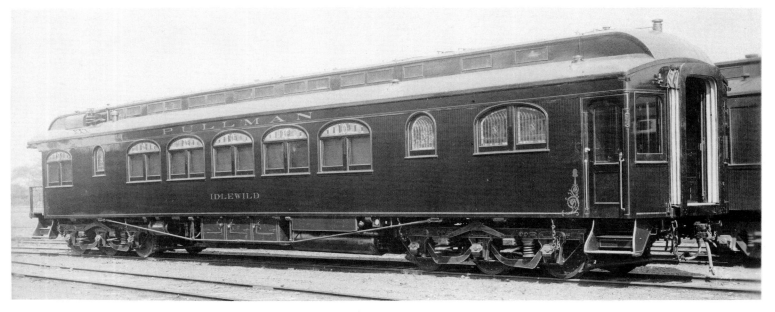

Figure I-24 The second Pullman private car to carry the name IDLEWILD was one of the flowers of the Pullman fleet. It is shown here in 1903 as it was substantially rebuilt from PRESIDENT. This private car was originally built in November 1876 and was sold to the American Smelting & Refining Company in March 1917. Smithsonian Institution Photograph P8000B.

CAR NAME	ST	CAR TYPE	CAR BUILDER (Owner)	DATE BUILT	PLAN	LOT
IL TROVATORE #106	P	7BovdB	(MannBCC)	/84	697	
ILCHESTER		12SecDrSr	PCW	9/9/99	1318E	2449
ILCHESTER		12SecDr	Pull Buf	4/28/08	1963C	B75
ILCHESTER #538	P	10SecB	B&O RR?		649	
ILIAD		10SecDrB	PCW	1/23/92	920B	1880
ILIKATO	P	16Ch4StsDr	B&S	7/82	3251	
ILINKA		12SecDrB	PCW	8/4/88	375E	1395
ILION		12Sec2Dr	PCW Det	1/14/88	537A	D334
ILLINOIS		OS	G&B/GMP	/66	7	Troy
ILLINOIS	N	14Ch13StsDrB	PCW	7-10/87	505	1376
ILLINOIS	P	BaggSmk	PCW	/99	1400	2410
ILLINOIS		20Ch9StsDrObs	PCW	11/14/99	1417G	2466
ILLINOIS		20Ch4StsObs	Pullman	7/16/05	2057	3217
ILLINOIS #84	P	12SecDr	H&H	/80-/82		2758
ILLINOIS #97		12SecDrB	(CRI&P RR)	12/15/79	204B	
ILLINOIS #106	P	Slpr	(IC RR)		1/228	
ILLINOIS #139	P	6Sec12ChSr	J&S (WS&PCC)	12/1/85	741B	
ILLINOIS #279	P	16SecO	Wagner	7/89	3063	Wg15
ILUS	N	12SecDrB	PCW	10/19/88	375F	1481
ILWACO	N	12SecDrB	Wagner	1/93	3073B	Wg83
IMBRAS	N	12SecDrB	Wagner	4/93	3073B	Wg91
IMERITIA		12SecDr	Pullman	9/12/00	1572	2585
IMOGEN		20Ch6StsB	PCW	10/1/84	167B	200
IMOLA SP-2541		10SecO	PCW	11/92	948A	W6
IMPERIAL	P	10SecDr	J&S (Pull/Sou)	/73	37	Wil10
IMPERIAL		Private	PCW	10/30/99	1418C	2451
IMPERIAL		12SecDr	Pullman	3/14/08	1963C	3600
INCA		12SecDr	PCW	10/25/90	784	1766
INCA	P	12SecDrB	Wagner	10/92	3070	Wg76
INDEPENDENCE		Private	Pullman	9/28/06	2203A	3382
INDIA		22Berths	PCW Det	7/28/76	48	D32
INDIA		12SecDrCpt	PCW	3/15/89	622A	1529
INDIA #96	P	28Ch10StsDr	Wagner	9/89	3125B	Wg18
INDIA #212	P	Parlor	PRR	7/28/73	29	Alt1
INDIANA		10Cpt	Pullman	1/23/02	1704	2758
INDIANA #17	P	12SecO	(WS&PCC)	6/30/73	833	
INDIANA #257	P	12SecO	G&B/GMP	/66	5	Troy
INDIANA #280		16SecO	Wagner	11/88	3063	Wg15
INDIANAPOLIS #510	N	OS	I&StLSCCo	7/71		
INDIANOLA	P	10SecB	Bowers, Dure	10/73	37	Wil11
INDIANOLA		12SecDr	Pullman	3/5/07	1963C	3525
INDIO	N	12SecDrCpt	Pullman	5/25/04	1985	3095

CAR NAME	ST	CAR TYPE	CAR BUILDER (Owner)	DATE BUILT	PLAN	LOT	CAR NAME	ST	CAR TYPE	CAR BUILDER (Owner)	DATE BUILT	PLAN	LOT
INDIO	R	10SecObs	PCW	11/17/88	445V	1510	IRIS		10SecDr	PCW	12/24/89	717B	1630
INDIO SP-48	P	10SecDrB	PCW	11/17/88	445H	1510	IRIS	P	12SecDrB	Wagner	10/92	3070	Wg76
INDUS		10SecDrB	PCW	1/20/90	725B	1629	IRMA		26ChDr	Pullman	7/27/04	1987A	3097
INDUS #95	P	28Ch10StsDr	Wagner	9/89	3125B	Wg18	IRMA SP-2542		10SecO	PCW	11/92	948A	W6
INELA		14SecO	PCW	3/2/93	1021	1969	IRONDALE		12SecDrB	PCW	2/23/97	1204D	2212
INEZ		10SecDr	PCW	12/30/89	717B	1630	IRONTON		12SecDr	Pullman	3/11/02	1581B	2818
INEZ	P	12SecDrB	Wagner	10/92	3070	Wg76	IROQUOIS		12SecO	PCW Det	1/15/81	91	D106
INFANTA		24StsDrSrB	Pullman	6/21/04	1988A	3086	IROQUOIS	?	10SecDrB	PCW	10-11/87	445E	1334
INGELOW		12SecDr	Pull Buf	4/8/08	1963C	B75	IROQUOIS		12SecDr	Pullman	9/28/09	1963E	3744
INGLEBY		26ChDrB	Pullman	6/28/07	1988B	3526	IROQUOIS #281	P	16SecO	Wagner	7/89	3063	Wg15
INGLENOOK		12SecDr	Pullman	12/28/06	2163	3460	IRVING		12SecDr	PCW	10/15/92	990	1931
INGLESIDE		12SecDr	Pullman	3/5/07	1963C	3525	IRVINGTON		12SecDr	PCW	9/26/98	1318B	2339
INGLEWOOD #282	P	16SecO	Wagner	6/89	3063	Wg15	IRVONA		12SecDr	Pullman	10/13/05	2049B	3232
INGOMAR #475	P	12SecDr	Wagner	1/92	3065C	Wg55	ISABEL	N	18Ch3Sts	Bowers, Dure	11/72	35G	11
INKERMAN #476	P	12SecDr	Wagner	1/92	3065C	Wg55	ISABEL		22ChDr	Pull Buf	6/12/02	1647C	B8
INOLA	N	6Sec8Ch	PCW Det	3/25/72	261	D10	ISABELLA		8SecLngObs	PCW	4/22/93	1032	1977
INTERLAKEN		12SecDr	Pullman	12/28/06	2163	3460	ISABELLA		20Ch4StsDrB	Pullman	6/13/01	1518E	2703
INTERNATLIONAL	?	Diner	?				ISABELLA	P	10Sec2DrB	Wagner	8/93	3074B	Wg97
INTERNATIONAL		10SecDrHotel	PCW Det	6/12/77	63	D45	ISAQUA		12SecDr	Pullman	5/23/02	1581B	2853
INTERVALE		12SecDr	Pull Buf	4/15/08	1963C	B75	ISARA 2565		14SecO	PCW Det	2/28/93	1021	D516
INTREPID	P	12SecO	PCW (Wagner)	6/85	3010	1070	ISAURIA		12SecDr	Pull Buf	10/26/03	1581F	B24
INVADER		12SecDr	Pullman	9/4/08	1963C	3646	ISELIN		12SecDr	Pullman	12/29/06	2163	3460
INVERARY		12SecDr	Pull Wilm	7/23/04	1581C	W16	ISIS #119	P	20Ch8StsDrB	Wagner	6/91	3126A	Wg51
INVERLAKE	N	7DrB Hotel SUV	Pull P-2491	10/10	3210	3811	ISLAND CREEK	P	BaggSmB SUV	H&H	/88	3219	
INVERNESS		12SecDr	Pull Buf	4/3/08	1963C	B75	ISLAND POND	P	10SecDr	Grand Trunk	9/72	31	Mon3
INVERNESS #477	P	12SecDr	Wagner	1/92	3065C	Wg55	ISLANDALE	P	7DrB Hotel SUV	Pull P-2491	10/10	3210	3811
INVERNESS #730	P	19Ch3StsB	H&H (B&O)		653A		ISLAY		10SecDrB	PCW	1/18/92	920B	1880
INVICTA	N	10SecDr	PC&StL RR	8/72	26A	Col3	ISLETA		10SecB	PCW	6/7/84	206	198
INWOOD		BaggClub	Pullman	8/15/05	2087	3220	ISLETA		10Sec2Dr SUV	Pullman	8/28/02	1746A	2925
IOLANDA		12SecDr	Pullman	8/6/08	1963C	3636	ISLEWORTH		12SecDr	Pullman	10/24/03	1581F	3025
IOLANTHE		Private	PCW	12/8/88	564	1494	ISLINGTON		12SecDr	Pullman	10/11/00	1581A	2588
IOMA	N	12SecDrCpt	Pullman	8/8/05	2088	3221	ISLYP #115	P		(WS&PCC)	7/83		
IONA		6SecCptSr	PCW Det	2/5/80	68	D54	ISMAIL	N	10SecDrB	(Wagner)	5/82	3046	
IONE #46	P	Parlor R-End	H&H (WS&PCC)	7/25/78	524?		ISMAILIA		12SecDr	Pullman	6/11/03	1581E	3010
IONIA		12SecDr	Pullman	6/19/03	1581F	3010	ISMAILIA	R	30Ch SUV	Pullman	6/11/03	3581E	3010
IONIA #145		10SecDr	H&H	10/70	19	Wil2	ISMENE		10SecDrB	PCW	11/3/91	920B	1866
IONIAN	P	10Sec2DrB	PCW (Wagner)	2/16/00	3079A	Wg134	ISOLDE		22ChDr	Pull Buf	6/17/02	1647C	B9
IOSCO	P	12SecDr	(Wagner)	/75	3008		ISTALINA	N	12SecDr	Pullman	10/11/07	2163A	3562
IOWA	P	SDR	C&NW RR	/69	12	FdL1	ISTLAN	R	10SecObs	PCW	6/12/88	445V	1414
IOWA	P	BaggSmk	PCW	/99	1400	2410	ISTLAN SP-64	P	10SecDrB	PCW	6/12/88	445F	1414
IOWA		10Cpt	Pullman	/02	1704	2758	ISTRIA	RN	16SecB	Pull Buf	12/16/02	1794	B□
IOWA #58	P	Slpr	(IC RR)				ISTROUMA		12SecDr SUV	Pullman	3/29/08	1963C	3600
IOWA #98	P		(CRI&P Assn)	12/15/79			ITALIA	S	12Sec	PCW Det	6/9/76	50	D30
IPSWICH	N	10SecDr	B&S	4/71	19F	Day5	ITALIA		16Sec	Pullman	10/24/01	1721	2766
IPSWICH #474	P	12SecDr	Wagner	1/92	3065C	Wg55	ITALY		12SecDr	PCW	6/7/87	473A	1323
IRELAND		12SecDr	PCW	6/7/87	473A	1323	ITALY	N	6SecDr2Sr	PCW	5/88	542A	1404
IRELAND	N	20Ch5StsB	PCW	?1/85	167A	?199	ITALY		12SecDr	Pullman	2/20/03	1581D	2975
IRENE		22ChDr	Pull Buf	6/10/02	1647C	B8	ITALY	R	30Ch SUV	Pullman	2/20/03	3581E	2975
IRENE #236	N	18Ch3Sts	Bowers, Dure	9/72	35G	1	ITASCA	P	OS	DetC&MCo	/71	17 1/2	D2
IREX		10SecDr	PCW	12/26/89	717B	1630	ITASCA	N	DDR	DetC&MCo	/71	17	D2
IREX	P	12SecDrB	Wagner	10/92	3070	Wg76	ITASKA #164	P		(WS&PCC)	9/24/87	426?	

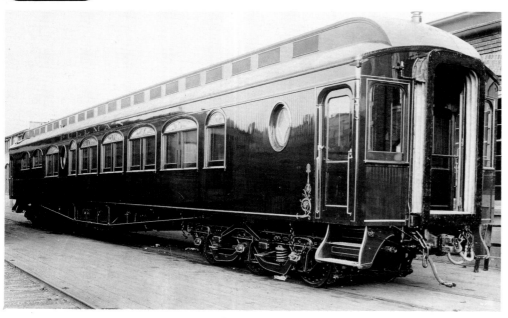

ITATA
Wood

Figure I-25 Pullman's Palace Car Company owned and operated several dining cars. ITURBIDE, built in 1887 as THE ITURBIDE, was rebuilt in the late 1890s to represent the cars that were being built at that time. Smithsonian Institution Photograph P6493. Courtesy of Robert Wayner.

CAR NAME	ST	CAR TYPE	CAR BUILDER (Owner)	DATE BUILT	PLAN	LOT
ITATA		10SecDrB	PCW	1/16/92	920B	1880
ITHACA	P	10SecO Tourist	(DL&W RR)	12/80	220	
ITHACA		12SecDr	Pull Buf	2/12/03	1581C	B13
ITURBIDE	N	Diner	PCW	6/7/87	378D	1324
IUKA #2518		14SecO	PCW Det	1/14/93	1014A	D510
IVAN		12SecDr	PCW	8/9/90	784	1707
IVANHOE		10SecDrB	PCW		445	
IVANHOE		12SecDr	Pullman	3/30/03	1581D	2983
IVANHOE #79	P	18Ch4StsDr	(Wagner)	/82	3112	
IVANHOE #123		10Sec	PFtW&C	2/70		FtW
IVANHOE #574	P	10SecDrB	J&S (B&O)	4/88	663	
IVEL		12SecDr	PCW	2/13/91	784B	1797
IVERNIA		12SecDrSr	Pullman	2/16/01	1581B	2638
IVORYDALE		12SecDr	Pull Buf	4/21/08	1963C	B75
IVRY	N	10Sec2Dr	Wagner	6/93	3074	Wg92
IXION	N	12SecDr	PCW	4-6/87	473A	1323
IZAAK WALTON	P	Private Hunting	(CH&D RR)	/73	71F	
J	P	OS	CB&Q/GMP	/67		Aur
JACINTO		12SecDr	Pullman	6/4/06	1963C	3360
JACKSON	P	OS	J&S (McComb)	/71	90?	
JACKSON		12SecDr	Pull Buf	2/4/04	1581F	B32

CAR NAME	ST	CAR TYPE	CAR BUILDER (Owner)	DATE BUILT	PLAN	LOT
JACKSON	RN	12ChCafeObs SUV	Pull	12/16/03	3968	3065
JACKSONVILLE		OS	B,P&Co	/61		Day
JACKSONVILLE	RN	8SecBObs	PCW Det	3/11/93	1021S	D516
JACQUES		12SecDr	Pull Buf	5/11/08	1963C	B76
JAFFA		12SecDr	Pullman	4/25/04	1963	3082
JALAPA	RN	10SecB (NG)	PCW Det	7/12/80	73F	D76
JALISCO		12SecDr	Pullman	4/26/04	1963	3082
JAMAICA	N	10SecDr	Grand Trunk	8/73	31B	Mon4
JAMAICA		12SecDr	Pullman	7/28/02	1581B	2852
JAMESBURG		12SecDr	Pullman	12/31/03	1581H	3053
JAMESTOWN		12SecDr	PCW Det	6/13/84	93D	D241
JAMESTOWN		12SecDr	Pullman	5/9/03	1581D	3004
JANELDA	P	28Ch3StsB	PCW	1/98	1167A	2268
JANESVILLE		12SecDr	Pullman	1/23/07	1963C	3448
JANISCH	P?	Private	(MannBCC)		6—	
JANUS	P	12SecDrB	Wagner	6/93	3073B	Wg82
JAPAN		12SecDrCpt	PCW	7/27/89	622A	1529
JAPAN	P	12SecDrB	Wagner	1/93	3073B	Wg82
JAPAN #144	P	10SecDr	C&NW Ry	10/70	19	FdL3
JAPETUS	N	12SecDr	Pullman	5/92	3065C	Wg68
JAPONICA		10SecDr2Cpt SUV	Pullman	11/22/09	2451	3747
JARALD		12SecDr	Pull Buf	5/2/08	1963C	B76
JARAMA		12SecDr	Pullman	7/15/08	1963C	3630
JARILLA		7Cpt2Dr	Pullman	10/29/02	1845	2926
JASCO		10SecO	PCW	8/3/92	948A	1944
JASON		10SecO	PCW	8/3/92	948A	1944
JASPER		12SecDr	Pull Buf	5/7/08	1963C	B76
JAVA	N	12SecDr	PCW	4-6/87	473A	1323
JAVA	N	12SecDr	PCW	5/28/92	784E	1915
JAVA	P	12SecDrB	Wagner	1/93	3073B	Wg82
JAVA #211	P	Parlor	PRR	11/72	29	Alt1
JAY GOULD E&A-3	P	Slpr	(E&A Assn)			
JEANNETTE		22ChDr	Pull Buf	6/18/02	1647C	B9
JEFFERSON	RN	Diner	PCW	12/31/90	783A	1705
JEFFERSON		BaggClubBarber	Pullman	4/6/04	2000	3026
JEFFERSON CITY	P	OS	MP RR	C:9/72	27	
JENA		14SecO	PCW Det	4/8/93	1021	D517
JERMYN		12SecDr	Pullman	3/24/08	1963C	3600
JERSEY		12SecDr	Pullman	4/16/03	1581D	2984
JERSEY SP-2544		10SecO	PCW	11/92	948A	W6
JERSEY CITY #284		10SecDr	PCW Det	4/21/73	26 3/4	D15
JESSAMINE		24ChDrSr	Pullman	7/20/04	1980A	3088
JESSAMY		28ChDr	Pullman	6/24/04	1986A	3087
JESSICA	P	18Ch3StsDrObs	(B&M RR)		1203	
JESSICA #130	N	Parlor (3'gauge)	(WS&PCC)	5/17 84	528?	
JEWEL		12SecDr	Pullman	7/15/08	1963C	3630
JEWETT		BaggBLibSmk	Pullman	4/24/07	2087	3534
JOANNA	P	26Ch3Sts	PCW	9/95	1153	2117
JOCAN	P	12SecDrB	Wagner	1/93	3073B	Wg82
JOCKEY CLUB #130	P	Parlor (3'gauge)	(WS&PCC)	5/17 84	528?	
JOHN A. LOGAN	N	29ChDr	Pullman	7/16/05	2061	3217
JOHN DAVENPORT	N	22Ch6StsDr	PCW	9/1/92	976A	1910

70

CAR NAME	ST	CAR TYPE	CAR BUILDER (Owner)	DATE BUILT	PLAN	LOT
JOHN ENDICOTT		BaggSmkBLib	Pullman	7/30/03	1945A	3020
JOHN WINTHROP	N	22Ch6StsDr	PCW	8/29/92	976A	1910
JOHNSTOWN		12SecDrCpt	Pullman	5/23/04	1985	3095
JOHNSTOWN #286		10SecDr	PCW Det	5/13/73	26 3/4	D15
JOLIET		12SecO	PCW Det	9/21/76	58	D42
JONQUIL	N	28Ch3Sts	PCW (Wagner)	7/16/86	3124	1200
JONQUIL SP-2543		10SecO	PCW	11/92	948A	W6
JOPLIN	N	10SecDr	PCW Det	2/29/76	52	D33
JOPLIN		12SecDr	Pullman	2/13/04	1963	3070
JOPPA	N	12SecDr	Wagner	2/92	3065C	Wg64
JOPPA SP-2545		10SecO	PCW	11/92	1024	W7
JORDAN		12SecDrB	PCW Det	1/4/89	375H	D378
JOSEPHINE	P	18Ch3StsDrObs	(B&M RR)		1203	
JOSEPHUS	N	14SecO	PCW (Wagner)	3/19/86	3018A	1191
JOSETTE	N	26ChDr	Pullman	5/29/07	1987B	3524
JOTHAM	N	10SecDr	(Wagner)	6/85	3048	
JOUNA		12SecDr	PCW Det	5/25/91	784D	D462
JOVIAN		14SecDr	Pullman	6/2/00	1523	2559
JUANALUSKA	N	12SecDrB	PCW	8/8/88	375E	1395
JUANALUSKA	P	8BoudB	J&S (UnionPCC)	/89	373	
JUANITA		28ChDr	Pullman	6/23/04	1986A	3087
JUDA		12SecDr	PCW	5/6/91	784D	1805
JUDEA	N	26ChB	PCW (Wagner)	7/20/86	3124A	1200
JUDITH		22Ch6StsDr	PCW	8/30/92	976A	1910
JUDITH	P	24ChDrB	(Big-4 RR)	/01	2378	
JULESBURG		12SecDr	PCW	8/29/82	93B	23
JULIA #43		8SecParlorRE	H&H (WS&PCC)	7/25/78		
JULIAN	P	10Sec2DrB	PCW (Wagner)	3/3/00	3079A	Wg135
JULIANA		22ChDr	Pull Buf	6/26/02	1647C	B9
JULIET		16Ch12StsDrB	PCW	12/11/88	614C	1508
JULIUS		12SecDr	Pull Buf	5/15/08	1963C	B76
JUMNA		14SecO	PCW Det	1/20/93	1014A	D510
JUNEAU		10SecDr2Cpt	Pullman	10/5/06	2078D	3398
JUNIATA	P	Slpr	J&S (WS&PCC)	6/84		Wil
JUNIATA	N	7CptLngObs	PCW	10/30/95	1159A	2126
JUNIATA	P	12SecDrB	Wagner	1/93	3073B	Wg82
JUNIATA #1	N	12SecO	J&S (WS&PCC)	6/73	833	
JUNIATA #134	P	DDR	PRR	9/70	17	Alt1
JUNIOR		10SecLibObs	Pullman	11/20/09	2351A	3748
JUNIPER		24ChDr	Pull Buff	4/10/06	2156	B51
JUNIUS	R	BaggSmkBLib			1706	
JUNO	N	16Ch6Sts	PCW Det	7/12/76	54E	D40
JUNO		27StsParlor	PCW Det	4/28/76	45	D31
JUNO		24ChDr	Pullman	6/29/05	1980A	3171
JUNO #18	P	20Ch10StsDr	(Wagner)	6/95	3109A	
JUNO #149	P	11Ch12StsDr	J&S	7/72	35	Wil1
JUPITER		27StsParlor	PCW Det	8/23/75	45	D25
JUPITER		14Ch18StsBObs	PCW	11/4/98	1279C	2354
JUPITER	P	12SecDrB	Wagner	1/93	3073B	Wg82
JUPITER #248	P	6SecCh	Taunton C.C.	3/71	29D	Tau1
JURA		10SecDrB	PCW	1/17/90	725B	1629

CAR NAME	ST	CAR TYPE	CAR BUILDER (Owner)	DATE BUILT	PLAN	LOT
JURGEN	N	10Sec2Dr	Wagner	6/93	3074	Wg92
JUSTINE		26Ch	Pullman	5/25/01	1646C1	2701
JUSTITIA		6CptLngObs	PCW	1/11/98	1299A	2283
JUSTUS		12SecDr	Pull Buf	5/19/08	1963C	B76
KAATERSKILL		8CptLibObs	Pullman	12/26/06	2260	3502
KAATERSKILL	P	12SecDr	(Wagner)	1/81	3022D	
KABEKONA	RN	12SecDr	PCW Det	6/24/84	1581C	D241
KAISER	RN	10SecB	PCW Det	7/16/73	37	D16
KAISER		12SecDrSr	PCW	5/27/99	1318C	2423
KALAMA		12SecDr	PCW Det	5/16/84	93D	D241
KALAMA	N	10SecDr	(Wagner)	/85	3053	
KALAMAZOO	P	OS	MichCent/GMP	8/1/67	27	Det
KALAMAZOO		12SecDr	Pullman	3/25/04	1963	3080
KALAMAZOO #125	P	12SecDr	(WS&PCC)	9/18/83	914	
KALIDA		12SecDr	Pull Buf	5/23/08	1963C	B77
KALITAN		12SecDr	Pullman	9/18/09	1963E	3744
KALKASKA #123	P	18Ch12Sts	(WS&PCC)	8/23/83	749C	
KALMAR		12SecDr	Pull Buf	5/28/08	1963C	B77
KALMIA		12SecDr	PCW	4/10/91	784C	1798
KALMIA	N	12SecDr	Pullman	10/26/05	2049B	3232
KALO		10SecDrB	PCW	6/6/91	920B	1808
KAMA		10SecDrB	PCW	1/14/90	725B	1629
KAMADERA	N	14SecO	B&O RR	10/80	657C	
KAMELA		10SecDrB	PCW	3/29/87	445C	1281
KAMELA	R	16Sec	PCW	3/29/87	1880	1281
KAMIAH		12SecDr	Pullman	1/30/03	1581C	2946
KAMOUR		10SecDrB	PCW	11/9/87	445E	1334
KAMOUR	RN	12SecDrB	PCW	4/9/84	1611	113
KAMOUR	N	12SecDrB	PCW?		1611	
KAMPALA	N	12SecDr	Pullman	12/27/06	2163	3460
KANATA	P	12SecDrB	Wagner	1/93	3073B	Wg88
KANAWHA		10SecB	B&O	4/30/73	37	B&O
KANAWHA		12SecDrSr	PCW	5/13/99	1318C	2423
KANAWHA #720	P	16Ch12StsB	B&S (B&O)		664	
KANDAHAR		12SecDr	Pullman	4/14/03	1581D	2984
KANKAKEE #97		Slpr	(IC RR)			
KANKAKEE #138	P	6Sec12ChB	J&S (WS&PCC)	11/23/85	741	
KANORADO		10Sec2Dr	PCW	5/10/93	995A	1942
KANSAS		10Cpt	Pullman	1/13/02	1704	2758
KANSAS CITY	P	SDR	CB&Q RR	/68		Aur
KANSAS CITY	P	12SecO	DetC&MCo	/71	5	D2
KANSAS CITY	N	20Ch9StsDrObs	PCW	11/15/99	1417H	2466
KANSAS CITY #315	P	8BoudB	J&S (MannBCC)	/89	373A	
KARA	P	12SecDrB	Wagner	1/93	3073B	Wg88
KARELIA		12SecDr	Pullman	6/25/03	1581F	3010
KARLA		14SecO	PCW	1/24/93	1017C	1962
KARNAK		14SecO	PCW Det	4/22/93	1021	D518
KARNAK #117	P	20Ch8StsDrB	Wagner	6/91	3126A	Wg51
KARONDA		12SecDr	Pullman	5/29/06	1963C	3360
KASHGAR		12SecDrB	PCW Det	10/5/88	375G	D358
KASHMIR		10SecDrB	PCW	5/24/84	149D	192

CAR NAME	ST	CAR TYPE	CAR BUILDER (Owner)	DATE BUILT	PLAN	LOT
KASHMIR	N	BaggClub	Pullman	8/16/05	2087	3220
KASHMIR	N	10Cpt	Wagner	6/96	3075A	Wg111
KASOTA	P	10SecO	(Wagner)		3002	
KASOTA	P	14Sec	B&S	8/93	3204	
KATAHDIN		12SecDr	Pull Buf	6/2/08	1963C	B77
KATHARINE		18Ch4StsDrB	PCW	6/24/93	1039A	1994
KATHLAMET	RN	12SecDr	PCW	9/30/82	1467A	22
KATHLEEN		18Ch6StsDrB	PCW Det	8/16/90	772D	D426
KATHLEEN	N	24ChDr	Pullman	7/5/05	1980A	3171
KATONAH	N	12SecDr	Pullman	10/24/05	2049B	3232
KATONAH	P	10SecO	(Wagner)	/86	3038	
KATRINA		20Ch5StsB	PCW	8/31/84	167A	199
KAWMO	RN	6SecCafeLng SUV	Pullman	1/14/08	3986	3588
KAZAN	N	20Ch4Sts	(Wagner)	9/86	3102	
KEARNEY		12SecDr	PCW Det	8/25/82	93B	D162
KEARNEY		12SecDrCpt	Pullman	10/5/05	2079	3204
KEARSARGE		12SecDr	Pull Buf	6/5/08	1963C	B77
KEARSARGE #206		Parlor	PCW Det	8/31/71	22	D7
KEATING		26ChDrB	Pullman	6/29/07	1988B	3526
KEDRON		12SecDrB	Pull Wilm	3/26/03	1611A	W9
KEDRON #375		10SecDrB	PCW Det	1/9/83	149	D176
KEENE		Parlor	PCW Det	2/15/72	22	D9
KEEWAYDIN		12SecDr	Pullman	1/23/03	1581C	2946
KEGONSA		12SecDr	Pullman	12/22/09	1963E	3766
KEMBLE	RN	Private	PCW Det	12/1/83	1929B	D212
KEMPTON	N	12SecDr	PCW	10/24/98	1318B	2340
KENABEEK		14SecDr	Pullman	5/24/06	2165	3352
KENDORA		12SecDr	Pull Buf	6/9/08	1963C	B77
KENESAW	P	12SecO?	J&S (W&A RR)	/72	4	
KENILWORTH		10SecDr	PCW Det	3/1/76	52	D33
KENILWORTH		6CptLngObs	Pullman	12/16/03	1299D	3065
KENILWORTH	P	10SecO Tourist	(Wagner)		3047	
KENILWORTH #568	P	12SecB	PCW	3/4/85	227	1034
KENMORE		Diner	PCW	11/20/90	662D	1777
KENMORE		12SecDr	Pullman	12/13/06	1963C	3448
KENNEBEC		12SecDr	PCW	12/4/83	93B	88
KENNEBEC		12SecDr	Pullman	4/25/07	1963C	3533
KENNEBEC #162	P	10SecDrB	J&S (WS&PCC)	9/24/87	663	
KENNESAW		12SecDr	Pullman	4/23/07	1963C	3533
KENNESAW	N	DDR	J&S (W&A RR)	/76	4A	Wil
KENNET		12SecDrB	PCW Det	1/4/88	375H	D378
KENNETH	N	16Ch2StsDrB	(Wagner)	6/85	3115B	
KENNEWICK		12SecDr	Pullman	1/20/03	1581C	2946
KENOMA #122	N		(WS&PCC)	7/83		
KENOSHA		10SecO (NG)	PCW Det	9/26/82	73A	D120
KENOSHA	P	12SecO	Wagner	2/90	3069	Wg63
KENSINGTON		12SecDr	Pullman	4/6/04	1963	3080
KENSINGTON	P	10SecDr	(Wagner)	/86	3043	
KENSINGTON #161	P	12SecO	J&S	2/71	18	Wil3
KENT	P	10SecDr	(Wagner)		3042	
KENT #162	N	10SecDrB	J&S (WS&PCC)	9/24/87	663E	
KENTON		12SecDr	Pull Buf	8/17/03	1581F	B21
KENTUCKY		10Cpt	Pullman	1/14/02	1704	2758
KENTUCKY #178	P	OS	PRR	10/71	27	Alt2
KENTUCKY #328	P	8BoudB	J&S (UnionPCC)	/89	373A	
KENWOOD		12SecO	PCW Det	3/7/71	5	D5
KENWOOD	N	12SecDr	PCW Det	5/16/84	93D	D241
KENWOOD	P	10SecDr	(Wagner)	/86	3043A	
KENWOOD	P	14SecDr SUV	Pull P-2069A	6/05	3207B	3187
KENWOOD	R	12SecO?	PCW Det	3/7/71	23F	D5
KENYON		12SecDr	Pullman	2/10/08	1963C	3598
KEOKUK	P	12SecO	DetC&MCo	/68	5?	Det
KEOKUK		12SecDr	Pullman	2/17/04	1963	3070
KEOKUK #19	N		(WS&PCC)	7/1/75		
KEOWEE #316	P	8BoudB	J&S (UnionPCC)	/89	373A	
KERRISTON	N	5CptLibObs	Pullman	8/14/05	2089	3222
KERRY		10SecO	PCW	8/1/92	948A	1944
KEWANEE		12SecDr	PCW Det	5/19/81	93A	D107
KEWANEE	RN	8SecBLibObs	PCW		1707	
KEYSTONE		12Sec	PCW Det	6/11/79	65	D53
KEYSTONE		12SecDr	Pullman	5/20/03	1581D	3004
KEYSTONE	P	10SecO	(Wagner)	8/85	3038A	
KEYSVILLE #171	R	12Ch38Sts SUV	Pull Buf	11/23/03	3581D	B25
KHARTOUM	N	10SecB	B&O RR?		649	
KHARTOUM		12SecDr	Pullman	4/13/03	1581D	2984
KHARTOUM	P	12SecDrB	Wagner	1/93	3073B	Wg88
KHEDIVE		12SecDrSr	PCW	5/26/99	1318C	2423
KHEDIVE	P	12SecDrB	Wagner	1/93	3073B	Wg88
KHEDIVE	N	10SecB	PCW Det	7/16/73	37	D16
KHIVA		12SecDrB	PCW	10/27/88	375F	1481
KHYBER		12SecDr	Pullman	2/2/03	1581D	2962
KIAMENSI #572	P	10SecDrB	J&S (B&O)	C:/88	663E	
KICKAPOO	N	14SecO	B&S	7/92	1593	
KILAMOX		14SecDr	Pullman	8/26/09	2438	3716
KILBOURN	N	26ChDr	Pullman	3/27/06	1987A	3335
KILBOURNE #73	P	12SecDr	H&H	/80-/82		
KILDARE		14SecO	PCW Det	5/13/93	1021	D525
KILL VAN KULL #122	P		(WS&PCC)	7/83		
KILLARNEY	P	12SecDr	(Wagner)	1/82	3022D	
KILLAWATS	N	10SecDr2Cpt SUV	Pullman	3/10/10	2471	3784
KILMORE		16Sec	Pullman	12/7/06	2184	3409
KINDERHOOK	P	12SecDr	(Wagner)	1/82	3022E	
KINGMAN		10SecDr2Cpt	Pullman	11/11/07	2078D	3583
KINGS MOUNTAIN	N	23ChDrLibObs	Pullman	8/28/06	2206	3388
KINGSLAND		16Ch6StsDrB	PCW Det	7/13/87	167F	D283
KINGSLAND	N	12SecDr	Pull Buf	6/22/08	1963C	B78
KINGSLEY		10SecLibObs	Pullman	11/23/09	2351A	3748
KINGSTON		10SecDr	Grand Trunk	3/71	19	Mon1
KINGSTON	N	12SecDrB	Wagner	2/90	3065	Wg32
KINGWOOD		OS	B&O RR	4/30/73	27 1/2	B&O2
KINGWOOD		12SecDr	Pullman	3/14/06	2163A	3333
KINKORA	N	OS	DetC&MCo	/73	27	Det

CAR NAME	ST	CAR TYPE	CAR BUILDER (Owner)	DATE BUILT	PLAN	LOT
KINKORA		12SecDr	Pullman	3/8/06	2163A	3333
KINNEY	N	12SecDrCpt	Pullman	5/24/04	1985	3095
KINSLEY		12SecDrB	PCW	9/7/88	375F	1480
KINZER		26ChDrB	Pullman	6/29/07	1988B	3526
KIOWA		10SecDrB	PCW	1/7/88	445E	1360
KIOWA	RN	12SecDr	PCW Det	6/13/84	1467A	D241
KIPLING		10SecLibObs	Pullman	11/26/09	2351A	3748
KIRKLAND	N	10SecDrB	PCW	4/3/84	149B	101
KIRKWOOD	N	10SecB	B&S	4/15/73	37	Day9
KIRKWOOD	P	10SecO	(Wagner)	/86	3038	
KISKIMINETAS #81	N	SlprParlor	(Woodruff)	6/1/80		
KISMET	P	26Ch3Sts	B&S	9/92	3260	
KISMET #125	N	12SecDr	(WS&PCC)	9/18/83	914A	
KIT CARSON		12SecDrB	PCW	2/11/89	375G	1557
KITTANNING		16Sec	Pullman	7/10/03	1721	3014
KITTANNING #280	P	10SecDr	J&S (PRR)	7/63	26	Wil9
KITTITAS	RN	12SecDr	PCW	9/19/82	1467A	22
KLAMATH	R	10SecObs	PCW	6/15/88	445V	1415
KLAMATH SP-65	P	10SecDrB	PCW	6/15/88	445F	1415
KLICKITAT	RN	12SecDr	PCW	3/6/83	1467A	184
KNICKERBOCKER		8CptLibObs	Pullman	12/26/06	2260	3502
KNIGHTSEN		10SecDr2Cpt	Pullman	11/15/07	2078D	3583
KNIGHTSVILLE	N	8CptLibObs	Pullman	12/26/06	2260	3502
KNOBEL		12SecDr	Pullman	3/26/02	1581B	2821
KNOLLSWOOD		12SecDr	Pullman	12/14/03	1581F	3043
KNOXVILLE	P	Slpr	(SouTranCo)			
KNOXVILLE		12SecDr	Pullman	2/16/04	1963	3070
KNOXVILLE #180	P	10SecDr	H&H	1/72	26	Wil6
KNUTSFORD		12SecDr	Pullman	2/10/08	1963C	3598
KOBE		12SecDrB	PCW	11/20/88	375G	1513
KOHINOOR		10SecLibObs	Pullman	12/1/09	2351A	3748
KOKOMO		BaggClub	Pullman	8/16/05	2087	3220
KOLIMA		10SecDrB	PCW	1/12/85	149D	1028
KOLIMA	N	8CptB	Wagner	6/89	3064	Wg17
KONIGSBERG		12SecDr	PCW	11/26/92	990	1941
KOOSKIA		12SecDr	Pullman	1/31/03	1581C	2946
KORDOFAN		12SecDr	Pullman	2/26/03	1581D	2975
KREMLIN		10SecDrB	PCW	11/22/83	149	84
KREMLIN		12SecDrSr	PCW	7/8/99	1318C	2424
KREMLIN	P	12SecDrB	Wagner	1/93	3073B	Wg88
KRESTREL		24ChDrB	Pullman	6/30/00	1518C	2553
KULON		10SecO	PCW	8/2/92	948A	1944
KUNA		10SecDrB	PCW	5/21/88	445F	1361
KUSHAQUA	P	12SecDrB	Wagner	1/93	3073B	Wg88
KUWANA		12SecDr	Pullman	12/23/09	1963E	3766
KWASIND	P	16Ch4StsB	B&S	9/82	3252B	
L'AFRICANA #132	P	7BoudB	(MannBCC)	/84	776	
L'ETOILE DU NORD #140	P	7BoudB	(MannBCC)	/84	776	
LA BRANCH	P	10SecDr	J&S (McComb)	/73	19	Wil
LA CROSSE		12SecDr	Pullman	3/28/04	1963	3080

CAR NAME	ST	CAR TYPE	CAR BUILDER (Owner)	DATE BUILT	PLAN	LOT
LA CROSSE #241		10SecB	J&S	6/73	37	Wil12
LA FAVORITA #104	P	7BoudB	(MannBCC)	/84	697	
LA GIOCONDA #124	P	7BoudB	(MannBCC)	/84	776	
LA GITANA	RN	6SecSr (NG)	PCW Det	11/4/82	155D	D173
LA GOLETA CP-8074	P	10Sec2Dr	PCW	4/14/93	995A	1942
LA GOLONDRINA	N	12ChCafeObs SUV	Pull	12/18/03	3968D	3065
LA GRANDE		10SecDrSr	PCW	6/21/90	784A	1715
LA HARPE	N	SlprParlor	(WS&PCC)	6/1/80		
LA HEROICA	RN	10SecB (NG)	PCW Det	8/24/80	73F	D76
LA JUNTA		10SecDr	PCW Det	3/25/79	59	D52
LA MUETTA #126	P	7BoudB	(MannBCC)	/84	776	
LA OTTO	N	SlprParlor	(WS&PCC)	6/1/80		
LA PALOMA	N	12ChCafeObs SUV	Pull	12/18/03	3968D	3065
LA PALOMA	RN	7SecSrB (NG)	PCW Det	9/26/82	1001	D120
LA REINE	N	6SecSr (NG)	PCW Det	11/18/82	155D	D173
LA REVE	RN	8SecB (NG)	PCW Det	11/18/82	155A	D173
LA ROSE	RN	8SecB (NG)	PCW Det	11/23/82	155A	D173
LA SALLE		12SecDr	PCW Det	9/30/80	91	D82
LA SONNAMBULA #103	P	7BoudB	(MannBCC)	/84	879?	
LA TRAVIATA #102	P	7BoudB	(MannBCC)	/83	697	
LA TROVATORE #106	P	7BoudB	(MannBCC)	/84	697	
LA VETA		10SecO (NG)	PCW Det	2/23/83	73A	D120
LA ZINGARA #130	P	7BoudB	(MannBCC)	/84	776	
LABRADOR		12SecDr	Pullman	7/30/02	1581B	2852
LACADIE	N	10SecDr	Grand Trunk	8/71	26E	Mon2
LACERTA	N	12SecDr	B&O (UnionPCC)		657B	
LACERTA	N	BaggBLibSmk	Pullman	4/24/07	2087	3534
LACEY	N	12SecDr	Pullman	3/28/04	1963	3080
LACHINE	P	10SecDr	Grand Trunk	6/72	31	Mon3
LACHINE #191	S	12SecDrB	PCW	3/19/86	93H	1192
LACHINE #460	P	12SecDrB	Wagner	4/91	3065	Wg50
LACHLAN		12SecDr	Pullman	9/29/09	1963E	3744
LACHLAN #341	N	10SecDrB	(UnionPCC)		565B	
LACINATA		16Sec SUV	Pullman	3/31/10	2184G	3808
LACKAWANNA	N	20Ch6StsDrObs	Pullman	6/16/04	1979A	3084
LACKAWAXEN	N	23ChDrLibObs	Pullman	8/28/06	2206	3388
LACLEDE		10SecDr	PCW Det	10/1/81	112A	D124
LACLEDE	N	12SecDr	Wagner	4/92	3065C	Wg66
LACOLLE	N	10SecDr	Grand Trunk	10/73	31B	Mon4
LACONIA		12SecDr	PCW	8/11/98	1318A	2313
LACONIA	P	12SecDr	Wagner	5/92	3065C	Wg85
LADOGA		10SecDrB	PCW	2-4/83	149	27
LADOGA	N	12SecDr	(Wagner)	8/82	3023B	
LADORE	N	10SecB	Bowers,Dure	5/31/73	37M	Wil11
LADOWA		12SecDrB	PCW Det	10/9/88	375G	D358
LADRONE		10SecDrB	PCW	11/22/83	149	84
LADRONE	N	12SecDr	Wagner	1/92	3065C	Wg84
LADY DAYTON #37	P		H&H (WS&PCC)	5/1/78	753	
LADY FRANKLIN #21	P	12SecO	H&H (WS&PCC)	6/14/71	736	
LADY WASHINGTON #22	P	12SecO	H&H (WS&PCC)	6/14/71	736	

CAR NAME	ST	CAR TYPE	CAR BUILDER (Owner)	DATE BUILT	PLAN	LOT
LAERTES		10Sec2Dr	PCW	3/24/96	1173A	2153
LAFAYETTE		10SecDr	TW&W Ry	/71	19	Tol1
LAFAYETTE		Diner	PCW	5/31/88	538D	1411
LAFAYETTE	P	10SecDr	(Wagner)	/84	3057C	
LAGARTO	N	10SecB	PCW Det	8/5/73	37	D16
LAGARTO	N	12SecDrCpt	Pullman	8/9/05	2088	3221
LAGONDA	P	12SecHotel	J&S (FEC RR)	1/99	2371A	
LAGOS	N	26ChB	Wagner	11/92	3125	Wg74
LAGUNA		10SecDr3Cpt	Pullman	4/26/06	2078C	3342
LAGUNA SP-66	P	10SecDrB	PCW	6/16/88	445F	1415
LAKE						
CHARLES M&LT-3	P	Slpr	(M&L RR)		37?	
PARK		12SecDr	PCW Det	6/20/84	93D	D241
PEPIN		12SecDrSr	PCW	8/7/86	375	1215
PLACID #521	P	12SecDrB	Wagner	8/92	3070	Wg72
SIDE #60	P	16Ch12StsDr	(Wagner)	8/86	3103	
VIEW	N	12SecO	(NP Assn)		18E	
VIEW #50	P	16Ch12StsDr	(Wagner)	7/90	3103	
VILLA WC-102	P	10SecDr	PCW Det	3/87	489	D294
LAKELAND		12SecDrB	Pullman	12/19/06	2142A	3405
LAKESIDE	P	12SecDrB	PCW	4/24/84	202	195
LAKESIDE		10SecDr2Cpt	Pullman	9/9/05	2078	3203
LAKETON		12SecDr	PCW	9/30/98	1318B	2339
LAKEVIEW		12SecDr	Pull Buf	6/13/08	1963C	B78
LAKEWOOD	N	12SecDr	PCW Det	7/26/82	93B	D162
LAKEWOOD #160	P	20Ch4StsB	J&S (WS&PCC)	6/27/87	711	
LAMAR		12SecDr	Pullman	10/29/07	1963C	3570
LAMBERTA		6CptDrLibObs	Pullman	3/13/02	1751B	2799
LAMIRA		12SecDrB	PCW	10/19/88	375F	1481
LAMIRA	N	BaggClub	Pullman	3/1/06	2136	3319
LAMIRA	N	14SecO	(Wagner)	/75	3052	
LAMMERMOOR #120	P	10Sec2Dr	PRR	8/70		WPhil
LAMMERMOOR #570	P	12SecB	PCW	3/4/85	227	1034
LAMPASAS		8SecSrB (NG)	PCW Det	11/18/82	155	D173
LAMPAZAS	N	8SecSrB (NG)	PCW Det	11/18/82	155	D173
LAMY		10SecDr	PCW	1/21/85	229	1032
LANARK		12SecDr	Pullman	7/31/02	1581B	2852
LANCASTER #171	P	6SecBCh	PRR	5/71	17S?	Alt1
LANCASTER #192	S	12SecDr	PCW	4/1/86	93H	1192
LANCASTER #553	P	14SecSr	H&H (B&O)	8/87	654	
LANCELOT	N	10SecDr	B&S	6/73	37	Day9
LANCELOT		12SecDr	Pullman	12/14/09	1963E	3766
LANDENBERG #554	P	14SecSr	H&H (B&O)	8/87	654	
LANDOVER		12SecDr	Pullman	5/23/08	1963C	3616
LANDRAIL		12SecDr	Pullman	9/5/08	1963C	3646
LANDSEER		12SecDr	Pullman	7/30/01	1581A	2727
LANGHAM		10SecDr2Cpt	Pullman	11/2/09	2451	3747
LANGHAM #140	RN	7BoudB	(MannBCC)	/84	776A	
LANGHAM #272	P	12SecDr	Wagner	1/89	3061B	Wg13
LANGHORNE		24ChDrB	Pull Buf	4/22/07	2278	B59
LANGLEY	N	12SecDr	PCW	4/26/92	784E	1899
LANNES	N	12SecDr	(Wagner)	8/86	3049A	
LANSDOWNE		28ChDr	Pullman	4/27/07	1986B	3527
LANSING		12SecDr	Pull Buf	6/29/04	1963	B38
LANTANA		10SecDr2Cpt	Pullman	12/22/08	2078E	3661
LANZO		14SecO	PCW	2/2/93	1017C	1962
LAONA	N	10SecDr	PCW	1/21/85	229	1032
LAPEER	RN	12SecO	B&S	7/31/74	114	Day
LAPEER		12SecDr	Pullman	3/25/04	1963	3080
LAPEER	P	12SecDrB	(Wagner)	9/86	3050A	
LAPLAND #166	RN	10SecDrB	(WS&PCC)	9/24/87	663F	
LAPORTE		12SecDr	Pullman	5/13/03	1581D	3004
LAPORTE	P	12SecDr	(Wagner)	/82	3021	
LAPWAI		12SecDr	Pullman	6/11/07	2271A	3458
LARABEE		28ChDr	Pullman	4/26/07	1986B	3527
LaRABIDA		Diner	PCW	4/23/93	1030	1976
LARAMIE	P	12SecO	DetC&MCo	/68	5	Det
LARAMIE		10Sec2Dr	PCW	9/24/96	1158A	2195
LARAMIE #461	P	12SecDrB	Wagner	4/91	3065	Wg50
LARCHMONT #102	P	20Ch8StsDrB	Wagner	7/90	3126	Wg36
LAREDO		10SecDrB	PCW	1/6/88	445E	1360
LAREDO	R	16Sec	PCW	1/6/88	1880	1360
LAREDO	P	12SecDrB	(Wagner)	9/87	3050A	
LARK		24ChDrB	Pullman	6/22/00	1518C	2552
LARKSPUR		12SecDrCpt	PCW	8/6/92	983	1919
LARWILL		12SecDrCpt	Pullman	8/9/05	2088	3221
LAS CRUCES		10SecDr	PCW Det	4/5/81	112	D149
LAS VEGAS		10SecDr	PCW Det	6/2/79	59	D63
LASSIE	N	28ChDr	Pullman	4/29/07	1986B	3527
LATAH		12SecDrSr	Pullman	3/16/01	1581A	2654
LATAKIA		12SecDr	Pullman	9/13/00	1572	2585
LATONIA	N	12SecO	(L&N RR)	/69	114D	
LATONIA		12SecDr	Pull Buf	4/27/03	1581C	B16
LATROBE		12SecDr	Pullman	1/27/03	1581D	2962
LAUDERDALE #273	P	12SecDr	Wagner	1/89	3061B	Wg13
LAUNCELOT		12SecDr	Pullman	3/21/10	1963E	3807
LAUNCELOT #231	P	4Sec5SrB	PCW (Wagner)	11/23/86	3045	1246
LAUREL	RN	14Sec	PCW	6/12/89	668F	1587
LAUREL	N	12SecO	(Wagner)	4/70	3010B	
LAUREL #10	P	Slpr	(SouTranCo)			
LAUREL #528	P	12SecO	B&O RR	/80	665	
LAURENTIA		8Sec3DrCpt SUV	Pullman	1/5/06	2119	3275
LAURETTE	N	28ChDr	Pullman	4/26/07	1986B	3527
LAURITA		20Ch6StsDr	PCW	6/26/91	928A	1826
LAUSANNE		10SecDr2Cpt	Pullman	12/22/08	2078E	3661
LAVACA	P	12SecO	(Pull/Sou)	/72	4	
LAVACA	N	10Sec2DrB	Wagner	12/99	3079A	Wg133
LAVACA	N	12SecDr	Wagner	4/92	3065C	Wg66
LAVENDER		24ChDrSr	Pullman	7/20/04	1980A	3088
LAVEROCK		26ChDr	Pullman	3/26/06	1987A	3335
LAVINIA		20Ch6StsB	PCW	1/22/85	167B	200
LAVINIA #732	P	17Ch11StsB	J&S (B&O)	6/29/87	686	2519

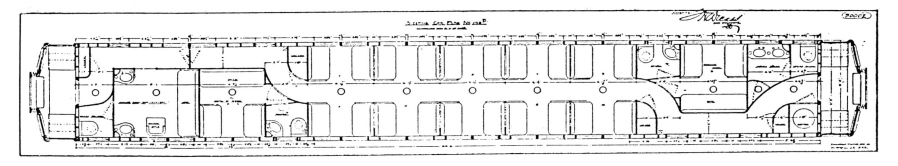

Figure I-26   Plan 1158B represents a minor modification of the original Plan 1158A. The 10-section, 2-drawing room car LARAMIE, built for service on the Union Pacific, carried this floor plan from prior to 1900 until sometime after 1906 when it was redesignated Plan 1158C. It was converted to Tourist Car 4720 in July 1917. The Newberry Library Collection.

CAR NAME	ST	CAR TYPE	CAR BUILDER (Owner)	DATE BUILT	PLAN	LOT	CAR NAME	ST	CAR TYPE	CAR BUILDER (Owner)	DATE BUILT	PLAN	LOT
LAWNDALE		12SecDr	Pullman	8/26/01	1581A	2756	LEHIGH #224	P	Parlor	J&S	9/72	35	Wil2
LAWNSIDE (#165)	RN	37Ch SUV	Pullman	12/17/03	3581P	3043	LEICESTER		Diner	PCW	11/21/90	662D	1777
LAWRENCE	P	36Ch SUV	Pull P-2494	3/11	3275	3821	LEICESTER		12SecDrSr	PCW	9/12/99	1318E	2449
LAWRENCE #48	N	OS	B&S	/70		Day	LEIGHTON		12SecDr	Pullman	10/11/00	1581A	2588
LAWRENCE #271	P	12SecDr	Wagner	1/89	3061B	Wg13	LEIPSIC		10SecDrB	PCW	6/28/84	149D	192
LE PARADIS		Private	PCW	8/10/85	247	1105	LEIPSIC	R	16Sec	PCW	6/28/84	1877A	192
LEADVILLE		10SecO (NG)	PCW Det	11/5/79	73	D61	LEIRE		12SecDr	PCW Det	6/3/91	784D	D462
LEADVILLE		12SecDr	PCW Denver	10/12/03	1581F	Den	LEITH		12SecDr	PCW	11/26/92	990	1941
LEAMINGTON #285	P	12SecDrB	Wagner	8/89	3065	Wg23	LEITRIM		16SecO	PCW	6/3/93	1040	1990
LEANDER		12SecDr	Pullman	5/15/08	1963C	3615	LELAH	N	OS	I&StLSCC	/71		
LEAVENWORTH		12SecDr	Pullman	10/29/07	1963C	3570	LELAH	N	20Ch3StsB	B&S (B&O)		653B	
LEAVENWORTH CTC-13	P		H&H (CTC)	/72			LELAND		Diner	PCW	11/9/89	662D	1641
LEAVITTSBURG #561	P	11SecDr	PCW	2/23/85	222	1033	LELAND		16Sec	Pullman	3/26/07	2184	3523
LEBANON	N	12SecDrSr	H&H (B&O)	? 8/87	654F		LELIA	P	10Sec	(OPLines)		47?	
LEBANON		12SecDr	Pull Buf	7/12/04	1963	B38	LEMNOS	N	10SecO	(Wagner)	/72	3001B	
LEBANON #189	S	12SecDr	PCW	3/19/86	93H	1192	LEMONT	N	10SecDrB	PCW	3/11/88	445F	1380
LEBANON #291	P	9CptB	Wagner	9/89	3066	Wg24	LEMOYNE		28ChDr	Pullman	4/26/07	1986B	3527
LEBO		12SecDr	PCW	5/14/91	784D	1805	LENA	R	12SecDrB	PCW Det	1/9/83	1611	D176
LEBRUN	N	14SecDr SUV	Pull P-1855	1/03	3207	2900	LENA #376		10SecDrB	PCW Det	1/9/83	149	D176
LECANTO		12SecDrB	PCW Det	9/8/88	375G	D346	LENAH	P	10Sec2DrB	Wagner	12/99	3079A	Wg133
LECANTO	N	12SecDr	Wagner	2/96	3077	Wg81	LENAXA	N	12SecDr	Pullman	1/24/01	1581C	2614
LEDA	N	10SecDr	Vermont Cent	5/72	31	StA2	LENNOX #190	S	12SecDr	PCW	3/19/86	93H	1192
LEDBURY		12SecDr	Pullman	12/15/09	1963E	3766	LENOLA		12SecDr	Pullman	3/9/06	2163A	3333
LEEDS		12SecDr	Pullman	9/25/03	1581F	3022	LENORE		14Ch13StsDrB	PCW	10/27/87	505	1376
LEFLORE	N	OS	PCW Det	12/12/70	27	D4	LENOVER		12SecDr	Pullman	3/9/06	2163A	3333
LEFLORE	N	12SecDr	PCW	11/1/98	1318B	2340	LENOX	N	12SecDrB	(Wagner)	10/80	3025	
LEGHORN	N	10SecDrB	Grand Trunk	11/10/81	445E	Mont8	LENOX #140	N	6Sec12ChSr	J&S (WS&PCC)	12/5/85	741B	
LEGHORN		12SecDr	Pull Buf	6/5/03	1581C	B18	LEO	P	Parlor	Taunton C.C.	3/71	29B	Tau1
LEGHORN	N	12SecO	(Wagner)	/85	3027D		LEO		27Sts Parlor	PCW Det	1/25/74	39	D20
LEGHORN	N	OS	Grand Trunk	10/31/72	27	Mon2	LEOLA	P	10Sec2DrB	Wagner	12/99	3079A	Wg133
LEHIGH		16Ch7StsDrObs	PCW	5/2/96	1193	2179	LEOMINSTER	P	26Ch SUV	Pull P-2071	6/05	3273	3189

CAR NAME	ST	CAR TYPE	CAR BUILDER (Owner)	DATE BUILT	PLAN	LOT
LEON		12SecDr	PCW	8/12/90	784	1707
LEONA	P	10Sec2DrB	Wagner	12/99	3079A	Wg133
LEONARDO		12Sec2Dr	Pullman	4/2/01	1580A	2646
LEONATA	N	10SecDrB	PCW	2-4/88	149	27
LEONATO		12Sec2Dr	Pullman	4/2/01	1580A	2646
LEONCITO		12SecDrB	PCW Det	8/14/88	375G	D346
LEONCITO	N	12SecDr	Wagner	3/93	3065C	Wg84
LEONIDAS	N	10SecB	(IC RR)		231	1/172
LEONIDAS	N	10SecDr	(Wagner)	/88	3057A	
LEONIDES	N	12SecDrSr	Pullman	5/24/07	2088	3553
LEONORA		18Ch5Sts	PCW Det	6/75	44	D24
LEONTES	N	8BoudB	J&S (UnionPCC)	/89	373?	
LEOTA	P	10Sec2DrB	Wagner	12/99	3079A	Wg133
LEOTI	N	10SecDrB	PCW Det	3/6/84	149B	D228
LERDO	N	10SecDr	PCW Det	9/9/81	112A	D112
LERDO	P	10Sec2DrB	Wagner	12/99	3079A	Wg133
LEROY	N	10SecDr	B&S	8/15/72	26A	Day8
LES HUGUENOTS #136	P	7BoudB	(MannBCC)	/84	776	
LESBOS	N	12SecDrB	Wagner	4/93	3073B	Wg91
LESLIE	N	26ChDr	Pullman	5/25/07	1987B	3524
LETITIA	P	28Ch3StsB	PCW	1/98	1167A	2268
LETITIA		18Ch4StsDrB	PCW	7/31/93	1039A	1995
LEVANA	P	22Ch6StsB	PCW	9/27/89	645	GT9
LEVANT	RN	12SecDr	Grand Trunk	2/71	21B	Mon1
LEVIS	N	10SecDr	Grand Trunk	6/72	31K	Mon3
LEWISTON	N	10SecDrB	B&S	3/17/70	149S	Day2
LEWISTON	RN	14Sec	PCW	5/8/90	668F	1709
LEWISTON		12SecDr	Pullman	11/1/00	1581A	2589
LEXINGTON	P	DDR	CB&Q RR	1/71	17	Aur
LEXINGTON		12SecDr	PCW	9/27/98	1318B	2339
LEYDEN		12SecDr	Pullman	7/27/01	1581A	2727
LEZEKA		10SecDr2Sr	Pullman	8/25/09	2078J	3721
LIBANUS #445		10SecDrB	PCW	9/25/85	240	1080
LIBERIA		16SecO	PCW	6/2/93	1040	1990
LIBERTAS		6CptLngObs	PCW	1/12/98	1299A	2283
LIBERTY		BaggClub	Pullman	3/1/06	2136	3319
LIBRA		14SecO	PCW Det	3/23/93	1021	D517
LIBURNIA		12SecDr	Pull Buf	1/4/04	1581F	B27
LICKING #176	P	10SecDr	PRR	9/71	21	Alt2
LIEGE		12SecDr	PCW	11/30/92	990	1941
LIFFEY		12SecDrB	PCW Det	12/28/88	375H	D378
LIFFEY		12SecDrB	PCW	12/6/95	1066	2130
LIGONIER	N	16SecB	Pull Cal	9/16/02	1794	C3
LIGONIER	N	12SecDr	Wagner	4/92	3065C	Wg66
LIGONIER #140	P	DDR	PRR	2/71	17	Alt1
LIGURIA		12SecDr	Pullman	6/17/03	1581E	3010
LIGURIA #408		12SecDrB	PCW	10/28/85	180	1122
LILA		12SecDr	PCW	8/11/90	784	1707
LILAC		18Ch6StsDrB	PCW	6/2/90	772D	1704
LILLE		14SecO	PCW Det	4/8/93	1021	D517

CAR NAME	ST	CAR TYPE	CAR BUILDER (Owner)	DATE BUILT	PLAN	LOT
LILLIAN		20Ch6StsDr	PCW	6/26/91	928A	1826
LILLIE San#3	P	OS	DetC&MCo			Det
LILY		10SecDr	PCW	12/24/89	717B	1630
LIMA		16Sec	Pullman	7/11/03	1721	3014
LIMA #266	P	10SecDr	J&S	11/72	26	Wil8
LIMERICK		16SecO	PCW	6/15/93	1040	1990
LIMON		12SecDr	Pullman	5/3/04	1963	3082
LIMOSA	N	12SecDrSr	Pullman	5/24/07	2088	3553
LINCOLN		12SecDr	PCW Det	8/21/82	93	D169
LINDELL	N	10SecB	PFtW&C	1/71	17	FtW2
LINDELL	N	12SecDr	Wagner	4/92	3065C	Wg66
LINDEN	NP	12SecDr	Wagner	6/93	3065C	Wg85
LINDEN CTC-91			(CentTrans)			
LINFIELD	P	14SecDr SUV	Pull P-2262	8/07	3207C	3498
LINGAYEN	N	10SecO	(Wagner)	/66	3003	
LINNET		24ChDrB	Pullman	6/15/00	1518C	2552
LINNIE	N	18Ch7StsB	(NYLE&W Assn)	/72	147	
LINWOOD		16Sec	Pullman	12/4/06	2184	3409
LIONEL		12SecDr	Pullman	3/23/10	1963E	3807
LIORNA	R	10SecObs	PCW	6/20/88	445V	1415
LIORNA	RN	6SecCafeLng SUV	Pullman	6/2/06	3986	
LIORNA SP-67	P	10SecDrB	PCW	6/20/88	445F	1415
LISBON		10SecDrB	PCW	11/6/83	149	80
LISKA		14SecO	PCW	2/24/93	1021	1969
LISZT		12SecDr	PCW	4/26/92	784E	1899
LITCHFIELD		14SecDr	PCW	12/27/99	1318J	2450
LITERNUM	N	12SecDr	Pull Buf	10/30/03	1581F	B24
LITHIA	R	10SecDrB	PCW	3/28/85	240	1037
LITHIA	R	12SecDr	PCW	3/28/85	1603	1037
LITHIA #439		10SecDrB	PCW	3/28/85	149E	1037
LITHUANIA		12SecDr	Pull Buf	10/30/03	1581F	B24
LITTLE FALLS		16Sec	Pullman	4/3/07	2184	3523
LITTLE ROCK	P	10SecDr	H&StJ RR	1/73	26	Han3
LITTLE ROCK	N	12SecDr	Pullman	7/13/08	1963C	3630
LITTLE ROCK #10	RN	10SecSrB	(WS&PCC)	6/5/73		
LIVADIA		12SecDr	PCW Det	1/21/82	93B	D135
LIVADIA		12SecDr	Pullman	11/20/01	1581A	2784
LIVERPOOL		12SecDrB	PCW	1/22/87	446A	1260
LIVERPOOL #5	P	22Ch4StsDr	(Wagner)	6/85	3108D	
LIVINGSTON		12SecDr	PCW Det	6/24/84	93D	D241
LIVINGSTON		12SecDrCpt	Pullman	4/6/04	2002	3028
LIVINGSTON #274	P	12SecDr	Wagner	1/98	3061B	Wg13
LIVIUS		BaggDiner	PCW	10/4/90	783A	1705
LIVIUS	P	BaggLibB	PCW	/98	1310A	2294
LIVONIA	N	12SecB	PCW Det	5/23/83	168	D195
LIVONIA	R	12SecDrB	PCW Det	5/23/83	1468	D195
LIVORNO		12SecDr	Pull Buf	10/28/02	1581C	B11
LLEWELLYN #232	P	4Sec5SrB	PCW (Wagner)	11/23/86	3045	1246
LOANDA	N	12SecO	B&O RR	C:/87	665	
LOBELIA		26Ch4StsDr	Pull Buf	7/12/00	1519C	B2
LOCARNO		16Sec	Pullman	6/7/09	2184D	3700

CAR NAME	ST	CAR TYPE	CAR BUILDER (Owner)	DATE BUILT	PLAN	LOT
LOCHIEL		14SecO	PCW Det	8/2/82	115	D155
LOCHIEL	N	12SecDr	Pullman	3/19/08	1963C	3604
LOCHIEL	P	12SecDr	Wagner	5/92	3065C	Wg85
LOCHINVAR		14SecO	PCW Det	10/27/81	115	D121
LOCHINVAR	P	14SecO	(Wagner)	/85	3019A	
LOCHINVAR E&A-4	N	Slpr	(E&A Assn)			
LOCK HAVEN #293		10SecDr	J&S	7/73	37	Wil12
LOCKPORT	P	12SecDrB	Wagner	11/92	3073	Wg75
LOCKSLEY		12SecDrSr	Pullman	2/14/01	1581B	2638
LOCUST POINT		12SecO	B&O (Union PCC)		657	
LOCUST POINT #504	P	12SecO	B&O RR		657	
LODI		10SecDr	PCW	1/10/90	717B	1630
LODORE		12SecDr	PullBuf	6/17/08	1963C	B78
LODOVICO		12Sec2Dr	Pullman	4/2/01	1580A	2646
LOGANSPORT	N	OS	PCW Det	11/13/71	27	D8
LOGANSPORT		12SecDrCpt	Pullman	5/24/04	1985	3095
LOHENGRIN		12SecDr	Pullman	3/19/08	1963C	3604
LOHENGRIN #115	P	7BoudB	(MannBCC)	/84	697	
LOIRET		16SecO	PCW	8/30/93	1040	2008
LOLITA	N	12SecDrB	PCW Det	6/7/82	93B	D154
LOLITA	R	12SecDrB	PCW Det	6/7/82	1468	D154
LOMBARDY	N	10SecDr	Grand Trunk	8/71	26E	Mon2
LOMBARDY		12SecDr	Pull Buf	8/27/03	1581F	B22
LOMOND		14SecO	PCW Det	8/2/82	115	D155
LONDON		12SecDr	Pullman	8/11/02	1581D	2852
LONDON	P	14SecO	(Wagner)	/76	3051A	
LONDON #37		OS	Grand Trunk	12/71	27	Mon2
LONE STAR	N	10SecDr	J&S	10/15/73	37	Wil12
LONG BRANCH		26ChDr	Pullman	5/29/07	1987B	3524
LONG BRANCH #161	P	18Ch7StsB	J&S (WS&PCC)	6/27/87	711A	
LONG BRANCH #241	P	SummerParlor	H&H (B&O)	7/31/74	40	Wil2
LONG ISLAND #155	P	22ChSr	J&S (WS&PCC)	5/12/87	516?	
LONGACRE		12SecDr	Pullman	2/8/08	1963C	3598
LONGFELLOW		12SecDr	PCW	10/14/92	990	1931
LONGFORD		16SecO	Pullman	5/4/05	2068	3177
LONGINUS		7CptLngObs	Pullman	6/16/04	1977B	3091
LONGMEADOW #101	P	20Ch8StsDrB	Wagner	7/90	3126	Wg36
LONGMONT		12SecDrB	PCW	2/12/89	375G	1557
LONGMONT	R	14SecDr	PCW	2/12/89	1882	1557
LONGOS	N	12SecDrB	Wagner	11/92	3073	Wg65
LONGVIEW	N	10SecB	PCW Det	3/27/72	26A	D10
LONGVIEW		16SecSr SUV	Pullman	8/31/04	2033A	3135
LONSDALE	P	26Ch3Sts	B&S	1/93	3260	
LOON LAKE #525	P	12SecDrB	Wagner	8/92	3070	Wg72
LORCA		12SecDr	PCW	12/11/92	990	1941
LORDSBURG	N	12SecDrCpt	Pullman	5/23/04	1985	3095
LORENZO		7Cpt2Dr	PCW	1/10/96	1065A	2140
LORENZO #117	N		(Woodruff)	7/83		
LORETTA	N	12SecO	(L&N RR)	/72	114	
LORNA		22Ch7StsDrObs	Pullman	6/17/04	2023	3126
LORNA	P	12SecDr	(Wagner)	10/80	3022F	
LORNE	N	10SecDr	Grand Trunk	9/72	31K	Mon3
LORRAINE		10SecDrB	PCW Det	9/8/83	149	D206
LORRAINE		12SecDr	Pullman	6/21/01	1581A	2726
LORRAINE	R	30Ch SUV	Pullman	6/21/01	3581E	2726
LOS						
ALAMOS SP-37	P	10SecDrB	PCW	2/19/87	445B	1298
ANGELES		14SecDr	Pull Buf	10/27/00	1582A	B3
ANGELES GH&SA-10	P	Slpr	(SP Co)	11/28/83		
GATOS CP-8071	P	10Sec2Dr	PCW	3/6/93	995A	1942
VEGAS	N	10SecDr	PCW Det	6/2/79	59	D63
LOTHAIRE	N	10Sec2Dr	Wagner	6/93	3074	Wg92
LOTHIAN	N	12SecDr	Wagner	5/88	3061B	Wg11
LOTHIAN #120	RN	7BoudB	(MannBCC)	/84	776A	
LOTUS	N	12SecDrB	PCW Det	9/8/83	1468	D206
LOTUS	R	12SecDrB	PCW Det	4/10/84	1468	D242
LOUISE	RN	27StsParlor	PCW Det	8/76	45	D31
LOUISE		20Ch6StsDr	PCW	7/2/91	928A	1826
LOUISIANA	N	12SecO	DetC&MCo	/68	5	Det
LOUISIANA	P	12SecO	(OPLines)	/72	4	
LOUISIANA		6CptLngObs	Pullman	4/6/04	2004	3030
LOUISIANA #286	P	12SecDrB	Wagner	8/89	3065	Wg23
LOUISIANA #329	P	8BoudB	J&S (UnionPCC)	/89	373?	
LOUISIANA M&LT-5	P	Sleeper	(M&LT RR)	9/28/83	37?	
LOUISVILLE	P	12SecO	(OPLines)	/71	5N	
LOUISVILLE		12SecDr	Pullman	7/27/03	1581F	3015
LOUISVILLE #179	P	OS	PRR	10/71	27	Alt2
LOUISVILLE CTC-87	R		PRR rblt/71			
LOUVAIN	P	14SecDr SUV	Pull P-2069A	5/06	3207B	3320
LOUVAIN #413		10SecDrB	PCW (UnionPCC)	11/3/83	149	D207
LOUVRE		14SecO	PCW Det	3/3/93	1021	D517
LOVELAND		16Sec	Pullman	11/27/06	2184	3409
LOVELAND	R	6SecCafeLng SUV	Pull	11/27/06	3986	3409
LOWELL	P	10SecDr	Vermont Cent	5/72	31	StA2
LOWELL		14SecDr	Pullman	12/30/01	1523F	2798
LOWELL		14SecDrO SUV	Pull	12/12	2673	4047
LOWNDES		12SecDr	Pullman	4/8/03	1581D	2984
LOYOLA		10SecDrB	PCW	11/3/91	920B	1866
LOYOLA	N	16SecB	Pull Cal	9/10/02	1794	C3
LOYOLA	R	12SecDrB	PCW	11/3/91	1940	1866
LOZIER SPofA-3	P	10Sec	(SPofA)	7/1/83	37L	
LUBECK		16SecO	PCW	8/10/93	1040	2008
LUCAN		14SecO	PCW	3/1/93	1021	1969
LUCANIA		Private	PCW	1/15/98	1273A	2251
LUCANIA		12SecDr	Pull Buf	6/25/08	1963C	B78
LUCCA		12SecDr	PCW	12/11/92	990	1941
LUCEITA	N	30Ch	PCW	4/27/98	1278A	2293
LUCENTIO		7Cpt2Dr	PCW	1/17/96	1065A	2143
LUCERIA		12SecDr	Pull Buf	11/6/03	1581F	B25
LUCERNE		12SecDr	PCW Det	12/15/88	635	D377
LUCERNE	N	20Ch8StsDrBObs	Wagner	11/92	3127	Wg73
LUCIA #137	RN	7BoudB	(MannBCC)	/84	776A	

CAR NAME	ST	CAR TYPE	CAR BUILDER (Owner)	DATE BUILT	PLAN	LOT
LUCIA de LAMMERMOOR #137	P	7BoudB	(MannBCC)	/84	776	
LUCIAN	P	10Sec2DrB	PCW (Wagner)	3/1/00	3079A	Wg135
LUCIDA		12SecDr	Pull Buf	6/29/08	1963C	B78
LUCIE		18Ch4StsDrB	PCW	7/1/93	1039A	1994
LUCILLE		20Ch6StsDr	PCW	7/2/91	928A	1826
LUCIN		10SecDr2Cpt	Pullman	9/7/05	2078	3203
LUCINA		28ChDr	Pullman	6/24/04	1986A	3087
LUCIPHENE	P	22Ch6StsDrB	B&S	2/00	2368	
LUCIUS	P	BaggLibB	PCW	/98	1310A	2294
LUCKNOW	P	12SecDrB	Wagner	11/92	3073	Wg75
LUCRETIA		18Ch4StsDrB	PCW	7/10/93	1039A	1994
LUCULLUS	N	Private	Wagner	11/88	3084	Wg16
LUDINGTON		16SecSr	Pullman	8/29/04	2033A	3135
LUDLOW		12SecDr	Pullman	3/27/03	1581D	2983
LUDLOW	P	12SecO Tourist	(Wagner)	3/86	3059	
LUELLA	N	16Ch3StsDrB	(Wagner)	4/85	3108B	
LUFKIN	N	10SecB	PCW	6/84	206	198
LUMEN		6CptDrLibObs	Pullman	2/25/02	1751B	2799
LUMEN		12SecDr SUV	Pullman	3/21/10	1963E	3807
LUMINARY		14SecO	PCW Det	4/26/93	1021	D518
LUNA		10SecDrB	PCW	5/14/87	445E	1292
LUNA	R	14Sec	PCW	5/14/87	445X	1292
LUPIA		12SecDr	Pull Buf	11/10/03	1581F	B25
LURAY	N	12SecDrB	Wagner	6/96	3073C	Wg109
LURAY #437		10SecDrB	PCW	4/1/85	149E	1037
LURLINE	P	12SecDr	(Wagner)	10/80	3022F	
LUSATIA		12SecDr	Pullman	9/14/00	1572	2585
LUSITANIA		12SecDr	Pull Buf	6/22/08	1963C	B78
LUTETIA	RN	16SecB	Pull Buf	12/20/02	1794	B
LUTZEN	N	14SecO	(Wagner)	/72	3020	
LUXEMBURG		12SecDr	Pullman	6/22/01	1581A	2726
LUXINIA	P	16Ch6StsDrB	PCW	8/85	3254	
LUXOR		14SecO	PCW Det	3/29/93	1021	D517
LUXOR #118	P	20Ch8StsDrB	Wagner	6/91	3126A	Wg51
LUZANO		10SecDrB	PCW Det	1/5/84	149	D223
LUZANO	R	16Sec	PCW Det	1/5/84	1877C	D223
LUZERNE #295		10SecDr	J&S		37	Wil12
LUZON		10Cpt	PCW	1/25/93	993A	1951
LUZON	P	12SecDrB	Wagner	12/98	3073E	Wg117
LYCIDAS #124	R	7BoudB	(MannBCC)	/84	776A	
LYCOMEDES		10Sec2Dr	PCW	7/6/96	1173A	2153
LYCOMING	N	7CptLngObs	PCW	10/30/95	1159E	2126
LYCON	N	10Sec2Dr	Wagner	6/93	3074	Wg92
LYDIA	N	17Ch12StsB	J&S (B&O)	7/2/87	686C	2520
LYDONIA	N	22Ch6Sts	PCW	8-9/92	976B	1910
LYLE #304	P	20Ch12StsB	(MannBCC)	/89	708	
LYLETE		10SecDrB	PCW	8/31/87	445E	1333
LYNCH #181	RN	12Ch36Sts SUV	Pull Buf	12/15/03	3581B	B26
LYNCHBURG #17	P	Slpr	(SouTranCo)			
LYNDEN	P	14SecO	(Wagner)	/76	3051A	
LYNDHURST		12SecDrB	PCW Det	7/26/88	375G	D345
LYNN #144	N	6Sec4Dr?	J&S (WS&PCC)	4/8/86	720?	
LYNWOOD	N	12SecDr	PCW	6/83	93B	38
LYONS		12SecDr	PCW	2/10/92	784E	1881
LYONTON		12SecDr	Pullman	10/18/00	1581A	2588
LYRA		12SecDr	PCW Det	3/10/91	784B	D460
LYRA	P	12SecDrB	Wagner	11/92	3073	Wg75
LYRATA	N	26ChDr	Pullman	3/26/06	1987A	3335
LYRAVINE	N	30Ch	Pullman	5/18/01	1646D	2700
LYSANDER	N	10SecO	(Wagner)	3/85	3039	
LYSTER	N	10SecDr	Grand Trunk	2/71	21D	Mon1
LYTTON		12SecDr	Pull Buf	8/18/03	1581F	B21
MABEL		18Ch4StsDrB	PCW	7/8/93	1039A	1995
MABEL		24ChDr	Pullman	7/5/05	1980A	3171
MaBELLE	N	14Ch13StsDrB	PCW	7/23/87	505	1376
MACAO		10SecDrB	PCW	4/9/84	149	113
MACAO	R	12SecDrB	PCW	4/9/84	1611	113
MACAULAY		10SecLibObs	Pullman	11/22/09	2351A	3748
MACEDON		12SecDr	Pullman	10/28/03	1581F	3025
MACEDONIA		12SecDr	Pullman	6/9/03	1581E	3010
MACKINAW		12SecDr	Pullman	5/23/03	1581D	3004
MACKINAW	P	12SecDr	Wagner-Jones	10/82	3023C	
MACKINAW #21	N	12SecO	H&H (WS&PCC)	6/14/71	736	
MACON		12SecDr	Pullman	2/19/04	1963	3070
MACUTA		12SecDrB	PCW	8/8/88	375E	1395
MACUTA	N	12SecDr	Wagner	5/88	3061B	Wg11
MADEIRA	P	12SecDr	H&H	7/88	1899	
MADELINE		16Ch12StsDrB	PCW	12/5/88	614C	1508
MADERA SPofA-2	P	10Sec	(SP Co)	9/8/84	37L	
MADERO		12SecDr	Pullman	9/22/06	1963C	3397
MADISON		12SecO	PCW Det	6/8/80	58	D74
MADISON	P	12SecDr	(Wagner)	9/83	3023D	
MADISON #76	P	12SecDr	H&H	/80-/82		
MADONNA		18Ch6StsDrB	PCW Det	8/16/90	772D	D426
MADRAS	N	10SecDr	B&S	7/31/74	37 1/4	Day
MADRAS	P	10Sec2DrB	Wagner	9/92	3071	Wg77
MADRID	P	12SecDr	(Wagner)	6/81	3022C	
MADRID	N	10SecDr	J&S	/73	37	Wil10
MADRIGAL		16Sec	Pullman	6/8/06	2184	3361
MADRINE		12SecDr	Pullman	9/15/09	1963E	3744
MADRONA		BaggLib	PCW	6/7/87	477	1325
MADRONA		12SecDr	Pullman	1/28/01	1581C	2614
MADRONA	N	15Ch5StsDr	(Wagner)	4/86	3106	
MAFEKING		12SecDr SUV	Pullman	5/4/04	1963	3082
MAGDEBURG #438	P	12SecDr	Wagner	11/90	3065B	Wg42
MAGDELIN		Diner	PCW	2/11/92	964A	1874
MAGELLAN	N	10SecDr	Grand Trunk	3/71	21-30C	Mon1
MAGELLAN	N	BaggClub	Pullman	1/22/07	2136A	3461

Figure I-27 The open platform 12-section drawing room MAGENTA #345 was lost overboard on the transfer boat "Maryland" in the New York Harbor in December 1888. It was built in March 1881 for General Service on the Pennsylvania Railroad at the Detroit Shops of Pullman's Palace Car Company. The Robert Wayner Collection.

CAR NAME	ST	CAR TYPE	CAR BUILDER (Owner)	DATE BUILT	PLAN	LOT
MAGENTA #277	P	10Cpt	Wagner	4/89	3062	Wg14
MAGENTA #345		12SecDr	PCW Det	3/18/81	93	D88
MAGGIORE		14SecO	PCW Det	4/18/93	1021	D518
MAGNET	N	10SecDr	B&S	8/72	26	Day8
MAGNET	RN	Private	Wagner	4/89	3062Q	Wg14
MAGNOLIA		12SecDrSr	PCW	12/10/98	1318C	2370
MAGNOLIA #11	P	Slpr	(SouTranCo)			
MAGNOLIA #33	P	14Ch5StsDr	(Wagner)	5/71	3106C	
MAGNUS		6CptDrLibObs	Pullman	3/2/03	1751B	2986
MAHAPA		12SecDr	Pullman	11/30/08	1963E	3659
MAHOPAC		12SecDr	Pullman	2/27/07	1963C	3525
MAHTOWA		12SecDr	Pullman	1/31/03	1581C	2946
MAID of the MIST E&A-33	P	Parlor	(E&A Assn)	/72		Elm0
MAIDA	N	10Ch14StsDr	PCW Det	8/7/72	35 1/2	D12
MAIDA		22ChDr	Pull Buf	6/28/02	1647C	B9
MAIDSTONE		16Sec	Pullman	6/2/09	2184D	3700
MAINE		12SecB	PCW Det	6/24/76	57	D41
MAINE	P	10SecO	(Wagner)	/67	3006	
MAINE	R	12SecDr	PCW		1603	
MAINTENON		12SecDr	Pull Buf	1/5/07	1963C	B56
MAITLAND	R	12SecDrB	PCW?		1611	
MAJESTIC	N	32StsB Parlor	PCW	7/99	1376B	2371

CAR NAME	ST	CAR TYPE	CAR BUILDER (Owner)	DATE BUILT	PLAN	LOT
MAJESTIC		6SecDr4Sr	PCW	2/6/91	887C	1788
MAJESTIC		12SecDr	Pullman	9/3/08	1963C	3646
MAJOR JOHN W.THOMAS	RN	18ChLngCafe SUV	Pullman	7/1/08	3968A	3623
MAJORIAN	N	12SecDrB	(Wagner)	6/87	3050A	
MALABAR	N	10Sec2DrB	Wagner	9/92	3071A	Wg77
MALACCA		10SecDrB	PCW	3/14/84	149	102
MALACCA #542	P	10Sec2DrB	Wagner	9/92	3071	Wg77
MALAGA	R	12SecDr	PCW Det	7/6/83	1602B	D193
MALAGA #392		12SecB	PCW Det	7/6/83	168	D193
MALAGA	RN	12SecDr	PCW Det		1602B	D41
MALAY	N	10SecDrB	PCW	10/30/86	445	1262
MALAY	R	14Sec	PCW	10/30/86	445X	1262
MALCOLM		12SecDr	Pull Buf	10/9/05	1963C	B47
MALDEN	N	OS	CB&Q RR	5/72	36	Hinc1
MALDEN		8SecSrB (NG)	PCW Det	11/23/82	155	D173
MALDEN		16Sec	Pullman	3/27/07	2184	3523
MALDIVE	P	14SecDr SUV	Pull P-2069A	5/06	3207B	3320
MALDWINA	N	30Ch	PCW	/98	1278A	2293
MALINCHE #146	N	6Sec4Dr	J&S (WS&PCC)	4/8/86	720H	
MALINTA		10SecDr2Cpt	Pullman	10/5/06	2078D	3398
MALINTZI	RN	10SecB (NG)	PCW Det	4/11/83	73F	D186
MALMO		10SecDrB	PCW	5/29/84	149D	192
MALTA		10SecDr	PCW Det	4/30/71	26	D6
MALTA	N	12SecB	PCW Det	5/18/83	168	D195
MALTA	R	12SecDrB	PCW Det	5/18/83	1468	D195
MALTA	P	10SecDr	(Wagner)	/85	3001C	
MALTRATA		12SecDr	Pullman	7/19/06	1963C	3385
MALVASIA		12SecDr	Pullman	9/15/00	1572	2585
MALVERN		12SecDr	Pullman	3/27/02	1581B	2821
MALVERN #315		10SecB	PRR	7/31/75	37	Alt
MALVOLIO		12SecDrCpt SUV	Pullman	5/24/04	1982	3090
MAMBRINO		12SecDrB	PCW Det	4/27/88	375G	D346
MAMBRINO	R	14SecDr	PCW Det	4/27/88	1882	D346
MAMELUKE	P	10Sec2DrB	Wagner	9/92	3071	Wg77
MANA		12SecDr	PCW Det	4/6/91	784D	D461
MANASQUAN		28ChDr	Pullman	4/29/07	1986B	3527
MANASSAS		10SecLibObs	Pullman	4/30/06	2176	3355
MANAYUNK		12SecDr	Pullman	5/23/08	1963C	3616
MANCHESTER		Parlor	PCW Det	2/15/72	22	D9
MANCHESTER		14SecDr	PCW	9/2/99	1318E	2447
MANCHESTER #267	P	12SecDr	Wagner	1/89	3061B	Wg13
MANCHURIA		12SecDr	Pullman	10/11/01	1581A	2765
MANDALAY	P	10Sec2DrB	Wagner	9/92	3071	Wg77
MANDAN		14SecO	PCW Det	7/5/82	93B	D151
MANDAN		12SecDr	Pullman	5/19/02	1581B	2853
MANDARIN		12SecDr	Pull Buf	9/29/06	1963C	B55
MANHATTAN		12Sec	PCW Det	7/31/79	65	D53
MANHATTAN		Diner	Pullman	12/23/01	1694	2762
MANHATTAN #116	P		(WS&PCC)	7/83		
MANILA		12SecDrSr	PCW	6/27/99	1318C	2424

CAR NAME	ST	CAR TYPE	CAR BUILDER (Owner)	DATE BUILT	PLAN	LOT
MANILA	P	12SecDrB	Wagner	12/98	3073E	Wg117
MANIPUR	N	10Sec2DrB	Wagner	9/92	3071A	Wg77
MANISTEE #104	N		B&O (WS&PCC)	5/19/82		
MANISTEE		OS	PCW Det	11/13/71	27	D8
MANISTEE		12SecDr	Pull Buf	6/22/04	1963	B38
MANITO		12SecDr	Pullman	5/26/02	1581B	2853
MANITOBA #62	P	12SecDr	B&S	/76-/78		
MANITOBA #288	P	12SecDrB	Wagner	8/89	3065	Wg23
MANITOU		10SecO (NG)	PCW Det	4/11/83	73A	D186
MANITOU	R	8SecSrB (NG)	PCW Det	4/11/83	1662A	D186
MANITOWOC		12SecDr	Pull Buf	9/15/06	1963C	B55
MANKATO	P	12SecDrB	PCW	4/18/84	202	195
MANKATO		16SecSr SUV	Pullman	8/31/04	2033A	3135
MANLIUS		10SecDrB	PCW	9/17/85	240	1080
MANOA #311	P	11Boud	(MannBCC)	/89	703	
MANOR #159	N	20Ch4StsB	J&S (WS&PCC)	6/27/87	711	
MANSET	N	12SecDr	Pullman	9/22/06	1963C	3397
MANSFIELD	N	10SecDr	J&S	11/71	26	Wil8
MANSFIELD		12SecDrCpt	Pullman	5/24/04	1985	3095
MANSFIELD #184	P	OS	PFtW&C	11/71	27	All1
MANSFIELD #268	P	12SecDr	Wagner	1/88	3061B	Wg13
MANTON	P	26Ch3Sts	B&S	3/93	3260	
MANTUA #136	P	DDR	PRR	11/70	17	Alt1
MANTURA	N	16Sec	Pullman	12/3/06	2184	3409
MANVILLE	P	26ChBarB SUV	Pull P-2265	8/07	3274	3497
MANZANO		7Cpt2Dr	Pullman	10/29/02	1845	2926
MAOLYN		12SecDr	Pullman	11/30/08	1963E	3659
MAPLETON		10SecDrCpt	PCW	5/23/89	668	1587
MAPLEWOOD	N	12SecB	PCW Det	7/2/83	168	D193
MAPLEWOOD		12SecDr	Pull Buf	1/30/07	1963C	B56
MARACAIBO		12SecDr	Pull Buf	4/30/04	1581F	B36
MARAJAH		12SecDr	Pullman	11/30/08	1963E	3659
MARANON		10SecDrB	PCW Det	1/5/84	149	D223
MARANON	R	12SecDrB	PCW Det	1/5/84	1468	D223
MARATHON	N	OS	GW Ry/GMP	8/1/67	13	Ham
MARATHON	N	16SecO	Wagner	7/89	3063	Wg15
MARATHON #50	S	12SecDr	PCW	8/26/86	93H	1205
MARCELINE		22ChDr	Pullman	7/9/01	1647C	2702
MARCELLA		16Ch12StsDrB	PCW	3/7/89	614B	1508
MARCELLUS		14SecDr	Pullman	2/27/00	1318J	2532
MARCHENA		BaggBSmkBarber	PCW	4/23/93	1029	1975
MARCHENA	N	BaggClub	Pullman	8/17/06	2136A	3386
MARCHIONESS	R	21Ch7StsDrObs	Pullman	7/25/05	2091	3219
MARCHIONESS		22Ch6StsDrObs	Pullman	7/25/05	2023	3219
MARCIAN		14SecDr	Pullman	6/2/00	1523	2559
MARCO		12SecDr	PCW	9/4/91	784C	1828
MARCUS		12SecDr	Pullman	7/15/08	1963C	3630
MARCUS		12SecDr	Pullman	3/26/10	1963E	3807
MARDIAN	N	7Cpt2Dr	PCW	1/17/96	1065A	2143
MARENGO		14SecO	PCW Det	5/27/93	1021	D525
MARENGO WC-104	N	10SecDr	CentCarCo	3/87	489	
MARFA		14SecO	PCW	2/24/93	1021	1969
MARGARET		18Ch4StsDrB	PCW	7/3/93	1039A	1994
MARGATE		14SecO	PCW Det	6/16/93	1041	D519
MARGUERITE		20Ch5StsB	PCW	8/23/84	167A	199
MARGUERITE #581	P	10SecDrB	J&S (B&O)	C:/88	663	
MARIANA	N	7Cpt2Dr	PCW	1/17/96	1065A	2143
MARIBEL	N	26ChDr	Pullman	5/28/07	1987B	3524
MARICOPA		10SecDr2Cpt	Pullman	9/11/05	2078	3203
MARIE		18Ch4StsDrB	PCW	6/27/93	1039A	1994
MARIETTA	P	OS	(PK&RSCC)			
MARIETTA		12SecDr	Pullman	9/26/07	1963C	W19
MARIETTA #79	P	17Ch2StsB	(Wagner)	8/85	3122	
MARIETTA #195	P	10SecDr	B&O RR	4/30/73	19	B&O1
MARIGOLD		12SecDrCpt	PCW	8/13/92	983	1919
MARILLA	N	26ChDrB	Pullman	6/28/07	1988B	3526
MARINA	RN	24Ch3Sts	PCW Det	5/24/82	1928B	D154
MARINETTE		10SecO	PCW	8/2/88	617	StL5
MARION	P	12SecDr	(Wagner)	11/82	3023A	
MARION #79	P	10SecDr	H&H	/81-/82	499	
MARION #273	P	10SecDr	Elmira Car W	2/73	26	Elm1
MARION F. #82	P	29Ch	H&H (WS&PCC)	2/20/80	750	
MARIPOSA	P	6Dr	Wagner	3/95	3076	Wg105
MARIPOSA SP-33	P	10SecDrB	PCW	2/25/87	445B	1298
MARIQUITA		Private	(Wagner)			
MARISSA	R	12SecDr	PCW		1603	
MARITANA		18Ch5Sts	PCW Det	6/75	44	D24
MARITANA #120	P	7BoudB	(MannBCC)	/84	776	
MARIUS		24ChDr	Pullman	7/3/05	1980A	3171
MARIUS		14SecDr	Pullman	2/24/00	1318J	2532
MARJORAM		24ChDrSr	Pullman	7/21/04	1980A	3088
MARJORIE		22Ch6StsDr	PCW	8/29/92	976A	1910
MARJORIE		32ChDr	Pullman	3/13/07	2274	3509
MARLBOROUGH		10SecDrB Hotel	PCW Det	5/20/76	56	D39
MARLBOROUGH		12SecDr	Pullman	8/24/07	1963C	W19
MARLBOROUGH	P	12SecDr	(Wagner)	/83	3023D	
MARLEY		16Sec	Pullman	11/28/06	2184	3409
MARMION		14SecO	PCW Det	10/27/81	115	D121
MARMION	P	14SecO	(Wagner)	11/85	3019A	
MARMORA	RN	10SecDr	Grand Trunk	2/71	19?	Mon1
MARMORA	N	12SecDr	(Wagner)	11/80	3022F	
MARNOHTAH		10secDr2Cpt SUV	Pullman	3/22/10	2471	3784
MAROPA		14SecDr	Pullman	9/2/09	2438	3716
MAROS		14SecO	PCW	1/30/93	1017C	1962
MARQUETTE	N	10SecDr	(Wagner)	/83	3057B	
MARQUETTE		12SecDr	Pull Buf	7/1/04	1963	B38
MARQUETTE #148	P	14SecO	H&StJ RR/GMP	/70	20	Han1
MARQUISE		12SecDr	Pullman	9/4/08	1963C	
MARS	RN	8Sec24Sts	Taunton C.C.	1/71	29I	Tau1
MARS		27StsParlor	PCW Det	10/26/75	45	D25

Figure I-28 MARYLAND, was one of the thrity-four cars built under the Mann "Boudoir" patents. It was delivered by Jackson & Sharp to the new Union Palace Car Company in 1888 and painted dark blue with gold trim, an ochre roof, and sported spoked wheels. Around 1900, along with its sisters, it underwent major rebuilding to a parlor car in the "ST." series. (Fig. I-37, page 112) The Robert Wayner Collection.

CAR NAME	ST	CAR TYPE	CAR BUILDER (Owner)	DATE BUILT	PLAN	LOT
MARS	P	Parlor	Taunton C.C.	3/71	29B	Tau1
MARSDEN		12SecDr	Pullman	12/10/08	1963E	3659
MARSEILLES		12SecDr	PCW	11/30/92	990	1941
MARSEILLES		12SecDr	Pullman	10/27/03	1581F	3025
MARSHALL		12SecDr	Pullman	10/29/07	1963C	3570
MARSHALLTOWN #11	N	12SecO	(WS&PCC)	6/5/73	694	
MARSHFIELD		12SecDr	Pull Buf	1/23/07	1963C	B56
MARSTON		16SecO	PCW	5/31/93	1040	1990
MARTELLO		10SecDrB	PCW	9/17/87	445E	1362
MARTHA		22ChDr	Pullman	6/20/02	1647F	2863
MARTHA #108	P	8BoudB	J&S (MannBCC)	/84	373B	
MARTIA		24ChDr	Pull Buf	3/15/07	2156	B58
MARTINSBURG		OS	B&O RR	4/30/73	27 1/2	B&O2
MARTIUS		14SecDr	Pullman	2/28/00	1318J	2532
MARULLUS		12SecDr	Pull Buf	5/6/10	1963E	B83
MARVEL		16Sec	Pullman	6/4/08	2184C	3623
MARY #79	P		(WS&PCC)	2/17/79		
MARY C. #41		8SecParlorRE	H&H (WS&PCC)	7/25/78		
MARY E. #44		Slpr	H&H (WS&PCC)	/78		
MARYLAND	P	12SecDr	(Wagner)	/83	3023D	
MARYLAND #326	P	8BoudB?	J&S (UnionPCC)	/89	373?	
MARYLAND	N	30ChObs	PCW	4/27/98	1278B	2299
MARYSVILLE	N	5CptLngObs	PCW	1/12/98	1299C	2283
MASCOTTE	RN	Private	PRR	8/72	29	Alt1
MASCOTTE		12SecDr	Pullman	3/30/08	1963C	3604
MASPETH		12SecDr	Pullman	12/10/08	1963E	3659
MASSACHUSETTS		12SecB	PCW Det	6/14/76	57	D41
MASSACHUSETTS	R	12SecDr	PCW?		1703	
MASSANIELLO #138	P	7BoudB	(MannBCC)	/84	776	
MASSASOIT		12SecDr	Pullman	3/26/08	1963C	3600
MASSAWEPIE #36	P	14Ch5StsDr	(Wagner)	5/86	3106A	
MASSENA	N	12SecDr	Wagner	6/93	3065C	Wg85
MASSILLON		16Sec	Pullman	7/13/03	1721	3014
MASSILLON #185		OS	PFtW&C	11/71	27	Al11
MATABON		12SecDr	Pullman	8/22/04	1963	3098
MATADAN #163	N	10SecDrB	(WS&PCC)	9/24/87	663	
MATADOR		12SecDr	Pullman	12/4/08	1963E	3659
MATAMORA	P	12SecDr	(Wagner)	6/81	3022G	
MATAMORAS	RN	8SecB (NG)	PCW Det	9/15/81	73D	D110
MATAMOROS	N	12SecDrSrB	PCW	/99	1319	2314
MATANZAS		12SecDr	Pullman	5/11/04	1963	3082
MATAQUA #82	N	22Ch10Sts	H&H (WS&PCC)	2/20/80	750E	
MATAWAN		12SecDr	Pullman	3/10/06	2163A	3333
MATCHLESS		16Sec	Pullman	6/9/06	2184	3361

CAR NAME	ST	CAR TYPE	CAR BUILDER (Owner)	DATE BUILT	PLAN	LOT
MATFIELD	P	36Ch SUV	Pull P-2263	8/07	3275	3492
MATILDA	P	20ChB	Wason & Co	8/87	1902	
MATTAPAN	P	32ChDr SUV	Pull P-2264	8/07	3277	3496
MATTERHORN		12SecDrB	PCW	5/13/84	191	132
MATTERHORN		32ChDr	Pullman	3/5/07	2274	3509
MATURA		12SecDr	Pullman	11/9/05	1963C	W17
MAUD	RN	27StsParlor	PCW Det	/77	45	D31
MAUD		18Ch4StsDrB	PCW	7/5/93	1039A	1994
MAUMEE VALLEY #418		14SecO	PCW	3/25/84	115	114
MAURETANIA		12SecDr	Pull Buf	1/16/07	1963C	B56
MAURITIUS		12SecDr	Pullman	12/19/02	1581B	2852
MAXIMUS		14SecDr	Pullman	2/24/00	1318J	2532
MAXINE		26Ch	Pullman	5/27/01	1646C1	2701
MAXON		16Sec	Pullman	3/26/07	2184	3523
MAYBROOK		12SecDr	Pullman	1/23/07	1963C	3448
MAYENCE	N	12SecO	Grand Trunk	12/71	27C	Mon2
MAYENNE	N	10Sec2DrB	Wagner	8/93	3074B	Wg97
MAYETTA		10SecDrB	PCW	2/15/88	445F	1369
MAYFAIR		12SecDr	Pullman	1/31/03	1581D	2962
MAYFIELD	P	12SecO	B&O (UnionPCC)		665F	
MAYFIELD		12SecDr	Pullman	9/7/01	1581A	2756
MAYFLOWER	P	OS	MC RR/GMP	/67		Det
MAYFLOWER			B&S	/69		Pit
MAYFLOWER		Private	Pullman	11/20/01	1672	2747
MAYFLOWER #141	N	Private	(MannBCC)	/84	684A	
MAYFLOWER E&A-26	P	Slpr	(E&A Assn)			
MAYSVILLE		12SecDr	Pullman	1/11/07	2163	3449
MAYWOOD		12SecDrB	PCW	9/8/88	375H	StL6
MAZARIN	N	12SecDr	Wagner	1/92	3065C	Wg55
MAZATLAN	P	6Sec4Dr	(WS&PCC)		741E	
MAZATLAN #465		12SecDrB	Wagner	8/91	3065	Wg53
MAZDA	RN	12SecDrCpt	PCW	4/23/93	1011E	1968
MAZO		12SecDr	PCW	9/13/90	784	1765
McGIBBON	P	10SecO	PCW (Wagner)	10/26/86	3035	1242
McGREGOR #72	P	25Ch2StsB	(Wagner)	6/82	3100A	
McGREGOR #105	P	10SecB	(IC RR)		231	
McHENRY	P	10SecO	PCW (Wagner)	10/20/86	3035	1242
McINTOSH	P	10SecO	PCW (Wagner)	11/2/86	3034	1242
McKENZIE	P	12SecO	PCW (Wagner)	10/27/86	3026A	1242
McPHERSON	P	12SecO	PCW (Wagner)	10/18/86	3026A	1242
MEACHAM LAKE #520	P	12SecDrB	Wagner	1/92	3070	Wg72
MEADOWBROOK		10SecDr2Cpt	Pullman	10/8/06	2078D	3398
MEADVILLE		12SecDr	Pullman	11/25/03	1581F	3025
MEANDER		12SecDr	Pullman	12/29/08	1963E	3659
MEATH		12SecDr	PCW	2/10/92	784E	1881
MECCA	N	10SecDr	Grand Trunk	9/72	31C	Mon3
MECCA	P	12SecDr	(Wagner)	6/86	3049A	
MECHAMA	P	14SecDr SUV	B&S	6/11	3207C	
MECHLENBERG #115	P	20Ch8StsDrB	Wagner	12/90	3126	Wg44
MECHLIN		10SecDrB	PCW	5/29/84	149D	192
MECHLIN	R	16Sec	PCW	5/29/84	1877E	192

CAR NAME	ST	CAR TYPE	CAR BUILDER (Owner)	DATE BUILT	PLAN	LOT
MEDEA		10SecDrB	PCW	7/8/85	240	1057
MEDEA	N	12SecDrB	Wagner	6/96	3073C	Wg109
MEDFIELD	P	20Ch6StsDr	Pull	2/91	3259A	
MEDFORD WC-51	N	12SecO	CentCarCo	2/84	532	
MEDINA #70	P	13Ch9StsDr	(Wagner)	8/84	3110	
MEDINAH	R	12SecDrB	PCW	10/6/88	1835	1481
MEDINAH	N	12SecDrB	PCW	10/6/88	375G	1481
MEDORA	RN	14Sec	PCW	2/14/90	668F	1649
MEDUSA		12SecDr	PCW Det	3/14/91	784D	D460
MEDWAY		10SecDrB	PCW Det	12/1/83	149	D212
MEDWAY	P	20Ch6StsDr	B&S	12/93	3262A	
MEGANTIC		16Sec	Pullman	6/2/06	2184	3361
MELAMPUS		12SecDr	Pullman	5/19/08	1963C	3615
MELBA	N	26ChDr	Pullman	5/28/07	1987B	3524
MELBOURNE	N	12SecDr	Pullman	3/26/04	1963	3080
MELBOURNE #82	P	12SecDr	H&H	/80-/82	197	
MELBOURNE #287	P	12SecDrB	Wagner	8/89	3065	Wg23
MELISANDE	N	6Dr	PCW	11/6/89	685D	1598
MELISSA	R	12SecDr	Pullman		1581C	
MELISSA #136	N	10SecDrB	(WS&PCC)	1/15/85	886D	
MELITA	N	10SecDrB	PCW	6/26/85	240	1057
MELROSE	N	10SecDrB	PCW Det	2/20/84	149	D227
MELROSE #91	P	12SecDrB	PCW (Wagner)	9/2/86	3049	1205
MELROSE #121	P	14SecO	PRR	9/70	69A	Pit
MELTON		12SecDr	Pullman	10/31/01	1581A	2784
MELTONA #116	N		(WS&PCC)	7/83		
MELUSINA		24ChDr	Pullman	7/11/05	1980A	3171
MELVERN	P		PRR	10/74		Alt
MEMNON		14SecO	PCW Det	6/28/82	115	D151
MEMNON	N	12SecDr	Wagner	4/88	3061B	Wg10
MEMPHIS	P	OS	(PK&RSCC)			
MEMPHIS		12SecDr	Pullman	12/19/02	1581B	2852
MEMPHIS #4	RN	10SecSrB	(WS&PCC)	6/30/73	759	
MENA	N	12SecDr	Pullman	1/27/02	1581A	2792
MENDELSSOHN		12SecDr	PCW	5/4/92	784E	1899
MENDOCINO		32ChDr	Pullman	3/5/07	2274	3509
MENDOTA		12SecDr	PCW Det	5/10/81	93A	D107
MENDOTA		12SecDr	Pullman	10/31/07	1963C	3570
MENDOTA	R	12SecDrB	PCW Det	5/10/81	1611	D107
MENDOZA	R	12SecDr	PCW		1603	
MENDOZA	N	12SecO	PCW (Wagner)	3/29/87	3026A	1264
MENGER		12SecDr	Pullman	3/6/08	1963C	3600
MENHADEN		12SecDr	Pullman	3/10/06	2163A	3333
MENLO	P	12SecDr	(Wagner)	7/86	3049A	
MENLO GH&SA-5	P	Slpr	(SP Co)	7/1/83	37?	
MENOKEN		10SecDrCpt	PCW	6/21/89	668	1587
MENOMINEE		10SecDr	PCW Det	11/13/72	26	D14
MENOMINEE		12SecDr	Pullman	2/20/04	1963	3070
MENOMINEE	P	10SecO	(Wagner)	/72	3001A	
MENOSHA		12SecDr	Pull Buf	9/22/06	1963C	B55
MENTONE #187	P	12SecSr	PCW (Wagner)	9/22/86	3049	1205
MENTOR	N	12SecO	B&O RR	/81	665	

CAR NAME	ST	CAR TYPE	CAR BUILDER (Owner)	DATE BUILT	PLAN	LOT
MERCED		10Sec2Dr	Pullman	8/29/02	1746A	2925
MERCED CP-37	P		B&S	10/3/83	37?	Day
MERCEDES	N	20Ch5StsB	PCW	8/8/84	167A	199
MERCEDES #467	P	12SecDrB	Wagner	8/91	3065	Wg53
MERCURY		27StsParlor	PCW Det	4/28/76	45	D31
MERCURY		14Ch18StsBObs	PCW	11/4/98	1279C	2354
MERCURY #216	P	8Sec24Sts	CB&Q RR	3/72	291	Aur2
MERCUTIO		7Cpt2Dr	PCW	1/25/96	1065A	2143
MEREDITH		12SecDrB	PCW Det	4/21/88	375E	D328
MERIDA		12SecDrB	PCW Det	10/27/88	375G	D358
MERIDA	R	14SecDr	PCW Det	10/27/88	1882	D358
MERIDEN		14SecDrO SUV	Pullman	12/12	2673	4047
MERIDIAN		12SecDr	Pullman	3/26/04	1963	3080
MERIDIAN	RN	12ChCafeObs SUV	Pullman	12/18/03	3968	3065
MERION		12SecDrSr	Pullman	1/14/01	1581A	2614
MERLIN		10SecDrB	PCW	2/11/84	149B	101
MERLIN	N	12SecDrB	Wagner	2/90	3065	Wg32
MERMAID		12SecDr	Pullman	9/15/09	1963E	3744
MERRIMAC	P	10SecDr	B&S	8/19/72	26	Day8
MERRIMAC #54	S	12SecSr	PCW	8/31/86	93H	1205
MERRIMAC #166	P	10SecDrB?	(Woodruff)	9/24/87	663?	
MERRIMACK		16Ch6StsDr	PCW?	6/93	1100	
MERSEY		10SecDrB	PCW Det	8/12/84	149D	D248
MERTON		14SecO	PCW Det	7/3/93	1041	D519
MERTON	N	BaggClub	Pullman	1/22/07	2136A	3461
MESA		10SecDrB	PCW	6/4/91	920B	1808
MESABA		12SecDr	Pullman	2/13/06	1963C	3316
MESILLA		12SecDrB	PCW Det	9/29/88	375G	D358
MESILLA	N	16SecO	Wagner	4/93	3072	Wg90
MESSENIA		12SecDr	Pull Buf	1/8/04	1581F	B27
MESSINA		12SecDr	PCW	6/83	93B	38
MESSINA #95	P	12SecSr	PCW (Wagner)	9/4/86	3049	1205
META		12SecDr	PCW Det	4/9/91	784D	D461
META #16	P	12SecO	(WS&PCC)	6/30/73	833	
METAPEDIA						
METAPEDIA		12SecDr	Pullman	10/22/06	1963C	3408
METAPONTO		Slpr	PCW Det	5/10/84	170	D222
METELLUS		14SecDr	Pullman	2/28/00	1318J	2532
METEOR		12SecO	PCW Det	7/30/81	114	D113
METEOR		12SecDr	Pullman	12/10/08	1963E	3659
METIS		10SecDrB	PCW	10/9/84	149D	1007
METON	N	12SecDr	Wagner	5/87	3061D	Wg3
METROPOLE		Diner	PCW	10/30/89	662D	1641
METROPOLE		16Sec	Pullman	6/7/06	2184	3361
METROPOLIS		12SecO	PCW Det	5/30/76	58	D42
METROPOLIS		12SecDr	Pullman	12/29/08	1963E	3659
METROPOLIS E&A-34	P	Parlor	(E&A Assn)			
METROPOLITAN		10SecDrHotel	PCW Det	8/2/72	33	D11
METROPOLITAN		12SecDr	Pull Buf	1/8/07	1963C	B56
METUCHEN		12SecDr	Pull Buf	10/6/06	1963C	B55

CAR NAME	ST	CAR TYPE	CAR BUILDER (Owner)	DATE BUILT	PLAN	LOT
METZ		12SecDr	PCW	5/26/92	784E	1915
MEXICANO		10SecB (NG)	PCW Det	9/17/80	73A	D76
MEXICO	N	10SecDr	PCW Det	10/5/81	112A	D124
MEXICO	N	10SecDrB	PCW Det	3/21/84	149B	D234
MIAMA	N	12SecO	PCW Det	5/10/82	114	D143
MIAMI	N	12SecO	PCW Det	5/10/82	114	D143
MIAMI	RN	8SecBLngObs	PCW Det	2/18/93	1021S	D516
MIAMI	P	12SecDr	(Wagner)	10/82	3023A	
MIAMI #172	P	10SecDr	PRR	6/71	21	Alt2
MIAMI CTC-93	R		PRR rblt/72			
MICAH	N	12SecDrB	Wagner	2/90	3065	Wg30
MICHIGAN		12SecO	B&S/GMP	/68	5	Day
MICHIGAN	R	12SecO	B&S/GMP	/68	119	Day
MICHIGAN		8SecCpt	PCW Det	10/82	129	D160
MICHIGAN	R	12SecDr	B&S/GMP	/68	1603	Day
MIDAS		12SecDr	PCW	10/24/90	784	1766
MIDAS	P	12SecDrB	(Wagner)	4/87	3050	
MIDDLEBORO	P	34Ch SUV	Pull P-1484	1/03	3265	2904
MIDDLEBURY	P	12SecO	(Wagner)	9/72	3027F	
MIDDLESEX	N	16SecB	J&S	1/99	2371B	
MIDLAKE		10SecDr2Cpt	Pullman	12/30/05	2078C	3287
MIDLAND		30Berths	PCW Det	1/25/74	38	D19
MIDLAND	N	12SecO	PCW Det	5/10/82	114	D143
MIDLAND		10SecDr	PCW	10/29/87	508B	1368
MIDLOTHIAN #571	P	12SecB	PCW	3/4/85	227	1034
MIDVALE		12SecDrB	PCW Det	4/7/88	375G	D346
MIDVILLE #176	R	12Ch38Sts SUV	Pullman	11/29/01	3581D	2784
MIFFLIN		12SecDrSr	Pullman	3/2/01	1581B	2654
MIGNON		12SecDr	Pullman	12/29/08	1963E	3659
MIGNON #112	P	7BoudB	(MannBCC)	/84	697	
MIGNONETTE		24ChDr	Pull Buf	4/25/06	2156	B52
MIGON		20Ch6StsB	PCW	1/22/85	167B	200
MIKADO	N	6Sec8Ch	PCW Det	3/25/72	261	D10
MIKADO	P	12SecDr	(Wagner)	6/81	3022C	
MIKAWA		12SecDr	Pullman	10/4/07	1963C	W19
MILAN		12Sec	PCW Det	6/19/75	49	D28
MILAN		12SecDr	PCW	8/14/90	784	1707
MILAN #315	P	10SecDr	(Wagner)	/75	3016	
MILANO		10SecDrB	PCW	12/20/87	445E	1360
MILANO	R	16Sec	PCW	12/20/87	1880A	1360
MILBURN		12SecDr	Pullman	10/26/03	1581F	3025
MILDRED	N	20Ch3StsB	B&S (B&O)		653B	
MILES CITY		12SecDr	PCW	9/30/82	93B	22
MILES CITY	RN	14Sec	PCW	6/5/89	668F	1587
MILETUS		10SecDrB	PCW	9/27/84	149D	1006
MILETUS	R	16Sec	PCW	9/27/84	1877A	1006
MILFORD	N	12SecDr	(Wagner)	3/86	3049	
MILFORD #290		10SecB	B&S	7/73	37	Day10
MILL PLAIN	P	BaggSmk	B&S	2/93	3213A	
MILLADORE		10SecDrB	PCW Det	6/2/88	445E	D336

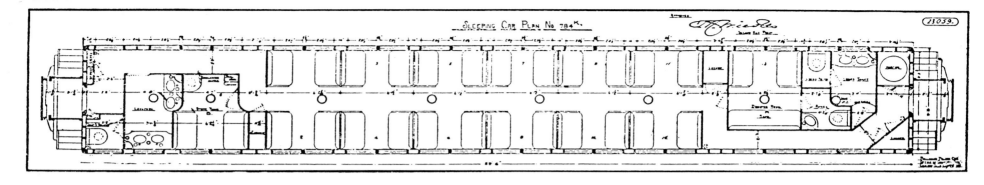

Figure I-29   Plan 784K, 12-section drawing room, smoking room, involved minor modifications of six cars from Plan 784C, prior to 1900. MIMORA, built in April 1891 under Lot 1798, was one of these cars. These same cars were later modified to Plan 784L, again involving only minor modification. The Newberry Library Collection.

CAR NAME	ST	CAR TYPE	CAR BUILDER (Owner)	DATE BUILT	PLAN	LOT
MILLBRAE		32ChDr	Pullman	3/6/07	2274	3509
MILLDALE	P	24Ch6Sts	B&S	12/93	3262	
MILLET		12SecDr	Pullman	8/6/01	1581A	2727
MILLICENT	N	24ChDr	Pull Buf	5/3/06	2156	B52
MILLICENT		24ChDr	Pull Buf	3/20/07	2156	B58
MILLIS	P	23ChDr SUV	Pull P-2070	6/05	3268	3188
MILLPORT	P	14SecDr SUV	Pull P-2262	8/07	3207C	3498
MILLSTON		10Sec3Cpt SUV	Pullman	11/2/05	2098	3236
MILLVILLE		24ChDrB	Pullman	6/8/05	1988A	3170
MILO		12SecDr	PCW	9/13/90	784	1765
MILSTEAD		10SecDr2Cpt	Pullman	6/12/08	2078D	3622
MILTIADES		12SecDrB	PCW	9/2/85	191	1074
MILTIADES	R	12SecDrB	PCW	9/2/85	1611	1074
MILTON		12SecDr	PCW	9/9/91	784C	1828
MILTON	N	6CptLibObs	Pullman	5/31/07	2089A	3533
MILTON	P	12SecO	(Wagner)	/85	3005	
MILWAUKEE	P	10SecO	G&B (PK & RSCC)	/68	47A	Troy
MILWAUKEE		12SecDr	Pull Buf	2/18/04	1581F	B34
MILWAUKEE WC-106	P	12SecO	CentCarCo	2/84	532	
MIMAS		10SecDrB	PCW	10/8/84	149D	1007
MIMAS	R	12SecDrB	PCW	10/8/84	1468	1007
MIMORA		12SecDr	PCW	4/11/91	784C	1798
MINDANAO		12SecDr	Pullman	8/22/04	1963	3098
MINDEN #439	P	12SecDr	Wagner	11/90	3065B	Wg42
MINDORO		12SecDr	Pullman	8/22/04	1963	3098
MINEOLA	P	24Ch5StsDr	(Wagner)	5/85	3108	
MINEOLA #90	P		B,D (WS&PCC)	6/7/81		
MINEOLA #232		Parlor	PCW Det	7/14/73	35 1/4	D17
MINERVA		27StsParlor	PCW Det	7/28/76	45	D31
MINERVA		20Ch5StsB	PCW Det	6/27/83	167	D203
MINETTE	N	26ChDr	Pullman	3/31/06	1987A	3335

CAR NAME	ST	CAR TYPE	CAR BUILDER (Owner)	DATE BUILT	PLAN	LOT
MINIDOKA		12SecDrCpt	Pullman	10/7/05	2079	3204
MINIKAHDA		14SecDr	Pullman	5/16/06	2165	3352
MINNEAPOLIS	P	12SecO	(NP Assn)		18D	
MINNEAPOLIS	P	14SecO	B&S	7/92	1593	
MINNEAPOLIS		12SecDr	Pull Buf	2/24/04	1581F	B34
MINNEAPOLIS #61	P	12SecDr	B&S	/76-/78		
MINNEAPOLIS WC-104	P	10SecDr	PCW Det	3/87	489	D294
MINNEDOSA		12SecDr	Pullman	10/20/06	1963C	3408
MINNEHAHA		28ChDr	Pullman	6/25/04	1986A	3087
MINNELUSA	N	12SecDr	Pull Buf	9/8/06	1963C	B55
MINNEOLA	N	6SecDr2Sr	PCW	5/23/88	542A	1404
MINNEQUA	P	10SecB	B&S	4/15/73	37	Day9
MINNEQUA		5CptLibObs	Pullman	8/14/05	2089	3222
MINNESELA	N	28ChDr	Pullman	6/25/04	1986A	3087
MINNESOTA		DDR	DetC&MCo	/70	17	Det
MINNESOTA	R	12SecDrB	PCW		1498	
MINNESOTA	N	7Cpt2Dr	Pullman	10/29/02	1845	2926
MINNESOTA #28	P		(CM&StP)			
MINNESOTA #269	P	12SecDr	Wagner	1/89	3061B	Wg13
MINNETONKA		12SecDr	Pull Buf	9/8/06	1963C	B55
MINNETONKA #70	P	12SecDr	H&H	/80-/82	197	
MINNEWASKA	N	10Ch15StsDr	J&S	11/72	35	Wil2
MINOCQUA		12SecDr	PCW Det	12/3/88	635	D377
MINOCQUA	N	12SecDr	Pullman	8/11/02	1581B	2852
MINORCA		10SecDrB	PCW	2/9/84	149B	101
MINOT	N	12SecB	PCW Det	7/31/83	168	D193
MINOTAUR		16Sec	Pullman	6/3/08	2184C	3623
MINTA	RN	12SecDr	PCW Det	5/20/84	1476A	D241
MINOTOLA		5CptLibObs	Pullman	8/16/05	2089	3222
MINTURN	N	6Dr	PCW	2/8/90	685D	1598

CAR NAME	ST	CAR TYPE	CAR BUILDER (Owner)	DATE BUILT	PLAN	LOT
MIRA		12SecDr	PCW	8/20/90	784	1708
MIRA	N	20Ch4Sts	(Wagner)	8/70	3101	
MIRABEAU		12SecDrB	PCW	12/3/88	375G	1513
MIRABEAU		12SecDr	Pullman	12/4/05	1963C	W17
MIRAGE		10SecDrB	PCW	11/4/87	445E	1334
MIRAMAR		10SecDrB	PCW	8/88	445E	
MIRAMON		12SecDr	Pullman	7/26/06	1963C	3385
MIRANDA		20Ch5StsB	PCW	8/8/84	167A	199
MIRANDA	P	22Ch6StsDrB	B&S	2/00	2368	
MIRIAM		22ChDr	Pull Buf	6/30/02	1647C	B9
MIRIAM #80	P	SlprParlor	(WS&PCC)	6/1/80		
MIRZA		12SecDr	PCW	8/15/90	784	1707
MISSISSIPPI	P	OS	(OPLines)	/72	27A	
MISSISSIPPI		10SecDr	H&H	9/72	26	Wil7
MISSISSIPPI	N	16SecB	PCW	6/ /93	1040J	1990
MISSISSIPPI	P	14SecO	B&S	/86	1593	
MISSISSIPPI		16Sec Tourist	Pullman	4/6/04	2008	3034
MISSISSIPPI	P	16SecO	(Wagner)	/67	3007	
MISSISSIPPI #44	P	12SecDr	B&S	/76-/78		
MISSISSIPPI #107	P	Slpr	(IC RR)	1/172		
MISSISSIPPI #137	P	SlprParlor?	(WS&PCC)	1/15/85	742?	
MISSOULA		12SecDr	PCW Det	3/21/83	93B	D184
MISSOULA		16Sec	Pullman	6/19/07	2247	3459
MISSOURI	P	OS	G,B&Co/GMP	/66	7	Troy
MISSOURI		8SecCpt	PCW Det	10/82	129	D160
MISSOURI		20Ch9StsDrObs	PCW	11/15/99	1417C	2466
MISSOURI	P	BaggSmk	PCW	/99	1400	2410
MISSOURI		20Ch4StsObs	Pullman	7/15/05	2057	3217
MISSOURI	P	16SecO	(Wagner)	/72	3029	
MISSOURI #99	P	12SecDrB	(CRI&P RR)	12/15/79	204	
MISSOURI #140	P	6Sec12ChSr	J&S (WS&PCC)	12/5/85	741B	
MISTLETOE		24ChDrSr	Pullman	7/21/04	1980A	3088
MISTRAL		10SecLibObs	Pullman	9/25/08	2351C	3634
MITANI	N	14SecDr	PCW	9/1/99	1318E	2447
MIZPAH	N	12SecDr	(Wagner)	6/86	3049A	
MOAPA		10SecDr3Cpt	Pullman	4/19/06	2078C	3342
MOBERLY		12SecDr	Pull Buf	4/6/03	1581C	B15
MOBILE	N	16SecO	Wagner	10/92	3072	Wg79
MOBILE	RN	12ChCafeObs SUV	Pull	12/16/03	3968	3065
MOBILE #99	P	Slpr	(IC RR)		2/176	
MOCHA		12SecDr	Pullman	5/4/04	1963	3082
MOCLIPS		12SecDr	Pullman	6/10/07	2271A	3458
MOCORITO	R	10SecObs	PCW	3/8/87	445W	1299
MOCORITO SP-45	P	10SecDrB	PCW	3/8/87	445B	1299
MOCTEZUMA	P	12SecO	(Pull/Sou)	/76	4A	
MODELLO	N	28ChDr	Pullman	4/9/06	1986A	3336
MODENA		12SecDr	PCW	12/2/82	93B	22
MODENA	P	14SecDr SUV	B&S	6/11	3207C	
MODESTO		10SecDr2Cpt SUV	Pullman	9/11/05	2078	3203
MODESTO CP-24	P	10SecDr	(CP Co)	7/30/84	37K	
MODOC #15	N		(WS&PCC)	9/28/72		

CAR NAME	ST	CAR TYPE	CAR BUILDER (Owner)	DATE BUILT	PLAN	LOT
MOHASKA		10SecDrB	PCW	5/31/89	623B	1566
MOHAVE		10SecDr	PCW Det	3/29/81	112	D149
MOHAWK	P	10SecDr	B&S	3/72	26	Day7
MOHAWK		OS	PCW Det	12/12/70	27	D4
MOHAWK	N	16SecO	Wagner	4/93	3072	Wg90
MOHAWK		Commissary	PCW Det	12/12/70	27	D4
MOHICAN		12SecDr	Pullman	12/9/07	1963C	3585
MOHONK	N	23ChDr	J&S	11/72	35	Wil2
MOJAVE		10Sec2Dr	Pullman	11/7/02	1746A	2928
MOLDAVIA		12SecDr	Pullman	10/7/01	1581A	2765
MOLINE		10SecDr	PCW Det	9/9/81	112A	D112
MOLINE		12SecDr	Pullman	2/19/04	1963	3070
MOLLENDO		12SecDr	Pullman	9/20/06	1963C	3397
MOMBASA	N	12SecDr	Pullman	10/10/07	2163A	3562
MOMENCE		16SecSr	Pullman	8/29/04	2033A	3135
MOMENCE #163	N	10SecDrB	(WS&PCC)	9/24/87	663E	
MONACO #143	N	6Sec4Dr	(WS&PCC)	4/8/86	720G	
MONACO #188	S	12SecDr	PCW	9/14/86	93H	1205
MONADNOCK		12SecDr	Pull Buf	12/30/05	1963C	B49
MONARCH		16Sec	Pullman	5/9/08	2184C	3623
MONCLOVA	N	12SecDrB	PCW	9/15/98	1319A	2314
MONCLOVA	R	12SecDrB	PCW	2/11/88	1566	1380
MONCLOVA M&LT-20	N	10SecDrB	PCW	2/11/88	445F	1380
MONCTON	P	SDR	B&S	9/74	27	Day11
MONCTON	N	16SecO	Wagner	4/93	3072	Wg90
MONDAMIN		12SecDr	Pullman	8/26/04	1581K	3118
MONERO		8SecDrB	PCW	7/26/88	556A	1463
MONGOLIA		12SecDr	Pullman	9/28/01	1581A	2765
MONIDA		10Sec2Dr	PCW	3/6/93	995A	1942
MONITOR	RN	Private	PCW Det	6/18/77	997	D44
MONITOR		12SecDr	Pullman	12/29/08	1963E	3659
MONITOR #263	P	10SecDr	B&S	8/72	26	Day8
MONMOUTH	P	12SecDr	(Wagner)	/83	3023D	
MONMOUTH #281	P	10SecDr	J&S (PRR)	2/73	26	Wil9
MONO SPofA-5	P	10Sec	(SPofA)	7/1/83	37L	
MONOCACY #520	P	12SecO	B&O RR	/81	665	
MONON	N	27ChDrObs	Pullman	6/8/08	2346	3624
MONONGAHELA		12SecDr	Pullman	4/25/07	1963C	3533
MONONGAHELA #136	P	10SecDrB	(WS&PCC)	1/15/85	886	
MONONGAHELA #297	P	10SecDr	Elmira Car W	8/73	37	Elm2
MONONGAHELA #524	P	12SecO	B&O RR	/81	665	
MONROE		12SecDr	Pullman	3/29/04	1963	3080
MONROE		Diner	Pullman	4/6/04	2001	3027
MONROE		10SecDr2Cpt	Pullman	6/13/08	2078D	3622
MONROEVILLE	R	10SecDr	PCW	2/23/85	222A	1033
MONROEVILLE #564	P	11SecDr	PCW	2/23/85	222	1033
MONROVIA		12SecDrSrB	PCW	2/28/99	1319A	2400
MONROVIA #29	N	12SecO?	G,B&Co/GMP	/68	27A	Troy
MONTALVO		32ChDr	Pullman	3/9/07	2274	3509
MONTANA	P	12SecDr	B&S/GMP	/68	12	Day
MONTANA	R	12SecO	B&S/GMP	/68	119B	Day

CAR NAME	ST	CAR TYPE	CAR BUILDER (Owner)	DATE BUILT	PLAN	LOT
MONTANA	P	12SecDr	(Wagner)	6/81	3022G	
MONTANO		12Sec2Dr	Pullman	4/3/01	1580A	2646
MONTAUK		12SecDr	Pullman	11/20/01	1581A	2784
MONTAUK	P	18Ch4StsDr	(Wagner)	4/85	3109	
MONTAUK #87	P		(WS&PCC)	6/7/81		
MONTCLAIR #270	P	12SecDr	Wagner	1/89	3061B	Wg13
MONTE CHRISTO		10SecB	PCW Det	8/5/73	37	D16
MONTEBELLO #522	P	12SecO	B&O RR	/81	665	
MONTECITO #139	N	6SecDrLibObs	J&S (WS&PCC)	12/1/85	741J	
MONTELLO		10SecDrB	PCW Det	6/2/88	445E	D336
MONTEMORA SP-68	P	10SecDrB	PCW	6/22/88	445F	1415
MONTEREY		10Sec2Dr	Pullman	11/4/02	1746A	2928
MONTEREY #464	P	10SecDrB	Wagner	8/91	3065	Wg53
MONTEREY CP-43	P	10SecDrB	PCW Det	12/8/83	149	D213
MONTESANO		Diner	PCW	1/21/90	662D	1659
MONTESANO	RN	12SecDr	PCW Det	7/5/82	1581C	D153
MONTEZUMA	N	12SecDr	PCW Det	9/1/82	93	D169
MONTEZUMA #466	P	12SecDrB	Wagner	8/91	3065	Wg53
MONTGOMERY	N	OS	(PK&RSCC)	/73	74	
MONTGOMERY	P	10SecB	Bowers, Dure	10/17/73	37	Wil11
MONTGOMERY		12SecDr	Pullman	3/19/04	1963	3080
MONTICELLO		12SecDr	Pullman	2/17/06	1963C	3316
MONTICELLO #20	P	OS	H&H	8/71	27	Wil5
MONTJOI		10SecDrB	PCW	8/11/87	445E	1333
MONTMORENCI	N	10SecB	B&S		37	Day11
MONTMORENCY #44	S		PCW	8/25/86	93H	1205
MONTOYA		10SecDr	PCW Det	3/29/81	112	D149
MONTPELIER	P	12SecDr	Vermont Cent	7/71	21B	StA1
MONTPELIER		10SecDr2Cpt	Pullman	9/12/05	2078	3203
MONTPELIER #47	S	12SecDr	PCW	8/28/86	93H	1205
MONTPELIER #163	P	10SecDrB	(WS&PCC)	9/24/87	663	
MONTREAL	P	DDR	Grand Trunk	10/70	17	Mon1
MONTREAL		12SecDr	PCW	12/3/92	990	1941
MONTREAL #8	P	Parlor	Vermont Cent	C:/71	8	StA
MONTREAL #52	P	12SecDr	PCW P-93H	9/3/86	3049	1205
MONTROSE		16Ch6StsDrB	PCW Det	6/29/87	167F	D283
MONTROSE #569	P	12SecB	PCW	3/4/85	227	1034
MONTVALE	N	11Ch12StsDr	H&H	6/72	35	1
MONTVALE	N	12SecDr	Pullman	2/17/06	1963C	3316
MOORHEAD		14SecO	PCW Det	6/28/82	115	D153
MOOSUP	P	34Ch SUV	Pull P-1996C	7/04	3265	3114
MOQUAH		12SecDr	Pullman	8/26/04	1581K	3118
MOQUI		10SecO (NG)	PCW Det	8/24/80	73A	D76
MORAN		12SecDr	Pullman	8/23/04	1963	3098
MORAVIA		12SecDr	Pullman	9/28/01	1581A	2765
MORAY	N	10SecDr	PCW Det	3/29/81	112F	D149
MOREA	N	10SecDr	Elmira Car W	1/73	26A	Elm1
MORELAND	RN	10SecDrB	PCW	6/18/84	149S	192
MORELAND		10SecDr2Cpt	Pullman	6/12/08	2078D	3622
MORELIA		10SecDrB	PCW	10/7/87	445E	1334
MORENA #104	S	12SecDr	PCW	9/10/86	93H	1205

CAR NAME	ST	CAR TYPE	CAR BUILDER (Owner)	DATE BUILT	PLAN	LOT
MORGAN						
CITY M&LT-2	P	Slpr	(M&LT RR)	5/18/83	37?	
MORNING						
STAR	P	OS	B&S	/67	27	
STAR E&A-1	P	Slpr	(E&A Assn)			
MORNINGSIDE		12SecDr	Pullman	12/21/05	1963C	3252
MOROCCO	N	6Sec2DrB	Grand Trunk	10/70	17J	Mon1
MOROCCO		12SecDr	Pullman	12/19/02	1581B	2852
MORPHEUS		12SecDr	Pullman	12/1/08	1963E	3659
MORRISTOWN		16Ch6StsDrB	PCW Det	7/9/87	167F	D283
MOSCA		7Cpt2Dr	Pullman	10/29/02	1845	2926
MOSCOW		10SecDrB	PCW	11/5/83	149	80
MOSCOW	R	16Sec	PCW	11/5/83	1877C	80
MOSELLE	R	12SecDrB	PCW Det	1/20/83	1468	D176
MOSELLE #377		10SecDrB	PCW Det	1/20/83	149	D176
MOTOSA	N	OS	PCW Det	11/13/71	27	D8
MOUND CITY	N	16SecO	PCW	5/88	543K	1427
MOUND CITY	RN	8SecBLibObs	PCW	5/12/88	1707	1377
MOUNDSVILLE #531	P	10Sec	B&O RR	?/87	658	
MOUNT AIRY	N	40Ch6StsDrBObs	Pull P-2495	12/11	3276	3822
MOUNT CLAIR #190	P	10SecDrB	B&O RR	4/30/73	19	B&O1
MOUNT HOOD		10SecDrB	PCW Det	2/11/84	149	D227
MOUNT SHASTA		10SecDrB	PCW Det	2/11/84	149	D227
MOUNT VERNON #19	P	OS	H&H	8/71	27	Wil5
MOUNTAIN						
CITY CTC-22	R		PRR	rblt/71		
MOUNTAINEER #2	P	Slpr	(SouTranCo)			
MOZART		14SecDr	PCW	8/31/88	600	W1
MT. ANTERO		10SecLibObs	Pullman	5/3/09	2351H	3688
MT. CLARE #505	P	12SecO	B&O RR	?/72	657	B&O1
MT. ELBERT		10SecLibObs	Pullman	5/3/09	2351H	3688
MT. HOOD	R	12SecDr	PCW		1603	
MT. MASSIVE		10SecLibObs	Pullman	5/3/09	2351H	3688
MT. MEENAHGA	N	17Ch12StsDr	J&S	9/72	35	Wil2
MT. OLYMPUS	R	12SecDr	PCW		1603	
MT. RAINIER	R	12SecDr	PCW		1603	
MT. SHAVANO		10SecLibObs	Pullman	5/5/09	2351H	3688
MT. ST. HELENS	R	12SecDr	PCW		1603	
MT. VERNON		12SecDr	Pullman	9/7/07	1963C	W19
MT. VERNON #517	P	12SecO	B&O RR		665	
MUIRTON		12SecDr	Pullman	9/14/01	1581A	2756
MULTNOMAH		12SecDr	Pullman	1/28/03	1581C	2946
MUNCIE		12SecDr	Pullman	9/26/03	1581F	3022
MUNICH		10SecDrB	PCW Det	9/8/83	149	D206
MUNICH	R	12SecDrB	PCW Det	9/8/83	1468	D206
MUNSTER		10SecDrB	PCW	11/26/83	149	84
MURAT		12SecDr	PCW Det	8/26/91	784D	D481
MURAT	N	12SecDr	Wagner	12/89	3067	Wg27
MURCIA		12SecDr	PCW	12/7/92	990	1941
MURFREESBORO	P	10SecDr	Bowers, Dure	10/73	37	Wil11
MURIEL		24ChDr	Pullman	6/30/05	1980A	3171

CAR NAME	ST	CAR TYPE	CAR BUILDER (Owner)	DATE BUILT	PLAN	LOT
MURILLO		12SecDr	Pullman	7/31/01	1581A	2727
MUSCAT		12SecDr	Pullman	7/20/03	1581F	3019
MUSCOGEE	R	12SecDrB	PCW	8/23/83	1611	47
MUSIDORA		28ChDr	Pullman	6/25/04	1986A	3087
MUSKEGON		12SecO	PCW Det	11/13/71	27C	D8
MUSKEGON		12SecDr	Pullman	2/19/04	1963	3070
MUSKEGON #103	RN		B&O (WS&PCC)	5/19/82	729A	
MUSKINGUM		12SecDr	Pullman	9/19/07	1963C	W19
MUSKINGUM #298	P	10SecDr	Elmira Car W	6/74	37	Elm2
MUSKODA		12SecDr	Pullman	5/29/02	1581B	2853
MUSKOGEE	R	12SecB	PCW	8/23/83	168	47
MUSKOGEE		10SecDr	PCW	8/23/83	112	47
MUSKOKA	N	10SecDr	Grand Trunk	10/73	31B	Mon4
MUSKOKA		12SecDr	Pullman	4/26/07	1963C	3533
MYHISANA	P	12SecSr SUV	Pull P-2488	10/10	3209	3809
myles STANDISH	N	22Ch6StsDr	PCW	8/30/92	976A	1910
MYOPIA		10SecDr2Cpt	Pullman	10/8/06	2078D	3398
MYRMIDON		12SecDr	Pullman	5/15/08	1963C	3615
MYRTLE		18Ch6StsDrB	PCW Det	6/30/90	772D	D430
MYSTIC		10SecLibObs	Pullman	9/25/08	2351C	3634
MYSTIC		Parlor	PCW Det	8/7/72	35-1/2	D12
NACHUSA		10SecDr2Cpt	Pullman	3/25/07	2078D	3528
NADA		12SecDr	PCW	5/1/91	784D	1805
NADURA		14SecO	PCW Det	1/20/93	1014A	D510
NAGANOOK	P	10SecDr	B&S	1/83	3200A	
NAGOYA		10SecDrB	PCW Det	9/2/84	149D	D248
NAGOYA	R	12SecDrB	PCW Det	9/2/84	1611	D248
NAHANT	P	10SecDrB	(Wagner)	5/82	3046	
NAHCOTTA	RN	12SecDr	PCW Det	3/9/83	1467A	D184
NAHMA		12SecDr	Pullman	1/27/03	1581C	2946
NAIAD	N	10SecDrB	PCW Det	4/1/84	149	D234
NAIAD	P	12SecDrB	Wagner	4/93	3073B	Wg91
NAINARI	N	12SecDr	Pullman	10/8/08	1963E	3649
NAIROBI	N	12SecDr SUV	Pullman	10/15/07	2163A	3562
NAMOUNA	P	12SecO	PCW (Wagner)	5/71	3010	1070
NAMPA		10SecDrSr	PCW	6/24/90	784A	1715
NAMUR		14SecO	PCW Det	1/14/93	1014A	D510
NANIWA		10SecLibObs	Pullman	8/31/08	2351C	3634
NANKIN	P	12SecDr	Wagner	6/93	3065C	Wg85
NANKING		10SecO	Grand Trunk		37F	
NANKING	N	10SecO	Grand Trunk	12/71	27F	Mon2
NANNAR	R	12SecDr			1602B	
NANON	N	12SecO	J&S (McComb)	12/70	18	Wil3
NANON	P	12SecDrB	Wagner	4/93	3073B	Wg91
NANTAHALA	N	12SecDrB	PCW	8/4/88	375E	1395
NANTAHALA	N	12SecDr	Pullman	3/9/06	2163A	3333
NANTASKET		12SecDr SUV	Pullman	1/26/07	1963C	3448
NANTASKET #145	N	6Sec4Dr?	J&S (WS&PCC)	4/8/86	720?	
NANTES		14Sec				
NANTES		10SecDrB	PCW	4/1/87	445C	1281
NANTES	R	16Sec	PCW	4/1/87	1880	1281
NANTES	P	12SecDr	(Wagner)	8/86	3049A	
NANTICOKE		12SecDrSr	Pullman	3/27/05	2049A	3169
NANTUCKET	P	34Ch SUV	Pull P-1484A	6/00	3265	2518
NAOMI		28ChDr	Pullman	6/29/04	1986A	3087
NAPA		10Sec2Dr	Pullman	11/5/02	1746A	2928
NAPA CP-40	P	Slpr	(SP Co)	7/1/83	37?	
NAPATA		10SecDrB	PCW	10/7/87	445E	1334
NAPIER		14SecO	PCW Det	5/13/93	1021	D525
NAPLES		10SecDrB	PCW Det	9/29/83	149	D206
NAPLES		12SecDr	PCW	8/4/98	1318A	2313
NAPLES	P	12SecDr	Wagner	6/93	3065C	Wg85
NAPOLEON		22Ch4StsDr	Pullman	4/6/04	2003	3029
NAPOLEON		12SecDrB	Pullman	2/23/06	2142	3318
NAPOMA	N	10SecDr	H&H	11/71	26 1/4	W6
NAPOMA	N	12SecDr	Pullman	1/5/07	2163	3460
NARADA		12SecDr	Pullman	9/15/09	1963E	3744
NARCISSUS		12SecDrCpt	PCW	8/19/92	983	1919
NARKA		10SecDrB	PCW	1/22/92	920B	1880
NARRAGANSETT	P	34Ch	J&S	/93	3263	
NARRO	N	12SecDrB	Wagner	10/92	3070	Wg76
NASBY 2569		14SecO	PCW Det	3/4/93	1021	D516
NASHLYN	N	12SecDr	Pullman	3/8/06	2163A	3333
NASHOTAH		12SecDrB	PCW	4/28/84	202	195
NASHOTAH	N	12SecDr	Pullman	3/8/06	2163A	3333
NASHUA	P	10SecDr	Vermont Cent	8/71	30	StA1
NASHUA		12SecDr	Pullman	7/24/03	1581F	3015
NASHUA	P	10SecDr	(Wagner)	6/85	3058	
NASHVILLE	P	OS	(PK&RSCC)			
NASHVILLE		12SecDr	Pull Buf	2/25/04	1581F	B34
NASSAU		10SecDrB	PCW	11/6/83	149	80
NASSAU	N	6CptDrLibObs	Pullman	3/13/02	1751B	2799
NASSAU	P	12SecDr	Wagner	7/88	3061B	Wg12
NASSAU #157	P	22ChSr	J&S (WS&PCC)	5/15/87	516?	
NASTURTIUM		24ChDr	Pull Buf	4/9/06	2156	B51
NATAL	N	16SecO	Wagner	4/93	3072	Wg90
NATAL	N	10SecB	(OPLines)	/72	75A	
NATALIE		18Ch6StsDrB	PCW Det	8/22/90	772D	D426
NATCHEZ	RN	10SecB	(OPLines)	/72	75A	
NATCHEZ		12SecDr	Pullman	7/29/03	1581F	3015
NATHALENA #38	P		H&H (WS&PCC)	5/1/78	570?	
NATICK		12SecDrB	Pullman	12/16/07	2326	3587
NATIONAL		Diner	PCW	10/30/89	662D	1641
NATIONAL		BaggClub	Pullman	3/1/06	2136	3319
NATIONAL PARK		12SecDr	PCW Det	3/9/83	93B	D184
NATRONA		26ChDr	Pullman	3/27/06	1987A	3335
NATUNA		12SecDrB	PCW	10/19/88	375F	1481
NAUGATUCK		14SecDrO SUV	Pull	12/12	2673	4047
NAUTILUS	N	10SecB	PFtW&C	1/71	14C	FtW2
NAUTILUS		12SecDr	Pullman	2/9/06	1963C	3316
NAVAJO		10SecO (NG)	PCW Det	9/17/80	73A	D76
NAVAJO	P	12SecDr	(Wagner)	8/86	3049A	

CAR NAME	ST	CAR TYPE	CAR BUILDER (Owner)	DATE BUILT	PLAN	LOT
NAVARETE		10SecDrB	PCW	3/13/88	445E	1361
NAVARIN	N	16SecO	Wagner	7/89	3063	Wg15
NAVARRE		10SecDrB	PCW Det	10/6/83	149	D206
NAVESINK	N	12SecDrSr	Pullman	3/23/05	2049A	3169
NAVITO		14SecO	PCW Det	7/1/93	1041	D519
NEBO CTC-103	R		O&M rblt/69			
NEBRA		10SecO	PCW	8/2/92	948A	1944
NEBRASKA		12SecO	B&S/GMP	/67	5	Day
NEBRASKA	P	BaggSmk	PCW	/99	1400	2410
NEBRASKA	N	7Cpt2Dr	Pullman	10/29/02	1845	2926
NEBRASKA		12SecDr	Pull Buf	11/29/05	1963C	B49
NEBRASKA	P	12SecO	(Wagner)	/67	3044	
NECEDAH		12SecDrSr	PCW	8/7/86	375	1215
NECEDAH		10Sec3Cpt	Pullman	11/4/05	2098	3236
NECHO	N	12SecDrB	Wagner	10/92	3070	Wg76
NECKAR	R	12SecDrB	PCW	11/10/84	1468	1017
NECKAR #431		10SecDrB	PCW	11/10/84	149D	1017
NEDDO		12SecDr	Pullman	3/29/08	1963C	3600
NEEDHAM	P	20Ch6StsDr	Pullman	2/91	3259A	
NEENAH WC- 43	P	12SecO	CentCarCo	2/84	532	
NEENAHA	N	20Ch8StsDrB	Wagner	9/90	3126E	Wg41
NEHALEM	N	14SecO	B&S	4/88	1593	
NELL GWYN	RN	12SecDr	PCW	4/83	1602B	37
NELSON		12SecDr	Pull Buf	2/1/04	1581F	B32
NEMADJI		12SecDr	Pullman	6/13/07	2271A	3458
NEMAUSA		20Ch5StsB	PCW Det	6/27/83	167	D203
NEMEHA		12SecDrB	PCW Det	4/14/88	375E	D328
NEMEHA	N	16Sec	Pullman	3/13/03	1894	2981
NEMESIS		10SecLibObs	Pullman	12/9/08	2351A	3663
NEODESHA		12SecDrB	PCW Det	5/17/88	375E	D328
NEOGA	N	26ChDr	Pullman	7/29/04	1987A	3097
NEOLA	P	12SecDrB	PCW	4/18/84	202	195
NEOLA	N	20Ch8StsDrB	Wagner	12/90	3126	Wg44
NEOSHO	N	12SecDr	Pullman	3/13/06	2163A	3333
NEOSHO CTC-14	R		Det rblt/74			
NEPAWIN		12SecDr	Pullman	5/24/02	1581B	2853
NEPENTHE	N	12SecDrSr	Pullman	3/1/01	1581B	2654
NEPESTA		12SecDrB	PCW Det	5/24/88	375E	D328
NEPONSET	RN	8SecBLibObs	PCW		1707	
NEPONSET	P	22Ch4Sts	(Wagner)	4/86	3123	
NEPTUNE		14Ch18StsBObs	PCW	11/4/98	1279C	2354
NEPTUNE #219	P	11Ch12StsDr	J&S	7/72	35	Wil1
NERCHA	P	14Sec	B&S	8/93	3204	
NERCHA 2517		14SecO	PCW Det	1/20/93	1014A	D510
NERCHO	N	20Ch8StsDrB	Wagner	6/91	3126A	Wg51
NEREID		18Ch6StsDrB	PCW Det	9/5/90	772D	D427
NEREUS	N	28Ch5StsDr	Wagner	9/89	3125B	Wg18
NERISSA	P	18Ch3StsDrObs	(B&M RR)		1203	
NERO		12SecDr	PCW	8/21/90	784	1708
NERO	P	12SecDrB	Wagner	4/93	3073B	Wg91
NESHAMINY		12SecDr	Pullman	10/11/07	2163A	3562
NESTOR	N	10SecO	(IC RR)		308A	
NESTOR	P	12SecDr	Wagner	6/93	3065C	Wg85
NETHERLAND		16Sec	Pullman	6/11/06	2184	3361
NEUSTADT		12SecDr	Pullman	7/22/03	1581F	3019
NEVA		10SecDrB	PCW	10/29/91	920B	1867
NEVA #378		10SecDrB	PCW Det	1/20/83	149	D176
NEVADA	P	12SecDr	Wagner	6/93	3065C	Wg85
NEVADA CP-25	P	10SecDr	(CP Co)	10/2/84	37K	
NEVADA SPC-8157	P	10Sec2Dr	PCW	3/11/99	1158	2376
NEVERSINK	N	11Ch12StsDr	J&S	7/72	35	Wil1
NEVERSINK E&A-19	P	Slpr	(E&A Assn)			
NEVIS		16SecO	PCW	8/30/93	1040	2008
NEW						
BEDFORD	P	16Ch23StsBObs	New Haven RR	7/02	3266	
CASTLE #193	P	12SecDr	PCW P-93H	4/16/86	3049	1192
ENGLAND		12SecB	PCW Det	7/14/76	57	D41
ENGLAND	P	40Ch6StsDrBObs	Pull P-2495	2/11	3276	3822
FOUNDLAND #194	P	12SecDr	PCW P-93H	4/16/86	3049	1192
HAMPSHIRE		12SecB	PCW Det	6/29/76	57	D41
HAMPSHIRE	P	40Ch6StsDrBObs*	Pull P-2495	2/11	3276	3822
HARTFORD	P	Bagg24Ch SUV	Pull P-2267	4/07	3278	3494
JERSEY	N	30ChObs	PCW	4/28/98	1278B	2299
MEXICO		10SecDr	PCW Det	7/5/81	112	D118
MEXICO		12SecDr	Pull Buf	12/5/05	1963C	B49
MEXICO #196	S	12SecDr	PCW	4/5/86	93H	1192
ORLEANS		14SecDr	Pull Buf	10/30/00	1582A	B3
ORLEANS #100	P	Slpr	(IC RR)			
ORLEANS M&LT-1	P	12SecO?	(M&LT RR)	3/21/85	114?	
PRESTON	P	BaggSm	B&S	2/93	3213A	
YORK	N	OS	CB&Q RR	/66	13	Aur
YORK		12SecDr	PCW	8/1/98	1318A	2313
YORK		27ChDrObs	Pullman	6/8/08	2346	3624
YORK			CNJ (SilverPC)	/66		
YORK		Private Parlor?	(WS&PCC)	/79		
YORK #66	P	25Ch2StsB	(Wagner)	11/87	3100A	
YORK #83	P	12SecDr	H&H	/80-/82		
ZEALAND		12SecDrCpt	PCW	4/1/89	622	1529
ZEALAND #195	P	12SecDr	PCW	4/1/86	3049	1192
NEWARK	P	10SecO Tourist	(DL&W RR)	5/79	220	
NEWARK	N	30Ch	PCW	4/27/98	1278A	2293
NEWARK	P	10SecO Tourist	(Wagner)		3037A	
NEWARK #177	P	10SecDr	PRR	9/71	21	Alt2
NEWAUKUM		14SecDr	Pullman	5/19/06	2165	3352
NEWBURG		10SecDrB	PCW Det	5/5/83	149	D194
NEWBURGH		12SecDr	Pullman	10/29/03	1581F	3025
NEWBURGH #90	P	12SecDr	H&H	/80-/82		
NEWCASTLE		26Ch6StsDr	PCW		1061	
NEWELL	N	12SecDr	Pullman	7/22/03	1581F	3019
NEWLAND		14SecDr	Pullman	8/1/00	1523E	2567
NEWPORT	R	Private	PCW Det	8/2/71	29K	D7
NEWPORT		12SecDr	Pullman	3/28/02	1581B	2821

CAR NAME	ST	CAR TYPE	CAR BUILDER (Owner)	DATE BUILT	PLAN	LOT
NEWPORT #86	P		B&S (WS&PCC)	3/1/81		
NEWPORT #214		Parlor	PCW Det	8/2/71	22	D7
NEWPORT E&A-38	P	Parlor	(E&A Assn)			
NEWPORT NEWS		10SecDr	PCW Det	7/19/82	112A	D152
NEWSTEAD		12SecDr	Pullman	9/28/03	1581F	3022
NEWTON		10SecDr	PCW Det	9/13/76	59	D43
NEWTON		12SecDr	Pull Buf	7/2/03	1581C	B19
NIAGARA		OS	GW Ry/GMP	C:6/67	7	Ham
NIAGARA	R	16Ch3StsDr	(E&A Assn)		539	
NIAGARA		12SecDr	Pullman	2/3/03	1581D	2962
NIAGARA #150	?	10SecSrB	J&S (WS&PCC)	6/22/86	503?	
NIAGARA E&A-37	P	Parlor	(E&A Assn)			
NIANTIC		12SecDr	Pullman	5/25/06	1963C	3360
NICARAGUA		7Cpt2Dr	PCW	12/3/94	1065A	2025
NICARAGUA #295	P	12SecDr	Wagner	12/89	3067	Wg27
NICARIA		12SecDr	Pull Buf	10/3/03	1581F	B23
NICE		12SecDr	PCW	2/10/92	784E	1881
NICKERSON		10SecDr2Cpt	Pullman	11/20/07	2078D	3583
NICOSIA	RN	16SecB	Pull Buf	12/15/02	1794	B
NIGHTINGALE	N	16SecB	PCW	6/ /93	1040J	1990
NILAND	N	12SecDrCpt	Pullman	5/26/04	1985	3095
NILE		12SecDr	PCW Det	4/18/91	784D	D461
NILES		12SecDr	Pullman	7/21/03	1581F	3019 3008B
NILUS	N	Parlor	PRR	7/6/72	29J	Alt1
NILUS		10SecDrB	PCW	4/29/85	240	1051
NIMBUS	N	12SecDr	Pullman	6/8/03	1581E	3010
NIMBUS	P	18Ch6StsB	Wason	7/88	3256	
NIMROD	RN	Private	PRR	7/6/72	29J	Alt1
NIMROD	P	14Sec	B&S	7/93	3204	
NINA	RN	21Ch1Sts	PCW	4/25/93	1251B	1985
NINA	P	12SecDrB	Wagner	4/93	3073B	Wg91
NINA #45	N	Sleeper	H&H (WS&PCC)	7/25/78	524?	
NINDARA	N	12SecDrB	Pullman	2/26/01	1611A	2614
NINEVEH		10SecDrB	PCW	6/26/85	240	1057
NINEVEH		6CptLngObs	Pullman	12/16/03	1299D	3065
NINGPO	P	12SecDr	Wagner	12/89	3065C	Wg85
NINITA	N	Slpr	(NYLE&W Assn)			
NINITA	N	20Ch8StsDrB	Wagner	9/90	3126E	Wg41
NIOBE		20Ch5StsB	PCW	6/17/84	167A	104
NIOBE	P	12SecDrB	Wagner	4/93	3073B	Wg91
NIOBRARA	P	12SecDr	H&H	/80-/82		
NIOBRARA		12SecDr	PCW Det	4/26/82	93B	D142
NIOTA	N	12SecDr	PCW Det	7/5/81	112	D118
NIOTAZE	N	24ChDrB	Pullman	6/9/05	1988A	3170
NIPHATES		14SecO	PCW	5/22/93	1021	1972
NIPHON #125	RN	7BoudB	(MannBCC)	/84	776A	
NIPIGON #294	P	12SecDr	Wagner	12/89	3067	Wg27
NIPPENO		10SecDr2Cpt	Pullman	3/25/07	2078D	3528
NIPPON		12SecDr	Pullman	3/20/03	1581D	2975
NIPPON	R	30Ch SUV	Pullman	3/20/03	3581E	2975
NIRVANA		12SecDr	Pullman	9/16/09	1963E	3744
NISQUALLY	RN	12SecDr	PCW	9/19/82	1467A	22
NOANK	P	34Ch	J&S	/93	3263	
NOCTURNE		10SecDrB	PCW Det	9/11/84	149D	D248
NOCTURNE	N	12SecDr	Pull Buf	6/20/07	1963C	B61
NODAWAY		16Sec	Pullman	3/27/07	2184	3523
NOGALES		12SecDr	Pullman	7/27/06	1963C	3385
NOKOMIS	RN	12SecDr	PCW Det	6/28/82	1581C	D153
NOKOMIS #92	P	26ChB	PCW (Wagner)	5/29/86	3124A	1194
NOLYNN	N	12SecO?	(L&N RR)	/69	114?	
NOMAD		10SecLibObs	Pullman	8/31/08	2351C	3634
NOME	N	12SecDrCpt	Pullman	5/24/04	1985	3095
NONPAREIL		12SecDr	Pullman	2/9/06	1963C	3316
NOORNA	N	12SecDr	PCW	8/10/98	1318A	2313
NOPALA		12SecDr	Pullman	9/20/06	1963C	3397
NOQUEBAY		10SecDrB	PCW	3/11/88	445F	1380
NOQUEBAY	N	26ChDr	Pullman	5/28/07	1987B	3524
NORBERT	P	24Ch6Sts	B&S	12/93	3262	
NORCROSS		10SecDr2Cpt	Pullman	6/13/08	2078D	3622
NORFOLK	R	12SecDr	PCW	3/15/85	1603	1037
NORFOLK #436		10SecDrB	PCW	3/15/85	149E	1037
NORMA		12SecDr	Pullman	3/17/08	1963C	3604
NORMA #107	P	7BoudB	(MannBCC)	/84	697	
NORMAHAL		12SecDr	Pullman	8/5/08	1963C	3636
NORMAL		DDR	PCW Det	/70	17	D3
NORMAN	N	SDR	(OPLines)			
NORMAN		10SecDr	PCW Det	3/15/73	26 3/4	D15
NORMAN		22Berths	PCW Det	6/2/76	48	D26
NORMAN		12SecDr	Pullman	5/22/03	1581D	3004
NORMANDIE		10SecDrB	PCW	9/17/87	445E	1362
NORMANDY		12SecDr	Pull Buf	8/31/03	1581F	B22
NORMANDY	R	30Ch SUV	Pull Buf	8/31/03	3581E	B22
NORMANDY #276	P	7CptObs	Wagner	4/89	3062A	Wg14
NORMANDY #552	P	14SecSr	H&H (B&O)	8/87	654	
NORMANIA		16Sec	Pullman	10/28/01	1721	2766
NOROTON	P	36Ch SUV	Pull P-2494	3/11	3275	3821
NORSEMAN		10SecLibObs	Pullman	12/9/08	2351A	3663
NORTH						
ADAMS	P	Bagg24Ch SUV	Pull P-2267	4/07	3278	3494
AMERICA #239		Parlor	PCW Det	11/22/71	34	D4
CREEK #45	P	16Ch4Sts	(Wagner)	7/85	3120	
DAKOTA		7Cpt2Dr	Pullman	10/29/02	1845	2926
PLATTE		10Sec2Dr	PCW	5/11/93	995A	1942
SHORE	P	27Ch8Sts2DrBObs	Pull P-1997	7/04	3270	3116
SHORE #55	P	20Ch6Sts	(Wagner)	5/85	3116	
STAR	P	12SecO	CB&Q RR	/68	5	Aur
YAKIMA	RN	14Sec	PCW	5/15/90	668F	1709
NORTH-WEST	P	12SecDr	B&S/GMP	/68	12	Day
NORTHBORO	P	34Ch SUV	Pull P-1484	1/03	3265	2904
NORTHBRIDGE	P	32ChDr SUV	Pull P-2494	3/11	3277	3821
NORTHERN		12SecDrB	Pullman	12/18/07	2326	3587

CAR NAME	ST	CAR TYPE	CAR BUILDER (Owner)	DATE BUILT	PLAN	LOT
NORTHERN CROWN #52	P	22Ch4Sts	(Wagner)	10/70	3121	
NORTHFORD	P	20Ch6StsDr	B&S	1/94	3262A	
NORTHLAND		12SecDr	Pullman	1/24/07	1963C	3448
NORTHMOOR		10SecLibObs	Pullman	12/1/09	2351A	3748
NORTHUMBER- LAND #293	P	12SecDr	Wagner	12/89	3067	Wg27
NORTHUP	N	12SecDr	Pullman	2/8/06	1963C	3316
NORTHWEST		16Sec	Pullman	2/15/07	2184	3508
NORWALK		12SecDr	Pullman	9/29/03	1581F	3022
NORWAY	R	12SecDr	PCW	3/14/91	896B	W5
NORWAY		16SecO	PCW	3/14/91	780	W5
NORWICH		12SecO	PCW Det	6/8/80	58	D74
NORWOOD	R	12SecB	PCW		168	
NORWOOD #310	P	10SecB	B&S (B&O)	7/31/75	37	Day
NOTTINGHAM		10SecDr2Cpt	Pullman	1/9/08	2078D	3594
NOVALIS	N	16SecB	Pull Buf	12/20/02	1794	B
NOVALIS #17	N	12SecO	(WS&PCC)	6/30/73	833	
NOVARA		16SecO	PCW	8/29/93	1040	2008
NOVARA	P	12SecDr	(Wagner)	8/86	3049A	
NOVELLA		28ChDr	Pullman	6/27/04	1986A	3087
NUBE BLANCA	RN	12SecDrB	Pullman	5/16/06	2165C	3352
NUBIA	N	12SecO	Grand Trunk	7/71	18C	Mon1
NUBIA		12SecDr	Pullman	2/24/03	1581D	2975
NUBIAN	P	10Sec2DrB	PCW (Wagner)	2/15/00	3079A	Wg134
NUECES	R	12SecDrB	PCW Det	3/22/82	1468	D141
NUEVITAS		12SecDr	Pullman	8/23/04	1963	3098
NUMA		10SecDr	PCW	1/10/90	717B	1630
NUMANTIA		16Sec	Pullman	10/21/01	1720A	2766
NUMIDIA		12SecDr	PCW Det	6/21/82	93B	D154
NUMIDIA		12SecDr	Pullman	10/2/01	1581A	2765
NUREMBURG #292	P	12SecDr	Wagner	12/89	3067	Wg27
NUSHKA		12SecDr	Pullman	8/27/04	1581K	3118
NYANZA		10SecDrB	PCW	2-4/83	149	27
NYANZA	N	12SecDr	Pullman	3/10/06	2163A	3333
NYANZA	P	12SecDr	Wagner	6/93	3065C	Wg85
NYDIA	N	26ChDr	Pullman	3/29/06	1987A	3335
NYDIA	N	17Ch2StsB	(Wagner)	8/85	3122	
NYDIA #45	P	Parlor R-End	H&H (WS&PCC)	7/25/78	524?	
NYMPHE		16Sec	Pullman	7/1/08	2184C	3627
NYSA		14SecO	PCW Det	8/3/92	1014A	D509
NYSSIA		20Ch6StsDr	PCW	7/27/91	928A	1827
OAK BLUFFS	P	BaggSmkB SUV	H&H	/89	3219	
OAK LAWN	P	BaggSmk	Pull	2/91	3215A	
OAK SUMMIT	P	BaggSmk	B&S	8/92	3213A	
OAKDALE		12SecDr	Pull Buf	12/21/03	1581F	B26
OAKDALE #160	N	20Ch4StsB	J&S (WS&PCC)	6/27/87	711	
OAKLAND		14SecDr	Pullman	8/3/00	1523E	2567
OAKLAND #154		10SecDr	J&S (B&O)	4/30/73	19	Wil1
OAKLAND #511	P	12SecO	B&O RR	11/80	657	
OAKLAND CP-41	P	10Sec?	(CP Co)	4/29/84	37?	
OAKLEY		12SecDr	Pullman	10/30/03	1581F	3025
OAKLYN (#166)	RN	37Ch SUV	Pullman	11/5/00	3581P	2589
OAKMONT		12SecDrB	PCW	9/12/98	1319A	2314
OAKRIDGE		12SecDr	Pullman	1/28/07	1963C	3448
OAKTON		12SecDr	Pullman	11/4/01	1581A	2784
OAKVILLE	N	12SecDrCpt	Pullman	5/24/04	1985	3095
OAKWOOD		14SecDr	Pull Buf	11/7/00	1582A	B3
OAKWOOD #321	P	8BoudB	G&B (UnionPCC)	/67	373A	
OBEDA	N	12SecDr	Pullman	3/8/06	2163A	3333
OBERLIN		12SecDr	Pull Buf	4/10/03	1581C	B15
OBERON		12SecDr	PCW Det	1/20/81	93	D88
OBERON #503	P	12SecDrB	Wagner	11/92	3073	Wg65
OBISPO		10Sec2Dr	Pullman	8/30/02	1746A	2925
OBRA		14SecO	PCW Det	2/3/93	1014A	D511
OCALA	N	10SecDrB	PCW	1/7/88	445F	1360
OCAMO		10SecO	PCW	8/1/92	948A	1944
OCCIA		30Ch	PCW	4/28/98	1278A	2293
OCCIDENT		8Sec4Dr	PCW	10/14/96	1210A	2202
OCCIDENTAL	P	12SecDr	(Wagner)	12/82	3023A	
OCCIDENTAL		10SecDrHotel	PCW Det	8/2/72	33	D11
OCEAN	S	22Berths	PCW	/75		
OCEAN #245	P	SummerParlor	H&H	7/73	40	Wil2
OCEAN GROVE #172	P	18Ch6StsDr	(WS&PCC)	6/23/88	730A	
OCEAN QUEEN E&A-31	P	Parlor	(E&A Assn)	/72		Elm0
OCEANIA #200	S	12SecDr	PCW	6/5/86	93H	1199
OCEANIC	RN	Private	PCW Det	6/12/77	63R	D45
OCEANICA		10SecDrHotel	PCW Det	5/28/72	33	D11
OCEANUS	N	20Ch4StsDr	(Wagner)	6/95	3109A	
OCEANUS	N	12SecDrB	Wagner	4/90	3065	Wg33
OCELLINA		30Ch	PCW	4/27/98	1278A	2293
OCHRA		10SecDrB	PCW	12/7/86	445C	1261
OCHRA	R	14Sec	PCW	12/7/86	445X	1261
OCMULGEE		10SecDr	PCW Det	3/22/82	112A	D141
OCONEE	P	10SecO	G&B (Pull/Sou)	/67	47A	Troy
OCONEE	R	8BoudB?	G&B (UnionPCC)		373	
OCONO		14SecDr	Pullman	8/31/09	2438	3716
OCONOMOWOC #69	P	12SecDr	H&H	/80-/82		
OCOSTA		10SecDrSr	PCW	8/20/91	937A	1838
OCRESIA		30Ch	PCW	4/28/98	1278A	2293
OCTAVIA		20Ch6StsB	PCW	1/22/85	167B	200
OCTAVIA		18Ch4StsDrB	PCW	7/14/93	1039A	1995
OCTAVIUS		12SecDr	Pull Buf	4/21/10	1963E	B83
OCTORARO		24ChDrB	Pullman	6/10/05	1988A	3170
ODAHMIN		12SecDr	Pullman	1/29/03	1581C	2946
ODANAH	P	10SecDr	(Wagner)	6/85	3048	
ODESSA		12SecDr	PCW	11/7/82	93B	22
ODESSA		12SecDr	Pull Buf	6/20/03	1581C	B18
ODETTE	N	12SecDr	Wagner	6/90	3065J	Wg34
ODIN #251	P	10SecDr	CochranCCo	5/72	26A	
OELWEIN		16SecSr	Pullman	8/30/04	2033A	3135

Figure I-30 OCEAN GROVE, a Woodruff parlor car built in 1888, was sold to Pullman's Palace Car Company in the following year and retained its original name throughout its life. The Robert Wayner Collection.

CAR NAME	ST	CAR TYPE	CAR BUILDER (Owner)	DATE BUILT	PLAN	LOT
OELWEIN #314	P	8BoudB	J&S (MannBCC)	/89	373A	Troy
OENONE		30Ch	PCW	4/27/98	1278A	2293
OGALALLA	RN	12SecDr	PCW	9/30/82	1467A	22
OGALALLA		12SecDr	PCW Det	4/19/82	93B	D142
OGDEN		10SecB (NG)	PCW Det	10/16/83	178	D211
OGDEN		12SecDr	Pullman	7/28/03	1581F	3015
OGDEN #27		10SecDr	H&StJ RR	/71	19	
OGDENSBURG	P	10SecDr	Vermont Cent	2/72	31	StA2
OGDENSBURG		16Sec	Pullman	2/12/07	2184	3508
OGEECHEE	R	10SecO	PCW Det (rblt)	1/20/75	47A	D27
OGEMA WC-106	N	12SecO	CentCarCo	2/84	532	
OGENAW	P	12SecDr	(Wagner)	/75	3027G	
OGILVIE		12SecDr	Pull Cal	6/18/03	1581D	C5
OGONTZ		12SecDr	PCW Det	1/9/91	784	D455
OGYGIA		30Ch	PCW	4/27/98	1278A	2293
OHIO		27Sts Parlor	PCW Det	/75		
OHIO		10Cpt	Pullman	/02	1704	2758
OHIO CTC-79	R		CC&ICC	rblt/70		
OHIO #2 #125		12SecO	PFtW&C/GMP	/69	5	FtW1
OHIO No.1	P	12SecO	E,G/GMP	/67	5	Troy
OIL CITY		16Sec	Pullman	7/14/03	1721	3014
OIL CITY E&A-10	P	Slpr	(E&A Assn)			
OILDOM	RN	6SecCafeLng SUV	Pull	6/1/09	3986	3700
OJIBWA		12SecDr	Pullman	5/15/02	1581B	2853
OKANAGON		12SecDr	PCW Det	6/6/84	93D	D241

CAR NAME	ST	CAR TYPE	CAR BUILDER (Owner)	DATE BUILT	PLAN	LOT
OKLAHOMA		12SecDr	Pull Buf	2/18/03	1581C	B13
OKLAHOMA	N	7Cpt2Dr	Pullman	10/29/02	1845	2926
OKONOKO #526	P	12SecO	B&O RR	/80	665	
OLAUS	N	12SecDrB	Wagner	1/93	3073B	Wg82
OLDENBURG #116	P	20Ch8StsDrB	Wagner	12/90	3126	Wg44
OLDFIELD	N	12SecDr	Pullman	1/28/02	1581A	2792
OLDHAM		14SecDr	PCW Det	1/14/93	1014A	D510
OLEAN		14SecO	PCW Det	8/3/92	1014A	D509
OLEANDER		24ChDrSr	Pullman	7/23/04	1980A	3088
OLEQUA		12SecDr	Pullman	6/12/07	2271A	3458
OLESA		14SecO	PCW	1/21/93	1017C	1962
OLGA		12SecDr	PCW	8/21/90	784	1708
OLGA #90	P	28Ch3Sts	PCW (Wagner)	4/24/86	3124	1194
OLINDO #135	P	10SecDrB	(WS&PCC)	1/15/85	886C	
OLIVE	P	26Ch3Sts	PCW	9/95	1153	2117
OLIVE BRANCH #13	P	Slpr	(SouTranCo)			
OLIVETTE	RN	Private	PCW Det	8/12/80	63K	D77
OLIVETTE	RN	6SecB36Sts	B&S	/71	17S	Day4
OLIVIA		20Ch6StsB	PCW	9/10/84	167B	200
OLPI		12SecDrB	PCW	8/25/88	375F	1480
OLYMPIA		10SecDrCpt	PCW	5/16/90	668A	1709
OLYMPIA		Private	PCW	10/21/99	1418C	2451
OLYMPIA #504	P	12SecDrB	Wagner	11/92	3073	Wg65
OLYMPIC #147	N	8SecBObs	J&S (WS&PCC)	5/17/86	720E	
OLYMPUS	N	OS	GW Ry/GMP	C:6/67	7	Ham

CAR NAME	ST	CAR TYPE	CAR BUILDER (Owner)	DATE BUILT	PLAN	LOT
OMAHA	P	14SecO	C&NW Ry/GMP	/66	20	Gal
OMAHA	S	14SecDr	Pullman	/00	1515B	2544
OMAHA		12SecDr	Pullman	5/1/02	1581B	2849
OMAHA	P	10SecO	(Wagner)	/66	3003	
OMAHA #63	P	12SecDr	(CRI&P Assn)	3/6/82	527	
OMAR		10SecDr	PCW	1/6/90	717B	1630
OMENA		7Cpt2Dr	PCW	9/14/94	1065A	2025
OMPHALE		30Ch	PCW	4/28/98	1278A	2293
ONALASKA #298	P	12SecDr	Wagner	12/89	3067	Wg27
ONECO		14SecO	PCW	3/13/93	1021	1969
ONEGA	N	10SecDr	Grand Trunk	3/71	21C	Mon1
ONEIDA	P	DDR	B&S	3/70	17	Day2
ONEIDA	N	16SecO	Wagner	4/93	3072	Wg90
ONEONTA		12SecDr	PCW Det	2/28/83	93B	D184
ONEONTA	R	12SecDrB	PCW Det	2/28/83	1468	D184
ONEOTA	P	14SecO	B&S	10/86	1593	
ONIAS	N	12SecDrB	Wagner	1/93	3073B	Wg82
ONONDAGA #297	P	12SecDr	Wagner	12/89	3067	Wg27
ONOTO		10SecLibObs	Pullman	12/17/09	2351A	3748
ONSET	P	34Ch	J&S	/93	3263	
ONTARIO	P	12SecO	B&S/GMP	/68	5	Day
ONTARIO	RN	12SecDr	B&S	7/71	17E	Day4
ONTARIO E&A-12	P	Slpr	(E&A Assn)			
ONTEORA		10SecDrB	PCW	5/28/89	623B	1567
ONWARD	N	12SecDrObs	Pullman	8/28/06	2194	3395
ONWENTSIA		12SecDr	Pullman	12/15/03	1581F	3043
ONYTES	N	12SecDrB	PCW Det	4/7/88	375G	D345
ONYX		12SecDr	PCW Det	4/2/92	784E	D497
OPACHEE		12SecDr	Pullman	1/23/03	1581C	2946
OPAL		12SecDrSr	PCW	4/4/99	1318C	2390
OPAL	N	12SecDr	Pullman	6/22/01	1581A	2726
OPAL	R	30Ch SUV	Pullman	6/22/01	3581E	2726
OPECHEE	N	12SecDr	Pullman	1/23/03	1581C	2946
OPHELIA		20Ch5StsB	PCW Det	8/15/83	167	D210
OPHIR		12SecDr	PCW Det	4/11/92	784E	D497
OPHIR #505	P	12SecDrB	Wagner	11/92	3073	Wg65
OPHITES	N	12SecDrB	PCW Det	9/8/88	375G	D346
OPORTO		10SecDrB	PCW	2/11/84	149B	101
OPORTO	N	16SecO	Wagner	10/92	3072	Wg79
OPORTO #197	S	14SecO?	PCW	5/29/86	93H	1199
OPPIUS	N	12SecDrB	PCW Det	9/1/88	375G	D346
ORANGE		18Ch6StsDrB	PCW Det	7/18/90	772D	D436
ORANGE						
COUNTY E&A-34	N	Parlor	(E&A Assn)			
ORAVIA		10SecLibObs	Pullman	12/10/09	2351A	3748
ORCHID		24ChDr	Pull Buf	4/11/06	2156	B51
ORCHIS		12SecDr	PCW	4/14/91	784C	1798
ORCOMA		16Sec	Pullman	5/27/09	2184D	3700
OREB		14SecO	PCW	5/25/93	1021	1972
OREBRO		10SecDrB	PCW Det	8/21/84	149D	D248
OREGON		10SecO	PCW Det	2/10/81	102	D91
OREGON	R	12SecDrB	PCW Det	2/10/81	1468	D91
OREGON	N	7Cpt2Dr	Pullman	10/29/02	1845	2926
ORESTES		12SecO	PCW Det	2/21/82	114	D145
ORESTES	N	12SecDrSr	Pullman	4/6/05	2049A	3169
OREVAL		10SecDrB	PCW	11/4/91	920B	1866
ORIANA		20Ch5StsB	PCW	6/17/84	167A	104
ORIENT	P	DDR	DetC&MCo	/70	17	D3
ORIENT		8Sec4Dr	PCW	10/14/96	1210A	2202
ORIENT #91	P		(WS&PCC)	7/9/81		
ORIENTAL		12SecDr	Pullman	1/29/02	1581A	2792
ORIENTAL #154	P	22ChSr	J&S (WS&PCC)	5/12/87	516?	
ORINDA		26ChDr	Pullman	7/29/04	1987A	3097
ORINOCO	P	12SecDr	PCW P-93H	6/11/86	3049	1199
ORINOCO #393		12SecB	PCW Det	7/31/83	168	D193
ORIOLE		10SecO	PCW Det	10/9/82	102	D170
ORIOLE		10SecDrSr	PCW	7/2/92	937B	1925
ORIOLE	N	26ChDrB	Pullman	6/28/07	1988B	3526
ORION		12SecDr	PCW Det	12/1/80	93	D88
ORION	P	26ChB	Wagner	11/92	3125	Wg74
ORISKANY		12SecDr	Pullman	2/28/07	1963C	3525
ORISON #441	P	12SecDrB	Wagner	12/90	3065	Wg43
ORISSA		12SecDrB	PCW	11/20/88	375G	1513
ORIZABA		10SecDrB	PCW	12/12/86	445C	1261
ORIZABA	R	16Sec	PCW	12/12/86	1880	1261
ORIZABA #440	P	12SecDrB	Wagner	12/90	3065	Wg43
ORKNEY	N	12SecDrB	Wagner	1/93	3073I	Wg83
ORKNEY #435		10SecDrB	PCW	11/8/84	149D	1017
ORLANDO		10SecDrB	PCW	8/4/87	445E	1333
ORLANDO		12SecDr	Pullman	5/14/08	1963C	3615
ORLANDO #506	P	12SecDrB	Wagner	11/92	3073	Wg65
ORLEANS	P	DDR	B&S	3/17/70	17	Day2
ORLEANS		12SecDr	PCW	11/26/92	990	1941
ORLINE	P	31Ch6Sts SU	(CI&L RR)	/12	2748	
ORMONDE	R	16Ch3StsDr	(E&A Assn)		539	
ORMONDE		12SecDr	Pull Wilm	2/18/03	1581C	W8
ORMONDE #507	P	12SecDrB	Wagner	10/92	3073	Wg65
ORMUS		10SecDrB	PCW	2/8/84	149B	101
OROFINO		10Sec2Dr	Pullman	9/2/02	1746A	2925
ORONO		10SecO	PCW	8/1/92	948A	1944
ORONSO		10SecLibObs	Pullman	12/11/09	2351A	3748
OROVILLE	N	5CptLngObs	PCW	1/11/98	1299C	2283
OROZEMBO	N	14SecDr	PCW	8/7/99	1318D	2447
ORPHEUS	N	20Ch3StsB	H&H (B&O)		653	
ORPHEUS	P	28Ch5Sts	Wagner	11/92	3125	Wg74
ORRVILLE		12SecDr	Pullman	2/7/08	2163A	3599
ORSINO		12Sec2Dr	Pullman	4/3/01	1580A	2646
ORTA		12SecDr	PCW Det	11/26/90	784	D449
ORTEGA SP-69		10SecDrB	PCW	6/23/88	445F	1415
ORTONA		10SecLibObs	Pullman	12/18/09	2351A	3748
ORVA		10SecDrB	PCW	12/10/89	725B	1629
OSACA	N	10SecDr	PCW Det	7/23/81	112	D118

CAR NAME	ST	CAR TYPE	CAR BUILDER (Owner)	DATE BUILT	PLAN	LOT
OSAGE		12SecDrB	PCW Det	4/7/88	375E	D328
OSAGE	P	12SecDr	(Wagner)	11/80	3022F	
OSBORNE	N	Diner	PCW	12/27/89	662D	1655
OSCAR J. DANIELS	N	16Sec	Pullman	6/3/09	2184D	3700
OSCAWANA #296	P	12SecDr	Wagner	12/89	3067	Wg27
OSCAWANA #296	R	12SecDr	Wagner	12/89	3098	Wg27
OSCEOLA		10SecDrB	PCW Wilm	2/12/89	149Y	W4
OSCEOLA	N	6SecDr2Sr	PCW	5/24/88	542A	1404
OSCOLA	P	12SecDr	(Wagner)	/75	3008	
OSCURA		7Cpt2Dr	Pullman	10/29/02	1845	2926
OSHKOSH	P	OS	C&NW Ry	/65	27	
OSHKOSH WC-105	R	10SecDrCafe	CentCarCo	2/82	532B	
OSHKOSH WC-105	P	12SecO	CentCarCo	2/82	532	
OSIRIS		12SecDr	PCW	1/20/91	784	1786
OSIRIS #120	P	20Ch8StsDrB	Wagner	6/91	3126A	Wg51
OSKALOOSA #6	N		(WS&PCC)	7/16/72		
OSPREY		24ChDrB	Pullman	6/2/00	1518C	2552
OSSEO		12SecDr	Pullman	6/14/07	2271A	3458
OSSIAN		12secO	PCW Det	4/3/82	114	D143
OSSIAN	P	26ChB	Wagner	11/92	3125	Wg74
OSSIPEE		26Ch6StsDr	PCW			1061
OSSIPEE #198	S	12SecDr	PCW (Wagner)	6/4/86	93H	1199
OSTEND		12SecDr	Pullman	12/19/02	1581B	2852
OSWEGATCHIE	P	12SecDr	(Wagner)	10/82	3023A	
OSWEGO	N	10SecDrB	PCW Det	5/3/83	149N	D194
OSWEGO		16Sec	Pullman	2/12/07	2184	3508
OSWEGO	N	12SecDrB	Wagner	1/92	3073B	Wg83
OSYKA #37	P	Slpr	(IC RR)			2/273
OTEO	RN	12SecDr	PCW	10/10/82	1467A	22
OTERO		10SecDr	PCW Det	10/24/79	59	D62
OTERO	P	28Ch5Sts	Wagner	11/92	3125	Wg74
OTHELLO		7Cpt2Dr	PCW	1/4/96	1065A	2140
OTHELLO #91	N		(WS&PCC)	7/9/81		
OTHMAN	N	12SecDrB	Wagner	10/92	3070	Wg76
OTHO	N	12SecDrB	Wagner	10/92	3070	Wg76
OTRANTO	R	10SecDrB	PCW	4/18/85	240	1034
OTRANTO #440		10SecDrB	PCW	4/18/85	149E	1037
OTSEGO		10SecDr	H&H	2/71	19	Wil2
OTSEGO		12SecDr	Pull Buf	7/2/04	1963	B38
OTSEGO #199	S	12SecDr	PCW (Wagner)	5/29/86	93H	1199
OTTAWA	P	DDR	Grand Trunk	10/31/70	17	Mon1
OTTAWA	P	12SecDr	(Wagner)	10/82	3023A	
OTTAWA #71	P		(CRI&P Assn)	5/5/80		
OTTAWA #78	N		(WS&PCC)	2/17/79		
OTTOMAN		12SecDr	Pullman	1/28/02	1581A	2792
OTTUMWA	N	12SecO	DetC&MCo	/71	5	D2
OTTUMWA		12SecDr	PCW Det	5/19/81	93A	D107
OTTUMWA	RN	8SecBLibObs	PCW			1707
OUITA	RN	12SecDr	DetC&MCo	3/70	17E	D3
OURAY		10SecO (NG)	PCW Det	4/11/83	73A	D186
OURAY	N	10SecB (NG)	PCW Det	10/16/83	178	D211
OURAY	P	12SecDr	(Wagner)	11/80	3022F	
OUTING	N	12SecDr	Pullman	12/29/06	2163	3460
OVERBROOK		12SecDrSr	Pullman	3/1/01	1581B	2654
OVERLAND		12SecDr	Pullman	12/31/03	1581H	3053
OVERTON		10SecDr2Cpt	Pullman	12/27/05	2078C	3287
OVID		12SecDr	PCW	8/22/90	784	1708
OVID	N	12SecDrB	Wagner	1/92	3073B	Wg83
OVIEDO	R	16Sec	PCW Det	11/10/83	1877	D207
OVIEDO #415	P	10SecDrB	PCW (Union PCC)	11/10/83	149	D207
OWASCO SP-2547		10SecO	PCW Wilm	11/92	1024	W7
OWATONNA	P	12SecDrB	PCW	4/28/84	202	195
OWEGO	P		(DL&W RR)			
OWEGO		10SecDrB	PCW Det	4/20/87	445C	D282
OWEGO	R	16Sec	PCW Det	4/20/87	1880	D282
OWOSSO	N	OS	G&B/D&MSCC	2/68		Troy
OXALIS		24ChDr	Pull Buf	6/9/06	2156	B53
OXFORD	P	10SecDr	B&S		26	Day7
OXFORD	P	10SecDr	J&S (McComb)	/72	26H	Wil
OXFORD		12SecDr	PCW Det	12/24/90	784	D455
OXFORD	P	10SecO	(Wagner)	/72	3001B	
OXMOOR	P	14SecO	L&N RR (OPL)	/78	214	
OXUS		12SecDr	PCW Det	5/2/91	784D	D461
OZARK		10SecDrB	PCW Det	1/12/84	149	D223
OZARK		12SecDr	Pull Wilm	2/19/03	1581C	W8
OZARK	RN	6SecCafeLng SUV	Pull	5/22/08	3986	3623
P.P.C.		Private	PCW Det	6/18/77	62	D44
PABLO	N	27ChDrObs	Pullman	6/8/08	2346	3624
PACHUCA	N	12SecDrB	PCW Det	2/10/81	1468	D91
PACIFIC	P	OS	CB&Q/GMP	/66	7C	Aur
PACIFIC	S	Diner	PCW	12/1/90	662D	1778
PACIFIC	N	8SecLngObs	PCW	4/22/93	1032A	1977
PACIFIC #43	P	12SecDr	B&S	/76-/78		
PACIFIC #64	P		(CRI&P Assn)	4/29/82		
PACIFIC E&A-36	P	Parlor	(E&A Assn)			
PACOLET #343	P	10SecDrB	(Union PCC)		565B	
PACTOLUS	N	12SecDr	Pullman	10/11/07	2163A	3562
PACTOLUS #134	RN	7BoudB	(MannBCC)	/84	776A	
PADDINGTON		12SecDr	Pullman	3/6/02	1581B	2818
PADGETT	N	12SecDr	Pull Buf	11/19/02	1581C	B11
PADUCAH	N	OS	PRR	8/70		
PADUCAH		12SecDr	Pullman	7/29/03	1581F	3015
PADUCAH #23	RN	4Sec12Ch	CNJ (WS&PCC)			
PADUCAH #23	N		(WS&PCC)	6/29/72		
PAGODA #123	RN	7BoudB	(MannBCC)	/84	776A	
PAHASKA	N	12SecDrSrB	PCW	2/28/99	1319A	2400
PAHKEE		10secDr2Cpt SUV	Pullman	3/15/10	2471	3784
PAISANO	N	BaggClub	Pullman	8/15/05	2087	3220
PAISLEY		12SecDr	Pullman	12/20/02	1581B	2852
PAKOTWE #112	RN	7BoudB	(MannBCC)	/84	697A	
PALACE		12SecDr SUV	Pullman	3/7/08	1963C	3600
PALACIO	N	BaggSmkB SUV	Pull P-2267	4/07	3278A	3494

93

CAR NAME	ST	CAR TYPE	CAR BUILDER (Owner)	DATE BUILT	PLAN	LOT
PALAMEDES		10Sec2Dr	PCW	5/21/96	1173A	2153
PALAMON	N	14SecO	PCW Det	12/9/70	20A	D4
PALAMON	N	12SecDrB	Wagner	11/92	3073	Wg75
PALATIAL		16Sec	Pullman	6/4/06	2184	3361
PALATINE	P	12SecDr	(Wagner)	10/82	3023A	
PALATKA		16Sec	Pullman	2/15/07	2184	3508
PALATKA #21		10SecO	(SF&W RR)	/75	22C	
PALATKA #30	P	20Ch4Sts	(Wagner)	8/70	3101	
PALAWAN		12SecDr	Pullman	8/24/04	1963	3098
PALERMO		10SecDrB	PCW	3/29/84	149B	101
PALERMO	P	12SecDr	B&S (Wagner)	8/82	3023A	
PALERMO		16Sec	Pullman	6/5/09	2184D	3700
PALESTINE	N	10SecB	Bowers, Dure	10/73	37	Wil11
PALESTINE		12SecDr	Pullman	2/28/03	1581D	2975
PALISADE		10SecDr2Cpt	Pullman	9/14/05	2078	3203
PALISADE CP-33	P	10Sec	(CP Co)	8/28/84	37L	
PALISADES #117	P		(WS&PCC)	7/83		
PALLAS		6CptDrLibObs	Pullman	3/13/02	1751B	2799
PALM BEACH	RN	8SecBLibObs	PCW Det	1/27/93	1014F	D511
PALMA		12SecDr	PCW	4/29/92	784E	1899
PALMARIA		12SecDr	Pullman	6/15/03	1581E	3010
PALMER #29	P	19Ch4StsDr	(Wagner)	6/85	3117A	
PALMETTO		10SecDrB	PCW Det	1/14/84	149	D224
PALMETTO	R	12SecDrB	PCW Det	1/14/84	1468	D224
PALMYRA		Saloon	H&StJ RR/GMP	/68	27	Han
PALMYRA		12SecDr	PCW Det	12/3/88	635	D377
PALMYRA		12SecDrSr	PCW	12/20/98	1318C	2370
PALMYRA	P	12SecDr	B&S (Wagner)	8/82	3023A	
PALO		14SecO	PCW	2/24/93	1021	1969
PALOS	P	16SecO	Wagner	3/93	3072	Wg89
PALOUSE		14SecDr	Pullman	5/23/06	2165	3352
PAMBRUN	N	12SecDr	Pullman	3/9/06	2163A	3333
PAMLICO			(B&O RR)	4/30/75		
PAMPLONA	N	12SecDr SUV	Pullman	8/23/04	1963	3098
PANA	N	SDR	J&S	3/71	27	Wil4
PANAMA	P	Slpr	(SouTranCo)			
PANAMA		OS	J&S	3/71	27	Wil4
PANAMA		7Cpt2Dr	PCW	9/28/94	1065A	2025
PANAMA #463	P	12SecDrB	Wagner	8/91	3065	Wg53
PANARIA		12SecDr	Pull Buf	11/4/03	1581F	B24
PANARIA	R	30Ch SUV	Pull Buf	11/4/03	3581E	B24
PANDITA #133	N	10SecDrB	J&S (WS&PCC)	1/15/85	886D	
PANDORA		20Ch3StsB	PCW Det	9/18/83	167A	D209
PANDORA #496	P	12SecDr	Wagner	2/92	3065C	Wg64
PANSY		18Ch6StsDrB	PCW	6/2/90	772D	1704
PANTHEON		16SecO	PCW	5/25/97	1232B	2224
PANTHEON #133	N	10SecDrB	J&S (WS&PCC)	1/15/85	886	
PANZA		14SecO	PCW	1/30/93	1017C	1962
PAOLI	P	16SecO	Wagner	3/93	3072	Wg89
PAOLI #27	P	OS	PRR	12/71	27	Alt2
PARADISE	RN	12SecDr	PCW	8/20/91	937E	1838
PARAGON		12SecDr	PCW Det	2/26/81	93	D88
PARAGON		12SecDr	Pullman	11/13/05	1963C	3240
PARAGOULD		12SecDr	Pull Buf	8/19/07	1963C	B65
PARAGUAY		7Cpt2Dr	PCW	12/6/94	1065A	2025
PARAGUAY		12SecDr	Pullman	12/24/02	1581B	2852
PARAISO		12SecDr	Pullman	4/10/03	1581D	2984
PARAISO SP-42	P	10SecDrB	PCW	3/12/87	445B	1299
PARANA	R	12SecDrB	PCW (Union PCC)	11/10/83	1468	D207
PARANA #414	P	10SecDrB	PCW (Union PCC)	11/10/83	149	D207

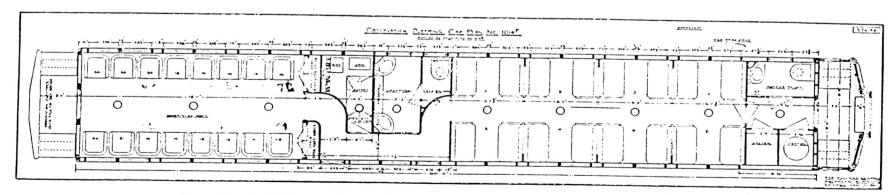

Figure I-31 Plan 1014F was used for two cars in 1901 when they were rebuilt from 14-section, Plan 1014A to 10-section library observation accommodations. One of these cars, PALM BEACH, shown on this page, was built as DIEPPE and was sold in February 1916 to Northwestern RR of South Carolina (ACL System). The Newberry Library Collection.

CAR NAME	ST	CAR TYPE	CAR BUILDER (Owner)	DATE BUILT	PLAN	LOT
PARIS	R	10SecDr	B&S	2/71	26H	Day5
PARIS	P	16SecO	Wagner	3/93	3072	Wg89
PARIS #187	P	10SecDr	B&S (PRR)	3/73	19	Day5
PARISIAN		Diner	PCW	11/20/90	662D	1777
PARKERSBURG		OS	B&O RR	4/30/73	27 1/2	B&O2
PARKERSBURG #529	P	12SecO	B&O RR	/80	665	
PARKLAND		12SecDr	Pullman	11/14/05	1963C	3240
PARKTON		10SecDrCpt	PCW	6/8/89	668	1587
PARMA	N	10SecB	PCW Det	9/5/73	37	D16
PARMA	N	12SecDrB	Wagner	8/92	3070	Wg71
PARMENIO	N	12SecDrB	Wagner	1/93	3073B	Wg88
PARNASSUS		12SecDrB	PCW	5/14/84	191	132
PARNASSUS		10SecDr2Cpt	Pullman	12/31/08	2078E	3661
PARRAL	N	12SecDrB	PCW	11/10/84	1468	1017
PARRAS	R	10SecObs	PCW	12/3/88	445W	1519
PARRAS IM-5		10SecDrB	PCW	12/3/88	445I	1519
PARSIFAL #135	P	7BoudB	(MannBCC)	/84	776A	
PARSIFAL		12SecDr	Pullman	3/18/08	1963C	3604
PARSONS	P	10SecDr	B&S	11/72	26	Day8
PARTHENIA		20Ch5StsB	PCW	11/24/83	167A	89
PARTHENON		16SecO	PCW	5/24/97	1232B	2224
PARTHENON	N	12SecDr	Pullman	3/18/08	1963C	3604
PARTHENON #500	P	12SecDr	Wagner	2/92	3065C	Wg64
PARTHIA	N	10SecB	PCW Det	9/5/73	37	D16
PARTHIA		12SecDrSr	PCW	10/5/99	1318E	2449
PARTHIA #40	P	18Ch4Sts	(Wagner)	4/86	3113	
PASADENA	N	12SecDr	PCW Det	8/25/82	93X	D162
PASCO		10SecDrCpt	PCW	6/6/89	668	1587
PASCOAG	P	36Ch SUV	Pull P-2263	8/07	3275	3492
PASHA	N	6Sec8Ch	B&S	12/71	26I	Day7
PASO						
ROBLES CP-8075		10Sec2Dr	PCW	5/2/93	995A	1942
ROBLES CP-8078		10Sec2Dr	PCW	11/12/95	1158	2125
PASSAIC	R	16Ch6Sts	J&S	9/72	54C	Wil2
PASSAIC		12SecDr	Pull Buf	4/5/04	1581G	B35
PASSAIC #144	P	6Sec4Dr?	J&S (WS&PCC)	4/8/86	720?	
PASSAIC #221	P	16Ch6Sts	J&S	9/72	35	Wil2
PASSAIC E&A-18	P	Sleeper	(E&A Assn)			
PASSUMPSIC		10SecDr	PCW Det	7/26/80	59	D89
PATAGONIA		7Cpt2Dr	PCW	9/6/94	1065A	2025
PATAPSCO	P	10SecB	B&S	7/31/74	37	
PATAPSCO		12SecDr	Pullman	5/16/08	1963C	3616
PATAPSCO #143	P	6Sec4Dr	(WS&PCC)	4/8/86	720?	
PATAPSCO #515	P	12SecO	B&O RR		665	
PATCHOGUE #89	P		(WS&PCC)	6/7/81		
PATERSON	N	12SecDrB	PCW	8/25/88	375G	1480
PATERSON		12SecDr SUV	Pullman	3/20/04	1963	3080
PATHFINDER	N	12SecDr	PCW	8/1/98	1318A	2313
PATIENCE		20Ch4StsDrB	Pull Buf	6/4/06	2177	B54
PATMOS #446		10SecDrB	PCW	9/19/85	240	1080
PATRELLA		20Ch4StsDrB	Pull Buf	6/12/06	2177	B54

CAR NAME	ST	CAR TYPE	CAR BUILDER (Owner)	DATE BUILT	PLAN	LOT
PATRICE		22ChDr	Pull Buf	7/2/02	1647C	B9
PATRICIAN		12SecDrB	Pullman	2/21/06	2142	3318
PATRIDA	N	26ChDrB	Pullman	6/25/07	1988B	3526
PATROCLUS		12SecDr	Pull Buf	5/18/10	1963E	B83
PATUXENT	P	12SecDr	B&S	8/93	3205	
PATUXENT #301		14SecO	B&S	10/73	37A	Day10
PATZCUARO	N	12SecDrSrB	PCW	2/28/99	1319A	2400
PATZCUARO	RN	8SecSrB (NG)	PCW Det	9/17/80	1594A	D76
PAUL SMITH #522	P	12SecDrB	Wagner	8/92	3070	Wg72
PAULA		14Ch13StsDrB	PCW	10/4/87	505	1376
PAULA	P	24ChDrB	(Big-4 RR)	/01	2378	
PAULINA		10SecDrB	PCW	11/12/87	445E	1334
PAULINA	R	12SecDr	PCW	11/12/87	1603	1334
PAULINE		30Ch	Pullman	5/20/01	1646D1	2700
PAULUS		14SecDr	Pullman	6/13/00	1523	2559
PAVIA	N	16SecO	Wagner	3/93	3072	Wg89
PAVILION		29Sts Parlor	PCW	2/93		
PAVONIA			(E&A Assn)	/76		
PAVONIA		12SecDr	Pullman	10/11/01	1581A	2765
PAVONIA	P	26ChB	(Wagner)	9/86	3124A	
PAWNEE	P	OS	CB&Q/GMP	/66		Aur
PAWNEE	P	12SecDr	(Wagner)	4/82	3022G	
PAWTUCKET	P	24ChB SUV	Pull P-1858	1/03	3269	2903
PAXICO		10SecDrB	PCW	2/17/88	445F	1369
PAXTON	N	12SecDr	PCW	10/3/98	1318B	2339
PAXTON		10SecDr2Cpt	Pullman	12/28/05	2078C	3287
PAYSON		10SecDr3Cpt	Pullman	4/28/06	2078C	3342
PEARL		12SecDrSr	PCW	4/7/99	1318C	2390
PEARLTON		12SecDr	Pullman	11/1/01	1581A	2784
PECHILI	N	12SecDrB	Wagner	12/90	3065	Wg43
PECOS	RN	10SecObs	PCW	3/8/87	445W	1299
PECOS		10SecB	PCW	6/5/84	206	198
PEEKSKILL	N	16SecO	PCW	5/88	543K	1427
PEERLESS		12Sec	PCW Det	7/26/79	65	D53
PEERLESS		16Sec	Pullman	6/4/06	2184	3361
PEERMONT		12SecDr	Pullman	12/29/06	2163	3460
PEGASUS	N	10Cpt	Wagner	1/94	3075	Wg104
PEGASUS #165	N		(WS&PCC)	9/24/87	426?	
PEKIN		12SecDr	PCW	3/12/91	784B	1801
PEKIN	N	12SecDrB	Wagner	11/92	3073	Wg75
PELEUS	N	12SecO	P&WB RR		18C	Wil3
PELHAM		12SecDrSr	Pullman	3/2/01	1581B	2654
PELICAN	P	10SecDr	(Wagner)	6/85	3048	
PELION		12Sec2Dr	PCW Det	1/2/88	537A	D334
PELONA		7Cpt2Dr	Pullman	10/29/02	1845	2926
PEMBA		12SecDr	Pull Buf	3/25/03	1581C	B14
PEMBERTON		28ChDr	Pullman	4/7/06	1986A	3336
PEMBINA		12SecDr	Pullman	1/21/03	1581C	2946
PEMBINA #66	P	12SecDr	B&S	/76-/78		
PEMBROKE		12SecDr	Pullman	11/13/01	1581A	2784
PENANG		10SecDrB	PCW	5/20/85	240A	1193

CAR NAME	ST	CAR TYPE	CAR BUILDER (Owner)	DATE BUILT	PLAN	LOT
PENARTH		12SecDr	Pull Buf	4/18/10	1963E	B83
PENASCO	N	12SecDr	Pullman	6/20/03	1581F	3010
PENASCO	R	30Ch SUV	Pullman	6/20/03	3581E	3010
PENBRYN		12SecDrCpt	Pullman	8/9/05	2088	3221
PENCADER		26ChDr	Pullman	3/31/06	1987A	3335
PENCOYD	P	14SecDr SUV	Pull P-2262	8/07	3207C	3498
PENDENNIS		12SecDr	Pullman	11/13/05	1963C	3240
PENDLETON		12SecDrCpt	Pullman	10/9/05	2079	3204
PENELOPE		18Ch4StsDrB	PCW	7/31/93	1039A	1995
PENELOPE #498	P	12SecDr	Wagner	2/92	3065C	Wg64
PENFIELD		14SecDr	PCW	12/28/99	1318J	2450
PENGUIN		24ChDrB	Pullman	6/29/00	1518C	2553
PENINSULA		10SecDr	PCW Det	11/19/72	26	D14
PENINSULA	P	10SecDr Tourist	(Wagner)	11/72	3012	
PENNINGTON		12SecDr	Pullman	9/5/07	1963C	3550
PENNSYLVANIA	N	30ChObs	PCW	4/28/98	1278B	2299
PENNSYLVANIA #124	P	12SecO	PFtW&C/GMP	/69	5	FtW1
PENOBSCOT		12SecDr	PCW	12/4/83	93B	88
PENOBSCOT	R	12SecDrB	PCW	12/4/83	1611	88
PENOBSCOT	P	10SecDrB	(Wagner)	5/81	3046	
PENOBSCOT #165	P		(WS&PCC)	9/24/87	426?	
PENOKEE #83	P		(WS&PCC)	7/1/80		
PENOKEE WC-103	P	10SecDr	PCW Det	3/87	489	D294
PENOLA		10SecDrB	PCW	6/18/87	445C	1322
PENRITH		12SecDr	Pullman	5/13/04	1963	3082
PENSACOLA	P	10SecDr	J&S (Pull/Sou)	/73	37	Wil10
PENSACOLA		12SecDr	Pullman	2/24/04	1963	3070
PENSEROSO		20Ch4StsDrB	Pull Buf	6/14/06	2177	B54
PENZANCE		12SecDr	Pullman	1/5/04	1581H	3053
PENZANCE #8	P	24Ch5StsDr	(Wagner)	6/85	3108	
PEONY		24ChDr	Pull Buf	4/27/06	2156	B52
PEORIA	P	12SecO	DetC&MCo	/68	5	Det
PEORIA	N	OS	PCW Det	11/13/71	27	D8
PEORIA	P	12SecDr	(Wagner)	10/82	3023A	
PEORIA	N		(WS&PCC)	2/17/79		
PEORIA #72	P		(CRI&P Assn)	1/24/80		
PEOSTA	N	Slpr	(IC RR)			2/127
PEPITA		20Ch4StsDrB	Pull Buf	6/9/06	2177	B54
PEQUEST		18Ch6StsDrB	PCW Det	7/17/90	772D	D436
PEQUOT	P	12SecDr	(Wagner)	4/81	3023A	
PERCIVALE		12SecDr	Pullman	3/22/10	1963E	3807
PERDITA		14Ch13StsDrB	PCW	10/3/87	505	1376
PERDIX		12SecDr	Pullman	1/5/07	2163	3460
PEREA		10SecDr	PCW Det	8/16/80	59	D78
PERFECTO	N	BaggClubB	Pullman	7/30/03	1945D	3020
PERHAM	RN	12SecDr	PCW	7/2/92	937E	1925
PERICLES		12SecO	PCW Det	2/21/82	114	D145
PERICLES	N	10Cpt	Wagner	1/94	3075	Wg104
PERIM		12SecDr	Pull Buf	3/20/03	1581C	B14
PERRILLA		7Cpt2Dr	Pullman	10/29/02	1845	2926
PERSEUS #449		10SecDrB	PCW	10/24/85	240	1080
PERSIA	P	10SecDr	C&NW Ry	7/31/71	19	FdL3
PERSIA		16SecO	PCW	2/7/91	780A	W5
PERSIA #12	P	16Ch8StsDrB	(Wagner)	6/85	3115C	
PERSIAN	P	10Sec2DrB	PCW (Wagner)	2/14/00	3079A	Wg134
PERTH		12SecDr	Pullman	7/24/03	1581F	3019
PERU	P	16SecO	Wagner	3/83	3072	Wg 89
PERU #28	N	12SecO	DetC&MCo/GMP	/69	5	Det
PERUGIA	N	10SecDr	Grand Trunk	10/31/72	31C	Mon3
PERUGIA		12SecDr	Pull Buf	1/24/02	1581A	B7
PESARO		12SecDr	Pull Buf	11/19/02	1581C	B11
PESCADERO SP-43	P	10SecDrB	PCW	3/8/87	445B	1299
PETERSBURG	RN	11SR	Wagner	9/89	1769C	Wg24
PETONIC #92	P		(WS&PCC)	7/9/81		
PETOSKEY		12SecDr	Pull Buf	6/27/04	1963	B38
PETRA		12SecDr	Pullman	3/20/03	1581D	2975
PETREL		10SecO	PCW Det	10/9/82	102	D170
PETREL	N	26ChDr	Pullman	5/25/07	1987B	3524
PETREL	P	12SecDrB	(Wagner)	6/86	3050	
PETRUCHIO		12Sec2Dr	Pullman	7/18/01	1580C	2742
PETUNIA		26Ch4StsDr	Pull Buf	7/9/00	1519C	B2
PEVERIL		14SecO	PCW Det	6/16/93	1041	D519
PEWAUKEE	P	12SecDrB	PCW	4/28/84	202	195
PHAETON #235	P	10Ch15StsDr	Bowers, Dure	9/72	35	Wil1
PHALARIS	N	18Ch6StsDrB	PCW Det	8/1/90	772D	D436
PHAON		14SecO	PCW Det	7/3/93	1041	D519
PHAROS		10SecDrB	PCW	10/16/84	149D	1007
PHAROS	P	12SecDr	(Wagner)	6/86	3049A	
PHEASANT		24ChDrB	Pullman	7/9/00	1518C	2553
PHENECIA		12SecDr	Pullman	6/20/03	1581F	3010
PHIDIAS		14SecO	PCW Det	5/6/93	1021	D518
PHILADELPHIA	N	26Ch3StsDr	PCW	4/29/98	1278C	2258
PHILADELPHIA #725	P	31StsB	B&S (B&O)	/85	653	
PHILADELPHIA CTC-81	R		PW&B rblt/73			
PHILARIO		12Sec2Dr	Pullman	7/20/01	1580C	2742
PHILENIA		20Ch4StsDrB	Pullman	6/15/01	1518E	2703
PHILETUS	N	12SecDrSrB	PCW	1/27/99	1319A	2400
PHILETUS	R	16Sec	PCW	10/15/85	1881	1080
PHILETUS #448		10SecDrB	PCW	10/15/85	240	1080
PHILISTIA		12SecDr	Pullman	6/12/03	1581E	3010
PHILLIS #396		20Ch5StsB	PCW	1/2/84	167A	90
PHILOMELA		26ChDr	Pullman	7/28/04	1987A	3097
PHOCION		12Sec2Dr	PCW Det	1/14/88	537A	D334
PHOEBE		22Ch6StsDr	PCW	9/1/92	976A	1910
PHOEBE		24ChDr	Pullman	7/6/05	1980A	3171
PHOENECIA		16Sec	Pullman	2/16/07	2184	3508
PHOENICIA	P	22Ch4Sts	(Wagner)	4/86	3123	
PHOENIX		10SecDrB	PCW	7/9/85	149D	1066
PHOENIX		12SecDr	Pullman	1/25/02	1581A	2792
PHOENIX #2		Diner	PCW	4/27/88	538D	1396
PHOTIUS	N	BaggBSmk	PCW	/99	1400A	2410

Figure I-32 Behind the partition with a lavishly-decorated transom can be seen section-like accommodations in the smoking room of PEVERIL, a 14-section car built in the Detroit Shops under Plan 1041. Smithsonian Institution Photograph P2517.

CAR NAME	ST	CAR TYPE	CAR BUILDER (Owner)	DATE BUILT	PLAN	LOT
PHRYGIA		12SecDr	Pullman	6/26/03	1581F	3010
PICADOR		12SecDr	Pullman	12/4/08	1963E	3659
PICARDY		12SecDr SUV	Pull Buf	9/2/03	1581F	B22
PICAYUNE		12SecDr	Pull Buf	5/2/10	1963E	B83
PICKWICK	RN	Private	PRR	6/10/72	29A	Alt1
PICKWICK		12SecDr	Pullman	8/12/08	1963C	3636
PICTON		12SecDr	Pull Buf	8/19/03	1581F	B21
PICTON	P	10SecB	B&S	9/74	37	Day11
PIEDMONT	P	Sleeper	(SouTranCo)			
PIEDMONT		12SecDrB	PCW	9/7/98	1319A	2314
PIEDMONT #192	P	10SecDr	B&O RR	4/30/73	19	B&O1
PIEMONTE		12Sec	PCW Det	6/9/76	50	D30
PIERCETON		12SecDr	Pullman	2/6/08	2163A	3599
PIERIDES		10Sec2Dr	PCW	5/23/96	1173A	2153
PIERMONTE	N	12SecDrSr	Pullman	4/8/05	2049A	3169
PILGRIM	P	26Ch3Sts	PCW Det	7/89	365	D391
PILGRIM		10SecDrB	PCW Det	12/28/82	149	D174
PILGRIM	N	Private	Wagner	11/28/88	3088	Wg16
PIMLICO		12SecDr	Pullman	7/25/03	1581F	3019
PIMLICO #513	P	12SecO	B&O RR		665	
PINAFORE		12SecDr	Pullman	3/23/08	1963C	3604
PINAFORE #10	P	24Ch5StsDr	(Wagner)	6/85	3108	
PINALENO		7Cpt2Dr	Pullman	10/29/02	1845	2926
PINDAR		12SecO	PCW Det	4/10/82	114	D143
PINDARUS #497	P	12SecDr	Wagner	2/92	3065C	Wg64
PINE ORCHARD	P	Bagg24Ch	Pull P-2267	4/07	3278	3494
PINECROFT	N	12SecDr	Pullman	6/15/03	1581E	3010
PINECROFT	R	30Ch SUV	Pullman	6/15/03	3581E	3010
PINEDA	N	12SecDr SUV	Pullman	10/16/07	2163A	3562
PINEHURST		16Sec	Pullman	6/9/06	2184	3361
PINON		8SecDrB	PCW	7/23/88	556A	1463
PINTA	P	6Sec8ChClub	(ATSF)		1757	
PINTA	P	16SecO	Wagner	3/93	3072	Wg89
PINZON	P	20Ch4StsDrB	Wagner	8/93	3128	Wg98
PIONEER		28ChDr	Pullman	4/6/06	1986A	3336
PIONEER #1	N	12SecO	GMP	/64	5	Chi
PIONEER #1	P	Slpr	(SouTranCo)			
PIQUA		16Sec	Pullman	7/15/03	1721	3014
PIQUA #292		10SecB	J&S	6/73	37	Wil12
PISA		12SecDr	PCW Det	11/21/90	784	D449
PISA	P	16SecO	Wagner	3/93	3072	Wg89
PISANIO		12SecDrCpt	Pullman	6/24/04	1982	3090
PISANO	N	10SecDr	B&S?	C:6/75	26H	Day5
PITTSBURG #205	P	10SecDr	H&H (B&O)	C:/71	26	Wil6
PITTSBURG		OS	CNJ (SilverPC)	/66		
PITTSBURGH		12SecDr	PCW	8/1/98	1318A	2313
PITTSFIELD		10SecDr2Cpt	Pullman	5/29/08	2078D	3622
PIZARRO	N	10SecDr	(Wagner)	/86	3043A	
PLACEDO		12SecDr	Pullman	9/21/06	1963C	3397
PLACIDIA		10SecDr2Cpt	Pullman	3/22/07	2078D	3528
PLAINFIELD #174	P	18Ch6StsDr	(WS&PCC)	6/23/88	730A	

CAR NAME	ST	CAR TYPE	CAR BUILDER (Owner)	DATE BUILT	PLAN	LOT
PLAISTED	N	12SecDr SUV	PCW	5/18/99	1318M	2423
PLAISTED	R	30Ch SUV	PCW	5/18/99	3581M	2423
PLANET		OS	G&B/GMP	/68		Troy
PLANET		27Sts Parlor	PCW Det	9/27/76	45	D31
PLANET	N	12SecDrB	Wagner	11/92	3073	Wg75
PLANET #256	P	OS	PC&StL RR	12/71	27	Col1
PLANKINTON		10SecDr2Cpt	Pullman	1/9/08	2078D	3594
PLAQUEMINE	N	12SecDr	Pullman	12/28/06	2163	3460
PLATO	N	16Ch6Sts	PCW Det	7/12/76	54E	D40
PLATO		10SecDrB	PCW	8/12/85	240	1069
PLATTE		12SecDrCpt	Pullman	9/30/05	2079	3204
PLATTE VALLEY		12SecO	B&S	12/70	5	Day
PLATTE VALLEY	R	12SecO	B&S	12/70	43	Day
PLATTSBURG		12SecDr	Pullman	1/24/07	1963C	3448
PLATTSMOUTH		12SecDr	PCW	8/10/82	93B	23
PLAZA		Diner	PCW	7/25/93	538D	2010
PLEASANT						
LAKE #524	P	12SecDrB	Wagner	8/92	3070	Wg72
PLEASANTON		12SecDr	Pull Buf	8/22/07	1963C	B65
PLEIADES		10SecDrB	PCW Det	2/29/84	149	D228
PLEIADES		12SecDr	Pull Wilm	3/21/03	1581C	W8
PLEIADES	N	20Ch4StsDr	(Wagner)	7/85	3109A	
PLEVNA	N	10SecDrB	PCW	11/22/83	149W	84
PLOVER		24ChDrB	Pullman	6/28/00	1518C	2553
PLUMAS		10SecDrB	PCW	5/27/89	623B	1567
PLUMOSA		16Sec SUV	Pullman	3/31/10	2184G	3808
PLUMSTEAD		12SecDr	Pullman	9/15/01	1581A	2756
PLUTUS	N	10SecB	(Wagner)	11/84	3036	
PLUTUS #79	N		(WS&PCC)	2/17/79		
PLYMOUTH		12SecDrCpt	Pullman	5/25/04	1985	3095
PLYMOUTH	P	12SecDr	(Wagner)	12/75	3027A	
PLYMOUTH #130	P	10SecB	PFtW&C	1/71	14C	FtW2
PLYMOUTH						
ROCK	P	10SecHotel	MC RR/GMP	/66	26	Det
ROCK	R	SDR	MC RR/GMP	/66	142A	Det
ROCK	N	Private	Pullman	4/6/04	2009	3035
ROCK E&A-29	P	Slpr	(E&A Assn)			
POCAHONTAS	P	OS	(OPLines)			
POCAHONTAS	RN	8SecBLibObs	PCW		1707	
POCANTICO		10SecDr2Cpt	Pullman	5/29/08	2078D	3622
POCASSET	P	10SecDrB	(Wagner)	5/81	3046	
POCATELLO		12SecDrB	PCW	2/8/89	375G	1557
POCATELLO	R	12SecDrB	Pull	2/8/89	1994	1557
POCHEQUET	P	OS	C&NW Ry	/65	27	Fdl
POCONO	P	Parlor	(DL&W RR)	4/75		
POCONO	N	16Ch7StsDrObs	PCW	5/2/96	1193	2179
PODOLIA		12SecDr	Pull Buf	11/23/03	1581F	B25
POINCIANA		24ChDr	Pull Buf	4/7/06	2156	B51
POINSETTIA		32ChDr	Pullman	3/8/07	2274	3509
POINT LEVI		OS	Grand Trunk	10/31/72	27	Mon2
POITIERS		14SecO	PCW Det	4/22/93	1021	D518
POKEGAMA		16Sec	Pullman	6/15/07	2247	3459
POLANA		6CptDrLibObs	Pullman	2/26/02	1751B	2799
POLAND	RN	12SecDr	PRR	5/71	17E	Alt1
POLAND	N	10SecDrB	PCW Det	12/29/83	149	D223
POLESTAR		12SecDr	Pullman	1/27/02	1581A	2792
POLILLO		12SecDr	Pullman	8/24/04	1963	3098
POLIO		12SecDr	PCW		784D	1801
POLONIUS #499	P	12SecDr	Wagner	2/92	3065C	Wg64
POLTAVA	N	12SecDrB	Wagner	1/93	3073B	Wg82
POLYNESIA	R	16Sec	PCW		1876	
POMARIA	N	12SecDrSr	PCW	5/13/99	1318C	2423
POMEROY		26ChDr	Pullman	3/27/06	1987A	3335
POMFRET	P	26Ch3Sts	B&S	1/93	3260	
POMMERN		12SecDr	Pullman	12/7/07	1963C	3585
POMONA	P	10Ch15StsDr	Bowers, Dure	9/72	35	Wil1
POMONA		26ChDr	Pullman	7/29/04	1987A	3097
POMONA	N	10SecO	(Wagner)	/86	3038	
POMPEII		14SecO	PCW Det	4/15/93	1021	D518
PONCE		12SecDr	Pullman	8/ /04	1963	3098
PONCE DE LEON	N	Diner	H&H	7/73	40C	Wil2
PONCELOT	N	12SecO	PCW Det	5/10/82	114	D143
PONCHA		8SecDrB	PCW	7/20/88	556A	1463
PONEMA	RN	12SecDr	PCW Det	7/5/82	1581C	D153
PONETO	N	12SecDr	Pullman	11/4/01	1581A	2784
PONSONBY		16Sec	Pullman	6/20/09	2184D	3700
PONTIAC	N	12SecO	G&B/D&MSCC	/68	5	Troy
PONTIAC	N	10SecDrB	PCW	11/16/86	445	1262
PONTIAC		29ChDr	Pullman	7/15/05	2061	3217
PONTIAC	P	10SecDrB	(Wagner)		3046	
PONTINUS	R	BaggSmkBLib	PCW		1706	
PONTINUS		10SecDr2Cpt SUV	Pullman	12/31/08	2078E	3661
PONTOTOC		10SecDr	PCW	8/22/83	112	47
PONTOTOC	R	12SecB	PCW	8/22/23	168	47
POPIEL	N	12SecDrB	Wagner	1/93	3073B	Wg82
PORT						
CHESTER	P	40Ch8StsDrObs	Pull P-2266A	4/07	3276	3493
CLYDE	N	40Ch6DinDrObs	Pull P-2266A	4/07	3276A	3493
COSTA	N	10SecDr2Cpt	Pullman	11/20/07	2078D	3583
HOPE	P	10SecDr	Grand Trunk	6/72	31	Mon3
JEFFERSON #97	P		(WS&PCC)	7/23/81		
MORRIS	P	Bagg24Ch	Pull P-2496	1/11	3278	3823
PORTA	N	20Ch8StsDrB	Wagner	6/91	3126A	Wg51
PORTAGE		12SecDrCpt	Pullman	5/25/04	1985	3095
PORTAGE E&A-46	P	Slpr	(E&A Assn)	/72		Elm0
PORTHOS	P	18Ch6StsB	B&S	11/88	3256A	
PORTIA		20Ch5StsB	PCW Det	8/15/83	167	D210
PORTIA	P	24ChDrB	(Big-4 RR)	/01	2378	
PORTIA	P	12SecDr	(Wagner)	6/86	3049A	
PORTLAND	P	6SecDrB	Grand Trunk	11/70	17	Mon1
PORTLAND		12SecDr	Pullman	5/16/03	1581D	3004
PORTLAND	P	12SecO	(Wagner)	/67	3044	

CAR NAME	ST	CAR TYPE	CAR BUILDER (Owner)	DATE BUILT	PLAN	LOT
PORTLAND SP-70	P	10SecDrB	PCW	6/26/88	445F	1415
PORTLEDGE	N	12SecDr	Pullman	12/7/07	1963C	3585
PORTOLA		12SecDr	Pull Buf	4/19/10	1963E	B83
PORTSMOUTH		12SecDr	Pullman	11/14/01	1581A	2784
PORTSMOUTH #327	N	8BoudB?	J&S (Union PCC)	/89	373?	
PORTUGAL	N	12SecDr	PCW	4/12/87	473A	1323
PORTUNUS		10SecDr2Cpt	Pullman	12/31/08	2078E	3661
POSTILION		12SecDrB	Pullman	2/24/06	2142	3318
POTOMAC		12SecDrSrB	PCW	1/27/99	1319B	2400
POTOMAC	N	BaggClub	Pullman	8/14/05	2087	3220
POTOMAC	P	12SecO	(Wagner)	9/72	3027C	
POTOMAC #148	P	6Sec4Dr?	J&S (WS&PCC)	5/17/86	720?	
POTOMAC #156	P	10SecDr	J&S (B&O)	4/30/73	19	Wil1
POTOMAC #514	P	12SecO	B&O RR		665	
POTOSI	P	10SecDr	H&StJ RR	2/73	26	Han3
POTOSI	P	12SecDrB	(Wagner)	6/86	3050	
POTSDAM		14SecO	PCW Det	5/20/93	1021	D525
POUGHKEEPSIE		16Sec	Pullman	2/13/07	2184	3508
POUGHKEEPSIE #36	P	16Ch5StsDr	(Wagner)	8/85	3106B	
POWHATAN		12SecDrSrB SUV	PCW	1/28/99	1319B	2400
PRAGUE		12SecDr	PCW	2/10/92	784E	1881
PRAIRIE QUEEN	P	DDR	DetC&MCo	/70	17	D3
PREMIER		BaggSmk	PCW	8/11/88	534C	1410
PREMIER	N	16Sec	Pullman	6/8/06	2184	3361
PRESCOTT		10SecDr	Grand Trunk	10/31/72	26	Mon2
PRESCOTT		12SecDr	Pullman	7/29/03	1581F	3015
PRESIDENT	P	OS Hotel	GW Ry/GMP	8/1/67	13	Ham
PRESIDENT		10SecDrBHotel	PCW Det	11/25/76	56	D39
PRESIDENT		Private	Pullman	4/6/04	2009	3035
PRESIDIO	RN	10SecObs	PCW	1/7/87	445W	1297
PRESTON	P	OS	H&H	2/72	27	Wil7
PRESTON	RN	16Sec	PCW Det	1/30/83	1877	D176
PRESTWICK	N	12SecDr	Pullman	12/16/03	1581F	3043
PRESWICK		12SecDr	Pullman	12/16/03	1581F	3043
PRETORIA		12SecDr	Pull Buf	1/19/04	1581F	B27
PRIMATE	N	5CptLibObs	Pullman	8/16/05	2089	3222
PRIMROSE	N	12SecDrSr	H&H (B&O)	? 8/87	654F	
PRINCE		26Sts Parlor	PCW	12/88		
PRINCE OF WALES	N	Parlor	PCW Det	/75		
PRINCE OF WALES	N	Parlor	PCW Det	3/1/74	39	D20
PRINCE REGENT		32Sts Parlor	PCW Det	6/17/95	45K	D508
PRINCESS	R	6SecB36Sts	DetC&MCo	7/73	17S	D3
PRINCESS		27Sts Parlor	PCW	12/88		
PRINCESS	P	DDR	DetC&MCo	7/73	17	D3
PRINCESS		22Berths	PCW Det	7/17/75	48	D26
PRINCESS	R	21Ch7StsDrObs	Pullman	7/25/05	2091	3219
PRINCESS		22Ch6StsDrObs	Pullman	7/25/05	2023	3219
PRINCESS ENA		32StsB Parlor	Pullman	1/06	1376C	3223
PRINCESS MARGARET		32Sts Parlor	PCW Det	10/93		D513
PRINCESS MARY		29Sts Parlor	PCW Det	6/17/95	45K	D508
PRINCESS OF WALES		36Sts Parlor	PCW	6/17/95	1149	2106

CAR NAME	ST	CAR TYPE	CAR BUILDER (Owner)	DATE BUILT	PLAN	LOT
PRINCESS PATRICIA		32Sts Parlor	Pullman	1/06	1376C	3223
PRINCETON		12SecDr	PCW	10/3/98	1318B	2339
PRINCETON		12SecDrB	Pullman	2/26/01	1611A	2614
PRINCETON	RN	10SecCpt	Pullman	5/16/06	2165D	3352
PRINCETON #104	P		B&O (WS&PCC)	5/19/82		
PRINCETON #160	P	12SecO	J&S	/71	18	Wil3
PRISCILLA		20Ch12StsB	PCW	5/27/89	618A	1514
PRISCILLA #65	P	28Ch4Sts	PCW (Wagner)	7/14/86	3124	1200
PRISTINA		20Ch4StsDrB	Pull Buf	6/6/06	2177	B54
PROBUS	R	BaggSmkBLib	PCW		1706	
PROCLUS		6CptDrLibObs	Pullman	3/3/03	1751B	2986
PROGRESS		10SecO (NG)	PCW Det	9/15/81	73A	D110
PROGRESS		12SecDrObs	Pullman	8/28/06	2194	3395
PROMETHEUS	P	12SecDr	Wagner	2/92	3065C	Wg64
PROMONTORY		10SecDr2Cpt	Pullman	9/20/05	2078	3203
PROMONTORY	P	12SecO	PFtW&C/GMP	/69	5	FtW 1
PROSERPINE		16Ch6StsB	PCW Det	7/14/76	54	D40
PROSERPINE		21Ch7StsDrObs	Pullman	10/7/05	2091	3228
PROSPECT	N	12SecDrB	(Wagner)	3/81	3050	
PROSPECT PARK #127	P	Parlor (3' gauge)	J&S (WS&PCC)	5/17/84	552?	
PROSPERO		14SecO	PCW Det	10/15/81	115	D125
PROSPERO		12SecDr	Pull Buf	5/3/10	1963E	B83
PROTEUS	N	12SecO	B&O RR		665	
PROTEUS		10SecDr2Cpt	Pullman	12/31/08	2078E	3661
PROVENCE		12SecDr	Pullman	5/12/08	1963C	3615
PROVIDENCE		12SecDr	Pull Buf	3/23/04	1581G	B35
PROVINCETOWN	P	34Ch SUV	Pull P-1484	1/03	3265	2904
PROVO		10SecB (NG)	PCW Det	10/23/83	178	D211
PROVO		12SecDr	Pullman	11/11/05	1963C	3240
PRUDENCE		18Ch4StsDrB	PCW	6/24/93	1039A	1994
PRUSSIA		16Sec	Pullman	10/22/01	1720A	2766
PSYCHE	N	Parlor	PCW Det	11/22/71	34C	D4
PSYCHE		26ChDr	Pullman	7/28/04	1987A	3097
PTOLEMY	N	12SecDrB	Wagner	1/93	3073B	Wg88
PUEBLA #462	P	12SecDrB	Wagner	8/91	3065	Wg53
PUEBLO		10SecDr	PCW Det	8/26/76	59	D43
PUENTE		10Sec2Dr	Pullman	10/31/02	1746A	2928
PUENTE O&C-8008		10SecDrB	PCW	5/12/88	445F	1377
PUERTO RICO	P	12SecDrB	Wagner	12/98	3073E	Wg117
PUGET SOUND		10SecDrB	PCW Det	2/20/84	149	D227
PUKWANA	N	12SecDrB	PCW	3/17/90	375J	1662
PULASKI	P	10SecB	Bowers, Dure	9/26/73	37	Wil11
PULASKI		10SecDrB	PCW Det	1/19/84	149	D224
PULASKI	R	16Sec	PCW	1/19/84	1877C	
PURDUE		12SecDr	Pullman	12/11/05	1963C	3252
PURITAN	P	26Ch3Sts	PCW Det	7/89	365	D391
PURITAN		10SecDrB	PCW Det	12/28/82	149	D174
PURITAN #81	P	28Ch4Sts	PCW (Wagner)	7/16/86	3124	1200
PUTNAM	N	14Ch5StsB	(DL&W RR)	4/75	211	
PUTNEY		12SecDr	Pullman	10/29/03	1581F	3025

99

Figure I-33 Princeton University leased this 10-section compartment car after it was rebuilt and steel-sheathed from a former Northern Pacific Association car, MINIKAHDA. The University used it during the summer months for geology field trips from 1926 until about 1940. Pullman Photograph 30068, the Arthur D. Dubin Collection.

CAR NAME	ST	CAR TYPE	CAR BUILDER (Owner)	DATE BUILT	PLAN	LOT
PUYALLUP		12SecDr	Pullman	5/20/02	1581B	2853
PYRAMID PARK		12SecDr	PCW	9/19/82	93B	22
PYRAMUS		12SecDr	Pullman	11/11/05	1963C	3240
PYRENEES		10SecDrB	PCW Det	3/6/84	149	D228
PYRRHUS	P	12SecDr	B&S	8/93	3205	
PYRRHUS	N	22Ch4StsDr	(Wagner)	6/85	3108D	
PYTHIA		26ChDr	Pullman	7/30/04	1987A	3097
PYTHON #237	P	18Ch3Sts	Bowers, Dure	11/72	35	Wil1
QUAKERESS	N	30ChObs	PCW	4/28/98	1278B	2299
QUANNA		16Sec	Pullman	5/27/09	2184D	3700
QUANTICO	N	OS	CB&Q RR	5/72	36B	Hinc1
QUANTICO		12SecDr	Pullman	5/18/08	1963C	3616
QUANTOCK		12SecDr	Pullman	12/21/09	1963E	3766
QUANTZINTECOMATZIN						
	LN	Diner	(NP RR)			
QUANTZINTECOMATZIN						
	N	Diner	CB&Q RR	12/9/71	34C	
QUEBEC		10SecDr	Grand Trunk	2/71	19	Mon1
QUEBEC		12SecDr	Pullman	3/16/03	1581D	2975
QUECHEE		12SecDrB	PCW	8/8/88	375E	1395
QUEEN		Parlor	PCW P-254	8/3/87		1071
QUEEN		16Ch6StsB	PCW Det	12/1/76	54	D40
QUEEN	N	26Ch3StsDr	PCW	4/25/98	1278C	2258
QUEEN CITY	RN	8SecBLibObs	PCW	6/26/88	1707	1415
QUEEN CITY	RN	5CptDrBLib	Pullman	3/3/03	1751G	2986
QUEEN CITY #270		10SecDr	PCW Det	10/7/72	26 1/2	D13
QUEENSLAND		12SecDr	Pullman	12/26/02	1581B	2852
QUEENSTOWN		12SecDrB	PCW	1/25/87	446A	1260
QUERETARO		10SecDrB	PCW Det	4/1/84	149B	D234
QUESTA		12SecDr	PCW Det	3/30/92	784E	D497
QUIBO		12SecDr	PCW	8/27/90	784	1708
QUIDNICK	P	26ChBarB SUV	Pull P-2265	8/07	3274	3497
QUIETO	P	12SecDr	B&S	8/93	3205	
QUIETO SP-53	P	10SecDrB	PCW	3/20/88	445F	1377
QUILDA	P	18Ch6StsB	New Haven RR	3/81	3250	
QUILON #124	N	12SecDr	(WS&PCC)	9/18/83	914A	
QUINCY	N	12SecDr	CB&Q/GMP	/67	13B	Aur
QUINCY	RN	12SecDr	CB&Q RR	C:12/68	12B	Aur
QUINEBAUG	P	23ChDr SUV	Pull P-2070	6/05	3268	3188
QUINTA SP-2546		10SecO	PCW	11/92	1024	W7
QUINTIUS	P	12Sec2Cpt SUV	Pull P-1854A	1/03	3206	2899
QUINTON		12SecDr	Pullman	10/13/00	1581A	2588
QUINTUS	R	BaggSmkBLib			1706	
QUIVETTE	P	12Sec2Cpt SUV	Pull P-1854A	1/03	3206	2899
QUIXOTE	P	18Ch6StsB	New Haven RR	3/81	3250	
QUOGUE #102	P		(WS&PCC)	8/5/81	465?	
R.C. FLOWERS #141	P	Private	(MannBCC)	/84	684	
RACELAND		12SecDr	Pullman	12/28/07	1963C	3586
RACINE	P	OS	G&B/GMP	/67		Troy
RACINE	P	12SecDr	(Wagner)	8/82	3023B	
RADCLIFFE	P	8Sec4Dr	Wagner	6/90	3068	Wg34
RADIANT		8Sec4SrB	PCW	5/1/91	894B	1807
RADNOR	P	12SecDr	Wagner	3/93	3065C	Wg84
RAGLAN		12SecDr	Pullman	12/26/02	1581B	2852
RAGO		12SecDr	PCW	12/19/90	784	1780
RAHULA		12SecDrB	PCW	10/31/88	375F	1481
RAHWAY	N	30Ch	PCW	4/28/98	1278A	2293
RAHWAY #231	P	Parlor	H&H	8/72	35 3/4	Wil1
RAILWAY AGE	P	Private	PCW	1/4/82	117	24
RAINBOW	RN	Private	PCW	1/28/88	1920	1361
RAJAH	N	SDR	B&S	12/71	26J	Day7
RAJAH	P	10Cpt	Wagner	6/96	3075A	Wg104

CAR NAME	ST	CAR TYPE	CAR BUILDER (Owner)	DATE BUILT	PLAN	LOT
RALEIGH #148	P	BaggBLib	J&S (WS&PCC)	5/17/86	720D	
RALEIGH #509	P	12SecDr	Wagner	4/92	3065C	Wg66
RAMA BLANCA	RN	12SecDrB	Pullman	5/16/06	2165C	3352
RAMAPO	N	14Ch13StsDrB	PCW	7/29/87	505	1376
RAMAPO E&A-15	P	Slpr	(E&A Assn)			
RAMBLER		10SecO (NG)	PCW Det	8/24/80	73A	D76
RAMBLER	N	Private	PCW	8/10/85	247E	1105
RAMBLER		12SecDr	Pull Buf	6/24/07	1963C	B62
RAMBLER	P	12SecO	(Wagner)	/85	3027E	
RAMBO		26ChDr	Pullman	3/26/06	1987A	3335
RAMESES #151	P	10SecDrB	J&S (WS&PCC)	7/6/86	696	
RAMILLIES	N	12SecDr	Wagner	5/92	3065C	Wg85
RAMLEH		14SecO	PCW Det	4/29/93	1021	D518
RAMOLA		12SecDrB	PCW	9/6/88	375F	1480
RAMONA #91	P	28Ch3Sts	PCW (Wagner)	5/25/86	3124	1194
RAMSGATE		12SecDr	Pullman	10/31/03	1581F	3025
RANDOLPH		12SecDr	Pullman	9/16/09	1963E	3744
RANDOLPH #103	P	12SecDr	(MannBCC)	/89	878	
RANGELEY	N	17Ch12StsB	J&S	6/29/87	686E	
RANGELEY	N	10secDr2Cpt SUV	Pullman	3/10/10	2471	3784
RANGELY #85	N		B&S (WS&PCC)	3/1/81		
RANGER	RN	Private	Wagner	9/92	3062-O	Wg78
RANGOON #167	N		(WS&PCC)	12/6/87	731?	
RANIER		16SecO	Pull Cal	/01	1637A	C2

CAR NAME	ST	CAR TYPE	CAR BUILDER (Owner)	DATE BUILT	PLAN	LOT
RANKIN	N	12SecDrCpt	Pullman	8/9/05	2088	3221
RANSCLEAVE	N	12SecDr	Pullman	12/21/09	1963E	3766
RANSOM		12SecDrB	PCW	8/27/88	375F	1480
RANSTON	P	22Ch6Sts	B&S	8/82	3253A	
RANTOUL		12SecDr	Pullman	1/5/04	1581H	3053
RAPHAEL		12SecDr	Pullman	8/12/01	1581A	2727
RAPIDAN	N	12SecO	B&O RR	/86	655	
RAPIDAN		12SecDrSr	PCW	5/17/99	1318C	2423
RAPIDAN	P	12SecO	(Wagner)	/85	3027D	
RAPIDAN #152	P	10SecSrB	J&S (WS&PCC)	7/7/86	401?	
RAPPAHANNOCK		12SecDrSr	PCW	5/20/99	1318C	2423
RAPPAHN-NOCK CTC-40	R		Det rblt/74			
RAPPAHN-NACK #151		SlprParlor	J&S (WS&PCC)	C:/80	401?	
RAQUETTE RIVER	P	10Sec2DrB	Wagner	9/92	3071A	Wg77
RARITAN		12SecDr	PCW	11/22/98	1318C	2359
RARITAN		12SecDr	Pullman	5/16/08	1963C	3616
RARITAN #167	P		(WS&PCC)	12/6/87	731?	
RARITAN #222	P	27Ch	J&S	9/72	35	Wil2
RARITAN #510	P	12SecDr	Wagner	4/92	3065C	Wg66
RATON		10SecDr	PCW	1/21/85	229	1032
RAVALLI	RN	12SecDr	PCW	8/20/91	937E	1838
RAVANEL	P	26Ch3Sts	B&S	9/92	3260	

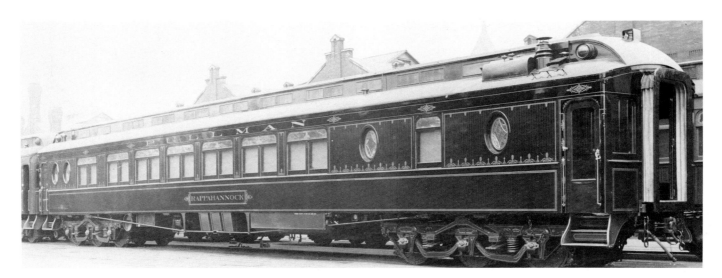

Figure I-34 The year 1889 was a good one for railroad car design as shown in the 12-section drawing room RAPPAHANNOCK, constructed for General Service on the Chesapeake & Ohio Railroad. Eight years later, it was destroyed by fire in the Wilmington Shops. Smithsonian Institution Photograph P4681.

CAR NAME	ST	CAR TYPE	CAR BUILDER (Owner)	DATE BUILT	PLAN	LOT
RAVEN		24ChDrB	Pullman	6/22/00	1518C	2552
RAVENDEN		12SecDrB	PCW Det	4/7/88	375E	D328
RAVENNA	N	14SecO	B&O RR	/80	657C	
RAVENNA #511	P	12SecDr	Wagner	4/92	3065C	Wg66
RAVENSWOOD	P	16Sec	B&S	12/82	3202	
RAWLINS		12SecO	B&S	12/70	5	Day
RAWLINS		10SecDr2Cpt	Pullman	9/21/05	2078	3203
RAWSON	N	12SecDr	PCW	8/ /98	1318A	2313
RAYMOND	RN	Diner	H&H	7/73	40C	Wil2
RAYMOND		12SecDr	Pull Buf	6/27/07	1963C	B62
RAYMOND	P	10SecDr	(Wagner)	/85	3009	
RAYNHAM	P	26ChBarB SUV	Pull P-2493	1/11	3274	3820
READING		12SecDr	Pullman	3/30/03	1581D	2983
READING	P	10SecDr	(Wagner)	/85	3009	
READING #7	N		(WS&PCC)	7/16/72		
READVILLE	P	24Ch3StsDr	B&S	3/93	3261	
REBEMAR	N	12SecDr	Pullman	10/10/07	2163A	3562
RECRUIT	N	24ChDrB	Pullman	6/6/05	1988A	3170
RED CLOUD		12SecDr	PCW Det	8/21/82	93	D169
RED CLOUD	R	12SecDr	PCW Det	8/21/82	1602B	D169
RED LODGE		10SecDrCpt	PCW	5/17/90	668A	1709
RED WING #68	P	12SecDr	H&H	/80-/82		
REDFIELD		14SecDr	PCW	12/22/99	1318J	2450
REDFIELD	R	12SecDr SUV	PCW	12/12/99	2296	2450
REDLAND	N	10SecDrB	PCW	2/11/84	149B	101
REDLAND	N	12SecDr	Pullman	3/21/10	1963E	3807
REDLAND	R	16Sec	PCW	2/11/84	1877C	101
REDONDO		10Sec2Dr	Pullman	11/1/02	1746A	2928
REDONDO #4		Diner	PCW	4/25/88	538D	1396
REDWING		24ChDrB	Pullman	7/10/00	1518C	2553
REDWOOD	N	12SecDr	PCW Det	5/19/81	93N	D107
REFECTORIO	N	BaggSmkB SUV	New Haven RR	11/04	3218B	
REGAL	N	14SecO	PCW Det	5/27/93	1021	D525
REGAL	P	10Cpt	Wagner	6/96	3075A	Wg111
REGENT	N	10SecB	PFtW&C	8/1/70	14C	FtW2
REGENT		12SecDrB	Pullman	12/17/07	2326	3587
REGENT	P	12SecDr	Wagner	2/96	3077	Wg81
REGGIO		12SecDr	Pull Buf	11/6/02	1581C	B11
REGINA		20Ch6StsDr	PCW	7/29/91	928A	1827
REGINA		12SecDrSr	PCW	12/17/98	1318C	2370
REGINALD		12SecDr	Pullman	12/23/07	1963C	3586
REGINALD #558	P	14SecSr	H&H (B&O)	? 8/87	654F?	
REGULUS		BaggDiner	PCW	12/31/90	783A	1705
REGULUS	N	12SecDr	Pullman	10/13/05	2049B	3232
REGULUS	N	10SecDr	(Wagner)	/84	3057C	
REINA	RN	24Ch3Sts	PCW Det	12/30/80	1931A	D88
REINDEER		12SecO	G,B&Co/GMP	/68	5	Troy
REITA		12SecDrB	PCW	11/17/88	375G	1513
RELAY		12SecDr	Pullman	12/21/07	1963C	3586
RELAY #503	P	14SecO	B&O RR	/80	657C	
RELIANCE		10SecDr3Cpt	Pullman	4/19/06	2078C	3342
REMBRANT		12SecDr	Pullman	8/13/01	1581A	2727
REMINGTON		12SecDr	Pullman	11/4/01	1581A	2784
REMUS	P	12SecDr	Wagner	3/93	3065C	Wg84
RENDCOMB		12SecDr	Pullman	2/8/08	2163A	3599
RENFREW		12SecDr	Pull Wilm	8/10/04	1581C	W16
RENNERT		12SecDrB	Pullman	12/16/07	2326	3587
RENO		10SecDr2Cpt	Pullman	9/21/05	2078	3203
RENO CP-34	P	10Sec	(CP Co)	10/31/84	37L	
RENOVO		16Sec	Pullman	7/16/03	1721	3014
RENOVO #186	P	OS	Phil&Erie RR	5/71	27	Erie1
RENSSELAER		12SecDr	Pullman	12/11/05	1963C	3252
RENWICK	N	12SecDr	PCW	4/29/92	784E	1899
REPUBLIC		12SecO	DetC&MCo	/68	5	Det
REPUBLIC		10SecDrB	PCW	4/3/88	445F	1377
REPUBLIC	RN	12SecDrCpt	PCW	4/23/93	1011D	1968
REPUBLIC		Private	Pullman	1/27/06	2110	3262
RESOLUTE		12SecDr	Pull Buf	7/1/07	1963C	B62
RESOLUTE	P	12SecO	(Wagner)	/85	3027D	
RESTORIA	P	8Cpt8Berth	Wagner	8/90	3203	
REVA		10SecDrB	PCW	12/10/89	725B	1629
REVELSTONE		16Sec	Pullman	5/17/09	2184D	3700
REVERE	P	12SecO	CB&Q RR	C:/70	17	Aur1
REVERE		12SecDr	Pullman	8/14/01	1581A	2727
REVERIE	N	12SecDrB	PCW	8/25/88	375G	1480
REVILLO	RN	10SecDrB	B&S	3/72	149S	Day7
REXFORD	N	10SecDrB	PCW Det	4/19/84	149D	D242
REXHAM		16SecO	Pull Cal	/01	1637A	C2
REYNOSA		10SecDrB	PCW	3/22/88	445E	1361
REYNOSA	N	12SecDrSr	Pullman	4/7/05	2049A	3169
RHAITIA		12SecDr	Pull Buf	1/15/04	1581F	B27
RHEIMS		12SecDr	PCW	5/28/92	784E	1915
RHINE		12SecDr	PCW Det	5/25/91	784D	D462
RHINECLIFF		8CptLibObs	Pullman	12/22/06	2260	3502
RHINELAND		14SecDr	Pullman	8/8/00	1523E	2567
RHODA	N	10SecB	PCW Det	9/2/73	37	D16
RHODA		24ChDr	Pullman	7/7/05	1980A	3171
RHODE ISLAND		12SecB	PCW Det	7/17/76	57	D41
RHODE ISLAND		12SecDr	Pull Buf	6/21/07	1963C	B62
RHODES	N	12SecDrB	Wagner	11/92	3073	Wg75
RHODESIA		12SecDr	Pullman	6/23/03	1581F	3010
RHODOPE	N	10Sec	B&O RR	?/87	658	
RHONE	P	12SecDr	Wagner	3/93	3065C	Wg84
RIALTO		12SecDrB	PCW Det	8/18/88	375G	D346
RIALTO	N	12SecDrCpt	PCW	4/23/93	1011D	1968
RIALTO	N	10SecDr	(Wagner)	/85	3033	
RICARDO	P	24Ch6Sts	Wason	7/88	3257A	
RICHELIEU		14SecO	PCW Det	8/9/82	115	D155
RICHELIEU		Diner	PCW	8/1/93	538D	2010
RICHELIEU		16Ch6StsDr	(B&M RR)	6/93	1100	
RICHELIEU	P	20Ch6Sts	B&S	7/89	3258A	
RICHELIEU #284	P	8CptB	Wagner	6/89	3064	Wg17

CAR NAME	ST	CAR TYPE	CAR BUILDER (Owner)	DATE BUILT	PLAN	LOT
RICHFIELD	P	14Ch5StsB	(DL&W RR)	4/75	211	
RICHFIELD	N	12SecDr	Pullman	10/20/05	2049B	3232
RICHLAND		10SecDr2Cpt	Pullman	10/8/06	2078D	3398
RICHMOND		12SecDr	Pullman	12/21/09	1963E	3766
RICHMOND #30	P	12SecO	G,B&Co/GMP	/68	5	Troy
RICHMOND #283	P	8CptB	Wagner	6/89	3064	Wg17
RICHTON		12SecDr	Pullman	11/3/00	1581A	2589
RICO	N	8SecSrB (NG)	PCW Det	10/30/82	155	D173
RIDEAU	N	12SecO	Grand Trunk	4/71	18C	Mon1
RIDEAU		12SecDr	Pullman	10/23/06	1963C	3408
RIDEAU	P	12SecO	(Wagner)	4/70	3010B	
RIDGEFIELD	P	24Ch3StsDr	B&S	3/92	3261	
RIDGELAND		14SecDr	Pullman	8/8/00	1523E	2567
RIDGELEY	N	10secDr2Cpt SUV	Pullman	3/9/10	2471	3784
RIDGEWOOD		21Ch6StsDrObs	Pullman	3/20/06	2091B	3334
RIENZI	N	10SecDrB	PCW	5/29/84	149M	192
RIENZI	RN	16Sec	PCW	5/29/84	1877	192
RIENZI	P	10SecDr	(Wagner)	/85	3053	
RIGA	P	12SecDr	Wagner	1/92	3065C	Wg84
RIGA #2520		14SecO	PCW Det	1/14/93	1014A	D510
RIGOLETTO		12SecDr SUV	Pullman	3/18/08	1963C	3604
RIGOLETTO #113	P	7BoudB	(MannBCC)	/84	697	
RIMMON	N	12SecDrSr	Pullman	3/1/01	1581B	2654
RIMOUSKI		10SecB	Grand Trunk	1/74	37	Mon5
RINALDO		12SecDrCpt SUV	Pullman	8/19/05	2103	3227
RINCON		10SecB	PCW	6/7/84	206	198
RINCON	R	12SecDrB	PCW	6/7/84	1611	198
RINGOLD		16SecO	Pull Cal	/01	1637A	C2
RIO AMAPA	N	10SecDrB	PCW	4/3/88	445F	1377
RIO CASALAPA	N	10SecDrB	PCW	6/18/87	445M	1322
RIO GRANDE		10SecDr	PCW Det	7/23/81	112	D118
RIO GRANDE #98	P	14SecO	(Wagner)	/69	3019	
RIO HONDO	RN	6SecCafeLng SUV	Pull	6/2/06	3986	3361
RIO JALTEPEC	N	10SecDrB	PCW Det	3/23/87	445C	D282
RIO TRINIDAD	N	10SecDrB	PCW Det	3/23/87	445C	D282
RIO VERDE	RN	6SecCafeLng SUV	Pull	4/3/07	3986	3523
RIO VISTA CP-36	P	10Sec	(CP Co)	10/13/84	37N	
RIONDA	N	12SecDr	Pullman	10/1/01	1581A	2765
RIPON		12SecDr	PCW	4/23/91	784D	1801
RIPON	P	12SecDr	Wagner	3/93	3065C	Wg84
RIPPOWAM	N	12SecDr	Pullman	12/12/05	1963C	3252
RIVA #241	P	Private	Wagner	4/87	3081	Wg1
RIVANNA		12SecDr	Pullman	11/30/01	1581A	2784
RIVANNA	P	8BoudB	J&S (Union PCC)		373A	
RIVARD		16Sec	Pullman	6/2/06	2184	3361
RIVER POINT	P	27Ch8Sts2DrBObs	New Haven RR	11/04	3270	
RIVERDALE		12SecDrB	PCW	8/19/96	1204C	2187
RIVERHEAD	N	12SecDr	Pullman	5/11/04	1963	3082
RIVERSIDE	P	Saloon	B&S (H&StJ RR)	/70		Day3
RIVERSIDE	N	12SecDr	B&O RR	/79	657B	
RIVERSIDE	P		(CRI&P Assn)	5/5/80		

CAR NAME	ST	CAR TYPE	CAR BUILDER (Owner)	DATE BUILT	PLAN	LOT
RIVER- SIDE SPC-8158	P	10Sec2Dr	PCW	3/15/99	1158	2376
RIVERTON		12SecDr	PCW Det	9/8/82	93	D169
RIVERTON	R	12SecDr	Pullman		1581C	
RIVERTON	P	12SecO	(Wagner)		3027D	
RIVIERA		BaggBSmkBarber	PCW	10/4/95	1155A	2122
RIVOLI		14SecO	PCW Det	4/8/93	1021	D517
RIZAH		14SecO	PCW	3/2/93	1021	1969
ROAMER		12SecDr	Pullman	9/16/09	1963E	3744
ROANOKE			B&O RR			
ROANOKE		12SecDrSrB	PCW	1/26/99	1319B	2400
ROANOKE #508	P	12SecDr	Wagner	4/92	3065C	Wg66
ROANOKE CTC-83	R	C	B&O rblt/75			
ROB ROY #74	P	25Ch2StsB	(Wagner)	6/82	3100B	
ROBERUAL		16Sec	Pullman	6/2/06	2184	3361
ROBIN		24ChDrB	Pullman	6/2/00	1518C	2552
ROCHEFORT		12SecDr	Pullman	11/2/03	1581F	3025
ROCHELLE		10SecDrB	PCW	12/5/87	445E	1360
ROCHELLE	P	12SecDr	(Wagner)	8/82	3023B	
ROCHESTER		14SecDr	PCW	9/2/99	1318E	2447
ROCK ISLAND		12SecDr	PCW Det	9/30/80	91	D82
ROCK ISLAND		12SecDr	Pull Buf	6/25/07	1963C	B62
ROCKAWAY		10SecDr2Cpt	Pullman	10/9/06	2078D	3398
ROCKAWAY #101	P	R-end	(WS&PCC)	8/5/81	465?	
ROCKAWAY #242	P	SummerParlor	H&H	7/73	40	Wil2
ROCKBRIDGE		10SecDrB	PCW	3/15/88	445B	1419
ROCKBRIDGE		12SecDr	Pullman	12/26/07	1963C	3586
ROCKBURN		26ChDr	Pullman	3/27/06	1987A	3335
ROCKDALE		16Sec	Pullman	6/2/06	2184	3361
ROCKET	RN	Private	Wagner	4/89	3062-O	Wg14
ROCKFORD		12SecDr	PCW Det	12/12/88	635	D377
ROCKFORD		12SecDr	Pullman	7/30/03	1581F	3015
ROCKFORD	R	30Ch SUV	Pullman	7/30/03	3581E	3015
ROCKHAM	N	12SecDrB	PCW	9/6/88	375G	1480
ROCKINGHAM		12SecDr	Pullman	3/6/07	1963C	3525
ROCKLAND		10SecDr	PCW Det	4/30/71	19A	D6
ROCKLAND		14SecDr	Pullman	7/31/00	1523E	2567
ROCKLEDGE	N	26ChDrB	Pullman	6/27/07	1988B	3526
ROCKLIFFE	N	16Ch12StsB	B&S (B&O)		664	
ROCKVILLE		12SecDrSr	Pullman	3/22/05	2049A	3169
ROCKWOOD		12SecDr	Pullman	10/30/07	1963C	3570
ROCKWOOD #533	P	10Sec	B&O RR	?/87	658	
ROCROY	N	20Ch8StsDrB	Wagner	6/91	3126A	Wg51
RODANO SP-71	P	10SecDrB	PCW	6/28/88	445F	1415
RODERICK		12SecDr	Pull Buf	10/18/05	1963C	B47
RODERIGO		12Sec2Dr	Pullman	7/21/01	1580C	2742
RODOSTO		12SecDr	Pullman	12/21/07	1963C	3586
ROGER WILLIAMS	N	22Ch6StsDr	PCW	9/1/92	976A	1910
ROGERO		12SecDrCpt	Pullman	6/24/04	1982	3090
ROGUE RIVER	N	10SecDr2Cpt	Pullman	11/15/07	2078D	3583
ROKEBY		12SecDr	Pull Buf	10/23/05	1963C	B47

CAR NAME	ST	CAR TYPE	CAR BUILDER (Owner)	DATE BUILT	PLAN	LOT
ROLAND	P	20Ch6Sts	Wagner	8/90	3258A	Wg40
ROLLA	N	10SecDr	PCW Det	6/16/81	112	D118
ROLLA	R	12SecDrB	PCW Det	6/16/81	1611	D118
ROLLINS		12SecDr	Pullman	12/12/05	1963C	3252
ROMANO		BaggSmk	H&H ?	? 8/72	534	
ROMANO #152	RN	10SecDrB	J&S (WS&PCC)	7/7/86	696	
ROMANOF		12SecDr	Pullman	6/25/01	1581A	2726
ROMANOF	R	30Ch SUV	Pullman	6/25/01	3581E	2726
ROMANTIC		12SecDr	Pullman	12/28/07	1963C	3586
ROME		12SecDr	PCW	12/23/90	784	1780
ROMEO		12Sec2Dr	Pullman	4/3/01	1580A	2646
ROMFORD	P	24Ch6StsB	B&S	1/94	3262B	
ROMULUS		10SecDrB	PCW	9/27/84	149D	1006
ROMULUS		BaggClubBarber	PCW	11/26/98	1372B	2365
RONCEVERTE		12SecDr	Pullman	1/12/07	2163	3449
RONDINELLA	N	12SecDr	Pullman	12/27/06	2163	3460
RONDOUT		12SecDr	Pullman	9/30/03	1581F	3022
RONKONKOMA #94	P	28Ch5StsDr	Wagner	9/89	3125B	Wg18
ROQUEFORT		12SecDr	Pullman	2/6/04	1581H	3053
ROSA BLANCA	RN	12SecDrB	Pullman	5/24/06	2165C	3352
ROSABELLE	N	26Ch3StsDr	PCW	4/27/98	1278C	2258
ROSALIE	N	20Ch3StsB	PCW Det	9/22/83	167A	D209
ROSALIND		20Ch6StsB	PCW	1/22/85	167B	200
ROSAMOND	P	18Ch3StsDrObs	(B&M RR)		1203	
ROSAMOND	N	27ChDrObs	Pullman	6/8/08	2346	3624
ROSARA #103	N	7BoudB	(MannBCC)	/83	879?	
ROSARIO		12SecDr	Pullman	12/31/07	1963C	3586
ROSCIUS	N	12SecO	PCW Det	11/13/71	27C	D8
ROSCIUS	P	12SecO	B&O (Union PCC)		665F	
ROSCIUS	N	BaggBLibSmk	Pullman	4/24/07	2087	3534
ROSEBURG	R	16Sec	PCW	6/2/85	1881	1048
ROSEBURG	N	10SecDr2Cpt	Pullman	10/5/06	2078D	3398
ROSEBURG O&C-8004	P	10SecDrB	PCW	6/2/85	240	1048
ROSEDALE		10SecDr	H&StJ RR	/71	19	Han2
ROSEDALE		12SecDrB	PCW	8/15/96	1204C	2187
ROSEDALE	P	10SecDr	(Wagner)	/88	3057A	
ROSELAND		10SecDrB	PCW	5/10/89	623B	1566
ROSELAWN		12SecDr	Pullman	12/23/07	1963C	3586
ROSELLE	N	10SecB	B&O RR?		649	
ROSEMARY		24ChDrSr	Pullman	7/22/04	1980A	3088
ROSEMERE		12SecDr	Pullman	10/23/06	1963C	3408
ROSEMONT		10SecDr	B&S	1/71	26	Day6
ROSEMONT	RN	SlprParlor	PRR		17	Alt1
ROSEMONT		12SecDrB	PCW	9/9/98	1319A	2314
ROSETTA #397		20Ch5StsB	PCW	1/8/84	167A	90
ROSEVILLE		10SecDr2Cpt	Pullman	9/29/05	2078	3203
ROSEWOOD	N	12SecDr	PCW Det	4/19/82	93X	D142
ROSITA		24ChDr	Pull Buf	4/3/07	2156	B58
ROSITA #109	N	Parlor	(WS&PCC)	/75	696	
ROSLIN #414	P	12SecDrB	Wagner	2/90	3065	Wg32
ROSLYN	N	6Sec2DrB	Grand Trunk	11/70	17J	3
ROSLYN #114	P		(WS&PCC)	7/83		
ROSSANO		12SecDrCpt SUV	Pullman	8/21/05	2103	3227
ROSSEAU		12SecDr	Pullman	12/26/07	1963C	3586
ROSSERNE	N	12SecDr	Pullman	11/30/01	1581A	2784
ROSSINI #330	P	10SecDrB	(UnionPCC)		565B	
ROSSLYN		12SecDrB	Pullman	12/16/07	2326	3587
ROSSMORE		10SecDrB Hotel	PCW Det	5/24/76	56	D39
ROSSMORE		12SecDr	Pullman	12/30/07	1963C	3586
ROSWELL	P	20Ch6Sts	Wagner	8/90	3258A	Wg40
ROTA		12SecDr	PCW	12/20/90	784	1780
ROTHSAY		12SecDr	PCW Det	4/7/92	784E	D497
ROTONDO	P	19Ch3Sts	B&S	7/82	3252	
ROTTERDAM		12SecDr	Pullman	5/17/04	1963	3082
ROUMANIA		10Cpt	PCW	2/8/93	993A	1951
ROUMANIA #512	P	12SecDr	Wagner	4/92	3065C	Wg66
ROUPHIA		12SecDr	Pull Buf	8/2/04	1963	B40
ROUSSEAU	N	12SecDr	Pullman	12/26/07	1963C	3586
ROVER		12SecO	PCW Det	3/7/71	5	D5
ROVER	RN	Private	Wagner	7/92	3062-O	Wg78
ROVER #16	P	Slpr	(SouTranCo)			
ROVER #43	N	Private	H&H (WS&PCC)	5/25/78	689	
ROWAYTON	P	36Ch SUV	Pull P-2494	3/11	3275	3821
ROWEN		12SecDr	PCW Det	2/5/92	784E	D488
ROWENA		20Ch5StsB	PCW Det	6/30/83	167	D203
ROWLAND		12SecDr	Pullman	3/12/06	2163A	3333
ROXANA		18Ch6StsDrB	PCW Det	9/5/90	772D	D427
ROXBURY #150	RN	BaggBLib?	J&S (WS&PCC)	5/17/86	720D	
ROXBURY #513	P	12SecDr	Wagner	4/92	3065C	Wg66
ROYAL	?					
ROYAL CITY	RN	5CptDrBLib	Pullman	2/25/02	1751G	2799
ROYALTON		10SecDrB	PCW	9/25/87	445E	1365
ROYALTON		12SecDr	Pull Buf	7/13/03	1581C	B19
ROYSTON	P	16Sec SUV	B&S	7/11	3212	
RUBENS		12SecDr	Pullman	7/29/01	1581A	2727
RUBICON	N	Parlor	PCW Det	2/5/72	22	D9
RUBICON	R	Parlor	PCW Det	2/5/72	29O	D9
RUBICON		12SecDr	Pullman	1/23/04	1581H	3053
RUBRIC	P	18Ch6StsB	B&S	11/88	3256	
RUBY	P	Commissary	C&NW Ry	/70	27	Chi3
RUBY		12SecDrSr	PCW	4/28/99	1318C	2390
RUBY #57	P	20Ch6StsSr	(Wagner)	6/96	3118	
RUDOLPH	P	18Ch6StsB	B&S	11/88	3256	
RUELLA	N	10SecDrB	PCW Det	8-11/80	63N	D77
RUELLA	N	12SecDrSr	Pullman	3/22/05	2049A	3169
RUFFORD		12SecDr	Pullman	12/23/09	1963E	3766
RUFFSDALE		12SecDr	Pullman	1/4/07	2163	3460
RUGBY	P	12SecDr	Wagner	3/93	3065C	Wg84
RUGBY #14	N		(WS&PCC)	9/28/72		
RUHE		14SecO	PCW Det	3/23/93	1021	D517
RUISSEAUMONT		12SecDr	Pullman	11/5/03	1581F	3025
RUMELIA		12SecDr	PCW	10/2/99	1318E	2449

CAR NAME	ST	CAR TYPE	CAR BUILDER (Owner)	DATE BUILT	PLAN	LOT
RUMSON #173	P	18Ch6StsDr	(WS&PCC)	6/23/88	730A	
RUNCORN CTC-54	R		(O&M RR)	rblt /70		
RUNNYMEDE		12SecDr	Pullman	1/25/04	1581H	3053
RUPICOLA	N	12SecO	B&O (Union PCC)	/80	665F	
RUSKIN		12SecDr	Pullman	7/25/01	1581A	2727
RUSSIA		10SecDr	B&S	/70	19	Day5
RUSSIA		12SecDr	PCW	6/7/87	473A	1323
RUSSIA	N	12SecDrB	(Wagner)	6/86	3050	
RUTGERS		12SecDr	Pullman	12/12/05	1963C	3252
RUTH		18Ch4StsDrB	PCW	6/24/93	1039A	1994
RUTHELLA		20Ch6StsDr	PCW	7/31/91	928A	1827
RUTHERFORD		21Ch6StsDrObs	Pullman	3/20/06	2091B	3334
RUTHVEN #2550		14SecO	PCW Det	2/18/93	1021	D516
RUTLAND	P	10SecDr	Vermont Cent	3/72	31	StA2
RUTLAND		12SecDr	Pullman	5/10/03	1581D	3004
RUTLAND	P	12SecDr	(Wagner)	/86	3027B	
RUTLEDGE		10SecDr2Cpt	Pullman	11/20/07	2078D	3583
RUVO		12SecDr	PCW	9/18/90	784	1765
RUXTON		12SecDr	Pullman	5/18/08	1963C	3616
RYDAL		24ChDrB	Pull Buf	4/29/07	2278	B59
RYDE		12SecDr	Pullman	3/13/06	2163A	3333
RYE BEACH		26Ch6StsDr	PCW		1061	
SAALE		14SecO	PCW	2/28/93	1021	1969
SABARA		12SecDr	PCW	1/23/91	784	1787
SABETHA		10SecDrB	PCW	2/17/88	445F	1369
SABINA #219	P	12SecDr	PCW	6/12/86	93H	1199
SABINE	R	12SecDrB	PCW	10/13/88	1566	1510

CAR NAME	ST	CAR TYPE	CAR BUILDER (Owner)	DATE BUILT	PLAN	LOT
SABINE	N	12SecDrCpt	Pullman	8/8/05	2088	3221
SABINE M&LT-1	N	12SecO	M&LT RR	3/21/85	114	
SABRINA #335	P	10SecDrB	(Union PCC)		565B	
SABULA #78	P	10SecDr	H&H	/81		
SACHEM	N	10SecDr	H&H	10/70	19	Wil2
SACHEM	N	12SecDr	Wagner	8/91	3065D	Wg59
SACO	N	10SecDr	B&S	4/71	19E	Day5
SACRAMENTO	N	12SecO	C&NW Ry	12/70	5	Fdl
SACRAMENTO #247	P	12SecDr	Wagner	4/88	3061B	Wg10
SADOWA	N	4Sec5SrB	PCW (Wagner)	11/23/86	3045	1246
SAFRANO		10SecDr2CptSUV	Pullman	11/19/09	2451	3747
SAG HARBOR #93	P		J&S (WS&PCC)	7/9/81		
SAGAMORE #248	P	12SecDr	Wagner	4/88	3061B	Wg10
SAGEM		14SecDr	Pullman	5/16/06	2165	3352
SAGEMERE		12SecDrB	Pullman	12/19/07	2326	3587
SAGINAW #249	P	12SecDr	Wagner	4/88	3061B	Wg10
SAGOULA		10SecDrB	PCW	5/31/89	623B	1566
SAGUENAY #64	P	28Ch3Sts	PCW (Wagner)	6/5/86	3124	1194
SAHARA		10SecDr3Cpt	Pullman	4/14/06	2078C	3342
SAHIBA	N	12SecDr	Pullman	12/28/06	2163	3460
SAHWA		12SecDr	Pullman	8/27/04	1581K	3118
SAKETO	N	SDR ?	J&S	3/71	27	Wil4
SALA	N	BaggSmB SUV	New Haven RR	6/11	3219A	
SALADIN	N	12SecDr	PCW P-93H	9/3/86	3049	1205
SALAMEA		12SecDr	Pullman	9/26/06	1963C	3397
SALAMIS		10SecDrB	PCW	9/26/84	149D	1006
SALANIO		12Sec2Dr	Pullman	7/17/01	1580C	2742

Figure I-35 The most popular railroad car plan (12-section drawing room) was used for the SABARA. The letterboard advertises its use on the Grand Trunk-Boston & Maine Route between Chicago and Boston. In November 1916, this car, along with six others, was sold to the Hillsboro & Northeastern RR in Wisconsin. Smithsonian Institution Photograph P1327.

CAR NAME	ST	CAR TYPE	CAR BUILDER (Owner)	DATE BUILT	PLAN	LOT
SALARINO		12SecDrCpt SUV	Pullman	6/24/04	1982	3090
SALEM	N	10SecDr2Cpt	Pullman	3/25/07	2078D	3528
SALEM O&C	P	10SecDrB	PCW	6/20/85	240	1048
SALERNO	N	10SecB	Grand Trunk	1/74	37M	Mon5
SALERNO		12SecDr	Pull Buf	1/24/03	1581C	B12
SALFORD		14SecO	PCW Det	4/12/93	1021	D517
SALIDA		10SecB (NG)	PCW Det	10/23/83	178	D211
SALINA		12SecDr	Pullman	5/3/02	1581B	2849
SALINA #221	S	12SecDr	PCW (Wagner)	6/10/86	93H	1199
SALINAS SP-49	P	10SecDrB	PCW	12/24/88	445H	1519
SALISBURY		12SecDr	Pull Buf	5/31/04	1963	B41
SALLUST	N	14Ch5StsDr	(Wagner)	5/71	3106C	
SALMON RIVER	P	10Sec2DrB	Wagner	9/92	3071A	Wg77
SALO		12SecDr	PCW	9/19/90	784	1765
SALOME		12SecDrB	PCW StL	2/8/88	375E	StL1
SALONA		12SecDr	PCW	1/24/91	784	1787
SALONICA		12SecDr	PCW	12/11/92	990	1941
SALT LAKE		10SecB (NG)	PCW Det	10/16/83	178	D211
SALT LAKE	RN	10SecDrSr	PCW	6/17/90	784R	1715
SALTILLO	N	8SecSrB (NG)	PCW Det	11/18/82	155	D173
SALTILLO		12SecDr	Pullman	7/21/06	1963C	3385
SALUDA	R	10SecO	PCW Det (rblt)	10/26/74	47	D27
SALUDA	N	23ChDrLibObs	Pullman	8/28/06	2206	3388
SALVADOR		10Cpt	PCW	1/30/93	993A	1951
SALVIA		24ChDr	Pull Buf	6/9/06	2156	B53
SALVIUS	R	BaggSmkBLib	PCW		1706	
SALVO	N	10SecB	J&S	6/73	37I	Wil12
SAMAR		12SecDr	Pullman	5/14/04	1963	3082
SAMARIA		14SecDr	PCW	8/9/99	1318D	2447
SAMARIA #207	P	Parlor	PRR	6/10/72	29	Alt1
SAMOA		10SecDrB	PCW	2/8/84	149B	101
SAMOA	P	14SecDr SUV	Pull P-2069A	5/06	3207B	3320
SAMOSET	N	10SecDrB	PCW	10/29/91	920G	1867
SAN						
ANDREAS	R	12SecDr	PCW	1/14/88	1603	1361
ANDREAS SP-54	P	10SecDrB	PCW	1/14/88	445E	1361
ANTONIO		14SecDr	Pull Buf	10/31/00	1582A	B3
ANTONIO GH&SA	P	Slpr	(SP Assn)			
ARDO CP-8073	P	10Sec2Dr	PCW	11/12/95	1158	2125
ARDO CP-8076	P	10Sec2Dr	PCW	4/12/93	995A	1942
BENITO		10SecDrB	PCW	1/11/87	445B	1297
BENITO	R	10SecObs	PCW	8/23/88	445V	1499
BENITO SP-29	P	10SecDrB	PCW	8/23/88	445F	1499
CARLOS		10SecO (NG)	PCW Det	8/22/81	73A	D108
DIEGO #257	P	12SecDr	Wagner	5/88	3061B	Wg11
DIEGO SPofA-4	P	Slpr	(SP Co)	7/1/83	37?	
FELIPE CP-8069		10Sec2Dr	PCW	7/21/93	995A	1942
FRANCISCO	N	12SecO	C&NW RR	12/70	5	Fdl
FRANCISCO		14SecDr	Pull Buf	10/27/00	1582A	B3
FRANCISCO #60	P	12SecDr	(CRI&P Assn)	7/20/80	527	
FRANCISCO #256	P	12SecDr	Wagner	5/88	3061B	Wg11
SAN						
GABRIEL	N	28Ch4StsDrObs	Pullman	1/16/06	2116A	3273
JACINTO	N	12SecO	C&NW Ry	2/28/71	4	
JUAN		10SecO (NG)	PCW Det	11/5/79	73	D61
JUAN		12SecDr	Pullman	8//04	1963	3098
LEANDRO	R	12SecDr	PCW	3/28/88	1603	1377
LEANDRO SP-47	P	10SecDrB	PCW	3/28/88	445F	1377
LORENZO	R	12SecDr	PCW	1/7/87	1603	1297
LORENZO SP-28	P	10SecDrB	PCW	1/7/87	445B	1297
LUCAS CP-8078	P	10Sec2Dr	PCW	4/15/93	995A	1942
MARCIAL		10SecDr	PCW Det	7/5/81	112	D118
MARCO #258	P	12SecDr	Wagner	5/88	3061B	Wg11
MARCOS SP-48		10SecDrB	PCW	2/24/88	445F	1380
MARCOS GH&SA		Slpr	(SP Co)		37?	
MATEO	R	10SecObs	PCW	12/25/86	445W	1297
MATEO SP-27	P	10SecDrB	PCW	12/25/86	445B	1297
MIGUEL SP-49		10SecDrB	PCW	1/28/88	445E	1361
PABLO CP-26	P	10SecDr	(CP Co)	10/17/84	37K	
PEDRO CP-46	P	10SecDrB	PCW Det	12/22/83	149	D213
RAFAEL	R	10SecObs	PCW	12/25/86	445W	1297
RAFAEL SP-26	P	10SecDrB	PCW	12/25/86	445B	1297
RAFAEL SP-8101		10SecDrB	PCW	12/25/86	445B	1297
RAMON CP-8072	P	10Sec2Dr	PCW	3/8/93	995A	1942
SALVADOR	P	22Ch4StsDrB	Wagner	8/93	3130	Wg96
VICENTE CP-8079	P	10Sec2Dr	PCW	5/6/93	995A	1942
SANDHURST		12SecDr	Pullman	10/2/03	1581F	3022
SANDOVAL		12SecDr	Pull Buf	11/14/05	1963C	B48
SANDUSKY		12SecDr	Pullman	10/30/07	1963C	3570
SANDUSKY #509	P	12SecO	B&O RR	11/80	657	
SANDY HOOK #243	P	SummerParlor	H&H (B&O)	7/30/74	40	Wil2
SANFORD	N	10SecDrB	PCW Det	3/26/84	149F	D234
SANGAMON		10SecDr	PCW Det	1/31/71	37 1/4	D51
SANGAMON		12SecDr	Pullman	9/7/07	1963C	3550
SANTA						
ANA		10SecDr	PCW	1/19/85	229	1032
ANITA		28Ch4StsDrObs	Pullman	1/16/06	2116	3273
BARBARA	R	12SecDr	PCW	12/27/86	1603	1297
BARBARA SP-25	P	10SecDrB	PCW	12/27/86	445B	1297
CLARA		10SecDrB	PCW Det	12/15/83	149	D213
CLARA		28Ch4StsDrObs	Pullman	1/16/06	2116	3273
CLARA CP-44	P	10SecDrB	PCW Det	12/15/83	149	D213
CRUZ	R	12SecDr	PCW	3/27/88	1603	1377
CRUZ SP-56	P	10SecDrB	PCW	3/27/88	445F	1377
FE		10SecDr	PCW Det	8/26/76	59	D43
HELENA CP-8074	P	10Sec2Dr	PCW	4/12/93	995A	1942
LUCIA IM-4 SP-52		10SecDrB	PCW	1/28/88	445E	1361
MARIA		22Ch6StsDr	PCW	4/25/93	1031	1978
MARIA	P	6Sec8ChClub	(ATSF)		1757	
MARIA SP-55	P	10SecDrB	PCW	2/11/88	445F	1380
MONICA		28Ch4StsDrObs	Pullman	1/16/06	2116	3273
PAULA CP-8068		10Sec2Dr	PCW	7/20/93	995A	1942

CAR NAME	ST	CAR TYPE	CAR BUILDER (Owner)	DATE BUILT	PLAN	LOT
SANTA						
RITA		10SecDrB	PCW Det	12/15/83	149	D213
RITA CP-45	P	10SecDrB	PCW Det	12/15/83	149	D213
ROSA SP-30	P	10SecDrB	PCW	1/10/87	445B	1297
SUSANA		28Ch4StsDrObs	Pullman	1/16/06	2116	3273
YSABEL CP-8077	P	10Sec2Dr	PCW	4/13/93	995A	1942
SANTAQUIN		10SecDr2Cpt	Pullman	3/23/07	2078D	3528
SANTEE		12SecDrSrB	PCW	1/25/99	1319B	2400
SANTEE #39		10SecO	PCW Det (rblt)	10/26/74	47	
SANTIAGO	R	12SecDrB	PCW Det	8/11/83	1468	D193
SANTIAGO #394		12SecB	PCW Det	8/11/83	168	D193
SANTIAGO #222		12SecDr	PCW (Wagner)	7/2/86	3049	1199
SANTOS		14SecO	PCW	5/18/93	1021	1972
SANTOS		12SecDr	Pullman	5/19/04	1963	3082
SAONA	N	10SecDr	J&S	2/73	26A	Wil9
SAONA	P	14SecDr SUV	Pull P-1855	1/03	3207	2900
SAPA		12SecDr	PCW	5/8/91	784D	1805
SAPPHIRE		12SecDrSr	PCW	4/17/99	1318C	2390
SAPPHIRE		16Sec	Pullman	1/5/09	2184D	3662
SAPPHO	P	Parlor	H&H	12/72	35	Wil1
SAPPHO	P	7CptObs	Wagner	9/92	3062A	Wg78
SAPPHO		26ChDr	Pullman	7/30/04	1987A	3097
SARACEN		16Sec	Pullman	5/17/09	2184D	3700
SARAGOSSA		12SecDr	Pullman	6/17/04	1963	3082
SARANAC #122	P	20Ch8StsDrB	Wagner	6/91	3126A	Wg51
SARANAC #146		10SecDr	H&H	10/70	19	Wil2
SARANO #130	RN	7BoudB	(MannBCC)	/84	776A	
SARASOTA	N	5CptDrBLibObs	Pullman	3/13/02	1751C	2799
SARATOGA	P	28Ch8StsDrBObs	Wagner	5/99	3131	Wg129
SARATOGA #215		Parlor	PCW Det	8/2/71	22	D7
SARATOGA #85	P		B&S (WS&PCC)	3/1/81		
SARATOGA #86	N		B&S (WS&PCC)	3/1/81		
SARDINIA		10Cpt	PCW	2/8/93	993A	1951
SARDINIA #49	P	15Ch5StsDr	(Wagner)	4/85	3106	
SARDIS		12SecDr	Pullman	3/18/03	1581D	2975
SARDIS	R	30Ch SUV	Pullman	3/18/03	3581E	2975
SARMATIA		12SecDr	PCW Det	1/28/82	93B	D135
SARMATIA	R	16Sec	PCW		1876	
SARNIA	R	16Sec	PCW		1876	
SARNIA #48	P	15Ch5StsDr	(Wagner)	4/85	3106	
SARNIA No.1		10SecDr	Grand Trunk	2/71	19	Mon1
SARNIA No.2		SDR	B&S	/70	14	Day4
SARNO	N	24ChDrB	Pullman	6/5/05	1988A	3170
SAROS	N	12SecDr	PCW P-93H	4/1/86	3049	1192
SATELLITE		12SecDr	Pullman	9/17/09	1963E	3744
SATILLA		10SecDr	PCW Det	3/8/82	112A	D141
SATSUMA		12SecDr	Pullman	1/28/04	1581H	3053
SATURN		27Sts Parlor	PCW Det	9/10/75	45	D25
SATURN		12SecDr	Pull Wilm	3/16/03	1581C	W8
SATURN #247		23ChDr	Taunton C.C.	5/71	29	Tau1
SATURNIA		12SecDr	Pullman	6/18/03	1581E	3010
SAUGATUCK	P	36Ch SUV	Pull P-2494	3/11	3275	3821
SAUKIN		12SecDrB	PCW	10/6/88	375F	1481
SAUSALITO SP-57	P	10SecDrB	PCW	4/6/88	445F	1377
SAVANNAH	P	OS	(PK&RSCC)			
SAVANNAH		10SecDr	PCW Det	2/28/82	112A	D141
SAVANNAH		12SecDrSrB	PCW	1/24/99	1319B	2400
SAVANNAH #79	P	10SecDr	H&H	/81	499	
SAVARIN		Diner	PCW	2/11/92	964A	1874
SAVOY		10SecDr	PCW	12/20/84	149D	1028
SAVOY	P	12SecDrB	Wagner	6/96	3073C	Wg109
SAXON	P		Bowers, Dure			
SAXON		10SecDr	PCW Det	3/15/73	26 3/4	D15
SAXON		22Berths	PCW Det	2/4/76	48	D26
SAXONIA		8Sec3DrCpt	Pullman	1/5/06	2119	3275
SAXONY		10Cpt	PCW	1/29/93	993A	1951
SAXONY	P	12SecDrB	Wagner	6/96	3073C	Wg109
SAYBROOK		BaggClub	Pullman	1/22/07	2136A	3461
SAYONA	N	12SecDr	Pullman	12/29/06	2163	3460
SCANDIA		12SecDrSr	PCW	6/7/99	1318C	2423
SCARBORO		12SecDr	Pullman	5/19/08	1963C	3616
SCARBOROUGH	P	12SecDr	(Wagner)	/74	3008D	
SCEPTRE		12SecDr	Pullman	1/30/02	1581A	2792
SCHELDT		12SecDr SUV	Pullman	1/26/04	1581H	3053
SCHENECTADY		12SecDr	(Wagner)	8/82	3023E	
SCHENLEY		12SecDr	Pullman	10/26/05	2049B	3232
SCHIEDAM		12SecDr	Pullman	1/28/04	1581H	3053
SCHILLER		12SecDr	PCW	10/14/92	990	1931
SCHUBERT		12SecDr	PCW	4/29/92	784E	1899
SCHUMANN		12SecDr	PCW	4/30/92	784E	1899
SCHUYLER		12SecDrB	Pullman	12/19/07	2326	3587
SCHUYLKILL	N	7CptLngObs	PCW	10/30/95	1159E	2126
SCHUYLKILL #134	P	10SecDrB	(WS&PCC)	1/15/85	886	
SCHUYLKILL #225	P	23ChDr	J&S	11/72	35	Wil2
SCHUYLKILL #726	P	31StsB	H&H (B&O)		653	
SCIO		12SecDr	PCW	2/21/91	784B	1797
SCIOTO		12SecDr SUV	Pullman	11/2/03	1581F	3025
SCIOTO #173	P	10SecDr	PRR	7/71	21	Alt2
SCIOTO CTC-96	R	Slpr	PRR RR	/72		
SCIPIO		10SecDrB	PCW	8/11/85	240	1069
SCIPIO	R	16Sec	PCW	8/11/85	1881	1069
SCOTIA	P	10SecDr	C&NW Ry	9/70	19	FdL3
SCOTIA		22Berths	PCW Det	4/28/76	48	D32
SCOTIA		12SecDrSr	PCW	12/13/98	1318C	2370
SCOTIA #7	P	16Ch8StsDrB	(Wagner)	7/71	3108C	
SCOTLAND		12Sec2Dr	PCW Det	12/88	537A	D334
SCOTLAND	P	14SecO	(Wagner)	/75	3052	
SCOTT		12SecDr	PCW	10/25/92	990	1931
SCOTTDALE		12SecDr	Pullman	10/14/05	2049B	3232
SCRANTON	P	Parlor	(DL&W RR)	5/76		
SCRANTON		16Sec	Pullman	2/13/07	2184	3508
SCRANTON	P	14SecO	(Wagner)	/75	3052	

CAR NAME	ST	CAR TYPE	CAR BUILDER (Owner)	DATE BUILT	PLAN	LOT
SCREVEN	N	12SecDr	Pullman	11/3/03	1581F	3025
SCYLLA		6CptDrLibObs	Pullman	3/13/02	1751B	2799
SCYTHIA		14SecO	PCW Det	12/29/75	53	D34
SCYTHIA		12SecDrSr	PCW	7/7/99	1318C	2424
SEA BREEZE	N	27ChDrObs	Pullman	6/8/08	2346	3624
SEABURY		12SecDr	Pull Buf	7/11/05	1963	B42
SEAFORD	N	10SecDrB	J&S (B&O)	4/88	663C	
SEASIDE	N	16Ch6Sts	PCW Det	12/1/76	54D	D40
SEASIDE	P	36Ch SUV	Pull P-2263	8/07	3275	3492
SEASIDE #244	P	SummerParlor	H&H (B&O)	7/30/74	40	Wil2
SEATTLE		12SecDr	Pullman	5/21/03	1581D	3004
SEATTLE SP-72	P	10SecDrB	PCW	6/29/88	445F	1415
SEAVIEW		12SecDr	Pullman	3/13/06	2163A	3333
SEAVIEW	P	Bagg16ChB	Pull P-1995	7/04	3272	3115
SEAWANHAKA #93	P	28Ch5Sts	Wagner	9/89	3125A	Wg18
SEBAGO	N	DDR	Grand Trunk	10/70	17J	Mon1
SECOR	N	6CptDrLibObs	Pullman	2/25/02	1751B	2799
SECURITY		10SecO (NG)	PCW Det	9/15/81	73A	D110
SEDALIA		10SecDr	PCW Det	3/27/72	26A	D10
SEDALIA		12SecO	PCW Det	11/13/71	27	D8
SEDALIA		12SecDr	Pullman	5/5/02	1581B	2849
SEDAN	N	12SecDr	(Wagner)	10/80	3022F	
SEDAN	P	12SecDr	B&S	8/93	3205	
SEGOVIA		12SecDr	Pullman	6/25/03	1581F	3010
SEGUIN	N	12SecDrCpt	Pullman	8/9/05	2088	3221
SEINE		12SecDr	PCW Det	9/25/91	784D	D470
SELIGMAN		10SecDr2Cpt	Pullman	11/11/07	2078D	3583
SELIM	N	12SecDrB	(Wagner)	9/86	3050A	
SELISH		12SecDr	Pullman	5/19/02	1581B	2853
SELKIRK #58	P	28Ch3Sts	PCW (Wagner)	5/29/86	3124	1194
SELVA		12SecDr	PCW	3/14/91	784B	1801
SELWYN		14SecO	PCW Det	5/31/93	1021	D525
SEMINOLE	P	OS	(Pull/Sou)			
SEMINOLE		12SecDr	PCW Det	12/31/81	93B	D135
SEMINOLE		16Sec	Pullman	2/14/07	2184	3508
SEMIRAMIDE #105	P	7BoudB	(MannBCC)	/84	297	
SEMIRAMIS #103	P	20Ch8StsDrB	Wagner	7/90	3126	Wg36
SENATOR	N	12SecDr	PCW	6/14/87	473A	1323
SENECA		16Ch7StsDrObs	PCW	2/1/02	1193B	2789
SENECA		16Ch7StsDrObs	PCW	5/2/96	1193	2179
SENECA #282	P	10SecDr	J&S (PRR)	2/73	26	Wil
SENEGAL		10SecDrB	PCW	4/3/84	149B	101
SENEGAL		12SecDr	Pullman	12/26/02	1581B	2852
SENIOR		10SecLibObs	Pullman	11/22/09	2351A	3748
SENORITA		20Ch5StsB	PCW Det	8/20/83	167	D210
SENTINEL	N	10SecB	H&StJ RR	10/72	26Y	Han3
SENTINEL		12SecDr	Pullman	8/6/08	1963C	3636
SEOUL		12SecDr	Pullman	6/17/04	1963	3082
SEPTIMIA		20Ch4StsDrB	Pullman	6/18/01	1518E	2703
SERAPIS		14SecO	PCW	8/31/88	611A	StL4
SERENA		10SecDrB	PCW	8/4/87	445E	1333
SERENA #30	P	17Ch2StsB	(Wagner)	8/85	3122	
SERGIUS	N	12SecDrB	Wagner	4/91	3065	Wg50
SERVIA		12SecDr	PCW Det	1/21/82	93B	D135

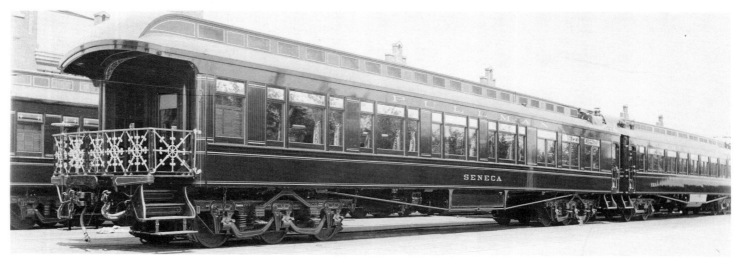

Figure I-36 In this photograph, the parlor observation, SENECA, graces the rear of Lehigh Valley's "Black Diamond" as the train appears ready for delivery to the railroad. This car was wrecked on the L&N in August 1901. Smithsonian Institution Photograph P3424.

CAR NAME	ST	CAR TYPE	CAR BUILDER (Owner)	DATE BUILT	PLAN	LOT
SERVIA #13	P	20Ch12StsDr	(Wagner)	5/74	3104	
SETON		12SecDr	Pullman	12/12/05	1963C	3252
SEVERANCE		12SecDr	Pullman	3/28/08	1963C	3600
SEVERN		12SecDrB	PCW	12/6/95	1066	2130
SEVERN #379		10SecDrB	PCW Det	2/2/83	149	D176
SEVERUS	N	12SecDrB	(Wagner)	4/87	3050	
SEVILLE		10SecDrB	PCW	3/29/84	149B	101
SEVILLE	RN	12SecDr	Pull Buf	7/9/03	1581C	B19
SEVILLE	N	12SecDrB	Wagner	8/91	3065K	Wg59
SEVRES		12SecDr	Pullman	7/26/03	1581F	3019
SEWARD		BaggClub	Pullman	8/21/06	2136A	3386
SEWARD	N	BaggClub	Pullman	3/1/06	2136	3319
SEWICKLEY		12SecDr SUV	Pullman	2/7/08	2163A	3599
SEYBO	N	10SecO	B&S	2/71	19A	Day5
SEYBO	N	12SecDr	Wagner	1/90	3067A	Wg28
SEYMOUR		12SecDr	Pullman	3/26/08	1963C	3600
SHABBONA	P	10SecDrB	(Wagner)	5/82	3046	
SHADYSIDE	P	26Ch3Sts	B&S	9/92	3260	
SHAFTESBURY		10SecDr2Cpt	Pullman	12/16/08	2078E	3661
SHAHALA		10SecDr2Sr	Pullman	9/1/09	2078J	3721
SHAKESPEARE		14SecDr	Pullman	12/30/01	1523F	2798
SHAMOKIN		12SecDr	Pullman	5/20/08	1963C	3616
SHAMROCK #5	N		(WS&PCC)	6/73		
SHAMROCK #75	P	25Ch2StsB	(Wagner)	6/85	3100B	
SHANGHAI		12SecDr	Pullman	6/17/04	1963	3082
SHANNON		12SecDrB	Pull Wilm	4/30/03	1611A	W9
SHANNON #380		10SecDrB	PCW Det	1/30/83	149	D176
SHANOMA		14SecDr	Pullman	8/31/09	2438	3716
SHARHA		10secDr2Cpt SUV	Pullman	3/18/10	2471	3784
SHARON		12SecDr	PCW Det	2/5/92	784E	D488
SHARON #67	P	25Ch2StsB	(Wagner)	10/81	3100B	
SHARON SPRINGS E&A-8	P	Slpr	(E&A Assn)			
SHASTA #412	P	12SecDrB	Wagner	2/90	3065	Wg32
SHASTA CP-35	P	Slpr	(SP Co)	7/1/83	37?	
SHASTA SPNM-8300	P	10Sec2Dr	PCW	3/16/99	1158	2376
SHASTA SPRINGS	N	10SecDr2Cpt	Pullman	10/8/06	2078D	3398
SHAWANO		10SecDrB	PCW	5/28/89	623B	1567
SHAWMONT		12SecDr	Pullman	3/15/06	2163A	3333
SHAWNEE #320	P	8BoudB?	(UnionPCC)		373?	
SHAWNEE #721	P	20Ch12StsB	B&S		678	
SHEBOYGAN #253	P	12SecDr	Wagner	5/88	3061B	Wg11
SHEERNESS		12SecDr	Pullman	10/5/03	1581F	3022
SHEFFIELD		14SecDr	PCW	12/16/99	1318J	2450
SHELBURNE		10SecDr2Cpt	Pullman	5/27/08	2078D	3622
SHELBURNE #31	P	6Sec8Ch	(Wagner)	6/85	3096	
SHELBY		12SecDr	Pullman	10/1/03	1581F	3022
SHELLEY		14SecDr	Pullman	12/30/01	1523F	2798
SHELTER		12SecDr SUV	Pullman	12/10/08	1963E	3659
SHELTER ISLAND #95	P		(WS&PCC)	7/23/81		
SHELTON	P	36Ch SUV	Pull P-2263	8/07	3275	3492
SHENANDOAH		10SecB	B&O	/73	37	B&O
SHENANDOAH		12SecDr	Pullman	11/27/01	1581A	2784
SHENANDOAH #146	P	6Sec4Dr?	J&S (WS&PCC)	4/8/86	720?	
SHENANDOAH #252	P	12SecDr	Wagner	4/88	3061B	Wg10
SHENANDOAH #523	P	12SecO	B&O RR	/81	657	
SHENANGO #251	P	12SecDr	Wagner	4/88	3061B	Wg10
SHERBOURN	P	36Ch SUV	Pull P-2263	8/07	3275	3492
SHERBROOKE		12SecO	Grand Trunk	11/71	27	Mon2
SHERBROOKE		12SecDr	Pull Buf	3/28/04	1581G	B35
SHERIDAN	N	10SecDrB	PCW	10/25/87	445E	1334
SHERIDAN #254	P	12SecDr	Wagner	5/88	3061B	Wg11
SHERMAN		10SecDr2Cpt	Pullman	9/23/05	2078	3203
SHERWOOD	N	12SecDr	PCW Det	4/12/82	93G	D142
SHERWOOD	P	14SecDr SUV	Pull P-1855	1/03	3207	2900
SHETLAND		10Cpt	PCW	2/9/93	993A	1951
SHIELA	N	22ChDr	Pull Buf	7/8/02	1647C	B9
SHIELDS	N	12SecDr	Pullman	12/26/02	1581B	2852
SHILOH		12SecDr	Pullman	12/26/02	1581B	2852
SHILOH	RN	12ChCafeObs SUV	Pull	12/18/03	3968	3065
SHIPKA		12SecDr	Pullman	5/13/04	1963	3082
SHIRLEY		12SecDr	Pullman	2/13/06	1963C	3316
SHOPIERE		10Sec3Cpt	Pullman	11/2/05	2098	3236
SHOREHAM	N	12SecDrB	PCW	11/24/88	375G	1513
SHOREHAM		Diner	PCW	7/26/93	538D	2010
SHOSHONE		12SecDrB	PCW	2/8/89	375G	1557
SHOSHONE		12SecDrCpt	Pullman	10/10/05	2079	3204
SHOTO		10secDr2Cpt SUV	Pullman	3/24/10	2471	3784
SHREVEPORT	N	10SecB	PCW	6/84	206	198
SHROPSHIRE	P	14SecDr SUV	Pull P-2262	8/07	3207C	3498
SHURTLEFF		12SecDr	Pullman	12/12/05	1963C	3252
SIAM		10SecDrB	PCW	12/20/84	149D	1028
SIAM	R	16Sec	PCW	12/20/84	1877B	1028
SIBERIA		12SecDr	Pullman	10/12/01	1581A	2765
SIBONEY		12SecDr	Pullman	8/ /04	1963	3098
SIBYL	P	12SecDrB	(Wagner)	4/87	3050	
SIBYLLA		26ChDr	Pullman	8/4/04	1987A	3097
SICILY	N	10SecDr	Grand Trunk	9/71	26F	Mon2
SICILY		12SecDr	Pull Buf	9/4/03	1581F	B22
SIDDARTHA #104	P	20Ch8StsDrB	Wagner	7/90	3126	Wg36
SIDNEY		12SecDr	PCW Det	8/25/82	93B	D162
SIDNEY		Diner	PCW	11/14/89	662D	1641
SIDONIA		12SecDr	Pullman	9/6/00	1572	2585
SIDONIA #209	P	Parlor	PRR	8/26/72	29	Alt1
SIERRA		10Sec2Dr	Pullman	11/3/02	1746A	2928
SIERRA CP-29	P	Slpr	(SP Co)	7/1/83	37?	
SIERRA BLANCA	RN	12SecDrB	Pullman	5/16/06	2165C	3352
SIESTA		10SecDrB	PCW Det	11/22/83	149	D208
SIGMA	N	12SecDr	CB&Q/GMP	/66	13B	Aur
SIGNET	RN	Private	Wagner	4/89	3062Q	Wg14
SILAO	N	10SecDr	PCW Det	3/8/82	112A	D141
SILESIA		12SecDr	PCW	6/83	93B	38
SILICA	N	10SecDrB	PCW	11/22/83	149	84
SILOAM		18Ch4StsDrB	PCW	7/13/93	1039A	1995
SILVANO		10Sec2Dr	Pullman	8/23/02	1746A	2925

CAR NAME	ST	CAR TYPE	CAR BUILDER (Owner)	DATE BUILT	PLAN	LOT
SILVER						
CITY		12SecB	PCW Det	8/15/76	58	D42
SPRING #534	P	12SecO	B&O RR	/86	655	
SILVERTON		12SecDr	PCW	10/24/98	1318B	2340
SILVIA		14Ch13StsDrB	PCW	10/5/87	505	1376
SILVIA #21	P	15Ch8StsDrB	(Wagner)	4/85	3115A	
SIMARA	P	20Ch6StsDr	B&S	7/89	3258	
SIMLA		12SecDr	Pullman	5/14/04	1963	3082
SIMODA		12SecDr	PCW	1/28/91	784	1787
SIMONIDES		10Sec2Dr	PCW	3/13/96	1173A	2153
SINCHON	P	16Sec SUV	B&S	7/11	3212	
SINCLAIR	N	14SecDr	Pullman	12/14/01	1523I	2798
SINGAPORE		10SecDrB	PCW	6/2/85	240A	1193
SINTON	N	6CptDrLibObs	Pullman	3/13/02	1751B	2799
SIOUX CITY	P	12SecDr	(Wagner)	4/84	3023A	
SIOUX CITY #101	P	10SecB	(IC RR)		231	
SIREN		16Sec	Pullman	6/3/09	2184D	3700
SIRIUS		12SecDr	PCW	1/10/91	784	1786
SIROCCO		16Sec	Pullman	6/3/09	2184D	3700
SIRONA		12SecDr	PCW	2/ /91	784B	D460
SIRONA		6CptDrLibObs	Pullman	2/27/02	1751B	2799
SISKIYOU	N	10SecDr2Cpt	Pullman	11/11/07	2078D	3583
SISSON	N	10SecDr2Cpt	Pullman	3/22/07	2078D	3528
SITKA	N	12SecO	H&H	2/72	27	Wil 7
SITKA		12SecDr	Pullman	7/26/03	1581F	3019
SIVA		12SecDr	PCW	1/8/91	784	1781
SKAGIT		12SecDr	Pullman	6/8/07	2271A	3458
SKALKAHO		16Sec	Pullman	3/21/01	1649A	2709
SKILLUTE		10secDr2Cpt SUV	Pullman	3/23/04	2471	3784
SKOKIE		12SecDr	Pullman	12/18/03	1581F	3043
SKYLARK	N	24ChDrB	Pullman	6/10/05	1988A	3170
SLAVONIA		12SecDr	Pullman	6/10/03	1581E	3010
SLIGO		14SecO	PCW	3/11/93	1021	1969
SMERDIS	N	16Ch12StsDr	(Wagner)	7/90	3103	
SMILAX		24ChDr	Pull Buf	6/7/06	2156	B53
SMYRNA		10SecDrB	PCW	2/9/84	149D	101
SNOHOMISH		14SecDr	Pullman	5/23/06	2165	3352
SNOQUALMIE		10SecDrSr	PCW	8/20/91	937A	1838
SNOWDEN		12SecDr	Pullman	1/29/04	1581H	3053
SNOWDEN #134	P	10SecDrB	(WS&PCC)	1/15/85	886	
SNOWDROP		24ChDrSr	Pullman	7/23/04	1980A	3088
SOBRAON	N	12SecDr	Wagner	1/89	3061B	Wg 13
SOCORRO		10SecDr	PCW Det	6/16/81	112	D118
SOCORRO		10Sec2Dr	Pullman	8/27/02	1746A	2925
SOCOTRA		12SecDr	Pullman	2/21/03	1581D	2975
SOCRATES		10Sec2Dr	PCW	2/26/96	1173A	2153
SOFALA		12SecDr	Pullman	2/23/03	1581D	2975
SOFIA		12SecDr	Pull Buf	5/11/03	1581C	B16
SOHO		12SecDr	PCW	2/21/91	784B	1797
SOKULK		14SecDr	Pullman	9/7/09	2438	3716
SOLANO	N	10Sec2Dr	Pullman	8/22/02	1746A	2925
SOLENT		14SecO	PCW	2/25/93	1021	1969
SOLFERINO	N	16SecO	(Wagner)	/72	3029	
SOLINUS		12SecDr	Pull Buf	4/25/10	1963E	B83
SOLITAIRE		16Sec	Pullman	1/5/09	2184D	3662
SOLOLA		12SecDr	PCW	1/30/91	784	1787
SOLWAY		10SecDrB	PCW Det	9/25/84	149D	D248
SOMERSET	N	6CptDrLibObs	Pullman	2/27/02	1751B	2799
SOMERSET #729	P	19Ch3StsB	H&H (B&O)		653A	
SOMERTON		24ChDrB	Pull Buf	4/17/07	2278	B59
SOMONAUK		10SecDr	PCW Det	10/1/81	112A	D124
SONATA		16Sec	Pullman	1/7/09	2184D	3662
SONDRIO		12SecDr	Pull Buf	11/16/05	1963C	B48
SONNAMBULA	P	8Cpt8Berth	Wagner	8/90	3203	
SONOMA CP-39	P	10SecB	(CP Co)	5/31/84	37M	
SONORA	N	12SecB	(OPLines)	/72	114A	
SONORA	R	12SecDrB	(OPLines)	/72	1468	
SONORA #537	P	10SecB	B&O RR?		649	
SOPHIA		22ChDr	Pull Buf	7/8/02	1647C	B9
SOPHOCLES		10Sec2Dr	PCW	3/14/96	1173A	2153
SOPHOMORE		10SecLibObs	Pullman	11/11/09	2351A	3748
SOPRIS	N	14SecO	PCW (Wagner)	4/19/86	3019A	1191
SOPRIS #167	N	15Ch4StsB	(WS&PCC)	12/6/87	731A	
SORANO	P	7BoudB	(MannBCC)	/84	776A	
SORIANO		12SecDr	Pull Buf	11/22/05	1963C	B48
SORREL		24ChDr	Pull Buf	6/7/06	2156	B53
SORRENTO		12SecDr	Pull Buf	1/29/03	1581C	B12
SOUDAN		10SecDrB	PCW	1/12/85	149D	1028
SOUDAN	N	12SecDr	Pullman	3/16/04	1963E	3080
SOUDAN #4	P	17Ch3StsDrB	(Wagner)	6/85	3108A	
SOUND BEACH	P	Bagg24Ch SUV	Pull P-2267	4/07	3278	3494
SOUND VIEW	P	40Ch6StsDrObs	Pull P-2266A	4/07	3276	3493
SOUTH						
DAKOTA	N	7Cpt2Dr	Pullman	9/25/02	1845	2926
PARK		10SecO (NG)	PCW Det	10/20/79	73	D61
SIDE #63	P	20Ch6Sts	(Wagner)	5/85	3116	
SIDE #112	P	Parlor	J&S (WS&PCC)	7/83	698?	
WILTON	P	BaggSmk	B&S	8/93	3213C	
SOUTHAMPTON #107	P	Parlor	(WS&PCC)	/73	731?	
SOUTHAMPTON	P	36Ch SUV	Pull P-2494	3/11	3275	3821
SOUTHBORO	P	36Ch SUV	Pull P-2263	8/07	3275	3492
SOUTHBRIDGE		10SecDr2Cpt	Pullman	11/8/09	2451	3747
SOUTHBURY	P	36Ch SUV	Pull P-2494	3/11	3275	3821
SOUTHEND		12SecDr	Pullman	10/3/03	1581F	3022
SOUTHERN	P	Diner	CB&Q RR	/68	16 Aur	
SOUTHERN	N	Diner	PCW	11/8/89	662D	1641
SOUTHEY		14SecDr	Pullman	12/30/01	1523F	2798
SOUTHINGTON	P	34Ch SUV	Pull P-1996C	7/04	3265	3114
SOUTHLAND	N	6CptDrLibObs	Pullman	3/13/02	1751B	2799
SOUTHRON	N	12SecDr	PCW	6/14/87	473A	1323
SOUTHWOLD		12SecDr	Pullman	12/20/09	1963E	3766
SOVEREIGN		12SecO	PCW Det	11/22/81	114	D122
SOVEREIGN		16Sec	Pullman	6/4/09	2184D	3700
SPAIN		12SecDr	PCW	6/7/87	473A	1323

CAR NAME	ST	CAR TYPE	CAR BUILDER (Owner)	DATE BUILT	PLAN	LOT
SPAIN		12SecDr	Pullman	3/13/03	1581D	2975
SPARLAND	N	22Ch6Sts	PCW	8-9/92	976B	1910
SPARLIN	N	12SecDr	Pull Buf	11/9/05	1963C	B48
SPARTA		12SecDr	Pull Buf	11/9/05	1963C	B48
SPARTAN	P	10SecDr	B&S	9/71	26	Day6
SPARTAN		12SecDr	PCW	11/16/98	1318C	2359
SPARTANBURG		10SecLibObs	Pullman	4/30/06	2176	3355
SPEEDWELL		12SecDr	Pullman	8/11/08	1963C	3636
SPENSER		12SecDr	PCW	10/28/92	990	1931
SPHINX	N	26ChDr	Pullman	5/29/07	1987B	3524
SPIREA		10SecDr2Cpt	Pullman	12/1/09	2451	3747
SPOKANE		12SecDr	PCW Det	3/31/83	93B	D184
SPOKANE		10SecDrCpt	PCW	2/11/90	668	1649
SPRINGDALE		12SecDr	Pullman	5/20/08	1963C	3616
SPRINGFIELD		12SecO	PCW Det	9/21/76	58	D42
SPRINGFIELD #1	P	14Sec	Wason/GMP	7/63	27	Spr
SPRINGPORT	N	12SecDr	Pullman	3/13/06	2163A	3333
SPRINGTON		12SecDr	Pullman	3/7/02	1581B	2818
SPUYTEN						
DUYVIL #65	P		(WS&PCC)	5/1/79		
ST.						
ADELE	RN	22Ch4Sts	(MannBCC)	/84	697C	
AGATHA	RN	20Ch4Sts	(MannBCC)	/84	776C	
ALBANS	RN	10SecDr	Vermont Cent	/71	30	StA1
ALBANS	P	14SecO	(Wagner)	/73	3020A	
ALDAN	RN	22Ch4StsB	(MannBCC)	/89	373I	
ALEXIS	RN	22Ch4StsB	J&S (Union PCC)	/89	373E	
ANDREW		Slpr	PCW Det	10/82	68A	D159
ANDREWS		12SecDr	Pullman	12/17/03	1581F	3043
ANGELE	RN	22Ch4Sts	(MannBCC)	/84	776F	
ANTHONY	RN	22Ch4StsB	J&S (Union PCC)	/89	373E	
ARMAND	RN	22Ch4StsB	(Union PCC)	/89	373E	
ARSENE	RN	22Ch4Sts	(MannBCC)	/84	776F	
AUBERT	RN	22Ch4Sts	(MannBCC)	/84	776F	
AUGUSTINE		10SecO	(ACL Assn)	6/82	185?	
AUGUSTINE	RN	8SecBLibObs	PCW Det	1/14/93	1014E	D510
BERNARD	RN	22Ch4StsB	J&S (Union PCC)	/89	373E	
CARVAN	RN	22Ch4Sts	(MannBCC)	/84	776F	
CECILE	RN	22Ch4Sts	(MannBCC)	/84	697C	
CELESTE	RN	22Ch4Sts	(MannBCC)	/84	697C	
CHARLES	P	Commissary	C&NW Ry	C:/70	4C	Chi2
CHARLES	N	Diner	PCW	1/10/90	662D	1778
CHARLES		Diner	Pullman	12/24/01	1694	2762
CLAIR		10SecDr	PCW Det	10/7/72	26 1/2	D13
CLAIR	RN	22Ch4StsDr	(MannBCC)	/89	703B	
CLAUDE	RN	22Ch4StsB	(Union PCC)		373E	
CLOUD	P	Commissary	C&NW Ry	C:/70	4C	Chi2
CLOUD		Diner	PCW	1/6/90	662D	1659
CLOUD	RN	22Ch5StsDr	(MannBCC)	/89	703C	
CROIX	RN	22Ch5StsDr	(MannBCC)	/89	703C	
CROIX		Diner	PCW	1/9/90	662D	1659

CAR NAME	ST	CAR TYPE	CAR BUILDER (Owner)	DATE BUILT	PLAN	LOT
ST.						
DENIS	P	CommissaryHotel	CB&Q RR	1/2/71		Aur3
DENIS	S	8SecCpt	PCW Det	6/30/83	130	D158
DENIS		Diner	Pullman	12/23/01	1694	2762
ELMO	RN	22Ch4StsB	J&S (Union PCC)	/89	373E	
FRANCIS	RN	20Ch4Sts	(MannBCC)	/84	776D	
GEORGE	P	12SecO	CB&Q RR	4/74	4C	
GEORGE		22Berth	PCW Det	6/28/75	48	D26
GEORGE #549	P	14SecSr	H&H (B&O)		654	
GERMAIN	RN	22Ch4Sts	(MannBCC)	/84	776F	
GRETNA	RN	22Ch4StsB	J&S (Union PCC)	/89	373I	
HELENA	RN	22Ch4Sts	(MannBCC)	/84	776F	
HELIER	RN	22Ch4Sts	(MannBCC)	/84	776F	
HYACINTHE	P	OS	Grand Trunk	12/71	27	Mon2
IGNACE	RN	22Ch4StsB	(MannBCC)	/89	373E	
ISIDORE	RN	22Ch4StsB	J&S (MannBCC)	/84	373G	
IVERS	RN	22Ch4Sts	J&S (MannBCC)	/84	776F	
JAMES	P	CommissaryHotel	CB&Q RR	1/2/71		Aur3
JAMES		Diner	PCW	7/25/93	538D	2010
JAMES	N	Diner	Pullman	12/23/01	1694	2762
JARVIS	RN	20Ch4Sts	(MannBCC)	/84	776D	
JEAN	RN	22Ch4StsB	G&B (Union PCC)	/67	373E	Troy
JOHN	P	10SecDr	Grand Trunk	7/73	31	Mon4
JOHNS		10SecO	(ACL Assn)		185A	
JOSEPH	P	SDR	CB&Q RR			Aur
JOSEPH	N	12SecDr	PCW	2/24/91	784B	1798
JOSEPH #312	P	11Boud	(MannBCC)	/89	703	
JULIEN	RN	22Ch4Sts	(MannBCC)	/84	697C	
LAWRENCE	P	OS	GWRy/GMP	C:7/67	7	Ham
LAWRENCE	RN	22Ch4StsB	(Union PCC)		373E	
LAWRENCE		Diner	PCW	11/8/89	662D	1641
LEONARDS #337	P	12SecDr	Wagner	3/88	3061	Wg9
LOUIS	S	8SecCpt	PCW Det	6/30/83	130	D158
LOUIS	N	20Ch9StsDrObs	PCW	11/14/99	1417H	2466
LOUIS	RN	16SecB	PCW	6/3/93	1040I	1990
LOUIS	P	10SecDr	(Wagner)	/83	3057B	
LOUIS	P	20Ch8StsDrBObs	Wagner	11/92	3127	Wg73
LOUIS #25		12SecO	(WS&PCC)	6/27/74	833?	
LOUIS #507	P	12SecO	B&O RR	/79	657	
LOUIS No.1	P	12SecO	DetC&MCo	/71	5	Det2
LOUIS No.2	N	SDR	I&StLSCCo		27	
LUCIEN	RN	35ChB	(MannBCC)	/84	776E	
MALO	RN	22Ch4Sts	(MannBCC)	/84	776F	
MARIE	RN	22Ch4StsB	J&S (Union PCC)	/89	373E	
MARTIN	RN	22Ch4Sts	(MannBCC)	/84	776F	
MARY'S #41	P	10SecB	(SF&W RR)	6/82	37	
MARYS		18Ch5Sts		?6/82	44 ?	?D24
MICHAELS	RN	22Ch4Sts	(MannBCC)	/84	697C	
MUNGO		Slpr	PCW Det	10/82	68A	D159
NICHOLAS	P	Commissary	C&NW Ry	2/28/71	4	Chi2
NICHOLAS		10SecDrHotel	PCW Det	6/12/77	63	D45

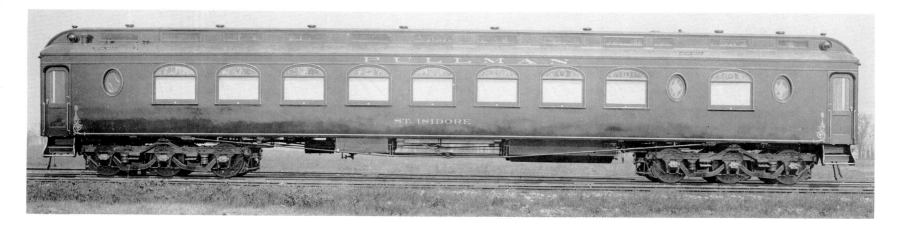

Figure I-37   The parlor car, ST. ISIDORE, is the 1900 version of the updated and rebuilt Mann Boudoir Car Company's MARTHA #108 built by Jackson & Sharp in 1884. Original Pullman Photograph 5179, courtesy of Robert Wayner.

CAR NAME	ST	CAR TYPE	CAR BUILDER (Owner)	DATE BUILT	PLAN	LOT
ST.						
NICHOLAS	RN	22Ch4Sts	(MannBCC)	/84	697D	
NORBERT	RN	22Ch4StsB	(Union PCC)		373E	
OMER	RN	22Ch4Sts	(MannBCC)	/84	776F	
PALLAS	RN	22Ch4StsB	(Union PCC)		373I	
PANCRAS	RN	22Ch4Sts	(MannBCC)	/84	776F	
PAUL	P	12SecO	(NP ASSN)		18D	
PAUL	RN	14SecO	I&StLSCCo	/71		
PAUL	P	12SecO	CentCarCo	2/84	532	
PAUL	P	14SecO	B&S	7/92	1593	
PAUL	N	12SecDr	Pull Buf	6/10/04	1963	B37
PAUL	P	20Ch8StsDrBObs	Wagner	10/92	3127	Wg73
PAUL #55	P	12SecDr	B&S	/76-/78		
PETERSBURG #338	P	12SecDr	Wagner	5/88	3061A	Wg9
PIERRE	RN	26Ch	(MannBCC)	/84	697D	
REGIS	RN	22Ch4Sts	(MannBCC)	/84	776F	
ROSE	RN	22Ch4Sts	(MannBCC)	/83	879A	
SABINE	RN	22Ch4StsB	J&S (Union PCC)	/89	373E	
SERVAN	RN	35ChB	(MannBCC)	/84	776E	
THOMAS	RN	26Ch	(MannBCC)	/84	697D	
VICTOR	RN	22Ch4StsB	(MannBCC)	/89	373E	
VINCENT #339	P	12SecDr	Wagner	5/88	3061A	Wg9
STAFFORD		16SecO	PCW	6/20/93	1040	1990
STALACTITE		16Sec	Pullman	7/1/08	2184C	3627
STAMFORD	P	24Ch6StsDr	(Wagner)	6/86	3100	
STANDARD		12SecDr	PCW Det	11/1/80	93	D88

CAR NAME	ST	CAR TYPE	CAR BUILDER (Owner)	DATE BUILT	PLAN	LOT
STANDISH		12SecDrSr	PCW	1/6/99	1318C	2379
STANFIELD		26ChDr	Pullman	3/31/06	1987A	3335
STANFORD	N	10SecDrB	PCW	12/20/85	149D	1028
STANHOPE		14SecO	PCW Det	5/24/93	1021	D525
STANLEY		10SecDr2Cpt	Pullman	12/21/08	2078E	3661
STANTON		12SecDr	PCW	9/9/91	784C	1828
STANTON		12SecDr	Pull Buf	7/9/03	1581C	B19
STANTON #25		10SecB	B&S	6/73	37	Day9
STANWICK		12SecDr	Pullman	10/10/07	2163A	3562
STANWOOD		12SecDr	PCW Det	5/7/81	93A	D92
STANWOOD		10SecDr2Cpt	Pullman	9/22/05	2078	3203
STARLIGHT	RN	Private	PCW	2/19/87	1920	1298
STARLING		24ChDrB	Pullman	6/22/00	1518C	2552
STARUCCA E&A-45	P	Slpr	(E&A Assn)			Elm0
STATE LINE	P	BaggBClub SUV	Pull P-1503B	/00	3216	2517
STATE of GEORGIA	P	DDR	J&S (W&A RR)	/76	4	Wil
STATEN						
ISLAND #550	P	14SecSr	H&H (B&O)		654	
STELLA	N	20Ch3StsB	PCW Det	9/18/83	167A	D209
STEPHANIE		22ChDr	Pull Buf	7/3/02	1647C	B9
STEPHANO		12Sec2Dr SUV	Pullman	4/4/01	1580A	2646
STEPNEY	P	34Ch SUV	Pull P-1996C	7/04	3265	3114
STERLING		12SecDr	PCW Det	4/12/82	93B	D142
STERLING		12SecDr	Pullman	5/2/02	1581B	2849
STERLING	N	12SecDrB	(Wagner)	11/82	3025	
STETTIN		12SecDr	Pullman	5/27/04	1963	3082

CAR NAME	ST	CAR TYPE	CAR BUILDER (Owner)	DATE BUILT	PLAN	LOT
STEVENSON	P	16Sec	Pullman	10/08	2355	3638
STILLASHA	N	10SecDr2Cpt SUV	Pullman	3/9/10	2471	3784
STILLWATER	P	14SecO	B&S	4/88	1593	
STILLWATER	N	14SecDr	Pullman	12/30/01	1523I	2798
STILLWATER	P	12SecDr	(Wagner)	8/82	3023B	
STILTON		12SecDr	Pullman	11/5/01	1581A	2784
STOCKBRIDGE	P	36Ch SUV	Pull P-2494	3/11	3275	3821
STOCKHOLM		12SecDr	Pull Buf	6/9/03	1581C	B18
STOCKTON		12SecDr SUV	Pull Buf	8/21/03	1581F	B21
STONEHAM	N	12SecO	B&O RR		665F	
STONELEIGH	N	10SecDr	PCW Det	7/19/82	112A	D152
STONELEIGH	N	6CptDrLibObs	Pullman	2/27/02	1751B	2799
STONINGTON		12SecDr	Pullman	11/21/00	1581A	2589
STONY CREEK	P	BaggSmk	Pull P-1991	4/04	3217	3093
STORM KING #119	P		(WS&PCC)	7/83		
STOUGHTON	P	32ChDr SUV	Pull P-2264	8/07	3277	3496
STOVROS	N	25Ch2StsB	(Wagner)	11/87	3100A	
STRADELLA #121	P	7BoudB	(MannBCC)	/84	776	
STRADELLA	N	14SecDr	(Wagner)	6/87	3050D	
STRALSUND #437	P	12SecDr	Wagner	11/90	3065B	Wg42
STRASBURG		16SecO	PCW	7/30/90	780	1706
STRASBURG	P	12SecDr	(Wagner)	3/86	3049	
STRATFORD		10SecDr	Grand Trunk	8/71	26	Mon2
STRATFORD		16SecO	PCW	7/31/90	780	1706
STRATHCONA		10SecDr2Cpt	Pullman	12/16/08	2078E	3661
STRATHROY	N	10SecB	Grand Trunk	1/74	37M	5
STRATHROY	P	12SecDr	(Wagner)	4/86	3049	
STREATOR		10SecDrB	PCW	3/13/88	445E	1361
STROBEL	N	12SecDrCpt	Pullman	8/8/05	2088	3221
STROUDSBURG			(DL&W Assn)			
STUART		12SecDr	Pullman	6/26/01	1581A	2726
STUTTGART #436	P	12SecDr	Wagner	11/90	3065B	Wg42
STUYVESANT		12SecDr	Pullman	2/28/07	1963C	3525
SUABIA		12SecDr	Pull Buf	5/5/03	1581C	B16
SUENO		12SecDr	PCW	8/28/90	784	1708
SUEZ		10SecDrB	PCW	1/14/90	725B	1629
SUFFIELD		14SecDr	PCW	12/18/99	1318J	2450
SUFFOLK	R	10SecLibObs	PCW Det	4/4/93	1021N	D517
SUFFOLK		14SecO	PCW Det	4/4/93	1021	D517
SUFFOLK	N	12SecDr	Pullman	5/27/04	1963	3082
SUGARLAND	N	12SecDrCpt	Pullman	8/8/05	2088	3221
SUISUN		12SecDr	Pullman	4/2/04	1963	3080
SUISUN CP-28	P	Slpr	(SP Co)	7/1/83	37?	
SULTAN		BaggSmk	PCW	8/16/88	534C	1410
SULTAN	N	16Sec	Pullman	2/13/07	2184	3508
SULTANA		20Ch5StsB	PCW Det	8/20/83	167	D210
SUMAC		24ChDr	Pull Buf	6/8/06	2156	B53
SUMATRA		10SecDrB	PCW	3/21/83	149	27
SUMATRA	N	10Cpt	PCW	1/5/93	993A	1951
SUMMIT		12SecO	B&S	12/70	5	Day
SUMMIT		12SecDrCpt	Pullman	10/11/05	2079	3204
SUMNER		12SecDr	Pullman	3/28/03	1581D	2983
SUNBEAM	RN	Private	PCW	6/23/88	1920	1415
SUNBURY	N	28Ch3Sts	PCW (Wagner)	6/22/86	3124	D256
SUNBURY #278	P	10SecDr	J&S (PRR)	2/73	26	Wil9
SUNFLOWER	N	Slpr	(IC RR)			2/273
SUNFLOWER		24ChDrSr	Pullman	7/23/04	1980A	3088
SUNNYSIDE	N	40Ch6StsDrObs	Pull P-2266A	4/07	3276	3493
SUNNYSIDE #82	P	6Sec10Ch	(Wagner)	8/70	3093	
SUNNYSIDE E&A-16	P	Slpr	(E&A Assn)			
SUNOL		10Sec2Dr	Pullman	11/1/02	1746A	2928
SUNSET	RN	Private	PCW	1/11/87	1920	1297
SUPERB		6Dr	PCW	11/6/89	685B	1598
SUPERIOR		10SecDr	PCW Det	11/19/72	26	D14
SUPERIOR	P	40Ch6StsDrBObs	Pull P-2495	2/11	3276	3822
SUPERIOR #77	P	6Sec10Ch	(Wagner)	4/66	3093A	
SURIGAO		12SecDr	Pullman	8/ /04	1963	3098
SURREY	P	12SecDrB	Wagner	6/96	3073C	Wg109
SURREY	R	30Ch SUV	Pull Buf	5/11/03	3581E	B16
SUSA		12SecDr	PCW	2/20/91	784B	1797
SUSANNE	N	22Ch4Sts	(Wagner)	10/70	3121	
SUSETTE	N	26ChDr	Pullman	5/25/07	1987B	3524
SUSQUEHANNA #135		10SecDrB	(WS&PCC)	1/15/85	886	
SUSQUEHANNA #545	P	10Sec	B&O RR	?/87	658	
SUSQUE-HANNA CTC-71	R		PRR rblt/70			
SUSQUE-HANNA E&A-21	P	Slpr	(E&A Assn)			
SUSSEX	P	12SecDrB	Wagner	6/96	3073C	Wg109
SUSSEX	P	26Ch3Sts	B&S	8/92	3260	
SUSSEX #310	P	11Boud	(MannBCC)	/89	703	
SUTHERLAND		10SecDr2Cpt	Pullman	12/26/05	2078C	3287
SUTTON		12SecDr	Pullman	3/8/02	1581B	2818
SUWANEE		16Sec	Pullman	2/16/07	2184	3508
SUWANNEE		10SecDr	PCW Det	3/15/82	112A	D141
SUWANNEE #320	P	8BoudB?	J&S (Union PCC)	/89	373	
SWALLOW		24ChDrB	Pullman	7/9/00	1518C	2553
SWAN		24ChDrB	Pullman	5/31/00	1518C	2552
SWANHILDA		10SecDr2Cpt	Pullman	12/22/08	2078E	3661
SWANINGTON #290	P	9CptB	Wagner	9/89	3066	Wg24
SWANNANOA	N	12SecDrB	PCW	8/8/88	375E	1395
SWANNANOA #336	P		(Union PCC)			
SWANSEA	N	10SecDr	Grand Trunk	10/31/72	31K	Mon3
SWANSEA #112	N		(WS&PCC)	7/83	698	
SWANTON		12SecDr	Pullman	11/6/01	1581A	2784
SWARTHMORE		12SecDr	Pullman	12/12/05	1963C	3252
SWASTIKA		16Sec	Pullman	7/1/08	2184C	3627
SWATARA CTC-86		Slpr	PFtW&StL RR			
SWATARA CTC-95	R	Slpr	PRR rblt/72			
SWEDEN		16SecO	PCW	3/13/91	780	W5
SWEDEN	R	12SecDr	PCW	3/13/91	896B	W5
SWEETWATER		12SecDr	PCW Det	4/19/82	93B	D142

CAR NAME	ST	CAR TYPE	CAR BUILDER (Owner)	DATE BUILT	PLAN	LOT
SWISSVALE		BaggClub	Pullman	1/22/07	2136A	3461
SYBARIS		Slpr	PCW Det	4/10/84	170	D222
SYBARIS		8SecLngObs	PCW	1/5/89	644B	1563
SYBARITA	P	12SecSr SUV	Pull P-2488	10/10	3209	3809
SYBIL #46	N	Sleeper	H&H (WS&PCC)	7/25/78	524?	
SYCAMORE	N	7CptLngObs	PCW	10/30/95	1159E	2126
SYCAMORE #102	P	12SecDr	(MannBCC)	/89	878	
SYDENHAM #146	N	6Sec4Dr?	J&S (WS&PCC)	4/8/86	720?	
SYDNEY		12SecDr	Pullman	3/16/04	1963	3080
SYLVAN		OS	Altoona	3/31/73	27	2
SYLVAN	N	12SecDr	Wagner	4/88	3061B	Wg10
SYLVANIA		14SecDr	PCW	9/1/99	1318E	2447
SYLVANUS	N	BaggBSmk	PCW	/99	1400A	2410
SYLVANUS #312	N	11Boud	(MannBCC)	/89	703	
SYLVIA	P	7BoudB	(MannBCC)	/84	776	
SYLVIA		24ChDr	Pull Buf	3/11/07	2156	B57
SYMPHONY		16Sec	Pullman	1/7/09	2184D	3662
SYRACUSE		12SecO	(DL&W RR)	5/79	235A	
SYRACUSE		12SecB	PCW Det	5/18/83	168	D195
SYRACUSE		12SecDr	Pullman	11/3/03	1581F	3025
SYRACUSE	RN	6CptDrLibObs	Pullman	10/29/02	1845F	2926
SYRDARIA		12SecDr	Pull Buf	11/23/03	1581F	B25
SYRIA	P	10SecDr	B&S	/71	26	Day6
SYRIA		12SecDr	Pullman	6/24/03	1581F	3010

CAR NAME	ST	CAR TYPE	CAR BUILDER (Owner)	DATE BUILT	PLAN	LOT
SYRINGA		24ChDrSr	Pullman	7/23/04	1980A	3088
SYRMIA		12SecDr	Pull Buf	11/15/05	1963C	B48
SYSTON		16SecO	PCW	6/16/93	1040	1990
T W PIERCE GH&SA	P	Slpr	(CP Co)		37?	
TABERNILLA	N	12SecDr	Pullman	10/19/05	2049B	3232
TABRIZ		12SecDr	PCW	7/9/92	784E	1915
TACITUS		10SecDrB	PCW	4/29/85	240	1051
TACOMA		12SecDr	PCW Det	5/20/84	93D	D241
TACOMA		12SecDr	Pull Buf	2/26/04	1581F	B34
TACONIC		12SecDr	Pullman	12/18/03	1581F	3043
TACONY		12SecDrCpt	Pullman	5/26/04	1985	3095
TAGUS	R	12SecDrB	PCW Det	2/15/83	1611	D176
TAGUS	P	16SecO	Wagner	10/92	3072	Wg79
TAGUS #381		10SecDrB	PCW Det	2/15/83	149	D176
TAHITA		12SecDrB	PCW	10/19/88	375F	1481
TAHOE		10Sec2Dr	Pullman	11/6/02	1746A	2928
TAHOE CP-38	P	Slpr	(SP Co)	7/1/83	37?	
TAKU	RN	12SecDr	PCW	10/10/82	1467B	22
TALAVERA		12SecDr	Pullman	4/9/03	1581D	2984
TALBOT		14SecO	PCW	5/25/93	1021	1972
TALEQUAH #108	P	20Ch8StsDrB	Wagner	9/90	3126E	Wg41
TALISMAN		12SecO	PCW Det	11/28/81	114	D123
TALISMAN	P	22Ch6Sts	B&S	7/82	3253B	
TALLADAGA		10SecDr	PCW	8/22/83	112	47

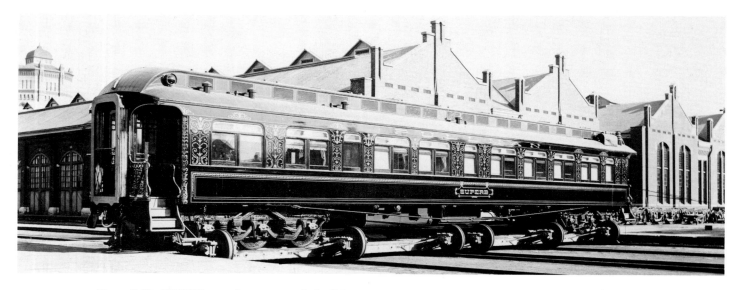

Figure I-38 SUPERB, an early narrow-vestibule all-drawing room car, stands on the transfer table in the early morning sun ready for delivery. This elaborately-decorated car was renamed MELISANDE in June 1911 and later was sold to the Chicago-based used railroad car dealer, Hotchkiss Blue & Company. Smithsonian Institution Photograph P630.

CAR NAME	ST	CAR TYPE	CAR BUILDER (Owner)	DATE BUILT	PLAN	LOT
TALLADAGA	R	12SecB	PCW	8/22/83	168	47
TALLAHASSEE	R	10SecO	PCW Det (rblt)	12/15/74	47	D27
TALLAHATCHIE #240	N	Slpr	(IC RR)			2/273
TALLAPOOSA		10SecDrB	PCW Det	1/19/84	149	D224
TALLAPOOSA	R	12SecDrB	PCW Det	1/19/84	1468	D224
TALLULAH		10SecDr	PCW Det	3/8/82	112A	D141
TALPA		12SecDr	PCW	3/18/91	784B	1801
TAMA		12SecDrCpt	Pullman	10/6/05	2079	3204
TAMAR		10SecDrB	PCW Det	11/24/83	149	D212
TAMAR		12SecDrB	Pull Wilm	4/11/03	1611A	W9
TAMAR	P	16SecO	Wagner	4/93	3072	Wg80
TAMARACK #266	P	12SecDr	Wagner	7/88	3061B	Wg12
TAMARINDO	N	14SecDr	Pull Buf	10/31/00	1582A	B3
TAMAYO	N	14SecDr	Pullman	6/11/00	1523G	2559
TAMEGA		10SecDrB	PCW	2/20/84	149B	101
TAMEGA	N	12SecDrSr	Pullman	3/28/05	2049A	3169
TAMERLANE		12SecDr	Pullman	9/17/09	1963E	3744
TAMORA		20Ch5StsB	PCW Det	6/30/83	167	D203
TAMPA		12SecDr	Pullman	4/4/04	1963	3080
TAMPA #22		10SecO	(SF&W RR)	7/75	22C	
TAMPA #29	P	20Ch4Sts	(Wagner)	8/70	3101	
TAMPICO	R	10SecDrB	PCW	4/10/85	240	1037
TAMPICO	R	12SecDr	PCW	4/10/85	1603	1037
TAMPICO #441		10SecDrB	PCW	4/10/85	149E	1037
TANCRED	RN	10SecDrB	PCW Det	4/18/93	1021H	D518
TANGA		12SecDr	Pull Buf	3/30/03	1581C	B14
TANGIER		14SecO	PCW Det	4/18/93	1021	D518
TANGIER	R	12SecDr	PCW	6/29/88	1603	1415
TANGIER #110	P	20Ch8StsDrB	Wagner	9/90	3126E	Wg41
TANGIER SP-73	P	10SecDrB	PCW	6/29/88	445F	1415
TANNHAUSER #116	P	7BoudB	(MannBCC)	/84	697	
TAOMA #225	S	12SecDr	PCW (Wagner)	6/26/86	93H	1199
TARASCON	N	14SecDr	Pullman	6/11/00	1523G	2559
TARASCON #112	P	20Ch8StsDrB	Wagner	12/90	3126	Wg44
TARBORO	N	12SecO	PCW Det	2/21/82	114	D145
TARENTUM		12SecDr	Pullman	10/16/05	2049B	3232
TARKIO		12SecDr	Pullman	12/15/05	1963C	3252
TARO		12SecDr	PCW	9/23/90	784	1765
TARPEIAN	N	12SecDrB	Wagner	8/91	3065	Wg53
TARQUIN		12SecDr	Pull Buf	6/18/07	1963C	B61
TARQUIN #106	RN	7BoudB	(MannBCC)	/84	697A	
TARRAFA		10SecO	PCW	8/2/92	948A	1944
TARRIEN	N	12SecDr	Pullman	2/17/06	1963C	3316
TARRYTOWN		10SecDr2Cpt	Pullman	5/29/08	2078D	3622
TARSA #2551		14SecO	PCW Det	2/23/93	1021	D516
TARSUS #447		10SecDrB	PCW	9/26/85	240	1080
TARTARY		14SecO	PCW	5/20/93	1021	1972
TARVIA	N	12SecDr	Pullman	3/7/06	2163A	3333
TASHMOO		10SecDr2Cpt	Pullman	11/8/09	3747	
TASMANIA		20Ch3StsB	PCW Det	10/4/83	167A	D209
TASMANIA #111	P	20Ch8StsDrB	Wagner	12/90	3126	Wg44
TASSO		12SecO	PCW Det	4/3/82	114	D143

CAR NAME	ST	CAR TYPE	CAR BUILDER (Owner)	DATE BUILT	PLAN	LOT
TAUNTON		14SecO	PCW Det	3/29/93	1021	D517
TAUNTON #263	P	12SecDr	Wagner	7/88	3061B	Wg12
TAURUS		12SecDr	PCW	1/9/91	784	1786
TAVARES		10SecDrB	PCW	2/23/88	445E	1361
TAVENER	N	12SecDr	Pullman	1/11/07	2163	3449
TAVETA		14SecO	PCW	3/2/93	1021	1969
TAVISTOCK #265	P	12SecDr	Wagner	7/88	3061B	Wg12
TAVORA		14SecO	PCW Det	6/13/93	1041	D519
TECHE M&LT-4	N	Slpr	(M&LT RR)	4/24/85	37?	
TECOAC	N	14SecDr	Pull Buf	9/23/00	1523D	B1
TECOLETE		10SecDr	PCW	1/21/85	229	1032
TECUMSEH	P	OS	CB&Q/GMP	/66		Aur
TECUMSEH		12SecDr	PCW Det	12/30/81	93B	D135
TECUMSEH		12SecDr	Pullman	9/18/09	1963E	3744
TEDWORTH		12SecDr	Pullman	12/18/09	1963E	3766
TEHAMA	N	10SecDr2Cpt	Pullman	12/16/08	2078E	3661
TEHAMA #413	P	12SecDrB	Wagner	2/90	3065	Wg32
TEHERAN	N	12SecO	Grand Trunk	6/71	18C	Mon1
TELFORD		12SecDr SUV	Pull Cal	8/ /04	1963	C9
TEMECULA		10SecDrB	PCW	12/5/87	445E	1361
TEMERAIRE		12SecDr	Pullman	5/11/08	1963C	3615
TEMESCAL		10SecDrB	PCW	12/15/87	445E	1360
TEMILPA	N	10SecDrB	PCW	8/88	445E	
TEMILPA	N	14SecDr	Pullman	12/30/01	1523I	2798
TEMPLAR		10SecDr	PCW Det	3/15/73	26 3/4	D15
TEMPLAR	N	10SecDrB	PCW	11/3/87	445E	1334
TEMPLETON		12SecDr	PCW	11/1/98	1318B	2340
TENANGO	N	14SecDr	Pullman	8/2/00	1523H	2567
TENEDOS	N	12SecDrB	Wagner	11/92	3073	Wg65
TENERIFFE #107	P	20Ch8StsDrB	Wagner	9/90	3126E	Wg41
TENINO		10SecDrCpt	PCW	2/14/90	668	1649
TENNESSEE	P	10SecO	PCW Det (rblt)	12/15/74	47	D27
TENNESSEE	N	4Sec12Ch	(WS&PCC)			
TENNESSEE		10Cpt	Pullman	1/15/02	1704	2758
TENNESSEE #20	N		(WS&PCC)	6/29/72		
TENNESSEE #24	P	10SecDr	H&H	4/72	26A	Wil7
TENNIS #134	N	10SecDrB	(WS&PCC)	1/15/85	886A	
TENNYSON		12SecDr	PCW	10/13/92	990	1931
TENSAS	P	SDR	J&S	5/73	37	Wil10
TEOCALCO	N	14SecDr	Pull Buf	9/27/00	1523D	B1
TERAMO		12SecDr	Pull Buf	2/4/03	1581C	B12
TERRE HAUTE	RN	8SecBLibObs	PCW			1707
TERRYVILLE	P	32Ch8Sts SUV	Pull P-1990	4/04	3271	3092
TERTULIA	N	BaggSmkB SUV	New Haven RR	6/11	3219A	
TEUTONIC		6SecDr4Sr	PCW	2/7/91	887C	1788
TEXANA	N	26ChDr	Pullman	3/27/06	1987A	3335
TEXARKANA		12SecDr	Pullman	5/6/02	1581B	2849
TEXAS		10Cpt	Pullman	1/16/02	1704	2758
TEXAS M&LT-6	P	Slpr	(SP Co)	3/86	37?	
THACKERAY		12SecDr	PCW	10/14/92	990	1931
THALES	N	12SecDr	(Wagner)	10/83	3023D	
THALIA		12SecDr	PCW Det	12/30/80	93	D88

THALIA
Wood

CAR NAME	ST	CAR TYPE	CAR BUILDER (Owner)	DATE BUILT	PLAN	LOT
THALIA		26ChDr	Pullman	8/4/04	1987A	3097
THALLOT		20Ch5StsB	PCW	11/24/83	167A	89
THAMES	S	Diner	PCW	11/27/89	662D	1641
THAMES	P	16SecO	Wagner	10/92	3072	Wg79
THASOS	N	12SecDrB	Wagner	11/92	3073	Wg65
THE ARUNDEL	S	32Sts Parlor	PCW	2/2/99	1376B	2371
THE CHICHESTER	S	32Sts Parlor	PCW	2/2/99	1376B	2371
THE ITURBIDE		Diner	PCW	6/7/87	378	1324
THE PALACE		Diner	PCW	6/7/87	378	1324
THE PONCE DE LEON		Diner	PCW	4/12/87	378	1324
THE QUEEN	RN	29Sts Parlor	PCW Det	1/25/74	39	D20
THE VENDOME		Diner	PCW	6/7/87	378D	1324
THEBES		14SecO	PCW Det	7/1/93	1041	D519
THEKLA		10SecDrB	PCW	11/9/91	920B	1866
THEKLA	N	26ChDr	Pullman	8/4/04	1987A	3097
THELMA	P	22Ch6StsDrB	B&S	2/00	2368	
THEMIS		26Ch3StsDr	PCW	4/28/98	1278C	2258
THEMIS	N	12SecDr	(Wagner)	11/83	3023D	
THEODORA		22Ch6StsDr	PCW	9/1/92	976A	1910
THEODORA	N	24Ch5StsDr	(Wagner)	6/86	3100	
THERAPIA		12SecDr	Pullman	6/16/03	1581E	3010
THERAPIA	R	30Ch SUV	Pullman	6/16/03	3581E	3010
THERESA		20Ch6StsDr	PCW	7/31/91	928A	1827
THESEUS		10SecDrB	PCW	9/27/84	149D	1006
THESEUS	R	16Sec	PCW	9/27/84	1877	1006
THESPIS		16SecO	Pull Cal	/01	1637B	C1
THESSALY		10SecDrB	PCW Det	9/25/84	149D	D248
THETIS		10SecDrB	PCW	10/13/84	149D	1007
THETIS	R	12SecDrB	PCW	10/13/84	1468	1007
THIBET		16SecO	Pull Cal	/01	1637B	C1
THISBE		14SecO	PCW Det	5/20/93	1021	D525
THISTLE	N	12SecO	H&H	2/72	27R	Wil7
THISTLE	N	12SecDrB	Wagner	8/92	3070	Wg71
THOMASTON	N	12SecDr	(Wagner)	11/80	3022F	
THOMASTON	P	32Ch8Sts SUV	Pull P-1990	4/04	3271	3092
THORNCLIFFE		12SecDr	Pullman	10/20/06	1963C	3408
THORNDALE		12SecDrCpt	Pullman	8/8/05	2088	3221
THORNDYKE		12SecDr	Pullman	3/26/08	1963C	3600
THORNTON	N	10SecDrB	PCW	10/23/86	445	1262
THORNTON	R	14Sec	PCW	10/23/86	445X	1262
THRACIA		12SecDr	PCW Det	1/28/82	93B	D135
THRALL	N	10SecDr2Cpt	Pullman	3/23/07	2078D	3528
THRUSH		24ChDrB	Pullman	6/4/00	1518C	2552
THULE 2552		14SecO	PCW Det	2/18/93	1021	D516
THURINGIA		12SecDr	Pull Buf	11/25/03	1581F	B25
THURIO		10SecDrB	PCW	11/5/91	920B	1866
THURMOND		12SecDr	Pullman	2/17/06	1963C	3316
THYRZA	N	14Sec	(Wagner)	4/87	3049C	
TIARA	RN	10SecB	Bowers, Dure	10/73	37U	Wil11
TIBER	R	12SecDrB	PCW Det	2/15/83	1468	D176
TIJER	P	16SecO	Wagner	4/93	3072	Wg90
TIBER #382		10SecDrB	PCW Det	2/15/83	149	D176

CAR NAME	ST	CAR TYPE	CAR BUILDER (Owner)	DATE BUILT	PLAN	LOT
TIBERIUS	N	BaggSmk	PCW	/99	1400	2410
TICAO		12SecDr	Pullman	8/ /04	1963	3098
TICINIUS #442		10SecDrB	PCW	8/21/85	240	1058
TICONDEROGA		Parlor	PCW Det	2/3/72	22	D9
TICONDEROGA	P	12SecDr	(Wagner)	11/80	3023F	
TICUMAN	N	14SecDr	Pullman	12/30/01	1523I	2798
TIDAL WAVE	P	12SecO	PCW (Wagner)	4/65	3010	1070
TIDANUM	N	12SecDr	Pullman	3/27/03	1581D	2983
TIFTON	P	12SecDrB	(GH&SF Ry)		2490	
TIGRANES	N	12SecDrB	Wagner	8/91	3065	Wg53
TIGRIS		14SecO	PCW	4/26/93	1021	1972
TIGRIS #224	P	12SecDr	PCW	6/11/86	3049	1199
TILLAMOOK		16Sec	Pullman	3/22/01	1649A	2709
TIMBERTON	N	12SecDr	Pullman	3/17/03	1581D	2975
TIMOR		12SecDr	Pullman	5/13/04	1963	3082
TIOGA	P	12SecDr	(Wagner)	4/82	3022H	
TIOGA	P	10SecO	(Pull/Sou)	/75	47A	
TIOSA	N	Private	J&S (WS&PCC)	7/25/78	689	
TIPPECANOE	P	12Ch24StsObs SU	(CI&L RR)	C:/11	2747	
TIPTON	N	10SecDr	PCW Det	3/1/76	52B	D33
TIPTON		BaggClub	Pullman	8/18/06	2136A	3386
TISONIA		10SecDrB	PCW	9/12/87	445E	1362
TISONIA	R	16Sec	PCW	9/12/87	1880C	1362
TITAN	N	10SecDr	PRR	2/71	26 3/4	Alt1
TITAN		12SecDr	PCW	11/30/98	1318C	2359
TITANIA		12SecDr	PCW Det	2/26/81	93	D88
TITANIA	R	12SecDrB	PCW Det	2/26/81	1611	D88
TITIAN		12SecDr	Pullman	8/8/01	1581A	2727
TITUS	R	BaggSmkBLib	PCW		1706	
TITUSVILLE #169	P	10SecDr	PRR	2/71	26 3/4	Alt1
TIVERTON		12SecDr	Pullman	3/11/02	1581B	2818
TIVOLI	N	12SecDrB	Wagner	11/92	3073A	Wg80
TOBOCA #416		10SecDrB	PCW	11/16/83	149	D207
TOCCOA #339	P	10SecDrB	(Union PCC)		565B	
TOKAY	N	10SecDr	PCW Det	7/23/81	112	D118
TOKIO		10SecDrB	PCW	4/8/84	149	113
TOKIO	P	16SecO	Wagner	10/92	3072	Wg79
TOLARE		10Sec2Dr	Pullman	11/3/02	1746A	2928
TOLEDO		10SecDr	TW&W Ry	6/30/75	19	Tol1
TOLEDO #264	P	12SecDr	Wagner	7/88	3061B	Wg12
TOLLESTON		12SecDr	Pullman	11/21/00	1581A	2589
TOLONO		10SecDr	TW&W RY	6/30/75	19	Tol1
TOLOSA	N	12SecDrB	Wagner	12/98	3073E	Wg117
TOLTEC		10SecO (NG)	PCW Det	6/10/80	73A	D76
TOLTEC	P	16SecO	Wagner	10/92	3072	Wg79
TOLUCA	N	8SecSrB (NG)	PCW Det	11/23/82	155	D173
TOLUCA	RN	14SecDr	PCW	9/26/84	1468	1006
TOLVA		14SecO	PCW Det	8/3/92	1014A	D509
TOMELLIN	N	14SecDr	Pullman	8/8/00	1523H	2567
TOMMY DODD E&A-28	P	Slpr	(E&A Assn)			
TONAWANDA		12SecDr	Pullman	2/27/07	1963C	3525
TONGA		12SecDr	PCW			

Figure I-39 Completed in June 1880, before the famous Calumet Shops were opened, TOLTEC, a 10-section open platform narrow gauge sleeper, rides the transfer table at the Pullman Car Works in Detroit. TOLTEC traveled The Scenic Line on the narrow gauge rails of the Denver & Rio Grande Railway until 1905 when it was sold to Mr. H. Schlacks who, in turn, sold it to the 3" gauge Uintah Railway. The Arthur D. Dubin Collection.

CAR NAME	ST	CAR TYPE	CAR BUILDER (Owner)	DATE BUILT	PLAN	LOT	CAR NAME	ST	CAR TYPE	CAR BUILDER (Owner)	DATE BUILT	PLAN	LOT
TONILITA	N	14SecDr	Pull Buf	9/21/00	1523D	B1	TOULON	P	16SecO	Wagner	4/93	3072	Wg90
TONOPAH	N	12SecDr	Pull Buf	11/25/03	1581F	B25	TOULOUSE		14SecO	PCW Det	5/20/93	1021	D525
TONOPAH	N	12SecDrB	(Wagner)	6/86	3050		TOURAINE		12SecDr	Pullman	5/9/08	1963C	3615
TONQUIN		10SecDrB	PCW	8/4/87	445E	1333	TOURIST		13SecO	J&S	6/82	187	
TOPAZ		12SecDr	PCW Det	2/5/92	784E	D488	TOURIST		12SecDr	Pull Buf	6/20/07	1963C	B61
TOPAZ	N	14Sec	PCW (Wagner)	7/2/86	3049B	1199	TOURMALINE		12SecDr	Pullman	12/9/07	1963C	3585
TOPEKA	P	10SecDr	H&StJ RR	10/72	26	Han3	TOURNAY	P	14SecDr SUV	Pull P-2069A	6/05	3207B	3187
TOPEKA		10Sec2Dr	PCW	3/4/93	995A	1942	TOURS		14SecO	PCW	5/19/93	1021	1972
TOPO CHICO		8SecSrB (NG)	PCW Det	11/92	1001	D514	TOWANDA	N	10SecDrB	PCW Det	2/29/84	149V	D228
TOPPENISH		14SecDr	Pullman	5/16/06	2165	3352	TOWANTIC	P	22Ch4Sts	B&S	8/82	3253	
TOREADOR		12SecDr	Pullman	12/4/08	1963E	3659	TOYAH	N	10SecDrB	PCW	11/6/83	149	80
TORGAN	N	12SecDr	(Wagner)	6/81	3022G		TRAFALGAR		12SecDr	Pullman	9/15/01	1581A	2756
TORINO		12SecDr	Pull Buf	10/24/02	1581C	B11	TRAJAN	N	12SecO	(Wagner)	/85	3005	
TORONTO	P	DDR	Grand Trunk	10/70	17	Mon1	TRANIO		10SecDrB	PCW	11/6/91	920B	1866
TORONTO		12SecDr	Pullman	7/30/03	1581F	3015	TRANSIT		22Berths	PCW Det	10/13/75	48	D26
TORQUAY		16SecO	Pullman	5/4/05	2068	3177	TRANSIT		12SecDr	Pull Buf	6/14/07	1963C	B61
TORREON		12SecDr	Pullman	7/27/03	1581F	3019	TRANSIT #262	N	10SecDr	H&H	9/72	26A	Wil7
TORREON	R	30Ch SUV	Pullman	7/27/03	3581E	3019	TRAVE		12SecDr	PCW Det	10/1/91	784D	D470
TORREON IM-6	P	10SecDrB	PCW	12/8/88	445H	1519	TRAVELER	RN	Slpr	(SouTranCo)			
TORRINGTON	P	32Ch8Sts SUV	Pull P-1990	4/04	3271	3092	TRAVELER		12SecDr	Pull Buf	6/6/07	1963C	B61
TOSCA		10SecDr	PCW	12/30/89	717B	1630	TRAVELER #209	P	Private	(Wagner)	/87	3080	
TOSCANA	S	12Sec	PCW Det	6/9/76	50	D30	TRAVIATA		12SecDr	Pullman	3/25/08	1963C	3604
TOSCANA	N	BaggClub	Pullman	8/17/05	2087	3220	TREMONT	P	Diner	CB&Q RR	/68	16	Aur
TOULON		10SecDrB	PCW	4/30/84	149D	192	TREMONT		Diner	PCW	11/22/90	662D	1777
TOULON	R	12SecDrB	PCW	4/30/84	1468	192	TREMONT		12SecDrB	PCW	9/9/98	1319A	2314

CAR NAME	ST	CAR TYPE	CAR BUILDER (Owner)	DATE BUILT	PLAN	LOT
TRENHOLME	N	10SecDr2Cpt SUV	Pullman	5/29/08	2078D	3622
TRENT		12SecDr	PCW Det	9/28/91	784D	D470
TRENT	N	16Sec	Pullman	1/11/08	2184	3588
TRENTON		12SecDr	Pullman	10/13/00	1581A	2588
TRENTON #157		12SecO	J&S	12/70	18	Wil3
TREVINO		10SecDrB	PCW	10/7/87	445E	1334
TREVISO		12SecDr	Pull Buf	10/31/02	1581C	B11
TREVORTON	N	12SecDrSr	Pullman	4/7/05	2049A	3169
TREVORTON #299	P	10SecB	B&S	9/5/73	37U	Day10
TREVOSE		24ChDrB	Pull Buf	5/18/07	2278	B59
TREYSA		14SecO	PCW Det	3/25/93	1021	D517
TRIANON		12SecDr SUV	Pullman	10/25/06	1963C	3408
TRIBES HILL		16Sec	Pullman	4/4/07	2184	3523
TRIDENT		12SecDr	Pullman	12/10/08	1963E	3659
TRIERMAIN		12SecDr	Pullman	12/6/07	1963C	3585
TRIESTE		12SecDr	Pullman	7/28/03	1581F	3019
TRIESTE #2523		14SecO	PCW Det	8/1/92	1014A	D509
TRILBY	P	22Ch6StsDrB	B&S	2/00	2368	
TRINCULO		12SecDrCpt	Pullman	6/24/04	1982	3090
TRINIDAD		12SecDrB	PCW	9/7/88	375F	1480
TRINIDAD	N	12SecDrB	PCW Det	5/2/84	1611A	D242
TRINIDAD #223	S	14SecO?	PCW (Wagner)	6/12/86	93H	1199
TRINITY	R	12SecDrB	PCW	1/28/88	1566	1361
TRIONAL	P	14SecDr SUV	B&S	6/11	3207C	
TRIPOLI		10SecDrB	PCW Det	10/6/83	149	D206
TRIPOLI #109	P	20Ch8StsDrB	Wagner	9/90	3126E	Wg41
TRISTRAM		12SecDr	Pullman	12/18/09	1963E	3766
TRITON	N	12SecDr	Wagner	1/89	3061B	Wg13
TRITON	N	12SecO	B&O RR		665F	
TRIUMPH		12SecDr	Pull Buf	6/11/07	1963C	B61
TRIUMPH O&C-8005	P	10SecDrB	PCW	12/18/86	240A	1138
TROJAN	P	10SecDr	B&S	9/71	26	Day6
TROJAN		12SecDr	PCW	11/17/98	1318C	2359
TROPHY	RN	14Sec	PCW	11/19/89	717F	1630
TROPIC	N	10SecDr	H&H	9/72	26A	Wil7
TROSACHS	N	Slpr	(NYLE&W Assn)		569	
TROSOA 2553		14SecO	PCW Det	2/25/93	1021	D516
TROUBADOUR		12SecDr	Pullman	12/4/08	1963E	3659
TROVATORE		12SecDr	Pullman	3/24/08	1963C	3604
TROY	N	DDR	PCW Det	/70	17K	D3
TROY	P	28Ch5StsDr	Wagner	11/92	3125B	Wg74
TRUCKEE		10SecDr2Cpt	Pullman	9/25/05	2078	3203
TRUCKEE CP-27	P	Slpr	(SP Co)	7/1/83	37?	
TRURO	P	10SecDr	Grand Trunk	8/73	31	Mon4
TRURO	N	14SecO	PCW Det	2/25/93	1021	D516
TRUXILLO	P	14Sec	B&S	7/93	3204	
TRYPHENA		8SecLngObs	PCW	5/15/90	644D	1668
TRYPHOSA		8SecLngObs	PCW	6/9/90	644D	1668
TUCKAHOE		12SecDr	Pullman	2/14/06	1963C	3316
TUCSON		10Sec2Dr	Pullman	8/25/02	1746A	2925
TUCSON GH&SA-8	P	Slpr	(CP Co)	5/24/84	37?	
TUDOR	P	16SecO	Wagner	4/93	3072	Wg90
TUDOR #102	RN	7BoudB	(MannBCC)	/83	697A	
TUERTO		7Cpt2Dr	Pullman	10/29/02	1845	2926
TUGALO #332	P	10SecDr	(Union PCC)		31	
TUGELA	N	10SecDr	Grand Trunk	6/72	31B	Mon3
TULA	N	10SecDr	PCW Det	2/28/82	112A	D141
TULANE		12SecDr	Pullman	12/15/05	1963C	3252
TULARE		10Sec2Dr	Pullman	11/3/02	1746A	2928
TULARE GH&SA-7	P	Slpr	(CP Co)	6/16/84	37?	
TULAROSA		7Cpt2Dr	Pullman	10/29/02	1845	2926
TULIP		18Ch6StsDrB	PCW	6/11/90	772D	1704
TULLAHOMA	R	10SecO	PCW Det (rblt)	2/13/75	47	D27
TUNBRIDGE		12SecDr	Pull Cal	8//04	1963	C9
TUNIS		10SecDrB	PCW	6/27/84	149D	192
TUNIS	R	12SecDrB	PCW	6/27/84	1468	192
TUNIS	P	16SecO	Wagner	4/93	3072	Wg90
TUPELO		12SecDr	Pull Cal	8//04	1963	C9
TUPPER LAKE #121	P	20Ch8StsDrB	Wagner	6/91	3126A	Wg51
TURENNE	N	12SecDr	(Wagner)	9/83	3023D	
TURIN		12Sec	PCW Det	6/19/75	49	D28
TURIN	N	12SecDr	Wagner	8/91	3067A	Wg28
TURKESTAN		12SecDr	Pullman	2/1/04	1581H	3053
TURKEY		12SecDr	Pullman	3/17/03	1581D	2975
TURLA		14SecO	PCW Det	1/27/93	1014A	D511
TURQUOISE		12SecDrSr	PCW	5/27/99	1318C	2390
TUSCALOOSA		10SecDrB	PCW Det	1/26/84	149	D224
TUSCANY	N	10SecDr	Grand Trunk	9/71	26H	Mon2
TUSCANY		12SecDr	Pull Buf	9/24/03	1581F	B22
TUSCARORA		12SecDrCpt	Pullman	5/26/04	1985	3095
TUSCARORA #175	P	10SecDr	PRR	8/71	21	Alt2
TUSCARORA CTC-94	R		PRR rblt/72			
TUSCOLA		16SecSr	Pullman	8/29/04	2033A	3135
TUSCOLA #317	N	8BoudB	(Union PCC)		373A	
TUSCUMBIA		12SecDr	Pull Buf	1/24/02	1581A	B7
TUSCUMBIA #37	R	10SecO	PCW Det (rblt)	2/13/75	47	D27
TUSHIPAH		10secDr2Cpt SUV	Pullman	3/16/10	2471	3784
TUSKEGEE	R	12SecB	PCW	8/83	168	47
TUSKOGEE		10SecDr	PCW	8/22/83	112	47
TUXEDO	N	14Ch13StsDrB	PCW	10/4/87	505	1376
TUXEDO		10SecDrB	PCW	11/1/87	445E	1334
TWEED		12SecDr	Pullman	9/17/01	1581A	2756
TWILIGHT	N	12SecO	PCW Det		5	Det2
TWILIGHT	RN	Private	PCW	1/31/88	1920	1361
TWILIGHT	N	16Ch3StsDrB	(Wagner)	7/71	3108C	
TWIN LAKES	P	BaggBClub SUV	Pull P-1503B	/00	3216	2517
TYBURNIA		12SecDr	Pullman	10/4/01	1581A	2765
TYLER		8SecSrB (NG)	PCW Det	11/18/82	155	D173
TYLOS #2514		14SecO	PCW Det	2/3/93	1014A	D511
TYNEDALE		12SecDr	Pullman	12/20/09	1963E	3766

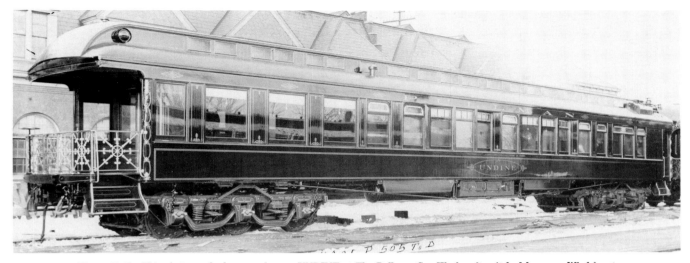

Figure I-40 This photograph shows parlor car UNDINE at The Pullman Car Works after it had been modified by the addition of an observation platform. It was built in 1887 for service in the New York, Lake Erie and Western Association as a parlor car. In 1913 it was sold to the Nevada Northern Railway. Smithsonian Institution Photograph P4001.

CAR NAME	ST	CAR TYPE	CAR BUILDER (Owner)	DATE BUILT	PLAN	LOT	CAR NAME	ST	CAR TYPE	CAR BUILDER (Owner)	DATE BUILT	PLAN	LOT
TYNEDALE #577	P	10SecDrB	J&S (B&O)	C:/88	663		ULYSSES		BaggDiner	PCW	6/26/91	783B	1845
TYPHON	N	10SecO	(Wagner)	/67	3006		ULYSSES		BaggClubBarber	PCW	12/3/98	1372B	2365
TYRE		14SecO	PCW Det	4/5/93	1021	D517	ULYSSES S. GRANT	N	20Ch4StsObs	Pullman	7/16/05	2057B	3217
TYRO	N	10SecB	B&S	9/73	17R	Day4	UMATILLA		10SecO	PCW Det	2/10/81	102	D91
TYROL		10SecDrB	PCW	3/8/84	149	102	UMATILLA	R	12SecDrB	PCW Det	2/10/81	1468	D91
TYROL	P	16SecO	Wagner	4/93	3072	Wg90	UMBRIA	RN	10SecDrB	Grand Trunk		240	
TYROL #226	S	14Sec?	PCW (Wagner)	7/8/86	93H	1199	UMBRIA #25	P	16Ch8StsDrB	(Wagner)	6/85	3115D	
TYROLEAN		12SecDr	PCW	11/29/98	1318C	2359	UMPYNA	RN	12SecDr	PCW Det	3/16/83	1467A	D184
TYRONE		12SecDrSr	Pullman	1/19/01	1581A	2614	UNAKA	N	12SecDrCpt	Pullman	8/8/05	2088	3221
TYRONE #135	P	DDR	PRR	10/70	17	Alt1	UNAKA SP-2549		10SecO	PCW	11/92	1024	W7
UARDA	P	24ChDrB	(Big-4)	/01	2378		UNAVISTA	P	16Ch6StsDrB	Pullman	8/85	3254	
UGANDA		18Ch6StsDrB	PCW Det	9/9/90	772D	D427	UNAWANDA	P	26Ch6Sts	New Haven RR	8/87	3255	
UGANDA	N	12SecDr	Pullman	10/15/07	2163A	3562	UNDINE		14Ch13StsDrB	PCW Det	9/7/87	505	D309
UKAMBA	N	12SecDr	Pullman	1/4/07	2163	3460	UNDINE	P	12SecDrB	Wagner	1/93	3073B	Wg83
UKASSA	N	12SecDr	Pullman	10/15/07	2163A	3562	UNICORN		12SecDr	Pullman	1/31/02	1581A	2792
UKRAINE		14SecO	PCW	6/2/93	1021	1972	UNIOLA		12SecDr	PCW	4/15/91	784C	1798
ULAMA	N	12SecDr	Pullman	4/24/07	2087	3534	UNION	P	12SecO	DetC&MCo	1/30/69	5	Det
ULAMA SP-2548		10SecO	PCW	11/92	1024	W7	UNION		10SecDr2Cpt	Pullman	9/26/05	2078	3203
ULIDIA		20Ch6StsDr	PCW	8/1/91	928A	1827	UNIQUE		12SecDr	Pullman	2/13/06	1963C	3316
ULINA		12SecDr	PCW Det	4/26/92	784E	D497	UNITAH		12SecDrB	PCW	2/13/89	375G	1557
ULRICA #86	P	28Ch3Sts	PCW (Wagner)	6/14/84	3124	D256	UNITAH	P	12SecDrB	Wagner	1/93	3073B	Wg83
ULSEAH		10SecDr2Sr	Pullman	8/22/09	2078J	3721	UNIVERSAL		12SecDr	Pullman	1/26/07	1963C	3448
ULSTER		16SecO	PCW	6/8/93	1040	1990	UNOME	N	24ChDrB	Pullman	6/3/05	1988A	3170
ULTIMA	P	26Ch6Sts	New Haven RR	8/87	3255		UPSAL		12SecDrSr	Pullman	3/1/01	1581B	2654
ULTONIA		14SecDr	PCW	8/28/99	1318E	2447	UPTON	N	10SecDrB	PCW	6/4/91	920B	1808
ULVA		10SecDr	PCW	12/9/89	717B	1630	URAL		10SecDrB	PCW	1/17/90	725B	1629

CAR NAME	ST	CAR TYPE	CAR BUILDER (Owner)	DATE BUILT	PLAN	LOT
URAL	P	12SecDrB	Wagner	1/93	3073B	Wg83
URANIA	P	12SecDrB	Wagner	1/93	3073B	Wg83
URANUS		12SecDr	PCW	1/22/91	784	1786
URBANA	P	12SecDrB	Wagner	1/93	3073B	Wg83
URBANA #253	P	10SecDr	PC&StL RR	8/72	26	Col3
URIEL		14SecO	PCW Det	4/5/93	1021	D517
URSA		12SecDr	PCW Det	11/18/90	784	D449
URSULA		18Ch6StsDrB	PCW Det	9/11/90	772D	D427
URSULA #87	P	Parlor	PCW (Wagner)	6/22/84	3124	D256
URUAPAN	N	12SecDrB	PCW		1498	
URUAPAN	RN	8SecSrB (NG)	PCW Det	2/23/83	1594A	D120
URUGUAY		14SecO	PCW Det	5/31/93	1021	D525
USONIAN	P	16Sec SUV	B&S	7/11	3212	
UTAH		10SecDr	PCW	10/27/87	508B	1368
UTAH	P	12SecDrB	Wagner	1/92	3073B	Wg83
UTAWANA		Parlor	(Wagner)	C:/75		
UTICA	P	10SecO Tourist	(DL&W RR)	12/79	220	
UTICA		12SecB	PCW Det	5/23/83	168	D195
UTICA		12SecDr	Pull Buf	4/18/03	1581C	B15
UTOPIA		20Ch3StsB	PCW Det	9/18/83	167A	D209
UTOPIA		BaggBSmkBarber	PCW	10/4/95	1155A	2122
UTRECHT		14SecO	PCW Det	4/12/93	1021	D517
UVADA		10SecDr3Cpt	Pullman	4/20/06	2078C	3342
UVALDE	RN	10SecObs	PCW	3/12/87	445W	1299
UVONIA	P	24Ch6Sts	Wason	7/88	3257	
UXBRIDGE		12SecDr	Pullman	10/6/03	1581F	3022
UXORA	P	14Sec	B&S	12/82	3201	
VACUNA		8SecLngObs	PCW	2/9/89	644B	1563
VACUNA		30ChDr	Pullman	7/2/04	1983	3089
VACUNA		28ChDr	Pullman	7/2/04	2024	3089
VALAIS	N	10SecDrB	PCW	11/14/84	149D	1017
VALAIS	R	16Sec	PCW	11/14/84	1877	1017
VALCOUR		14SecO	PCW	5/13/93	1021	1972
VALDA		10SecDrB	PCW	12/10/89	725B	1629
VALDIVIA	RN	16SecB	Pull Cal	9/20/02	1794	C3
VALDIVIA	P	12SecSr SUV	Pull P-2488	10/10	3209	3809
VALDOSTA		12SecDr	Pull Buf	2/16/03	1581C	B13
VALENCE		14SecO	PCW Det	3/31/93	1021	D517
VALENCIA		16Sec	Pullman	10/16/01	1720A	2766
VALENCIA #55		12SecDr	PCW Det	6/21/82	93B	D154
VALENCIA #229	S	16SecO?	PCW (Wagner)	6/18/86	93H	1199
VALENTIN	R	Diner SUV	PCW	2/11/92	1981	1874
VALENTIN		Diner	PCW	2/11/92	964A	1874
VALENTINE		24Ch2StsDrObs	Pull Buf	5/4/01	1634A	B4
VALENZA		14SecO	PCW Det	5/20/93	1021	D525
VALERIA		18Ch4StsDrB	PCW	7/18/93	1039A	1995
VALERIA #40	P		H&H (WS&PCC)	5/1/78	570?	
VALERIAN		12SecDr	Pullman	5/4/01	1581A	2726
VALHALLA		12SecDr	Pullman	5/14/08	1963C	3615
VALIANT	N	22Ch6Sts	PCW	8-9/92	976B	1910
VALIANT	P	12SecHotel	J&S	1/99	2371A	

CAR NAME	ST	CAR TYPE	CAR BUILDER (Owner)	DATE BUILT	PLAN	LOT
VALIANT	P	22Ch2Sts	B&S	7/84	3252A	
VALKYR		18Ch6StsDrB	PCW Det	9/9/90	772D	D427
VALKYRIE		30ChDr	Pullman	7/2/04	1983	3089
VALKYRIE		28ChDr	Pullman	7/2/04	2024	3089
VALLECITO SP-46	P	10SecDrB	PCW	12/19/88	445H	1519
VALLEJO		10Sec2Dr	Pullman	8/26/02	1746A	2925
VALLEJO GH&SA-11	P	Slpr	(SP Co)	7/1/83	37?	
VALLEY CITY		12SecDr	PCW Det	6/27/84	93D	D241
VALLEY CITY	RN	14Sec	PCW	2/5/90	668F	1649
VALLEY FALLS	N	12SecDrSr	Pullman	3/30/05	2049A	3169
VALLEY FALLS #536	P	10SecB	B&O RR?		649	
VALMONT		10SecDr2Cpt	Pullman	12/29/05	2078C	3287
VALMORE		10SecLibObs	Pullman	8/24/08	2351C	3634
VALPA #2554		14SecO	PCW Det	2/18/93	1021	D516
VALPARAISO		10SecO	B&S	12/72	26K	8
VALPARAISO #515	P	12SecDrB	Wagner	8/92	3070	Wg71
VALSPAR	N	12SecDr	Pullman	1/5/07	2163	3460
VANADIS		12SecDr	Pullman	8/5/08	1963C	3636
VANCEBORO		12SecDr	Pull Buf	7/7/04	1963	B38
VANCOUVER		10SecDrB	PCW Det	2/23/84	149	D228
VANCOUVER		12SecDrB	PCW	8/22/88	375F	1496
VANCOUVER #230	S	16SecO?	PCW	6/26/86	93H	1199
VANDALIA		12SecDrSr	PCW	12/12/98	1318C	2370
VANDALIA #137	P	6SecBCh	PRR	11/70	17S?	Alt1
VANDYKE		12SecDr	Pullman	8/6/01	1581A	2727
VANEGAS	N	12SecDrB	Pull Wilm	3/30/03	1611F	W9
VANESSA		20Ch11StsObsB	Pullman	7/7/04	1984	3096
VANITIE	N	22Ch6Sts	PCW	8-9/92	976B	1910
VAQUEROS		12SecDr	Pullman	9/28/06	1963C	3397
VARDO #2555		14SecO	PCW Det	2/25/93	1021	D516
VARENNES		12SecDrSr	Pullman	2/8/01	1581B	2638
VARZO #2512		14SecO	PCW Det	8/1/92	1014A	D509
VASHTI		20Ch11StsObsB	Pullman	7/7/04	1984	3096
VASONA		10Sec2Dr	Pullman	10/30/02	1746A	2928
VASQUEZ	N	12SecO	(Wagner)	/85	3027E	
VASSAR	N	10SecDr	H&H	4/72	26	7
VASSAR		12SecDr	Pullman	12/18/05	1963C	3252
VASSAR	N	20Ch6Sts	(Wagner)	5/74	3104	
VAUBAN	N	12SecDrB	Wagner	8/91	3065	Wg53
VEGA		12SecDr	PCW	1/9/91	784	1781
VEGA	P	12SecDrB	Wagner	11/92	3073	Wg75
VELASCO	R	10SecObs	PCW	8/31/87	445V?	1333
VELASCO		10SecDrB	PCW	8/31/87	445E	1333
VELASCO	P	10SecB	(Wagner)	11/84	3036	
VELETA		12SecDr SUV	Pullman	7/28/06	1963C	3385
VELMAR	N	Parlor	(NYLE&W Assn)			
VELMER	N	12SecDrB	(Wagner)	5/85	3050	
VELOCIPEDE		15Ch17StsDr	CB&Q RR	/68	6	Aur
VENADITO	N	10SecDrB	PCW	1/28/88	445E	1361
VENADO	P	10SecB	(Wagner)	9/84	3036	
VENCEDOR		12SecDr	Pullman	5/12/08	1963C	3615

Figure I-41 The interior of VENEZUELA represents extraordinary craftsmanship in marquetry, incorporating an intricate design of leaves and flowers resembling morning glories. Even the window armrests are covered in a plush upholstery. The car was originally built as a 10-section drawing room buffet but was later downgraded to a utility 16-section in 1902. Smithsonian Institution Photograph P-2471.

CAR NAME	ST	CAR TYPE	CAR BUILDER (Owner)	DATE BUILT	PLAN	LOT
VENDOME #516	P	12SecDrB	Wagner	8/92	3070	Wg71
VENETIA		12SecDr	PCW Det	2/4/82	93B	D144
VENETIAN		12SecDr	PCW	11/21/98	1318C	2359
VENEZUELA		10SecDrB	PCW	7/10/84	149D	193
VENEZUELA	R	16Sec	PCW	7/10/84	1877	193
VENEZUELA #514	P	12SecDrB	Wagner	8/92	3070	Wg71
VENICE		10SecDrB	PCW Det	9/29/83	149	D206
VENICE	N	10SecDrCpt	Wagner	6/93	3074F	Wg93
VENICE	R	16Sec	PCW Det	9/29/83	1877C	D206
VENICE #517	P	12SecDrB	Wagner	8/92	3070	Wg71
VENTA		14SecO	PCW Det	2/4/93	1014A	D511
VENTNOR		6CptLngObs	Pullman	12/18/03	1299D	3065
VENTURA		14SecO	PCW Det	5/11/93	1021	D525
VENUS		27Sts Parlor	PCW Det	5/3/76	45	D25
VENUS	P	12SecDrB	Wagner	11/92	3073	Wg75
VENUS #217	P	6SecCh	CB&Q RR	3/72	29D	Aur2
VENUSTA		10SecDr2Cpt	Pullman	11/11/09	2451	3747
VERA		10SecDrB	PCW	12/10/89	725B	1629
VERA CRUZ	N	10SecDrB	PCW Det	4/24/84	149D	D242
VERBENA		26Ch4StsDr	Pull Buf	7/12/00	1519C	B2
VERDI		12SecDr	PCW	4/29/92	784E	1899
VERDI	P	10Cpt	Wagner	1/94	3075	Wg104
VERDUGO		10Sec2Dr	Pullman	10/31/02	1746A	2928
VERDUN		14SecO	PCW Det	4/12/93	1021	D517
VERITAS		6CptLngObs	PCW	2/3/98	1299A	2283
VERMONT	P	DDR	Ohio Falls	12/70	17	Jef
VERMONT	N	DDR	PCW Det	7/31/73	17	D2
VERMONT		10Cpt	Pullman	1/11/02	1704	2758
VERMONT #208	P	12SecO	(Wagner)	12/6/93	3015	
VERNDALE		10SecDrCpt	PCW	5/23/89	668	1587
VERNIA		10SecDrB	PCW	11/11/91	920B	1866
VERNIA	R	12SecDr	PCW	11/11/91	1497	1886
VERNON 2556		14SecO	PCW Det	2/14/93	1021	D516
VERNONDALE	N	12SecDr	Pullman	9/7/07	1963C	W19
VEROCHIO	P	12SecDr	PCW P-93H	9/2/86	3049	
VERONA	N	OS	DetC&MCo	7/71		
VERONA #518	P	12SeCDrB	Wagner	8/92	3070	Wg71
VERONICA		26Ch4StsObsB	Pullman	8/6/04	1993B	3111
VERSAILLES		12SecDrSr	PCW	7/7/99	1318C	2424
VERSAILLES #227	S	16SecO?	PCW (Wagner)	6/19/86	93H	1199
VERSAILLES #258	P	OS	CB&Q RR	5/72	36	Hinc1
VESPASIAN		12SecDr	Pullman	5/7/01	1581A	2726
VESPER	N	OS	C&NW Ry	C:2/71	27	2
VESPER	P	10Cpt	Wagner	1/94	3075B	Wg104
VESTA		27Sts Parlor	PCW Det	7/28/76	45	D31
VESTA		22ChDr	Pull Buf	7/12/02	1647C	B9
VESTA #16	P	20Ch10StsDr	(Wagner)	7/85	3109A	
VESTA #150	P	27Ch	J&S	7/72	35	Wil1
VETUS	R	BaggSmkBLib			1706	
VEVAY		12SecDr	PCW Det	8/18/91	784D	D481

CAR NAME	ST	CAR TYPE	CAR BUILDER (Owner)	DATE BUILT	PLAN	LOT
VEYTIA		10SecDrB	PCW	5/7/87	445E	1292
VEYTIA	R	16Sec	PCW	5/7/87	1880	1292
VICEREINE		24Ch2StsDrObs	Pull Buf	4/24/01	1634A	B4
VICEROY	P	OS	GW Ry/GMP	/65	13	Ham
VICEROY		10SecDrB Hotel	PCW Det	5/20/76	56	D39
VICEROY	N	BaggClubBarber	Pullman	4/6/04	2000	3026
VICEROY	P	Private	(Wagner)	11/13/95	3082	
VICKERY	N	12SecDrSrB	PCW	9/12/98	1319B	2314
VICKSBURG		12SecDr	Pullman	3/27/03	1581D	2983
VICTOR		12SecDr	PCW Det	12/30/81	93B	D135
VICTOR		12SecDrB	PCW	8/16/88	375F	1480
VICTORIA	RN	Diner	PCW	/76		
VICTORIA		Hotel	Great Western	/65		Ham
VICTORIA		27Sts Parlor	PCW Det	3/1/74	39	D20
VICTORIA		Diner	PCW	10/13/88	538D	1411
VICTORIA #88	P	Parlor	PCW (Wagner)	6/22/84	3124	D256
VICTORIA #23	P		(WS&PCC)	6/29/72		
VICTORIA #76	P		(WS&PCC)	7/1/79	551?	
VICTORINE		24Ch2StsDrObs	Pull Buf	4/29/01	1634A	B4
VIDA		14Ch13StsDrB	PCW	10/27/87	505	1376
VIDALIA		10SecDrB	PCW	10/29/87	445E	1334
VIDETTE	N	10SecDr	J&S (McComb)	/73	19H	Wil
VIDETTE	P	10Cpt	Wagner	1/94	3075	Wg104
VIENNA		10SecDr	H&H	7/31/71	19	Wil2
VIENNA		12SecDr	Pull Buf	6/15/03	1581C	B18
VIENNA	P	10SecDr	(Wagner)	/85	3057C	
VIENTO		10SecDrB	PCW	4/5/87	445C	1281
VIGAN		12SecDr	Pullman	8/04	1963	3098
VIGILANT	N	20Ch9StsDrObs	PCW	11/15/99	1417G	2466
VIGILANT		10SecLibObs	Pullman	8/24/08	2351C	3634
VIGO		12SecDr	PCW	2/19/91	784B	1797
VIKING #2567		14SecO	PCW Det	3/9/93	1021	D516
VILAINE		14SecO	PCW	5/16/93	1021	1972
VILAINE	N	12SecDr	Pull Buf	6/15/03	1581C	B18
VILANO #2568		14SecO	PCW Det	3/11/93	1021	D516
VILLA BLANCA	RN	12SecDrB	Pullman	5/24/06	2165C	3352
VILLETTE		14SecO	PCW	5/24/93	1021	1972
VILLISCA	RN	8SecBLibObs	PCW		1707	
VINARA		12SecDr	Pullman	7/28/06	1963C	3385
VINCENNES		12SecDrSr	Pullman	2/5/01	1581B	2638
VINCENNES #250	P	10SecDr	CochranCCo	6/30/72	26A	
VINCENTIO		7Cpt2Dr	PCW	1/17/96	1065A	2143
VINELAND		14SecDr	Pullman	8/2/00	1523E	2567
VINITA		10SecDr	H&StJ RR	/71	19A	Han2
VINITA	N	12SecDrCpt	Pullman	8/8/05	2088	3221
VINTON	N	10SecDrB	PCW	3/22/88	445E	1361
VINTON	R	16Sec	PCW	3/22/88	1880B	1361
VIOLA	N	10SecB	B&S	9/73	17R	Day4
VIOLA	P	15Ch8StsDrB	(Wagner)	4/85	3115A	
VIOLET		18Ch6StsDrB	PCW	6/11/90	772D	1704
VIRGENE		24ChDr	Pullman	7/7/05	1980A	3171
VIRGIL		12SecO	PCW Det	5/10/82	114	D143
VIRGIL #519	P	12SecDrB	Wagner	8/92	3070	Wg71
VIRGILIA		26ChDr	Pullman	8/4/04	1987A	3097
VIRGINIA	?	15SecO	NC&StL	rblt /72	114G	
VIRGINIA		12SecDr	Pullman	10/1/01	1581A	2765
VIRGINIA		8Sec4Cpt SUV	Pullman	2/6/06	2144	3300
VIRGINIA #1	P	12SecO	(PK&RSCC)	/72	114	
VIRGINIA #194	P	10SecDr	B&O RR	4/30/73	19	B&O1
VIRGINIA #323	P	8BoudB	J&S (Union PCC)	/89	373	
VIRGO		14SecO	PCW	5/20/93	1021	1972
VIRGO	P	12SecDrB	Wagner	11/92	3073	Wg75
VIROQUA		12SecDrSr	PCW	8/7/86	375	1215
VISCOUNT	N	6CptDrLibObs	Pullman	2/25/02	1751B	2799
VISKA		10SecO	PCW	8/3/92	948A	1944
VISTULA		12SecDr	Pull Buf	8/9/04	1962	B40
VISTULA #228	S	16SecO?	PCW	6/19/86	93H	1199
VITERBO		12SecDr	Pull Buf	11/10/02	1581C	B11
VIVERA		10SecDrB	PCW	11/12/91	920B	1866
VIVIAN #40	N		H&H (WS&PCC)	5/1/78	753A	
VIVIAN #89	P	28Ch3Sts	PCW (Wagner)	6/25/84	3124	D256
VIXEN	N	26ChDr	Pullman	3/26/06	1987A	3335
VIZIER	N	6CptDrLibObs	Pullman	3/13/02	1751B	2799
VLADIMIR		12SecDr	Pullman	5/6/01	1581A	2726
VLADIMIR	R	30Ch SUV	Pullman	5/6/01	3581E	2726
VLADISH	N	12SecDrB	Wagner	/98	3073Q	Wg110
VOLANTE	N	20Ch12StsB	B&S		678B	
VOLANTE	P	10Cpt	Wagner	1/94	3075D	Wg104
VOLGA		12SecDr	Pull Buf	8/6/04	1963	B40
VOLGA #383		10SecDrB	PCW Det	2/26/83	149	D176
VOLHYNIA		12SecDr	Pull Buf	10/9/03	1581F	B23
VOLTA		10Sec2Dr	Pullman	11/3/02	1746A	2928
VOLTAIRE		12SecDr	PCW	10/13/92	990	1931
VOLTAIRE	N	12SecDr	PCW?		1963Y	
VOLTURNO		12SecDr	Pull Buf	2/9/03	1581C	B12
VOLUNTEER	N	20Ch9StsDrObs	PCW	11/14/99	1417G	2466
VOLUNTEER		10SecLibObs	Pullman	8/25/08	2351C	3634
VOLUSIA		SDR	PRR	3/73	28A	2
VOLUSIA	RN	10SecDrB	Phil&Erie RR		149S	
VRANA		14SecO	PCW	4/25/93	1021	1972
VULCAN		27StsParlor	PCW Det	/77	45	D31
VULCAN	RN	8Sec24Sts	PRR	9/72	29E	Alt1
VULCAN		12SecDr	PCW	11/29/98	1318C	2359
WABANA	N	12SecDr	PCW	/98	1318M	2313
WABANSE	RN	10SecDrB	PCW	2/23/85	222C	1033
WABASH		10SecDr	TW&W Ry	6/30/76	19	Tol1
WABASH	P	10SecDr Tourist	(Wagner)		3054	
WABASH #145	P	6Sec4Dr?	J&S (WS&PCC)	4/8/86	720?	
WABENO		14SecDr	Pullman	5/24/06	2165	3352
WACHESAW	RN	6CptLngObs	Pullman	12/18/03	1299E	3065
WACO	RN	OS	I&StLSCCo	7/71	27 3/4	
WACONDA	P	12SecDr	(Wagner)	3/82	3022G	

CAR NAME	ST	CAR TYPE	CAR BUILDER (Owner)	DATE BUILT	PLAN	LOT	CAR NAME	ST	CAR TYPE	CAR BUILDER (Owner)	DATE BUILT	PLAN	LOT
WADENA		10SecDrCpt	PCW	2/17/90	668	1649	WAMEGA	N	10SecDr	Grand Trunk	5/73	31B	Mon4
WADENA		10SecDrSr	PCW	7/2/92	937B	1925	WAMEGA	N	BaggClub	Pullman	8/21/06	2136A	3386
WAGRAM		14SecO	PCW Det	4/18/93	1021	D518	WAMPUM		12SecDrCpt	Pullman	8/9/05	2088	3221
WAHSATCH		10SecDr2Cpt	Pullman	9/27/05	2078	3203	WANATAH #184	P	10SecDr	J&S	11/72	26	Wil8
WAHSATCH #127	P	12SecO	PFtW&C/GMP	/69	5	FtW 1	WANATAH		5CptLibObs	Pullman	8/17/05	2089	3222
WALDECK	P	14Sec	B&S	7/93	3204		WANATAH		6CptLibObs	Pullman	5/31/07	2089A	3533
WALDEMAR	N	32StsB Parlor	PCW	2/2/99	1376B	2371	WANDERER	RN	Private	J&S	7/31/73	35E	Wil1
WALDEMAR	R	12SecDrB	PCW	2/25/87	1566	1298	WANDERER	N	8SecBLibObs	PCW		1707B	
WALDORF		Diner	PCW	7/31/93	538D	2010	WANDERER #93	P	Private	(Wagner)		3097	
WALDORF		12SecDr	Pullman	9/3/08	1963C	3646	WANDERMEER	P	7Dr	Pullman	1/06	2120	3268
WALES		14SecO	PCW Det	5/24/93	1021	D525	WANDRILLE		16Sec SUV	Pullman	4/4/10	2184G	3808
WALKILL E&A-47	P	Slpr	(E&A Assn)	/72		Elm0	WAPATO		14SecDr	Pullman	5/18/06	2165	3352
WALLA WALLA		10SecO	PCW Det	2/24/81	102	D91	WAPITIA #323	P	8BoudB	J&S (Union PCC)		373A	
WALLABOUT #120	P		(WS&PCC)	7/83			WARBANO		10SecDrB	PCW	5/9/89	623B	1566
WALLINGFORD		10SecDrB	PCW	9/21/87	445E	1365	WARCAHTO		10SecDr2Sr	Pullman	9/4/09	2078J	3721
WALLINGFORD	N	BaggClub	Pullman	8/18/06	2136A	3386	WARCHAPA		10secDr2Cpt SUV	Pullman	3/9/10	2471	3784
WALLINGTON		12SecDrSr	Pullman	3/30/05	2049A	3169	WAREHAM	P	34Ch SUV	Pull P-1484A	6/00	3265	2518
WALLOWA		12SecDr	Pullman	5/28/02	1581B	2853	WARFIELD		14SecDr	PCW	12/23/99	1318J	2450
WALLULA		10SecO	PCW Det	2/24/81	102	D91	WARHAM	N	12SecDr	(Wagner)	10/82	3023A	
WALPOLE	P	26ChBarB SUV	Pull P-2265	8/07	3274	3497	WARNCLIFFE #555	P	14SecSr	H&H (B&O)	8/87	654	
WALSINGHAM #401	P	12SecDr	Wagner	1/90	3067A	Wg28	WARRINGTON #402	P	12SecDr	Wagner	1/90	3067A	Wg28
WALTHAM		14SecO	PCW	6/2/93	1021	1972	WARRIOR		12SecDr	Pullman	9/5/08	1963C	3646
WALTHAM #168	R	30Ch SUV	Pull Buf	10/28/02	3581E	B11	WARSAW		12SecDr	PCW	8/21/90	784	1708
WALTON		Diner	Pullman	12/24/01	1694	2762	WARSAW E&A-48	P	Sleeper	(E&A Assn)	/72		Elm0

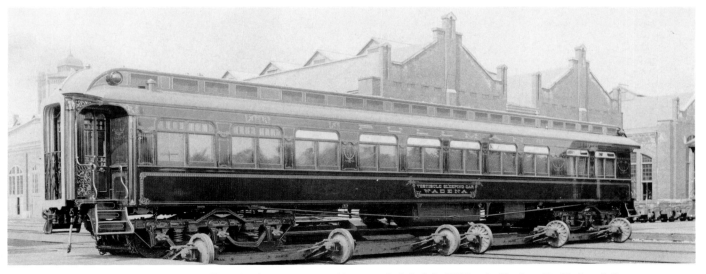

Figure I-42 WADENA, a 10-section drawing room smoking room, built in July 1892 for the Northern Pacific Association, was later rebuilt along with all its sister cars to 12-section drawing room accommodations and renamed PERHAM. It continued in NP Association service until 1915. Smithsonian Institution Photograph P-1957, courtesy of Robert Wayner.

CAR NAME	ST	CAR TYPE	CAR BUILDER (Owner)	DATE BUILT	PLAN	LOT
WARWICK		12SecDr	PCW	8/22/90	784	1708
WARWICK	RN	12SecDrCpt SUV	Pullman	8/17/05	2103D	3227
WARWICK #39	P	14SecO	PCW (Wagner)	4/19/86	3019A	1191
WARWICK E&A-49	P	Slpr	(E&A Assn)	/75		
WASANTA #33	P	/72	(E&A Assn)			Elm0
WASANTA	N	12SecDr	Pullman	3/12/06	2163A	3333
WASHINGTON	P	10SecB	B&S	9/15/73	37	Day10
WASHINGTON	N	26Ch3StsDr	PCW	4/27/98	1278C	2258
WASHINGTON #516	P	12SecO	B&O RR	/80	665	
WASHITA	P	12SecDr	(Wagner)	/82	3022H	
WATCH HILL	P	16Ch23StsBObs	Pull P-1486A	6/00	3266	2519
WATER GAP	P	10SecO Tourist	(DL&W RR)	12/80	220	
WATERBURY	P	16SecO	H&H	7/88	1901	
WATERFORD	N	12SecDr	B&O RR	/79	657B	
WATERFORD #403	P	12SecDr	Wagner	8/91	3067A	Wg28
WATERLOO	P	10SecDr	(Wagner)	/85	3033	
WATERLOO #115	N	22Ch4Sts	(MannBCC)	/84	697C	
WATERVILLE		10SecDrB	PCW Det	3/23/87	445C	D282
WATHENA	N	BaggClub	Pullman	8/15/06	2136A	3386
WATHERA		10SecDrB	PCW	2/16/88	445F	1369
WATSEKA		12SecDr	Pull Buf	2/28/03	1581C	B13
WATSEKA #103	P		B&O (WS&PCC)	5/19/82	729	
WATUPPA	P	26Ch SUV	Pull P-2071	6/05	3273	3189
WATURUS	P	10Cpt SUV	B&S	12/07	3208	
WAU WINET		10SecDr	PCW Det	9/23/82	112A	D171
WAUKESHA		12SecDr	Pullman	7/31/03	1581F	3015
WAUKESHA #74	P	12SecDr	H&H	/80-/82	197	
WAUKESHA WC-100	P	12SecO	CentCarCo	2/86	532	
WAUNAKEE		10Sec3Cpt	Pullman	11/1/05	2100	3238
WAUNETA		10SecLibObs	Pullman	7/22/09	2351A	3715
WAUPACA WC-44	R	10SecDrCafe	CentCarCo	2/82	532B	
WAUPACA WC-44	P	12SecO	CentCarCo	2/82	532	
WAUREGAN	P	26Ch3Sts	B&S	2/93	3260	
WAUSAUKEE		10SecDrB	PCW	3/8/88	445F	1380
WAUTAUGA	R	12SecB	PCW	8/22/83	168	47
WAUTAUGA		10SecDr	PCW	8/22/83	112	47
WAUWINET		12SecDr	Pullman	11/11/01	1581A	2784
WAVELAND		14SecDr	Pullman	8/10/00	1523E	2567
WAVELAND		16Sec	Pullman	12/4/06	2184	3409
WAVERLEY			(DL&W Assn)			
WAVERLEY #122	P	10Sec	PFtW&C	2/70		
WAVERLEY #566	P	12SecB	PCW	3/4/85	227	1034
WAVERLY	RN	12SecDrSr	PCW Det	3/30/92	784E	D497
WAWONA	R	8SecBLibObs	PCW	3/29/87	1707	1299
WAWONA		10SecLibObs	Pullman	7/27/09	2351A	3715
WAWONA SP-44	P	10SecDrB	PCW	3/29/87	445B	1299
WAYFARER	P	10Cpt SUV	Pull P-2489	10/10	3208	3810
WAYLAND		16Sec	Pullman	6/7/06	2184	3361
WAYMART	N	12SecDr	Pullman	3/28/02	1581B	2821
WAYNE	P	10SecB	B&S	7/31/71	17	Day4
WAYNE		12SecDr	Pullman	5/19/03	1581D	3004
WAYNE	P	12SecDr	(Wagner)	/75	3008B	
WAYNE #172	R	30Ch SUV	Pullman	5/07/01	3581E	2726
WEATHERSFIELD	P	36Ch SUV	Pull P-2263	8/07	3275	3492
WEBOTUCK #480	P	12SecDr	Wagner	8/91	3065D	Wg59
WEBSTER	N	10SecB	PFtW&C/GMP	8/15/75	14D	FrW2
WEBSTER		12SecDr	Pullman	5/7/02	1581B	2849
WEEHAWKEN		12SecDr	Pullman	7/29/03	1581F	3019
WEIMAR		8Sec10Ch3Sts	PCW	4/29/88	568A	StL3
WELAKA	R	10SecO	PCW Det (rblt)	1/20/75	47	D27
WELAKA #34	P	14Ch5StsDr	(Wagner)	7/71	3106C	
WELDON	P	Composite	(IC RR)			
WELDON		12SecDrB	Pullman	12/18/06	2142A	3405
WELLAND		12SecDrB	Pullman	1/13/02	1611A	2787
WELLAND	P	14SecO	(Wagner)		3030	
WELLAND #453		10SecDrB	PCW	12/30/85	240A	1144
WELLESLEY		12SecDr	PCW Det	1/13/91	784	D455
WELLESLEY #482	P	12SecDr	Wagner	8/91	3065D	Wg59
WELLINGTON	RN	Diner	PCW	6/26/91	783C	1845
WELLINGTON #483	P	12SecDr	Wagner	8/91	3065D	Wg59
WENDELL		12SecDrB	Pullman	12/18/07	2326	3587
WENDOVER	N	6CptLngObs	PCW	2/3/98	1299C	2283
WENONAH		5CptLibObs	Pullman	8/15/05	2089	3222
WENTWORTH	N	12SecDr	Pullman	3/29/04	1963	3080
WENTWORTH #728	P	19Ch3StsB	H&H (B&O)		653A	
WESLEY	N	12SecDr	PCW	11/21/90	784	1767
WEST BADEN	P	12Ch24StsObs SU	(Cl&L RR)	C:/11	2747	
WEST END		10SecDrHotel	PCW Det	6/6/77	63	D45
WEST FARMS	P	Bagg24Ch	Pull P-2496	1/11	3278	3823
WEST HAVEN	P	40Ch6StsDrObs	Pull P-2266A	4/07	3276	3493
WEST POINT		10SecDrB	PCW Det	5/3/83	149	D194
WEST POINT		16Sec	Pullman	4/3/07	2184	3523
WESTANA	P	10Cpt SUV	Pull P-2489	10/10	3208	3810
WESTBROOK		16Sec	Pullman	6/12/06	2184	3361
WESTCHESTER		12SecDr	Pullman	10/27/06	1963C	W18
WESTDALE	P	20Ch3StsB	B&S	2/93	3260A	
WESTERLY		12SecDr	Pullman	10/31/07	1963C	3570
WESTERLY #147	P	14SecO	CB&Q RR	10/70	20	
WESTERN #203	P	DayParlor	CB&Q RR	4/71	29	Aur1
WESTERN WORLD	P	Hotel	MC RR/GMP	/66	26X	Det
WESTERNER	N	8SecBLibObs	PCW	5/27/88	1707	1377
WESTFIELD		12SecDr	Pullman	9/9/01	1581A	2756
WESTFIELD #141		DDR	PCW Det	10/20/70	26A	D4
WESTFORD #15	P	12SecB	PCW (Wagner)	4/17/86	3018B	1191
WESTGARTH	P	26Ch3Sts	B&S	8/92	3260	
WESTHAMPTON	P	10Cpt SUV	B&S	12/07	3208	
WESTLAKE	N	12SecDr	PCW Det	4/18/91	784D	D461
WESTLAKE		12SecDr2Sr	Pullman	11/5/09	2447	3745
WESTMEATH		12SecDr	Pullman	12/17/09	1963E	3766
WESTMINSTER	RN	DDR Hotel	PFtW&C	8/15/70	14	FtW2
WESTMINSTER	P	Commissary	CB&Q RR			Aur1
WESTMINSTER	N	10Cpt SUV	Pull P-2489	10/10	3208	3810

CAR NAME	ST	CAR TYPE	CAR BUILDER (Owner)	DATE BUILT	PLAN	LOT
WESTMINSTER #245	P	12SecDr	Wagner	12/87	3061	Wg6
WESTMINSTER #276	P	10SecDrB Hotel	DetC&MCo	/73	27	Det
WESTMORELAND #244	P	12SecDr	Wagner	12/87	3061A	Wg6
WESTMOUNT		12SecDr	Pullman	10/25/06	1963C	3408
WESTON		12SecDr	Pullman	11/7/01	1581A	2784
WESTOVER		12SecDr	Pullman	2/13/06	1963C	3316
WESTPHALIA		12SecDr	Pullman	6/27/03	1581F	3010
WESTPORT		16SecO	Pullman	5/2/05	2068	3177
WESTVILLE		12SecDr	Pullman	2/1/04	1581H	3053
WETUMPKA		10SecDr	PCW	8/27/83	112	47
WETUMPKA	R	12SecB	PCW	8/83	168	47
WEXFORD	N	10SecDrB	PCW Det	4/30/84	149D	D242
WEXFORD	R	16Sec	PCW Det	4/30/84	1877	D242
WEYANOKE		12SecDr	Pullman	2/13/06	1963C	3316
WEYBURN	RN	12SecDrSr	PCW Det	2/5/92	784E	D488
WEYMOUTH		12SecDr	Pull Buf	6/15/04	1963	B41
WHEATON		12SecDr	Pullman	11/9/01	1581A	2784
WHEELING		12SecDr	Pull Buf	4/8/04	1581G	B35
WHEELING #151	P	10SecDr	J&S (B&O)	4/30/73	19	Wil1
WHEELING #527	P	12SecO	B&O RR	/80	665	
WHILEAWAY	P	7Dr	Pullman	1/06	2120	3268
WHITCOMB	N	12SecDr	PCW Det	2/14/82	93B	D144
WHITCOMB	N	12SecDr	PCW Det	3/31/83	93X	D184
WHITEHALL	P	Parlor	Vermont Cent	/71	8	StA
WHITELAND		16Sec	Pullman	11/30/06	2184	3409
WHITFIELD		14SecDr	PCW	12/20/99	1318J	2450
WHITFORD		12SecDrSr	Pullman	5/24/07	2088	3553
WHITFORD		12SecDrCpt	Pullman	8/8/05	2088	3221
WHITING		12SecDr	Pullman	2/6/08	2163A	3599
WHITMAN		10SecLibObs	Pullman	11/27/09	2351A	3748
WHITMORE	N	7CptLngObs	PCW	10/25/95	1159E	2126
WHITNEY		BaggClub	Pullman	8/17/06	2136A	3386
WHITTENDON		24ChDrB SUV	Pull	12/12	2672	4046
WHITTIER		14SecDr	Pullman	12/30/01	1523F	2798
WICHITA	RN	12SecDr	B&S	5/70	17E	Day 3
WICHITA		12SecDr	Pullman	10/30/07	1963C	3570
WICKFORD	P	34Ch SUV	Pull P-1484A	6/00	3265	2518
WICKHAM	N	12SecDrB	PCW Det	9/29/88	375G	D358
WICKLOW		16SecO	PCW	6/10/93	1040	1990
WIESDABEN		12SecDr	Pullman	2/3/04	1581H	3053
WILBURTHA		12SecDr	Pullman	3/7/06	2163A	3333
WILD ROSE #12	P	Slpr	(SouTranCo)			
WILDMERE		16Sec	Pullman	6/15/06	2184	3361
WILDWOOD	P	Private	PCW	7/9/83	117	24
WILDWOOD #431	P	8SecDr	Wagner	8/90	3068	Wg37
WILHELMINE		20Ch4StsDrB	Pullman	6/13/01	1518E	2703
WILLAMETTE	P	6Sec3CptObs	Pullman	10/08	2354	3637
WILLAPA		10SecDrSr	PCW	8/21/91	937A	1838
WILLARD		Diner	Pullman	12/23/01	1694	2762
WILLEWAH		10secDr2Cpt SUV	Pullman	3/17/10	2471	3784
WILLIAM BRADFORD	N	22Ch6StsDr	PCW	8/30/92	976A	1910

CAR NAME	ST	CAR TYPE	CAR BUILDER (Owner)	DATE BUILT	PLAN	LOT
WILLIAM BREWSTER	N	22Ch6StsDr	PCW	8/29/92	976A	1910
WILLIAM TELL #125	P	7BoudB	(MannBCC)	/84	776	
WILLIAMSBURG	R	12SecDr	PCW	4/83	1602B	37
WILLIAMSBURG #385		10SecDr	PCW	4/83	112A	37
WILLIAMSPORT	R	12SecDrB	PCW		1611	
WILLIAMSPORT #277	P	10SecDr	Phil&Erie RR		26	Ren2
WILLIAMSTOWN #164	RN	37Ch SUV	Pullman	5/10/01	3581P	2726
WILLIMANTIC		24ChDrB SUV	Pull	12/12	2672	4046
WILLISTON #32	P	14SecO	PCW (Wagner)	4/20/86	3019A	1191
WILLOMERE	N	12SecDr	Pullman	2/26/04	1963	3070
WILLOWEMOC	N	27Ch	J&S	9/72	35	Wil2
WILLOWGATE	N	12SecDr	Pullman	2/3/04	1581H	3053
WILMERDING		12SecDrSr	Pullman	1/24/01	1581A	2614
WILMINGTON	N	30 Ch	PCW	4/28/98	1278A	2293
WILMINGTON #162	P	12SecO	J&S	2/71	18	Wil3
WILMINGTON #563	P	11SecDr	PCW	2/23/85	222	1033
WILMORE		12SecDrSr	Pullman	1/28/01	1581A	2614
WILTON	N	10SecDr	H&StJ RR	2/72	26	Han3
WILTSHIRE		12SecDr	Pullman	11/4/03	1581F	3025
WIMBERLY		12SecDr	Pullman	12/7/07	1963C	3585
WIMBLEDON		12SecDr	Pullman	1/28/03	1581D	2962
WINCHESTER		14SecDr	PCW	9/1/99	1318E	2447
WINCHESTER #311	P	10SecB	B&S (B&O)	7/31/75	37	Day
WINCHESTER #400	P	12SecDr	Wagner	1/90	3067A	Wg28
WINDERMERE	P	32ChDr SUV	Pull P-2264	8/07	3277	3496
WINDERMERE #243	P	12SecDr	Wagner	12/87	3061	Wg6
WINDSOR		10SecDrB Hotel	PCW Det	5/20/76	56	D39
WINDSOR	RN	Diner	PCW Det	3/15/74	39	D20
WINDSOR		Diner	PCW	5/31/88	538D	1411
WINDSOR		12SecDrSr	PCW	6/29/99	1318C	2424
WINDSOR	P	12SecDrB	(Wagner)	11/82	3025	
WINETKA	R	12SecO	PCW Det	3/7/71	23B	D5
WINETKA		12SecO	PCW Det	3/7/71	5	D5
WINFIELD		14SecDr	PCW	12/21/99	1318J	2450
WINGATE	RN	12SecDrSr	PCW Det	8/26/91	784E	D481
WINIFRED		18Ch4StsDrB	PCW	7/12/93	1039A	1995
WINIFRED	N	27ChDrObs	Pullman	6/8/08	2346	3624
WINIFRED #81	P	SlprParlor	(WS&PCC)	6/1/80		
WINNEBAGO		10SecLibObs	Pullman	7/27/09	2351A	3715
WINNECONNE	N	12SecDrCpt	Pullman	8/9/05	2088	3221
WINNEMAC		10SecLibObs	Pullman	7/22/09	2351A	3715
WINNEMUCCA		10SecDr2Cpt	Pullman	9/28/05	2078	3203
WINNETKA	P	10SecDrB	(Wagner)	3/82	3046A	
WINNIPAUK	P	36Ch SUV	Pull P-2263	8/07	3275	3492
WINNIPEG		12SecDr	PCW Det	7/26/82	93B	D162
WINNIPEG		12SecDr	Pullman	7/31/03	1581F	3015
WINNISQUAM	N	12SecDr	Pullman	3/14/06	2163A	3333
WINONA	P	DDR	DetC&MCo	7/31/73	17	D2
WINONA		12SecDr	Pullman	2/26/04	1963	3070
WINONA	P	12SecDr	(Wagner)	4/84	3023G	
WINONA #67	P	12SecDr	H&H	/80-/82	197	

CAR NAME	ST	CAR TYPE	CAR BUILDER (Owner)	DATE BUILT	PLAN	LOT
WINSLOW		12SecDr	Pullman	1/25/07	1963C	3448
WINSTEDCK		14SecDrO SUV	Pull	12/12	2673	4047
WINSTON		12SecDrB	Pullman	12/20/06	2142A	3405
WINTHROP	N	12SecDrSr	PCW	1/5/99	1318C	2379
WISCONSIN	P	DDR	DetC&MCo	/72	17	D2
WISCONSIN	N	7Cpt2Dr	Pullman	9/26/02	1845	2926
WISCONSIN #27	P		(PCM&StP)			
WISCONSIN #481	P	12SecDr	Wagner	8/91	3065D	Wg59
WISSAHICKON		12SecDr	Pullman	11/3/06	1963C	W18
WISSAHICKON #229	P	Parlor	H&H	8/72	35 3/4	Wil1
WISTAR	N	10SecDr	J&S	11/72	26A	Wil8
WISTERIA		12SecDrCpt	PCW	8/17/92	983	1919
WITMER		12SecDr	Pullman	3/8/06	2163A	3333
WIZARD		12SecDr	Pullman	9/8/08	1963C	3646
WOLLASTON		12SecDr	Pullman	12/19/03	1581F	3043
WOLSEY		12SecDr	PCW	8/20/90	784	1708
WOLVERTON		12SecDr	Pullman	10/18/00	1581A	2588
WONEWOC		10Sec3Cpt	Pullman	11/2/05	2100	3238
WOOD RIVER #2253	P	BaggSmk	Pull	2/91	3215	
WOODBINE		10SecB	B&S		37	Day9
WOODBINE	N	28Ch3Sts	PCW (Wagner)	6/22/86	3124	D256
WOODBOURNE		24ChDrB	Pull Buf	5/14/07	2278	B59
WOODBRIDGE		12SecDr	Pullman	3/14/06	2163A	3333
WOODBURY		24ChDrB	Pullman	6/3/05	1988A	3170
WOODFORD		12SecDr	Pull Buf	11/19/01	1581A	B6
WOODLAND	P	12SecO	(L&N RR)	/72	114	
WOODLAND		12SecDrSr	Pullman	2/8/01	1581B	2638
WOODLAWN		12SecDr	Pullman	12/19/03	1581F	3043
WOODMONT	P	34Ch SUV	Pull P-1484	1/03	3265	2904
WOODS HOLE	P	BaggSmkB SUV	New Haven RR	11/04	3218	
WOODSIDE		16Sec	Pullman	6/14/06	2184	3361
WOODSTOCK	P	DDR	Ohio Falls	12/70	17	Jef
WOODSTOCK	R	12SecDrB	PCW		1498	
WOODVILLE		12SecDrSr	Pullman	2/12/01	1581B	2638
WOOLWICH		14SecO	PCW	4/26/93	1021	1972
WOONSOCKET		24ChDrB SUV	Pull	12/12	2672	4046
WORCESTER		12SecDrSr	PCW	9/11/99	1318E	2449
WRENTHAM	P	26Ch SUV	Pull P-2071	6/05	3273	3189
WYANDOTTE	P	12SecDrB	(Wagner)	11/80	3024	
WYEVALE	N	10SecDr	Grand Trunk	5/73	31K	Mon4
WYLTWYCK	?	12SecO	B&O RR	C:/87	665F	
WYNDHAM		14SecO	PCW Det	5/13/93	1021	D525
WYNNEWOOD		6CptLngObs	Pullman	12/18/03	1299D	3065
WYNOOCHE	N	14SecO	B&S	4/88	1593	
WYOMING	P	DDR	CB&Q RR	1/71	17	Aur
WYOMING	R	12SecDrB	PCW		1611	
WYOMING	N	7Cpt2Dr	Pullman	10/4/02	1845	2926
WYUTA		10Sec2Dr	PCW	5/2/93	995A	1942
XANTHO		18Ch6StsDrB	PCW Det	9/11/90	772D	D427
XANTHUS	R	BaggSmkBLib	PCW		1706	
XENIA		12SecDr	Pullman	11/18/01	1581A	2784
XENIA #174	P	10SecDr	PRR	7/71	21	Alt2

CAR NAME	ST	CAR TYPE	CAR BUILDER (Owner)	DATE BUILT	PLAN	LOT
XERXES	N	12SecDr	(Wagner)	8/86	3049A	
XOCHITL	RN	10SecB (NG)	PCW Det	2/23/83	73F	D120
YADKIN #333	P	10SecDrB	(Union PCC)		565B	
YAKIMA	N	10SecDr	H&StJ RR	3/71	19A	Han 2
YAKIMA	N	12SecDrB	Wagner	8/92	3070	Wg71
YALE		12SecDrB	Pullman	2/25/01	1611A	2614
YALE	P	12SecDrB	Wagner	11/92	3073A	Wg80
YALE #164	N	10SecDrB?	(Woodruff)	9/24/87	663?	
YALU		12SecDr	Pull Buf	8/12/04	1963	B40
YAMATA	N	28ChDr	Pullman	4/6/06	1986A	3336
YAMPAH		10SecDr2Sr	Pullman	8/28/09	2078J	3721
YANKEE	N	23Ch2StsDrLibO*	Pullman	8/28/06	2206	3388
YANKTON		12SecDr	Pull Buf	7/17/03	1581C	B19
YANTIC	RN	10SecDrB	PCW	2/23/85	222E	1033
YANTIC	P	12SecDrB	Wagner	11/92	3073A	Wg80
YAQUI		12SecDr	PCW Det	5/30/91	784D	D462
YARMOUTH		12SecDr	Pull Buf	4/14/03	1581C	B15
YARROW #384		10SecDrB	PCW Det	3/3/83	149	D176
YAZOO		12SecDr	Pullman	6/2/04	1963	3082
YEDDO		10SecDrB	PCW	5/22/85	240A	1193
YEDDO	P	12SecDrB	Wagner	11/92	3073A	Wg80
YELLOWSTONE		12SecDr	PCW Det	4/26/82	93B	D142
YELLOWSTONE		12SecDr	Pullman	1/29/07	1963C	3448
YELLOWSTONE #41	RN	Private	H&H (WS&PCC)	7/25/78	689	
YENTOI		12SecDr	Pullman	11/16/08	1963E	3659
YEOMAN	RN	6CptLngObs	Pullman	12/16/03	1299E	3065
YOKOHAMA		10SecDrB	PCW	5/20/85	240A	1193
YOLANDA		12SecDr	Pullman	2/8/06	1963C	3316
YOLANDE		26ChDr	Pullman	8/2/04	1987A	3097
YONKERS		12SecDr	Pullman	2/26/07	1963C	3525
YORK		12SecDr	PCW	2/20/91	784B	1797
YORK	P	12SecDrB	Wagner	11/92	3073A	Wg80
YORKLYN #578	P	10SecDrB	J&S (B&O)	C:/88	663	
YORKSHIRE		12SecDr	Pullman	7/19/03	1581F	3019
YORKTOWN		12SecDr	Pullman	9/1/03	1581F	3019
YOSEMITE	P	DDR	B&S	5/70	17	Day3
YOSEMITE	S	Diner	PCW	12/1/90	662D	1778
YOSEMITE		12SecDr	Pullman	11/15/01	1581A	2784
YOSEMITE #43	RN	Private	H&H (WS&PCC)	7/25/78	689	
YOUCONE		10SecDr2Sr	Pullman	8/30/09	2078J	3721
YOUGHIOGHENY #521	P	12SecO	B&O RR	/81	665	
YOUGHIOGHENY #150		6Sec4Dr?	J&S (WS&PCC)	5/17/86	720?	
YOUNG AMERICA		Commissary?	C&NW	C:/70	Comm1	
YOUNGSTOWN		12SecDr	Pullman	1/25/07	1963C	3448
YOUNGSTOWN	P	10SecO Tourist	(Wagner)		3037A	
YSIDORA		10SecDrB	PCW	1/6/88	445E	1360
YSIDORA	N	16Ch2StsDrB	(Wagner)	6/85	3115D	
YUBA	N	SDR	DetC&MCo	/72	12D	Det
YUBA	P	12SecDrB	Wagner	11/92	3073A	Wg80
YUCATAN	N	10SecDr	PCW Det	3/22/82	112A	D141
YUCCA		10SecDrB	PCW	5/14/87	445E	1292
YUHO		12SecDr	PCW Det	5/7/91	784D	D461

CAR NAME	ST	CAR TYPE	CAR BUILDER (Owner)	DATE BUILT	PLAN	LOT
YUKON		12SecDr	PCW Det	5/27/91	784D	D462
YUMA	N	12SecDr2Sr	Pullman	11/3/09	2447	3745
YUMA	P	12SecDrB	Wagner	11/92	3073A	Wg80
YUMA GH&SA-6	P	Slpr	(GH&SA RR)	5/8/84	37?	
YVONNE	N	15Ch3StsDrB	(Wagner)	4/85	3115A	
ZACA #2522		14SecO	PCW Det	1/30/93	1014A	D511
ZACATECAS		10SecDrB	PCW Det	3/21/84	149B	D234
ZACATECAS	R	12SecDrB	PCW	3/21/84	1468	234
ZAMA	N	7CptObs	Wagner	4/89	3062A	Wg14
ZAMBA #2558		14SecO	PCW Det	2/14/93	1021	D516
ZAMBESI		14SecO	PCW Det	5/17/93	1021	D525
ZANDRIE		16Sec (SUV)	Pullman	3/31/10	2184G	3808
ZANESVILLE		10SecB	B&O RR	/73	37	B&O
ZANESVILLE		12SecDr	Pullman	4/5/04	1963	3080
ZANESVILLE #518	P	12SecO	B&O RR		665	
ZANONI	P	10SecDr	(Wagner)	5/85	3053	
ZANOW SP-2508		10SecO	PCW	8/1/92	948A	1944
ZANTE		14SecO	PCW Det	3/25/93	1021	D517
ZANZIBAR		12SecDrB	PCW	10/13/88	375F	1481
ZANZIBAR	R	14SecDr	PCW	10/13/88	1882	1481
ZARA		12SecDr	PCW	6/17/92	784E	1915
ZARA	P	12SecDrB	(Wagner)	6/86	3050A	
ZEALAND		12SecDr	Pullman	11/7/03	1581F	3025
ZEDA		12SecDr	PCW	5/6/91	784D	1805
ZELANDA	R	12SecDr	PCW	6/30/88	1603	1415
ZELANDA SP-74	P	10SecDrB	PCW	6/30/88	445F	1415
ZELDA #2566		14SecO	PCW Det	3/4/93	1021	D516
ZELIENOPLE #560	P	11SecDr	PCW	2/23/85	222	1033
ZEMAR	N	12SecDr	Pullman	3/28/03	1581D	2983
ZEMAR	R	30Ch SUV	Pullman	3/28/03	3581E	2983
ZENDA		10SecDr3Cpt	Pullman	4/13/06	2078C	3342
ZENITH		12SecDrB	PCW	2/23/97	1204D	2212
ZENO		12SecDr	PCW	2/21/91	784B	1797
ZENO	P	12SecDrB	(Wagner)	5/85	3050	
ZENOBIA			(Wagner)	?5/85		
ZENOBIA	N	15Ch17StsDr	DetC&MCo	7/74	6	
ZENOBIA	N	22ChDr	Pull Buf	7/13/02	1647C	B9
ZENTA	N	10SecDr	(Wagner)	/75	3016	
ZEPHYR	N	30 Ch	PCW	4/27/98	1278A	2293
ZERAH	N	12SecDrB	Wagner	4/93	3073B	Wg91
ZERBA	N	10SecDrSr	PCW Det	3/29/81	112F	D149
ZERBINO	N	12SecO	B&O RR		657	
ZERBINO	N	12SecDrB	(Wagner)	6/86	3050A	
ZETA	P	12SecDrB	(Wagner)	5/83	3050	
ZETA		12SecDr	PCW	12/24/90	784	1780
ZETES		10Sec2Dr	PCW	3/26/96	1173A	2153
ZEUS		12SecDr	PCW	1/12/91	784	1781
ZEYLA		12SecDrB	PCW Det	10/13/88	375G	D358
ZIMARA		14SecO	PCW Det	5/11/93	1021	D525
ZIMRI	P	12SecDr	PCW P-93H	4/18/86	3049	1192
ZIMRIDA	N	12SecDrSrB	PCW	9/9/98	1319B	2314
ZINGARA	P	16Sec	B&S	12/82	3202	
ZORAYDA	N	12SecDrB	PCW Det	12/28/82	1611	D174
ZORITA #2557		14SecO	PCW Det	2/23/93	1021	D516
ZULEIKA		26ChDr	Pullman	8/3/04	1987A	3097
ZUNI		10SecO (NG)	PCW Det	6/10/80	73A	D76
ZUNI		7Cpt2Dr	Pullman	10/29/02	1845	2926
ZURICH		8Sec10Ch3Sts	PCW	4/19/88	568A	StL3
ZURICH	N	12SecDrB	Wagner	5/83	3050	

Figure I-43 This 12-section drawing room buffet, ZENITH, was one of two cars built in Lot 2212. It spent its entire life in general service until 1917 when it was converted to Tourist Car 4471 for World War I troop service. Smithsonian Institution Photograph P-3623A.

THE STANDARD CAR PRODUCTION

Steel, Standardization, Modernization And Mobilization, 1910-1931

Figure I-44 ALBION COLLEGE, a 10-section, 2-double-bedroom compartment "Betterment" car, was rebuilt in September 1935 from CABAZON. It received its two-tone gray paint scheme in January 1956 and is shown here in Kansas City, Missouri in October 1964, the month and year it was sold for scrap. The Robert Wayner Collection.

CAR NAME	ST	CAR TYPE	BUILT	PLAN	LOT
A.S. BUFORD	N	10Sec2Dr	9-10/25	3584A	4899
ABBAKAN		12SecDr	2-5/21	2410F	4612
ABBIE		26ChDr	4-5/13	2416A	4136
ABBOTT		12SecDr	10-12/16	2410F	4431
ABELARD		12SecDr	12/10-11	2410	3866
ABERDEEN		12SecDr	1-3/17	2410F	4450
ABERNATHY		12SecDr	10/22-23	2410H	4647
ABIGAIL		28ChDr	4-6/25	3416	4864
ABIGAIL ADAMS		28ChDr	1-2/27	3416A	6032
ABINGDON	N	12SecDr	5-8/25	3410A	4868
ABIQUA		12SecDr	10/22-23	2410H	4647
ABRONDA		12SecDr	2-3/15	2410B	4318
ABSECON		28ChDr	11/18/10	2418	3862
ABYSSINIA		12SecDr	8-9/20	2410F	4574
ACADEMY		26ChDr	4-5/17	2416C	4492
ACADIA		12SecDr	8-10/26	3410A	4945
ACADIA UNIVERSITY	RN	12Sec2Dbr	5-6/17	4046	4497
ACAPULCO NdeM	SN	8SecDrBLngObs	12/18	4030	4489
ACAPULCO NdeM-100	SN	12SecDr	2-5/25	3410	4845
ACAYUCAN NdeM	SN	8SecDr2Cpt	3/29	3979A	6237
ACCOTINK		12SecDr	8-9/20	2410F	4574
ACELGA		12SecDr	1-3/14	2410A	4249
ACHILLES		12SecDr	8-9/20	2410F	4574
ACKERSON		12SecDr	1-2/15	2410B	4311
ACKLAND		12SecDr	2-3/15	2410B	4318
ACME		12SecDr	1-2/22	2410F	4625
ACROPOLIS		12SecDr	2-5/16	2410E	4367
ACTIUM		12SecDr	8-9/20	2410F	4574
ACTOPAN		12SecDr	8-9/20	2410F	4574
ACULZINGO NdeM-101	SN	14Sec	11-12/28	3958A	6212
ACUSHNET		32ChDr	8/27	3917A	6078
ACUSHNET	R	24ChDr10StsLng	8/27	3917C	6078
AD-VANTAGES #1	LN	Exhibit (Rexall)	6/16	2916A	4389
ADAIR		BaggBClubSmkBarber	7/23	2951	4698
ADAMS		28ChDr	5-7/26	3416	4958
ADAMSBORO		12SecDr	9/28/10	2410	3813
ADAMSDALE		12SecDr	2-5/25	3410	4845
ADANA		12SecDr	8-9/20	2410F	4574
ADELAIDE		26ChDr	3-4/14	2416A	4264
ADELINA PATTI		30ChDr	1/27	3418A	6034
ADELPHI		12SecDr	12/10-11	2410	3866
ADENMOOR		12SecDr	4-5/13	2410	4106
ADIRONDACK		12SecDr	12/12-13	2410	4076
ADJUTANT		10SecLngObs	12/15	2521B	4353
ADJUTANT		BaggBClubSmk	2-4/09	2136C	3660
ADLON		12SecDr	12/10-11	2410	3866
ADMIRAL		10SecLngObs	12/15	2521B	4353
ADMIRAL		BaggBClubSmk	2-4/09	2136C	3660
ADMIRAL DEWEY		12SecDr	1-2/29	3410B	6220
ADOREA	N	30ChDr	11/12	2668	4051

CAR NAME	ST	CAR TYPE	BUILT	PLAN	LOT
ADRIAN		BaggBClubSmkBarber	7/23	2951	4698
ADRIAN ISELIN	N	10Sec2Dr	9-10/25	3584A	4899
ADRIANOPLE		12SecDr	2-3/13	2410	4105
ADRIATHA		12SecDr	9-10/13	2410A	4215
ADRIENNE	LN	Lecture (Rexall)	6/16	2916C	4389
ADVANCE		Private (GS)	3/11	2502	3848
ADWALTON		12SecDr	10-12/15	2410D	4338
AFRICA		12SecDr	8-9/20	2410F	4574
AFTON CANYON		3CptDrLngBSunRm	1/30	3975F	6337
AGASSIZ		12SecDr	2-5/25	3410	4845
AGATHA	N	26ChDr	5/12	2416	3986
AGERATUM		28ChDr	5-7/26	3416	4958
AGNES		26ChDr	5/23	2416D	4691
AGOSTA		10SecDr2Cpt	5/13	2585A	4141
AGRICOLA		12SecDr	8-9/20	2410F	4574
AGUA PRIETA NdeM	SN	8SecDrBLngObs	12/18	4030	4489
AGUSTIN LARA	N	10SecDrCpt	8/29	3973A	6273
AGUSTIN LARA FCP-202	SN	10SecDr2Cpt	8-10/23	3585C	4725
AIKEN	N	12SecDr	1-3/14	2410A	4249
AILANTHUS		12SecDr	12/10-11	2410	3866
AILSA		16Sec	12/14-15	2412C	4304
AIMEE		26ChDr	4/12	2416	3950
AINSWORTH		12SecDr	2-5/21	2410F	4612
AIREY		12SecDr	11/18/10	2410	3864
AJAX		12SecDr	5-8/18	2410F	4540
AJUSCO NdeM	SN	10SecDrCpt	9-10/17	4031	4525
AKELEY		12SecDr	4-5/11	2410	3893
AKISTA		16Sec	12/14-15	2412C	4304
AKRON		12SecDr	10/4/10	2410	3813
ALABAMA	N	10SecLngObs	10-12/26	3521A	4998
ALABAMA	N	3CptDrLngBSunRm	6/29	3975C	6262
ALABAMA COLLEGE	RN	10Sec2DbrCpt	7/21/10	4042A	3800
ALADDIN		24ChDrB	1/14	2417A	4239
ALAMANCE		10SecLngObs	10/11	2521	3926
ALAMO		12SecDr	8-10/26	3410A	4945
ALAMOS NdeM	SN	8SecDrBLngObs	12/18	4030	4489
ALANA		12SecDr	3-6/20	2410F	4565
ALANSON		12SecDr	2-4/24	3410	4762
ALASTAIR		18ChBLngObs	4/16	2901	4365
ALAZON		12SecDr	12/27-28	3410B	6127
ALBA		12SecDr	10-12/20	2410F	4591
ALBANY		12SecDr	1-2/30	3410B	6351
ALBERENE		12SecDr	4-5/13	2410	4106
ALBERT FINK	N	12SecDr	4-6/14	2410A	4271
ALBERT GALLATIN		12SecDr	8-12/25	3410A	4894
ALBERT LEA		12SecDr	8-9/12	2410	4035
ALBION		12SecDr	9/29/10	2410	3813
ALBION COLLEGE	RN	10Sec2DbrCpt	6/13	4042A	4159
ALBURGH		12SecDr	3-6/11	2410	3881
ALCIDAS		12SecDr	8-9/20	2410F	4574

CAR NAME	ST	CAR TYPE	BUILT	PLAN	LOT	CAR NAME	ST	CAR TYPE	BUILT	PLAN	LOT
ALCONA		12SecDr	2-3/13	2410	4105	ALFRED NOBEL		14Sec	10-11/26	3958	6012
ALDBURN		26ChDr	7/23/10	2416	3804	ALGECIRAS		12SecDr	1/16	2410E	4354
ALDEBARAN		12SecDr	8-9/20	2410F	4574	ALGODON		12SecDr	5-7/13	2410	4149
ALDERNEY		12SecDr	8-9/20	2410F	4574	ALGONQUIN CLUB		8SecBLngSunRm	10/30	3989C	6393
ALDHAM		12SecDr	10/28/10	2410	3860	ALGONQUIN PARK		8SecDr2Cpt	5-6/29	3979A	6261
ALDRICH		12SecDrCpt	10-12/21	2411C	4624	ALGONQUIN PASS		8SecDr2Cpt	2-3/30	3979A	6353
ALEJANDRA FCP-216	SN	10SecDr2Cpt	8-10/23	3585	4725	ALHAMBRA		12SecDr	2-3/27	3410A	6055
ALEPPO		12SecDr	3-6/20	2410F	4565	ALICANTE		12SecDr	10-12/15	2410D	4338
ALESIA		12SecDr	5-8/18	2410F	4540	ALIDA		12SecDr	1-2/15	2410B	4311
ALETHIA		26ChDr	6/11	2416	3902	ALINDA		24ChDrB	6-7/14	2417B	4265
ALEXANDER B. ANDREWS	N	10Sec2Dr	9-10/25	3584A	4899	ALKIRE		12SecDr	5-7/13	2410	4149
ALEXANDER B. ANDREWS	N	12SecDr	2-5/25	3410	4845	ALLAIRE		24ChDrB	12/10	2417	3863
ALEXANDER BELL		14Sec	10-11/26	3958	6012	ALLEDONIA		26ChDr	5/23	2416D	4691
ALEXANDER GRIGGS		14Sec	6/30	3958A	6373	ALLEGHENY		12SecDr	10/11/10	2410	3814
ALEXANDER GRIGGS	R	6Sec6Dbr	6/30	4084B	6373	ALLEGRA		26ChDr	6/11	2416	3902
ALEXANDER H. STEPHENS		BaggBSmkBarber	10-11/25	3951	4885	ALLEGRIPPUS		12SecDr	10/11/10	2410	3814
ALEXANDER HAMILTON		32ChObs	9-10/24	3420	4785	ALLEMAGNE		12SecDr	7-8/13	2410A	4192
ALEXANDER HAMILTON		12ChDrBLngSunRm	9/30	4002C	6385	ALLENBURY		12SecDr	9-10/15	2410D	4327
ALEXANDER HENRY		8SecDr2Cpt	5-6/29	3979A	6261	ALLENDALE		26ChDr	7/2/10	2416	3804
ALEXANDER RAMSEY		8SecDr2Cpt	5-6/29	3979A	6261	ALLENE		26ChDr	3/13	2416	4107
ALEXANDER SPOTSWOOD	N	8SecDr2Cpt	10-12/28	3979A	6205	ALLENHURST		26ChDr	3/13	2416	4107
ALEXANDER SPOTSWOOD	N	10SecDrCpt	4-6/27	3973C	6043	ALLENPORT		12SecDr	11/22/10	2410	3864
ALEXANDRIA		12SecDr	4/26	3410A	4943	ALLENWOOD		26ChDr	5/12	2416	3986
ALEXANDRIA BAY		3Sbr2CptDrBLngSunR	11/29	3991	6276	ALLIANCE		12SecDr	5-8/18	2410F	4540
ALFARATA		12SecDr	5-6/23	2410H	4699	ALLOWAY		12SecDrCpt	6/27/10	2411	3800
ALFONSO		12SecDr	2-3/15	2410B	4318	ALLSWORTH		26ChDr	7/29/10	2416	3804
ALFONSO ORTIZ						ALMA		26ChDr	3-4/14	2416A	4264
TIRADO FCP-233	SN	10SecDr2Cpt	9-10/26	3585G	4996	ALMADEN	N	12SecDr	8-12/25	3410A	4894

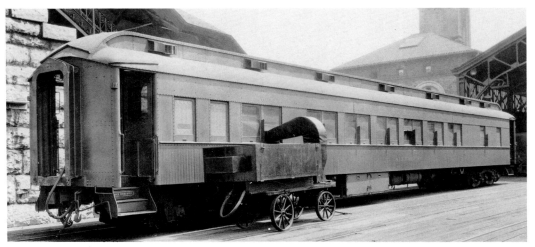

Figure I-45 This circa 1930 photograph, taken at The B&O Mount Royal Station in Baltimore, Maryland, shows the 12-section drawing room ALLENBURY built in 1915 with simulated wood steel sheathing. The contraption was a 1930s solution to cooling a car, prior to occupancy, that had been sitting in the sun. The Author's Collection.

CAR NAME	ST	CAR TYPE	BUILT	PLAN	LOT
ALMADEN		12SecDr	6-9/26	3410A	4969
ALMANZA		12SecDr	10-12/15	2410D	4338
ALMENARA		12SecDr	10-12/15	2410D	4338
ALMIDOR		12SecDr	5-7/13	2410	4149
ALOQUIN		12SecDr	5-6/17	2410F	4497
ALPHA		28ChDr	5-7/26	3416	4958
ALPHA		28ChDr	9/28/10	2410	3813
ALPINE BLUEBELL		14Sec	3-4/30	3958A	6357
ALPINE BUTTERCUP		14Sec	3-4/30	3958A	6357
ALPINE CLOVER		14Sec	3-4/30	3958A	6357
ALPINE PINK		14Sec	3-4/30	3958A	6357
ALPLAND		12SecDr	2-5/16	2410E	4367
ALSACE		12SecDr	2-3/15	2410B	4318
ALSACIENNE		12SecDr	3-6/15	2410B	4319
ALSINA		24ChDrB	6-7/14	2417B	4265
ALSUMA		16Sec	6-7/20	2412F	4568
ALTAMONT		12SecDr	2-3/27	3410A	6055
ALTERTON		12SecDr	7-8/13	2410A	4192
ALTHEA		24ChDrB	6-7/14	2417B	4265
ALTHEEN		12SecDr	3-6/20	2410F	4565
ALTHOM		12SecDr	6-9/26	3410A	4969
ALTMAR		12SecDr	4-5/11	2410	3893
ALTOONA		12SecDr	11/19/10	2410	3864
ALVARETTA		26ChDr	4/12	2416	3950
ALVEDO		12SecDr	10-12/20	2410F	4591
ALVERNO		12SecDr	11-12/13	2410A	4234
ALVERTON		12SecDr	10/10/10	2410	3814
ALVINSTON		12SecDr	4-5/11	2410	3893
ALVINSTON	N	12SecDr	2-3/15	2410B	4318
ALZBETA		26ChDr	4-5/16	2416C	4366
AMADO NERVO NdeM-103	SN	12SecDr	8-12/25	3410A	4894
AMALFI		12SecDr	7/13	2410A	4179
AMANDA		28ChDr	4-6/25	3416	4864
AMARGOSA		10SecDr2Cpt	10-11/13	2585A	4218
AMARYLLIS		28ChDr	5-6/24	3416	4761
AMASA		12SecDr	10-11/11	2410	3936
AMATLAN NdeM-104	SN	8SecDr2Cpt	10-12/28	3979A	6205
AMAWALK		12SecDr	10-11/11	2410	3936
AMBASSADOR		12SecDr	4/26	3410A	4943
AMBLESIDE		16Sec	6-7/13	2412B	4160
AMBLUCO		12SecDr	2-4/24	3410	4762
AMBOY		BaggBClubSmkBarber	7/23	2951	4698
AMBRIDGE		BaggBClubSmk	9/12/10	2415	3834
AMBROSE		12SecDr	10/22-23	2410H	4647
AMBROSE LIGHT		36Ch	12/29-30	3916B	6319
AMELIA		28ChDr	8/24	3416	4801
AMENTA		12SecDr	10-11/11	2410	3936
AMERENE		12SecDr	1-2/22	2410F	4625
AMERICA		12SecDr	8-9/20	2410F	4574
AMERICAN FORD		10SecLngObs	9-10/14	2521A	4292

CAR NAME	ST	CAR TYPE	BUILT	PLAN	LOT
AMERICAN FORD	R	10Sec12ChObs	9-10/14	2521J	4292
AMERICAN LEGION	N	14Sec	10-11/29	3958A	6285
AMERICAN REVOLUTION	RN	8SecBLngObs	1-5/18	4020	4531
AMERICAN UNIVERSITY	RN	12Sec2Dbr	7-9/17	4046	4515
AMES		28ChDr	5-7/26	3416	4958
AMESBURY		12SecDr	1-2/22	2410F	4625
AMESTOY		12SecDr	1-3/14	2410A	4249
AMHERST COLLEGE	RN	10Sec2DbrCpt	5-6/16	4042B	4386
AMIDAS		12SecDr	11-12/13	2410A	4234
AMIDON		12SecDr	1-2/22	2410F	4625
AMINA		24ChDrB	6-7/14	2417B	4265
AMIRANTE		12SecDr	12/12-13	2410	4076
AMITY		26ChDr	4-5/17	2416C	4492
AMLIN		12SecDr	2-3/14	2410B	4318
AMNERIS		26ChDr	3/13	2416	4107
AMON G. CARTER	N	12SecDr	1-2/29	3410B	6220
AMPERSAND		12SecDr	9-10/15	2410D	4327
AMSBRY		26ChDr	3/13	2416	4107
AMSDEN		12SecDr	4-5/11	2410	3893
AMSTERDAM		12SecDr	1-2/30	3410B	6351
AMYLUM		12SecDr	12/10-11	2410	3866
ANACONDA		12SecDr	4-6/24	3410	4763
ANACORTES		12SecDr	4-6/24	3410	4763
ANANDALE		10SecDr2Cpt	6-7/15	2585B	4328
ANAPRA		12SecDr	4-5/13	2410	4106
ANATOK		12SecDr	1-2/16	2410E	4356
ANBURY		12SecDr	1-3/14	2410A	4249
ANCIETTE		26ChDr	4-5/16	2416C	4366
ANCON		12SecDr	1-3/14	2410A	4249
ANCRAM		12SecDr	2-3/15	2410B	4318
ANDALUSIA		12SecDr	12/10-11	2410	3866
ANDERSON		26ChDr	7/2/10	2416	3804
ANDES		12SecDr	10/22-23	2410H	4647
ANDOVER		12SecDr	4-5/11	2410	3893
ANDRADE		12SecDr	1-2/16	2410E	4356
ANDREAS		12SecDr	7-8/15	2410D	4322
ANDRES					
QUINTANA ROO NdeM	SN	12SecDr	2-5/25	3410	4845
ANDREW CARNEGIE		14Sec	7/30	3958A	6376
ANDREW JACKSON		12SecDr	4/26	3410A	4943
ANDREW JOHNSON	N	12SecDr	2-5/25	3410	4845
ANDREW PICKENS		10Sec2Dr	9-10/25	3584A	4899
ANDREW PICKENS		10Sec2Dr	9-10/29	3584B	6284
ANDREW PICKENS	R	10SecDr2Dbr	9-10/29	4074D	6284
ANDREW SQUIRE		14Sec	3-4/30	3958A	6357
ANDREW W. MELLON	N	CptDrBObsLng	9-10/27	3975V	6076
ANDREW W. MELLON	N	8SecDr2Cpt	2-3/30	3979A	6353
ANDRICO		24ChDrB	1/14	2417A	4239
ANDRUSA		12SecDr	12/10-11	2410	3866
ANEMONE	N	12SecDr	7-8/15	2410D	4322

CAR NAME	ST	CAR TYPE	BUILT	PLAN	LOT
ANERLEY		12SecDr	8-9/20	2410F	4574
ANGEL'S CAMP	N	Dining Lounge	10/28	3969A	6011
ANGELINE		18ChBLngObs	4/16	2901	4365
ANGELO		12SecDr	1-2/22	2410F	4625
ANGELUS		12SecDr	1-2/15	2410B	4311
ANGERONA		12SecDr	4-5/13	2410	4106
ANGOLA		12SecDr	10-12/20	2410F	4591
ANGORA		28ChDr	11/18/10	2418	3862
ANIELA		26ChDr	4-5/16	2416C	4366
ANITA		12SecDr	1-2/22	2410F	4625
ANJOU		12SecDr	8-9/20	2410F	4574
ANN ARBOR		12SecDrCpt	3/14	2411A	4269
ANN McGINTY	N	12SecDr	1-3/25	3410	4844
ANNABEL		28ChDr	6-7/27	3416A	6087
ANNADORE		12SecDr	4-5/13	2410	4106
ANNE BAILEY	N	14Sec	9-10/25	3958	4869
ANSELIN'S TOWER	RN	8SecDr3Dbr	9-10/25	4090D	4899
ANSONIA		32ChDr	8/27	3917A	6078
ANSONIA	R	24Chdr10StsLng	8/27	3917C	6078
ANSPACH		12SecDr	7-8/13	2410A	4192
ANTARES		12SecDr	2-5/21	2410F	4612
ANTHONY WAYNE		12SecDr	5-8/25	3410A	4868
ANTHRACITE		26ChDr	4-5/13	2416A	4136
ANTHRACITE CLUB	N	8SecBLngSunRm	8-9/29	3989U	6274
ANTIETAM		12SecDr	7-8/12	2410	4014
ANTIGUA		12SecDr	2-5/16	2410E	4367
ANTILLES		10SecDr2Cpt	11-12/11	2585	3942
ANTIOPE	N	24ChDrB	6-7/14	2417B	4265
ANTIPHON		12SecDrCpt	10-12/21	2411C	4624
ANTLER PEAK		10Sec2Dr	12/26-27	3584A	6031
ANTLERS		12SecDr	2-3/13	2410	4105
ANTOINETTE		26ChDr	3/13	2416	4107
ANTONIO CASO NdeM-106	N	8sec4Dbr	10-11/12	4022B	4067
ANTRIM		12SecDr	8-9/20	2410F	4574
ANTWERP		12SecDr	3-4/11	2410	3881
APITON		12SecDr	12/10-11	2410	3866
APOLLO		12SecDr	6-9/26	3410A	4969
APPIAN		12SecDr	3-6/20	2410F	4565
APPLEBY		12SecDr	1-2/16	2410E	4356
APPOMATTOX	N	10SecLngObs	11-12/25	3521A	4923
APPOMATTOX COUNTY	N	10SecLngObs	11-12/25	3521A	4923
APPONAUG		36Ch	10/24	3916	4802
APTHORP		12SecDr	7-8/14	2410A	4277
APTHORPE HOUSE		13Dbr	8/30	3997A	6392
AQUA		28ChDr	11/19/10	2418	3862
AQUEDUCT		12SecDr	5-6/17	2410F	4497
AQUIDIBAN		12SecDr	10-12/15	2410D	4338
AQUITANIA		12SecDr	5-7/13	2410	4149
ARABELLA		28ChDr	4-6/25	3416	4864
ARANSAS		12SecDr	8-10/26	3410A	4945
ARAPAHOE		12SecDr	7-8/28	3410B	6184
ARATUS		12SecDrCpt	10-12/21	2411C	4624
ARCADE		BaggBClubSmk	9/12/10	2415	3834
ARCANUM		12SecDr	2-3/15	2410B	4318
ARCH PEAK		10Sec2Dr	12/28-29	3584B	6213
ARCHBOLD		12SecDr	1-4/12	2410	3949
ARCHER		10SecDr2Cpt	10-11/13	2585A	4218
ARCHIBALD GUTHRIE		14Sec	6/30	3958A	6373
ARCHIBALD GUTHRIE	R	6Sec6Dbr	6/30	4084B	6373
ARCHON		12SecDr	8-9/20	2410F	4574
ARCTURUS		12SecDr	2-5/21	2410F	4612
ARDARA		26ChDr	3/13	2416	4107
ARDATH		12SecDr	1-3/14	2410A	4249
ARDELAN		12SecDr	1-2/16	2410E	4356
ARDENHEIM		12SecDr	10/25/10	2410	3860
ARDMORE		12SecDr	10/22/10	2410	3860
ARDOYNE		12SecDr	9-10/13	2410A	4215
ARDROSSAN		12SecDr	2-5/16	2410E	4367
ARDSLEY CLUB		8SecBObsLng	2-3/29	3989	6229
ARDWICK		26ChDr	7/18/10	2416	3804
ARENA		16Sec	2/13	2412	4101
ARENOSA		12SecDr	12/10-11	2410	3866
ARESTO		12SecDr	12/10-11	2410	3866
ARGENSOLA		16Sec	6-7/20	2412F	4568
ARGO		12SecDr	1-2/22	2410F	4625
ARGONNE		12SecDr	3-6/20	2410F	4565
ARGUS		12SecDr	2-3/15	2410B	4318
ARGYLE		7Cpt2Dr	12/11	2522	3943
ARIEL		12SecDr	5-8/18	2410F	4540
ARION		12SecDr	8-9/20	2410F	4574
ARISTINE		26ChDr	4/12	2416	3950
ARISTON		12SecDr	12/10-11	2410	3866
ARISTOTLE		Sec Dr Cpt	10-12/21	2411C	4624
ARKALON		10SecDr2Cpt	1-2/13	2585	4091
ARKLOW		12SecDr	10-12/20	2410F	4591
ARLETTA		24ChDrB	6-7/14	2417B	4265
ARLINGHAM		12SecDr	10/26/10	2410	3860
ARLINGHAM		26ChDr	5/23	2416D	4691
ARLINGTON		12SecDr	11/14/10	2410	3860
ARLINGTON		12SecDr	4/26	3410A	4943
ARMADA		12SecDr	2-5/16	2410E	4367
ARMAGH		12SecDr	8-9/20	2410F	4574
ARMANDAVE		12SecDr	2-3/13	2410	4105
ARMENIA		12SecDr	2-5/16	2410E	4367
ARMILDIA		26ChDr	4/12	2416	3950
ARMILLA		26ChDr	4-5/16	2416C	4366
ARMINGTON		12SecDr	4-6/24	3410	4763
ARMISTA		26ChDr	4/12	2416	3950
ARMISTICE		12SecDr	3-6/20	2410F	4565
ARMITAGE		12SecDr	2-3/13	2410	4105

CAR NAME	ST	CAR TYPE	BUILT	PLAN	LOT
ARMORY HILL	N	16ChBLngObs	1-2/27	3961G	6035
ARMORY SQUARE	N	BaggBSmkBarber	12/26-27	3951C	6022
ARMSMEAR		12SecDr	7-8/14	2410A	4277
ARMSTRONG		12SecDr	2-5/16	2410E	4367
ARMY-NAVY CLUB	N	BaggBSmkBarber	10-11/25	3951B	4885
ARNAZ		12SecDr	1-3/14	2410A	4249
ARNOLD		12SecDrCpt	6/25/10	2411	3800
ARONA		12SecDr	2-5/25	3410	4845
AROYA		10SecDr2Cpt	11-12/11	2585	3942
ARRAS		12SecDr	8-9/20	2410F	4574
ARRIAN		12SecDrCpt	10-12/21	2411C	4624
ARRINGDALE		12SecDr	4-5/13	2410	4106
ARRINGTON		10SecDr2Cpt	1-2/13	2585	4091
ARROWHEAD		10SecDrCpt	4-6/27	3973	6043
ARROWSMITH		12SecDr	1-4/12	2410	3949
ARSENAL TOWER	RN	8SecDr3Dbr	1-3/25	4090F	4844
ARTCRAFT		16Sec	6-7/20	2412F	4568
ARTEMISA		12SecDr	1-2/15	2410B	4311
ARTHINGTON		12SecDr	2-3/13	2410	4105
ARTHUR		12SecDrCpt	6/21/10	2411	3800
ARTHUR A. DENNY		8SecDr2Cpt	5-6/29	3979A	6261
ARTHUR BRISBANE	N	8SecDr2Cpt	2-3/30	3979A	6353
ARTIGAN		12SecDr	1-2/16	2410E	4356
ARTSUL		12SecDr	3-6/20	2410F	4565
ARUNDEL		12SecDr	5-6/17	2410F	4497
ARVONIA		10SecDr2Cpt	1-2/13	2585	4091
ASBURY PARK	RN	30ChBLibObs	4/16	2901A	4365
ASCALON		10SecDr2Cpt	10-11/13	2585A	4218
ASH BELT	RN	8SecSbr6RmtDbr	3/27	4174	6054
ASH FORK		10SecLngObs	9-10/14	2521A	4292
ASH FORK	RN	8SecSbr6RmtDbr	9-10/25	4174	4869
ASHBOURNE		12SecDr	10-12/26	3410A	6023
ASHBRIDGE		12SecDr	9-10/15	2410D	4327
ASHBY		12SecDr	10-12/20	2410F	4591
ASHBY'S GAP		8SecDr2Cpt	12/29-30	3979A	6334
ASHCOM		12SecDr	2-5/25	3410	4845
ASHEVILLE	N	Private (GS)	3/11	2502	3848
ASHIPPUN		14Sec	8-10/28	3958A	6181
ASHLAND		12SecDr	12/10-11	2410	3866
ASHLAND	RN	10ChRestLng	6-7/13	4018	4160
ASHLAND COUNTRY CLUB	RN	8SecRestLngObs	1-5/18	4026B	4531
ASHLEY	N	12SecDr	12/10-11	2410	3866
ASHTON		12SecDr	1-2/16	2410E	4356
ASHWOOD		12SecDr	3-4/11	2410	3881
ASPASIA		12SecDr	4-5/13	2410	4106
ASPEN		12SecDr	12/27-28	3410B	6127
ASPERN		12SecDr	1-2/22	2410F	4625
ASPINOOK		12SecDr	11-12/13	2410A	4234
ASSEMBLY HALL		4CptLngObs	4/26	3960	4944
ASTORIA		12SecDr	5-8/18	2410F	4540

CAR NAME	ST	CAR TYPE	BUILT	PLAN	LOT
ASTRA		12SecDr	12/12-13	2410	4076
ASTRAKAN		12SecDr	8-9/20	2410F	4574
ASTRAKHAN	N	12SecDr	8-9/20	2410F	4574
ASUNCION NdeM	SN	6Sec6Dbr	10-12/21	4060	4624
ATASCADERO		16Sec	6-7/20	2412F	4568
ATCO		12SecDr	5-6/23	2410H	4699
ATGLEN		12SecDrCpt	6/15/10	2411	3800
ATHLETIC CLUB		8SecBLngSunRm	8-9/29	3989A	6274
ATHLONE		12SecDr	8-9/20	2410F	4574
ATHOL SPRINGS		12SecDrCpt	5/17	2411C	4494
ATHRIN		12SecDr	12/10-11	2410	3866
ATKINSON		12SecDr	4-5/11	2410	3893
ATLANTIC CITY		26ChDr	9-11/22	2416D	4649
ATLAS		12SecDr	8-9/20	2410F	4574
ATOKA		12SecDr	7/13	2410A	4179
ATOYAC NdeM-232	N	10SecDr2Dbr	5-6/25	4074C	4843
ATTAKAPA TRIBE	RN	6SecDr4Dbr	9-10/29	4092	6284
ATTERCLIFFE		12SecDr	1-4/12	2410	3949
ATTICA		12SecDr	8-9/20	2410F	4574
ATWOOD		12SecDr	9/27/10	2410	3813
AUBURN		10SecDr2Cpt	11-12/11	2585	3942
AUBURN		32ChDr	5/26	3917	4957
AUBURN	R	24ChDr10StsLng	5/26	3917E	4957
AUBURN	N	24ChDr10StsLng	6/30	3917E	4803
AUBURNDALE		16Sec	10-11/14	2412C	4296
AUCKLAND		12SecDr	8-9/20	2410F	4574
AUDENREID		12SecDr	7-8/13	2410A	4192
AUDRAIN		12SecDr	4-5/13	2410	4106
AUDREY		12SecDr	1-3/17	2410F	4450
AUDREY		28ChDr	6-7/30	3416B	6372
AUDUBON		12SecDr	10-12/16	2410F	4431
AUDWIN		12SecDr	2-5/16	2410E	4367
AUGUST A. BUSCH	N	12SecDr	8-12/25	3410A	4894
AUGUSTA	N	12SecDr	11-12/13	2410A	4234
AUGUSTA	RN	17StsRestLng	6/11	4019F	3902
AUGUSTA	N	10ChRestLng	6-7/13	4018	4160
AUGUSTIN					
LARA FCP-202	SN	10SecDr2Cpt	8-10/23	3585	4725
MELGAR NdeM	SN	10Sec3Dbr	3/14	3411	4269
AULANDER		12SecDr	9-10/15	2410D	4327
AULDEARN		12SecDr	1/10/16	2410E	4354
AURANIA		26ChDr	4-5/16	2416C	4366
AUREOLA		26ChDr	6/11	2416	3902
AURORA		26ChDr	4-5/13	2416A	4136
AUSABLE		16Sec	12/14-15	2412C	4304
AUSABLE PASS		8SecDr2Cpt	2-3/30	3979A	6353
AUSTELL		12SecDr	9-10/13	2410A	4215
AUSTINBURG		12SecDr	10/3/10	2410	3813
AUTOMEDON		12SecDr	9-10/15	2410D	4327
AUVERGNE	N	12SecDr	11-12/13	2410A	4234

Figure I-46 BALSAM FIR was built in May 1930 as a 10-section drawing room compartment sleeper. In December 1948 it was sold to the Union Pacific Railroad where it served until placed in government storage at Tooele Ordnance Depot, Utah, in November 1958. Upon withdrawal from government storage in June 1962, it was rebuilt as UP Dynamometer Car 903001, and in 1964 it was renumbered UP-210. George Cockle Photograph.

CAR NAME	ST	CAR TYPE	BUILT	PLAN	LOT
AVALANCHE PASS		8SecDr2Cpt	2-3/30	3979A	6353
AVALANCHE PEAK	N	10Sec2Dr	9-10/29	3584B	6284
AVALON		14Sec	7/29	3958A	6271
AVENA		28ChDr	4/12	2418	3951
AVENEL		24ChDrB	1/14	2417A	4239
AVENING		12SecDr	2-5/16	2410E	4367
AVERIL		12SecDr	3-6/15	2410B	4319
AVIGNON		12SecDr	12/10-11	2410	3866
AVILLA		12SecDr	7/13	2410A	4179
AVOCA		12SecDr	7/13	2410A	4179
AVON		12SecDr	1-2/22	2410F	4625
AVONDALE		26ChDr	9-11/22	2416D	4649
AVONMORE		BaggBClubSmk	7/13	2415	4158
AYLESBURY		12SecDr	2-3/15	2410B	4318
AYLESWORTH		12SecDr	2-5/25	3410	4845
AYLMER		12SecDr	4-5/11	2410	3893
AZALEA		28ChDr	5-7/26	3416	4958
AZUREA		16Sec	5-6/16	2412F	4386
B'NAI B'RITH	RN	10ChCafeLng	6/11	4019H	3902
BAALBEC		12SecDr	3-6/20	2410F	4565
BABETTE		26ChDr	4-5/13	2416A	4136
BABRIUS		12SecDrCpt	10-12/21	2411C	4624
BACK BAY		36Ch	8/27	3916A	6077
BACK BAY	R	22Ch12StsBObsLng	8/27	4040	6077
BACKUS	N	12SecDr	9/28/10	2410	3813
BADEN		26ChDr	4-5/17	2416C	4492

CAR NAME	ST	CAR TYPE	BUILT	PLAN	LOT
BADGER STATE	N	10SecDr2Cpt	11-12/11	2585	3942
BAFFIN BAY		3Cpt2DrLngObs	11-12/26	3959	6015
BAGDAD		26ChDr	4-5/17	2416C	4492
BAGGOTT	N	12SecDr	10/26/10	2410	3860
BAHAMA		7Cpt2Dr	11/13	2522A	4222
BAHIA					
ASCENSION NdeM	SN	8SecDLng	12/17-18	4025V	4528
CHACAHUA NdeM	SN	8SecDLng	11/12	4025J	4049
ESPIRITU SANTO NdeM	SN	8SecDLng	10/13	4025U	4202
MAGDALENA NdeM	SN	8SecDLng	9-10/14	4025U	4292
BAILEY		12SecDr	2-5/16	2410E	4367
BAINBRIDGE		26ChDr	8/3/10	2416	3804
BAIRD		28ChDr	11/19/10	2418	3862
BAKER		32Ch	11/10/10	2419	3861
BAKERSFIELD		12SecDrCpt	6/13	2411A	4159
BAKERTON		12SecDr	5-6/17	2410F	4497
BALA		28ChDr	5-7/26	3416	4958
BALAKLAVA		12SecDr	10-12/15	2410D	4338
BALBOA		3Cpt2DrLngObs	11-12/26	3959	6015
BALCARRES		12SecDr	2-5/16	2410E	4367
BALD EAGLE		28ChDr	6-7/27	3416A	6087
BALD PEAK		10Sec2Dr	12/26-27	3584A	6031
BALDROMMA		12SecDr	8-9/20	2410F	4574
BALDUR	N	12SecDr	12/16/10	2410	3864
BALDWIN		12SecDrCpt	6/24/10	2411	3800
BALFOUR		16Sec	12/17-18	2412F	4528

CAR NAME	ST	CAR TYPE	BUILT	PLAN	LOT	CAR NAME	ST	CAR TYPE	BUILT	PLAN	LOT
BALLARAT		12SecDr	2-3/13	2410	4105	BARRETT		12SecDr	2-5/16	2410E	4367
BALLARD	N	12SECDR	11/14/10	2410	3860	BARRON		16Sec	12/17-18	2412F	4528
BALLINA		12SecDr	10-11/11	2410	3936	BARRYTOWN	N	6CptLngObs	8/6/10	2413	3802
BALLSTON		16Sec	10-11/14	2412C	4296	BARTHOLOMEW PENROSE		14Sec	7/30	3958A	6376
BALSAM FIR		10SecDrCpt	5/30	3973A	6358	BARTLETT		16Sec	12/17-18	2412F	4528
BALSAMO		12SecDr	10-12/20	2410F	4591	BARTON		12SecDr	10-12/16	2410F	4431
BALTIMORE		26ChDr	5/12	2416	3986	BARTRAM OAK		8SecDr2Cpt	2-3/30	3979A	6353
BALTUSROL CLUB		8SecBLngSunRm	6-7/30	3989B	6362	BASEL	N	12SecDr	10/22/10	2410	3814
BALYMENA		12SecDr	3-6/15	2410B	4319	BASILMAR		12SecDr	2-3/13	2410	4105
BALZAC	N	12SecDr	4/29/10	2410	3772	BASKIN		12SecDr	11-12/13	2410A	4234
BAMBROUGH		12SecDr	5-7/13	2410	4149	BASS LAKE		10SecDrCpt	10-12/29	3973A	6300
BAMFORD		24ChDrB	1/14	2417A	4239	BASSETT	N	12SecDr	9/30/10	2410	3813
BANAVIE		12SecDr	7-8/13	2410A	4192	BATES PEAK		10Sec2Dr	9-10/29	3584B	6284
BANDELIER		12SecDr	10-12/16	2410F	4431	BATESVILLE		12SecDr	1-4/12	2410	3949
BANDURA		12SecDr	1-3/17	2410F	4450	BATH		12SecDr	8-9/20	2410F	4574
BANKERS' CLUB		8SecBLngSunRm	8-9/29	3989A	6274	BATTLE CREEK		10Cpt	1/12	2505	3969
BANNOCK		12SecDr	4/30	3410B	6360	BATTLEMENT TOWER	RN	8SecDr3Dbr	12/26-27	4090D	6031
BANNOCK PEAK		10Sec2Dr	12/28-29	3584B	6213	BAXTER		12SecDr	2-5/16	2410E	4367
BANNON		12SecDr	10-11/11	2410	3936	BAY CITY		12SecDrCpt	8-9/11	2411	3922
BANTRY		16Sec	6-7/13	2412B	4160	BAY HEAD		28ChDr	6-7/27	3416A	6087
BARBERTON		12SecDr	5/4/10	2410	3772	BAY POND		10Cpt	1/12	2505	3969
BARBERTON		12SecDr	2-4/24	3410	4762	BAY POND	R	5CptLngObs	1/12	2505B	3969
BARBOURSVILLE	N	12SecDr	1-3/14	2410A	4249	BAY STATE		BaggBSmkBarber	10-11/25	3951	4885
BARCLAY		28ChDr	11/21/10	2418	3862	BAYBERRY		16Sec	4/17	2412F	4484
BARDORA		12SecDr	3-6/20	2410F	4565	BAYBORO		12SecDr	3-6/20	2410F	4565
BARKER		12SecDr	5-8/18	2410F	4540	BAYHURST	N	12SecDr	11/14/10	2410	3860
BARKSDALE		12SecDr	10-12/20	2410F	4591	BAYLISS	N	12SecDr	10/12/10	2410	3813
BARKSDALE	RN	10Sec3Dbr	2-5/25	4087B	4845	BAYONNE		16Sec	4/17	2412F	4484
BARLETT TOWER	RN	8SecDr3Dbr	5-6/25	4090	4846	BAYVIEW		24ChDrObs	4/14	2819A	4263
BARLOW		16Sec	12/17-18	2412F	4528	BAYWOOD	N	12SecDr	11/23/10	2410	3864
BARMOUTH		26ChDr	5/12	2416	3986	BAZANTAH		12SecDr	10/14	2410B	4293
BARNARD		12SecDr	4-5/11	2410	3893	BEACH HAVEN		12SecDrCpt	7/7/10	2411	3800
BARNARD COLLEGE	RN	10Sec2DbrCpt	7/1/10	4042A	3800	BEACH RIDGE		12SecDrCpt	3/11	2411	3800
BARNEGAT		BaggBClubSmk	7/13	2415	4158	BEACON		10SecDr2Cpt	11/17	2585D	4527
BARNESBORO		28ChObs	8/24/10	2421	3806	BEACON HILL	N	16ChBLngObs	1-2/27	3961G	6035
BARNESLEY		12SecDrCpt	6/13	2411A	4159	BEACON TOWER	RN	8SecDr3Dbr	5-6/25	4090E	4846
BARNESTON		26ChDr	7/26/10	2416	3804	BEALLSVILLE		12SecDr	6-9/26	3410A	4969
BARNESVILLE		10SecDr2Cpt	10-11/13	2585A	4218	BEAR OAK		8SecDr2Cpt	7-8/30	3979A	6377
BARNETTS		12SecDr	9/28/10	2410	3813	BEARMOUTH		10SecDrCpt	4-6/27	3973	6043
BARNETTS		12SecDr	10-12/16	2410F	4431	BEARTOWN		26ChDr	7/25/10	2416	3804
BARNEVELD		BaggBClubSmk	1/12	2415	3946	BEASLEY	N	12SecDr	11/5/10	2410	3860
BARNSLEY		12SecDr	6-9/26	3410A	4969	BEATRICE		26ChDr	3-4/14	2416A	4264
BARNSTABLE BAY		32ChDr	12/29-30	3917B	6318	BEATTY		12SecDrCpt	5/17	2411C	4494
BARODA		12SecDr	4-5/11	2410	3893	BEAUDESERT	N	12SecDr	5/11/10	2410	3773
BARON DE KALB		12SecDr	4/26	3410A	4943	BEAUDETTE		12SecDr	1-3/14	2410A	4249
BARON ROCHAMBEAU	N	12SecDr	2-4/24	3410	4762	BEAUFORT		12SecDr	2-3/27	3410A	6055
BARON VON STEUBEN	N	12SecDr	2-4/24	3410	4762	BEAULIEU		12SecDr	5-7/13	2410	4149
BARONESS		26ChDr	3-4/14	2416A	4264	BEAUMORIS		16Sec	6-7/13	2412B	4160
BARREE		26ChDr	4-5/17	2416C	4492	BEAUREGARD		12SecDr	2-5/21	2410F	4612
BARREN HILL		BaggBSmkBarber	12/26-27	3951C	6022	BEAUVAIS		16Sec	4/17	2412F	4484

CAR NAME	ST	CAR TYPE	BUILT	PLAN	LOT
BEAVER		12SecDr	6-9/26	3410A	4969
BEAVER FALLS		12SecDr	10/3/10	2410	3813
BEAVERDALE		28ChObs	8/25/10	2421	3806
BECKET		BaggBClubSmkBarber	4-5/17	2951	4483
BECKLEY		36Ch	6/16	2916	4389
BECKWITH		12SecDr	1-4/12	2410	3949
BEDFORD		12SecDr	9/29/10	2410	3813
BEDINGTON		12SecDr	2-4/24	3410	4762
BEDMINSTER	N	12SecDr	5/2/10	2410	3772
BEECH CREEK		10Sec2Dr	1/12	2584	3947
BEECH FOREST	RN	12RmtDrDbr2Sbr	6-7/17	4157A	4503
BEECH GROVE	RN	12RmtDrDbr2Sbr	6-7/17	4157	4503
BEECH PARK	RN	12RmtDrDbr2Sbr	10/22-23	4157B	4647
BEECH WOODS	RN	12RmtDrDbr2Sbr	10/22-23	4157A	4647
BEECHER		12SecDr	6-7/16	2410F	4385
BEECHFIELD		26ChDr	3/13	2416	4107
BEECHWOOD		BaggBClubSmkBarber	2-3/12	2602	3948
BEEKMAN		12SecDr	2-5/16	2410E	4367
BEESON		BaggBClubSmkBarber	4-5/17	2951	4483
BEETHOVEN		12SecDr	7-8/15	2410D	4322
BEETHOVEN		6Cpt3Dr	11-12/24	3523A	4833
BEGONIA		28ChDr	5-7/26	3416	4958
BELDENS		12SecDr	4-5/11	2410	3893
BELFAST		12SecDr	2-4/24	3410	4762
BELFORT		12SecDr	1/16	2410E	4354
BELGIUM	N	12SecDr	11/12/10	2410	3860
BELGRADE		16Sec	5-6/13	2412B	4150
BELINDA		26ChDr	3/13	2416	4107
BELKNAP		12SecDr	4-5/11	2410	3893
BELL PEAK		10Sec2Dr	12/28-29	3584B	6213
BELL TOWER	RN	8SecDr3Dbr	5-6/25	4090	4846
BELLAMY		12SecDrCpt	10-12/21	2411C	4624
BELLBROOK		12SecDr	6-7/16	2410F	4385
BELLE		26ChDr	3-4/14	2416A	4264
BELLE BRIDGE		12SecDrCpt	3/14	2411A	4269
BELLE CENTRE		10Cpt	1/12	2505	3969
BELLE FOURCHE		14SecCpt	10/12	2634	4040
BELLE HAVEN		28ChDr	6-7/27	3416A	6087
BELLE ISLE		12SecDrCpt	7/18/10	2411	3800
BELLE PLAINE		14SecCpt	10/12	2634	4040
BELLE VERNON		12SecDrCpt	8-9/11	2411	3922
BELLEAU WOOD		7Dr	12/20	2583B	4621
BELLEFLOWER		10SecDr2Cpt	12/15	2585C	4351
BELLEFONTAINE		BaggBClubSmk	1/12	2415	3946
BELLEFONTE		26ChDr	3/13	2416	4107
BELLEROSE		12SecDr	7-8/13	2410A	4192
BELLEVUE		16Sec	4/17	2412F	4484
BELLEVUE	N	12SecDr	8-10/26	3410A	4945
BELLHURST		12SecDr	5-6/17	2410F	4497
BELLINI		6Cpt3Dr	11-12/24	3523A	4833

CAR NAME	ST	CAR TYPE	BUILT	PLAN	LOT
BELLONIA		12SecDr	3-6/20	2410F	4565
BELLWOOD		12SecDr	11/15/10	2410	3860
BELLWOOD		26ChDr	9-11/22	2416D	4649
BELMAR		26ChDr	3/13	2416	4107
BELNORD		12SecDr	5-7/13	2410	4149
BELSENA		28ChDr	11/19/10	2418	3862
BELTON		12SecDr	4-6/24	3410	4763
BELTRAMI		12SecDr	2-5/21	2410F	4612
BELVIDERE		16Sec	6/2/10	2412	3801
BELVONA		12SecDr	12/10-11	2410	3866
BEMENT		12SecDr	2-5/21	2410F	4612
BEMIS	N	12SecDr	10/4/10	2410	3813
BENBURB		12SecDr	8-9/20	2410F	4574
BENCHOR		7Cpt2Dr	12/11	2522	3943
BENEDICT		16Sec	12/17-18	2412F	4528
BENGAL		12SecDr	8-9/20	2410F	4574
BENGIES		28ChDr	11/21/10	2418	3862
BENICIA		LngBDorm(5Sec)	7-8/27	3981	6061
BENITO		12SecDr	12/10-11	2410	3866
BENJAMIN					
FRANKLIN		30ChDr	9-10/24	3418	4783
FRANKLIN		30ChDr	8-9/30	4000A	6384
GRUBB HUMPHREYS		10Sec2Dr	9-10/25	3584A	4899
BENJAMIN RUSH		12SecDr	4/26	3410A	4943
BENMORE		12SecDr	10-12/20	2410F	4591
BENNETT		26ChDr	7/12/10	2416	3804
BENNING		6CptLngObs	8/8/10	2413	3802
BENNING	R	5CptLngObs	8/8/10	2413A	3802
BENTALON		16Sec	12/17-18	2412F	4528
BENTHAM		12SecDr	5-7/13	2410	4149
BENTLEY		6CptLngObs	8/10/10	2413	3802
BENTLEYVILLE		12SecDr	6-9/26	3410A	4969
BENTON		32Ch	11/10/10	2419	3861
BENTONVILLE		12SecDr	9/30/10	2410	3813
BENTONVILLE		12SecDr	10/22-23	2410H	4647
BENWOOD	N	12SecDr	11/26/10	2410	3860
BENZINGER		12SecDr	5-6/17	2410F	4497
BERESINA		12SecDr	10-12/15	2410D	4338
BERGAMO		12SecDr	6-7/17	2410F	4503
BERGENFIELD		12SecDr	4-6/14	2410A	4271
BERGERAC		16Sec	12/17-18	2412F	4528
BERGHEIM	N	12SecDr	11/9/10	2410	3860
BERKELEY		12SecDr	8-10/26	3410A	4945
BERLIN		12SecDr	11/17/10	2410	3864
BERMUDA		16Sec	4/17	2412F	4484
BERNADOTTE		12SecDr	8-9/20	2410F	4574
BERNE		12SecDr	8-9/20	2410F	4574
BERNETTA		24ChDrB	6-7/14	2417B	4265
BERNWOOD		12SecDr	1-3/17	2410F	4450
BEROSUS		12SecDrCpt	10-12/21	2411C	4624

CAR NAME	ST	CAR TYPE	BUILT	PLAN	LOT
BERRIEN	N	12SecDr	10/24/10	2410	3814
BERRYVILLE	N	12SecDr	5-8/25	3410A	4868
BERTHA		28ChDr	4-6/25	3416	4864
BERTHOLD		12SecDr	4-6/24	3410	4763
BERTHOUD PASS		8SecDr2Cpt	12/29-30	3979A	6334
BERTILLON		12SecDr	10-12/16	2410F	4431
BERTRAND		12SecDr	12/10-11	2410	3866
BERWICK		16Sec	12/17-18	2412F	4528
BERWINDALE		26ChDr	7/23/10	2416	3804
BERWYN		16Sec	4/17	2412F	4484
BESCO		28ChDr	5-7/26	3416	4958
BESSIE		26ChDr	4/12	2416	3950
BETHALTO		16Sec	2-3/11	2412	3878
BETHEL		36Ch	10/24	3916	4802
BETHEVAN		12SecDr	6-9/26	3410A	4969
BETHSHAN	N	12SecDr	10/14/10	2410	3813
BETSY ROSS		12SecDr	8-12/25	3410A	4894
BETTINA		26ChDr	4/12	2416	3950
BETTSVILLE		12SecDr	6-9/26	3410A	4969
BETZWOOD		26ChDr	3/13	2416	4107
BEVAN	N	12SecDr	11/9/10	2410	3860
BEVERLY		16Sec	4/17	2412F	4484
BEVIER		12SecDr	7-8/15	2410D	4322
BEVINGTON		12SecDr	1-3/14	2410A	4249
BEXAR		12SecDr	8-10/26	3410A	4945
BEXLEY		12SecDr	12/12-13	2410	4076
BICKMORE		12SecDr	1-3/17	2410F	4450
BICKNELL		12SecDr	2-5/16	2410E	4367
BIDDEFORD	N	12SecDr	7-8/15	2410D	4322
BIDDLE		26ChDr	4-5/17	2416C	4492
BIENVILLE	RN	10ChRestLng	8/13	4056	4195
BIG ROCK		10SecDrCpt	4-6/27	3973	6043
BIG WALNUT		28ChDr	6-7/27	3416A	6087
BIGELOW		16Sec	12/17-18	2412F	4528
BIGLER		BaggBClubSmkBarber	4-5/17	2951	4483
BIJOU		12SecDr	7-8/15	2410D	4322
BILLERICA		16Sec	12/17-18	2412F	4528
BILLINGSWOOD		12SecDr	10/22-23	2410H	4647
BILLINGTON		16Sec	12/17-18	2412F	4528
BILLOP		16Sec	12/17-18	2412F	4528
BINGHAM		12SecDr	10-11/11	2410	3936
BINSTEAD		12SecDrCpt	3/11	2411	3880
BIRCHWOOD		12SecDr	1-4/12	2410	3949
BIRDSBORO		26ChDr	7/9/10	2416	3804
BIRKENDALE		12SecDr	3-6/15	2410B	4319
BISBEE	N	12SecDr	9/16/10	2410	3794
BISCAY	N	12SecDr	10/14/10	2410	3813
BISHOP QUARTER	N	6Cpt3Dr	10-12/25	3523A	4922
BISMA-REX #3	LN	Exhibit (Rexall)	6/16	2916B	4389
BISON		12SecDr	4-6/24	3410	4763

CAR NAME	ST	CAR TYPE	BUILT	PLAN	LOT
BISON PEAK		10Sec2Dr	12/26-27	3584A	6031
BISSELL	N	12SecDr	4/12/10	2410	3771
BIZET		6Cpt3Dr	11-12/24	3523A	4833
BLACK BUTTES		4Cpt2DrLngObs	3/13	2703	4100
BLACK DIAMOND		24ChLngSunRm	11/27	3984	6092
BLACK HAWK		10SecDrCpt	4-6/27	3973	6043
BLACK HILLS	N	10SecDr2Cpt	11-12/11	2585	3942
BLACK OAK		8SecDr2Cpt	7-8/30	3979A	6377
BLACK RIVER		14Sec	8-10/28	3958A	6181
BLACK ROCK		12SecDrCpt	3/11	2411	3880
BLACKBURN		6CptLngObs	8/4/10	2413	3802
BLACKSTOCKS		12SecDr	10-12/26	3410A	6023
BLACKWELL		12SecDr	4-6/14	2410A	4271
BLADEN		16Sec	12/17-18	2412F	4528
BLAIRSVILLE		12SecDr	11/14/10	2410	3860
BLAIRSVILLE		12SecDr	10/22-23	2410H	4647
BLAKELEY		26ChDr	7/26/10	2416	3804
BLANCA PEAK		10Sec2Dr	12/28-29	3584B	6213
BLANCHARD		12SecDr	4-5/11	2410	3893
BLANCHE		28ChDr	6-7/27	3416A	6087
BLANDFORD		16Sec	12/17-18	2412F	4528
BLANDIN		12SecDr	1-2/15	2410B	4311
BLANTON		12SecDr	4-5/13	2410	4106
BLASDELL		12SecDr	10-11/11	2410	3936
BLAUMOND		16Sec	5-6/16	2412F	4386
BLAUVELT		12SecDr	4-5/11	2410	3893
BLEDSOE		16Sec	12/17-18	2412F	4528
BLEEKER		16Sec	12/17-18	2412F	4528
BLENCATHRA		12SecDr	11-12/14	2410B	4297
BLISSFIELD		12SecDr	10/3/10	2410	3813
BLOIS	N	12SecDr	10/19/10	2410	3814
BLOOMSBURG		12SecDr	5-6/17	2410F	4497
BLOOMVILLE		12SecDr	9/30/10	2410	3813
BLOSSVALE		12SecDr	10-11/11	2410	3936
BLOUNT		16Sec	12/17-18	2412F	4528
BLUE BIRD LAKE		10SecDrCpt	3-5/30	3973A	6338
BLUE GENTIAN		8SecDr2Cpt	4-5/30	3979A	6359
BLUE GRASS		26ChB	4/14	2416B	4264
BLUE RAPIDS		4Cpt2DrLngObs	3/13	2703	4100
BLUE RIDGE		BaggBSmkBarber	12/26-27	3951C	6022
BLUE STONE		28ChDr	6-7/27	3416A	6087
BLUEBELL		28ChDr	5-7/26	3416	4958
BLUEFIELD	N	12SecDr	5-8/25	3410A	4868
BLUEFIELD	N	10Sec3Dbr	3/11	3411	3880
BLUESTEM		12SecDr	4-6/24	3410	4763
BLYTHESWOOD		12SecDr	4-6/14	2410A	4271
BOARDMAN		12SecDr	4-5/11	2410	3893
BOCA DEL RIO NdeM-107	SN	12SecDr	1-2/30	3410B	6351
BOCA GRANDE		10SecLngObs	11/13	2521	4221
BOGOTA		12SecDr	8-9/20	2410F	4574

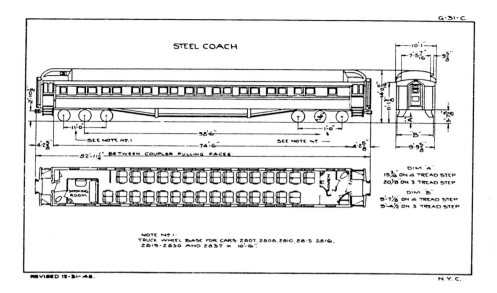

G-31-C

STEEL COACH

N.Y.C.

REVISED 12-31-48.

Figure I-47 In 1941 and 1942 Pullman sold 325 parlor cars, most of which went to the railroads for conversion to coaches. Among the 47 purchased for this purpose by the New York Central, BLANCHE was 1 of the 15 Plan 3416A parlors on the list. It carried the NYC number 2817, had Pullman mechanical air conditioning, and a seating capacity of 60. The Author's Collection.

CAR NAME	ST	CAR TYPE	BUILT	PLAN	LOT
BOGOTA NdeM	SN	6Sec6Dbr	10-12/16	4084	4431
BOHEMIAN		12SecDr	12/12-13	2410	4076
BOK TOWER	RN	8SecDr3Dbr	12/24-25	4090	4836
BOLLING		16Sec	12/17-18	2412F	4528
BOLTON		36Ch	6/16	2916	4389
BOLTONIA		28ChDr	5-7/26	3416	4958
BON AIR	N	12SecDr	1-3/14	2410A	4249
BON AIR	N	12SecDr	10/14/10	2410	3814
BONANZA		12SecDr	8-9/20	2410F	4574
BONAPARTE		12SecDr	8-9/20	2410F	4574
BONBRIGHT		12SecDr	5-7/13	2410	4149
BONCHURCH		12SecDr	7/13	2410A	4179
BONDSVILLE		12SecDr	4-6/14	2410A	4271
BONDURANT		12SecDr	4-5/13	2410	4106
BONIFACE		12SecDr	10-22/23	2410H	4647
BONIFAY		12SecDr	9-10/13	2410A	4215
BONITA		12SecDr	8-9/20	2410F	4574
BONIVARD		12SecDr	12/12-13	2410	4076
BONLEE		12SecDr	1-3/14	2410A	4249
BONNEAU		12SecDr	11-12/13	2410A	4234

CAR NAME	ST	CAR TYPE	BUILT	PLAN	LOT
BONNEVAL		12SecDr	8-9/12	2410	4035
BONNIEWOLD	N	12SecDr	10/21/10	2410	3814
BONNYMEADE		12SecDr	1/14	2410A	4240
BONSAL	N	12SecDr	10/28/10	2410	3860
BONSALL	N	12SecDr	10/28/10	2410	3860
BONSECOUR		12SecDr	5-7/13	2410	4149
BOONE		10SecDr2Cpt	10-11/12	2585	4067
BOONTON	N	12SecDr	11/9/10	2410	3860
BOOTHBAY	N	12SecDr	5/5/10	2410	3773
BOOTHWYN		12SecDr	1-3/17	2410F	4450
BOPPARD		12SecDr	7-8/13	2410A	4192
BOQUILLA	N	12SecDr	11/16/10	2410	3860
BORDENTOWN		6CptLngObs	8/11/10	2413	3802
BORDLEY		16Sec	12/17-18	2412F	4528
BORDULAC		12SecDr	8-9/12	2410	4035
BOREAS RANGE		10SecDrCpt	5/30	3973A	6358
BORGIA		24ChDrB	6-7/14	2417B	4265
BORIS	N	12SecDr	8/29/10	2410	3794
BORNEO		12SecDr	8-9/20	2410F	4574
BORSIPPA	N	12SecDr	10/31/10	2410	3860
BORTON		12SecDr	9/15/10	2410	3794
BOSCOBEL	N	12SecDr	5/26/10	2410	3794
BOSPORUS		12SecDr	6-7/17	2410F	4503
BOSSERT	N	12SecDr	10/26/10	2410	3814
BOSSLER		12SecDr	5-6/17	2410F	4497
BOSTON		Private (GS)	1/14	2502A	4211
BOSTON BAY		32ChDr	12/29-30	3917B	6318
BOSTON COLLEGE	RN	10Sec2DbrCpt	6/13	4042A	4159
BOSTON LIGHT		36Ch	12/29-30	3916B	6319
BOSTON LIGHT	R	28Ch10StsLng	12/29-30	3916D	6319
BOTETOURT	N	10SecLngObs	11-12/25	3521A	4923
BOTETOURT COUNTY	N	10SecLngObs	11-12/25	3521A	4923
BOUDINOT		12SecDr	10-12/16	2410F	4431
BOUFFANT		16Sec	5-6/16	2412F	4386
BOUGAINVILLE		16Sec	12/17-18	2412F	4528
BOULDER		16Sec	2/13	2412	4101
BOUND BROOK		24ChDrObs	4/14	2819A	4263
BOURNE		12SecDr	10-22/23	2410H	4647
BOUVINES	N	12SecDr	10/24/10	2410	3814
BOWDIL		28ChDr	5-7/26	3416	4958
BOWDOIN		16Sec	12/17-18	2412F	4528
BOWERSTON		12SecDr	9/29/10	2410	3813
BOWIE		28ChDr	11/21/10	2418	3862
BOWMAN		16Sec	12/17-18	2412F	4528
BOWMAN LAKE		10SecDr2Cpt	5-6/24	3585	4770
BOYERO		10SecDr2Cpt	11-12/11	2585	3942
BOYLAN	N	12SecDr	6/10/10	2410	3794
BOYLE		12SecDr	10-22/23	2410H	4647
BOYNE		12SecDr	2-5/21	2410F	4612
BOYNTON		6CptLngObs	8/8/10	2413	3802

CAR NAME	ST	CAR TYPE	BUILT	PLAN	LOT
BOZEMAN	N	12SecDr	10/28/10	2410	3860
BRACEVILLE		12SecDr	4-5/11	2410	3893
BRACKEN		16Sec	10/13	2412B	4216
BRADBURY		12SecDr	10-12/16	2410F	4431
BRADDOCK		12SecDr	1-4/12	2410	3949
BRADENVILLE		12SecDr	1/14	2410A	4240
BRADFORD		36Ch	10/24	3916	4802
BRADGATE		16Sec	4/17	2412F	4484
BRADLEY		12SecDr	4-6/14	2410A	4271
BRADNER	N	12SecDr	11/17/10	2410	3864
BRADSHAW		12SecDr	4/30	3410B	6360
BRADSTREET		16Sec	12/17-18	2412F	4528
BRADWELL		12SecDr	11-12/13	2410A	4234
BRADY		16Sec	12/17-18	2412F	4528
BRAE BURN CLUB		8SecBObsLng	2-3/29	3989	6229
BRAHMS		6Cpt3Dr	11-12/24	3523A	4833
BRAIDWOOD		16Sec	4/17	2412F	4484
BRAINERD		16Sec	4/17	2412F	4484
BRAMCOTE		26ChDr	3/13	2416	4107
BRAMHALL		16Sec	4/17	2412F	4484
BRANCHPORT		12SecDr	5-6/17	2410F	4497
BRANDON		12SecDr	1-3/14	2410A	4249
BRANDRETH		BaggBClubSmk	1/12	2415	3946
BRANSTOCK		16Sec	4/17	2412F	4484
BRANT BEACH		24ChDrB	3/14	2417A	4250
BRATTLEBORO		12SecDr	7-8/14	2410A	4277
BRAXTON		16Sec	12/17-18	2412F	4528
BRAXTON BRAGG	N	12SecDr	7-8/12	2410	4014
BRAY		12SecDr	10/22-23	2410H	4647
BRAZIL		12SecDr	5/27/10	2410	3794
BRAZIL		26ChDr	4-5/17	2416C	4492
BRAZNELL		12SecDr	5-6/17	2410F	4497
BRAZOS		12SecDr	8-10/26	3410A	4945
BRAZOS PEAK		10Sec2Dr	12/28-29	3584B	6213
BREDA		12SecDr	5-8/18	2410F	4540
BREED'S HILL		6SbrBLngSunRm	12/29	3994B	6291
BREMEN		12SecDr	6-9/26	3410A	4969
BREMOND		12SecDr	1-2/16	2410E	4356
BRENAU COLLEGE	N	10Sec2DbrCpt	12/16-17	4042	4443
BRENFORD		26ChDr	7/12/10	2416	3804
BRENTA		12SecDr	12/10-11	2410	3866
BRENTHORP		12SecDr	2-3/13	2410	4105
BRENTWOOD		12SecDr	10-12/26	3410A	6023
BRESLIN TOWER	RN	8SecDr3Dbr	5-6/25	4090	4846
BRET HARTE		12SecDr	1-2/29	3410B	6220
BRETONA		12SecDr	12/27-28	3410B	6127
BRETTON WOODS		12SecDr	7-8/14	2410A	4277
BREVARD		12SecDr	9/28/10	2410	3813
BREWERTON		12SecDr	1-4/12	2410	3949
BRIANA		12SecDr	10/22-23	2410H	4647

CAR NAME	ST	CAR TYPE	BUILT	PLAN	LOT
BRIARWOOD		16Sec	4/17	2412F	4484
BRICKNER	N	12SecDr	6/13/10	2410	3794
BRIDESBURG		26ChDr	9-11/22	2416D	4649
BRIDGE TOWER	RN	8SecDr3Dbr	8-12/25	4090F	4894
BRIDGEDALE		12SecDr	1-3/14	2410A	4249
BRIDGEHILL		16Sec	4/17	2412F	4484
BRIDGEPORT		12SecDr	10/4/10	2410	3813
BRIDGER		16Sec	2/13	2412	4101
BRIDGER RANGE		10SecDrCpt	5/30	3973A	6358
BRIDGEVILLE		26ChDr	7/29/10	2416	3804
BRIDLESPUR CLUB	RN	8SecRestObsLng	9-10/14	4025A	4292
BRIER HILL		10Sec2Dr	1/12	2584	3945
BRIERCREST		12SecDr	11-12/13	2410A	4234
BRIGANTINE		12SecDr	7-8/13	2410A	4192
BRIGHTER TOWER	RN	8SecDr3Dbr	5-6/25	4090	4846
BRIGHTSIDE		12SecDr	5-6/17	2410F	4497
BRIGHTWOOD		12SecDr	4-5/11	2410	3893
BRIGSTOCK TOWER	RN	8SecDr3Dbr	5-6/25	4090	4846
BRILHART		26ChDr	5/12	2416	3986
BRILLION		16Sec	9-10/12	2412	4037
BRIMFIELD		12SecDr	4-5/11	2410	3893
BRIMSON		12SecDr	4-5/13	2410	4106
BRINDISI		12SecDr	8-9/20	2410F	4574
BRINDLEY		16Sec	12/17-18	2412F	4528
BRINGHURST		12SecDr	10/1/10	2410	3813
BRINK HAVEN		28ChDr	6-7/27	3416A	6087
BRINKER		26ChDr	4-5/17	2416C	4492
BRINKLOW		16Sec	4/17	2412F	4484
BRIXHAM		12SecDr	10/22-23	2410H	4647
BRIXWORTH TOWER	RN	8SecDr3Dbr	5-6/25	4090	4846
BROAD RIPPLE	N	28ChObs	3/14	2421A	4251
BROAD RIVER		BaggBSmkBarber	12/26-27	3951C	6022
BROADACRE		12SecDr	6-9/26	3410A	4969
BROADLAND		16Sec	4/17	2412F	4484
BROADMOOR		16Sec	9-10/12	2412	4037
BROADSTAIRS		12SecDr	10/22-23	2410H	4647
BROADWATER		10SecDr2Cpt	10-11/13	2585A	4218
BROADWAY LIMITED	N	CptDrLngBSunRm	9-10/27	3975G	6076
BROCKLEY		12SecDr	7-8/15	2410D	4322
BROCKMAN		16Sec	4/17	2412F	4484
BROCKWAY		12SecDr	10-12/26	3410A	6023
BRODHEAD		12SecDr	10/22-23	2410H	4647
BROKAW		BaggBClubSmkBarber	4-5/17	2951	4483
BROMHOLM		12SecDr	2-3/13	2410	4105
BRONSON		12SecDr	4-5/11	2410	3893
BRONXVILLE		12SecDr	4-5/11	2410	3893
BROOK CLUB	RN	8SecRestObsLng	10/13	4025A	4202
BROOK FOREST		12SecDr	11-12/29	3410B	6299
BROOKBRIDGE		12SecDr	7/13	2410A	4179
BROOKFORD		16Sec	4/17	2412F	4484

CAR NAME	ST	CAR TYPE	BUILT	PLAN	LOT	CAR NAME	ST	CAR TYPE	BUILT	PLAN	LOT
BROOKHURST	N	12SecDr	9/7/10	2410	3794	BUNGALOW	N	12SecDr	9/23/10	2410	3794
BROOKS		12SecDrCpt	10-12/21	2411C	4624	BUNKER HILL		10Sec2Dr	5/13	2584	4140
BROOKSIDE		BaggBClubSmkBarber	2-3/12	2602	3948	BUNOLA		12SecDr	4-6/14	2410A	4271
BROOKSVALE		34Ch	8/13	2764A	4195	BUNYAN		6Cpt3Dr	8/24	3523	4804
BROOKVIEW		6CptLngObs	8/11/10	2413	3802	BURCHELL	N	12SecDr	5/4/10	2410	3772
BROTHER JONATHAN	N	12SecDr	8-12/25	3410A	4894	BURDETTE		12SecDr	9-10/15	2410D	4327
BROUGHAM		16Sec	12/17-18	2412F	4528	BURDICK		12SecDr	1-4/12	2410	3949
BROUGHTON		12SecDr	1-3/14	2410A	4249	BUREAU		12SecDrCpt	6/13	2411	4151
BROUSSARD		12SecDrCpt	6/13	2411A	4159	BURGESS		12SecDr	4-5/11	2410	3893
BROUSSARD	R	24ChLng	6/13	4009	4159	BURGNER		28ChDr	5-6/24	3416	4761
BROWNELL		12SecDr	4-5/11	2410	3893	BURKHART		12SecDr	6-9/26	3410A	4969
BROWNFIELD		12SecDr	1/14	2410A	4240	BURLAND		12SecDr	10-12/16	2410F	4431
BROWNFIELD	N	12SecDr	4/26	3410A	4943	BURLEIGH		12SecDr	10-12/16	2410F	4431
BROWNHELM		12SecDr	4-6/14	2410A	4271	BURLINGAME		12SecDr	10/22-23	2410H	4647
BROWNING		16Sec	4/17	2412F	4484	BURLINGTON		16Sec	10/13	2412B	4217
BROWNLOW		16Sec	4/17	2412F	4484	BURLINGTON BAY	N	4Cpt2DrLngObs	3/13	2703	4100
BROWNSTOWN		12SecDr	10/3/10	2410	3813	BURLINGTON BRIDGE		12ChBLngSunRm	6/26	3964	4965
BROWNSTOWN		12SecDr	10-12/16	2410F	4431	BURLINGTON HOUSE		12ChBLngSunRm	6/26	3964	4965
BROWNSVILLE		12SecDr	2-5/16	2410E	4367	BURLINGTON HOUSE	R	16ChBLngSunRm	6/26	3964B	4965
BRUANTI		16Sec	6-7/20	2412F	4568	BURLINGTON LIGHT		12ChBLngSunRm	6/26	3964	4965
BRUCEVILLE		12SecDr	3/17/10	2410	3769	BURLINGTON LIGHT	R	16ChBLngSunRm	6/26	3964D	4965
BRUMAIRE		12SecDr	1-2/16	2410E	4356	BURLINGTON ROUTE		12ChBLngSunRm	6/26	3964	4965
BRUMLEY		16Sec	5-6/13	2412B	4150	BURLINGTON ROUTE	R	16ChBLngSunRm	6/26	3964B	4965
BRUNDAGE		16Sec	4/17	2412F	4484	BURNABY		12SecDr	2-5/21	2410F	4612
BRUNER	N	12SecDr	10/8/10	2410	3813	BURNHAM		12SecDr	12/12-13	2410	4076
BRUNSWICK		16Sec	4/17	2412F	4484	BURNLEY		16Sec	12/17-18	2412F	4528
BRUSSELS		16Sec	6-7/13	2412B	4160	BURNS		6Cpt3Dr	11-12/23	3523	4726
BRYANT	N	12SecDr	5/20/10	2410	3794	BURNSIDE		12SecDr	12/12-13	2410	4076
BRYCE CANYON		3CptDrLngBSunRm	1/30	3975F	6337	BURR OAK		10Cpt	1/12	2505	3969
BRYN MAWR		6CptLngObs	4/11	2413	3879	BURRAGE		36Ch	6/16	2916	4389
BRYN MAWR COLLEGE	RN	10Sec2DbrCpt	11-12/22	4042	4648	BURRELL		24ChDrB	1/14	2417A	4239
BRYSON		12SecDr	4-5/13	2410	4106	BURROUGHS		16Sec	12/17-18	2412F	4528
BUCCANEER		10SecDr2Cpt	12/15	2585C	4351	BURRWOOD	N	12SecDr	12/13/10	2410	3864
BUCEPHALUS		12SecDr	7/13	2410A	4179	BUSHARD		12SecDr	1-3/14	2410A	4249
BUCHANAN		12SecDr	10/12/10	2410	3813	BUSHNELL		16Sec	2/13	2412	4101
BUCHANAN	N	12SecDr	5-8/25	3410A	4868	BUSHROD		26ChDr	4-5/17	2416C	4492
BUCHARIA		16Sec	6-7/20	2412F	4568	BUSHWICK		12SecDr	7-8/13	2410A	4192
BUCKEYE		12SecDr	4-6/14	2410A	4271	BUSKIRK	N	12SecDr	11/8/10	2410	3860
BUCKEYE STATE		BaggBSmkBarber	10-11/25	3951	4885	BUSTLETON		12SecDrCpt	6/23/10	2411	3800
BUCKHORN		12SecDr	6-9/26	3410A	4969	BUSWELL	N	12SecDr	11/14/10	2410	3860
BUCKLAND		12SecDr	10-11/11	2410	3936	BUTLER		12SecDr	3/18/10	2410	3794
BUCKNER		12SecDr	10-12/16	2410F	4431	BUTLER UNIVERSITY	RN	12Sec2Dbr	7-9/17	4046	4515
BUCYRUS		12SecDrCpt	5/17	2411C	4494	BUTTERCUP		28ChDr	5-7/26	3416	4958
BUENOS AIRES NdeM-110	SN	6Sec6Dbr	4-6/24	4084A	4763	BUTTERWORTH		12SecDr	10/4/10	2410	3813
BUFFALO		25ChDrObs	12/12	2669A	4053	BUTTONWOOD		12SecDr	11/25/10	2410	3864
BUFFALO CLUB		8SecBLngSunRm	6-7/30	3989B	6362	BUTTRICK		16Sec	12/17-18	2412F	4528
BUFFINGTON		BaggBClubSmk	1/12	2415	3946	BUXAR		12SecDr	1/16	2410E	4354
BUFORD		16Sec	2/13	2412	4101	BUZZARDS BAY		36Ch	8/27	3916A	6077
BULGARIA		12SecDr	8-9/20	2410F	4574	BUZZARDS BAY	R	22Ch12StsBObsLng	8/27	4040	6077
BULLOCH	N	12SecDr	9/13/10	2410	3794	BYERS	N	12SecDr	3/10/10	2410	3769

Figure I-48   In March 1952, the CALEXICO, shown here standing in San Francisco, appears to be in excellent condition. In February 1958, it was placed in Government storage at Pueblo Ordnance Depot, Colorado, where it remained until July 1962 when it was sold for scrap to Midwest Steel.   Wilber C. Whittaker photograph.

CAR NAME	ST	CAR TYPE	BUILT	PLAN	LOT
BYERTON		12SecDr	3-6/20	2410F	4565
BYWATER		12SecDr	2-5/21	2410F	4612
BYZANTINE		12SecDr	5-7/13	2410	4149
CABAZON		12SecDrCpt	6/13	2411A	4159
CABELL		16Sec	1-5/18	2412F	4531
CABEZON PEAK		10Sec2Dr	9-10/29	3584B	6284
CABILDO		12SecDr	8-10/26	3410A	4945
CABIN CREEK		12SecDr	2-3/13	2410	4105
CABO					
CATOCHE NdeM-111	SN	14Sec	7/30	3958A	6376
CORRIENTES NdeM-112	SN	14Sec	8-10/28	3958A	6181
SAN LUCAS NdeM-113	SN	12SecDr	2-5/25	3410	4845
CABORCA FCS-BC	L	10SecDr2Cpt	12/25-26	3585K	4933
CABOT		12SecDr	10/22-23	2410H	4647
CABRA		16Sec	1-5/18	2412F	4531
CACAHUAMILPA NdeM	SN	8SecDr2Cpt	8-9/29	3979A	6283
CACHALOT		12SecDr	3-6/15	2410B	4319
CADBURY		12SecDr	1-3/14	2410A	4249
CADDO		12SecDr	8-10/26	3410A	4945
CADESIA		16Sec	1/16	2412E	4352
CADET		12SecDrCpt	6/13	2411	4151
CADET		BaggBClubSmk	2-4/09	2136C	3660
CADIZ		16Sec	4/17	2412F	4484
CADMUS		12SecDr	8-9/20	2410F	4574
CADWALLADER					
C. WASHBURN	N	6Sbr2DbrObsLng	8/28	3974D	6183
CAESAR RODNEY		30ChDr	9-10/24	3418	4783

CAR NAME	ST	CAR TYPE	BUILT	PLAN	LOT
CAHOKIA		12SecDr	10-12/26	3410A	6023
CAJON		7Cpt2Dr SBS	11/11	2522	3941
CALABAR		10SecDrCpt	4-6/27	3973	6043
CALAFIA		12SecDr	8-10/26	3410A	4945
CALAIS		12SecDr	6-7/16	2410F	4385
CALAMARES		12SecDr	1-3/14	2410A	4249
CALAMUS		10SecDr2Cpt	10-11/12	2585	4067
CALAVERAS		12SecDr	8-10/26	3410A	4945
CALCASIEU		12SecDr	6-7/17	2410F	4503
CALCUTTA		14SecDr	8/13	2762	4193
CALDWELL	N	12SecDr	5/12/10	2410	3773
CALEB STRONG		30ChDr	2-3/30	4000	6324
CALEDONIA		12SecDr	10/25/10	2410	3860
CALENA		26ChDr	4-5/16	2416C	4366
CALENDULA		12SecDr	2-3/13	2410	4105
CALERA		14SecDr	8/13	2762	4193
CALEXICO		10SecDr2Cpt	1-2/13	2585	4091
CALGARY		16Sec	1-5/18	2412F	4531
CALHOUN		12SecDr	1-2/16	2410E	4356
CALKINS	N	16Sec	4/17	2412F	4484
CALLA LILY		34Ch	4/26	3419	4956
CALLISON		10SecDr2Cpt	1-2/13	2585	4091
CALLISTO		12SecDr	1-2/22	2410F	4625
CALLOWAY		16Sec	1-5/18	2412F	4531
CALNEVA		12SecDr	5-7/13	2410	4149
CALVERT		26ChDr	3/13	2416	4107
CALVERTON		24ChDrB	12/10	2417	3863

CAR NAME	ST	CAR TYPE	BUILT	PLAN	LOT
CALVERTON	R	26StsLngBDr	12/10	2417J	3863
CALVIN		12SecDr	4-5/11	2410	3893
CALVIN COOLIDGE	N	12SecDr	1-2/29	3410B	6220
CALYDON		12SecDr	8-9/20	2410F	4574
CALZONA		10SecDr2Cpt	8-9/13	2585A	4201
CAMAGUEY		23StsBLng	11/26	3967	4979
CAMALOA		12SecDr	2-5/21	2410F	4612
CAMARGO		12SecDr	5-7/13	2410	4149
CAMBRAY		12SecDrCpt	6/13	2411A	4159
CAMBRIDGE		12SecDr	11/21/10	2410	3864
CAMDEN		26ChDr	5/12	2416	3986
CAMERON PASS		8SecDr2Cpt	12/29-30	3979A	6334
CAMERONIA		12SecDr	5-7/13	2410	4149
CAMMACK		12SecDr	10-11/11	2410	3936
CAMP BRAGG		10SecDr2Cpt	11-12/21	2585D	4628
CAMP BULLIS		10SecDr2Cpt	11-12/21	2585D	4628
CAMP CARLIN		10SecDr2Cpt	11-12/21	2585D	4628
CAMP CODY		10SecDr2Cpt	11-12/21	2585D	4628
CAMP DEVENS		10SecDr2Cpt	11-12/21	2585D	4628
CAMP DIX		10SecDr2Cpt	11-12/21	2585D	4628
CAMP EUSTIS		10SecDr2Cpt	11-12/21	2585D	4628
CAMP FORREST		10SecDr2Cpt	11-12/21	2585D	4628
CAMP FREMONT		10SecDr2Cpt	11-12/21	2585D	4628
CAMP FUNSTON		10SecDr2Cpt	11-12/21	2585D	4628
CAMP GLENN		10SecDr2Cpt	11-12/21	2585D	4628
CAMP KNOX		10SecDr2Cpt	11-12/21	2585D	4628
CAMP LEE		10SecDr2Cpt	11-12/21	2585D	4628
CAMP LEWIS		10SecDr2Cpt	11-12/21	2585D	4628
CAMP MABRY		10SecDr2Cpt	11-12/21	2585D	4628
CAMP McCOY		10SecDr2Cpt	11-12/21	2585D	4628
CAMP MEIGS		10SecDr2Cpt	11-12/21	2585D	4628
CAMP MERRITT		10SecDr2Cpt	11-12/21	2585D	4628
CAMP MILLS		10SecDr2Cpt	11-12/21	2585D	4628
CAMP PIKE		10SecDr2Cpt	11-12/21	2585D	4628
CAMP POLK		10SecDr2Cpt	11-12/21	2585D	4628
CAMP SEVIER		10SecDr2Cpt	11-12/21	2585D	4628
CAMP UPTON		10SecDr2Cpt	11-12/21	2585D	4628
CAMP VAIL		10SecDr2Cpt	11-12/21	2585D	4628
CAMP WHEELER		10SecDr2Cpt	11-12/21	2585D	4628
CAMPANULA		10SecDr2Cpt	12/15	2585C	4351
CAMPBELL		12SecDrCpt	7/7/10	2411	3800
CAMPBELL WALLACE	N	10Sec2Dr	9-10/25	3584A	4899
CAMPECHE		4Cpt2DrLngObs	9/16	2950	4423
CAMPECHE NdeM-114	SN	8SecDr2Cpt	12/29-30	3979A	6334
CAMPERDOWN		12SecDr	10-12/15	2410D	4338
CAMPODONICO FCP-218	SN	10SecDr2Cpt	12/25-26	3585A	4933
CAMPUS		16Sec	2/13	2412	4101
CAMROSE		12SecDr	11-12/13	2410A	4234
CAMULOS		12SecDr	8-9/20	2410F	4574
CANAAN		36Ch SBS	1/13	2691	4056

CAR NAME	ST	CAR TYPE	BUILT	PLAN	LOT
CANADA		12SecDr	8-9/20	2410F	4574
CANAJOHARIE		12SecDr	4-6/14	2410A	4271
CANANDAIGUA		12SecDr	11/10/10	2410	3860
CANATHA		7Cpt2Dr	12/11	2522	3943
CANDACE		26ChDr	11/12	2416	4052
CANDELARIA FC-DS	SN	8Sec5Dbr	2-3/15	4036E	4318
CANETA		12SecDr	1-3/17	2410F	4450
CANFIELD		12SecDr	4-5/11	2410	3893
CANISTEO		16Sec	4/17	2412F	4484
CANNA		28ChDr	5-7/26	3416	4958
CANNES		12SecDr	6-7/16	2410F	4385
CANNON		26ChDr	4-5/17	2416C	4492
CANON CITY		10SecLngObs	9-10/14	2521A	4292
CANONBURY TOWER	RN	8SecDr3Dbr	5-6/25	4090	4846
CANONICUS		12SecDr	7-8/15	2410D	4322
CANONSBURG		12SecDr	10/15/10	2410	3813
CANOPIC		12SecDr	5-7/13	2410	4149
CANTERBURY		12SecDr	11-12/13	2410A	4234
CANTERBURY CLUB	RN	8SecRestObsLng	9-10/14	4025A	4292
CANTINFLAS FCP-225	SN	10SecDr2Cpt	8-10/23	3585	4725
CANTON		12SecDr	5-6/23	2410H	4699
CANTRALL		16Sec	6-7/20	2412F	4568
CANUTE		12SecDr	8-9/20	2410F	4574
CANVEY		12SecDr	2-5/21	2410F	4612
CANWOOD		12SecDr	8-9/20	2410F	4574
CAPE ALAVA		10SecDr2Cpt	11-12/22	2585D	4648
CAPE ANN		10SecDr2Cpt	11-12/22	2585D	4648
CAPE ARAGO		10SecDr2Cpt	11-12/22	2585D	4648
CAPE CHARLES		10SecDr2Cpt	11-12/22	2585D	4648
CAPE COD		10SecDr2Cpt	11-12/22	2585D	4648
CAPE FEAR		10SecDr2Cpt	11-12/22	2585D	4648
CAPE FERRELO		10SecDr2Cpt	11-12/22	2585D	4648
CAPE FLATTERY		10SecDr2Cpt	11-12/22	2585D	4648
CAPE FORTUNAS		10SecDr2Cpt	11-12/22	2585D	4648
CAPE HENRY		10SecDr2Cpt	11-12/22	2585D	4648
CAPE KIWANDA		10SecDr2Cpt	11-12/22	2585D	4648
CAPE LOOKOUT		10SecDr2Cpt	11-12/22	2585D	4648
CAPE MAY		26ChDr	5/12	2416	3986
CAPE MEARES		10SecDr2Cpt	11-12/22	2585D	4648
CAPE NEDDICK		10SecDr2Cpt	11-12/22	2585D	4648
CAPE POGE		10SecDr2Cpt	11-12/22	2585D	4648
CAPE PORPOISE		10SecDr2Cpt	11-12/22	2585D	4648
CAPE ROMANO		10SecDr2Cpt	11-12/22	2585D	4648
CAPE ROZIER		10SecDr2Cpt	11-12/22	2585D	4648
CAPE SABLE		10SecDr2Cpt	11-12/22	2585D	4648
CAPE SAN BLAS		10SecDr2Cpt	11-12/22	2585D	4648
CAPHAR		7Cpt2Dr	12/11	2522	3943
CAPIDON		12SecDr	2-5/21	2410F	4612
CAPIRA		12SecDr	11-12/13	2410A	4234
CAPISTRANO		12SecDr	9-10/15	2410D	4327

CAR NAME	ST	CAR TYPE	BUILT	PLAN	LOT
CAPITOL ARMS		3CptDrLngBSunRm	7/29	3975C	6275
CAPITOL BRIDGE		BaggBSmkBarber	10/25	3951	4885
CAPITOL CITY		BaggBSmkBarber	10/25	3951	4885
CAPITOL CITY		3CptDrLngBSunRm	7/5/29	3975C	6275
CAPITOL COURIER		3CptDrLngBSunRm	7/6/29	3975C	6275
CAPITOL ESCORT		3CptDrLngBSunRm	7/10/29	3975C	6275
CAPITOL GARDEN		BaggBSmkBarber	10-11/25	3951	4885
CAPITOL HEIGHTS	N	10SecLngObs	11-12/25	3521A	4923
CAPITOL HILL	N	BaggBSmkBarber	10-11/25	3951	4885
CAPITOL HOME		BaggBSmkBarber	10-11/25	3951	4885
CAPITOL PARK	N	10SecLngObs	11-12/25	3521A	4923
CAPITOL ROAD	N	10SecLngObs	12/12/25	3521A	4923
CAPITOL SQUARE		3CptDrLngBSunRm	7/9/29	3975C	6275
CAPITOL VIEW	N	10SecLngObs	11-12/25	3521A	4923
CAPRICHO	N	10SecDrCpt	8/29	3973A	6273
CAPRICHO FCP-204	SN	10SecDr2Cpt	5-6/24	3585	4770
CAPRICORN		4Cpt2DrLngObs	9/16	2950	4423
CAPTAIN		6Sbr7StsCafeLng	8/27	3976	6058
CAPTAIN JOHN SUTTER	N	12SecDr	2-4/24	3410	4762
CAPTIVA		12SecDr	12/12-13	2410	4076
CAPULET		12SecDr	1-2/16	2410E	4356
CARA NOME #4	LN	Exhibit (Rexall)	6/16	2916B	4389
CARACAS NdeM	SN	6Sec6Dbr	12/12-13	4084C	4076
CARBERY		12SecDr	11-12/14	2410B	4297
CARDINAL BONZANO	N	10Cpt	8/16/10	2505	3833
CARDINAL HAYES	N	6Cpt3Dr	10-12/25	3523A	4922
CARIBBEAN		4Cpt2DrLngObs	9/16	2950	4423
CARIBOU PASS		8SecDr2Cpt	2-3/30	3979A	6353
CARLETON		12SecDr	8-10/26	3410A	4945
CARLETON CLUB	RN	8SecRestObsLng	6-7/23	4025H	4690
CARLISLE		12SecDr	12/12-13	2410	4076
CARLOTTA	N	26ChDr	7/2/10	2416	3804
CARNEGIE		12SecDr	1/27/10	2410	3723
CARNFORTH		12SecDr	5-7/13	2410	4149
CAROL BOARDMAN	N	8SecDr2Cpt	12/29-30	3979A	6334
CARPENTER		6CptLngObs	8/5/10	2413	3802
CARPENTER'S HALL		4CptLngObs	10/25	3960	4889
CARRANZA	N	10SecDrCpt	8/29	3973A	6273
CARRANZA FCP-200	SN	10SecDr2Cpt	12/23-24	3585	4728
CARRINGTON		12SecDr	10-11/11	2410	3936
CARROLLTON		12SecDr	10/28/10	2410	3860
CARROTHERS		12SecDr	10/1/10	2410	3813
CARROW		12SecDr	3-6/20	2410F	4565
CARSON-NEWMAN COLLEGE	N	10SecDr2Cpt	1-2/24	3585	4743
CARTER LAKE		10SecDrCpt	10-12/29	3973A	6300
CARTER'S GROVE	N	6SecDr2Dbr2Cpt	6/11	2547C	3892
CARTHAGE		12SecDr	2-5/21	2410F	4612
CARVER		12SecDr	5-8/18	2410F	4540
CARVILLE		12SecDr	2-4/24	3410	4762
CARYL		12SecDr	1-4/12	2410	3949

CAR NAME	ST	CAR TYPE	BUILT	PLAN	LOT
CASA BLANCA		10SecLngObs	9-10/14	2521A	4292
CASA MONICA		10Sec2Dr	2/13	2584	4092
CASCADES	RN	8Rmt3Dbr2Dr	12/26-27	4176	6031
CASCO		12SecDr	5-8/18	2410F	4540
CASHMERE		12SecDr	4-6/24	3410	4763
CASPAR		12SecDr	8-9/20	2410F	4574
CASSADAGA		12SecDr	1-4/12	2410	3949
CASSELTON		12SecDr	4-6/24	3410	4763
CASSIUS M. CLAY	N	12SecDr	8-10/26	3410A	4945
CASSOPOLIS		12SecDr	2-5/21	2410F	4612
CASSVILLE		12SecDr	4-6/14	2410A	4271
CASTALIA		12SecDr	11/18/10	2410	3860
CASTINE		12SecDr	10-12/26	3410A	6023
CASTLE CREST	RN	8Sec5Dbr	9-10/17	4036J	4525
CASTLE GATE		10SecLngObs	9-10/14	2521A	4292
CASTLE PEAK		10Sec2Dr	12/26-27	3584A	6031
CASTLE RANGE	RN	8Sec5Dbr	1-3/17	4036J	4450
CASTLE ROCK		4Cpt2DrLngObs	3/13	2703	4100
CASTLE TOWER	RN	8SecDr3Dbr	8-12/25	4090F	4894
CASTLEFORD		12SecDr	6-7/17	2410F	4503
CASTLETON		12SecDr	1-2/30	3410B	6351
CASTLINE		12SecDr	4-5/11	2410	3893
CASTOR		12SecDr	2-5/21	2410F	4612
CASWELL		12SecDr	1-2/15	2410B	4311
CATALINA		LngBDorm(5Sec)	7-8/27	3981	6061
CATARACT		12SecDr	4-5/11	2410	3893
CATAWBA		12SecDr	4-5/11	2410	3893
CATAWISSA		26ChDr	7/27/10	2416	3804
CATEMACO NdeM	SN	14Sec	9-10/25	3958	4869
CATHARPIN		12SecDr	4-5/13	2410	4106
CATHEDRAL PEAK		10Sec2Dr	12/26-27	3584A	6031
CATHERINE		28ChDr	6-7/27	3416A	6087
CATHMAID		12SecDr	11-12/13	2410A	4234
CATLIN		16Sec	6-7/20	2412F	4568
CATORCE	N	14SecDr	8/13	2762	4193
CATSKILL VALLEY		DrSbrBLngObs	4/29	3988	6221
CAVENDISH		12SecDr	4-5/11	2410	3893
CAWDOR		12SecDr	1-3/17	2410F	4450
CAWNPORE		12SecDr	10-12/15	2410D	4338
CAYENNE		12SAecDr	8-9/20	2410F	4574
CAZADERO		10SecDr2Cpt	1-2/13	2585	4091
CEBORUCO FCP-226	SN	10SecDr2Cpt	12/25-26	3585G	4933
CECELIA	N	26ChDr	7/26/10	2416	3804
CECIL		12SecDr	5-6/23	2410H	4699
CEDAR FALLS		12SecDr	8-9/12	2410	4035
CEDAR GROVE		10Sec2Dr	1/12	2584	3945
CEDAR LANE		10SecDr2Cpt	7-8/23	2585D	4707
CEDAR LEDGE		10SecDr2Cpt	7-8/23	2585D	4707
CEDAR MOUNTAIN		BaggBSmkBarber	12/26-27	3951C	6022
CEDAR PASS		8SecDr2Cpt	12/29-30	3979A	6334

CAR NAME	ST	CAR TYPE	BUILT	PLAN	LOT
CEDAR POINT		4Cpt2DrLngObs	3/13	2703	4100
CEDAR RAPIDS	N	10Sec3Dbr	3/11	3411C	3880
CEDAR RIVER		10Sec2Dr	5/13	2584	4140
CEDAR SPRINGS		10SecDr2Cpt	7-8/23	2585D	4707
CEDARTOWN		12SecDr	6-7/16	2410F	4385
CEDRIC		12SecDr	12/10-11	2410	3866
CELESTE		28ChDr	4-6/25	3416	4864
CELIA		28ChDr	6-7/30	3416B	6372
CELINDA		24ChDrB	6-7/14	2417B	4265
CEMONTON		12SecDr	1-4/12	2410	3949
CENDRILLON		12SecDr	2-3/13	2410	4105
CENTABELLA		8SecDr2Cpt	10-12/28	3979A	6205
CENTACORRA		8SecDr2Cpt	10-12/28	3979A	6205
CENTACRE		8SecDr2Cpt	8-9/29	3979A	6283
CENTAGA		8SecDr2Cpt	10-12/28	3979A	6205
CENTALASKA		8SecDr2Cpt	10-12/28	3979A	6205
CENTALBA		8SecDr2Cpt	10-12/28	3979A	6205
CENTALPINA		8SecDr2Cpt	10-12/28	3979A	6205
CENTAMIA		8SecDr2Cpt	10-12/28	3979A	6205
CENTAMURA		8SecDr2Cpt	10-12/28	3979A	6205
CENTANDA		8SecDr2Cpt	10-12/28	3979A	6205
CENTANOVA		8SecDr2Cpt	10-12/28	3979A	6205
CENTAQUA		8SecDr2Cpt	10-12/28	3979A	6205
CENTARCH		8SecDr2Cpt	10-12/28	3979A	6205
CENTASCA		8SecDr2Cpt	10-12/28	3979A	6205
CENTASH		8SecDr2Cpt	10-12/28	3979A	6205
CENTAURORA		8SecDr2Cpt	10-12/28	3979A	6205
CENTAVERA		8SecDr2Cpt	10-12/28	3979A	6205
CENTAVIA		8SecDr2Cpt	10-12/28	3979A	6205
CENTAVILLA		8SecDr2Cpt	10-12/28	3979A	6205
CENTAVISTA		8SecDr2Cpt	10-12/28	3979A	6205
CENTAVON		8SecDr2Cpt	3/29	3979A	6237
CENTAXIA		8SecDr2Cpt	10-12/28	3979A	6205
CENTAYA		8SecDr2Cpt	10-12/28	3979A	6205
CENTBROOK		8SecDr2Cpt	10-12/28	3979A	6205
CENTBURNE		8SecDr2Cpt	10-12/28	3979A	6205
CENTCALLE		8SecDr2Cpt	10-12/28	3979A	6205
CENTCAMPO		8SecDr2Cpt	10-12/28	3979A	6205
CENTCREST		8SecDr2Cpt	10-12/28	3979A	6205
CENTCROFT		8SecDr2Cpt	8-9/29	3979A	6283
CENTDALE		8SecDr2Cpt	8-9/29	3979A	6283
CENTDORE		8SecDr2Cpt	10-12/28	3979A	6205
CENTDOYA		8SecDr2Cpt	10-12/28	3979A	6205
CENTELLO		8SecDr2Cpt	8-9/29	3979A	6283
CENTELM		8SecDr2Cpt	10-12/28	3979A	6205
CENTENNIAL		12SecDr	7/13	2410A	4179
CENTER PARK		12SecDrCpt	3/11	2411	3880
CENTERBURG		12SecDr	1-4/12	2410	3949
CENTERPOINT		12SecDr	10/4/10	2410	3813
CENTERPOINT	N	12SecDr	10/14/10	2410	3814

CAR NAME	ST	CAR TYPE	BUILT	PLAN	LOT
CENTERVILLE	N	30ChDr	9-10/24	3418	4783
CENTESIMA		8SecDr2Cpt	10-12/28	3979A	6205
CENTESSA		8SecDr2Cpt	10-12/28	3979A	6205
CENTFAUN		8SecDr2Cpt	10-12/28	3979A	6205
CENTFORD		8SecDr2Cpt	3/29	3979A	6237
CENTGARDE		8SecDr2Cpt	10-12/28	3979A	6205
CENTGATE		8SecDr2Cpt	10-12/28	3979A	6205
CENTGLEN		8SecDr2Cpt	10-12/28	3979A	6205
CENTHAM		8SecDr2Cpt	3/29	3979A	6237
CENTHILL		8SecDr2Cpt	10-12/28	3979A	6205
CENTHOLM		8SecDr2Cpt	8-9/29	3979A	6283
CENTIDA		8SecDr2Cpt	10-12/28	3979A	6205
CENTILLA		8SecDr2Cpt	10-12/28	3979A	6205
CENTLAWN		8SecDr2Cpt	10-12/28	3979A	6205
CENTLEA		8SecDr2Cpt	10-12/28	3979A	6205
CENTLOCH		8SecDr2Cpt	10-12/28	3979A	6205
CENTLONA		8SecDr2Cpt	10-12/28	3979A	6205
CENTLOW		8SecDr2Cpt	8-9/29	3979A	6283
CENTLYN		8SecDr2Cpt	3/29	3979A	6237
CENTMOOR		8SecDr2Cpt	3/29	3979A	6237
CENTMORA		8SecDr2Cpt	10-12/28	3979A	6205
CENTMOUNT		8SecDr2Cpt	10-12/28	3979A	6205
CENTNOME		8SecDr2Cpt	10-12/28	3979A	6205
CENTOAK		8SecDr2Cpt	10-12/28	3979A	6205
CENTOLIO		8SecDr2Cpt	10-12/28	3979A	6205
CENTONDRIA		8SecDr2Cpt	10-12/28	3979A	6205
CENTONIA		8SecDr2Cpt	10-12/28	3979A	6205
CENTOSA		8SecDr2Cpt	8-9/29	3979A	6283
CENTPAULO		8SecDr2Cpt	10-12/28	3979A	6205
CENTQUORA		8SecDr2Cpt	10-12/28	3979A	6205
CENTRAIL		8SecDr2Cpt	10-12/28	3979A	6205
CENTRAL AMERICA		3Cpt2DrLngObs	8-10/25	3959	4886
CENTRAL AVENUE		3Cpt2DrLngObs	8-10/25	3959	4886
CENTRAL BRIDGE		3Cpt2DrLngObs	8-10/25	3959	4886
CENTRAL BUTTE		3Cpt2DrLngObs	8-10/25	3959	4886
CENTRAL CITY		3Cpt2DrLngObs	8-10/25	3959	4886
CENTRAL COLLEGE		3Cpt2DrLngObs	8-10/25	3959	4886
CENTRAL FALLS		3Cpt2DrLngObs	8-10/25	3959	4886
CENTRAL FERRY		3Cpt2DrLngObs	8-10/25	3959	4886
CENTRAL GROVE		3Cpt2DrLngObs	8-10/25	3959	4886
CENTRAL HOUSE		3Cpt2DrLngObs	8-10/25	3959	4886
CENTRAL JUNCTION		3Cpt2DrLngObs	8-10/25	3959	4886
CENTRAL LAKE		3Cpt2DrLngObs	8-10/25	3959	4886
CENTRAL MILLS		3Cpt2DrLngObs	8-10/25	3959	4886
CENTRAL MINE		3Cpt2DrLngObs	8-10/25	3959	4886
CENTRAL MOUNTAINS		3Cpt2DrLngObs	9-10/26	3959	4971
CENTRAL PARK		3Cpt2DrLngObs	8-10/25	3959	4886
CENTRAL PLAINS		3Cpt2DrLngObs	9-10/26	3959	4971
CENTRAL PLATEAU		3Cpt2DrLngObs	9-10/26	3959	4971
CENTRAL POINT		3Cpt2DrLngObs	9-10/26	3959	4971

Figure I-49   CENTRAL PARK was 1 of 15 3-compartment, 2-drawing room / lounge observation cars built in 1925 for "The Twentieth Century Limited." It is fortunate for rail enthusiasts that this car is in the custody of The National Railway Historical Society, Lake Shore Chapter, who will, we hope, extend its life well into the future. A Howard Ameling Photograph.

CAR NAME	ST	CAR TYPE	BUILT	PLAN	LOT	CAR NAME	ST	CAR TYPE	BUILT	PLAN	LOT
CENTRAL PROVINCES		3Cpt2DrLngObs	8-10/25	3959	4886	CHADDS FORD		24ChDrB	3/14	2417A	4250
CENTRAL ROUTE		3Cpt2DrLngObs	8-10/25	3959	4886	CHADRON		12SecDr	8-9/12	2410	4035
CENTRAL SQUARE		3Cpt2DrLngObs	8-10/25	3959	4886	CHADWICK		16Sec	10/13	2412B	4216
CENTRAL STATE		3Cpt2DrLngObs	8-10/25	3959	4886	CHAFFEE		12SecDr	1-3/17	2410F	4450
CENTRAL VALLEY		3Cpt2DrLngObs	8-10/25	3959	4886	CHAGRES		12SecDr	5-7/13	2410	4149
CENTRAL VILLAGE		3Cpt2DrLngObs	9-10/26	3959	4971	CHALDEA		12SecDr	3-6/20	2410F	4565
CENTRANGE		8SecDr2Cpt	10-12/28	3979A	6205	CHALMETTE		12SecDr	8-10/26	3410A	4945
CENTREVILLE		26ChDr	7/20/10	2416	3804	CHALONS		7Cpt2Dr	11/17	2522B	4529
CENTRIDGE		8SecDr2Cpt	10-12/28	3979A	6205	CHAMANADE		12SecDr	2-3/15	2410B	4318
CENTROCK		8SecDr2Cpt	8-9/29	3979A	6283	CHAMBERS		12SecDrCpt	7/1/10	2411	3800
CENTROSS		8SecDr2Cpt	3/29	3979A	6237	CHAMBERSBURG		12SecDr	5-6/23	2410H	4699
CENTSALVA		8SecDr2Cpt	10-12/28	3979A	6205	CHAMIER		16Sec	1-5/18	2412F	4531
CENTSHIRE		8SecDr2Cpt	8-9/29	3979A	6283	CHAMITA		26ChDr	6/11	2416	3902
CENTSOMA		8SecDr2Cpt	10-12/28	3979A	6205	CHAMP CLARK	N	12SecDr	4-6/21	2410F	4613
CENTSPUR		8SecDr2Cpt	3/29	3979A	6237	CHAMPOTON NdeM-115	SN	8SecDr2Cpt	12/29-30	3979A	6334
CENTSTAR		8SecDr2Cpt	10-12/28	3979A	6205	CHAMPOTON FC-DS	SN	8Sec5Dbr	6/13	4036A	4159
CENTSYLVIA		8SecDr2Cpt	10-12/28	3979A	6205	CHANCELLOR LIVINGSTON	N	12SecDr	2-5/25	3410	4845
CENTVERN		8SecDr2Cpt	3/29	3979A	6237	CHANCELLOR LIVINGSTON	R	10SecDr2Dbr	2-5/25	4074E	4845
CENTWAY		8SecDr2Cpt	10-12/28	3979A	6205	CHANDA		12SecDr	10-12/15	2410D	4338
CENTWELL		8SecDr2Cpt	8-9/29	3979A	6283	CHANDLER		7Cpt2Dr	11/17	2522B	4529
CENTWICK		8SecDr2Cpt	3/29	3979A	6237	CHANDOS		12SecDr	8-9/20	2410F	4574
CENTWOOD		8SecDr2Cpt	10-12/28	3979A	6205	CHANTILLY		12SecDr	4-5/13	2410	4106
CERVANTES		6Cpt3Dr	11-12/23	3523	4726	CHANTREY		7Cpt2Dr	11/17	2522B	4529
CESSNA		12SecDr	5-6/23	2410H	4699	CHANUTE		10SecDr2Cpt	8-9/13	2585A	4201
CEYLON		12SecDr	4-5/11	2410	3893	CHAPEL HILL		10SecLngObs	12/15	2521B	4362
CHABERT		12SecDr	6-7/16	2410F	4385	CHAPIN		12SecDr	3-4/11	2410	3881
CHACAMAX FC-DS	N	8Sec5Dbr	3/11	4036B	3880	CHAPMAN		12SecDr	4-5/13	2410	4106

CAR NAME	ST	CAR TYPE	BUILT	PLAN	LOT
CHAPPAQUA		12SecDr	1-4/12	2410	3949
CHARFORD		7Cpt2Dr	12/11	2522	3943
CHARING CROSS		16Sec	10-11/14	2412C	4296
CHARLECOTE		12SecDr	6-7/17	2410F	4503
CHARLES A. BROADWATER		8SecDr2Cpt	5-6/29	3979A	6261
CHARLES CARROLL		30ChDr	9-10/24	3418	4783
CHARLES CARROLL					
OF CARROLLTON	N	BaggBSmkBarber	10-11/25	3951B	4885
CHARLES E. PERKINS		8SecDr2Cpt	12/29-30	3979A	6334
CHARLES H. COSTER	N	10Sec2Dr	9-10/25	3584A	4899
CHARLES PINCKNEY		12SecDr	5-8/25	3410A	4868
CHARLES RIVER		18ChBLngObs	6/16	2918	4391
CHARLES THOMPSON	N	12SecDr	1-3/25	3410	4844
CHARLES THOMSON	N	12SecDr	1-3/25	3410	4844
CHARLES THOMSON	R	10SecDr2Dbr	1-3/25	4074E	4844
CHARLESTON		12SecDr	12/10-11	2410	3866
CHARLESWORTH		12SecDr	4-5/11	2410	3893
CHARLOTTE		26ChDr	3/13	2416	4107
CHARLOTTESVILLE		12SecDr	11-12/13	2410A	4234
CHARLTON		12SecDr	3-4/11	2410	3881
CHARMION		26ChDr	4-5/16	2416C	4366
CHARTER OAK		18ChBLngObs	6/16	2918	4391
CHASE		26ChDr	4-5/17	2416C	4492
CHASELAND		12SecDr	10/5/10	2410	3813
CHASM FALLS		12SecDrCpt	3/11	2411	3880
CHASM LAKE		10SecDrCpt	10-12/29	3973A	6300
CHATAYA		12SecDr	11-12/13	2410A	4234
CHATHAM		10SecDr2Cpt	11/17	2585D	4527
CHATSWORTH		12SecDr	7/13	2410A	4179
CHATTANOOGA		12SecDr	9-10/13	2410A	4215
CHATTAROY		12SecDr	2-5/21	2410F	4612
CHATTOLANEE		12SecDr	2-4/24	3410	4762
CHATWOLD		12SecDr	1-3/17	2410F	4450
CHAUCER		6Cpt3Dr	11-12/23	3523	4726
CHAUDIER		12SecDr	1-3/14	2410A	4249
CHAUMONT		6CptLngObs	8/13/10	2413	3802
CHAUNCEY		12SecDr	1-4/12	2410	3949
CHAUTAUQUA		12SecDr	4-6/14	2410A	4271
CHAUTAUQUA	N	12SecDr	10-12/26	3410A	6023
CHAUVIN		12SecDr	11-12/13	2410A	4234
CHAVEZ		12SecDr	1-3/17	2410F	4450
CHELTONIA		16Sec	6-7/20	2412F	4568
CHEMUNG		10SecDr2Cpt	11-12/11	2585	3942
CHENANGO		16Sec	4/17	2412F	4484
CHENEQUA		8SecDr2Cpt	7-8/27	3979	6052
CHEPSTOW		12SecDr	7-8/13	2410A	4192
CHERAW		10SecDr2Cpt	8-9/13	2585A	4201
CHERBOURG		12SecDr	8-9/20	2410F	4574
CHERIBON		12SecDr	2-3/13	2410	4105
CHEROKEE	N	12SecDr	1-3/14	2410A	4249

CAR NAME	ST	CAR TYPE	BUILT	PLAN	LOT
CHERRYVALE		12SecDr	7-8/14	2410A	4277
CHERYL		30ChDr	11/12	2668	4051
CHESANING		12SecDr	1-4/12	2410	3949
CHESAPEAKE	N	12SecDr	1-3/14	2410A	4249
CHESHIRE		12SecDr	8-9/20	2410F	4574
CHESILHURST		28ChDr	11/22/10	2418	3862
CHESNAY		12SecDr	5-7/13	2410	4149
CHESTER		26ChDr	5/12	2416	3986
CHESTERFIELD		12SecDr	4-5/11	2410	3893
CHESTERWOOD		12SecDr	6-7/17	2410F	4503
CHESTNUT HILL		26ChDr	8/6/10	2416	3804
CHESWOLD		10Cpt	4/13	2505	4120
CHETOPA		12SecDr	10/14	2410B	4293
CHETUMAL NdeM-116	SN	12SecDr	2-5/25	3410	4845
CHEVALIER		10SecLngObs	12/15	2521B	4353
CHEVRON		12SecDr	11-12/14	2410B	4297
CHEVY CHASE		12SecDr	4/26	3410A	4943
CHEWACLA	P	12SecDr	2/21	2410G	4598
CHEYENNE		12SecDr	5-8/18	2410F	4540
CHEYNEY		12SecDr	6-9/26	3410A	4969
CHICAGO		Private (GS)	1/14	2492A	4210
CHICAGO CLUB		8SecBLngSunRm	6-7/30	3989B	6362
CHICKASHA		12SecDr	2-5/21	2410F	4612
CHICKIES		12SecDrCpt	7/1/10	2411	3800
CHICONTEPEC NdeM-118	SN	12SecDr	2-5/25	3410	4845
CHIEF AMERICAN HORSE		10SecDrCpt	3-5/30	3973A	6338
CHIEF ARLEE		10SecDrCpt	3-5/30	3973A	6338
CHIEF BAPTISTA		10SecDrCpt	3-5/30	3973A	6338
CHIEF BIG MEDICINE		10SecDrCpt	3-5/30	3973A	6338
CHIEF BIG WHITE		10SecDrCpt	3-5/30	3973A	6338
CHIEF BLACK BEAR		10SecDrCpt	3-5/30	3973A	6338
CHIEF BLACK FOOT		10SecDrCpt	3-5/30	3973A	6338
CHIEF CHARLOT		10SecDrCpt	3-5/30	3973A	6338
CHIEF COMCOMLY		10SecDrCpt	3-5/30	3973A	6338
CHIEF FIRE HEART		10SecDrCpt	3-5/30	3973A	6338
CHIEF GALL		10SecDrCpt	3-5/30	3973A	6338
CHIEF GARRY		10SecDrCpt	3-5/30	3973A	6338
CHIEF GOLIAH		10SecDrCpt	3-5/30	3973A	6338
CHIEF GOOD LANCE		10SecDrCpt	3-5/30	3973A	6338
CHIEF IRON TAIL		10SecDrCpt	3-5/30	3973A	6338
CHIEF JOSEPH		10SecDrCpt	3-5/30	3973A	6338
CHIEF KAMIAKIN		10SecDrCpt	3-5/30	3973A	6338
CHIEF LITTLE RAVEN		10SecDrCpt	3-5/30	3973A	6338
CHIEF LITTLE SHELL		10SecDrCpt	3-5/30	3973A	6338
CHIEF MANY HORNS		10SecDrCpt	3-5/30	3973A	6338
CHIEF PINE SHOOTER		10SecDrCpt	3-5/30	3973A	6338
CHIEF RED CLOUD		10SecDrCpt	3-5/30	3973A	6338
CHIEF RED LEAF		10SecDrCpt	3-5/30	3973A	6338
CHIEF RED SHIELD		10SecDrCpt	3-5/30	3973A	6338
CHIEF RED THUNDER		10SecDrCpt	3-5/30	3973A	6338

CAR NAME	ST	CAR TYPE	BUILT	PLAN	LOT
CHIEF RED TOMAHAWK		10SecDrCpt	3-5/30	3973A	6338
CHIEF RED WOLF		10SecDrCpt	3-5/30	3973A	6338
CHIEF ROCKY BEAR		10SecDrCpt	3-5/30	3973A	6338
CHIEF RUSHING EAGLE		10SecDrCpt	3-5/30	3973A	6338
CHIEF SEATTLE		10SecDrCpt	3-5/30	3973A	6338
CHIEF SITTING BULL		10SecDrCpt	3-5/30	3973A	6338
CHIEF SPOTTED TAIL		10SecDrCpt	3-5/30	3973A	6338
CHIEF STANDING BUFFALO		10SecDrCpt	3-5/30	3973A	6338
CHIEF SWORD		10SecDrCpt	3-5/30	3973A	6338
CHIEF TAHOLAH		10SecDrCpt	3-5/30	3973A	6338
CHIEF VICTOR		10SecDrCpt	3-5/30	3973A	6338
CHIEF WOLF ROBE		10SecDrCpt	3-5/30	3973A	6338
CHIEFTAIN		6SbrCafeLng	1/30	3976A	6313
CHIGWELL		12SecDr	2-5/21	2410F	4612
CHILAKELA		12SecDr	4/15	2410C	4309
CHILAKO		12SecDr	2-5/21	2410F	4612
CHILDWOLD		16Sec	10-11/14	2412C	4296
CHILHOWIE	N	12SecDr	5-8/25	3410A	4868
CHILI		12SecDr	1-2/22	2410F	4625
CHILLICOTHE		10SecDr2Cpt	12/16-17	2585D	4443
CHILMARK		34Ch	8/13	2764A	4195
CHILMINAR		12SecDr	8-9/20	2410F	4574
CHILOQUIN		10SecDr2Cpt	1-2/13	2585	4091
CHIMES TOWER	RN	8SecDr3Dbr	5-6/25	4090	4846
CHINA		10SecDr2Cpt	11-12/11	2585	3942
CHINNEBY		12SecDr	2-5/21	2410F	4612
CHINOOK		12SecDr	4-6/24	3410	4763
CHIPANA		12SecDr	7/13	2410A	4179
CHIPOLA		12SecDr	3-6/20	2410F	4565
CHIPPENDALE		12SecDr	12/12-13	2410	4076
CHIPPEWA		14Sec	8-10/28	3958A	6181
CHIQUITA		26ChDr	4-5/16	2416C	4366
CHISPA		12SecDr	7/13	2410A	4179
CHITTENANGO		12SecDr	4-5/11	2410	3893
CHLORIS		26ChDr	3-4/14	2416A	4264
CHOCORUA		12SecDr	7-8/14	2410A	4277
CHOCTAW TRIBE	RN	6SecDr4Dbr	9-10/29	4092	6284
CHOKOSNA		16Sec	10-11/14	2412C	4296
CHOLULA NdeM-119	SN	7Cpt2Dr	11/13	2522A	4222
CHOUDRANT		12SecDr	9-10/13	2410A	4215
CHRISTIAL		26ChDr	4-5/16	2416C	4366
CHRISTIANA		28ChDr	11/21/10	2418	3862
CHRISTOPHER COLUMBUS		12SecDr	8-12/25	3410A	4894
CHRISTOPHER GADSDEN		12SecDr	8-12/25	3410A	4894
CHRISTOPHER WREN		14Sec	10-11/26	3958	6012
CHRYSALIS		12SecDr	5-7/13	2410	4149
CHURCH TOWER	RN	8SecDr3Dbr	8-12/25	4090F	4894
CHURCHILL		12SecDr	4-5/11	2410	3893
CHURCHVILLE		12SecDr	1-2/30	3410B	6351
CICERO		12SecDr	4-6/14	2410A	4271

CAR NAME	ST	CAR TYPE	BUILT	PLAN	LOT
CICOLA		12SecDr	5-8/18	2410F	4540
CIMARRON		10SecDr2Cpt	1-2/13	2585	4091
CINALOA		12SecDr	8-9/20	2410F	4574
CINCINNATI	N	12SecDr	1-3/14	2410A	4249
CINNABAR		12SecDr	8-9/20	2410F	4574
CIRCLEVILLE		12SecDr	4/29/10	2410	3772
CIRCUMNAVIGATORS CLUB	RN	8SecRestLngObs	12/17-18	4025	4528
CITY CLUB	N	8SecBLngSunRm	8-9/29	3989K	6274
CITY OF ABERDEEN		LngBBarberObs	7-8/27	3982	6053
CITY OF BELLINGHAM		LngBBarberObs	7-8/27	3982	6053
CITY OF BUTTE		LngBBarberObs	7-8/27	3982	6053
CITY OF DANVILLE	N	25ChDrObs	4/27	3957K	6065
CITY OF DECATUR (Wab)	N	27ChDr	1/27	3957K	6038
CITY OF EVERETT		LngBBarberObs	7-8/27	3982	6053
CITY OF LAFAYETTE (Wab)	N	25ChDrObs	1/27	3957A	6038
CITY OF PORTLAND		LngBBarberObs	7-8/27	3982	6053
CITY OF SEATTLE		LngBBarberObs	7-8/27	3982	6053
CITY OF SPOKANE		LngBBarberObs	7-8/27	3982	6053
CITY OF TACOMA		LngBBarberObs	7-8/27	3982	6053
CITY OF WABASH	N	25ChDrObs	4/27	3957B	6065
CIUDAD DE					
CHIHUAHUA NdeM	SN	8Sec5Dbr	6/13	4036B	4159
GUADALAJARA NdeM	SN	8sec5Dbr	6/25/10	4036	3800
GUANAJUATO NdeM	SN	8Sec5Dbr	8/9/11	4036	3922
JALAPA NdeM	SN	8Sec5Dbr	3/14	4036B	4269
LEON NdeM	N	8Sec5Dbr	7/16/10	4036	3800
MONTERREY NdeM	SN	8Sec5Dbr	3/11	4036B	3880
MORELIA NdeM	SN	8Sec5Dbr	7/2/10	4036B	3800
TOLUCA NdeM	SN	8Sec5Dbr	3/11	4036B	3880
TORREON NdeM	SN	8Sec5Dbr	6/23/10	4036A	3800
CIVRAY		12SecDr	8-9/20	2410F	4574
CLAGGETT		12SecDr	2-5/21	2410F	4612
CLAGHORN		12SecDrCpt	6/13	2411A	4159
CLAIBORNE		12SecDr	10-12/16	2410F	4431
CLAIRTON		26ChDr	8/1/10	2416	3804
CLAN CAMERON	RN	10Sec3Dbr	6/27/10	3411	3800
CLAN CAMPBELL	RN	10Sec3Dbr	6/18/10	3411	3800
CLAN CHATTAN	RN	10Sec3Dbr	5/17	3411	4494
CLAN DONALD	RN	10Sec3Dbr	5/17	3411	4494
CLAN GORDON	RN	10Sec3Dbr	5/17	3411	4494
CLAN GREGOR	RN	10Sec3Dbr	5/17	3411	4494
CLARA BARTON		28ChDr	1-2/27	3416A	6032
CLARA MORRIS		26ChDrObs	1/27	3957A	6038
CLARE		24ChDrB	5/13	2417A	4137
CLAREMONT		10Cpt	4/13	2505	4120
CLARENDON		12SecDr	2-3/27	3410A	6055
CLARIDGE		26ChDr	5/12	2416	3986
CLARK		12SecDrCpt	10-12/21	2411C	4624
CLARKES GAP		8SecDr2Cpt	12/29-30	3979A	6334
CLARKSBURG		12SecDr	7-8/12	2410	4014

CAR NAME	ST	CAR TYPE	BUILT	PLAN	LOT
CLARKSVILLE		12SecDr	10/6/10	2410	3813
CLARNIE		12SecDr	6-7/16	2410F	4385
CLASON		12SecDr	6-7/16	2410F	4385
CLAUDIA		28ChDr	8/24	3416	4801
CLAVERACK		12SecDr	3-4/11	2410	3881
CLAYMONT		24ChDrB	1/14	2417A	4239
CLAYMORE		12SecDr	5-7/13	2410	4149
CLAYPOOL		12SecDr	4-5/11	2410	3893
CLAYSBURG		12SecDr	5-6/17	2410F	4497
CLAYVILLE		12SecDr	2-5/16	2410E	4367
CLEARBROOK		12SecDr	2-4/24	3410	4762
CLEARVIEW		26ChDr	5/12	2416	3986
CLEBOURNE		12SecDr	6-7/17	2410F	4503
CLEMATIS		28ChDr	5-6/24	3416	4761
CLEMENCEAU		12SecDr	2-5/21	2410F	4612
CLEMENT		12SecDr	1-2/16	2410E	4356
CLEMENTE					
OROZCO FCP-220	SN	10SecDr2Cpt	9-10/26	3585A	4996
CLEMONS		12SecDr	10-11/11	2410	3936
CLEON		12SecDr	8-9/20	2410F	4574
CLERMONT		12SecDr	5-6/17	2410F	4497
CLEVELAND		16Sec	12/18	2412X	4489
CLEVELAND CLUB		8SecBLngSunRm	6-7/30	3989B	6362
CLIFFORD		12SecDr	10/15/10	2410	3813
CLIFFWOOD		26ChDr	3/13	2416	4107
CLIFTON FORGE		12SecDr	12/10-11	2410	3866
CLIFTON SPRINGS		16Sec	10-11/14	2412C	4296
CLINTON		12SecDr	1-2/16	2410E	4356
CLIPPARD		12SecDr	4-5/13	2410	4106
CLIPSTONE		12SecDr	7-8/15	2410D	4322
CLIVEDEN		12SecDr	6-7/16	2410F	4385
CLOCK TOWER	RN	8SecDr3Dbr	5-6/25	4090	4846
CLODION		12SecDr	1-2/22	2410F	4625
CLONTARF		12SecDr	8-9/20	2410F	4574
CLOTHO	N	26ChDr	4-5/16	2416C	4366
CLOUD PEAK		10Sec2Dr	12/28-29	3584B	6213
CLOVER ACRES	RN	8Sec5Dbr	7/8/10	4036B	3800
CLOVER BANK	RN	8Sec5Dbr	7/13/10	4036A	3800
CLOVER BASIN	RN	8Sec5Dbr	2-3/15	4036E	4318
CLOVER BAY	RN	8Sec5Dbr	6/21/10	4036A	3800
CLOVER BED	RN	8Sec5Dbr	2-3/15	4036E	4318
CLOVER BLOOM	RN	8Sec5Dbr	6/22/10	4036A	3800
CLOVER BLOSSOM	RN	8Sec5Dbr	9/12	4036B	4036
CLOVER BLUFF	RN	8Sec5Dbr	7/2/10	4036B	3800
CLOVER BROOK	RN	8Sec5Dbr	7/6/10	4036A	3800
CLOVER CAMP	RN	8Sec5Dbr	1-2/22	4036F	4625
CLOVER CAPE	RN	8Sec5Dbr	10-11/13	4036C	4218
CLOVER CASTLE	RN	8Sec5Dbr	1-2/13	4036C	4091
CLOVER CHALET	RN	8Sec5Dbr	1-2/13	4036C	4091
CLOVER CITY	RN	8Sec5Dbr	6/13	4036B	4151

CAR NAME	ST	CAR TYPE	BUILT	PLAN	LOT
CLOVER CLIFF	RN	8Sec5Dbr	5/13	4036C	4141
CLOVER COLONY	RN	8Sec5Dbr	8-9/20	4036I	4574
CLOVER COLORS	RN	8Sec5Dbr	2-5/21	4036F	4612
CLOVER CORNERS	RN	8Sec5Dbr	6-7/16	4036I	4385
CLOVER COURT	RN	8Sec5Dbr	7/15/10	4036B	3800
CLOVER CREEK	RN	8Sec5Dbr	3/14	4036A	4269
CLOVER CREST	RN	8Sec5Dbr	6/28/10	4036A	3800
CLOVER CROFT	RN	8Sec5Dbr	6/13	4036A	4159
CLOVER CROP	RN	8Sec5Dbr	10-12/15	4036I	4338
CLOVER CROWN	RN	8Sec5Dbr	10-12/16	4036I	4431
CLOVER DALE	RN	8Sec5Dbr	7/1/10	4036A	3800
CLOVER DELL	RN	8Sec5Dbr	6/13	4036B	4159
CLOVER DOWNS	RN	8Sec5Dbr	2-5/21	4036F	4612
CLOVER FALLS	RN	8Sec5Dbr	1-2/13	4036C	4091
CLOVER FARM	RN	8Sec5Dbr	7/9/10	4036A	3800
CLOVER FIELD	RN	8Sec5Dbr	6/23/10	4036A	3800
CLOVER FLATS	RN	8Sec5Dbr	4-6/21	4036I	4613
CLOVER GAP	RN	8Sec5Dbr	3/11	4036G	3880
CLOVER GARDEN	RN	8Sec5Dbr	3/11	4036B	3880
CLOVER GARLAND	RN	8Sec5Dbr	6/28/10	4036G	3800
CLOVER GATE	RN	8Sec5Dbr	3/11	4036B	3880
CLOVER GEM	RN	8Sec5Dbr	6/13	4036G	4159
CLOVER GLADE	RN	8Sec5Dbr	6/13	4036A	4159
CLOVER GLEN	RN	8Sec5Dbr	6/13	4036A	4159
CLOVER GLOSS	RN	8Sec5Dbr	7/19/10	4036G	3800
CLOVER GLOW	RN	8Sec5Dbr	6/13	4036G	4159
CLOVER GRANGE	RN	8Sec5Dbr	6/28/10	4036G	3800
CLOVER GREENS	RN	8Sec5Dbr	9/12	4036G	4036
CLOVER GROVE	RN	8Sec5Dbr	7/14/10	4036G	3800
CLOVER GULLY	RN	8Sec5Dbr	3/11	4036G	3880
CLOVER HARVEST	RN	8Sec5Dbr	4-6/14	4036E	4271
CLOVER HAVEN	RN	8Sec5Dbr	6/13	4036B	4151
CLOVER HEIGHTS	RN	8Sec5Dbr	7/22/10	4036A	3800
CLOVER HIGHLANDS	RN	8Sec5Dbr	3/11	4036B	3880
CLOVER					
HIGHLANDS FC-DS	L	8Sec5Dbr	3/11	4036B	3880
CLOVER HILL	RN	8Sec5Dbr	3/11	4036A	3880
CLOVER HOLLOW	RN	8Sec5Dbr	6/29/10	4036A	3800
CLOVER HOME	N	8Sec5Dbr	3/11	4036G	3880
CLOVER ISLE	RN	8Sec5Dbr	3/11	4036B	3880
CLOVER KNOLL	RN	8Sec5Dbr	3/14	4036B	4269
CLOVER LAKE	RN	8Sec5Dbr	10-11/12	4036C	4067
CLOVER LAND	RN	8Sec5Dbr	6/24/10	4036A	3800
CLOVER LANE	RN	8Sec5Dbr	6/27/10	4036B	3800
CLOVER LAWN	RN	8Sec5Dbr	8-9/11	4036B	3922
CLOVER LEAF	RN	8Sec5Dbr	3/11	4036B	3880
CLOVER LODGE	RN	8Sec5Dbr	6/28/10	4036A	3800
CLOVER MANOR	RN	8Sec5Dbr	3/14	4036B	4269
CLOVER MANTLE	RN	8Sec5Dbr	6/13	4036B	4159
CLOVER MEADOW	RN	8Sec5Dbr	6/13	4036A	4159

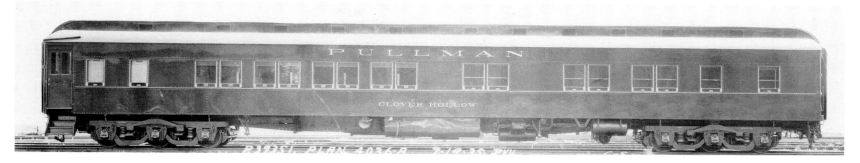

Figure I-50   This Pullman photograph taken July 19, 1935, shows CLOVER HOLLOW as it came out of the shops in fresh paint after being rebuilt from DOWLIN. It was streamlined in 1938 for service on The B&O "Capitol Limited," sold to the B&O in 1948, and withdrawn from Pullman lease in 1962. Pullman Photograph 39351, the Robert Wayner Collection.

CAR NAME	ST	CAR TYPE	BUILT	PLAN	LOT
CLOVER MILL	RN	8Sec5Dbr	3/11	4036B	3880
CLOVER MOUND	RN	8Sec5Dbr	10-12/15	4036I	4338
CLOVER MOUNTAIN	RN	8Sec5Dbr	3/14	4036B	4269
CLOVER NEST	RN	8Sec5Dbr	4-6/14	4036E	4271
CLOVER NOOK	RN	8Sec5Dbr	10-12/15	4036I	4338
CLOVER PARK	RN	8Sec5Dbr	3/11	4036A	3880
CLOVER PASSAGE	RN	8Sec5Dbr	10-12/15	4036I	4338
CLOVER PASTURE	RN	8Sec5Dbr	9-10/13	4036E	4215
CLOVER PATH	RN	8Sec5Dbr	8-9/11	4036B	3922
CLOVER PATH FC-DS	L	8Sec5Dbr	8-9/11	4036B	3922
CLOVER PLAINS	RN	8Sec5Dbr	6/13	4036B	4159
CLOVER PLAINS FC-DS	L	8Sec5Dbr	6/13	4036B	4159
CLOVER PLANT	RN	8Sec5Dbr	5-8/18	4036F	4540
CLOVER PLANT FCS-BC	SN	8Sec5Dbr	5-8/18	4036F	4540
CLOVER PLATEAU	RN	8Sec5Dbr	6/13	4036B	4151
CLOVER PLOT	RN	8Sec5Dbr	9-10/20	4036F	4590
CLOVER POINT	RN	8Sec5Dbr	3/11	4036B	3880
CLOVER POND	RN	8Sec5Dbr	6/13	4036B	4151
CLOVER PORT	RN	8Sec5Dbr	6/21/10	4036A	3800
CLOVER PRAIRIE	RN	8Sec5Dbr	3/11	4036B	3880
CLOVER QUEEN	RN	8Sec5Dbr	1-3/14	4036E	4249
CLOVER REST	RN	8Sec5Dbr	6/13	4036B	4159
CLOVER RIDGE	RN	8Sec5Dbr	6/24/10	4036A	3800
CLOVER ROAD	RN	8Sec5Dbr	3/11	4036A	3880
CLOVER ROW	RN	8Sec5Dbr	2-5/21	4036I	4612
CLOVER RUN	RN	8Sec5Dbr	3/11	4036B	3880
CLOVER SHORE	RN	8Sec5Dbr	3/14	4036B	4269
CLOVER SLOPE	RN	8Sec5Dbr	6/16/10	4036B	3800
CLOVER SPRAY	RN	8Sec5Dbr	2-5/16	4036I	4367
CLOVER SPRINGS	RN	8Sec5Dbr	8-9/11	4036A	3922
CLOVER STATE	RN	8Sec5Dbr	9-10/15	4036I	4327

CAR NAME	ST	CAR TYPE	BUILT	PLAN	LOT
CLOVER SUMMIT	RN	8Sec5Dbr	6/13	4036B	4159
CLOVER TERRACE	RN	8Sec5Dbr	8-9/11	4036A	3922
CLOVER TOWN	RN	8Sec5Dbr	3-6/20	4036F	4565
CLOVER TRAIL	RN	8Sec5Dbr	6/21/10	4036A	3800
CLOVER TRAIL FCS-BC	SN	8Sec5Dbr	6/21/10	4036A	3800
CLOVER VALE	RN	8Sec5Dbr	6/13	4036B	4159
CLOVER VALLEY	RN	8Sec5Dbr	6/13	4036B	4159
CLOVER VELDT	RN	8Sec5Dbr	10-12/20	4036F	4591
CLOVER VIEW	RN	8Sec5Dbr	8-9/11	4036A	3922
CLOVER VILLA	RN	8Sec5Dbr	3-6/20	4036F	4565
CLOVER VISTA	RN	8Sec5Dbr	3/14	4036B	4269
CLOVER WALK	R	8Sec5Dbr	12/10-11	4036E	3866
CLOVER WAY	RN	8Sec5Dbr	4-6/14	4036E	4271
CLOVER WOODS	RN	8Sec5Dbr	3/11	4036B	3880
CLOVER WREATH	RN	8Sec5Dbr	3-4/17	4036F	4485
CLOVERLY		10SecDr2Cpt	10-11/13	2585A	4218
CLOVIS		12SecDr	6-7/16	2410F	4385
CLUB					
CHURUBUSCO NdeM-362	SN	8secDLng	6/21/10	4025C	3801
JUANACATLAN NdeM-362	N	8SecDLng	6/21/10	4025C	3801
TZARARACUA NdeM	N	8secRestLngObs	7-8/11	4025C	3912
VERDE NdeM	SN	8secDLng	7-8/11	4025C	3912
CLYBOURN		12SecDr	2-3/13	2410	4105
CLYDE		BaggBClubSmkBarber	4-5/17	2951	4483
CLYMER		26ChDr	2-3/14	2416A	4248
CLYSMA		7Cpt2Dr	12/11	2522	3943
COACH DORM 2781 (UP)	RL	30StsCoach 6Sec	/25	4178	4890
COACH DORM 2783 (UP)	RL	30StsCoach 6Sec	/25	4178	4890
COACH DORM 2784 (UP)	RL	22StsCoach 8Sec	/25	4177	4890
COACH DORM 2785 (UP)	RL	22StsCoach 8Sec	/25	4177	4890
COACH DORM 2786 (UP)	RL	22StsCoach 8Sec	/25	4177	4890

CAR NAME	ST	CAR TYPE	BUILT	PLAN	LOT
COACH SLEEPER I	RN	45Sts (45Berths)	9-11/22	4094	4649
COACH SLEEPER II	RN	45Sts (45Berths)	9-11/22	4094	4649
COACH SLEEPER III	RN	42Sts (42Berths)	9-11/22	4094A	4649
COACH SLEEPER IV	RN	42Sts (42Berths)	9-11/22	4094A	4649
COALDALE	N	12SecDr	5-8/25	3410A	4868
COALPORT		26ChDr	4-5/17	2416C	4492
COALRIDGE		12SecDr	1/14	2410A	4240
COATEPEC NdeM-121	SN	14Sec	8-10/28	3958A	6181
COATESVILLE		BaggBClubSmkBarber	8/17/10	2414A	3803
COATZACOALCOS FC-DS	SN	8Sec5Dbr	3/11	4036B	3880
COBURN		12SecDrCpt	6/28/10	2411	3800
COCHECO		12SecDr	7-8/14	2410A	4277
COCHITI		7Cpt2Dr SBS	11/11	2522	3941
COCHRAN		6CptLngObs	8/9/10	2413	3802
COCKSPUR		12SecDr	8-9/20	2410F	4574
COCROFT		12SecDr	8-9/20	2410F	4574
CODY		16Sec	4/17	2412F	4484
COFRE DE PEROTE NdeM	SN	10SecDrCpt	4-6/27	3973C	6043
COHOES		12SecDr	10-11/11	2410	3936
COIT TOWER	RN	8SecDr3Dbr	5-6/25	4090	4846
COKATO		12SecDr	8-9/20	2410F	4574
COKEBURG		12SecDr	5-6/17	2410F	4497
COLADERO		12SecDr	12/10-11	2410	3866
COLD BROOK		10Sec2Dr	1/12	2584	3945
COLD SPRING		12SecDrCpt	7/22/10	2411	3800
COLE		6Cpt3Dr	7/25	3523A	4887
COLEGROVE		12SecDr	5-6/17	2410F	4497
COLENSO		12SecDr	4-5/13	2410	4106
COLERIDGE		6Cpt3Dr	11-12/23	3523	4726
COLETTA		26ChDr	4/12	2416	3950
COLFAX		10SecDr2Cpt	11-12/11	2585	3942
COLGATE		12SecDr	5-6/17	2410F	4497
COLISEUM	RN	Recreation	2-3/15	3966	4318
COLLAMER		12SecDr	3-4/11	2410	3881
COLLEGE OF NEW ROCHELLE	N	12SecDr	2-5/25	3410	4845
COLLIER		12SecDr	5-6/17	2410F	4497
COLLINGDALE		12SecDr	1-3/17	2410F	4450
COLLINGSWOOD		26ChDr	7/16/10	2416	3804
COLLINGTON		12SecDr	4/23/10	2410	3770
COLLINS		12SecDrCpt	7/6/10	2411	3800
COLOMBIA NdeM	SN	12SecDr	4-6/14	2410A	4271
COLONEL LINDBERGH	N	5CptLngObs	8/8/10	2413A	3802
COLONEL LINDBERGH	N	3Cpt2DrLngLibObs	12/29-30	3950G	4744
COLONIAL		24ChDrSunRm	9/30	4002B	6385
COLONIE CLUB		8SecBObsLng	2-3/29	3989	6229
COLONNADE		BaggBClubSmkBarber	2-3/12	2602	3948
COLOPHON		12SecDr	8-9/20	2410F	4574
COLORADO		10SecDr2Cpt	1-2/13	2585	4091
COLTER PEAK		10Sec2Dr	12/26-27	3584A	6031

CAR NAME	ST	CAR TYPE	BUILT	PLAN	LOT
COLUMBIA BASIN	N	10SecLngObs	9-10/14	2521A	4292
COLUMBIA BLUFFS	N	10SecLngObs	9-10/14	2521A	4292
COLUMBIA BRIDGE	N	10SecLngObs	12/15	2521B	4353
COLUMBIA CANYON	N	10SecLngObs	10/13	2521	4202
COLUMBIA CLUB		8SecBLngSunRm	8-9/29	3989A	6274
COLUMBIA COLLEGE	RN	10Sec2DbrCpt	11-12/22	4042	4648
COLUMBIA COUNTY	N	10SecLngObs	12/11-12	2521	3944
COLUMBIA CREST	N	10SecLngObs	11/12	2521	4049
COLUMBIA GORGE		10SecDrCpt	5/30	3973A	6358
COLUMBIA LAKE	N	10SecLngObs	12/11-12	2521	3944
COLUMBIA RIVER	N	10SecLngObs	10/11	2521	3926
COLUMBIAN		24ChDrSunRm	9/30	4002B	6385
COLUMBINE		28ChDr	5-6/24	3416	4761
COLUMBUS		10SecDrCpt	4-6/27	3973	6043
COLUSA		12SecDr	12/27-28	3410B	6127
COMANCHE		16Sec	4/17	2412F	4484
COMANCHE TRIBE	RN	6SecDr4Dbr	9-10/29	4092	6284
COMETA		12SecDr	8-9/20	2410F	4574
COMITAN NdeM-123	SN	12SecDr	5-8/25	3410A	4868
COMMANDANT		10SecLngObs	12/15	2521B	4353
COMMANDANT		BaggBClubSmk	2-4/09	2136C	3660
COMMANDER		6Sbr7StsCafeLng	8/27	3976	6058
COMMANDER-IN-CHIEF	RN	8SecBLngObs	12/17-18	4020	4528
COMMODORE		6SbrCafeLng	1/30	3976A	6313
COMMODORE		BaggBClubSmk	2-4/09	2136C	3660
COMMONER		12SecDr	2-3/15	2410B	4318
COMO		12SecDr	1-2/22	2410F	4625
COMPTON	N	16Sec	5-6/16	2412F	4386
COMSTOCK		12SecDr	4-5/11	2410	3893
CONCHITA		26ChDr	4-5/16	2416C	4366
CONDIT		12SecDr	4-5/11	2410	3893
CONDORCET		12SecDr	6-7/17	2410F	4503
CONDRON		26ChDr	3/13	2416	4107
CONEMAUGH		12SecDrCpt	6/17/10	2411	3800
CONESTOGA		26ChDr	3/13	2416	4107
CONEWAGO		12SecDr	10/27/10	2410	3860
CONGERS		12SecDr	4-5/11	2410	3893
CONGRESS		12SecDr	1-2/15	2410B	4311
CONGRESS HALL		4CptLngObs	10/25	3960	4889
CONKLIN		12SecDr	8-9/20	2410F	4574
CONLOGUE		16Sec	2-3/11	2412	3878
CONNEAUT		8CptLng	4/11	2540	3879
CONNEAUT	R	7CptBLng	4/11	2540F	3879
CONNECTICUT	N	DrSbrB16StsLng	4/29	3988D	6221
CONNING TOWER	N	8SecDr3Dbr	5-6/25	4090	4846
CONOMO		12SecDr	7/13	2410A	4179
CONOVER		12SecDr	9-10/15	2410D	4327
CONOWINGO		26ChDr	7/27/10	2416	3804
CONRAD		BaggBClubSmkBarber	4-5/17	2951	4483
CONSHOHOCKEN		12SecDrCpt	6/28/10	2411	3800

CAR NAME	ST	CAR TYPE	BUILT	PLAN	LOT
CONSORT		12SecDr	11-12/14	2410B	4297
CONSTANTINE		12SecDr	4-6/14	2410A	4271
CONSUL		12SecDr	12/12-13	2410	4076
CONTINENTAL		14SecDr	8/13	2762	4193
CONTINENTAL CONGRESS	RN	8SecDrLngObs	12/17-18	4024	4528
CONTINENTAL HALL		4CptLngObs	4/26	3960	4944
CONTRA COSTA		10Sec2Dr	2/13	2584	4092
CONVERSE		12SecDr	5/27/10	2410	3794
CONVOY	RN	16ChCafe	10-11/14	4012	4296
CONWAY		12SecDr	1-2/16	2410E	4356
COOPER		16Sec	1-5/18	2412F	4531
COOSA	P	12SecDr SBS	4/17	2410G	4454
COPAKE		12SecDr	1-4/12	2410	3949
COPPS HILL	N	16ChBLngObs	9/25	3961G	4910
COPTOS		12SecDr	8-9/20	2410F	4574
COQUILLE		12SecDr	1-2/16	2410E	4356
CORA		26ChDr	3-4/14	2416A	4264
CORAL BAY	N	2DrCptLibObs	12/29-30	3950	4744
CORAL BEACH	N	3Cpt2DrLngLibObs	12/29-30	3950C	4744
CORAL CANYON	N	3Cpt2DrLngLibObs	12/29-30	3950C	4744
CORAL COVE	N	3Cpt2DrLngLibObs	12/29-30	3950C	4744
CORAL ISLE	N	3Cpt2DrLngLibObs	12/29-30	3950E	4744
CORAL KEYS	N	3Cpt2DrLngLibObs	12/29-30	3950C	4744
CORAL POINT	N	3Cpt2DrLngLibObs	12/29-30	3950H	4744
CORAL REEF	N	3Cpt2DrLngLibObs	12/29-30	3950H	4744
CORAL STRAITS	N	3Cpt2DrLngLibObs	12/29-30	3950H	4744
CORAL STRAND	N	3Cpt2DrLngLibObs	12/29-30	3950C	4744
CORANTO		16Sec	12/14-15	2412C	4304
CORAZON		12SecDr	12/10-11	2410	3866
CORBETT		6CptLngObs	8/9/10	2413	3802
CORBIN		16Sec	1-5/18	2412F	4531
CORCORAN		12SecDr	4/26	3410A	4943
CORDAVILLE		12SecDr	4-6/14	2410A	4271
CORDELIA		28ChDr	8/24	3416	4801
CORDILLERA		4Cpt2DrLngObs	9/16	2950	4423
CORDOVA		12SecDr	2-3/13	2410	4105
CORFU		12SecDr	8-9/20	2410F	4574
CORINTH		12SecDr	2-3/13	2410	4105
CORLETT		10SecDr2Cpt	10-11/12	2585	4067
CORLISS		16Sec	12/14-15	2412C	4304
CORMENAIS		12SecDr	9-10/15	2410D	4327
CORNELIA		28ChDr	4-6/25	3416	4864
CORNELIUS HENDRICKSON	N	12ChDrBLngSunRm	9/30	4002C	6385
CORNHUSKER		12SecDr	11-12/29	3410B	6299
CORONADO		12SecDr	8-10/26	3410A	4945
COROT		6Cpt3Dr	7/25	3523A	4887
CORRAL		16Sec	2/13	2412	4101
CORREGIDORA DE QUERETARO NdeM	SN	12SecDr	2-5/25	3410	4845
CORRY		12SecDr	11/15/10	2410	3860

CAR NAME	ST	CAR TYPE	BUILT	PLAN	LOT
CORRY		26ChDr	4-5/17	2416C	4492
CORSAIR		16Sec	12/14-15	2412C	4304
CORSO		12SecDr	3-6/20	2410F	4565
CORTEZ		16Sec	1-5/18	2412F	4531
CORTLAND		16Sec	2/13	2412	4101
CORVETTE		12SecDr	6-7/16	2410F	4385
CORWITH		12SecDr	4-5/11	2410	3893
CORYDALIS		16Sec	1/16	2412E	4352
CORYDON		7Cpt2Dr	11/17	2522B	4529
CORYELL		12SecDr	4-5/11	2410	3893
COS COB		32ChDr	8/27	3917A	6078
COSDEN		16Sec	5-6/16	2412F	4386
COSHOCTON		16Sec	10/13	2412B	4216
COSMOPOLITAN		12SecDr	2-3/13	2410	4105
COSMOS		28ChDr	5-7/26	3416	4958
COSMOS CLUB	N	BaggBSmkBarber	10-11/25	3951B	4885
COSTA RICA NdeM	SN	12SecDr	1-4/12	2410I	3949
COSTELLA PEAK		10Sec2Dr	12/28-29	3584B	6213
COSTILLA		CptDrLngBSunRm	10/27	3975	6047
COTTENHAM		12SecDr	11-12/13	2410A	4234
COTTESMORE		12SecDr	11-12/14	2410B	4297
COTTON LAND	RN	8SecBCoach	6/13	4052	4159
COTTON LAND	R	8SecDinLng	6/13	4052A	4159
COTTON VALLEY	RN	8SecBCoach	3/11	4052	3922
COTTON VALLEY	R	8SecDinLng	3/11	4052A	3922
COTTONDALE		12SecDr	9-10/13	2410A	4215
COULSON		12SecDr	10/6/10	2410	3813
COULTER		12SecDr	1-2/16	2410E	4356
COUNCIL BLUFFS		10SecDr2Cpt	10-11/13	2585A	4218
COUNCIL GROVE		10SecLngObs	9-10/14	2521A	4292
COUNCIL OAK		8SecDr2Cpt	2-3/30	3979A	6353
COUNT DE ROCHAMBEAU	N	12SecDr	2-4/24	3410	4762
COUNTESS		30ChDr	9/25	3418	4911
COUNTRY CLUB		8SecBLngSunRm	8-9/29	3989A	6274
COURAGEOUS	RN	17StsRestLng	3-4/14	4019B	4264
COURIER	RN	16ChCafe	10-11/14	4012	4296
COURLAND		7Cpt2Dr	12/11	2522	3943
COURLANDER		12SecDr	12/12-13	2410	4076
COURTIER		10SecLngObs	12/15	2521B	4353
COVALLEN		12SecDr	5-7/13	2410	4149
COVE		12SecDr	2-4/24	3410	4762
COVEDALE		26ChDr	2-3/14	2416A	4248
COVENTRY		7Cpt2Dr	11/17	2522B	4529
COVESVILLE		12SecDrCpt	3/11	2411	3880
COVINGTON		12SecDr	10/29/10	2410	3860
COWANSBURG		12SecDr	2-5/16	2410E	4367
COWDRAY		14SecDr	8/13	2762	4193
COWLEY		28ChDr	11/22/10	2418	3862
COWPENS BATTLEGROUND	N	10SecLibLng	1-2/28	3521B	6128
COWPER		6Cpt3Dr	8/24	3523	4804

CAR NAME	ST	CAR TYPE	BUILT	PLAN	LOT
COXSACKIE		12SecDr	4-5/11	2410	3893
CRAGSTON		12SecDr	7/13	2410A	4179
CRAIGHURST		12SecDr	5-7/13	2410	4149
CRAIGIE HOUSE		13Dbr	6-7/30	3997A	6314
CRAIGLEITH		12SecDr	11-12/13	2410A	4234
CRAIGMOOR		12SecDr	7/13	2410A	4179
CRAMPTON'S GAP		8SecDr2Cpt	12/29-30	3979A	6334
CRANCH		12SecDrCpt	10-12/21	2411C	4624
CRANDALL		12SecDr	4-5/11	2410	3893
CRANDON		12SecDr	8-9/12	2410	4035
CRANMOOR		12SecDr	5-7/13	2410	4149
CRASSUS		12SecDr	9-10/15	2410D	4327
CRATINUS		12SecDrCpt	10-12/21	2411C	4624
CRAWFORD		12SecDr	10-12/16	2410F	4431
CRAYLAND		12SecDr	5-7/13	2410	4149
CREEK CLUB	RN	8SecRestObsLng	10/13	4025A	4202
CREIGHTON UNIVERSITY	N	12Sec2Dbr	7-9/17	4046	4515
CREOLE		10SecDr2Cpt	11-12/11	2585	3942
CRESCENT CITY		BaggClubBSmk	3/25	3415A	4848
CRESHEIM		12SecDr	10/12/10	2410	3814
CRESSON		12SecDr	10/29/10	2410	3860
CRESTWOOD		12SecDr	3-4/11	2410	3881
CRESWELL		24ChDrB	12/10	2417	3863
CREVANT		12SecDr	10-12/15	2410D	4338
CRILLON		12SecDr	7-8/13	2410A	4192
CRISFIELD		12SecDr	7-8/13	2410A	4192
CRISMAN		12SecDr	4-5/11	2410	3893
CRITERION		12SecDr	2-3/15	2410B	4318
CRITTENDEN		12SecDr	3-4/11	2410	3881
CROCKETT		12SecDr	10-12/16	2410F	4431
CROCUS		28ChDr	5-7/26	3416	4958
CROFTON		12SecDr	11-12/13	2410A	4234
CROMER		12SecDr	1-3/14	2410A	4249
CROMWELL		12SecDr	12/12-13	2410	4076
CROOKSTON		12SecDr	4-6/24	3410	4763
CROOKSVILLE		12SecDr	2-4/24	3410	4762
CROSBY		28ChDr	11/21/10	2418	3862
CROSSETT		12SecDr	10-12/16	2410F	4431
CROSSLEY		12SecDr	2-4/24	3410	4762
CROSSMAN PEAK		10Sec2Dr	9-10/29	3584B	6284
CROTON FALLS		12SecDrCpt	3/11	2411	3880
CROTON LAKE		12SecDrCpt	3/14	2411A	4269
CROWLEY		10SecDr2Cpt	5/13	2585A	4141
CROWN POINT		10SecDrCpt	5/30	3973A	6358
CROYDON		28ChDr	4-6/25	3416	4864
CROYLAND		26ChDr	7/30/10	2416	3804
CROZIER		12SecDr	2-5/21	2410F	4612
CRUCERO		10SecDr2Cpt	10-11/13	2585A	4218
CRUGERS		16Sec	2-3/11	2412	3878
CRUSADER		12SecDr	3-6/15	2410B	4319
CRUSADER ROSE	N	12SecDr	8-12/25	3410A	4894
CRYSANTHEMUM	N	25ChDrObs	12/12	2669A	4053
CRYSLER		12SecDr	4-5/11	2410	3893
CRYSTAL BAY		3Cpt2DrLngObs	11-12/29	3959B	6316
CRYSTAL BEACH		3Cpt2DrLngObs	4-5/28	3959A	6125
CRYSTAL BLUFF		3Cpt2DrLngObs	11-12/29	3959B	6316
CRYSTAL BROOK		3Cpt2DrLngObs	11-12/29	3959B	6316
CRYSTAL CAVE		3Cpt2DrLngObs	4-5/28	3959A	6125
CRYSTAL FALLS		3Cpt2DrLngObs	4-5/28	3959A	6125
CRYSTAL GORGE		3Cpt2DrLngObs	11-12/29	3959B	6316
CRYSTAL MOUNT		3Cpt2DrLngObs	11-12/29	3959B	6316
CRYSTAL PEAK		3Cpt2DrLngObs	4-5/28	3959A	6125
CRYSTAL POINT		3Cpt2DrLngObs	11-12/29	3959B	6316
CRYSTAL RIDGE		3Cpt2DrLngObs	11-12/29	3959B	6316
CRYSTAL RIVER		3Cpt2DrLngObs	4-5/28	3959A	6125
CRYSTAL ROCK		3Cpt2DrLngObs	4-5/28	3959A	6125
CRYSTAL SPRINGS		3Cpt2DrLngObs	4-5/28	3959A	6125
CRYSTAL VIEW		3Cpt2DrLngObs	11-12/29	3959B	6316
CUAUHTEMOC NdeM	SN	12SecDr	2-4/24	3410	4762
CUBA		23StsBLng	11/26	3967	4979
CUDWORTH		12SecDr	1-2/16	2410E	4356
CUERNA		12SecDr	1-3/14	2410A	4249
CULBERTSON		12SecDr	5-8/18	2410F	4540
CULEBRA		12SecDr	10-12/26	3410A	6023
CULIACAN NdeM-124	SN	12SecDr	5-8/25	3410A	4868
CULLEN		16Sec	1-5/18	2412F	4531
CULLISON		10SecDr2Cpt	1-2/13	2585	4091
CULLODEN		14SecDr	8/13	2762	4193
CULVER		12SecDr	5/10/10	2410	3773
CULVERMERE		BaggBClubSmkBarber	2-3/12	2602	3948
CUMBERLAND CLUB		8SecBLngSunRm	10/30	3989C	6393
CUMBERLAND ROAD		BaggBSmkBarber	12/26-27	3951C	6022
CUMMINGS		16Sec	1-5/18	2412F	4531
CUNAKA		12SecDr	8-9/20	2410F	4574
CUPOLA		12SecDr	2-5/16	2410E	4367
CURRY		28ChDr	11/21/10	2418	3862
CURTIN		32Ch	11/17/10	2419	3861
CURTIS BAY		28ChDr	6-7/27	3416A	6087
CURWEN		12SecDr	7-8/15	2410D	4322
CURWENSVILLE		12SecDr	2-4/24	3410	4762
CURWOOD		12SecDr	8-9/20	2410F	4574
CURZON		16Sec	4/17	2412F	4484
CUSHING		16Sec	1-5/18	2412F	4531
CUSHMAN		12SecDr	10/22-23	2410H	4647
CUSTER		7Cpt2Dr	9/13	2522A	4208
CUSTIS		16Sec	1-5/18	2412F	4531
CUTHBERT		12SecDrCpt	6/13	2411A	4159
CUTLER		12SecDr	8-9/20	2410F	4574
CUTTYHUNK		12SecDr	10/22-23	2410H	4647
CUVINA		26ChDr	4-5/16	2416C	4366

CAR NAME	ST	CAR TYPE	BUILT	PLAN	LOT
CUYAMA		16Sec	1-5/18	2412F	4531
CUYLERVILLE		12SecDrCpt	6/13	2411A	4159
CYGNUS		12SecDr	8-9/20	2410F	4574
CYLBURN		26ChDr	7/15/10	2416	3804
CYMRIC		12SecDr	5-7/13	2410	4149
CYNTHIA		12SecDr	3-6/15	2410B	4319
CYNWYD		16Sec	10/13	2412B	4216
CYPRIAN		12SecDr	2-3/15	2410B	4318
CYPRUS		12SecDr	6-7/16	2410F	4385
CYRIL		12SecDr	5-8/18	2410F	4540
CYRUS FIELD		14Sec	10-11/26	3958	6012
CYRUS H. JENKS		LngBBarberSunRm	5-6/29	3990	6249
CYRUS H. McCORMICK		14Sec	3-4/30	3958A	6357
CYRUS NORTHROP		8SecDr2Cpt	12/29-30	3979A	6334
D-100		Dining Room	10/28	3969	6011
D. WILLIS JAMES		LngBBarberSunRm	5-6/29	3990	6249
DA COSTA		28ChDr	6-7/27	3416A	6087
DAGGETT		10SecDr2Cpt	8-9/13	2585A	4201
DAGMAR		12SecDr	3-6/15	2410B	4319
DAGSBORO		12SecDr	2-5/16	2410E	4367
DAHLIA		28ChDr	5-7/26	3416	4958
DAHLONEGA		12SecDr	10-12/26	3410A	6023
DAKOMING		12SecDr	9-10/13	2410A	4215
DALBERG		12SecDrCpt	6/13	2411A	4159
DALE CREEK		4Cpt2DrLngObs	3/13	2703	4100
DALE SUMMIT	RN	10Sec4PvtSec	9-10/12	2412H	4037
DALEBORO	RN	10Sec4PvtSec	5-6/16	2412H	4386
DALEBROOK	RN	10Sec4PvtSec	4/17	2412H	4484
DALECREST	RN	10Sec4PvtSec	6-7/20	2412H	4568
DALEFORD	RN	10Sec4PvtSec	12/14-15	2412H	4304
DALEGROVE	RN	10Sec4PvtSec	10-11/14	2412H	4296
DALEHURST	RN	10Sec4PvtSec	12/14-15	2412H	4304
DALEMEAD	RN	10Sec4PvtSec	1-5/18	2412H	4531
DALEMONT	RN	10Sec4PvtSec	10/13	2412H	4216
DALEPARK	RN	10Sec4PvtSec	10/13	2412H	4217
DALEROSE	RN	10Sec4PvtSec	12/14-15	2412H	4304
DALESBURG	RN	10Sec4PvtSec	5-6/16	2412H	4386
DALESHIRE	RN	10Sec4PvtSec	12/17-18	2412H	4528
DALESIDE	RN	10Sec4PvtSec	6-7/13	2412H	4160
DALEVIEW	RN	10Sec4PvtSec	12/14-15	2412H	4304
DALEVILLE	RN	10Sec4PvtSec	6-7/13	2412H	4160
DALHOUSIE UNIVERSITY	RN	12Sec2Dbr	5-6/17	4046	4497
DALTON		12SecDr	7/13	2410A	4179
DALZELL		12SecDr	9-10/15	2410D	4327
DAMAR		16Sec	2/13	2412	4101
DAMON		12SecDr	8-9/20	2410F	4574
DAMOSEL		26ChDr	4-5/16	2416C	4366
DANA		12SecDrCpt	10-12/21	2411C	4624
DANBRIDGE		12SecDr	11-12/13	2410A	4234
DANBURY		12SecDr	4-5/11	2410	3893

CAR NAME	ST	CAR TYPE	BUILT	PLAN	LOT
DANIEL BOONE		12SecDr	1-2/29	3410B	6220
DANIEL MORGAN	N	12SecDr	6-7/17	2410F	4503
DANIEL WEBSTER		12SecDr	5-8/25	3410A	4868
DANTE		6Cpt3Dr	11-12/23	3523	4726
DANTON		12SecDr	4-5/13	2410	4106
DANVILLE		12SecDr	10/10/10	2410	3813
DARBY		12SecDr	2-5/25	3410	4845
DARCIA		28ChDr	4/12	2418	3951
DARDANELLES		12SecDr	8-9/20	2410F	4574
DARENT		26ChDr	3/13	2416	4107
DARETOWN		12SecDr	5-6/17	2410F	4497
DARIEN		36Ch	10/24	3916	4802
DARIUS WELLS		14Sec	3/27	3958	6054
DARKWATER		26ChDr	7/25/10	2416	3804
DARLOW		16Sec	2/13	2412	4101
DARTMOOR		16Sec	6-7/13	2412B	4160
DARTMOUTH		12SecDr	10/22-23	2410H	4647
DATHEMA		7Cpt2Dr	12/11	2522	3943
DAUNTLESS		12SecDr	12/12-13	2410	4076
DAUPHIN		32Ch	11/18/10	2419	3861
DAVENPORT		12SecDr	11-12/29	3410B	6299
DAVID C. SHEPARD		LngBBarberSunRm	5-6/29	3990	6249
DAVID FARRAGUT	N	12SecDr	2-5/25	3410	4845
DAVID LIVINGSTONE		Private	7/27	3972	6037
DAVID MOSSOM	RN	10SecDrCpt	9-10/17	4031	4525
DAVID SINTON		14Sec	3-4/30	3958A	6357
DAVIDA		26ChDr	6/11	2416	3902
DAVIDSON COLLEGE	RN	10Sec2DbrCpt	6/13	4042A	4159
DAVIS		12SecDrCpt	10-12/21	2411C	4624
DAVY CROCKETT	N	Private (GS)	9/17	2502C	4490
DAYLESFORD		26ChDr	8/3/10	2416	3804
DAYSVILLE		12SecDr	3-4/11	2410	3881
DAYTON		10SecDr2Cpt	5/13	2585A	4141
DAYTONA		12SecDr	10/22-23	2410H	4647
DE COVERLEY		8SecDr2Cpt	8-9/29	3979A	6283
DE FOREST		8SecDr2Cpt	8-9/29	3979A	6283
DE LANCEY		8SecDr2Cpt	8-9/29	3979A	6283
DE LONG		8SecDr2Cpt	8-9/29	3979A	6283
DE PEYSTER		8SecDr2Cpt	8-9/29	3979A	6283
DE WITT		8SecDr2Cpt	8-9/29	3979A	6283
DE WOLF		8SecDr2Cpt	8-9/29	3979A	6283
DE YOUNG		8SecDr2Cpt	8-9/29	3979A	6283
DEAN LAKE		10SecDr2Cpt	11-12/27	3585B	6123
DEANS		24ChDrB	4/16	2417C	4364
DEANWOOD		26ChDr	5/12	2416	3986
DEARBORN		12SecDr	12/12-13	2410	4076
DEBORAH		28ChDr	4-6/25	3416	4864
DEEMSTER		12SecDr	3-6/20	2410F	4565
DEEP SOUTH	RN	8SecDr3Dbr	1/25	4090B	4843
DEER CREEK		10Sec2Dr	5/13	2584	4140

Figure I-51 The Pullman private car, DAVID LIVINGSTONE, was temporarily disguised for use in one of the United States Government's Defense Special Trains. These trains carried government officials throughout the country to influence and assist business in undertaking defense contracts. Pullman Photograph 45599, taken October 31, 1941 from the Arthur D. Dubin Collection.

CAR NAME	ST	CAR TYPE	BUILT	PLAN	LOT	CAR NAME	ST	CAR TYPE	BUILT	PLAN	LOT
DEER LODGE		10SecDrCpt	4-6/27	3973	6043	DELLWOOD		16Sec	2-3/11	2412	3878
DEER TRAIL		4Cpt2DrLngObs	3/13	2703	4100	DELMAR		12SecDr	8-9/20	2410F	4574
DEERHURST		12SecDr	3-6/15	2410B	4319	DELORA		24ChDrB	6-7/14	2417B	4265
DEFENDER	RN	17StsRestLng	5/12	4019A	3986	DELORME		12SecDr	7-8/13	2410A	4192
DEFOE		6Cpt3Dr	8/24	3523	4804	DELPHINIUM		28ChDr	5-7/26	3416	4958
DEKKAN		12SecDr	8-9/20	2410F	4574	DELPHOS		12SecDrCpt	5/17	2411C	4494
DEL MONTE		10Sec2Dr	2/13	2584	4092	DELTA		BaggBClubSmkBarber	9/24	2951C	4805
DELABOLE		12SecDr	7-8/13	2410A	4192	DEMASBY		16Sec	5-6/16	2412F	4386
DELACOUR		12SecDr	1-2/16	2410E	4356	DEMOSTHENES		12SecDrCpt	10-12/21	2411C	4624
DELACROIX		12SecDrCpt	6/13	2411A	4159	DENDRON		12SecDr	4-5/13	2410	4106
DELAFIELD	N	12SecDr	4/8/10	2410	3770	DENHOLM		24ChDrB	12/10	2417	3863
DELAGOA		12SecDr	8-9/20	2410F	4574	DENVER		CptDrLngBSunRm	10/27	3975	6047
DELAIR		28ChDr	11/22/10	2418	3862	DENVER TOWER	RN	8SecDr3Dbr	4-5/26	4090E	4961
DELANCO		24ChDrB	12/10	2417	3863	DEPEW		BaggBClubSmkBarber	4-5/17	2951	4483
DELANSON		12SecDr	7/16	2410F	4412	DERRICK		BaggBClubSmkBarber	4-5/17	2951	4483
DELAPLAINE		10SecDr2Cpt	6-7/15	2585B	4328	DERRINGER		26ChDr	4-5/17	2416C	4492
DELAUNAY		12SecDr	9-10/15	2410D	4327	DERRY		24ChDrB	12/10	2417	3863
DELAWANNA		12SecDr	7-8/12	2410	4014	DERVISH		12SecDr	3-6/20	2410F	4565
DELAWARE		12SecDr	4-5/11	2410	3893	DERWOOD		12SecDr	4/30	3410B	6360
DELCAMBRE		12SecDrCpt	6/13	2411A	4159	DES MOINES		10SecDrCpt	4-6/27	3973	6043
DELCO		16Sec	6-7/20	2412F	4568	DES MOINES CLUB		8SecBLngSunRm	8-9/29	3989A	6274
DELECTO		12SecDr	5-8/18	2410F	4540	DES PLAINES		8SecDr2Cpt	8-9/29	3979A	6283
DELFTHAVEN		12SecDr	10-12/26	3410A	6023	DESDEMONA		28ChDr	8/24	3416	4801
DELGADA		12SecDr	12/10-11	2410	3866	DESERET		12SecDr	10-12/26	3410A	6023
DELIA		26ChDr	3/13	2416	4107	DESIERTO		7Cpt2Dr	11/17	2522B	4517
DELL LAKE		10SecDr2Cpt	11-12/27	3585B	6123	DESIREE		28ChDr	4-6/25	3416	4864
DELLVALE		12SecDr	1-3/14	2410A	4249	DETROIT	N	5CptLngObs	8/13/10	2413A	3802

CAR NAME	ST	CAR TYPE	BUILT	PLAN	LOT
DETROIT CLUB		8SecBLngSunRm	10/30	3989C	6393
DEUEL		LngBDorm(5Sec)	7-8/27	3981	6061
DEVAULT		12SecDr	5/12/10	2410	3773
DEVEREUX		16Sec	6-7/20	2412F	4568
DEVILS LAKE		10SecDr2Cpt	5-6/24	3585	4770
DEVILS TOWER	RN	8SecDr3Dbr	5-6/25	4090	4846
DEVON	N	12SecDr	7-8/13	2410A	4192
DEWART		12SecDr	11/26/10	2410	3864
DEWEY		12SecDr	4-6/14	2410A	4271
DEXTER		12SecDr	3-6/20	2410F	4565
DIANA		24ChDrB	6-7/14	2417B	4265
DIAS		28ChDr	5-7/26	3416	4958
DIASCUND		12SecDr	11-12/13	2410A	4234
DICKENS		6Cpt3Dr	11-12/23	3523	4726
DICKINSON		12SecDr	7/16	2410F	4412
DICTATOR		12SecDr	9-10/15	2410D	4327
DIEGO RIVERA FCP-217	SN	10SecDr2Cpt	12/23-24	3585C	4728
DIGHTON ROCK		BaggBSmkBarber	12/26-27	3951C	6022
DILLERVILLE		12SecDrCpt	7/9/10	2411	3800
DILLONVALE		16Sec	10-11/14	2412C	4296
DILLSBURG		12SecDr	2-4/24	3410	4762
DIMONDALE		12SecDr	10-11/11	2410	3936
DINANT		12SecDr	1-2/16	2410E	4356
DINWIDDIE	N	10SecLngObs	10-12/26	3521A	4998
DINWIDDIE COUNTY	N	10SecLngObs	10-12/26	3521A	4998
DIOGENES		12SecDrCpt	10-12/21	2411C	4624
DIPLOMAT		25ChDrObs	4/27	3957B	6065
DISCOVERER	RN	17StsRestLng	3-4/14	4019A	4264
DIVERSEY		12SecDr	2-3/13	2410	4105
DIXIANA		7Cpt2Dr	11/13	2522A	4222
DIXIE BAY		3CptDrLngBSunRm	12/28-29	3975C	6217
DIXIE HOME		3CptDrLngBSunRm	12/28-29	3975C	6217
DIXIE LAND		3CptDrLngBSunRm	12/28-29	3975C	6217
DIXIE SPRINGS		3CptDrLngBSunRm	1/30	3975F	6337
DIXIE TRAIL		3CptDrLngBSunRm	12/28-29	3975C	6217
DIXON		24ChDrB	4/16	2417C	4364
DOBBS FERRY		12SecDrCpt	3/11	2411	3880
DODGE		12SecDr	10/22-23	2410H	4647
DODGE CITY		10SecLngObs	12/16-17	2521C	4442
DODGEVILLE		36Ch	10/24	3916	4802
DODONA		12SecDr	8-9/20	2410F	4574
DODSON		BaggBClubSmkBarber	4-5/17	2951	4483
DOLA		12SecDr	2-5/25	3410	4845
DOLLY MADISON		16ChBLngObs	1-2/27	3961A	6035
DOLORES		32Ch	6/30	3916C	6363
DOLORES					
HIDALGO NdeM-134	SN	8SecDr3Dbr	5-6/25	4090A	4843
DOLORITA		26ChDr	3-4/14	2416A	4264
DOLPHIN		12SecDr	7-8/13	2410A	4192
DOLPHIN		12SecDr	5-8/18	2410F	4540

CAR NAME	ST	CAR TYPE	BUILT	PLAN	LOT
DOME ROCK		10Sec2Dr	2/13	2584	4092
DOMINGO		7Cpt2Dr	11/17	2522B	4517
DOMINICA		12SecDr	12/10-11	2410	3866
DONAJI NdeM	SN	6SecDr2Cpt2Dbr	2/16	2547C	4363
DONALD McKAY		34Ch	2/30	4001	6325
DONCASTER		12SecDr	7-8/13	2410A	4192
DONIZETTI		6Cpt3Dr	11-12/24	3523A	4833
DONNER LAKE	N	BaggDormKitchen	10/28	3970A	6010
DONOHOE		12SecDr	5-6/17	2410F	4497
DONWELL		12SecDr	1-3/14	2410A	4249
DORA		28ChDr	5-6/24	3416	4761
DORADO		12SecDr	5-8/18	2410F	4540
DORCHESTER		12SecDr	10-12/26	3410A	6023
DOREMUS	N	12SecDrCpt	6/13	2411A	4159
DORKING		16Sec	6-7/20	2412F	4568
DORRANCE		10SecDr2Cpt	11-12/11	2585	3942
DORSEY		BaggBClubSmkBarber	4-5/17	2951	4483
DOSARIS		12SecDr	3-6/15	2410B	4319
DOTAME		12SecDr	4/15	2410C	4309
DOUBLE PEAK		10Sec2Dr	12/28-29	3584B	6213
DOUGLAS FIR		10SecDrCpt	5/30	3973A	6358
DOUTHITT		12SecDr	3-6/20	2410F	4565
DOVER		12SecDr	2-3/27	3410A	6055
DOVER BAY	RN	6DbrBLng	7/23	4015A	4698
DOVER CASTLE	RN	6DbrBLng	9/24	4015A	4805
DOVER CLIFFS	R	6DbrBLng	9/24	4015A	4805
DOVER FORT	R	6DbrBLng	9/24	4015A	4805
DOVER HARBOR	RN	6DbrBLng	7/23	4015A	4698
DOVER HILL	RN	6DbrBLng	7/23	4015A	4698
DOVER PATROL	R	6DbrBLng	9/24	4015A	4805
DOVER PLAINS		12SecDrCpt	8-9/11	2411	3922
DOVER PLAINS	RN	6DbrBLng	7/23	4015A	4698
DOVER STRAIT	R	6DbrBLng	9/24	4015A	4805
DOVRAY		12SecDr	9-10/15	2410D	4327
DOWAGIAC		12SecDr	1-4/12	2410	3949
DOWLIN		12SecDrCpt	6/29/10	2411	3800
DOWNINGTOWN		26ChDr	7/28/10	2416	3804
DRACON		12SecDr	7-8/15	2410D	4322
DRAKE		12SecDrCpt	10-12/21	2411C	4624
DRAKE UNIVERSITY	N	12Sec2Dbr	10-12/26	4046B	6023
DRAPER		16Sec	5-6/16	2412F	4386
DREADNOUGHT		10SecDr2Cpt	12/15	2585C	4351
DRESDEN		12SecDr	2-4/24	3410	4762
DRESHER		26ChDr	7/14/10	2416	3804
DRIFTWOOD		26ChDr	3/13	2416	4107
DRISCOLL		12SecDr	5-6/17	2410F	4497
DROMEDARY TOWER	RN	8SecDr3Dbr	5-6/25	4090	4846
DROMIO		12SecDr	1-2/22	2410F	4625
DROMORE		12SecDr	8-9/20	2410F	4574
DRUID PEAK		10Sec2Dr	12/26-27	3584A	6031

CAR NAME	ST	CAR TYPE	BUILT	PLAN	LOT
DRURY COLLEGE	N	10Sec2DbrCpt	12/17-18	4042B	4528
DU CHAILLU		3Cpt2DrLngObs	11-12/26	3959	6015
DU PAGE		LngBDorm(5Sec)	7-8/27	3981	6061
DUANE		12SecDr	4-6/14	2410A	4271
DUBOIS		26ChDr	4-5/17	2416C	4492
DUCHESS		26ChDr	3-4/14	2416A	4264
DUDLEY		Bagg24Ch	6/16	2915	4388
DUDLEY	R	BaggBSmk	6/16	2915E	4388
DUFFERIN		12SecDr	4-6/14	2410A	4271
DUFFIELD		12SecDr	1-2/16	2410E	4356
DULCIBEL		26ChDr	4-5/16	2416C	4366
DULCINEA		24ChDrB	5/17	2417D	4493
DULUTH		12SecDr	5-8/18	2410F	4540
DULVERTON		12SecDr	12/10-11	2410	3866
DUMAS	N	12SecDr	7-8/13	2410A	4192
DUMAS		6Cpt3Dr	11-12/23	3523	4726
DUMONT		12SecDr	4-5/11	2410	3893
DUNBAR		26ChDr	8/4/10	2416	3804
DUNCANNON		26ChDr	3/13	2416	4107
DUNCANNON	N	12SecDr	5-8/25	3410A	4868
DUNCIAD		16Sec	6-7/20	2412F	4568
DUNDALE		12SecDr	11/26/10	2410	3860
DUNDALE		12SecDr	2-5/25	3410	4845
DUNDALK		12SecDr	2-5/21	2410F	4612
DUNDAS VALLEY	N	4Cpt2DrLngObs	3/13	2703	4100
DUNDEE		12SecDr	8-9/20	2410F	4574
DUNE PARK		12SecDrCpt	5/17	2411C	4494
DUNGARVIN		12SecDrCpt	6/13	2411A	4159
DUNGENESS		12SecDr	3-6/15	2410B	4319
DUNKIRK		12SecDr	1-2/30	3410B	6351
DUNLAP		28ChDr	11/22/10	2418	3862
DUNLAY		12SecDrCpt	6/13	2411A	4159
DUNLIN		12SecDr	12/10-11	2410	3866
DUNLO		28ChDr	11/22/10	2418	3862
DUNNING		12SecDr	2-5/25	3410	4845
DUNREITH		12SecDr	4-5/11	2410	3893
DUNSANY		12SecDr	2-5/21	2410F	4612
DUNSTAN		12SecDr	12/10-11	2410	3866
DUPONT		12SecDr	10/7/10	2410	3813
DUPONT CIRCLE		4CptLngObs	4/26	3960	4944
DUQUESNE CLUB	RN	8SecRestObsLng	9-10/14	4025B	4292
DURER		6Cpt3Dr	7/25	3523A	4887
DURHAM		34Ch	8/13	2764A	4195
DUROYCE		12SecDr	1-2/16	2410E	4356
DURWARD		12SecDr	1-3/14	2410A	4249
DUSTANIA		16Sec	6-7/20	2412F	4568
DUWAMISH		12SecDr	4/15	2410C	4309
DWIGHT		12SecDr	3-6/20	2410F	4565
DYERSBURG		12SecDr	10/22-23	2410H	4647
DYERSVILLE		12SecDrCpt	9/12	2411	4036

CAR NAME	ST	CAR TYPE	BUILT	PLAN	LOT
DYKEMANS		12SecDr	4-5/11	2410	3893
DYSART		26ChDr	2-3/14	2416A	4248
E.C.CARRUTH	N	12SecDr	4-6/24	3410	4763
EAGLE BAY		BaggBSmkBarber	10-11/25	3951	4885
EAGLE BLUFF		BaggBSmkBarber	10-11/25	3951	4885
EAGLE CANYON		BaggBSmkBarber	10-11/25	3951	4885
EAGLE CLIFF		BaggBSmkBarber	10-11/25	3951	4885
EAGLE CREEK		BaggBSmkBarber	10-11/25	3951	4885
EAGLE FORD		BaggBSmkBarber	10-11/25	3951	4885
EAGLE GORGE		BaggBSmkBarber	10-11/25	3951	4885
EAGLE GROVE		BaggBSmkBarber	10-11/25	3951	4885
EAGLE HARBOR		BaggBSmkBarber	10-11/25	3951	4885
EAGLE HEIGHTS		BaggBSmkBarber	10-11/25	3951	4885
EAGLE HILL		BaggBSmkBarber	10-11/25	3951	4885
EAGLE ISLAND		BaggBSmkBarber	10-11/25	3951	4885
EAGLE LAKE		BaggBSmkBarber	10-11/25	3951	4885
EAGLE MOUNTAIN		BaggBSmkBarber	10-11/25	3951	4885
EAGLE PASS		10Sec2Dr	2/13	2584	4092
EAGLE PEAK		BaggBSmkBarber	10-11/25	3951	4885
EAGLE POINT		BaggBSmkBarber	10-11/25	3951	4885
EAGLE POND		BaggBSmkBarber	10-11/25	3951	4885
EAGLE RAPIDS		BaggBSmkBarber	10-11/25	3951	4885
EAGLE RIVER		BaggBSmkBarber	10-11/25	3951	4885
EAGLE ROCK		BaggBSmkBarber	10-11/25	3951	4885
EAGLE SPRING		BaggBSmkBarber	10-11/25	3951	4885
EAGLE VALLEY		BaggBSmkBarber	10-11/25	3951	4885
EAGLESMERE		12SecDr	1-3/14	2410A	4249
EAGLEVILLE		12SecDr	10/6/10	2410	3813
EARLVILLE		12SecDr	6-9/26	3410A	4969
EAST AKRON		12SecDr	5-8/25	3410A	4868
EAST ALBURGH		12SecDr	5-8/25	3410A	4868
EAST ALEXANDER		12SecDr	5-8/25	3410A	4868
EAST ALHAMBRA		12SecDr	5-8/25	3410A	4868
EAST ALLEGHANY		12SecDr	5-8/25	3410A	4868
EAST ALLENTOWN		12SecDr	5-8/25	3410A	4868
EAST ALLIANCE		12SecDr	5-8/25	3410A	4868
EAST ALTON		12SecDr	5-8/25	3410A	4868
EAST ANGUS		12SecDr	5-8/25	3410A	4868
EAST APPLEGATE		12SecDr	5-8/25	3410A	4868
EAST ARCADIA		12SecDr	5-8/25	3410A	4868
EAST ASCOT		12SecDr	5-8/25	3410A	4868
EAST AUBURN		12SecDr	5-8/25	3410A	4868
EAST AURORA		12SecDr	5-8/25	3410A	4868
EAST BANGOR		12SecDr	5-8/25	3410A	4868
EAST BANK		12SecDr	5-8/25	3410A	4868
EAST BARRE		12SecDr	5-8/25	3410A	4868
EAST BATAVIA		12SecDr	5-8/25	3410A	4868
EAST BERGEN		12SecDr	5-8/25	3410A	4868
EAST BERKSHIRE		12SecDr	5-8/25	3410A	4868
EAST BERLIN		12SecDr	5-8/25	3410A	4868

CAR NAME	ST	CAR TYPE	BUILT	PLAN	LOT
EAST BERNARD		12SecDr	5-8/25	3410A	4868
EAST BETHANY		12SecDr	5-8/25	3410A	4868
EAST BIGGS		12SecDr	5-8/25	3410A	4868
EAST BILLERICA		12SecDr	5-8/25	3410A	4868
EAST BLACKSTONE		12SecDr	5-8/25	3410A	4868
EAST BRADFORD		12SecDr	5-8/25	3410A	4868
EAST BRAINTREE		12SecDr	5-8/25	3410A	4868
EAST BRANCH		12SecDr	5-8/25	3410A	4868
EAST BREWSTER		12SecDr	5-8/25	3410A	4868
EAST BRIDGER		12SecDr	5-8/25	3410A	4868
EAST BRIGHTON		12SecDr	5-8/25	3410A	4868
EAST BROOKFIELD		12SecDr	5-8/25	3410A	4868
EAST BROOKSIDE		12SecDr	5-8/25	3410A	4868
EAST BROUGHTON		12SecDr	5-8/25	3410A	4868
EAST BUFFALO		12SecDr	5-8/25	3410A	4868
EAST BURLINGTON		12SecDr	5-8/25	3410A	4868
EAST BUSKIRK		12SecDr	5-8/25	3410A	4868
EAST BUTLER		12SecDr	5-8/25	3410A	4868
EAST BYARS		12SecDr	5-8/25	3410A	4868
EAST CADIZ		12SecDr	5-8/25	3410A	4868
EAST CAIRO		12SecDr	5-8/25	3410A	4868
EAST CAMBRIDGE		12SecDr	5-8/25	3410A	4868
EAST CANDIA		12SecDr	5-8/25	3410A	4868
EAST CANTON		12SecDr	5-8/25	3410A	4868
EAST CARNEGIE		12SecDr	5-8/25	3410A	4868
EAST CHARLEROI		12SecDr	5-8/25	3410A	4868
EAST CHARLOTTE		12SecDr	5-8/25	3410A	4868
EAST CHATHAM		12SecDr	5-8/25	3410A	4868
EAST CHESTER		12SecDr	5-8/25	3410A	4868
EAST CHICAGO		12SecDr	5-8/25	3410A	4868
EAST CLAREMONT		12SecDr	5-8/25	3410A	4868
EAST CLARENCE		12SecDr	5-8/25	3410A	4868
EAST CLARIDON		12SecDr	5-8/25	3410A	4868
EAST CLAYTON		12SecDr	5-8/25	3410A	4868
EAST CLINTON		12SecDr	5-8/25	3410A	4868
EAST COLUMBIA		12SecDr	5-8/25	3410A	4868
EAST CONCORD		12SecDr	5-8/25	3410A	4868
EAST COOPER		12SecDr	5-8/25	3410A	4868
EAST CORNING		12SecDr	5-8/25	3410A	4868
EAST CREEK		12SecDrCpt	7/19/10	2411	3800
EAST CREIGHTON	N	12SecDr	5-8/25	3410A	4868
EAST CUMMINSVILLE		12SecDr	5-8/25	3410A	4868
EAST DAYTON		12SecDr	5-8/25	3410A	4868
EAST DEFIANCE		12SecDr	5-8/25	3410A	4868
EAST DOVER		12SecDr	5-8/25	3410A	4868
EAST END		12SecDr	5-8/25	3410A	4868
EAST FALLS		12SecDr	5-8/25	3410A	4868
EAST GRAFTON		12SecDr	5-8/25	3410A	4868
EAST LEXINGTON		12SecDr	5-8/25	3410A	4868
EAST MADISONVILLE		12SecDr	5-8/25	3410A	4868
EAST MONROE		12SecDr	5-8/25	3410A	4868
EAST NEWARK		12SecDr	5-8/25	3410A	4868
EAST NORWOOD		12SecDr	5-8/25	3410A	4868
EAST OSWEGO		12SecDr	5-8/25	3410A	4868
EAST PALMYRA		12SecDr	5-8/25	3410A	4868
EAST PEMBROKE		12SecDr	5-8/25	3410A	4868
EAST ROCHESTER		12SecDr	5-8/25	3410A	4868
EAST SIDE		12SecDr	5-8/25	3410A	4868
EAST SOMERVILLE		12SecDr	5-8/25	3410A	4868
EAST SPARTA		12SecDr	5-8/25	3410A	4868
EAST STEUBEN		12SecDr	5-8/25	3410A	4868
EAST SYRACUSE		12SecDr	5-8/25	3410A	4868
EAST TOLEDO		12SecDr	5-8/25	3410A	4868
EAST VIEW		12SecDrCpt	3/14	2411A	4269
EAST WILLIAMSON		12SecDr	5-8/25	3410A	4868
EAST YOUNGSTOWN		12SecDr	5-8/25	3410A	4868
EASTAMPTON		36Ch	10/24	3916	4802
EASTER LILY		34Ch	4/26	3419	4956
EASTERN STAR		12SecDr	10/22-23	2410H	4647
EASTWOOD		16Sec	10/13	2412B	4217
EATON		28ChDr	5-7/26	3416	4958
EAU CLAIRE		12SecDr	7-8/28	3410B	6184
ECHO		12SecDr	12/27-28	3410B	6127
ECHO CANYON		3CptDrLngBSunRm	6/29	3975C	6262
ECHO LAKE		10SecDr2Cpt	11-12/27	3585B	6123
ECHO PEAK		10Sec2Dr	12/26-27	3584A	6031
ECHOTA		12SecDr	4-5/11	2410	3893
ECHOVILLE		12SecDr	3-6/20	2410F	4565
ECKENRODE		12SecDr	5-6/17	2410F	4497
ECKFORD		12SecDr	1-4/12	2410	3949
ECKSTEIN NORTON	N	12SecDr	4-5/11	2410	3893
ECLIPSE		16Sec	12/14-15	2412C	4304
ECONOMY		12SecDr	10/14/10	2410	3814
EDDINGTON		12SecDrCpt	6/13	2411A	4159
EDDYVILLE		12SecDr	3-6/20	2410F	4565
EDELLYN		12SecDr	2-5/21	2410F	4612
EDEN		26ChDr	4-5/17	2416C	4492
EDENBURG		12SecDr	10/10/10	2410	3813
EDENWOLD		12SecDr	9-10/13	2410A	4215
EDGAR ALLEN POE		12SecDr	1-2/29	3410B	6220
EDGECLIFF		12SecDrCpt	6/13	2411A	4159
EDGECOT		12SecDr	10-12/15	2410D	4338
EDGEHILL		12SecDr	10-12/15	2410D	4338
EDGEMERE		16Sec	10/13	2412B	4217
EDGERLY		12SecDrCpt	6/13	2411A	4159
EDGERTON		BaggClubBSmk	11/24	3415	4808
EDGEWOOD		12SecDr	12/27-28	3410B	6127
EDGEWORTH		12SecDr	11/9/10	2410	3860
EDISON		10SecDr2Cpt	5/13	2585A	4141
EDISONVILLE		12SecDr	6-9/26	3410A	4969

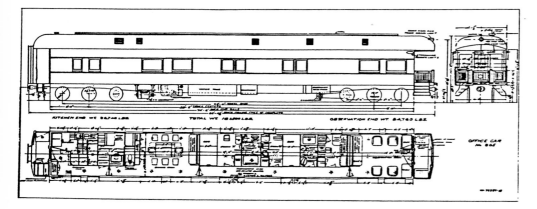

Figure I-52 The 12-section drawing room EAST NEWARK was wrecked on the B&O in 1944. In the settlement with Pullman, B&O took posession of the car and rebuilt it to B&O Business Car 905, shown here in a clearance diagram from the Author's Collection.

CAR NAME	ST	CAR TYPE	BUILT	PLAN	LOT
EDITH		28ChDr	4-6/25	3416	4864
EDLAM		26ChDr	5/23	2416D	4691
EDMINSTER		10SecDr2Cpt	6-7/15	2585B	4328
EDMORE		12SecDr	3-6/20	2410F	4565
EDMUND CARTWRIGHT		14Sec	3/27	3958	6054
EDMUND HALLEY		14Sec	10-11/26	3958	6012
EDMUND W. PETTUS		10Sec2Dr	9-10/25	3584A	4899
EDMUND W. PETTUS		10Sec2Dr	9-10/29	3584B	6284
EDUMIA		12SecDr	8-9/20	2410F	4574
EDWARD C. MARSHALL	N	10Sec2Dr	9-10/25	3584A	4899
EDWARD CARY WALTHALL		10Sec2Dr	9-10/25	3584A	4899
EDWARD CARY WALTHALL		10Sec2Dr	9-10/29	3584B	6284
EDWARD HOPKINS	N	BaggClubBSmk	10/24	3415	4786
EDWARD PREBLE		12SecDr	1-2/29	3410B	6220
EDWARD RUTLEDGE		28ChDr	4-6/25	3416	4864
EDWARD WINSLOW	N	BaggClubBSmk	10/24	3415	4786
EDWARDSVILLE		12SecDr	4-6/14	2410A	4271
EDWINA		32Ch	6/30	3916C	6363
EGERIA		12SecDr	2-3/15	2410B	4318
EGG HARBOR		12SecDrCpt	6/28/10	2411	3800
EIDOLON		16Sec	12/14-15	2412C	4304
EIFFEL TOWER	RN	8SecDr3Dbr	5-6/25	4090	4846
EILEEN		28ChDr	6-7/30	3416B	6372
EL					
ARELLANO		10SecLngObs	10/13	2521	4202
CANEY		12SecDr	8-9/20	2410F	4574
CIBOLA		10SecLngObs	10/13	2521	4202

CAR NAME	ST	CAR TYPE	BUILT	PLAN	LOT
EL					
DIAZ		10SecLngObs	10/13	2521	4202
DORADO NdeM	SN	8SecDrBLngObs	12/18	4030	4489
DUQUE JOB NdeM	SN	12SecDr	2-3/27	3410A	6055
DUQUE JOB NdeM	SN	8SecDr3Dbr	8-12/25	4090F	4894
JARAMILLO		10SecLngObs	10/13	2521	4202
MALDONADO		10SecLngObs	10/13	2521	4202
MIXTECO NdeM	SN	10SecDr2Cpt	8-9/13	2583D	4201
MONTE		10SecDr2Cpt	11-12/11	2585	3942
MOSCOSO		10SecLngObs	10/13	2521	4202
NARVAEZ		10SecLngObs	10/13	2521	4202
NEVADO NdeM	N	10SecDrCpt	9-10/17	4031	4525
NIGROMANTE NdeM	SN	12SecDr	2-4/24	3410	4762
NORTE		10SecDr2Cpt	11-12/11	2585	3942
OCCIDENTE		10SecDr2Cpt	11-12/11	2585	3942
ONATE		10SecLngObs	10/13	2521	4202
ORIENTE		10SecDr2Cpt	11-12/11	2585	3942
PENSADOR					
MEXICANO NdeM-224	N	14Sec	9-10/25	3958	4869
PIPILA NdeM	SN	12SecDr	10-12/23	3410	4724
QUIVIRA		10SecLngObs	10/13	2521	4202
SALVADOR NdeM	SN	12SecDr	10-12/15	2410i	4338
SALVADOR NdeM-142	SN	6Sec6Dbr	1-3/17	4084	4450
SUD		10SecDr2Cpt	10-12/11	2585	3942
TOTONACA NdeM	SN	10SecDr2cpt	10-12/20	2585D	4592
ULLOA		10SecLngObs	10/13	2521	4202
ZAPOTECO NdeM	SN	10SecDr2Cpt	11-12/21	2585D	4628
ZARCO NdeM	SN	12SecDr	8-10/26	3410A	4945
ELANETTE		28ChDr	4/12	2418	3951
ELATMA		12SecDr	2-5/21	2410F	4612
ELBEN		12SecDr	9/23/10	2410	3794
ELBERON		16Sec	5-6/13	2412B	4150
ELBERT H. GARY		14Sec	7/30	3958A	6376
ELBERT PEAK		10Sec2Dr	12/26-27	3584A	6031
ELBRIDGE GERRY		28ChDr	4-6/25	3416	4864
ELBURN		10SecDr2Cpt	10-11/12	2585	4067
ELDON		12SecDr	12/10-11	2410	3866
ELDORA		24ChDrB	6-7/14	2417B	4265
ELDORA	R	26StsLngBDr	6-7/14	2417J	4265
ELDRED		12SecDr	5/11/10	2410	3773
ELDRED		12SecDr	5/11/10	2410	3773
ELDRIDGE		12SecDr	12/10-11	2410	3866
ELECTRA		26ChDr	6/11	2416	3902
ELFREDA		26ChDr	4-5/13	2416A	4136
ELINDA		24ChDrB	4/16	2417C	4364
ELISHA GRAY		14Sec	3/27	3958	6054
ELITE		12SecDr	2-3/13	2410	4105
ELIZABETH		28ChDr	11/23/10	2418	3862
ELIZABETH					
BROWNING		28ChDr	1-2/27	3416A	6032

CAR NAME	ST	CAR TYPE	BUILT	PLAN	LOT	CAR NAME	ST	CAR TYPE	BUILT	PLAN	LOT
ELIZABETH						ELPHINSTONE		12SecDr	1-3/14	2410A	4249
CADY STANTON	N	28ChDr	4-6/25	3416	4864	ELRAMA		26ChDr	4-5/17	2416C	4492
ELIZABETHTOWN		12SecDr	9/16/10	2410	3794	ELSADOR		16Sec	12/14-15	2412C	4304
ELK PASS		8SecDr2Cpt	2-3/30	3979A	6353	ELSBERRY		16Sec	5-6/13	2412B	4150
ELKHART		16Sec	10-11/14	2412C	4296	ELSINORE		12SecDr	6-7/17	2410F	4503
ELKHART VALLEY		DrSbrBLngObs	4/29	3988	6221	ELSMERE		16Sec	1-5/18	2412F	4531
ELKHORN		12SecDr	12/27-28	3410B	6127	ELSWICK		12SecDr	12/10-11	2410	3866
ELKINGTON		10SecDr2Cpt	12/15	2585C	4351	ELTONBURG		12SecDr	2-5/16	2410E	4367
ELKLAND		16Sec	9/11	2412	3923	ELVEDEN		12SecDr	7/13	2410A	4179
ELKS CLUB	N	16ChB15StsLng	4/30	3996C	6312	ELVIN		12SecDr	1-2/15	2410B	4311
ELKTON		26ChDr	4-5/17	2416C	4492	EMALINDA		26ChDr	3-4/14	2416A	4264
ELKVIEW		BaggBClubSmk	9/7/10	2415	3834	EMBASSY		25ChDrObs	4/27	3957B	6065
ELLEN		28ChDr	6-7/27	3416A	6087	EMBRUN		12SecDr	10-11/11	2410	3936
ELLENDALE		12SecDr	2-5/25	3410	4845	EMBURY		12SecDr	6-7/16	2410F	4385
ELLENSBURG	N	12SecDr	2-5/16	2410E	4367	EMERALD		32ChDr	8/27	3917A	6078
ELLERY		12SecDr	2-3/15	2410B	4318	EMERALD BAY	RN	8Sec4Dbr	5-6/17	4022	4497
ELLINGTON		12SecDr	6-7/16	2410F	4385	EMERALD BEACH	RN	8Sec4Dbr	6-7/17	4022	4503
ELM BROOK	RN	12Rmt2Sbr3Dbr	2-5/25	4158	4845	EMERALD BORDER	RN	8Sec4Dbr	6-7/20	4022C	4568
ELM CITY	RN	12Rmt2Sbr3Dbr	4-6/24	4158	4763	EMERALD BROOK	RN	8Sec4Dbr	6-7/17	4022	4503
ELM CREEK	RN	12Rmt2Sbr3Dbr	6-8/24	4158	4764	EMERALD CHASM	RN	8Sec4Dbr	7-9/17	4022A	4515
ELM CREST	RN	12Rmt2Sbr3Dbr	2-5/25	4158	4845	EMERALD CREEK	RN	8Sec4Dbr	7-9/17	4022A	4515
ELM FARM	RN	12Rmt2Sbr3Dbr	2-5/25	4158	4845	EMERALD DELL	RN	8Sec4Dbr	7-9/17	4022A	4515
ELM FIELD	RN	12Rmt2Sbr3Dbr	8-12/25	4158	4894	EMERALD FALLS	RN	8Sec4Dbr	7-9/17	4022	4515
ELM FOREST	RN	12Rmt2Sbr3Dbr	6-7/17	4158	4503	EMERALD GLADE	RN	8Sec4Dbr	7-9/17	4022A	4515
ELM GLEN	RN	12Rmt2Sbr3Dbr	6-8/24	4158	4764	EMERALD GLEN	RN	8Sec4Dbr	7-9/17	4022A	4515
ELM GROVE	RN	12Rmt2Sbr3Dbr	6-7/17	4158	4503	EMERALD GORGE	RN	8Sec4Dbr	7-9/17	4022A	4515
ELM HEIGHTS	RN	12Rmt2Sbr3Dbr	6-8/24	4158	4764	EMERALD GROVE	RN	8Sec4Dbr	1-2/13	4022B	4091
ELM HILL	RN	12Rmt2Sbr3Dbr	2-5/25	4158	4845	EMERALD HILL	RN	8Sec4Dbr	1-2/13	4022B	4091
ELM LANE	RN	12Rmt2Sbr3Dbr	6-8/24	4158	4764	EMERALD ISLE	RN	8Sec4Dbr	5-6/17	4022A	4497
ELM LAWN	RN	12Rmt2Sbr3Dbr	8-12/25	4158	4894	EMERALD LAKE	RN	8Sec4Dbr	7-9/17	4022	4515
ELM LEAF	RN	12Rmt2Sbr3Dbr	4-6/24	4158	4763	EMERALD LAWN	RN	8Sec4Dbr	8-9/13	4022D	4201
ELM LODGE	RN	12Rmt2Sbr3Dbr	2-5/25	4158	4845	EMERALD LODGE	RN	8Sec4Dbr	8-9/13	4022D	4201
ELM MANOR	RN	12Rmt2Sbr3Dbr	2-5/25	4158	4845	EMERALD PARK	RN	8Sec4Dbr	12/16-17	4022B	4443
ELM PARK	RN	12Rmt2Sbr3Dbr	10/22-23	4158	4647	EMERALD PASS		8SecDr2Cpt	7-8/30	3979A	6377
ELM POINT	RN	12Rmt2Sbr3Dbr	2-5/25	4158	4845	EMERALD RAPIDS	RN	8Sec4Dbr	12/15	4022B	4351
ELM QUEEN	RN	12Rmt2Sbr3Dbr	1-3/25	4158	4844	EMERALD SEA	RN	8Sec4Dbr	12/15	4022B	4351
ELM REST	RN	12Rmt2Sbr3Dbr	6-8/24	4158	4764	EMERALD SPRING	RN	8Sec4Dbr	12/15	4022B	4351
ELM RIDGE	RN	12Rmt2Sbr3Dbr	1-3/25	4158	4844	EMERALD STREAM	RN	8Sec4Dbr	5-6/17	4022A	4497
ELM SPRINGS	RN	12Rmt2Sbr3Dbr	2-5/25	4158	4845	EMERALD SUMMIT	RN	8Sec4Dbr	10-11/12	4022B	4067
ELM TRAIL	RN	12Rmt2Sbr3Dbr	4-6/24	4158	4763	EMERALD TRAIL	RN	8Sec4Dbr	9-10/17	4022A	4525
ELM WOODS	RN	12Rmt2Sbr3Dbr	10/22-23	4158	4647	EMERALD VALE	RN	8Sec4Dbr	12/15	4022B	4351
ELMBANK		16Sec	1-5/18	2412F	4531	EMERALD WATERS	RN	8Sec4Dbr	1-5/18	4022C	4531
ELMER		26ChDr	4-5/17	2416C	4492	EMERALD WAVES	RN	8Sec4Dbr	5-6/16	4022C	4386
ELMETA		26ChDr	6/11	2416	3902	EMIGSVILLE		12SecDr	3/17/10	2410	3794
ELMINGTON		12SecDr	4-5/13	2410	4106	EMIGSVILLE		12SecDr	5-6/17	2410F	4497
ELMIRA COLLEGE	RN	10Sec2DbrCpt	12/17-18	4042B	4528	EMILY		26ChDr	3-4/14	2416A	4264
ELMO		14Sec	8-10/28	3958A	6181	EMLEN HOUSE		13Dbr	6-7/30	3997A	6314
ELMSFORD		12SecDr	11-12/14	2410B	4297	EMLENTON		26ChDr	7/28/10	2416	3804
ELOISE		26ChDr	4-5/16	2416C	4366	EMLITA		28ChDr	4/12	2418	3951
ELOTA FCP-228	SN	10SecDr2Cpt	5/13	2585E	4141	EMMA ABBOTT		16ChBLngObs	1-2/27	3961B	6035

EMMONS
Steel

CAR NAME	ST	CAR TYPE	BUILT	PLAN	LOT
EMMONS		12SecDr	4-6/14	2410A	4271
EMPIRE STATE		25ChDrObs	12/12	2669A	4053
EMPIRE STATE		BaggBSmkBarber	10-11/25	3951	4885
EMPRESS		30ChDr	9/25	3418	4911
EMPRESS CATHERINE		20ChLngLibObs	4/25	3956	4849
EMPRESS JOSEPHINE		20ChLngLibObs	4/25	3956	4849
EMPRESS VICTORIA		20ChLngLibObs	4/25	3956	4849
EMRICK		12SecDr	11-12/13	2410A	4234
EMRYTRA		16Sec	12/14-15	2412C	4304
EMSWORTH		12SecDr	10/7/10	2410	3813
ENCORSE		12SecDr	4-5/11	2410	3893
ENDERSLAKE		12SecDr	3-6/20	2410F	4565
ENDICOTT		16Sec	5-6/13	2412B	4150
ENGADINE		16Sec	6-7/20	2412F	4568
ENGINEERS CLUB		13ChRestLng	12/29	3992	6289
ENGLAND		12SecDr	8-9/20	2410F	4574
ENNERDALE		12SecDrCpt	6/29/10	2411	3800
ENOCH ENSLEY	N	12SecDr	7-8/12	2410	4014
ENOLA		28ChDr	11/22/10	2418	3862
ENON		12SecDr	4-6/14	2410A	4271
ENSENADA FCS-BC	SN	10SecDr2Cpt	5-6/24	3585	4770
ENSIGN		12SecDrCpt	6/13	2411	4151
ENTERPRISE	RN	17StsRestLng	5/12	4019A	3986
ENTRIKEN		12SecDr	5/28/10	2410	3794
ENVOY	N	12SecDr	6-10/21	2410F	4614
EPHRAIM McDOWELL		8SecDr4Dbr	6/30	4003	6339
EPICURUS		12SecDrCpt	10-12/21	2411C	4624
EPIRUS		12SecDr	2-5/16	2410E	4367
EPWORTH		12SecDr	5-8/18	2410F	4540
ERIE		12SecDr	4-6/14	2410A	4271
ERIEVILLE		16Sec	9/11	2412	3923
ERLIN		12SecDr	10-11/11	2410	3936
ERMINIE		26ChDr	4-5/16	2416C	4366
ERMITA		26ChDr	3/13	2416	4107
ERNSTON		12SecDr	4/4/10	2410	3769
ERSKINE		12SecDr	7-8/13	2410A	4192
ERVINA		26ChDr	3-4/14	2416A	4264
ESCALON		10SecDr2Cpt	8-9/13	2585A	4201
ESCANABA		12SecDr	5-8/18	2410F	4540
ESCARCEGA FCS-BC	N	8Sec5Dbr	6/21/10	4036A	3800
ESCONDIDO		10SecDr2Cpt	8-9/13	2585A	4201
ESCORT		12SecDr	12/10-11	2410	3866
ESKDALE		12SecDr	9-10/15	2410D	4327
ESMAR		12SecDr	7/13	2410A	4179
ESMOND		16Sec	1-5/18	2412F	4531
ESOPUS		12SecDr	10-11/11	2410	3936
ESPANOLA		12SecDr	2-3/27	3410A	6055
ESPLANADE		12SecDr	7-8/13	2410A	4192
ESPYVILLE		12SecDr	6-9/26	L3410A	4969
ESSEX		12SecDr	10-12/26	3410A	6023

CAR NAME	ST	CAR TYPE	BUILT	PLAN	LOT
ESTEPAR		12SecDr	2-5/21	2410F	4612
ESTES PARK		12SecDr	11-12/29	3410B	6299
ESTEVAN		12SecDr	12/10-11	2410	3866
ESTRELLA		12SecDr	8-10/26	3410A	4945
ESTRELLA					
BLANCA NdeM-137	SN	12Sec2Dbr	5-6/17	4046A	4497
DE ORIENTE NdeM-138	SN	12Sec2Dbr	8-9/20	4046A	4574
DEL NORTE NdeM-139	SN	12Sec2Dbr	8-9/20	4046A	4574
DEL SUR NdeM-140	SN	12Sec2Dbr	8-9/20	4046A	4574
POLAR NdeM-141	SN	12Sec2Dbr	8-9/20	4046A	4574
ETHAN ALLEN		12SecDr	5-8/25	3410A	4868
ETHELIND		28ChDr	4/12	2418	3951
ETHELWYN		30ChDr	11/12	2668	4051
ETHERLEY		12SecDr	3-6/20	2410F	4565
ETHIOPIA		12SecDr	2-5/21	2410F	4612
ETNA		16Sec	6-7/20	2412F	4568
ETOWAH	P	12SecDr SBS	4/17	2410G	4454
ETOWAH		12SecDr	3-6/20	2410F	4565
ETTERICK		16Sec	6-7/20	2412F	4568
EUCLID		16Sec	1-5/18	2412F	4531
EUDORA		24ChDrB	4/16	2417C	4364
EUNICE		28ChDr	4-6/25	3416	4864
EUPHEMIA		26ChDr	4/12	2416	3950
EUPOLIS		12SecDrCpt	10-12/21	2411C	4624
EURASIA		12SecDr	12/10-11	2410	3866
EUREKA		12SecDr	8-10/26	3410A	4945
EURETTA		26ChDr	3-4/14	2416A	4264
EURIPIDES		12SecDrCpt	10-12/21	2411C	4624
EURYMEDON		12SecDr	10-12/15	2410D	4338
EUTERPE		12SecDr	3-6/20	2410F	4565
EVADNE		12SecDr	7/13	2410A	4179
EVANGELINE		28ChDr	4-6/25	3416	4864
EVANSTON		12SecDr	7-8/28	3410B	6184
EVARTS		12SecDr	5-8/18	2410F	4540
EVENDALE		12SecDr	4-5/11	2410	3893
EVENING STAR		12SecDr	10/22-23	2410H	4647
EVENTIDE	RN	16DuSr	7/23	4029	4698
EVERETT		7Cpt2Dr	11/13	2522A	4222
EVERGREEN		12SecDr	2-3/13	2410	4105
EVERMOORE		16Sec	6-7/20	2412F	4568
EVERSON		26ChDr	8/1/10	2416	3804
EVINGTON		12SecDrCpt	3/11	2411	3880
EVONA		12SecDr	7/13	2410A	4179
EXCELSIOR SPRINGS	RN	10Sec3Dbr	1-3/25	4087B	4844
EXETER		16Sec	1-5/18	2412F	4531
EXLINE		12SecDr	10-11/11	2410	3936
EXMOOR CLUB		8SecBObsLng	2-3/29	3989	6229
EXPLORER		8SecDr2Cpt	5-6/29	3979A	6261
EXPORT		BaggBClubSmkBarber	8/23/10	2414A	3803
EXTON		12SecDr	2-5/21	2410F	4612

Steel

CAR NAME	ST	CAR TYPE	BUILT	PLAN	LOT
EYERSGROVE		12SecDr	6-9/26	3410A	4969
EYLAU		12SecDr	1-2/22	2410F	4625
FADLADEEN		12SecDr	2-5/21	2410F	4612
FAENZA		12SecDr	10-12/15	2410D	4338
FAIR DEAL #1	LN	6Cpt3Dr	7/25	3523A	4887
FAIR DEAL #2	LN	6Cpt3Dr	11-12/23	3523	4726
FAIR HILL		12SecDr	9/16/10	2410	3794
FAIRBANKS		12SecDr	10/5/10	2410	3813
FAIRBROOK		12SecDr	11/16/10	2410	3860
FAIRBURY		12SecDrCpt	6/13	2411	4151
FAIRCHANCE		16Sec	6/7/10	2412	3801
FAIRCHILD		12SecDr	5-8/18	2410F	4540
FAIRFAX		12SecDr	5-8/18	2410F	4540
FAIRFAX HARRISON	N	12SecDr	1-3/25	3410	4844
FAIRFIELD		12SecDr	2-5/21	2410F	4612
FAIRHAVEN		36Ch	10/24	3916	4802
FAIRLAND		12SecDr	5-6/11	2410	3903
FAIRLAWN		12SecDr	10-11/11	2410	3936
FAIRMONT		12SecDr	9-10/15	2410D	4327
FAIRMOUNT PARK		8SecDr2Cpt	5-6/29	3979A	6261
FAIRVIEW		12SecDr	9-10/15	2410D	4327
FAIRVILLE		26ChDr	7/30/10	2416	3804
FAIRWEATHER		12SecDr	11-12/13	2410A	4234
FAITH		28ChDr	8/24	3416	4801
FAITHORN		12SecDr	3-6/20	2410F	4565
FALCONSTANE		12SecDr	7/13	2410A	4179
FALKIRK		12SecDr	10-12/15	2410D	4338
FALL RIVER		18ChBLngObs	6/16	2918	4391
FALLON		12SecDr	12/27-28	3410B	6127
FALLS CITY		BaggClubBSmk	3/25	3415A	4848
FALLS CREEK		24ChDrB	3/14	2417A	4250
FALLS VIEW		12SecDrCpt	8-9/11	2411	3922
FALLSINGTON		26ChDr	3/13	2416	4107
FALLSTON		12SecDr	5-6/11	2410	3903
FALMOUTH		12SecDr	8-9/20	2410F	4574
FALSTAFF		12SecDr	2-5/21	2410F	4612
FALSTRIA		12SecDr	7-8/15	2410D	4322
FAMILY CLUB		8SecBLngSunRm	3/30	3989B	6349
FAMOUS		10SecLngObs	12/15	2521B	4353
FANETTE		28ChDr	4/12	2418	3951
FANEUIL		12SecDr	5-6/11	2410	3903
FANEUIL HALL		4CptLngObs	10/25	3960	4889
FARADAY		12SecDr	2-3/13	2410	4105
FARIBAULT		10SecDrCpt	4-6/27	3973	6043
FARLANE		12SecDr	11-12/14	2410B	4297
FARMDALE		12SecDr	4/28/10	2410	3771
FARMVILLE	N	12SecDr	5-8/25	3410A	4868
FARNESE		16Sec	6-7/20	2412F	4568
FARNHAM		8CptLng	4/11	2540	3879
FARNHURST		28ChObs	3/14	2421	4251

CAR NAME	ST	CAR TYPE	BUILT	PLAN	LOT
FARQUHAR		12SecDr	10-12/15	2410D	4338
FARRAGUT SQUARE	N	BaggBSmkBarber	12/26-27	3951C	6022
FARRALONE		12SecDr	8-10/26	3410A	4945
FARRINGTON		12SecDr	4/2/10	2410	3770
FARWELL		12SecDr	8-9/12	2410	4035
FASANO		12SecDr	8-9/20	2410F	4574
FASSETT		26ChDr	7/30/10	2416	3804
FATHER MARQUETTE	N	6Cpt3Dr	10-12/25	3523A	4922
FAULTLESS		10SecLngObs	12/15	2521B	4353
FAUNCE		12SecDr	3-4/17	2410F	4485
FAWCETT GAP		8SecDr2Cpt	12/29-30	3979A	6334
FAYETTEVILLE		12SecDr	2-4/24	3410	4762
FEARLESS		10SecLngObs	12/15	2521B	4353
FEDERAL		Private (GS)	1/18/10	2492	3812
FEDERAL HALL		4CptLngObs	10/25	3960	4889
FELDMERE	N	12SecDr	4-6/14	2410A	4271
FELIPE					
PESCADOR FCP-211	SN	10SecDr2Cpt	12/23-24	3585	4728
FELLSMERE		Bagg20Ch	8/13	2765	4196
FELTON		12SecDrCpt	6/28/10	2411	3800
FENESTRA		12SecDr	7-8/15	2410D	4322
FENNIMORE		12SecDr	4-5/13	2410	4106
FENNS		28ChDr	5-7/26	3416	4958
FENRIS		12SecDr	1-3/14	2410A	4249
FERDINAND MAGELLAN		Private	8/29	3972B	6246
FERENBAUGH		12SecDr	4-6/14	2410A	4271
FERGUS		12SecDr	4-6/14	2410A	4271
FERINTOSH		12SecDr	9-10/13	2410A	4215
FERMO		12SecDr	8-9/20	2410F	4574
FERN BANK		10Sec2Dr	1/12	2584	3945
FERN GLEN		28ChDr	6-7/27	3416A	6087
FERN LAKE		10SecDr2Cpt	11-12/27	3585B	6123
FERNANDO MONTES					
DE OCA NdeM-143	N	8SecDr2Cpt	10-12/28	3979A	6205
FERNDELL		16Sec	6-7/20	2412F	4568
FERNE		26ChDr	3-4/14	2416A	4264
FERNLEY		10SecDr2Cpt	11-12/11	2585	3942
FERNWOOD		26ChDr	3/13	2416	4107
FERRON		12SecDr	7/16	2410F	4412
FERRY FARM	N	12SecDr	10-12/23	3410	4724
FERRY POINT		10SecLngObs	12/16-17	2521C	4442
FIELD		6Cpt3Dr	8/24	3523	4804
FIELDON		16Sec	6-7/20	2412F	4568
FIELDSBORO		12SecDr	5-6/17	2410F	4497
FIFESHIRE		12SecDr	2-5/21	2410F	4612
FIGART		12SecDr	6-9/26	3410A	4969
FILLMORE		12SecDr	11/23/10	2410	3864
FILSON		12SecDr	6-9/26	3410A	4969
FINGAL		12SecDr	8-9/20	2410F	4574
FINLEY		12Sec Dr	2-5/25	3410	4845

Figure I-53 FERDINAND MAGELLAN, shown here as built in August 1929, is perhaps the best known Pullman standard car because it was sold to the United States Government in November 1942 for use by the President of the United States (F.D.R.). During its use by the President, all lettering was removed. FERDINAND MAGELLAN is being maintained in immaculate and operating condition by the Gold Coast Museum in Florida. Pullman Photograph 33771, The Robert Wayner Collection.

CAR NAME	ST	CAR TYPE	BUILT	PLAN	LOT
FIORGYN		12SecDr	7-8/15	2410D	4322
FIR CREST	RN	6Sec4Rmt4Dbt	9-10/25	4179	4899
FIR FOREST	RN	6Sec4Rmt4Dbr	12/26-27	4179	6031
FIR GARDENS	RN	6Sec4Rmt4Dbr	12/26-27	4179	6031
FIR GLEN	RN	6Sec4Rmt4Dbr	9-10/25	4179	4899
FIR HILLS	RN	6Sec4Rmt4Dbr	9-10/25	4179	4899
FIR LAKE	RN	6Sec4Rmt4Dbr	9-10/25	4179	4899
FIR PARK	RN	6Sec4Rmt4Dbr	12/24-25	4179	4836
FIR PASS	RN	6Sec4Rmt4Dbr	12/24-25	4179	4836
FIR PEAK	RN	6Sec4Rmt4Dbr	9-10/25	4179	4899
FIR RANGE	RN	6Sec4Rmt4Dbr	9-10/25	4179	4899
FIR RAPIDS	RN	6Sec4Rmt4Dbr	12/25-26	4179	4932
FIR SLOPE	RN	6Sec4Rmt4Dbr	9-10/25	4179	4899
FIR SPRINGS	RN	6Sec4Rmt4Dbr	9-10/25	4179	4899
FIR SUMMIT	RN	6Sec4Rmt4Dbr	12/25-26	4179	4932
FIR TERRACE	RN	6Sec4Rmt4Dbr	12/25-26	4179	4932
FIR TRAIL	RN	6Sec4Rmt4Dbr	12/25-26	4179	4932
FIR VALLEY	RN	6Sec4Rmt4Dbr	12/25-26	4179	4932
FIR VIEW	RN	6Sec4Rmt4Dbr	12/25-26	4179	4932
FIR WOODS	RN	6Sec4Rmt4Dbr	12/25-26	4179	4932
FIR ZONE	RN	6Sec4Rmt4Dbr	12/25-26	4179	4932
FIRST AID	LN	Dorm (Rexall)	1-5/18	2412F	4531
FIRST CITIZEN	N	8SecDr2Cpt	3/29	3979A	6237
FISHERS		12SecDr	3-4/17	2410F	4485
FITCH		6Cpt3Dr	8/24	3523	4804
FITHIAN		12SecDr	5-6/11	2410	3903
FITZGERALD		12SecDr	3-4/17	2410F	4485
FLAGDALE		12SecDr	6-9/26	3410A	4969
FLAGG		12SecDr	1-3/14	2410A	4249
FLAGLER		12SecDr	11-12/13	2410A	4234
FLAGSTAFF		7Cpt2Dr	9/13	2522A	4208

CAR NAME	ST	CAR TYPE	BUILT	PLAN	LOT
FLAMBEAU		10SecDrCpt	4-6/27	3973	6043
FLAMSTED		12SecDr	12/12-13	2410	4076
FLANDERS		12SecDr	1-2/15	2410B	4311
FLAT ROCK		10Sec2Dr	1/12	2584	3945
FLATBUSH		12SecDr	7-8/13	2410A	4192
FLAVIA		28ChDr	6-7/27	3416A	6087
FLAXTON		12SecDr	8-9/12	2410	4035
FLEETLINE	N	10SecDrCpt	4-6/27	3973	6043
FLEETNOR	N	10SecDrCpt	4-6/27	3973C	6043
FLEETON	N	10SecDrCpt	4-6/27	3973C	6043
FLEETVILLE	N	10SecDrCpt	4-6/27	3973	6043
FLEETWOOD	N	10SecDrCpt	4-6/27	3973	6043
FLEMINGTON		12SecDr	4/6/10	2410	3770
FLETCHER		12SecDr	5-6/11	2410	3903
FLEURY		12SecDr	2-5/21	2410F	4612
FLINT		12SecDr	4-6/14	2410A	4271
FLODDEN		12SecDr	2-5/21	2410F	4612
FLORENCE		26ChDr	5/12	2416	3986
FLORENCE NIGHTINGALE		16ChBLngObs	1-2/27	3961A	6035
FLORILLA		24ChDrB	6-7/14	2417B	4265
FLORIMAL		12SecDr	12/10-11	2410	3866
FLORIMUND		12SecDr	2-3/13	2410	4105
FLORISTON		10SecDr2Cpt	1-2/13	2585	4091
FLOSSMOOR CLUB		8SecBObsLng	2-3/29	3989	6229
FLOTOW		6Cpt3Dr	11-12/24	3523A	4833
FLOYD		12SecDr	10/22-23	2410H	4647
FLOYD RIVER		10SecDrCpt	8/28	3973A	6185
FLUSHING		16Sec	6-7/13	2412B	4160
FLYING CLOUD		12ChDrBLngSunRm	2-3/30	4002	6326
FLYING CLOUD(2nd)	N	12ChDrBLngSunRm	7/30	4002	6364
FLYING FISH		12ChDrBLngSunRm	2-3/30	4002	6326
FLYING FISH(2nd)	N	12ChDrBLngSunRm	2-3/30	4002	6326
FOLCROFT		26ChDr	5/12	2416	3986
FOLKSTONE		16Sec	6-7/13	2412B	4160
FOLLANSBEE		12SecDr	5-6/23	2410H	4699
FOLSOM		12SecDr	7/16	2410F	4412
FOLWELL		12SecDr	5-6/17	2410F	4497
FONITA		12SecDr	12/10-11	2410	3866
FONSO		12SecDr	5-7/13	2410	4149
FONTANET		8CptLng	4/11	2540	3879
FONTANET	R	7CptBLng	4/11	2540F	3879
FONTENELLE		12SecDr	11-12/29	3410B	6299
FONTENOY		12SecDr	2-5/16	2410E	4367
FORD CITY		28ChDr	6-7/27	3416A	6087
FORDHAM UNIVERSITY	RN	12Sec2Dbr	7-9/17	4046	4515
FORDNEY		12SecDr	5-6/11	2410	3903
FORDWICK		12SecDr	7-8/12	2410	4014
FOREMOST		10SecLngObs	12/15	2521B	4353
FOREST LAWN		10Sec2Dr	1/12	2584	3945
FOREST PARK		10SecLngObs	12/15	2521B	4362

CAR NAME	ST	CAR TYPE	BUILT	PLAN	LOT
FORESTPORT		12SecDr	3-4/11	2410	3881
FORGRAY		12SecDr	2-5/16	2410E	4367
FORMIGNY		12SecDr	1/16	2410E	4354
FORMOSA		12SecDr	8-9/20	2410F	4574
FORNOVO		12SecDr	1/16	2410E	4354
FORQUER		12SecDr	2-3/13	2410	4105
FORREST		12SecDr	3-6/20	2410F	4565
FORREST HILLS		36Ch	8/27	3916A	6077
FORREST LAKE		4Cpt2DrLngObs	3/13	2703	4100
FORSYTH		12SecDr	3-6/20	2410F	4565
FORT ADAMS		10SecDr2Cpt	10-12/20	2585D	4592
FORT AMADOR		10SecDr2Cpt	10-12/20	2585D	4592
FORT ANDREWS		10SecDr2Cpt	10-12/20	2585D	4592
FORT BANKS		10SecDr2Cpt	10-12/20	2585D	4592
FORT BLISS		10SecDr2Cpt	10-12/20	2585D	4592
FORT CANBY		10SecDr2Cpt	10-12/20	2585D	4592
FORT CARROLL		10SecDr2Cpt	10-12/20	2585D	4592
FORT CASEY		10SecDr2Cpt	10-12/20	2585D	4592
FORT CLARK		10SecDr2Cpt	10-12/20	2585D	4592
FORT CRALO MANSION		13Dbr	8/30	3997A	6392
FORT CROOK		10SecDr2Cpt	10-12/20	2585D	4592
FORT DADE		10SecDr2Cpt	10-12/20	2585D	4592
FORT DAVIS		10SecDr2Cpt	10-12/20	2585D	4592
FORT DEARBORN		6SbrLngSunRm	8/28	3974A	6183
FORT DEARBORN	R	6Sbr2DbrObsLng	8/28	3974D	6183
FORT DODGE		10SecDr2Cpt	10-12/20	2585D	4592
FORT ERIE		10Sec2Dr	1/12	2584	3945
FORT EUSTIS	N	6SecDr2Dbr2Cpt	6/11	2547C	3892
FORT FOSTER		10SecDr2Cpt	10-12/20	2585D	4592
FORT GAINES		10SecDr2Cpt	10-12/20	2585D	4592
FORT GETTY		10SecDr2Cpt	10-12/20	2585D	4592
FORT GIBBON		10SecDr2Cpt	10-12/20	2585D	4592
FORT GRANT		10SecDr2Cpt	10-12/20	2585D	4592
FORT GREBLE		10SecDr2Cpt	10-12/20	2585D	4592
FORT GREENE		10SecDr2Cpt	10-12/20	2585D	4592
FORT HEATH		10SecDr2Cpt	10-12/20	2585D	4592
FORT HILL	N	40Ch8StsBDrObs	8/13	2766D	4197
FORT HUNT		10SecDr2Cpt	10-12/20	2585D	4592
FORT HUNTER		10Sec2Dr	5/13	2584	4140
FORT JAY		10SecDr2Cpt	10-12/20	2585D	4592
FORT KENT		10SecDr2Cpt	10-12/20	2585D	4592
FORT KEOGH		10SecDr2Cpt	10-12/20	2585D	4592
FORT LEE	N	14Sec	9-10/25	3958	4869
FORT LEVETT		10SecDr2Cpt	10-12/20	2585D	4592
FORT LISCUM		10SecDr2Cpt	10-12/20	2585D	4592
FORT LOGAN		10SecDr2Cpt	10-12/20	2585D	4592
FORT LYNN		10SecDr2Cpt	10-12/20	2585D	4592
FORT MADISON		10SecLngObs	9-10/14	2521A	4292
FORT MASON		10SecDr2Cpt	10-12/20	2585D	4592
FORT MEADE		10SecDr2Cpt	10-12/20	-2585D	4592

CAR NAME	ST	CAR TYPE	BUILT	PLAN	LOT
FORT MICHIE		10SecDr2Cpt	10-12/20	2585D	4592
FORT MILEY		10SecDr2Cpt	10-12/20	2585D	4592
FORT MIMS		10SecDr2Cpt	10-12/20	2585D	4592
FORT MONROE	N	12SecDr	1-3/14	2410A	4249
FORT MOTT		10SecDr2Cpt	10-12/20	2585D	4592
FORT MYER		10SecDr2Cpt	10-12/20	2585D	4592
FORT OMAHA		6SbrLngSunRm	8/28	3974A	6183
FORT OMAHA	R	65br2DbrObsLng	8/28	3974D	6183
FORT PICKENS		10SecDr2Cpt	10-12/20	2585D	4592
FORT PIERCE		10SecLngObs	11/13	2521	4221
FORT PORTER		10SecDr2Cpt	10-12/20	2585D	4592
FORT RILEY		10SecDr2Cpt	10-11/13	2585A	4218
FORT SILL		10SecDr2Cpt	10-12/20	2585D	4592
FORT SLOCUM		10SecDr2Cpt	10-12/20	2585D	4592
FORT SNELLING		10SecDr2Cpt	10-12/20	2585D	4592
FORT SPRING		12SecDr	2-3/13	2410	4105
FORT STEVENS		10SecDr2Cpt	10-12/20	2585D	4592
FORT STRONG		10SecDr2Cpt	10-12/20	2585D	4592
FORT TAYLOR		10SecDr2Cpt	10-12/20	2585D	4592
FORT TERRY		10SecDr2Cpt	10-12/20	2585D	4592
FORT THOMAS		10SecDr2Cpt	10-12/20	2585D	4592
FORT TRAVIS		10SecDr2Cpt	10-12/20	2585D	4592
FORT UNION	RN	8SecDrBLngObs	12/18	4030	4489
FORT WARD		10SecDr2Cpt	10-12/20	2585D	4592
FORT WASHINGTON	RN	8SecDrBLngObs	12/18	4030	4489
FORT WOOD		10SecDr2Cpt	10-12/20	2585D	4592
FORT WORDEN		10SecDr2Cpt	10-12/20	2585D	4592
FORT WORTH		12SecDr	10/22-23	2410H	4647
FORTIN NdeM	SN	14Sec	7/30	3958A	6376
FORTUNATA		6Sec5Cpt	6/11	2547	3892
FORTVILLE		12SecDr	9-10/11	2410	3924
FOSSTON		16Sec	2/13	2412	4101
FOSTER		12SecDr	10/22-23	2410H	4647
FOSTORIA		10SecDr2Cpt	10-11/13	2585A	4218
FOUNTAIN		12SecDr	2-3/13	2410	4105
FOURNET		12SecDrCpt	6/13	2411A	4159
FOWLER		12SecDr	5-6/11	2410	3903
FOX HILLS CLUB		8SecBObsLng	2-3/29	3989	6229
FOX OAKS		8SecDr2Cpt	2-3/30	3979A	6353
FOX POINT		18ChBLngObs	6/16	2918	4391
FOX RIDGE		12SecDrCpt	7/15/10	2411	3800
FOX RIVER		10SecDrCpt	8/28	3973A	6185
FOXBURG		12SecDr	4/12/10	2410	3771
FOXGLOVE		28ChDr	5-7/26	3416	4958
FOXMEAD		12SecDr	11-12/13	2410A	4234
FOXON		Bagg20Ch	8/13	2765	4196
FRADEL		12SecDr	1-3/14	2410A	4249
FRAMINGHAM		36Ch SBS	1/13	2691	4056
FRANCE		12SecDr	8-9/20	2410F	4574
FRANCES		28ChDr	8/24	3416	4801

CAR NAME	ST	CAR TYPE	BUILT	PLAN	LOT
FRANCES BURNETT		28ChDr	1-2/27	3416A	6032
FRANCES E. WILLARD	N	28ChDr	5-7/26	3416	4958
FRANCINA		26ChDr	3-4/14	2416A	4264
FRANCIS HOPKINSON		12SecDr	4/26	3410A	4943
FRANCIS KEY		12SecDr	5-8/25	3410A	4894
FRANCIS LYNDE STETSON	N	10Sec2Dr	9-10/25	3584A	4899
FRANCIS MARION		10Sec2Dr	9-10/25	3584A	4899
FRANCIS MARION		10Sec2Dr	9-10/29	3584B	6284
FRANCIS SCOTT KEY	N	12SecDr	8-12/25	3410A	4894
FRANCIS T. NICHOLLS		10Sec2Dr	9-10/25	3584A	4899
FRANCIS T. NICHOLLS		10Sec2Dr	9-10/29	3584B	6284
FRANCISCO JAVIER					
MINA NdeM-144	N	8SecDr2Cpt	10-12/28	3979A	6205
FRANCISCO					
MARQUEZ NdeM-145	N	8SecDr2Cpt	10-12/28	3979A	6205
ZARCO NdeN-146	SN	12SecDr	5-8/25	3410A	4868
FRANCOIS					
XAVIER MARTIN		10Sec2Dr	9-10/29	3584B	6284
XAVIER MARTIN		10Sec2Dr	9-10/25	3584A	4899
FRANKENTROST		12SecDr	4-6/14	2410A	4271
FRANKFORD		12SecDr	2-5/16	2410E	4367
FRANKLIN		BaggBClubSmkBarber	2-3/12	2602	3948
FRANKLIN SQUARE		6Cpt3Dr	7/25	3523A	4887
FRANKTON		12SecDr	10/6/10	2410	3813
FRANKTON		26ChDr	4-5/17	2416C	4492
FRAZER		12SecDrCpt	6/21/10	2411	3800
FREDERICK		12SecDr	12/10-11	2410	3866
FREDERICK MUHLENBERG	N	12SecDr	2-3/27	3410A	6055
FREDERICKTOWN		12SecDr	2-4/24	3410	4762
FREDONIA		12SecDr	5-6/11	2410	3903
FREEDOM		12SecDr	10/11/10	2410	3814
FREEHOLD		26ChDr	2-3/14	2416A	4248
FREEHOLD	N	30ChDr	9-10/24	3418	4783
FREEMAN		12SecDrCpt	7/8/10	2411	3800
FREMONT		12SecDr	8-10/26	3410A	4945
FREMONT PASS		8SecDr2Cpt	12/29-30	3979A	6334
FREMONT'S PEAK		10Sec2Dr	12/26-27	3584A	6031
FRENCHTOWN		12SecDr	10-12/26	3410A	6023
FRENIER		12SecDr	9-10/13	2410A	4215
FRESNO		12SecDr	8-10/26	3410A	4945
FRIARS CLUB		13ChRestLng	12/29	3992	6289
FRIENDSHIP		12SecDr	2-4/24	3410	4762
FRIENDSHIP	N	Private	8/27	3972	6037
FRODIN	N	12SecDr	4-6/14	2410A	4271
FRONTENAC		10SecDrCpt	4-6/27	3973	6043
FRUGALITY		12SecDr	2-4/24	3410	4762
FRUITLAND		12SecDr	1-2/15	2410B	4311
FRUITVALE		12SecDr	1-2/15	2410B	4311
FRYE		28ChDr	4-6/25	3416	4864
FUCINO		12SecDr	1-2/22	2410F	4625

CAR NAME	ST	CAR TYPE	BUILT	PLAN	LOT
FULFORD		12SecDr	10-12/15	2410D	4338
FULLER E. CALLAWAY	RN	10SecDr2Dbr	9-10/29	4074D	6284
FULTON CHAIN		10Sec2Dr	1/12	2584	3947
FULTONHAM		12SecDr	5-6/11	2410	3903
FULTONVILLE		12SecDr	10-11/11	2410	3936
FUNDY		12SecDr	8-9/20	2410F	4574
FUSAN		12SecDr	11-12/14	2410B	4297
GADSDEN		12SecDr	5-7/13	2410	4149
GADSHILL		12SecDr	3-6/20	2410F	4565
GAIL BORDEN	N	14Sec	9-10/25	3958	4869
GAINFORD		12SecDr	9-10/13	2410A	4215
GAINSBOROUGH		6Cpt3Dr	7/25	3523A	4887
GALATA TOWER	RN	8SecDr3Dbr	5-6/25	4090	4846
GALAX	N	12SecDr	5-8/25	3410A	4868
GALENA		12SecDr	10/5/10	2410	3813
GALESBURG		10SecDr2Cpt	12/16-17	2585D	4443
GALETON		16Sec	2/13	2412	4101
GALEWOOD		12SecDr	1-3/14	2410A	4249
GALIBON		12SecDr	4-6/21	2410F	4613
GALICIA		12SecDr	8-9/20	2410F	4574
GALINARA		12SecDr	2-5/21	2410F	4612
GALLANT		10SecLngObs	12/15	2521B	4353
GALLATIN CANYON	N	3CptDrLngBSunRm	6/29	3975U	6262
GALLATZIN		BaggBClubSmk	7/13	2415	4158
GALLINAS		12SecDr	1-3/17	2410F	4450
GALLUP		7Cpt2Dr	9/13	2522A	4208
GALT		24ChDrB	4/16	2417C	4364
GALTIER		14Sec	6/28	3958A	6174
GALVA		12SecDr	4-6/21	2410F	4613
GAMBIER		24ChDrB	1/14	2417A	4239
GAMBOA		12SecDr	7/13	2410A	4179
GAME COCK		20ChBLng	2/30	3999	6327
GANANOQUE		12SecDr	2-5/21	2410F	4612
GANISTER		12SecDr	12/15/10	2410	3864
GANNETT		10SecDr2Cpt	11-12/11	2585	3942
GANNETT PEAK		10Sec2Dr	12/28-29	3584B	6213
GANYMEDE		12SecDr	7-8/15	2410D	4322
GAP		28ChDr	11/23/10	2418	3862
GARDA		12SecDr	7-8/13	2410A	4192
GARDEN ARBOR	RN	8Sec5Dbr	6/25/10	4036	3800
GARDEN BROOK	RN	8Sec5Dbr	6/28/10	4036	3800
GARDEN CANYON	RN	8Sec5Dbr	6/15/10	4036	3800
GARDEN CITY		10SecDr2Cpt	10/12	2585	4039
GARDEN DALE	R	8Sec5Dbr	3/11	4036	3880
GARDEN DELL	RN	8Sec5Dbr	3/11	4036	3880
GARDEN GLADE	RN	8Sec5Dbr	6/13	4036	4159
GARDEN GLEN	RN	8Sec5Dbr	6/13	4036	4159
GARDEN GROVE	RN	8Sec5Dbr	6/29/10	4036	3800
GARDEN HEIGHTS	RN	8Sec5Dbr	6/13	4036	4159
GARDEN HILL	RN	8Sec5Dbr	3/11	4036	3880

CAR NAME	ST	CAR TYPE	BUILT	PLAN	LOT
GARDEN HOME	RN	8Sec5Dbr	6/30/10	4036	3800
GARDEN ISLE	RN	8Sec5Dbr	6/22/10	4036	3800
GARDEN LANE	RN	8Sec5Dbr	3/11	4036	3880
GARDEN STATE	RN	8Sec5Dbr	7/16/10	4036	3800
GARDEN VALLEY	RN	8Sec5Dbr	8-9/11	4036	3922
GARDENCOURT	N	12SecDr	5-6/11	2410	3903
GARDERE		12SecDr	11-12/13	2410A	4234
GARDONE		12SecDr	12/12-13	2410	4076
GARETH		16Sec	6-7/20	2412F	4568
GARFIELD		12SecDr	5-8/18	2410F	4540
GARLAND		28ChDr	4-6/25	3416	4864
GARMAN		12SecDr	11/26/10	2410	3864
GARNET		12SecDr	1-2/22	2410F	4625
GARRISON CLUB		8SecBLngSunRm	10/30	3989C	6393
GARTHBY		12SecDr	2-3/15	2410B	4318
GARTHOWEN		12SecDr	2-5/21	2410F	4612
GARWAY		26ChDr	3/13	2416	4107
GARY		24ChDrB	4/16	2417C	4364
GASCONADE		12SecDr	4-5/13	2410	4106
GASPARD		12SecDr	4-6/21	2410F	4613
GASTON		12SecDr	11-12/13	2410A	4234
GATESIDE		12SecDr	1-3/14	2410A	4249
GAULEY	N	12SecDr	1-3/14	2410A	4249
GAVARNIE		12SecDr	2-5/16	2410E	4367
GAYLORD		16Sec	7-8/11	2412	3912
GAYTHORNE		12SecDr	5-7/13	2410	4149
GEDDES		12SecDr	3-4/17	2410F	4485
GEN PEDRO ANTONIO					
de LEON NdeM	SN	12SecDr	2-4/24	3410	4762
GENDOVA		12SecDr	12/10-11	2410	3866
GENERAL					
A.H. TERRY		14Sec	6/30	3958B	6373
A.H. TERRY	R	6Sec6Dbr	6/30	4084B	6373
ANAYA NdeM	SN	12SecDr	2-5/25	3410	4845
BENTEEN		8SecDr2Cpt	5-6/29	3979A	6261
CANBY		8SecDr2Cpt	5-6/29	3979A	6261
CARR		10SecLibLng	1-2/30	3521C	6323
CHAFFEE		10SecLibLng	1-2/30	3521C	6323
CROOK		10SecLibLng	1-2/30	3521C	6323
CUSTER		8SecDr2Cpt	5-6/29	3979A	6261
E.S. GODFREY		14Sec	6/30	3958A	6373
E.S. GODFREY	R	6Sec6Dbr	6/30	4084A	6373
EARLY		10SecLibLng	1-2/30	3521C	6323
EWELL		10SecLibLng	1-2/30	3521C	6323
FORREST		10SecLibLng	1-2/30	3521C	6323
FREMONT		8SecDr2Cpt	5-6/29	3979A	6261
GIBBON		8SecDr2Cpt	5-6/29	3979A	6261
HANCOCK		10SecLibLng	1-2/30	3521C	6323
HILL		10SecLibLng	1-2/30	3521C	6323
HOWARD		10SecLibLng	1-2/30	3521C	6323

CAR NAME	ST	CAR TYPE	BUILT	PLAN	LOT
GENERAL					
KEARNY		10SecLibLng	1-2/30	3521C	6323
LAFAYETTE		12SecDr	8-12/25	3410A	4894
LAWTON		10SecLibLng	1-2/30	3521C	6323
LEAVENWORTH		8SecDr2Cpt	5-6/29	3979A	6261
LONGSTREET		10SecLibLng	1-2/30	3521C	6323
MERRILL		8SecDr2Cpt	5-6/29	3979A	6261
MERRITT		8SecDr2Cpt	5-6/29	3979A	6261
MILES		8SecDr2Cpt	5-6/29	3979A	6261
MILLS		10SecLibLng	1-2/30	3521C	6323
OBREGON	N	10SecDrCpt	8/29	3973A	6273
OBREGON FCP-201	SN	10SecDr2Cpt	12/25-26	3585A	4933
PICKETT		10SecLibLng	1-2/30	3521C	6323
POLK		10SecLibLng	1-2/30	3521C	6323
SCHOFIELD		10SecLibLng	1-2/30	3521C	6323
SCOTT		8SecDr2Cpt	5-6/29	3979A	6261
SHERIDAN		8SecDr2Cpt	5-6/29	3979A	6261
SHERMAN		8SecDr2Cpt	5-6/29	3979A	6261
STONEMAN		10SecLibLng	1-2/30	3521C	6323
STUART		10SecLibLng	1-2/30	3521C	6323
SULLY		8SecDr2Cpt	5-6/29	3979A	6261
W.C.BROWN		8SecDr2Cpt	5-6/29	3979A	6261
GENESSEE VALLEY	N	SbrDrBLngObs	7/30	3988A	6356
GENESSEE VALLEY		SbrDrBLngObs	7/30	3988A	6356
GENEVA		8CptLng	4/11	2540	3879
GENEVA	R	7CptBLng	4/11	2540F	3879
GENEVIEVE		28ChDr	8/24	3416	4801
GENOA		12SecDr	8-9/20	2410F	4574
GENTIAN		28ChDr	5-7/26	3416	4958
GENTILLY		12SecDr	7/16	2410F	4412
GEORGE ARMISTEAD		12SecDr	4/26	3410A	4943
GEORGE B. HARRIS		8SecDr2Cpt	12/29-30	3979A	6334
GEORGE B. WINSHIP		8SecDr2Cpt	5-6/29	3979A	6261
GEORGE F. BAKER		LngBBarberSunRm	5-6/29	3990	6249
GEORGE MASON		12SecDr	5-8/25	3410A	4868
GEORGE PEABODY	N	12SecDr	8-10/26	3410A	4945
GEORGE POINDEXTER		10Sec2Dr	9-10/25	3584A	4899
GEORGE POINDEXTER		10Sec2Dr	9-10/29	3584A	6284
GEORGE POINDEXTER	R	10SecDr2Dbr	9-10/29	4074D	6284
GEORGE ROGERS CLARK		14Sec	3-4/30	3958A	6357
GEORGE STEPHENSON		14Sec	3/27	3958	6054
GEORGE W. CABLE	N	12SecDr	2-5/25	3410	4845
GEORGE W. HOLDREGE		8SecDr2Cpt	12/29-30	3979A	6334
GEORGE WARING		14Sec	10-11/26	3958	6012
GEORGE WASHINGTON		12SecDr	8-12/25	3410A	4894
GEORGE WASHINGTON		20ChBLng	9/30	3999A	6383
GEORGE WESTINGHOUSE	N	8SecDr2Cpt	7-8/30	3979A	6377
GEORGE WYTHE		10Sec2Dr	9-10/25	3584A	4899
GEORGE WYTHE		10Sec2Dr	9-10/29	3584B	6284
GEORGE WYTHE	N	8SecDr2Cpt	8-9/29	3979A	6283

CAR NAME	ST	CAR TYPE	BUILT	PLAN	LOT
GEORGETOWN UNIVERSITY	RN	12Sec2Dbr	7-9/17	4046	4515
GEORGETTE		26ChDr	3-4/14	2416A	4264
GEORGIA		12SecDr	3-6/20	2410F	4565
GEORGIAN BAY		3Sbr2CptDrBLngSunR	11/29	3991	6276
GERANDO		12SecDr	12/10-11	2410	3866
GERLANE		10SecDr2Cpt	8-9/13	2585A	4201
GERMANIA		12SecDr	11/9/10	2410	3860
GERMANTOWN		12SecDr	10/28/10	2410	3860
GERTHA		12SecDr	7-8/15	2410D	4322
GERTRUDE		28ChDr	4-6/25	3416	4864
GERVILLE		26ChDr	4/12	2416	3950
GETTYSBURG		12SecDr	10/7/10	2410	3813
GIANT BEND	RN	12SecDr	3-6/20	2410F	4565
GIANT CAVE	RN	12SecDr	3-6/20	2410F	4565
GIANT CITY	N	12SecDr	11-12/13	2410A	4234
GIANT CRAG	N	12SecDr	11/14/10	2410	3860
GIANT DIKE	N	12SecDr	11-12/13	2410A	4234
GIANT DOME	N	12SecDr	12/10-11	2410	3866
GIANT ECHO	N	12SecDr	2-3/13	2410	4105
GIANT ELM	N	12SecDr	5-8/18	2410F	4540
GIANT GLEN	N	12SecDr	4-5/11	2410	3893
GIANT GLOW	N	12SecDr	10-12/15	2410D	4338
GIANT HILL	N	12SecDr	9-10/11	2410	3924
GIANT MILL	N	12SecDr	12/20-21	2410F	4611
GIANT OAK	N	12SecDr	4-6/14	2410A	4271
GIANT PALM	N	12SecDr	9-10/13	2410A	4215
GIANT PARK	N	12SecDr	5/2/10	2410	3772
GIANT PASS	N	12SecDr	4-6/14	2410A	4271
GIANT PEAK	RN	12SecDr	1-3/14	2410A	4249
GIANT PINE	N	12SecDr	1-3/17	2410F	4450
GIANT REEF	N	12SecDr	7-8/13	2410A	4192
GIANT ROCK	N	12SecDr	1-4/12	2410	3949
GIANT STAR	N	12SecDr	12/10-11	2410	3866
GIANT VIEW	N	12SecDr	6/13/10	2410	3794
GIANT WOOD	N	12SecDr	4-6/14	2410A	4271
GIBSONBURG		12SecDr	6-9/26	3410A	4969
GIBSONIA		12SecDr	3-6/20	2410F	4565
GIBSONTON		12SecDr	4-6/14	2410A	4271
GIDEON		12SecDr	8-9/20	2410F	4574
GIFFORD		12SecDr	12/10-11	2410	3866
GILBERT		12SecDr	12/12-13	2410	4076
GILBERTON		BaggBClubSmk	7/13	2415	4158
GILCREST		16Sec	2/13	2412	4101
GILDER		12SecDrCpt	10-12/21	2411C	4624
GILIA	N	12SecDr	11-12/14	2410B	4297
GILLESPIE		12SecDr	3-4/17	2410F	4485
GILLETT		12SecDr	10-12/16	2410F	4431
GILMAN		12SecDr	5-6/11	2410	3903
GILMORE		12SecDr	1-2/15	2410B	4311
GILOLO		12SecDr	4-6/21	2410F	4613

CAR NAME	ST	CAR TYPE	BUILT	PLAN	LOT
GILPIN		16Sec	5-6/16	2412F	4386
GILROY		12SecDr	1-2/16	2410E	4356
GIOTTO'S TOWER	RN	8SecDr3Dbr	5-6/25	4090	4846
GIRDLESTONE		16Sec	6-7/20	2412F	4568
GIRDLETREE		26ChDr	7/28/10	2416	3804
GIRVAN		12SecDr	1-2/15	2410B	4311
GIRVESEND		12SecDr	1-3/17	2410F	4450
GLACIER		12SecDr	4-6/24	3410	4763
GLACIER PARK	RN	2Cpt3Dr2PvtRm	10-12/25	3962	4922
GLACIER PARK	R	4Dr4Cpt	10-12/25	3962A	4922
GLACIER PEAK		10Sec2Dr	12/28-29	3584B	6213
GLADE		28ChDr	4-6/25	3416	4864
GLADIATOR		12SecDr	8-9/20	2410F	4574
GLADIOLUS		28ChDr	5-6/24	3416	4761
GLADSTONE		12SecDr	1-3/14	2410A	4249
GLADWIN		16Sec	7-8/11	2412	3912
GLAMIS		12SecDr	7-8/15	2410D	4322
GLASBY		16Sec	10-11/14	2412C	4296
GLASGOW		12SecDr	4-6/21	2410F	4613
GLASSBORO		12SecDrCpt	7/8/10	2411	3800
GLASSMERE		28ChDr	11/23/10	2418	3862
GLASSPORT		12SecDr	4-6/14	2410A	4271
GLEASON		16Sec	6-7/20	2412F	4568
GLEN ADAIR		6Cpt3Dr	8-10/28	3523C	6182
GLEN ADELAIDE		6Cpt3Dr	7-8/26	3523A	4970
GLEN AIRLIE		6Cpt3Dr	7-8/26	3523A	4970
GLEN ALADALE		6Cpt3Dr	7-8/26	3523A	4970
GLEN ALICE		6Cpt3Dr	8-10/28	3523C	6182
GLEN ALLEN		6Cpt3Dr	1-2/28	3523C	6124
GLEN ALMOND		6Cpt3Dr	7-8/26	3523A	4970
GLEN ALPINE		10SecLngObs	2/24/11	2521	3865
GLEN ALTA		6Cpt3Dr	10-12/25	3523A	4922
GLEN ANNA		6Cpt3Dr	8-10/28	3523C	6182
GLEN ARAY		6Cpt3Dr	7-8/26	3523A	4970
GLEN ARBOR		6Cpt3Dr	10-12/25	3523A	4922
GLEN ARCH		6Cpt3Dr	8-10/28	3523C	6182
GLEN ARDEN		6Cpt3Dr	10-12/29	3523C	6290
GLEN ASHDALE		6Cpt3Dr	7-8/26	3523A	4970
GLEN ATHOL		6Cpt3Dr	8-10/28	3523C	6182
GLEN AUBREY		6Cpt3Dr	8-10/28	3523C	6182
GLEN AYR		6Cpt3Dr	8-10/28	3523C	6182
GLEN BAY		6Cpt3Dr	10-12/25	3523A	4922
GLEN BEACH		6Cpt3Dr	8-10/28	3523C	6182
GLEN BLAIR		6Cpt3Dr	1-2/28	3523C	6124
GLEN BROOK		6Cpt3Dr	10-12/25	3523A	4922
GLEN CAMBRIA		6Cpt3Dr	1/30	3523C	6341
GLEN CAMPSIE		6Cpt3Dr	7-8/26	3523A	4970
GLEN CANYON		6Cpt3Dr	10-12/25	3523A	4922
GLEN CASTLE		6Cpt3Dr	10-12/25	3523A	4922
GLEN CEDAR		6Cpt3Dr	10-12/25	3523A	4922

Figure I-54   GLEN BLAIR, a 6-compartment, 3-drawing room car, remained in Pullman ownership until June 1967 when it was sold to Darby Wood Products of Hagerstown, Maryland. This premier accommodation car was parked at Brunswick, Maryland (as shown above) for about 13 years. Time has taken its toll. Photograph by Dave J. Spanagel in October 1983, the Howard Ameling Collection.

CAR NAME	ST	CAR TYPE	BUILT	PLAN	LOT	CAR NAME	ST	CAR TYPE	BUILT	PLAN	LOT
GLEN CITY		6Cpt3Dr	8-10/28	3523C	6182	GLEN FEE		6Cpt3Dr	7-8/26	3523A	4970
GLEN CLIFF		6Cpt3Dr	10-12/25	3523A	4922	GLEN FERRY		6Cpt3Dr	10-12/25	3523A	4922
GLEN CRAG		6Cpt3Dr	10-12/25	3523A	4922	GLEN FLORA		6Cpt3Dr	10-12/25	3523A	4922
GLEN CREEK		6Cpt3Dr	10-12/25	3523A	4922	GLEN FORGE		6Cpt3Dr	10-12/29	3523C	6290
GLEN CREST		6Cpt3Dr	10-12/29	3523C	6290	GLEN FRAZER		6Cpt3Dr	10-12/29	3523C	6290
GLEN CRYSTAL		6Cpt3Dr	8-10/28	3523C	6182	GLEN GARDNER		6Cpt3Dr	8-10/28	3523C	6182
GLEN DALE		6Cpt3Dr	10-12/25	3523A	4922	GLEN GARRY		6Cpt3Dr	10-12/25	3523A	4922
GLEN DEE		6Cpt3Dr	7-8/26	3523A	4970	GLEN GORDON		6Cpt3Dr	1-2/28	3523C	6124
GLEN DELTA		6Cpt3Dr	10-12/29	3523C	6290	GLEN GROVE		6Cpt3Dr	10-12/25	3523A	4922
GLEN DEVON		6Cpt3Dr	10-12/25	3523A	4922	GLEN HAVEN		10SecLngObs	12/15	2521B	4362
GLEN DOCHART		6Cpt3Dr	7-8/26	3523A	4970	GLEN HAZEL		24ChDrB	3/14	2417A	4250
GLEN DOLL		6Cpt3Dr	7-8/26	3523A	4970	GLEN HOLLOW		6Cpt3Dr	10-12/25	3523A	4922
GLEN DOUGLAS		6Cpt3Dr	7-8/26	3523A	4970	GLEN HOPE		6Cpt3Dr	7-8/26	3523A	4970
GLEN DOWER		6Cpt3Dr	1-2/28	3523C	6124	GLEN HURON		6Cpt3Dr	1-2/28	3523C	6124
GLEN EAGLES		6Cpt3Dr	10-12/25	3523A	4922	GLEN IRWIN		6Cpt3Dr	10-12/29	3523C	6290
GLEN ECHO		6Cpt3Dr	7-8/26	3523A	4970	GLEN ISLAND		6Cpt3Dr	10-12/25	3523A	4922
GLEN ELDER		6Cpt3Dr	10-12/25	3523A	4922	GLEN KARN		6Cpt3Dr	10-12/29	3523C	6290
GLEN ELLYN		6Cpt3Dr	10-12/25	3523A	4922	GLEN LAKE		6Cpt3Dr	10-12/25	3523A	4922
GLEN ELM		6Cpt3Dr	8-10/28	3523C	6182	GLEN LAWN		6Cpt3Dr	10-12/29	3523C	6290
GLEN ESK		6Cpt3Dr	7-8/26	3523A	4970	GLEN LOCH		28ChDr	11/23/10	2418	3862
GLEN EWEN		6Cpt3Dr	10-12/29	3523C	6290	GLEN LODGE		6Cpt3Dr	10-12/25	3523A	4922
GLEN EYRE		6Cpt3Dr	10-12/25	3523A	4922	GLEN LOMA LINDA	N	6Cpt3Dr	1-2/28	3523C	6124
GLEN FALLS		6Cpt3Dr	10-12/25	3523A	4922	GLEN LUCE		6Cpt3Dr	7-8/26	3523A	4970
GLEN FALLS		6Cpt3Dr	8-10/28	3523C	6182	GLEN MAIN		6Cpt3Dr	7-8/26	3523A	4970
GLEN FARM		6Cpt3Dr	10-12/29	3523C	6290	GLEN MAJOR		6Cpt3Dr	1-2/28	3523C	6124

CAR NAME	ST	CAR TYPE	BUILT	PLAN	LOT	CAR NAME	ST	CAR TYPE	BUILT	PLAN	LOT
GLEN MANOR		6Cpt3Dr	10-12/25	3523A	4922	GLEN TURRET		6Cpt3Dr	8-9/26	3523A	4974
GLEN MASSAN		6Cpt3Dr	8-9/26	3523A	4974	GLEN VALLEY		6Cpt3Dr	10-12/25	3523A	4922
GLEN MAWR		6Cpt3Dr	8-10/28	3523C	6182	GLEN VIEW CLUB	RN	8SecRestObsLng	11/13	4025D	4221
GLEN MEADAIL		6Cpt3Dr	8-9/26	3523A	4974	GLEN VISTA		6Cpt3Dr	10-12/25	3523A	4922
GLEN MEADOW		6Cpt3Dr	10-12/25	3523A	4922	GLEN WILLOW		6Cpt3Dr	10-12/25	3523A	4922
GLEN MORGAN		6Cpt3Dr	1/30	3523C	6341	GLEN WILTON		12SecDr	2-3/13	2410	4105
GLEN MORISTON		6Cpt3Dr	8-9/26	3523A	4974	GLENARM		12SecDr	3-6/20	2410F	4565
GLEN MUICK		6Cpt3Dr	8-9/26	3523A	4974	GLENARTNEY		7Cpt2Dr	12/29/10	2522	3867
GLEN NESS		6Cpt3Dr	8-9/26	3523A	4974	GLENAVON		12SecDr	1-3/14	2410A	4249
GLEN NEVIS		6Cpt3Dr	10-12/25	3523A	4922	GLENBURNIE		10SecDr2Cpt	10-11/13	2585A	4218
GLEN NORMAN		6Cpt3Dr	10-12/29	3523C	6290	GLENCAIRN		BaggBClubSmk	9/14/10	2415	3834
GLEN OAK		6Cpt3Dr	10-12/25	3523A	4922	GLENCOE		12SecDr	7/16	2410F	4412
GLEN ONOKO		6Cpt3Dr	8-10/28	3523C	6182	GLENDORA		10SecDr2Cpt	1-2/13	2585	4091
GLEN ORCHARD		6Cpt3Dr	10-12/25	3523A	4922	GLENELLA		12SecDr	1-3/14	2410A	4249
GLEN OSBORNE		28ChDr	6-7/27	3416A	6087	GLENFIELD		12SecDr	10/7/10	2410	3813
GLEN PARK		10Sec2Dr	1/12	2584	3945	GLENFINLAS		7Cpt2Dr	12/29/10	2522	3867
GLEN PASS		6Cpt3Dr	8-10/28	3523C	6182	GLENFORD		12SecDr	5-6/11	2410	3903
GLEN PLACID		6Cpt3Dr	8-10/28	3523C	6182	GLENGYLE		7Cpt2Dr	12/30/10	2522	3867
GLEN POND		6Cpt3Dr	8-10/28	3523C	6182	GLENHURST		BaggBClubSmkBarber	2-3/12	2602	3948
GLEN POOL		6Cpt3Dr	10-12/29	3523C	6290	GLENLYON		26ChDr	5/12	2416	3986
GLEN RAE		6Cpt3Dr	8-9/26	3523A	4974	GLENOLDEN		26ChDr	5/12	2416	3986
GLEN RAPIDS		6Cpt3Dr	10-12/25	3523A	4922	GLENOVER		12SecDr	6-9/26	3410A	4969
GLEN RICHEY		10Sec2Dr	5/13	2584	4140	GLENROCK		12SecDrCpt	3/11	2411	3880
GLEN RIDGE		6Cpt3Dr	10-12/25	3523A	4922	GLENROCK		24ChDrB	1/14	2417A	4239
GLEN RIO		6Cpt3Dr	10-12/29	3523C	6290	GLENSHANE		12SecDr	2-5/21	2410F	4612
GLEN RIVER		6Cpt3Dr	10-12/25	3523A	4922	GLENSIDE		BaggBClubSmkBarber	2-3/12	2602	3948
GLEN ROAD		6Cpt3Dr	10-12/25	3523A	4922	GLENTARKIN		7Cpt2Dr	12/30/10	2522	3867
GLEN ROBERTS		6Cpt3Dr	1/30	3523C	6341	GLENVAR		12SecDr	1-2/15	2410B	4311
GLEN ROGERS		6Cpt3Dr	1/30	3523C	6341	GLENVIEW		12SecDr	9-10/15	2410D	4327
GLEN ROSA		6Cpt3Dr	8-9/26	3523A	4974	GLENVILLE		10SecDr2Cpt	1-2/13	2585	4091
GLEN ROY		6Cpt3Dr	8-9/26	3523A	4974	GLENWOOD		16Sec	10/13	2412B	4217
GLEN SADDELL		6Cpt3Dr	8-9/26	3523A	4974	GLOAMING	N	32ChObs	8/19/10	2420H	3805
GLEN SANNOX		6Cpt3Dr	8-9/26	3523A	4974	GLORIA	P	29Ch SBS	8/18	3352	4536
GLEN SHIEL		6Cpt3Dr	8-9/26	3523A	4974	GLOUCESTER		12SecDr	2-3/27	3410A	6055
GLEN SPRINGS		6Cpt3Dr	10-12/25	3523A	4922	GLOVER GAP		8SecDr2Cpt	12/29-30	3979A	6334
GLEN SPRUCE		6Cpt3Dr	8-10/28	3523C	6182	GLOXINIA		28ChDr	5-7/26	3416	4958
GLEN STOCKDALE		6Cpt3Dr	8-9/26	3523A	4974	GLYNDEN		12SecDr	12/29/10	2410	3864
GLEN STRAE		6Cpt3Dr	8-9/26	3523A	4974	GOBELIN		12SecDr	3-6/20	2410F	4565
GLEN STREAM		6Cpt3Dr	10-12/25	3523A	4922	GODIVA		26ChDr	4-5/13	2416A	4136
GLEN SUMMIT		6Cpt3Dr	8-9/26	3523A	4974	GODWIN		12SecDrCpt	10-12/21	2411C	4624
GLEN SUNSET		6Cpt3Dr	8-10/28	3523C	6182	GOETHALS		12SecDr	11-12/13	2410A	4234
GLEN SUTTON		6Cpt3Dr	8-10/28	3523C	6182	GOETHE		6Cpt3Dr	8/24	3523	4804
GLEN TANA		6Cpt3Dr	8-9/26	3523A	4974	GOLD RUN	N	12SecDr	6-8/24	3410C	4764
GLEN TARBERT		6Cpt3Dr	8-9/26	3523A	4974	GOLDEN BEACH	N	3Cpt2DrLngObs	4-5/28	3959A	6125
GLEN TARSAN		6Cpt3Dr	8-9/26	3523A	4974	GOLDEN CAVE	N	3Cpt2DrLngObs	4-5/28	3959A	6125
GLEN TAY		6Cpt3Dr	1-2/28	3523C	6124	GOLDEN CITY		2DrCptLibObs	12/29-30	3950	4744
GLEN TERRACE		6Cpt3Dr	8-10/28	3523C	6182	GOLDEN CITY		3Cpt2DrLngObs	11-12/29	3959B	6316
GLEN TILT		6Cpt3Dr	8-9/26	3523A	4974	GOLDEN CITY	R	3Cpt2DrLngLibObs	12/29-30	3950C	4744
GLEN TORRIDON		6Cpt3Dr	8-9/26	3523A	4974	GOLDEN CITY	R	3Cpt3DrLngObs	11-12/29	3959H	6316
GLEN TRAIL		6Cpt3Dr	10-12/25	3523A	4922	GOLDEN CREEK		2DrCptLibObs	12/29-30	3950	4744

CAR NAME	ST	CAR TYPE	BUILT	PLAN	LOT
GOLDEN CREEK	R	3Cpt2DrLngLibObs	12/29-30	3950C	4744
GOLDEN FALLS	N	3Cpt2DrLngObs	4-5/28	3959A	6125
GOLDEN GATE		2DrCptLibObs	12/29-30	3950	4744
GOLDEN GATE		3Cpt2DrLngObs	11-12/29	3959B	6316
GOLDEN GATE	R	3Cpt2DrLngLibObs	12/29-30	3950C	4744
GOLDEN GATE	R	3Cpt3DrLngObs	11-12/29	3959H	6316
GOLDEN GROVE		2DrCptLibObs	12/29-30	3950	4744
GOLDEN GROVE		3Cpt2DrLngObs	11-12/29	3959B	6316
GOLDEN GROVE	R	3Cpt2DrLngLibObs	12/29-30	3950C	4744
GOLDEN GROVE	R	3Cpt3DrLngObs	11-12/29	3959H	6316
GOLDEN HILL		2DrCptLibObs	12/29-30	3950	4744
GOLDEN HILL	R	3Cpt2DrLngLibObs	12/29-30	3950C	4744
GOLDEN HORN		2DrCptLibObs	12/29-30	3950	4744
GOLDEN HORN		3Cpt2DrLngObs	11-12/29	3959B	6316
GOLDEN HORN	R	3Cpt2DrLngLibObs	12/29-30	3950C	4744
GOLDEN HORN	R	3Cpt3DrLngObs	11-12/29	3959H	6316
GOLDEN LAKE		2DrCptLibObs	12/29-30	3950	4744
GOLDEN LAKE		3Cpt2DrLngObs	11-12/29	3959B	6316
GOLDEN PEAK	N	3Cpt2DrLngObs	4-5/28	3959A	6125
GOLDEN ROCK	N	3Cpt2DrLngObs	4-5/28	3959A	6125
GOLDEN SPRINGS	N	3Cpt2DrLngObs	4-5/28	3959A	6125
GOLDEN STATE		2DrCptLibObs	12/29-30	3950	4744
GOLDEN STATE		3Cpt2DrLngObs	11-12/29	3959B	6316
GOLDEN STATE	R	3Cpt2DrLngLibObs	12/29-30	3950C	4744
GOLDEN STATE	R	3Cpt3DrLngObs	11-12/29	3959H	6316
GOLDEN STREAM		2DrCptLibObs	12/29-30	3950	4744
GOLDEN STREAM		3Cpt2DrLngObs	11-12/29	3959B	6316
GOLDEN STREAM	R	3Cpt2DrLngLibObs	12/29-30	3950C	4744
GOLDEN STREAM	R	3Cpt3DrLngObs	11-12/29	3959H	6316
GOLDEN VALLEY		2DrCptLibObs	12/29-30	3950	4744
GOLDEN VALLEY	R	3Cpt2DrLngLibObs	12/29-30	3950C	4744
GOLDSBORO	RN	21StsRestLng	6/11	4019J	3902
GOLDSMITH		12SecDr	12/10-11	2410	3866
GOLIAD		12SecDr	8-10/26	3410A	4945
GOMERA		12SecDr	2-3/15	2410B	4318
GONZALES	N	10SecDr2Cpt	11/17	2585D	4527
GONZALES BOCANEGRA FCP-209	SN	10SecDr2Cpt	9-10/26	3585A	4996
GONZALO CURIEL FCP-215	SN	10SecDr2Cpt	1-2/24	3585	4743
GOODFIELD		12SecDr	5-6/11	2410	3903
GOODISON		12SecDr	1-4/12	2410	3949
GOODLAND		12SecDr	10/19/10	2410	3813
GOODLETT		12SecDr	10-12/16	2410F	4431
GOODRICH		12SecDr	3-4/17	2410F	4485
GORDONVILLE		12SecDrCpt	7/12/10	2411	3800
GORGAS		12SecDr	4-6/21	2410F	4613
GORGEOUS ROSE	N	12SecDr	8-12/25	3410A	4894
GORLITZ		12SecDr	1-2/16	2410E	4356
GORMAN		12SecDr	8-9/20	2410F	4574
GOSHEN		BaggBClubSmkBarber	9/24	2951C	4805
GOSSETT		12SecDr	4-6/14	2410A	4271
GOTHA		12SecDr	4-6/21	2410F	4613
GOTHAM	N	25ChDrObs	12/12	2669A	4053
GOTHAM	R	24ChDrObs	12/12	2669C	4053
GOTHARD		12SecDr	10-12/15	2410D	4338
GOTHIC PEAK		10Sec2Dr	12/26-27	3584A	6031
GOTHIC TOWER	RN	8SecDr3Dbr	5-8/25	4090F	4868
GOULA		12SecDr	4-6/21	2410F	4613
GOUNOD		6Cpt3Dr	11-12/24	3523A	4833
GOUVERNEUR		12SecDr	3-4/11	2410	3881
GOUVERNEUR MORRIS		30ChDr	8-9/30	4000A	6384
GOUVERNEUR MORRIS		34ChDr	10/24	3419	4784
GOVERNOR CLINTON	N	28ChDr	4-6/25	3416	4864
GOVERNOR JAY	N	28ChDr	4-6/25	3416	4864
GOVERNOR NELSON	N	12SecDr	2-5/25	3410	4845
GOVERNOR NELSON	R	10SecDr2Dbr	2-5/25	4074E	4845
GOVERNOR SEWARD	N	34Ch	4/26	3419	4956
GOVERNOR SEYMOUR	N	34Ch	4/26	3419	4956
GRACCHUS		16Sec	6-7/20	2412F	4568
GRACE		32Ch	6/30	3916C	6363
GRAHAM		12SecDr	10-12/16	2410F	4431
GRAMATAN		12SecDr	2-3/13	2410	4105
GRAND CANYON		10SecLngObs	9-10/14	2521A	4292
GRAND CANYON	N	3CptDrLngBSunRm	6/29	3975U	6262
GRAND CROSSING		12SecDrCpt	5/17	2411C	4494
GRAND ISLAND		4Cpt2DrLngObs	3/13	2703	4100
GRAND LAKE		10SecDrCpt	3-5/30	3973A	6338
GRAND SUMMIT		10SecLngObs	12/16-17	2521C	4442
GRAND VALLEY		10SecLngObs	9-10/14	2521A	4292
GRAND VIEW		10SecLngObs	12/15	2521B	4362
GRANDE GAP	N	8SecDr2Cpt	12/29-30	3979A	6334
GRANICUS		12SecDr	10-12/15	2410D	4338
GRANITE PEAK		10Sec2Dr	9-10/29	3584B	6284
GRANLIDEN		12SecDr	11-12/14	2410B	4297
GRANOGUE		12SecDr	7-8/13	2410A	4192
GRANSBY		16Sec	5-6/16	2412F	4386
GRANTLEY		12SecDr	11/25/10	2410	3864
GRAPEVILLE		12SecDr	5-6/17	2410F	4497
GRASPAN		12SecDr	1/16	2410E	4354
GRASSELLI		12SecDr	4-6/14	2410A	4271
GRATWICK		12SecDr	3-4/11	2410	3881
GRATZTOWN		12SecDr	2-4/24	3410	4762
GRAVAN		12SecDr	1-2/15	2410B	4311
GRAVOSA		16Sec	6-7/13	2412B	4160
GRAY		6Cpt3Dr	8/24	3523	4804
GRAY GABLES		BaggBClubSmk	6/16	2919	4392
GRAY OAK		8SecDr2Cpt	7-8/30	3979A	6377
GRAYBILL		12SecDr	2-4/24	3410	4762
GRAYFORD		12SecDr	10/12/10	2410	3813

CAR NAME	ST	CAR TYPE	BUILT	PLAN	LOT
GRAYFORD		12SecDr	10-12/16	2410F	4431
GRAYLING		12SecDr	1-4/12	2410	3949
GRAYMERE		12SecDr	10-12/26	3410A	6023
GRAYMONT		12SecDr	12/12-13	2410	4076
GRAYNOOK	N	12SecDr	10-12/15	2410D	4338
GRAYSON		12SecDr	5-6/11	2410	3903
GRAYTOWN		16Sec	2-3/11	2412	3878
GRAYVILLE		12SecDr	5-6/11	2410	3903
GREAT BARRINGTON		32ChDr	8/27	3917A	6078
GREAT BEAR		2CptDrLngBLibObs	6/26	3953A	4955
GREAT BEND		12SecDrCpt	3/14	2411A	4269
GREAT CHIEF	N	2CptDrLngBLibObs	5/24	3953	4782
GREAT CIRCLE		2CptDrLngBLibObs	5/24	3953	4782
GREAT DIVIDE		2CptDrLngBLibObs	5/24	3953	4782
GREAT FALLS		2CptDrLngBLibObs	5/24	3953	4782
GREAT LAKES		2CptDrLngBLibObs	5/24	3953	4782
GREAT NECK		2CptDrLngBLibObs	5/24	3953	4782
GREAT NORTHERN		2CptDrLngBLibObs	5/24	3953	4782
GREAT PLAINS		2CptDrLngBLibObs	5/24	3953	4782
GREAT REPUBLIC		34Ch	2/30	4001	6325
GREAT SPIRIT		2CptDrLngBLibObs	5/24	3953	4782
GREEN BANK		12SecDr	8-12/25	3410A	4894
GREEN BAY		10SecDr2Cpt	10/12	2585	4039
GREEN BOTTOM		12SecDr	8-12/25	3410A	4894
GREEN FOREST	N	12SecDr	8-12/25	3410A	4894
GREEN ISLAND		12SecDrCpt	3/14	2411A	4269
GREEN LODGE		BaggBClubSmk	6/16	2919	4392
GREEN PEAK		10Sec2Dr	12/26-27	3584A	6031
GREEN RIDGE		12SecDr	8-12/25	3410A	4894
GREEN SPRING		12SecDr	8-12/25	3410A	4894
GREEN SPRINGS		12SecDrCpt	6/30/10	2411	3800
GREEN TREE		12SecDrCpt	7/2/10	2411	3800
GREEN VALLEY	N	12SecDr	8-12/25	3410A	4894
GREEN'S RUN		12SecDr	8-12/25	3410A	4894
GREENBRIER		6Sec5Cpt	2/16	2547A	4363
GREENBRIER	R	6SecDr2Cpt2Dbr	2/16	2547C	4363
GREENBUSH		36Ch	8/27	3916A	6077
GREENCASTLE		26ChDr	5/23	2416D	4691
GREENDALE		12SecDr	3-4/11	2410	3881
GREENLEAF		12SecDr	4-5/13	2410	4106
GREENOCK		12SecDr	1-4/12	2410	3949
GREENS FARMS		36Ch	8/27	3916A	6077
GREENSAND		12SecDr	2-5/16	2410E	4367
GREENSBURG		10Cpt	8/19/10	2505	3833
GREENSWARD		12SecDr	1-3/17	2410F	4450
GREENVILLE		12SecDr	3/30/10	2410	3769
GREENWAY		16Sec	2-3/11	2412	3878
GREENWICH		12SecDr	2-3/27	3410A	6055
GREENWICH BAY		32ChDr	12/29-30	3917B	6318
GREENWOOD		12SecDr	10/13/10	2410	3813

CAR NAME	ST	CAR TYPE	BUILT	PLAN	LOT
GREER		28ChDr	4-6/25	3416	4864
GRENOBLE		12SecDr	1-2/15	2410B	4311
GRENORE		16Sec	6-7/13	2412B	4160
GRENSHAW		12SecDr	2-3/13	2410	4105
GRETNA		26ChDr	5/23	2416D	4691
GREYSTONE		16Sec	9/11	2412	3923
GRIDIRON CLUB		BaggBSmkBarber	9-10/26	3951C	4972
GRIFFITH		12SecDr	10-12/16	2410F	4431
GRIMSEL PASS		8SecDr2Cpt	7-8/30	3979A	6377
GRINNELL		12SecDrCpt	6/13	2411	4151
GROSEILLERS		12SecDr	2-5/25	3410	4845
GROSSE POINTE	N	8SbrObsLng	4/30	4005A	6340
GROVE		28ChDr	4-6/25	3416	4864
GROVEDALE		6CptLngObs	8/6/10	2413	3802
GROVELAND		12SecDr	2-3/13	2410	4105
GROVERTOWN		12SecDr	10/12/10	2410	3814
GRUNEWALD		12SecDr	1-2/16	2410E	4356
GRYPHON		12SecDr	1-2/16	2410E	4356
GUADALUPE					
VICTORIA NdeM-152	SN	12SecDr	5-8/25	3410A	4868
GUAM		12SecDr	4-6/21	2410F	4613
GUANTANAMO		12SecDr	7/13	2410A	4179
GUARDIAN	RN	17StsRestLng	5/12	4019A	3986
GUASAVE NdeM-151	SN	12SecDr	8-12/25	3410A	4894
GUATEMALA		12SecDr	1-2/22	2410F	4625
GUATEMALA NdeM	SN	12SecDr	10/22-23	2410I	4647
GUAYMAS NdeM-153	SN	12SecDr	5-8/25	3410A	4868
GUDRUN		12SecDr	1/16	2410E	4354
GUELATAO NdeM	SN	8SecDr2Cpt	4-5/30	3979A	6359
GUERDON		12SecDr	12/10-11	2410	3866
GUERNSEY		12SecDr	8-9/20	2410F	4574
GUILDERLAND		12SecDr	5-6/11	2410	3903
GUILLERMO					
PRIETO NdeM-154	SN	12SecDr	2-5/25	3410	4845
GULF PARK COLLEGE	RN	10Sec2DbrCpt	12/16-17	4042	4443
GULFCREST		16Sec	6-7/20	2412F	4568
GULL LAKE		10SecDrCpt	10-12/29	3973A	6300
GULLIVER		12SecDr	7-8/13	2410A	4192
GUNNISON	N	12SecDr	1-2/16	2410E	4356
GUNSIGHT LAKE		10SecDr2Cpt	5-6/24	3585	4770
GUTY CARDENAS FCP-221	SN	10SecDr2Cpt	6-7/25	3585A	4888
GUYASUTA		26ChDr	8/5/10	2416	3804
GUYENCOURT		12SecDr	7-8/13	2410A	4192
GWENDOLYN		28ChDr	6-7/30	3416B	6372
GWIN		12SecDr	8-10/26	3410A	4945
GWLADYS		30ChLng	11/27	3983	6091
GYPSUM		12SecDr	3-4/17	2410F	4485
H.D. NEWCOMB	N	12SecDr	9-10/11	2410	3924
HACKENSACK		12SecDr	7-8/12	2410	4014
HACKETT		12SecDr	9-10/17	2410F	4525

170

Figure I-55   GREEN SPRING, a 12-section drawing room built in 1925 in Lot 4894, Plan 3410A, was stream-styled in 1938 for service on the B & O "Limited" trains.   It was sold to the B & O in 1948 and between January 1959 and April 1960 it was in government storage at the Louisville Medical Depot, Kentucky.   It is shown here in October 1948 in Elizabeth, New Jersey.   The Robert Wayner Collection.

CAR NAME	ST	CAR TYPE	BUILT	PLAN	LOT	CAR NAME	ST	CAR TYPE	BUILT	PLAN	LOT
HADDONFIELD		12SecDr	3/29/10	2410	3770	HAMILL		12SecDr	7/16	2410F	4412
HADLEY		12SecDr	5-6/11	2410	3903	HAMILTON		12SecDr	10/29/10	2410	3860
HADLOCK		16Sec	1-5/18	2412F	4531	HAMILTON COLLEGE	RN	10Sec2DbrCpt	6/13	4042A	4159
HADLYM		36Ch	6/16	2916	4389	HAMLET		12SecDr	11-12/13	2410A	4234
HADLYME	N	36Ch	6/16	2916	4389	HAMLIN		12SecDr	3-4/11	2410	3881
HADRIAN		12SecDr	2-5/16	2410E	4367	HAMMOND		12SecDr	9-10/17	2410F	4525
HAGAR		26ChDr	6/11	2416	3902	HAMMONTON		26ChDr	7/29/10	2416	3804
HAGERSTOWN		12SecDr	4/21/10	2410	3770	HAMPTON ROADS	N	BaggBSmk	12/26-27	3951G	6022
HAINES		28ChDr	4-6/25	3416	4864	HANCOCK		16Sec	1-5/18	2412F	4531
HAINESPORT		26ChDr	7/22/10	2416	3804	HANDLEY		12SecDr	1-3/14	2410A	4249
HAKONE		12SecDr	9-10/15	2410D	4327	HANGING ROCK		BaggBSmkBarber	12/26-27	3951C	6022
HALDANE		6Sec5Cpt	6/11	2547	3892	HANMER		12SecDr	1-3/14	2410A	4249
HALDEMAN		16Sec	1-5/18	2412F	4531	HANNAH		24ChDrB	1/14	2417A	4239
HALETHORPE		12SecDr	5-6/17	2410F	4497	HANOVER		12SecDr	1-4/12	2410	3949
HALF MOON		BaggBSmkBarber	12/26-27	3951C	6022	HANOVER		36Ch	10/24	3916	4802
HALFORD		16Sec	2/13	2412	4101	HANSARD		12SecDr	2-5/16	2410E	4367
HALIDON		12SecDr	10-12/15	2410D	4338	HANSFORD		12SecDr	1-3/14	2410A	4249
HALIFAX		36Ch	6/16	2916	4389	HANSTED		12SecDr	4-6/21	2410F	4613
HALLAM		16Sec	1-5/18	2412F	4531	HAPPY HOLLOW CLUB		4Sbr2CptDrLng	12/29	3995	6301
HALLECK		10SecDr2Cpt	11-12/11	2585	3942	HAPPY HOLLOW CLUB	R	2Sbr2Dbr2CptDrLng	12/29	3995C	6301
HALMAR		12SecDr	9-10/17	2410F	4525	HARBISON		12SecDrCpt	6/13	2411A	4159
HALSTED		12SecDr	9-10/17	2410F	4525	HARCUM		16Sec	1-5/18	2412F	4531
HALVAN		12SecDr	4-6/21	2410F	4613	HARDING		12SecDr	12/13/10	2410	3864
HAMBLEDON		12SecDr	11-12/14	2410B	4297	HARDWICK		12SecDrCpt	8/23	2411C	4697
HAMERTON		12SecDr	12/12-13	2410	4076	HARGWEN		12SecDr	2-5/16	2410E	4367

CAR NAME	ST	CAR TYPE	BUILT	PLAN	LOT
HARISTA		12SecDr	12/10-11	2410	3866
HARKNESS		12SecDr	7/16	2410F	4412
HARLISS		12SecDr	5-8/18	2410F	4540
HARMAN		32Ch	11/17/10	2419	3861
HARMARVILLE		26ChDr	8/1/10	2416	3804
HARMONY		12SecDr	5/5/10	2410	3773
HARMOZIA		12SecDr	4-6/21	2410F	4613
HARPER		16Sec	2/13	2412	4101
HARPERS FERRY		28ChObs	3/14	2421	4251
HARPSWELL		12SecDr	4-5/13	2410	4106
HARRIET		28ChDr	6-7/27	3416A	6087
HARRIET BEECHER STOWE		30ChDr	1/27	3418A	6034
HARRINGTON		26ChDr	7/19/10	2416	3804
HARRISBURG		BaggBClubSmk	7/13	2415	4158
HARRISON		26ChDr	8/3/10	2416	3804
HARRISON LAKE		10SecDr2Cpt	5-6/24	3585	4770
HARRODSBURG		12SecDr	10-12/26	3410A	6023
HARSHMAN		12SecDr	1-4/12	2410	3949
HARTE		12SecDrCpt	10-12/21	2411C	4624
HARTFORD		12SecDr	5-6/17	2410F	4497
HARTLAND		10SecDrCpt	4-6/27	3973	6043
HARTLEY		26ChDr	8/5/10	2416	3804
HARTMAN		12SecDr	5-6/11	2410	3903
HARTNEY		12SecDr	2-5/16	2410E	4367
HARVARD CLUB	RN	8SecRestObsLng	9-10/14	4025D	4292
HARVEY		12SecDr	3-6/24	3410	4763
HARVEY		12SecDr	10/15/10	2410	3813
HARWICH		12SecDr	12/10-11	2410	3866
HASBROUCK HOUSE		13Dbr	6-7/30	3997A	6314
HASELTON		12SecDr	10/15/10	2410	3814
HASKELL		12SecDr	7/16	2410F	4412
HASLEMERE		12SecDr	12/12-13	2410	4076
HASLETT		12SecDr	9-10/17	2410F	4525
HASTINGS		12SecDrCpt	8/23	2411C	4697
HATFIELD		12SecDr	9-10/17	2410F	4525
HATTERAS		12SecDr	12/12-13	2410	4076
HATTON		12SecDr	9-10/17	2410F	4525
HAULTAIN		12SecDr	11-12/13	2410A	4234
HAVANA		12SecDr	5-7/13	2410	4149
HAVANA		23StsBLng	11/26	3967	4979
HAVELOCK		12SecDrCpt	8/23	2411C	4697
HAVENS		12SecDr	5-8/18	2410F	4540
HAVERFORD		12SecDrCpt	3/11	2411	3880
HAVERSTRAW		16Sec	6/21/10	2412	3801
HAVINGTON		12SecDr	1-2/15	2410B	4311
HAVOLI		12SecDr	3-6/15	2410B	4319
HAVRE		12SecDr	4-6/24	3410	4763
HAWARDEN		12SecDr	9-10/17	2410F	4525
HAWKINS		12SecDr	2-5/25	3410	4845
HAWLEY		12SecDr	4/30	3410B	6360
HAWORTH		12SecDr	3-4/11	2410	3881
HAWSTONE		26ChDr	9-11/22	2416D	4649
HAWTHORNE		16Sec	10/13	2412B	4217
HAWTHORNE	R	12Sec4EnclSec	10/13	2412J	4217
HAWTREY		16Sec	7-8/11	2412	3912
HAYNES		12SecDr	9-10/17	2410F	4525
HAYSVILLE		12SecDr	2-4/24	3410	4762
HAYWARD		16Sec	1-5/18	2412F	4531
HAZEL DELL		28ChDr	6-7/27	3416A	6087
HAZELHURST		12SecDrCpt	3/11	2411	3880
HAZELTINE		16Sec	2/13	2412	4101
HAZEN		10SecDr2Cpt	11-12/11	2585	3942
HEATHCOTE		12SecDr	3-6/20	2410F	4565
HEATHFIELD		12SecDr	5-7/13	2410	4149
HEBRIDES		12SecDr	8-9/20	2410F	4574
HEBRON		12SecDr	1/14	2410A	4240
HECTOR		12SecDr	11-12/13	2410A	4234
HEDDA		24ChDrB	6-7/14	2417B	4265
HEDGELY		12SecDr	10-12/15	2410D	4338
HEDRICK		12SecDr	9-10/17	2410F	4525
HEIDELBERG COLLEGE	RN	10Sec2DbrCpt	2-3/11	4042B	3878
HEISLEY		12SecDr	1-4/12	2410	3949
HELEN HUNT JACKSON		34Ch	1/27	3419A	6033
HELENA		12SecDr	4/15	2410C	4309
HELENA MODJESKA		26ChDrObs	1/27	3957A	6038
HELIANTHUS		28ChDr	5-7/26	3416	4958
HELICON		16Sec	6-7/20	2412F	4568
HELLESPONT		12SecDr	4-6/21	2410F	4613
HELMETTA		26ChDr	4-5/16	2416C	4366
HEMLOCK		12SecDrCpt	6/28/10	2411	3800
HENEDINE		12SecDr	2-3/15	2410B	4318
HENEFER		10SecDr2Cpt	11-12/11	2585	3942
HENLEY		12SecDr	1-3/14	2410A	4249
HENLOPEN		12SecDr	4-6/21	2410F	4613
HENRIETTA		26ChDr	3-4/14	2416A	4264
HENRY BESSEMER		14Sec	3/27	3958	6054
HENRY CLAY		12SecDr	8-12/25	3410A	4894
HENRY D. McDANIEL		10Sec2Dr	9-10/25	3584A	4899
HENRY H. SIBLEY		8SecDr2Cpt	5-6/29	3979A	6261
HENRY KNOX		12SecDr	4/26	3410A	4943
HENRY LAURENS		12SecDr	4/26	3410A	4943
HENRY M. RICE		8SecDr2Cpt	5-6/29	3979A	6261
HENRY STANLEY		Private	7/27	3972	6037
HENRY TIMROD	N	12SecDr	1-3/25	3410	4844
HENRY W. CORBETT		14Sec	6/30	3958A	6373
HENRY W. GRADY	RN	10Sec4PvtSec	2-3/11	2412H	3878
HENRY W. GRADY		10Sec2Dr	9-10/25	3584A	4899
HENRY W. GRADY		14Sec	10-11/29	3958A	6285
HENRY W. LONGFELLOW		12SecDr	1-2/29	3410B	6220
HENRY W. WATTERSON		14Sec	3-4/30	3958A	6357

172

CAR NAME	ST	CAR TYPE	BUILT	PLAN	LOT
HENRY WATTERSON	N	14Sec	3-4/30	3958A	6357
HENRYVILLE		12SecDr	10/14/10	2410	3813
HENRYVILLE		12SecDr	6-9/26	3410A	4969
HENSHAW		12SecDr	4-6/21	2410F	4613
HEPBURN		12SecDr	9-10/17	2410F	4525
HEPLER		12SecDr	2-4/24	3410	4762
HERALD	N	16Sec	7-8/11	2412	3912
HERALD SQUARE		6Cpt3Dr	10-12/29	3523C	6290
HERCULES		12SecDr	12/12-13	2410	4076
HEREFORD	N	12SecDr	1-4/12	2410	3949
HERKEND		12SecDr	4-6/21	2410F	4613
HERKIMER		12SecDr	3-6/20	2410F	4565
HERMENEGILDO					
GALEANA NdeM	SN	12SecDr	10-12/23	3410	4724
HERMINIE		26ChDr	7/21/10	2416	3804
HERMOSA		12SecDr	1-3/14	2410A	4249
HERMOSILLO NdeM	SN	12SecDr	5-8/25	3410A	4868
HERNDON		28ChDr	11/23/10	2418	3862
HEROE					
DE NACOZARI NdeM	SN	12SecDr	1-3/25	3410	4844
HEROES DE					
CHAPULTEPEC NdeM	SN	12SecDr	10-12/26	3410A	6023
HEROICA					
VERACRUZ NdeM	SN	8SecDr3Dbr	5-6/25	4090A	4843
HERRICK		12SecDr	9-10/11	2410	3924
HERRON		12SecDr	9-10/17	2410F	4525
HERSHEY		10SecDr2Cpt	10-11/12	2585	4067
HESHBON		12SecDr	10/28/10	2410	3860
HESIOD		12SecDrCpt	10-12/21	2411C	4624
HESKETH		12SecDr	11-12/14	2410B	4297
HESPERUS		12SecDr	7-8/15	2410D	4322
HESS		12SecDr	2-4/24	3410	4762
HESTER		26ChDr	5/23	2416D	4691
HEVERLY		12SecDr	2-5/25	3410	4845
HEWLETT		12SecDr	1-3/14	2410A	4249
HIAWATHA		8SecDr2Cpt	7-8/27	3979	6052
HIBBARD		12SecDr	10/8/10	2410	3813
HIBISCUS		28ChDr	5-7/26	3416	4958
HIBISCUS	R	10Ch36StsLng	5-7/26	4093	4958
HICKTON		12SecDr	10/13/10	2410	3813
HIDDEN LAKE		10SecDr2Cpt	5-6/24	3585	4770
HIGGINSON		12SecDrCpt	10-12/21	2411C	4624
HIGH BRIDGE		12SecDrCpt	8-9/11	2411	3922
HIGH GATE		10Sec2Dr	1/12	2584	3945
HIGH GROVE		10SecLngObs	12/16-17	2521C	4442
HIGH POINT		10SecLngObs	2/25/11	2521	3865
HIGH ROCK		10SecLngObs	12/15	2521B	4362
HIGH TOWER	RN	8SecDr3Dbr	5-6/25	4090F	4846
HIGHLAND COUNTRY CLUB	RN	8SecRestLngObs	7-8/11	4025C	3912
HIGHLAND FALLS		10Cpt	3/14	2505A	4252
HIGHLAWN		12SecDr	5-6/11	2410	3903
HIGHMORE		16Sec	12/14-15	2412C	4304
HIGHSPIRE		26ChDr	7/26/10	2416	3804
HIGHTSTOWN		12SecDr	5-6/23	2410H	4699
HIGHWOOD		12SecDr	5-6/11	2410	3903
HILDRETH		12SecDr	9-10/17	2410F	4525
HILGER		16Sec	1-5/18	2412F	4531
HILL TOWER	RN	8SecDr3Dbr	5-8/25	4090F	4868
HILLCREST		BaggBClubSmkBarber	2-3/12	2602	3948
HILLCREST CLUB	RN	8SecRestObsLng	9-10/14	4025A	4292
HILLGROVE		12SecDr	5-6/11	2410	3903
HILLIARD		16Sec	9/11	2412	3923
HILLISBURG		12SecDr	5-6/11	2410	3903
HILLMAN		12SecDr	12/10-11	2410	3866
HILLS		28ChDr	5-7/26	3416	4958
HILLSDALE COLLEGE	RN	10Sec2DbrCpt	6/13	4042A	4151
HILLSIDE		12SecDrCpt	3/11	2411	3880
HILLSVILLE		12SecDr	10/10/10	2410	3813
HILONIAN		12SecDr	2-5/16	2410E	4367
HILTON		16Sec	6/21/10	2412	3801
HILYARD		12SecDrCpt	6/13	2411A	4159
HIMROD		12SecDr	5-6/17	2410F	4497
HINCKLEY		12SecDr	3-4/11	2410	3881
HINDOSTAN		12SecDr	6-7/17	2410F	4503
HINDSBORO		12SecDr	10/14/10	2410	3813
HINGHAM		36Ch	6/16	2916	4389
HINTON	N	12SecDr	4/18/10	2410	3770
HIPPOCRATES		12SecDrCpt	10-12/21	2411C	4624
HIRAMSBURG		12SecDr	6-9/26	3410A	4969
HISTRION		12SecDr	12/10-11	2410	3866
HOBART		12SecDr	3/16/10	2410	3794
HOBART COLLEGE	RN	10Sec2DbrCpt	3/14	4042A	4269
HOBLYN		12SecDr	4-6/21	2410F	4613
HOBOKEN		12SecDr	7-8/12	2410	4014
HOGARTH		6Cpt3Dr	7/25	3523A	4887
HOIMA		12SecDr	11-12/14	2410B	4297
HOLBEIN		6Cpt3Dr	7/25	3523A	4887
HOLBORN		10SecDr2Cpt	1-2/13	2585	4091
HOLCOMB		12SecDr	5-6/11	2410	3903
HOLDER TOWER	RN	8SecDr3Dbr	5-6/25	4090	4846
HOLDERNESS		12SecDr	11-12/14	2410B	4297
HOLLAND		12SecDrCpt	8/23	2411C	4697
HOLLIDAY		12SecDrCpt	8/23	2411C	4697
HOLLIDAYSBURG		28ChDr	11/26/10	2418	3862
HOLLINS		28ChDr	11/25/10	2418	3862
HOLLISTON		12SecDr	3-4/11	2410	3881
HOLLY BEACH		26ChDr	8/2/10	2416	3804
HOLLY SPRINGS		10SecLngObs	12/15	2521B	4362
HOLLYHOCK		28ChDr	5-7/26	3416	4958
HOLLYWOOD BEACH	N	10SecDr2Cpt	7-8/23	2585E	4707

CAR NAME	ST	CAR TYPE	BUILT	PLAN	LOT	CAR NAME	ST	CAR TYPE	BUILT	PLAN	LOT
HOLMES		12SecDrCpt	10-12/21	2411C	4624	HOWELL		12SecDr	5/6/10	2410	3773
HOLMESDALE		12SecDr	4-6/14	2410A	4271	HOWICK		12SecDr	4-6/21	2410F	4613
HOLMESVILLE		12SecDr	10/14/10	2410	3813	HOWSON		12SecDr	4-6/21	2410F	4613
HOLQUIST		12SecDr	9-10/17	2410F	4525	HOYLAKE		12SecDr	11-12/14	2410B	4297
HOLSON		12SecDr	1-3/14	2410A	4249	HUATUSCO NdeM-155	SN	8SecDr2Cpt	2-3/30	3979A	6353
HOLTENAU		12SecDr	11-12/13	2410A	4234	HUBBARD		12SecDr	9-10/17	2410F	4525
HOLTON		16Sec	2/13	2412	4101	HUBBELL		12SecDr	10-12/16	2410F	4431
HOLTWOOD		12SecDr	5-6/17	2410F	4497	HUBERTA		24ChDrB	5/17	2417D	4493
HOLYHEAD		16Sec	6-7/13	2412B	4160	HUDSON		16Sec	6/21/10	2412	3801
HOLYOKE		12SecDr	5-8/18	2410F	4540	HUDSON LAKE		12SecDrCpt	5/17	2411C	4494
HOMELAND		12SecDr	12/12-13	2410	4076	HUDSON RIVER	N	26ChDrObs	5/25	3957	4862
HOMER		26ChDr	4-5/17	2416C	4492	HUDSON VALLEY		DrSbrBLngObs	4/29	3988	6221
HOMESTEAD	N	6Sec5Cpt	6/11	2547	3892	HUESCA		12SecDr	10-12/15	2410D	4338
HOMESTEAD	R	6SecDr2Cpt2Dbr	6/11	2547C	3892	HUEY		28ChDr	5-7/26	3416	4958
HOMEWOOD		10SecDr2Cpt	12/15	2585C	4351	HUGO		6Cpt3Dr	11-12/23	3523	4726
HOMEWORTH		12SecDr	10/14/10	2410	3813	HUGONARD		12SecDr	9-10/13	2410A	4215
HONDO		12SecDrCpt	6/13	2411A	4159	HULDAH		26ChDr	6/11	2416	3902
HONDURAS		12SecDr	6-7/17	2410F	4503	HULTON		12SecDr	3/23/10	2410	3794
HONDURAS NdeM	SN	12SecDr	4-6/21	2410F	4613	HUMASTON		12SecDr	5-6/11	2410	3903
HONEOYE		16Sec	6/21/10	2412	3801	HUMBIRD		14Sec	8-10/28	3958A	6181
HONESDALE		12SecDr	7/16	2410F	4412	HUMBOLDT		12SecDr	1-3/17	2410F	4450
HONEY BROOK		26ChDr	7/27/10	2416	3804	HUMELA		12SecDr	12/10-11	2410	3866
HONEYSUCKLE		28ChDr	5-6/24	3416	4761	HUMMEL		12SecDr	6/10/10	2410	3794
HONNEDAGA		12SecDr	3-4/11	2410	3881	HUMPHRY DAVY		14Sec	10-11/26	3958	6012
HONOR LEGION	RN	32StsBSunRm	6/26	3964E	4965	HUNNEWELL		12SecDr	9-10/15	2410D	4327
HONORA		32Ch	6/30	3916C	6363	HUNT		6Cpt3Dr	8/24	3523	4804
HOOD		6Cpt3Dr	8/24	3523	4804	HUNTER		12SecDr	5-6/17	2410F	4497
HOOSIER STATE		BaggBSmkBarber	10-11/25	3951	4885	HUNTERS PASS		8SecDr2Cpt	2-3/30	3979A	6353
HOOVER		12SecDr	4/5/10	2410	3769	HUNTINGDON COLLEGE	N	10Sec2DbrCpt	2-3/11	4042B	3878
HOPATCONG		12SecDr	7-8/12	2410	4014	HUNTINGTON		12SecDr	10/28/10	2410	3860
HOPEDALE		12SecDr	5-6/11	2410	3903	HUNTINGTON		12SecDr	11-12/13	2410A	4234
HOPEWELL		26ChDr	7/25/10	2416	3804	HUNTSVILLE		12SecDrCpt	3/11	2411	3880
HOPI TRIBE	RN	6SecDr4Dbr	9-10/29	4092	6284	HURDLAND		10SecDr2Cpt	12/16-17	2585D	4443
HORATIO		12SecDr	2-5/25	3410	4845	HURDWAR		12SecDr	4-6/21	2410F	4613
HORNELL	N	26ChDr	2-3/14	2416A	4248	HURLOCK		26ChDr	5/12	2416	3986
HORNINGFORD		6CptLngObs	8/12/10	2413	3802	HURON	N	10SecDrCpt	8/28	3973A	6185
HORNSBY		12SecDr	4-6/14	2410A	4271	HURON VALLEY		DrSbrBLngObs	4/29	3988	6221
HORSHAM		12SecDr	11-12/14	2410B	4297	HURWORTH		12SecDr	9-10/15	2410D	4327
HORTENSE		28ChDr	4-6/25	3416	4864	HUTCHINSON		12SecDr	4/30	3410B	6360
HORTON		12SecDr	10-12/16	2410F	4431	HYACINTH		28ChDr	5-6/24	3416	4761
HOSMER		12SecDr	1-3/17	2410F	4450	HYDE PARK		12SecDr	9/22/10	2410	3794
HOT SPRINGS	N	12SecDr	1-3/14	2410A	4249	HYNDMAN		12SecDr	4/7/10	2410	3770
HOT SPRINGS	N	6Sec5Cpt	6/11	2547	3892	HYNER		12SecDr	12/17/10	2410	3864
HOT SPRINGS	R	6SecDr2Cpt2Dbr	6/11	2547C	3892	HYPERIDES		12SecDrCpt	10-12/21	2411C	4624
HOUTZDALE		12SecDr	5-6/17	2410F	4497	HYPERION		12SecDr	11-12/13	2410A	4234
HOUTZDALE		6CptLngObs	8/6/10	2413	3802	HYTHE		12SecDr	5-8/18	2410F	4540
HOVEY		12SecDrCpt	10-12/21	2411C	4624	IBERIA		12SecDr	8-10/26	3410A	4945
HOWARD		26ChDr	8/2/10	2416	3804	IBERVILLE		12SecDr	8-10/26	3410A	4945
HOWARD LAKE		10SecDr2Cpt	5-6/24	3585	4770	IBSEN		6Cpt3Dr	11-12/24	3523A	4833
HOWE		6Cpt3Dr	8/24	3523	4804	IBYCUS		12SecDrCpt	10-12/21	2411C	4624

CAR NAME	ST	CAR TYPE	BUILT	PLAN	LOT	CAR NAME	ST	CAR TYPE	BUILT	PLAN	LOT
ICHABOD		12SecDr	8-9/20	2410F	4574	INDEPENDENCE SQUARE		6Cpt3Dr	7-8/26	3523A	4970
IDAHO		10SecDr2Cpt	1-2/13	2585	4091	INDIAN CREEK		BaggClubBSmk	11/24	3415	4839
IDALIA		24ChDrB	6-7/14	2417B	4265	INDIAN FALLS		BaggClubBSmk	11/24	3415	4839
IDAMAR		16Sec	9/11	2412	3923	INDIAN HEAD		BaggClubBSmk	11/24	3415	4839
IDAVILLE		12SecDr	6-9/26	3410A	4969	INDIAN HILL		BaggClubBSmk	11/24	3415	4839
IDEAL		Private (GS)	3/11	2503	3847	INDIAN POINT		BaggClubBSmk	11/24	3415	4839
IDEAL ROSE	N	12SecDr	8-12/25	3410A	4894	INDIAN RIDGE		BaggClubBSmk	11/24	3415	4839
IDLEWILD		26ChDr	9-11/22	2416D	4649	INDIAN RIVER		BaggClubBSmk	11/24	3415	4839
IDLEWOOD		12SecDr	11/17/10	2410	3864	INDIAN ROCK		BaggClubBSmk	11/24	3415	4839
IGNACIO						INDIAN SPRINGS		BaggClubBSmk	11/24	3415	4839
ALDAMA NdeM-156	N	14Sec	9-10/25	3958	4869	INDIAN TRAIL		BaggClubBSmk	11/24	3415	4839
ALLENDE NdeM	SN	12SecDr	10-12/23	3410	4724	INDIANA HARBOR		12SecDrCpt	5/17	2411C	4494
DE LA LLAVE NdeM-157	N	10Sec3Dbr	8-9/11	3411	3922	INDIANAPOLIS		12SecDr	10/17/10	2410	3813
M. ALTAMIRANO NdeM	N	12SecDr	6-8/24	3410	4764	INDIANAPOLIS					
RAMIREZ FCP-208	SN	10SecDr2Cpt	9-10/26	3585A	4996	ATHLETIC CLUB	N	8SecB9StsLng	2-3/29	3989J	6229
RAYON NdeM-159	N	8SecDr2Cpt	7-8/30	3979A	6377	INDIRA		12SecDr	1-2/15	2410B	4311
ZARAGOZA NdeM	SN	12SecDr	1-3/25	3410	4844	INGALLS		12SecDr	1-4/12	2410	3949
IGRAINE		12SecDr	1-2/16	2410E	4356	INGHAMS		12SecDr	5-6/11	2410	3903
ILDERTON		12SecDr	1-2/16	2410E	4356	INGLEBY		26ChDr	5/12	2416	3986
ILESFORD		12SecDr	2-3/15	2410B	4318	INGLEHART		12SecDr	2-3/13	2410	4105
ILION		16Sec	10-11/14	2412C	4296	INGLEHOME		10SecLngObs	2/24/11	2521	3865
ILLINOIS	N	14Sec	3-4/30	3958A	6357	INGLENOOK		12SecDr	11/9/10	2410	3860
ILMA		26ChDr	6/11	2416	3902	INLAND EMPIRE		10SecDrCpt	5/30	3973A	6358
IMLAY		10SecDr2Cpt	10-11/12	2585	4067	INNSBRUCK		16Sec	6-7/13	2412B	4160
IMLER		26ChDr	9-11/22	2416D	4649	INSTRUCTION CAR NO.1	N	DrSbrBLngObs	4/29	3988K	6221
IMOLA		12SecDr	7/13	2410A	4179	INSTRUCTION CAR NO.2	N	BaggBSmk	10-11/25	3951I	4885
IMPERATOR		12SecDr	5-7/13	2410	4149	INSTRUCTION CAR NO.3	N	BaggBSmk	9-10/26	3951G	4972
INAVALE		12SecDr	9-10/15	2410D	4327	INSURGENTE					
INCA		12SecDr	4-6/21	2410F	4613	PEDRO MORENO NdeM	N	12SecDr	8-10/26	3410A	4945
INDEPENDENCE HALL		4CptLngObs	10/25	3960	4889	INTERBROOK	N	16Sec	6-7/13	2412B	4160

Figure I-56   Pullman Photograph 42714, taken July 18, 1939, shows INDIAN POINT at Calumet, fresh from the paint shop after stream-styling by Pullman for PRR service. This car was sold to PRR in October 1942 and became their Baggage-Passenger 5192. The Arthur D. Dubin Collection.

CAR NAME	ST	CAR TYPE	BUILT	PLAN	LOT
INTERLACHEN CLUB	RN	10SecRestLngObs	12/17-18	4027	4528
INTERLAKEN		12SecDr	11/5/10	2410	3860
INTREPID		16Sec	5-6/13	2412B	4150
INVERCLYDE		12SecDr	11-12/13	2410A	4234
INVERMAY		12SecDr	11-12/13	2410A	4234
INVERSNAID		12SecDr	7-8/13	2410	4192
INVINCIBLE		12SecDr	2-3/15	2410B	4318
INWOOD		12SecDr	2-4/24	3410	4762
INWOOD CLUB		8SecBObsLng	2-3/29	3989	6229
INYO		12SecDr	8-10/26	3410A	4945
IONA		12SecDr	3/18/10	2410	3794
IOWA RIVER		10SecDrCpt	8/28	3973A	6185
IPSUS		12SecDr	8-9/20	2410F	4574
IPSWICH		12SecDr	2-3/27	3410A	6055
IRANCOY		12SecDr	4-5/13	2410	4106
IRELAND		12SecDr	8-9/20	2410F	4574
IRENE	N	25ChDrObs	12/12	2669A	4053
IREX		12SecDr	4-6/21	2410F	4613
IRIS		26ChDr	6/11	2416	3902
IRON MOUNTAIN		10SecDrCpt	4-6/27	3973	6043
IRONDALE		16Sec	6-7/20	2412F	4568
IRONMASTER		12SecDr	12/12-13	2410	4076
IRONSHIRE		12SecDr	4/18/10	2410	3770
IRONSHIRE		12SecDr	10-12/16	2410F	4431
IRVINETON		12SecDr	4/22/10	2410	3771
IRVING		12SecDrCpt	10-12/21	2411C	4624
IRVONA		BaggBClubSmk	9/12/10	2415	3834
IRWIN		26ChDr	4-5/17	2416C	4492
IRWIN	N	12SecDr	5-8/25	3410A	4868
ISAAC HULL		12SecDr	8-12/25	3410A	4894
ISAAC I. STEVENS		8SecDr2Cpt	5-6/29	3979A	6261
ISAAC NEWTON		14Sec	10-11/26	3958	6012
ISAAC SHELBY	N	10SecDrCpt	4-6/27	3973	6043
ISAAC SHELBY	RN	10SecDr2Dbr	1/25	4074C	4843
ISELIN		BaggBClubSmk	9/13/10	2415	3834
ISHMONIE		12SecDr	8-9/20	2410F	4574
ISKANDER		6Sec5Cpt	6/11	2547	3892
ISLA DE					
COZUMEL NdeM-161	SN	14Sec	3/27	3958	6054
LOBOS NdeM-162	SN	14Sec	10-11/26	3958	6012
ISLA SACRIFICIOS					
NdeM-163	SN	12SecDr	2-3/27	3410A	6055
ISLA					
TIBURON NdeM-154	SN	12SecDr	8-12/25	3410A	4894
ISLA DEL					
CARMEN NdeM-160	SN	14Sec	7/30	3958A	6376
ISLA MARIA					
CLEOFAS NdeM	SN	12SecDr	8-12/25	3410A	4894
MADRE NdeM-165	SN	12SecDr	2-5/25	3410	4845
MAGDALENA NdeM	SN	12SecDr	10-12/26	3410A	6023

CAR NAME	ST	CAR TYPE	BUILT	PLAN	LOT
ISLAND CHALET		10SecDrCpt	8/29	3973A	6273
ISLAND CHARM		10SecDrCpt	8/29	3973A	6273
ISLAND CHIEF		10SecDrCpt	8/29	3973A	6273
ISLAND GLADE		10SecDrCpt	8/29	3973A	6273
ISLAND HOME		10SecDrCpt	8/29	3973A	6273
ISLAND KNIGHT		10SecDrCpt	8/29	3973A	6273
ISLAND NYMPH		10SecDrCpt	8/29	3973A	6273
ISLAND PALACE		10SecDrCpt	8/29	3973A	6273
ISLAND PARK		8SecDr2Cpt	5-6/29	3979A	6261
ISLAND PEER		10SecDrCpt	8/29	3973A	6273
ISLAND QUEEN		10SecDrCpt	8/29	3973A	6273
ISLAND REGAL		10SecDrCpt	8/29	3973A	6273
ISLAND REST		10SecDrCpt	8/29	3973A	6273
ISLAND ROSE		10SecDrCpt	8/29	3973A	6273
ISLAND ROYAL		10SecDrCpt	8/29	3973A	6273
ISLAND VALE		10SecDrCpt	8/29	3973A	6273
ISLAS REVILLAGIGEDO					
NdeM-166	SN	12SecDr	2-5/25	3410	4845
ISLESBORO		12SecDr	7/13	2410A	4179
ISLIP		12SecDr	8-9/20	2410F	4574
ISOBEL		28ChDr	6-7/27	3416A	6087
ISOLITA		12SecDr	3-6/15	2410B	4319
ISONDEGA	P	12SecDr	2/21	2410G	4598
ISPAHAN		12SecDr	8-9/20	2410F	4574
ISRAEL PUTNAM		12SecDr	4/26	3410A	4943
ISTOPOGA		12SecDr	4-6/21	2410F	4613
ITASCA		12SecDr	7/13	2410A	4179
ITHANELL		12SecDr	2-3/15	2410B	4318
IVADEL		12SecDr	1-2/15	2410B	4311
IVAN		12SecDr	4-6/21	2410F	4613
IVANHOE		12SecDr	4-6/21	2410F	4613
IVANHOE CLUB	RN	8SecRestObsLng	10/13	4025A	4202
IVANPAH		10SecDr2Cpt	10-11/13	2585A	4218
IVETO		12SecDr	1-2/15	2410B	4311
IVY HILL		12SecDr	12/14/10	2410	3864
IVY ROCK		12SecDr	6/1/10	2410	3794
IVYSIDE		12SecDr	5-6/17	2410F	4497
IXION		12SecDr	8-9/20	2410F	4574
IXTACCIHUATL	N	8Sec5Dbr	10-12/15	4036I	4338
IXWORTH		12SecDr	8-9/20	2410F	4574
J. FINLEY WILSON	N	12SecDr	2-4/24	3410	4762
JACELIA		12SecDr	3-6/20	2410F	4565
JACK JOUETT	N	12SecDr	1-3/25	3410	4844
JACK JOUETT	R	10SecDr2Dbr	1-3/25	4074E	4844
JACKSONVILLE		12SecDr	10/22-23	2410H	4647
JACQUELIN		24ChDrB	4/13	2417	4108
JACQUES CARTIER		3Cpt2DrLngObs	11-12/26	3959	6015
JACQUINET		12SecDr	3-6/20	2410F	4565
JAIME NUNO FCP-214	SN	10SecDr2Cpt	12/23-24	3585	4728
JAMES B. EADS		14Sec	3/27	3958	6054

CAR NAME	ST	CAR TYPE	BUILT	PLAN	LOT
JAMES BAINES		34Ch	2/30	4001	6325
JAMES BUCHANAN	N	12SecDr	8-12/25	3410A	4894
JAMES CRAIK	RN	10SecDrCpt	5-7/18	4031	4540
JAMES DWIGHT DANA		14Sec	3/27	3958	6054
JAMES E. OGLETHORPE	N	12SecDr	8-10/26	3410A	4945
JAMES FENIMORE COOPER		12SecDr	1-2/29	3410B	6220
JAMES GADSDEN	N	10Sec2Dr	9-10/25	3584A	4899
JAMES GUTHRIE	N	12SecDr	7/13	2410A	4179
JAMES HARROD	RN	10SecDrCpt	9-10/17	4031	4525
JAMES J. HILL		LngBBarberSunRm	5-6/29	3990	6249
JAMES K. POLK	N	30ChBLibObs	4/16	2901A	4365
JAMES LANE ALLEN	RN	10SecDr2Dbr	9-10/25	4074B	4899
JAMES LAWRENCE		12SecDr	5-8/25	3410A	4868
JAMES LOGAN		12SecDr	5-8/25	3410A	4868
JAMES LONGSTREET	N	12SecDr	1-4/12	2410	3949
JAMES MADISON		BaggClubBSmk	10/24	3415	4786
JAMES MADISON		30ChDr	8-9/30	4000A	6384
JAMES MARSHALL	N	12SecDr	2-5/25	3410	4845
JAMES MONROE		12SecDr	4/26	3410A	4943
JAMES N. GLOVER		8SecDr2Cpt	5-6/29	3979A	6261
JAMES OTIS		12SecDr	5-8/25	3410A	4868
JAMES P. ESPY		14Sec	3/27	3958	6054
JAMES RUMSEY	N	12SecDr	5-8/18	2410F	4540
JAMES RUSSELL LOWELL		12SecDr	1-2/29	3410B	6220
JAMES W. SLOSS	N	12SecDr	1-2/15	2410B	4311
JAMES W. WILSON	N	10Sec2Dr	9-10/25	3584A	4899
JAMES WATT		14Sec	10-11/26	3958	6012
JAMES WHITCOMB RILEY		12SecDr	1-2/29	3410B	6220
JAMES WILSON		28ChDr	4-6/25	3416	4864
JAMESINA		26ChDr	4/12	2416	3950
JAMESTOWN		12SecDr	3/15/07	1963D	3402
JANE		24ChDrB	5/13	2417A	4137
JANE AUSTEN		28ChDr	1-2/27	3416A	6032
JANICE		24ChDrB	4/13	2417	4108
JAPAN		12SecDr	8-9/20	2410F	4574
JARBONET		12SecDr	6-7/16	2410F	4385
JASMINE		28ChDr	5-7/26	3416	4958
JASONBY		12SecDr	7-8/15	2410D	4322
JASTROW		12SecDr	7/16	2410F	4412
JATHNIEL		16Sec	12/14-15	2412C	4304
JAVA		12SecDr	12/20-21	2410F	4611
JEANETTE		32Ch	6/30	3916C	6363
JEFFERSON		12SecDr	11-12/14	2410B	4297
JEFFERSON CLUB	RN	8SecRestObsLng	10/13	4025A	4202
JEFFERSONVILLE		12SecDr	6-9/26	3410A	4969
JEKYL ISLAND		BaggBSmkBarber	12/26-27	3951C	6022
JELLICO		12SecDr	4-6/21	2410F	4613
JEMEZ		7Cpt2Dr SBS	11/11	2522	3941
JENA		12SecDr	4-6/21	2410F	4613
JENERA		12SecDr	5-6/11	2410	3903

CAR NAME	ST	CAR TYPE	BUILT	PLAN	LOT
JENISON		12SecDr	4-6/21	2410F	4613
JENNINGS		12SecDr	2-5/25	3410	4845
JENNY LIND		16ChBLngObs	1-2/27	3961B	6035
JERDA		26ChDr	4-5/16	2416C	4366
JEROME		12SecDr	1-4/12	2410	3949
JERSEY CITY		12SecDr	2/19/10	2410	3769
JERSEY SHORE	RN	42StsBLng	6/11	4023	3902
JERSEYTOWN	N	12SecDr	8-12/25	3410A	4894
JERSEYTOWN		12SecDr	6-9/26	3410A	4969
JESSALYN		26ChDr	3-4/14	2416A	4264
JESSICA		28ChDr	6-7/30	3416B	6372
JESSIE		26ChDr	4-5/16	2416C	4366
JESUP		12SecDr	9-10/13	2410A	4215
JESUS					
GARCIA FCP-210	SN	10SecDr2Cpt	9-10/26	3585A	4996
JESUS GONZALEZ					
ORTEGA NdeM-167	N	8SecDr2Cpt	7-8/30	3979A	6377
JETHRO WOOD		14Sec	10-11/26	3958	6012
JEWETT		24ChDrB	1/14	2417A	4239
JINNISTAN		12SecDr	4-6/21	2410F	4613
JITOMIR		12SecDr	4-6/21	2410F	4613
JOAN MANNING	N	9CptOffice(Rexall)	8/26/10	2505D	3833
JOANNA		28ChDr	6-7/30	3416B	6372
JOAQUIN MILLER	RN	3Cpt4DrBarber	6/26	3962B	4895
JOAQUIN MILLER	R	4Cpt4Dr	6/26	3962C	4895
JOBSTOWN		12SecDr	9/17/10	2410	3794
JOEL CHANDLER HARRIS		3Cpt2DrLngObs	8-10/25	3959	4886
JOHN					
ADAMS		BaggClubBSmk	10/24	3415	4786
ADAMS		20ChBLng	9/30	3999A	6383
BARRY		12SecDr	8-12/25	3410A	4894
C. CALHOUN	N	12SecDr	5-8/25	3410A	4868
CALHOUN		12SecDr	5-8/25	3410A	4868
DICKINSON		12SecDr	8-12/25	3410A	4894
EAGER HOWARD		12SecDr	1-2/29	3410B	6220
ENDICOTT	N	20ChBLng	9/30	3999A	6383
ERICSSON		14Sec	10-11/26	3958	6012
FINCH	N	12SecDr	2-5/25	3410	4845
FOX JR.	N	12SecDr	8-10/26	3410A	4945
GREENLEAF WHITTIER		12SecDr	1-2/29	3410B	6220
H. INMAN	N	10Sec2Dr	9-10/25	3584A	4899
HANCOCK		34ChDr	10/24	3419	4784
HANCOCK		30ChDr	8-9/30	4000A	6384
HUNT MORGAN	N	12SecDr	1-4/12	2410	3949
JACOB ASTOR		8SecDr2Cpt	5-6/29	3979A	6261
JAMES AUDUBON	N	12SecDr	8-10/26	3410A	4945
JAMES AUDUBON		14Sec	3/27	3958	6054
JAY		12SecDr	8-12/25	3410A	4894
KINZIE		3Cpt2DrLngObs	11-12/26	3959	6015
L. HELM	N	12SecDr	2-5/25	3410	4845

CAR NAME	ST	CAR TYPE	BUILT	PLAN	LOT
JOHN					
LANGDON		30ChDr	2-3/30	4000	6324
M. FORBES		8SecDr2Cpt	12/29-30	3979A	6334
M. MOREHEAD	RN	10Sec4PvtSec	12/14-15	2412H	4304
M. MOREHEAD		14Sec	10-11/29	3958A	6285
M. MOREHEAD		10Sec2Dr	9-10/25	3584A	4899
MARSHALL		3Cpt2DrLngObs	8-10/25	3959	4886
McLOUGHLIN		8SecDr2Cpt	5-6/29	3979A	6261
MILLER		14Sec	6/30	3958A	6373
MILLER	R	6Sec6Dbr	6/30	4084B	6373
MORTON		12SecDr	4/26	3410A	4943
PAULDING		12SecDr	1-2/29	3410B	6220
RUTLEDGE		3Cpt2DrLngObs	8-10/25	3959	4886
S. BARBOUR	N	10Sec2Dr	9-10/25	3584A	4899
SEVIER	N	12SecDr	2-5/25	3410	4845
SLIDELL	RN	10Sec4PvtSec	1-5/18	2412H	4531
SLIDELL		10Sec2Dr	9-10/25	3584A	4899
SLIDELL		14Sec	10-11/29	3958A	6285
SMITH		3Cpt2DrLngObs	11-12/26	3959	6015
STEVENS		14Sec	3/27	3958	6054
T. MORGAN	RN	10Sec4PvtSec	9-10/12	2412H	4037
T. MORGAN		10Sec2Dr	9-10/25	3584A	4899
T. MORGAN		14Sec	10-11/29	3958A	6285
T. MORGAN	N	10Sec4PvtSec	12/14-15	2412H	4304
TROTWOOD MOORE	RN	10SecDr2Dbr	9-10/25	4074B	4899
W. ARRINGTON	N	12SecDr	2-5/25	3410	4845
W. JOHNSTON	N	10Sec2Dr	9-10/25	3584A	4899
WILKES	N	32ChDr	8/27	3917A	6078
WINTHROP		12SecDr	5-8/25	3410A	4868
WITHERSPOON	N	10SecDrCpt	4-6/27	3973C	6043
WITHERSPOON	N	8SecDr2Cpt	10-12/28	3979A	6205
JOHNATHAN CLUB		8SecBLngSunRm	3/30	3989B	6349
JOHNETTA		24ChDrB	1/14	2417A	4239
JOHNSONBURG		12SecDr	3/24/10	2410	3794
JOHNSTOWN		12SecDr	11/22/10	2410	3860
JOLIET		12SecDrCpt	6/13	2411	4151
JONATHAN TRUMBULL		12SecDr	1-2/29	3410B	6220
JONQUIL		26ChDr	4-5/16	2416C	4366
JOPPA		12SecDr	12/20-21	2410F	4611
JORALEMON		12SecDr	2-5/16	2410E	4367
JORDAN		12SecDr	2-5/16	2410E	4367
JORGE NEGRETE FCP-231	SN	10SecDr2Cpt	9-10/25	3585A	4996
JORULLO NdeM	SN	10SecDrCpt	9-10/17	4031	4525
JOSE VASCONCELOS					
NdeM-168	N	8SecDr3Dbr	8-12/25	4090F	4894
JOSEPH BRYAN	N	10Sec2Dr	9-10/25	3584A	4899
JOSEPH D. OLIVER	N	12SecDr	8-12/25	3410A	4894
JOSEPH E. SIRRINE	N	12SecDr	2-5/25	3410	4845
JOSEPH LISTER		8SecDr4Dbr	6/30	4003	6339
JOSEPH REED		12SecDr	4/26	3410A	4943

CAR NAME	ST	CAR TYPE	BUILT	PLAN	LOT
JOSEPH WHEELER	N	10SecLibLng	1-2/28	3521B	6128
JOSEPH WHEELER		BaggBSmkBarber	10-11/25	3951	4885
JOSEPHINE		26ChDr	3/13	2416	4107
JOSEPHINE LAKE		10SecDr2Cpt	5-6/24	3585	4770
JOSHUA BARNEY		12SecDr	4/26	3410A	4943
JOSSELYN		10SecDr2Cpt	11-12/11	2585	3942
JOURDANTON		12SecDr	2-5/16	2410E	4367
JOVE		12SecDr	4-6/21	2410F	4613
JOVITA		12SecDr	1-3/14	2410A	4249
JOYCE		26ChDr	4-5/16	2416C	4366
JOYLAND		12SecDr	12/20-21	2410F	4611
JUAN A.					
MATEOS NdeM-169	N	12SecDr	10-12/23	3410	4724
JUAN DE LA					
BARRERA NdeM-170	N	8sec4Dbr	7-9/17	4022A	4515
JUAN					
ESCUTIA NdeM-171	SN	8SecDr3Dbr	5-6/25	4090	4846
JUAN RUIZ DE					
ALARCON NdeM-172	N	8SecDr2Cpt	2-3/30	3979A	6353
JUANA		12SecDr	8-10/26	3410A	4945
JUANA DE ASBAJE NdeM	SN	8SecDr2Cpt	10-12/28	3979A	6205
JUCHITECO NdeM	SN	10SecDr2Cpt	10-12/20	2585D	4592
JUDGE SPENCER	N	12SecDr	8-12/25	3410A	4894
JUDITH		26ChDr	4-5/16	2416C	4366
JUDSON		12SecDr	5/5/10	2410	3773
JUDSONIA		10SecDr2Cpt	6-7/15	2585B	4328
JULESBURG		10SecDr2Cpt	11-12/11	2585	3942
JULIA		28ChDr	6-7/27	3416A	6087
JULIA WARD HOWE		28ChDr	1-2/27	3416A	6032
JULIER PASS		8SecDr2Cpt	7-8/30	3979A	6377
JULIET		28ChDr	8/24	3416	4801
JUMEL		12SecDr	6-7/16	2410F	4385
JUMNA		12SecDr	12/20-21	2410F	4611
JUNIATA		12SecDr	10/11/10	2410	3814
JUNIN		16Sec	1/16	2412E	4352
JUNIUS		12SecDr	7/16	2410F	4412
JUNKIN		12SecDr	2-5/16	2410E	4367
JUSTITIA		24ChDrB	5/17	2417D	4493
JUSTO SIERRA NdeM-173	N	8SecDr3Dbr	5-6/25	4090	4846
JUTLAND		12SecDr	12/20-21	2410F	4611
JYPOOR		12SecDr	4-6/21	2410F	4613
K-100		BaggDormKitchen	10/28	3970	6010
KALIQUE		12SecDr	11-12/13	2410A	4234
KALISPELL		12SecDr	4-6/21	2410F	4613
KALKASKA		12SecDr	9-10/11	2410	3924
KALMONT		12SecDr	1-2/15	2410B	4311
KALO		12SecDr	5-8/18	2410F	4540
KAMA		12SecDr	12/20-21	2410F	4611
KAMADERA		12SecDr	7/13	2410A	4179
KAMERLING		12SecDr	2-3/13	2410	4105

Figure I-57   In 1958, The National Railroad of Mexico acquired ONALUSKA PASS from the New York Central. In June 1959, the name JUAN RUIZ DE ALARCON, No. 172, was applied to the car after the interim use of the designation, MGRS-77. The Wilber C. Whittaker Collection.

CAR NAME	ST	CAR TYPE	BUILT	PLAN	LOT	CAR NAME	ST	CAR TYPE	BUILT	PLAN	LOT
KAMSIN		16Sec	6-7/20	2412F	4568	KAW TRIBE	RN	6SecDr4Dbr	12/24-25	4092	4836
KANASKAT		12SecDr	4/15	2410C	4309	KAYLOR		12SecDr	2/10/10	2410	3769
KANAWHA		6Sec5Cpt	2/16	2547A	4363	KAYMOOR		12SecDr	1-2/15	2410B	4311
KANAWHA	R	6SecDr2Cpt2Dbr	2/16	2547C	4363	KEARTON		12SecDr	11-12/14	2410B	4297
KANE		12SecDr	4/11/10	2410	3771	KEATING		26ChDr	5/12	2416	3986
KANKAKEE		12SecDr	10/22-23	2410H	4647	KEATS		12SecDr	10/22-23	2410H	4647
KANSAS CITY CLUB	RN	10SecRestLngObs	1-5/18	4027	4531	KEEFER		12SecDr	5-6/17	2410F	4497
KANTISHNA		16Sec	10-11/14	2412C	4296	KEENBROOK		10SecDr2Cpt	10-11/13	2585A	4218
KANTLEEK (Rexall)	LN	Dynamo-Storage	9/10/10	2415B	3834	KEENOY		12SecDr	4-6/21	2410F	4613
KANTY		28ChDr	4-6/25	3416	4864	KEGOMIC		12SecDr	4-6/21	2410F	4613
KANZANA		12SecDr	10/14	2410B	4293	KEIGHTLEY		12SecDr	10/22-23	2410H	4647
KAPLAN		12SecDr	9-10/17	2410F	4525	KEINATH		26ChDr	11/12	2416	4052
KARA		12SecDr	5-8/18	2410F	4540	KEITH		12SecDr	12/27-28	3410B	6127
KARAMAC		BaggBClubSmkBarber	2-3/12	2602	3948	KELSO		12SecDr	5-8/18	2410F	4540
KARENZA		12SecDr	12/12-13	2410	4076	KELTON		12SecDr	3/10/10	2410	3769
KARLDON		12SecDr	1-2/15	2410B	4311	KEMPER		12SecDr	9-10/17	2410F	4525
KARNER		16Sec	2-3/11	2412	3878	KENARDEN		12SecDr	1-3/17	2410F	4450
KARNS		28ChDr	4-6/25	3416	4864	KENASTON		12SecDr	1-3/14	2410A	4249
KASAN		12SecDr	7-8/15	2410D	4322	KENDALL		12SecDr	3-4/11	2410	3881
KASKASKIA		12SecDr	10-12/26	3410A	6023	KENDALLVILLE		16Sec	10-11/14	2412C	4296
KASOAG		12SecDr	4-6/14	2410A	4271	KENDLETON		12SecDr	9-10/17	2410F	4525
KATALLA		16Sec	10-11/14	2412C	4296	KENDRICK		12SecDrCpt	5/17	2411C	4494
KATAMA		12SecDr	11-12/13	2410A	4234	KENDRIK PEAK		10Sec2Dr	9-10/29	3584B	6284
KATE DOUGLAS WIGGIN		30ChDr	1/27	3418A	6034	KENFIELD		12SecDr	1-2/15	2410B	4311
KAUKAUNA		12SecDr	1-2/22	2410F	4625	KENILWORTH		12SecDr	10/14/10	2410	3814
KAVIRONDO		12SecDr	9-10/15	2410D	4327	KENNARD		16Sec	7-8/11	2412	3912

CAR NAME	ST	CAR TYPE	BUILT	PLAN	LOT
KENNARD	R	12Sec4EnclSec	7-8/11	2412K	3912
KENNERDELL		12SecDr	6/13/10	2410	3794
KENNETH		12SecDr	2-3/15	2410B	4318
KENOSHA		12SecDr	5-8/18	2410F	4540
KENOVA		12SecDr	7-8/12	2410	4014
KENSETT		12SecDr	2-5/16	2410E	4367
KENTLAND		12SecDr	10/8/10	2410	3813
KENTLAND	N	34ChDr	10/24	3419	4784
KENTUCKY	N	10SecLngObs	10-12/26	3521J	4998
KENTUCKY	N	3CptDrLngBSunRm	6/29	3975C	6262
KENTUCKY HOME	RN	8SecDrBLngObs	12/18	4030	4489
KENVIL		26ChDrObs	12/12	2669	4053
KERLIN		12SecDr	9-10/17	2410F	4525
KERMESS		12SecDr	9-10/17	2410F	4525
KERN		12SecDr	8-10/26	3410A	4945
KERRMOOR		12SecDr	1-4/12	2410	3949
KERSEY		12SecDr	9-10/17	2410F	4525
KERSTON		12SecDr	9-10/17	2410F	4525
KESWICK		12SecDr	12/10-11	2410	3866
KESWICK	N	12SecDr	10-12/23	3410	4724
KETNER		12SecDr	9-10/17	2410F	4525
KEUPER		12SecDr	2-5/16	2410E	4367
KEWANNA		26ChDr	9-11/22	2416D	4649
KEY LARGO	N	36Ch	10/24	3916	4802
KEY WEST		10SecLngObs	11/13	2521	4221
KEYMAR		24ChDrB	1/14	2417A	4239
KEYSTONE BANKS	RN	12Rmt2Sbr3Dbr	2-5/25	4158A	4845
KHYBER PASS		8SecDr2Cpt	7-8/30	3979A	6377
KICKAPOO		12SecDr	4-6/21	2410F	4613
KILBOURN		14Sec	4/27	3958	6041
KILBOURN	R	6Sec6Dbr	4/27	4084B	6041
KILCULLEN TOWER	RN	8SecDr3Dbr	1/25	4090	4843
KILDARE		16Sec	7-8/11	2412	3912
KILKENNY		12SecDr	4-6/21	2410F	4613
KILLARNEY		12SecDr	4-6/21	2410F	4613
KILLARNEY ROSE	N	12SecDr	8-12/25	3410A	4894
KIMBALL		16Sec	9/11	2412	3923
KIMBELL	N	12SecDr	2-5/25	3410	4845
KIMBERLEY		12SecDr	12/12-13	2410	4076
KIMBOLTON		12SecDr	5-6/23	2410H	4699
KIMBROUGH		12SecDr	9-10/15	2410D	4327
KIMLAR		16Sec	5-6/16	2412F	4386
KINARD		10SecDr2Cpt	6-7/15	2585B	4328
KINCAID		12SecDr	4-6/21	2410F	4613
KINDERHOOK		16Sec	10-11/14	2412C	4296
KINDERSLEY		12SecDr	2-3/15	2410B	4318
KING COLLEGE	N	10SecDr2Cpt	1-2/24	3585	4743
KINGS BRIDGE		12SecDrCpt	7/13/10	2411	3800
KINGS CROWN		8SecDr2Cpt	4-5/30	3979A	6359
KINGS MOUNTAIN		24ChDrObs	4/14	2819	4263

CAR NAME	ST	CAR TYPE	BUILT	PLAN	LOT
KINGSBURY		12SecDrCpt	6/13	2411A	4159
KINGSCUT		12SecDr	11-12/13	2410A	4234
KINGSDALE		12SecDr	5/13/10	2410	3773
KINGSMILL		12SecDr	5-6/11	2410	3903
KINGSTON		32ChDr	6/16	2917	4390
KINGSVILLE		12SecDr	1-2/30	3410B	6351
KINGWOOD		12SecDr	12/14/10	2410	3864
KINKORA		BaggBClubSmk	9/7/10	2415	3834
KINNELON	N	16Sec	9/11	2412	3923
KINNIKINNIK	N	12SecDr	11-12/13	2410A	4234
KINPORT		12SecDr	11/26/10	2410	3864
KINROSS		12SecDr	1-3/14	2410A	4249
KINTYRE		12SecDr	4-6/21	2410F	4613
KINZER		26ChDr	3/13	2416	4107
KIOWA TRIBE	RN	6SecDr4Dbr	4-5/26	4092	4961
KIOWAY		12SecDr	4-6/21	2410F	4613
KIPTON		12SecDr	10-11/11	2410	3936
KIRBEKAN		12SecDr	10-12/15	2410D	4338
KIRKLAND		12SecDr	10-12/16	2410F	4431
KIRKMAN		12SecDr	9-10/17	2410F	4525
KIRKVILLE		16Sec	2-3/11	2412	3878
KIRKWOOD		12SecDr	9/22/10	2410	3794
KIRVAN		12SecDr	9-10/15	2410D	4327
KISCADDIN		12SecDr	4-6/21	2410F	4613
KISMET		12SecDr	1-2/15	2410B	4311
KISSIMMEE		7Cpt2Dr	11/13	2522A	4222
KISSINGEN		12SecDr	10-12/15	2410D	4338
KIT CARSON		CptDrLngBSunRm	10/27	3975	6047
KITCHAWAN		12SecDr	5-6/11	2410	3903
KITCHENER		12SecDr	9-10/17	2410F	4525
KITCHI GAMMI CLUB	RN	8SecRestObsLng	6-7/23	4025H	4690
KITTANHAM		12SecDr	10-22-23	2410H	4647
KITTANNING		26ChDr	3/13	2416	4107
KITTREDGE		12SecDr	9-10/17	2410F	4525
KITTSON		14Sec	6/28	3958A	6174
KITTSON	R	6Sec6Dbr	6/28	4084B	6174
KIWANIS CLUB		13ChRestLng	12/29	3992	6289
KLAMATHON		12SecDr	9-10/13	2410A	4215
KLENZO #1	LN	Lecture (Rexall)	6/16	2916C	4389
KLIMWOC		12SecDr	2-3/15	2410B	4318
KNEELAND		12SecDr	4-6/14	2410A	4271
KNICKERBOCKER		12SecDr	10-12/26	3410A	6023
KNICKERBOCKER CLUB		8SecBLngSunRm	6-7/30	3989B	6362
KNOLLWOOD		12SecDr	10-12/26	3410A	6023
KNOLLWOOD CLUB		8SecBObsLng	2-3/29	3989	6229
KNOWLTON		12SecDr	5/25/10	2410	3794
KNOWLTON		12SecDr	10-12/16	2410F	4431
KOHLER		12SecDr	2-4/24	3410	4762
KOKOMO		16Sec	10/13	2412B	4216
KOOTENAI		12SecDr	4-6/21	2410F	4613

CAR NAME	ST	CAR TYPE	BUILT	PLAN	LOT
KOTONIA		12SecDr	5-7/13	2410	4149
KRAMARGE		12SecDr	11-12/13	2410A	4234
KRANTWOOD		16Sec	12/14-15	2412C	4304
KREAMER		12SecDr	6/14/10	2410	3794
KREETAN		12SecDr	9-10/15	2410D	4327
KUBAC		12SecDr	3-6/20	2410F	4565
KULON		12SecDr	11-12/13	2410A	4234
KUSTRIN		12SecDr	10-12/15	2410D	4338
KWASIND		12SecDr	3-6/15	2410B	4319
KYLEMORE		12SecDr	11-12/13	2410A	4234
L.L. SHREVE	N	12SecDr	2-5/25	3410	4845
L.Q.C. LAMAR		BaggBSmkBarber	10-11/25	3951	4885
LA ADELITA NdeM-175	SN	6SecDr2Cpt2Dbr	2/16	2547C	4363
LA ANTIGUA NdeM-176	SN	14Sec	6/30	3958A	6373
LA BOHEME		10SecDrCpt	10-12/29	3973A	6300
LA CROSSE		14Sec	4/27	3958	6041
LA CROSSE	R	6Sec6Dbr	4/27	4084B	6041
LA FONTAINE		10SecDrCpt	10-12/29	3973A	6300
LA FONTAINE CLUB		8SecBLngSunRm	8-9/29	3989A	6274
LA FONTAINE CLUB	RN	8SecRestLngObs	1-5/18	4025	4531
LA FOURCHE		12SecDrCpt	6/13	2411A	4159
LA GIOCONDA		10SecDrCpt	10-12/29	3973A	6300
LA GRANGE		7Cpt2Dr	10/12	2522	4038
LA JUNTA		10SecLngObs	9-10/14	2521A	4292
LA PLATA		10SecLngObs	9-10/14	2521A	4292
LA REINE		10SecDrCpt	10-12/29	3973A	6300
LA SALLE		12SecDrCpt	6/13	2411	4151
LA VERGNE		7Cpt2Dr	10/12	2522	4038
LACHINE		12SecDr	4-6/21	2410F	4613
LACKAWANNA		26ChDr	4-5/13	2416A	4136
LACLEDE		12SecDr	9-10/17	2410F	4525
LACONIA		12SecDr	10-12/26	3410A	6023
LACOTA		16Sec	9/11	2412	3923
LACOUR		12SecDr	9-10/17	2410F	4525
LADELLE		12SecDr	9-10/17	2410F	4525
LADNER		12SecDr	9-10/17	2410F	4525
LADOGA		12SecDr	3-6/15	2410B	4319
LADRILLO		7Cpt2Dr	11/17	2522B	4517
LADUE		12SecDr	9-10/17	2410F	4525
LAFAYETTE SQUARE		6Cpt3Dr	7-8/26	3523A	4970
LAFLIN		12SecDr	7/16	2410F	4412
LAGARTO		12SecDr	5-8/18	2410F	4540
LAGO					
AZUL	RN	8SecRest6StsObsLng	9-10/14	4025J	4292
CHICO	RN	8SecRest6StsObsLng	9-10/14	4025J	4292
DE CHALCO NdeM-177	SN	8SecDr2Cpt	10-12/28	3979A	6205
DE CHAPALA NdeM-178	SN	8Sec4Dbr	7-9/17	4022A	4515
DE TEXCOCO NdeM-182	SN	8SecDr2Cpt	10-12/28	3979A	6205
MANSO	RN	8SecRestLngObs	11/12	4025J	4049
PATZCUARO NdeM	SN	14Sec	9-10/25	3958	4869

CAR NAME	ST	CAR TYPE	BUILT	PLAN	LOT
LAGO					
SECO	RN	8SecRest6StsLngObs	12/11-12	4025J	3944
VERDE	RN	8SecRestLngObs	11/12	4025J	4049
LAGONDA		12SecDr	1-2/15	2410B	4311
LAGUNA DE					
TAMIAHUA NdeM-179	SN	12SecDr	6-8/24	3410	4764
TERMINOS NdeM-180	SN	12SecDr	8-12/25	3410A	4894
LAGUNA DEL					
CARMEN NdeM-181	SN	12SecDr	2-5/25	3410	4845
LAHAD		12SecDr	11-12/14	2410B	4297
LAHOMA		12SecDr	2-3/15	2410B	4318
LAIDLAW		12SecDr	9-10/17	2410F	4525
LAKE AINSLIE		10SecDr2Cpt	12/23-24	3585	4728
LAKE ALBERT		10SecDr2Cpt	10/24	3585A	4817
LAKE ALEXANDER		10SecDr2Cpt	10/24	3585A	4817
LAKE ANDREW		10SecDr2Cpt	1-2/24	3585	4743
LAKE ANN		10SecDr2Cpt	12/23-24	3585	4728
LAKE APOPKA		10SecDr2Cpt	8-10/23	3585	4725
LAKE ARIANA		10SecDr2Cpt	8-10/23	3585	4725
LAKE ARROW		10SecDr2Cpt	12/25-26	3585A	4933
LAKE ARTHUR		10SecDr2Cpt	8-10/23	3585	4725
LAKE ASHLEY		10SecDr2Cpt	10/24	3585A	4817
LAKE ATWOOD		10SecDr2Cpt	12/25-26	3585A	4933
LAKE AUBURN		10SecDr2Cpt	5-6/24	3585	4770
LAKE AUGUSTA		10SecDr2Cpt	5-6/24	3585	4770
LAKE BARON		10SecDr2Cpt	10/24	3585A	4817
LAKE BEAR		10SecDr2Cpt	6-7/25	3585A	4888
LAKE BELANONA		10SecDr2Cpt	9-10/26	3585A	4996
LAKE BENNETT		10SecDr2Cpt	12/23-24	3585	4728
LAKE BENTON		10SecDr2Cpt	12/23-24	3585	4728
LAKE BERNARD		10SecDr2Cpt	9-10/26	3585A	4996
LAKE BILLIOT		10SecDr2Cpt	8-10/23	3585	4725
LAKE BLUEFIELD		10SecDr2Cpt	9-10/26	3585A	4996
LAKE BLUFF		10SecDr2Cpt	12/23-24	3585	4728
LAKE BONAPARTE		10SecDr2Cpt	12/23-24	3585	4728
LAKE BORGNE		10SecDr2Cpt	9-10/26	3585A	4996
LAKE BRUIN		10SecDr2Cpt	6-7/25	3585A	4888
LAKE BRYANT		10SecDr2Cpt	12/23-24	3585	4728
LAKE BUFFUM		10SecDr2Cpt	12/23-24	3585	4728
LAKE BUTLER		10SecDr2Cpt	5-6/24	3585	4770
LAKE CADILLAC		10SecDr2Cpt	6-7/25	3585A	4888
LAKE CAILLOU		10SecDr2Cpt	8-10/23	3585	4725
LAKE CALHOUN		10SecDr2Cpt	8-10/23	3585	4725
LAKE CALUMET		10SecDr2Cpt	12/23-24	3585	4728
LAKE CAROLINE		10SecDr2Cpt	6-7/25	3585A	4888
LAKE CARY		10SecDr2Cpt	12/23-24	3585	4728
LAKE CATHERINE		10SecDr2Cpt	12/25-26	3585A	4933
LAKE CHABOT		10SecDr2Cpt	1-2/24	3585	4743
LAKE CHAMPLAIN		10SecDr2Cpt	8-10/23	3585	4725
LAKE CHAPLIN		10SecDr2Cpt	12/25-26	3585A	4933

CAR NAME	ST	CAR TYPE	BUILT	PLAN	LOT	CAR NAME	ST	CAR TYPE	BUILT	PLAN	LOT
LAKE CHARLES		10SecDr2Cpt	8-10/23	3585	4725	LAKE FRENCH		10SecDr2Cpt	6-7/25	3585A	4888
LAKE CHARM		10SecDr2Cpt	8-10/23	3585	4725	LAKE FRENCH FCS-BC	SN	10SecDr2Cpt	6-7/25	3585A	4888
LAKE CHELAN		10SecDr2Cpt	1-2/24	3585	4743	LAKE GAGE		10SecDr2Cpt	5-6/24	3585	4770
LAKE CHICOT		10SecDr2Cpt	8-10/23	3585	4725	LAKE GARDNER		10SecDr2Cpt	12/23-24	3585	4728
LAKE CHILDS		10SecDr2Cpt	9-10/26	3585A	4996	LAKE GARFIELD		10SecDr2Cpt	1-2/24	3585	4743
LAKE CLARK		10SecDr2Cpt	12/23-24	3585	4728	LAKE GENEVA		7CptLngObs	10/12	2635	4041
LAKE CLAY		10SecDr2Cpt	6-7/25	3585A	4888	LAKE GENEVA		10SecDr2Cpt	12/23-24	3585	4728
LAKE CLEALUM		10SecDr2Cpt	1-2/24	3585	4743	LAKE GENTRY		10SecDr2Cpt	10/24	3585A	4817
LAKE CLEAR		12SecDrCpt	3/11	2411	3880	LAKE GEORGE		10SecDr2Cpt	8-10/23	3585	4725
LAKE COLVILLE		10SecDr2Cpt	12/25-26	3585A	4933	LAKE GILES		10SecDr2Cpt	6-7/25	3585A	4888
LAKE COMO		10SecDr2Cpt	8-10/23	3585	4725	LAKE GIRARD		10SecDr2Cpt	5-6/24	3585	4770
LAKE CONBOY		10SecDr2Cpt	9-10/26	3585A	4996	LAKE GOODWIN		10SecDr2Cpt	1-2/24	3585	4743
LAKE CONE		10SecDr2Cpt	10/24	3585A	4817	LAKE GRANDIN		10SecDr2Cpt	12/25-26	3585A	4933
LAKE CONLIN		10SecDr2Cpt	12/25-26	3585A	4933	LAKE GREGORY		10SecDr2Cpt	12/25-26	3585A	4933
LAKE CONSTANCE		10SecDr2Cpt	9-10/26	3585A	4996	LAKE GRIFFIN		10SecDr2Cpt	8-10/23	3585	4725
LAKE CONWAY		10SecDr2Cpt	5-6/24	3585	4770	LAKE GROVE		10SecDr2Cpt	6-7/25	3585A	4888
LAKE COWAN		10SecDr2Cpt	12/25-26	3585A	4933	LAKE GUANO		10SecDr2Cpt	9-10/26	3585A	4996
LAKE CREEK		10SecDr2Cpt	9-10/26	3585A	4996	LAKE HAMILTON		10SecDr2Cpt	6-7/25	3585A	4888
LAKE CRYSTAL		10SecDr2Cpt	8-10/23	3585	4725	LAKE HANCOCK		10SecDr2Cpt	8-10/23	3585	4725
LAKE CUSHMAN		10SecDr2Cpt	12/23-24	3585	4728	LAKE HARNEY		10SecDr2Cpt	8-10/23	3585	4725
LAKE DIAS		10SecDr2Cpt	9-10/26	3585A	4996	LAKE HARRIET		10SecDr2Cpt	8-10/23	3585	4725
LAKE DICKEY		10SecDr2Cpt	9-10/26	3585A	4996	LAKE HARRIS		10SecDr2Cpt	8-10/23	3585	4725
LAKE DISSTON		10SecDr2Cpt	9-10/26	3585A	4996	LAKE HASSET		10SecDr2Cpt	10/24	3585A	4817
LAKE DORR		10SecDr2Cpt	9-10/26	3585A	4996	LAKE HAYDEN		10SecDr2Cpt	6-7/25	3585A	4888
LAKE DRUMMOND		10SecDr2Cpt	12/23-24	3585	4728	LAKE HAZEN		10SecDr2Cpt	9-10/26	3585A	4996
LAKE DUANE		10SecDr2Cpt	9-10/26	3585A	4996	LAKE HELEN		10SecDr2Cpt	12/23-24	3585	4728
LAKE DUMONT		10SecDr2Cpt	12/25-26	3585A	4933	LAKE HENDERSON		10SecDr2Cpt	12/23-24	3585	4728
LAKE DUNMORE		10SecDr2Cpt	12/23-24	3585	4728	LAKE HENDRICKS		10SecDr2Cpt	10/24	3585A	4817
LAKE EATON		10SecDr2Cpt	5-6/24	3585	4770	LAKE HENRY		10SecDr2Cpt	12/23-24	3585	4728
LAKE EDWARD		10SecDr2Cpt	5-6/24	3585	4770	LAKE HIAWATHA		10SecDr2Cpt	9-10/26	3585A	4996
LAKE ELEANOR		10SecDr2Cpt	12/25-26	3585A	4933	LAKE HOPATCONG		10SecDr2Cpt	12/23-24	3585	4728
LAKE ELEANOR FCS-BC	SN	10SecDr2Cpt	12/25-26	3585A	4933	LAKE HUNTLEY		10SecDr2Cpt	5-6/24	3585	4770
LAKE ELIZABETH		10SecDr2Cpt	12/23-24	3585	4728	LAKE HURON		10SecDr2Cpt	12/23-24	3585	4728
LAKE ELMO		10SecDr2Cpt	8-10/23	3585	4725	LAKE IRVIN		10SecDr2Cpt	12/23-24	3585	4728
LAKE EMMA		10SecDr2Cpt	6-7/25	3585A	4888	LAKE ISABELLA		10SecDr2Cpt	1-2/24	3585	4743
LAKE ERIE		10SecDr2Cpt	12/23-24	3585	4728	LAKE ITASCA		10SecDr2Cpt	1-2/24	3585	4743
LAKE ERNEST		10SecDr2Cpt	10/24	3585A	4817	LAKE JACKSON		10SecDr2Cpt	6-7/25	3585A	4888
LAKE EUGENE		10SecDr2Cpt	8-10/23	3585	4725	LAKE JAMES		10SecDr2Cpt	5-6/24	3585	4770
LAKE EUNICE		10SecDr2Cpt	12/23-24	3585	4728	LAKE JESSIE		10SecDr2Cpt	10/24	3585A	4817
LAKE EUSTIS		10SecDr2Cpt	8-10/23	3585	4725	LAKE JESSUP		10SecDr2Cpt	8-10/23	3585	4725
LAKE FELICITY		10SecDr2Cpt	8-10/23	3585	4725	LAKE JOE		10SecDr2Cpt	12/23-24	3585	4728
LAKE FERGUSON		10SecDr2Cpt	1-2/24	3585	4743	LAKE JULIAN		10SecDr2Cpt	6-7/25	3585A	4888
LAKE FLORENCE		10SecDr2Cpt	6-7/25	3585A	4888	LAKE KEGONSA		10SecDr2Cpt	12/23-24	3585	4728
LAKE FOREST		7CptLngObs	10/12	2635	4041	LAKE KENNEDY		10SecDr2Cpt	9-10/26	3585A	4996
LAKE FOREST		10SecDr2Cpt	12/23-24	3585	4728	LAKE KERR		10SecDr2Cpt	9-10/26	3585A	4996
LAKE FORTUNA		10SecDr2Cpt	8-10/23	3585	4725	LAKE KEZAR	N	10SecDr2Cpt	6-7/25	3585A	4888
LAKE FRANCIS		10SecDr2Cpt	12/25-26	3585A	4933	LAKE KLUANE		10SecDr2Cpt	9-10/26	3585A	4996
LAKE FRANKLIN		10SecDr2Cpt	12/25-26	3585A	4933	LAKE LABERGE		10SecDr2Cpt	9-10/26	3585A	4996
LAKE FREDERICK		10SecDr2Cpt	12/25-26	3585A	4933	LAKE LAPOURDE		10SecDr2Cpt	8-10/23	3585	4725
LAKE FREMONT		10SecDr2Cpt	12/23-24	3585	4728	LAKE LATHAM		10SecDr2Cpt	5-6/24	3585	4770

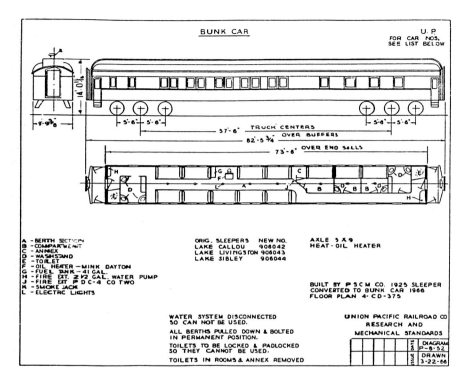

Figure I-58 In 1966, after 40 years of active service, the Union Pacific Bunk Car 906043 was converted from the former 10-section drawing room, 2-compartment car, LAKE LIVINGSTON. These work train cars were normally painted silver with black lettering. The Author's Collection.

CAR NAME	ST	CAR TYPE	BUILT	PLAN	LOT
LAKE LAURA		10SecDr2Cpt	10/24	3585A	4817
LAKE LAWRENCE		10SecDr2Cpt	5-6/24	3585	4770
LAKE LILLIAN		10SecDr2Cpt	12/23-24	3585	4728
LAKE LINDSAY		10SecDr2Cpt	5-6/24	3585	4770
LAKE LIVINGSTON		10SecDr2Cpt	12/25-26	3585A	4933
LAKE LONG		10SecDr2Cpt	8-10/23	3585	4725
LAKE LOUISE		10SecDr2Cpt	12/25-26	3585A	4933
LAKE LOWRY		10SecDr2Cpt	9-10/26	3585A	4996
LAKE LUCERNE		10SecDr2Cpt	12/25-26	3585A	4933
LAKE MAHOPAC		10SecDr2Cpt	12/23-24	3585	4728
LAKE MAITLAND		10SecDr2Cpt	8-10/23	3585	4725
LAKE MAJELLA		10SecDr2Cpt	8-10/23	3585	4725
LAKE MANITOBA		10SecDr2Cpt	12/25-26	3585A	4933
LAKE MANSFIELD		10SecDr2Cpt	9-10/26	3585A	4996
LAKE MARION		10SecDr2Cpt	6-7/25	3585A	4888
LAKE MARTHA		10SecDr2Cpt	12/23-24	3585	4728

CAR NAME	ST	CAR TYPE	BUILT	PLAN	LOT
LAKE MARY		10SecDr2Cpt	8-10/23	3585	4725
LAKE MASON		10SecDr2Cpt	1-2/24	3585	4743
LAKE MAUREPAS		10SecDr2Cpt	8-10/23	3585	4725
LAKE MELVILLE		10SecDr2Cpt	12/25-26	3585A	4933
LAKE MENDOTA		10SecDr2Cpt	8-10/23	3585	4725
LAKE MERRITT		10SecDr2Cpt	8-10/23	3585	4725
LAKE MICHIGAN		10SecDr2Cpt	12/23-24	3585	4728
LAKE MILTONA		10SecDr2Cpt	1-2/24	3585	4743
LAKE MINNETONKA		10SecDr2Cpt	12/23-24	3585	4728
LAKE MITCHELL		10SecDr2Cpt	5-6/24	3585	4770
LAKE MONONA		10SecDr2Cpt	5-6/24	3585	4770
LAKE MONROE		10SecDr2Cpt	8-10/23	3585	4725
LAKE MOORE		10SecDr2Cpt	12/25-26	3585A	4933
LAKE MOREAU		10SecDr2Cpt	8-10/23	3585	4725
LAKE NAHWATEL		10SecDr2Cpt	10/24	3585A	4817
LAKE NATCHEZ		10SecDr2Cpt	8-10/23	3585	4725
LAKE NICARAGUA		10SecDr2Cpt	12/25-26	3585A	4933
LAKE NIPIGON		10SecDr2Cpt	12/25-26	3585A	4933
LAKE NORRIS		10SecDr2Cpt	10/24	3585A	4817
LAKE OKEECHOBEE		10SecDr2Cpt	8-10/23	3585	4725
LAKE OLIVER		10SecDr2Cpt	5-6/24	3585	4770
LAKE ONOTA		10SecDr2Cpt	10/24	3585A	4817
LAKE ONTARIO		10SecDr2Cpt	12/23-24	3585	4728
LAKE ORDWAY		10SecDr2Cpt	5-6/24	3585	4770
LAKE OWEN		10SecDr2Cpt	12/23-24	3585	4728
LAKE PARK	N	12SecDrCpt	7/20/10	2411	3800
LAKE PARKER		10SecDr2Cpt	12/23-24	3585	4728
LAKE PEARL		10SecDr2Cpt	8-10/23	3585	4725
LAKE PEIGNEU		10SecDr2Cpt	8-10/23	3585	4725
LAKE PELICAN		10SecDr2Cpt	12/23-24	3585	4728
LAKE PHALEN		10SecDr2Cpt	6-7/25	3585A	4888
LAKE PICKETT		10SecDr2Cpt	12/25-26	3585A	4933
LAKE PIERCE		10SecDr2Cpt	12/25-26	3585A	4933
LAKE PIGEON		10SecDr2Cpt	8/25	3585A	4898
LAKE PLACID		10SecDr2Cpt	12/23-24	3585	4728
LAKE PLEASANT		10SecDr2Cpt	12/23-24	3585	4728
LAKE PONTCHARTRAIN		10SecDr2Cpt	8-10/23	3585	4725
LAKE POYGAN		10SecDr2Cpt	10/24	3585A	4817
LAKE PRESTON		10SecDr2Cpt	12/23-24	3585	4728
LAKE QUINALT		10SecDr2Cpt	10/24	3585A	4817
LAKE RAMON		10SecDr2Cpt	9-10/26	3585A	4996
LAKE ROLAND		10SecDr2Cpt	12/25-26	3585A	4933
LAKE RONKONKOMA		10SecDr2Cpt	8/25	3585A	4898
LAKE RUSSELL		10SecDr2Cpt	8/25	3585A	4898
LAKE SALVADOR		10SecDr2Cpt	8-10/23	3585	4725
LAKE SANFORD		10SecDr2Cpt	12/23-24	3585	4728
LAKE SARAH		10SecDr2Cpt	10/24	3585A	4817
LAKE SCOTT		10SecDr2Cpt	8/25	3585A	4898
LAKE SEDGWICK		10SecDr2Cpt	9-10/26	3585A	4996
LAKE SELBY		10SecDr2Cpt	8/25	3585A	4898

CAR NAME	ST	CAR TYPE	BUILT	PLAN	LOT
LAKE SHERIDAN		10SecDr2Cpt	1-2/24	3585	4743
LAKE SIBLEY		10SecDr2Cpt	10/24	3585A	4817
LAKE STEARNS		10SecDr2Cpt	5-6/24	3585	4770
LAKE SUNAPEE		10SecDr2Cpt	12/23-24	3585	4728
LAKE SUPERIOR		10SecDr2Cpt	12/23-24	3585	4728
LAKE SUTHERLAND		10SecDr2Cpt	9-10/26	3585A	4996
LAKE SYLVAN		10SecDr2Cpt	8/25	3585A	4898
LAKE TAHOE		10SecDr2Cpt	8-10/23	3585	4725
LAKE TAPPS		10SecDr2Cpt	9-10/26	3585A	4996
LAKE TERRELL		10SecDr2Cpt	5-6/24	3585	4770
LAKE THISTED		10SecDr2Cpt	10/24	3585A	4817
LAKE THOMPSON		10SecDr2Cpt	12/23-24	3585	4728
LAKE TRACY		10SecDr2Cpt	5-6/24	3585	4770
LAKE TRAVERSE		10SecDr2Cpt	12/23-24	3585	4728
LAKE UNION		10SecDr2Cpt	12/25-26	3585A	4933
LAKE VALE		10SecDr2Cpt	8-10/23	3585	4725
LAKE VERRET		10SecDr2Cpt	8-10/23	3585	4725
LAKE VICTOR		10SecDr2Cpt	8-10/23	3585	4725
LAKE VIEW		10SecDr2Cpt	12/23-24	3585	4728
LAKE VINEYARD		10SecDr2Cpt	8-10/23	3585	4725
LAKE VISTA		10SecDr2Cpt	1-2/24	3585	4743
LAKE WACCAMAW		10SecDr2Cpt	10/24	3585A	4817
LAKE WANITAH		10SecDr2Cpt	1-2/24	3585	4743
LAKE WASHINGTON		10SecDr2Cpt	9-10/26	3585A	4996
LAKE WAUBESA	N	10SecDr2Cpt	9-10/26	3585A	4996
LAKE WEIR		10SecDr2Cpt	8-10/23	3585	4725
LAKE WHATCOM		10SecDr2Cpt	9-10/26	3585A	4996
LAKE WHITE		10SecDr2Cpt	12/25-26	3585A	4933
LAKE WHITEWOOD		10SecDr2Cpt	12/23-24	3585	4728
LAKE WHITNEY		10SecDr2Cpt	12/25-26	3585A	4933
LAKE WILLIAMS		10SecDr2Cpt	12/23-24	3585	4728
LAKE WILSON		10SecDr2Cpt	5-6/24	3585	4770
LAKE WINDER		10SecDr2Cpt	9-10/26	3585A	4996
LAKE WINNEBAGO		10SecDr2Cpt	12/23-24	3585	4728
LAKE WINNIPEG		10SecDr2Cpt	12/25-26	3585A	4933
LAKE WINTHROP		10SecDr2Cpt	9-10/26	3585A	4996
LAKE WOODRUFF		10SecDr2Cpt	5-6/24	3585	4770
LAKE WORTH		10SecDr2Cpt	8-10/23	3585	4725
LAKE YUKON		10SecDr2Cpt	12/25-26	3585A	4933
LAKE ZURICH		10SecDr2Cpt	1-2/24	3585	4743
LAKEMONT		12SecDr	5/9/10	2410	3773
LAKEMONT		12SecDr	5-6/17	2410F	4497
LAKETON		16Sec	10/13	2412B	4216
LAKEVILLE		12SecDr	9/23/10	2410	3794
LAKEVILLE		12SecDr	6-9/26	3410A	4969
LAKEWOOD		12SecDr	7/16	2410F	4412
LAKOTA		12SecDr	9-10/15	2410D	4327
LALIVE		26ChDr	4/12	2416	3950
LAMANDA		12SecDr	1-3/14	2410A	4249
LAMARQUE		12SecDr	6-7/17	2410F	4503

CAR NAME	ST	CAR TYPE	BUILT	PLAN	LOT
LAMATH		12SecDr	11-12/13	2410A	4234
LAMBERTVILLE		12SecDr	6/11/10	2410	3794
LAMBETH		12SecDr	9-10/13	2410A	4215
LAMBS CLUB		13ChRestLng	12/29	3992	6289
LAMBSON		12SecDr	11/25/10	2410	3864
LAMOILLE		12SecDr	7-8/14	2410A	4277
LAMOKIN		12SecDr	9/13/10	2410	3794
LAMONT		12SecDrCpt	9/12	2411	4036
LAMORNA		12SecDr	2-5/16	2410E	4367
LAMOURE		10SecDr2Cpt	1-2/13	2585	4091
LAMPASAS		12SecDr	6-7/17	2410F	4503
LANCASTER		16Sec	2-3/11	2412	3878
LANCASTER CLUB		4Sbr2CptDrLng	12/29	3995	6301
LANCASTER CLUB	R	2Sbr2Dbr2CptDrLng	12/29	3995C	6301
LANDENBERG		12SecDr	9/24/10	2410	3794
LANDER		LngBDorm(5Sec)	7-8/27	3981	6061
LANDERNAU		12SecDr	9-10/13	2410A	4215
LANDISVILLE		12SecDr	9/16/10	2410	3794
LANDOVER		24ChDrB	4/13	2417	4108
LANDRO		12SecDr	7/13	2410A	4179
LANDSEER		6Cpt3Dr	7/25	3523A	4887
LANE		28ChDr	5-7/26	3416	4958
LANESVILLE		36Ch	6/16	2916	4389
LANGDON		12SecDr	8/24/10	2410	3794
LANGLEY		12SecDr	1-2/22	2410F	4625
LANGSIDE		16Sec	1/16	2412E	4352
LANIER		12SecDr	1-3/14	2410A	4249
LANKFORD		12SecDr	10-12/16	2410F	4431
LANRICK		7Cpt2Dr	1/4/11	2522	3867
LANSDALE		12SecDr	7-8/12	2410	4014
LANSDOWNE		26ChDr	5/12	2416	3986
LANTERN TOWER	RN	8SecDr3Dbr	12/26-27	4090D	6031
LANZO	N	12SecDr	7-8/13	2410A	4192
LAON		12SecDr	12/20-21	2410F	4611
LAPHAM		12SecDr	6-7/16	2410F	4385
LAPLACE		12SecDr	6-7/17	2410F	4503
LARABEE		26ChDr	5/12	2416	3986
LARAMIE		12SecDr	1-2/22	2410F	4625
LARAND		12SecDr	1-3/14	2410A	4249
LARCHLAND		16Sec	5-6/13	2412B	4150
LARCHMONT		32ChDr	6/16	2917	4390
LAREDO		12SecDr	1-3/17	2410F	4450
LARGO					
DE CHAPALA NdeM-178	N	8Sec4Dbr	7-9/17	4022A	4515
LARIAT CREST	N	10SecDrCpt	10-12/29	3973A	6300
LARIAT RANGE	N	10SecDrCpt	3-5/30	3973A	6338
LARIMER		12SecDr	9/24/10	2410	3794
LARISTAN		12SecDr	4-6/21	2410F	4613
LARKIN		12SecDr	1-2/22	2410F	4625
LARKSPUR		28ChDr	5-6/24	3416	4761

CAR NAME	ST	CAR TYPE	BUILT	PLAN	LOT
LARNED		7Cpt2Dr	9/13	2522A	4208
LARONE		12SecDr	4-5/13	2410	4106
LARPENTEUR		12SecDr	2-5/25	3410	4845
LARUE		12SecDr	2-5/25	3410	4845
LARWILL		12SecDr	11/15/10	2410	3860
LAS PLUMAS		10SecLngObs	9-10/14	2521A	4292
LAS TRAMPAS		10Sec2Dr	2/13	2584	4092
LAS VEGAS	T	10SecDr2Cpt	10-11/13	2585A	4218
LASCAR		12SecDr	1-2/22	2410F	4625
LASHBURN		12SecDr	1-3/14	2410A	4249
LASHMET		10SecDr2Cpt	8-9/13	2585A	4201
LASITA		12SecDr	11-12/13	2410A	4234
LASSEN		16Sec	4/17	2412F	4484
LATANACHE		12SecDr	6-7/17	2410F	4503
LATANIER		12SecDr	1-2/15	2410B	4311
LATCHA		12SecDr	6-9/26	3410A	4969
LATHAM		12SecDr	1-2/22	2410F	4625
LATHROP		10SecDr2Cpt	10-11/12	2585	4067
LATIMER		16Sec	7-8/11	2412	3912
LATOURELLE		12SecDr	2-5/16	2410E	4367
LATROBE		12SecDr	12/27-28	3410B	6127
LAUDECK		12SecDr	11-12/13	2410A	4234
LAUDERDALE		12SecDr	12/12-13	2410	4076
LAUGHLIN		12SecDr	6-9/26	3410A	4969
LAUNFAL		12SecDr	11-12/14	2410B	4297
LAURA		28ChDr	6-7/27	3416A	6087
LAURANA		12SecDr	3-6/20	2410F	4565
LAUREL		12SecDr	1-2/22	2410F	4625
LAUREL ART	N	8SecDr2Cpt	4-5/30	3979A	6359
LAUREL BAY		8SecDr2Cpt	4-5/30	3979A	6359
LAUREL BRANCH		8SecDr2Cpt	4-5/30	3979A	6359
LAUREL CREEK		8SecDr2Cpt	4-5/30	3979A	6359
LAUREL CREST		8SecDr2Cpt	4-5/30	3979A	6359
LAUREL DALE		8SecDr2Cpt	4-5/30	3979A	6359
LAUREL GAP		8SecDr2Cpt	4-5/30	3979A	6359
LAUREL GLEN		8SecDr2Cpt	4-5/30	3979A	6359
LAUREL GROVE		8SecDr2Cpt	4-5/30	3979A	6359
LAUREL HEIGHTS		8SecDr2Cpt	4-5/30	3979A	6359
LAUREL HILL		8SecDr2Cpt	4-5/30	3979A	6359
LAUREL PARK		8SecDr2Cpt	4-5/30	3979A	6359
LAUREL PASS		8SecDr2Cpt	4-5/30	3979A	6359
LAUREL RIDGE		8SecDr2Cpt	4-5/30	3979A	6359
LAUREL ROAD		8SecDr2Cpt	4-5/30	3979A	6359
LAUREL SUMMIT		8SecDr2Cpt	4-5/30	3979A	6359
LAUREL TERRACE		8SecDr2Cpt	4-5/30	3979A	6359
LAUREL VALLEY		8SecDr2Cpt	4-5/30	3979A	6359
LAUREL VIEW		8SecDr2Cpt	4-5/30	3979A	6359
LAUREL WOOD		8SecDr2Cpt	4-5/30	3979A	6359
LAURELTON		12SecDr	10-12/26	3410A	6023
LAURIN		12SecDr	10-11/11	2410	3936

CAR NAME	ST	CAR TYPE	BUILT	PLAN	LOT
LAUVETA		12SecDr	9-10/15	2410D	4327
LAVA LAKE		10SecDr2Cpt	11-12/27	3585B	6123
LAVACA		12SecDr	8-10/26	3410A	4945
LAVAL UNIVERSITY	RN	12Sec2Dbr	5-6/17	4046	4497
LAVALLETTE		12SecDr	5/19/10	2410	3794
LAVENHAM		12SecDr	11-12/13	2410A	4234
LAVEROCK		26ChDr	5/12	2416	3986
LAVIELLE		12SecDr	1-2/16	2410E	4356
LAVISTA		12SecDr	1-2/15	2410B	4311
LAWMAKER		12SecDr	4/26	3410A	4943
LAWN		12SecDr	5-6/23	2410H	4699
LAWSONHAM		12SecDr	4/27/10	2410	3772
LAWTON		12SecDr	1-4/12	2410	3949
LAWYERS CLUB		13ChRestLng	12/29	3992	6289
LAYLAND		12SecDr	6-9/26	3410A	4969
LAYTON		12SecDr	4-6/14	2410A	4271
LE GRAND		10SecDrCpt	10-12/29	3973A	6300
LE ROI		10SecDrCpt	10-12/29	3973A	6300
LE SEUER		14Sec	6/28	3958A	6174
LEAH		24ChDrB	5/13	2417A	4137
LEAMAN PLACE		12SecDr	9/1/10	2410	3794
LEANING TOWER	RN	8SecDr3Dbr	1/25	4090	4843
LEATHERWOOD		12SecDr	5-6/17	2410F	4497
LEBO		12SecDr	5-8/18	2410F	4540
LECLAIRE		12SecDr	2-3/13	2410	4105
LECOMPTON		10SecDr2Cpt	8-9/13	2585A	4201
LEDDRA		12SecDr	7-8/15	2410D	4322
LEDFORD		16Sec	7-8/11	2412	3912
LEDWITH		12SecDr	9-10/13	2410A	4215
LEDY		28ChDr	5-7/26	3416	4958
LEDYARD		12SecDr	11-12/13	2410A	4234
LEECHBURG		12SecDr	9/20/10	2410	3794
LEESPORT		12SecDr	5/20/10	2410	3794
LEETSDALE		12SecDr	2-4/24	3410	4762
LEHIGH		12SecDr	7/13	2410A	4179
LEIF ERICSON		3Cpt2DrLngObs	11-12/26	3959	6015
LEIGHTON		12SecDr	4-6/21	2410F	4613
LEINSTER		12SecDr	4-6/21	2410F	4613
LEIPSIC		12SecDr	7-8/13	2410A	4192
LEITRIM		12SecDr	12/20-21	2410F	4611
LELIA		26ChDr	3-4/14	2416A	4264
LEMAY		10SecDr2Cpt	11-12/11	2585	3942
LEMERT		12SecDr	5-6/11	2410	3903
LEMLEY		12SecDr	9-10/17	2410F	4525
LEMNOS		12SecDr	12/10-11	2410	3866
LEMOYNE		24ChDrB	4/13	2417	4108
LENAPE		10SecDr2Cpt	11-12/11	2585	3942
LENAWEE		12SecDr	12/20-21	2410F	4611
LENDORA		28ChDr	4/12	2418	3951
LENNIG		12SecDr	5-6/17	2410F	4497

CAR NAME	ST	CAR TYPE	BUILT	PLAN	LOT	CAR NAME	ST	CAR TYPE	BUILT	PLAN	LOT
LENOLA		12SecDr	10/21/10	2410	3814	LIMA NdeM	SN	6Sec6Dbr	10-12/21	4060	4624
LENORE		28ChDr	5-6/24	3416	4761	LIMAVILLE		12SecDr	6-9/26	3410A	4969
LENOVER		12SecDrCpt	3/11	2411	3880	LIMEDALE		12SecDr	4/13/10	2410	3771
LENOX		12SecDr	1-2/15	2410B	4311	LIMEDALE		12SecDr	2-4/24	3410	4762
LEOLYN		28ChDr	5-7/26	3416	4958	LIMERICK		12SecDr	3-6/20	2410F	4565
LEONA VICARIO NdeM	SN	12SecDr	8-12/25	3410A	4894	LIMITAR		12SecDr	11-12/14	2410B	4297
LEONTINE		12SecDr	7-8/15	2410D	4322	LIMOGES		12SecDr	12/20-21	2410F	4611
LEONVILLE		12SecDr	6-7/17	2410F	4503	LINARO		12SecDr	4-6/21	2410F	4613
LEPANTO		12SecDr	1/16	2410E	4354	LINCOLN		16Sec	10/13	2412B	4217
LERDO						LINCOLN MEMORIAL		34Ch	8/30	4001A	6386
DE TEJADA FCP-229	SN	10SecDr2Cpt	8-10/23	3585F	4725	LINCOLNDALE		12SecDr	4-6/14	2410A	4271
LERIDA		12SecDr	10-12/15	2410D	4338	LINDA ROSA		10SecLngObs	12/16-17	2521C	4442
LESNO		12SecDr	1/16	2410E	4354	LINDEN		12SecDr	6-9/26	3410A	4969
LETCHWORTH		12SecDr	5-6/17	2410F	4497	LINDENWOLD		24ChDrB	1/14	2417A	4239
LETTER GAP		8SecDr2Cpt	12/29-30	3979A	6334	LINDLEY		10SecDr2Cpt	5/13	2585A	4141
LEVANT		12SecDr	1-3/17	2410F	4450	LINDSAY		12SecDr	2-5/16	2410E	4367
LEVEREDGE		12SecDr	6-7/16	2410F	4385	LINESVILLE		12SecDr	10/8/10	2410	3813
LEVINGS		12SecDr	10-11/11	2410	3936	LINLITHGO		12SecDr	1-4/12	2410	3949
LEWALD		12SecDr	2-3/15	2410B	4318	LINN		28ChDr	5-6/24	3416	4761
LEWES		12SecDr	9/3/10	2410	3794	LINOMA LAKE		10SecDrCpt	10-12/29	3973A	6300
LEWIS		12SecDr	2-5/25	3410	4845	LINTON		12SecDr	12/10-11	2410	3866
LEWISBURG		12SecDr	9/27/10	2410	3794	LINVILLE		12SecDr	2-3/13	2410	4105
LEWISTOWN		12SecDr	2-5/16	2410E	4367	LINWOOD		10SecDr2Cpt	10-11/13	2585A	4218
LEXINGTON		12SecDr	11/11/10	2410	3860	LIONS CLUB	N	16ChB15StsLng	4/30	3996C	6312
LEXINGTON	N	12SecDr	2-3/27	3410A	6055	LISBON		12SecDr	6-7/16	2410F	4385
LEXINGTON	RN	10StsRestLng	12/14-15	4018	4304	LISMULLEN		12SecDr	4-6/21	2410F	4613
LIBERIA		12SecDr	12/20-21	2410F	4611	LISZT		6Cpt3Dr	11-12/24	3523A	4833
LIBERTY		12SecDrCpt	6/13	2411A	4159	LITA		26ChDr	3-4/14	2416A	4264
LIBERTY BELL		BaggBSmkBarber	10-11/25	3951	4885	LITCHFIELD		12SecDr	4-6/24	3410	4763
LIBERTY BOYS		BaggBSmkBarber	10-11/25	3951	4885	LITTLE ARCH	RN	14Sec	9-10/15	4061B	4327
LIBERTY CAP		BaggBSmkBarber	9-10/26	3951C	4972	LITTLE BANK	RN	14Sec	3-4/11	4061A	3881
LIBERTY COLORS		BaggBSmkBarber	9-10/26	3951C	4972	LITTLE BAY	RN	14Sec	3-6/20	4061	4565
LIBERTY COUNTY		BaggBSmkBarber	10-11/25	3951	4885	LITTLE BEAM	RN	14Sec	1-4/12	4061A	3949
LIBERTY GROVE		BaggBSmkBarber	10-11/25	3951	4885	LITTLE BEAR	RN	14Sec	4/28/10	4061	3771
LIBERTY HALL		4CptLngObs	10/25	3960	4889	LITTLE BELL	RN	14Sec	5/26/10	4061	3794
LIBERTY HILL		BaggBSmkBarber	9-10/26	3951C	4972	LITTLE BEND	RN	14Sec	12/10-11	4061A	3866
LIBERTY ISLAND		BaggBSmkBarber	10-11/25	3951	4885	LITTLE BLUE	RN	14Sec	12/10-11	4061	3866
LIBERTY PARK		BaggBSmkBarber	10-11/25	3951	4885	LITTLE BOW	RN	14Sec	2-5/21	4061	4612
LIBERTY POLE		BaggBSmkBarber	10-11/25	3951	4885	LITTLE CAMP	RN	14Sec	1-2/22	4061	4625
LIBERTY STREET		BaggBSmkBarber	9-10/26	3951C	4972	LITTLE CAVE	RN	14Sec	7/16	4061B	4412
LIBERTY TREE		BaggBSmkBarber	10-11/25	3951	4885	LITTLE CITY	RN	14Sec	9/27/10	4061	3794
LIBRA		12SecDr	12/20-21	2410F	4611	LITTLE CRAG	RN	14Sec	2-5/21	4061	4612
LIDDONFIELD		6CptLngObs	8/12/10	2413	3802	LITTLE DALE	RN	14Sec	11/12/10	4061	3860
LIGGETT		12SecDr	2-4/24	3410	4762	LITTLE DIKE	RN	14Sec	11-12/13	4061A	4234
LIGHTNING		30ChDr	2-3/30	4000	6324	LITTLE DOME	RN	14Sec	12/10-11	4061	3866
LIGONIER		BaggBClubSmkBarber	9/24	2951C	4805	LITTLE ECHO	RN	14Sec	9-10/11	4061A	3924
LILAC		26ChDr	5/23	2416D	4691	LITTLE ELM	RN	14Sec	3-6/20	4061	4565
LILAH		26ChDr	4-5/16	2416C	4366	LITTLE FARM	RN	14Sec	4/11/10	4061	3771
LILLIAN		26ChDr	5/23	2416D	4691	LITTLE FERN	RN	14Sec	9/1/10	4061	3794
LILLY		26ChDr	9-11/22	2416D	4649	LITTLE FIR	RN	14Sec	7-8/12	4061A	4014

CAR NAME	ST	CAR TYPE	BUILT	PLAN	LOT	CAR NAME	ST	CAR TYPE	BUILT	PLAN	LOT
LITTLE FORK	RN	14Sec	9/28/20	4061	3813	LITTLE SPUR	RN	14Sec	7/13	4061	4179
LITTLE FOX	RN	14Sec	10/17/10	4061	3814	LITTLE STAR	RN	14Sec	4-5/13	4061	4106
LITTLE GAP	RN	14Sec	10/11/10	4061	3814	LITTLE SWAN	RN	12Sec	2-5/21	4061	4612
LITTLE GATE	RN	14Sec	11/16/10	4061	3860	LITTLE TOWN	RN	14Sec	6/14/10	4061A	3794
LITTLE GLEN	RN	14Sec	12/10-11	4061A	3866	LITTLE VIEW	RN	14Sec	3-6/20	4061	4565
LITTLE GLOW	RN	14Sec	7/13	4061	4179	LITTLE VINE	RN	14Sec	10/20/10	4061	3814
LITTLE GULF	RN	14Sec	2-3/15	4061	4318	LIVE OAK		10SecLngObs	11/13	2521	4221
LITTLE HILL	RN	14Sec	10/20/10	4061	3814	LIVERMORE		12SecDr	8/29/10	2410	3794
LITTLE HOME	RN	14Sec	3-6/15	4061	4319	LIVERPOOL		12SecDr	5-6/17	2410F	4497
LITTLE HORN	RN	14Sec	11/11/10	4061	3860	LIVINGSTON		12SecDr	4-6/21	2410F	4613
LITTLE ISLE	RN	14Sec	7-8/15	4061B	4322	LIVONIA		12SecDr	3-6/20	2410F	4565
LITTLE KAW	RN	14Sec	11/12/10	4061	3860	LLOYDVILLE		12SecDr	6-9/26	3410A	4969
LITTLE MESA	RN	14Sec	5/25/10	4061A	3794	LOCH AWE		10SecDr2Cpt	6-7/25	3585A	4888
LITTLE MILL	RN	14Sec	5-8/18	4061	4540	LOCH DOON		10SecDr2Cpt	6-7/25	3585A	4888
LITTLE OAK	RN	14Sec	12/10-11	4061A	3866	LOCH EARN		10SecDr2Cpt	6-7/25	3585A	4888
LITTLE PALM	RN	14Sec	10/11/10	4061	3814	LOCH FYNE		10SecDr2Cpt	6-7/25	3585A	4888
LITTLE PARK	RN	14Sec	4/20/10	4061	3771	LOCH KATRINE		10SecDr2Cpt	6-7/25	3585A	4888
LITTLE PASS	RN	14Sec	10/17/10	4061	3814	LOCH LEVEN		10SecDr2Cpt	6-7/25	3585A	4888
LITTLE PEAK	RN	14Sec	12/10-11	4061	3866	LOCH LINNHE		10SecDr2Cpt	6-7/25	3585A	4888
LITTLE PINE	RN	14Sec	12/10-11	4061	3866	LOCH LOMOND		10SecDr2Cpt	6-7/25	3585A	4888
LITTLE POND	RN	14Sec	11/17/10	4061	3864	LOCH MAREE		10SecDr2Cpt	8/25	3585A	4898
LITTLE PORT	RN	14Sec	12/10-11	4061	3866	LOCH NESS		10SecDr2Cpt	8/25	3585A	4898
LITTLE REEF	RN	14Sec	12/10-11	4061A	3866	LOCH TARBERT		10SecDr2Cpt	8/25	3585A	4898
LITTLE ROAD	R	14Sec	8-9/20	4061	4574	LOCH TAY		10SecDr2Cpt	8/25	3585A	4898
LITTLE ROCK	RN	14Sec	2-5/16	4061	4367	LOCHARD		7Cpt2Dr	1/5/11	2522	3867
LITTLE ROSE	RN	14Sec	4/22/10	4061	3771	LOCHINVAR	N	12SecDr	10-12/15	2410D	4338

BECAME S-10c

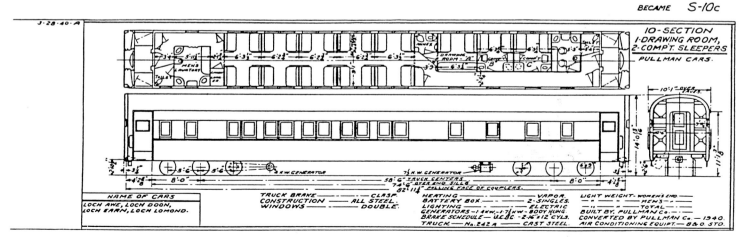

Figure I-60 All 11 of the remaining standard cars with the prefix "LOCH" were sold to the B&O in 1948. (LOCH LINNHE was wrecked on the ATSF in 1945.) Four of these cars received stream-styling in 1940 for service on the "Capitol and National Limited's" as indicated in the diagram. The Author's Collection.

Figure I-59 **B & O LOCH TAY carries the designation X-4566 in this October 1982 photograph by the author taken at Gassaway, West Virginia. This car was scrapped by the railroad during 1986.**

CAR NAME	ST	CAR TYPE	BUILT	PLAN	LOT
LOCHMOOR CLUB		8SecBObsLng	2-3/29	3989	6229
LOCK BERLIN		12SecDrCpt	7/20/10	2411	3800
LOCK HAVEN		12SecDr	9/20/10	2410	3794
LOCK SPRINGS	RN	10Sec3Dbr	2-5/25	4087B	4845
LOCKLAND		12SecDr	5-6/11	2410	3903
LOCKPORT		10SecDr2Cpt	11/17	2585D	4527
LOCKRIDGE		16Sec	5-6/13	2412B	4150
LOCKWOOD		12SecDr	10/20/10	2410	3814
LOCUST FOREST	RN	8SecSbr6RmtDbr	9-10/25	4175	4869
LOCUST WOODS	RN	8SecSbr6RmtDbr	9-10/25	4175	4869
LODEMA		12SecDr	2-3/13	2410	4105
LODI		12SecDr	12/20-21	2410F	4611
LOGAN SQUARE		6Cpt3Dr	7-8/26	3523A	4970
LOGANSPORT		12SecDr	11/16/10	2410	3860
LOGANSPORT		12SecDr	5-6/23	2410H	4699
LOGSTOWN		12SecDr	2-3/27	3410A	6055
LOIRE		12SecDr	12/20-21	2410F	4611
LOIS		26ChDr	3-4/14	2416A	4264
LOLITA		26ChDr	4-5/16	2416C	4366
LOLO PASS		8SecDr2Cpt	12/29-30	3979A	6334
LOMA VISTA		10Sec2Dr	2/13	2584	4092
LOMBARD		12SecDr	12/27-28	3410B	6127
LOMETA		10SecDr2Cpt	8-9/13	2585A	4201
LONDON		12SecDrCpt	5/17	2411C	4494
LONDON TOWER	RN	8SecDr3Dbr	9-10/25	4090D	4899
LONE LAKE		10SecDr2Cpt	11-12/27	3585B	6123
LONE PEAK		10Sec2Dr	12/28-29	3584B	6213
LONE STAR		12SecDr	10/22-23	2410H	4647
LONE STAR CLUB	RN	8SecRestLngObs	2/25/11	4025A	3865
LONG BEACH		10Sec2Dr	2/13	2584	4092
LONG BRANCH		24ChDrB	12/10	2417	3863
LONG CLIFF		10SecDr2Cpt	7-8/23	2585D	4707
LONG ISLAND		BaggBSmkBarber	12/26-27	3951C	6022
LONG PINE		12SecDr	7-8/28	3410B	6184
LONG'S PEAK		10Sec2Dr	12/26-27	3584A	6031
LONGDALE		12SecDr	11-12/13	2410A	4234
LONGFELLOW		12SecDrCpt	10-12/21	2411C	4624
LONGMONT	N	12SecDr	7-8/13	2410A	4192
LONGPORT		12SecDr	5/24/10	2410	3794
LONGWOOD		12SecDr	1-4/12	2410	3949
LONGWORTH		12SecDr	1-2/16	2410E	4356
LONSDALE		36Ch	6/16	2916	4389
LOOKOUT MOUNTAIN		BaggBSmkBarber	12/26-27	3951C	6022
LOOMIS		12SecDr	6-7/16	2410F	4385
LORANGER		12SecDr	7/11	2410	3913
LORCA		12SecDr	12/20-21	2410F	4611
LORD CORNWALLIS	N	12SecDr	2-4/24	3410	4762
LORD FAIRFAX	N	12SecDr	2-4/24	3410B	4762
LORD MOUNT STEPHEN		LngBBarberSunRm	5-6/29	3990	6249
LORD STRATHCONA		LngBBarberSunRm	5-6/29	3990	6249
LORELEI		12SecDr	7-8/13	2410A	4192
LORENDA		24ChDrB	6-7/14	2417B	4265
LORETTA		28ChDr	6-7/27	3416A	6087
LORILLARD		12SecDr	6-7/16	2410F	4385

CAR NAME	ST	CAR TYPE	BUILT	PLAN	LOT
LORING		12SecDr	11-12/13	2410A	4234
LORRISON		12SecDr	11-12/13	2410A	4234
LOS ANGELES	N	Private (GS)	3/11	2503	3847
LOS ANGELES		12SecDr	8-10/26	3410A	4945
LOSTANT		16Sec	7-8/11	2412	3912
LOTHARIO		12SecDr	3-6/20	2410F	4565
LOTHIAN		12SecDr	1-2/16	2410E	4356
LOTOS CLUB	RN	8SecRestObsLng	10/13	4025D	4202
LOUDON		12SecDr	10-12/15	2410D	4338
LOUDON PARK		28ChDr	6-7/27	3416A	6087
LOUDONVILLE		26ChDr	5/23	2416D	4691
LOUIS FRONTENAC		3Cpt2DrLngObs	11-12/26	3959	6015
LOUIS PASTEUR		14Sec	10-11/26	3958	6012
LOUISA ALCOTT		28ChDr	1-2/27	3416A	6032
LOUISE		26ChDr	5/23	2416D	4691
LOUISVILLE	RN	10ChRestLng	7-8/11	4018	3912
LOVENIA		12SecDr	1-2/15	2410B	4311
LOVERNA		12SecDr	1-2/15	2410B	4311
LOVETT(1st)		12SecDr	2-5/16	2410E	4367
LOVETT(2nd)	N	12SecDr	2-5/16	2410E	4367
LOVINGTON		12SecDr	5/2/10	2410	3772
LOVINGTON		26ChDr	9-11/22	2416D	4649
LOWDEN		12SecDr	10-12/16	2410F	4431
LOWLAND		10SecDr2Cpt	10-11/13	2585A	4218
LOWRY	N	12SecDr	5-8/25	3410A	4868
LOYAL LEGION	RN	32StsBSunRm	6/26	3964E	4965
LOYALHANNA		12SecDr	9/7/10	2410	3794
LOYALTON		12SecDr	11-12/14	2410B	4297
LOYOLA	N	14Sec	9-10/25	3958	4869
LOYOLA	N	14Sec	10/26	3958	4997
LUANNA		26ChDr	3-4/14	2416A	4264
LUBEC		16Sec	6-7/20	2412F	4568
LUBIN		12SecDr	7/13	2410A	4179
LUCASTON		12SecDr	11/26/10	2410	3864
LUCCA		12SecDr	12/20-21	2410F	4611
LUCKETT		12SecDr	5-6/17	2410F	4497
LUCKNOW		12SecDrCpt	5/17	2411C	4494
LUCRETIA		28ChDr	6-7/27	3416A	6087
LUCRETIA MOTT	N	28ChDr	5-7/26	3416	4958
LUDLOW		10SecDr2Cpt	8-9/13	2585A	4201
LUELLA		24ChDrB	4/13	2417	4108
LUGANO		12SecDr	5-7/13	2410	4149
LUIS G. URBINA NdeM	N	14Sec	9-10/25	3958	4869
LUJANE		12SecDr	7-8/15	2410D	4322
LUMBERTON		12SecDr	9/20/10	2410	3794
LUNDY'S LANE		BaggBSmkBarber	12/26-27	3951C	6022
LUNSFORD		12SecDr	3-6/20	2410F	4565
LURAY		12SecDr	12/20-21	2410F	4611
LURLINE		12SecDr	1-2/16	2410E	4356
LURTON		12SecDr	10-12/16	2410F	4431
LUSIAD		12SecDr	3-6/20	2410F	4565
LUTHER BURBANK		14Sec	10-11/26	3958	6012
LUTHERVILLE		12SecDr	5-6/17	2410F	4497
LUTON		12SecDr	7-8/15	2410D	4322
LUVENA		12SecDr	7-8/15	2410D	4322
LUZON		12SecDr	3-6/20	2410F	4565
LYCEUM	RN	Recreation	4-6/21	3966	4613
LYCIDAS		12SecDr	7/13	2410A	4179
LYCOMING		12SecDr	10-12/26	3410A	6023
LYCURGUS		12SecDrCpt	10-12/21	2411C	4624
LYDIA		26ChDr	3-4/14	2416A	4264
LYDICK		12SecDr	1-4/12	2410	3949
LYKENS		12SecDr	5/23/10	2410	3794
LYMAN		12SecDr	10-12/16	2410F	4431
LYNAFIELD		12SecDr	7-8/14	2410A	4277
LYNCHBURG		12SecDr	12/10-11	2410	3866
LYNDALE		12SecDr	2-3/13	2410	4105
LYNDONVILLE		12SecDr	3-4/11	2410	3881
LYONETT		12SecDr	1-4/12	2410	3949
LYONS		12SecDr	12/20-21	2410F	4611
LYONS FALLS		12SecDrCpt	3/14	2411A	4269
LYRA		12SecDr	12/20-21	2410F	4611
LYSANDER		12SecDr	9-10/15	2410D	4327
LYTHMORE		16Sec	9/11	2412	3923
MABYN		12SecDr	7/13	2410A	4179
MACARIA		24ChDrB	5/17	2417D	4493
MACATAWA		10SecDr2Cpt	11/17	2585D	4527
MACHIAS		12SecDr	7-8/12	2410	4014
MacINTYRE RANGE		10SecDrCpt	5/30	3973A	6358
MACKAY		10SecDr2Cpt	11/17	2585D	4527
MACKENZIE		10SecDr2Cpt	11/17	2585D	4527
MACKENZIE		12SecDr	2-5/25	3410	4845
MACKINAC		12SecDr	10-12/26	3410A	6023
MACKSVILLE		12SecDr	2-5/25	3410	4845
MACOMB HOUSE		13Dbr	8/30	3997A	6392
MADAGASCAR		12SecDr	12/20-21	2410F	4611
MADAWASKA		12SecDr	11-12/13	2410A	4234
MADELINE		28ChDr	6-7/27	3416A	6087
MADERA		12SecDr	9/2/10	2410	3794
MADISON		12SecDr	10/13/10	2410	3814
MADISON SQUARE		6Cpt3Dr	10-12/29	3523C	6290
MADLEY		12SecDr	2-5/25	3410	4845
MADRAS		16Sec	5-6/16	2412F	4386
MADRID		10SecDr2Cpt	5/13	2585A	4141

CAR NAME	ST	CAR TYPE	BUILT	PLAN	LOT
MAGDALA		10SecLngObs	12/11-12	2521	3944
MAGDALEN TOWER	RN	8SecDr3Dbr	5-6/25	4090	4846
MAGENTA		12SecDr	1-3/14	2410A	4249
MAGIC CITY		8SbrObsLng	4/30	4005	6340
MAGNOLIA		12SecDr	10/31/10	2410	3860
MAGNOLIA		26ChDr	9-11/22	2416D	4649
MAHAFFEY		12SecDr	5-6/17	2410F	4497
MAHALIA		12SecDr	9-10/15	2410D	4327
MAHASKA		12SecDrCpt	6/13	2411	4151
MAHOMET		12SecDr	4-6/14	2410A	4271
MAHONING		12SecDr	9/27/10	2410	3794
MAHONING					
VALLEY COUNTRY CLUB	RN	8SecCptRestLngObs	1-5/18	4026A	4531
MAIDEN ROCK		16Sec	9-10/12	2412	4037
MAINE		12SecDr	6-10/21	2410F	4614
MAINE	N	SbrDrB16StsLng	7/30	3988D	6356
MAINVILLE		12SecDr	7-9/17	2410F	4515
MAITLAND		12SecDr	2-5/25	3410	4845
MAIWAND		12SecDr	10-12/15	2410D	4338
MAKENA		10SecLngObs	12/11-12	2521	3944
MAKETEWAH COUNTRY CLUB	RN	8SecRestLngObs	7-8/11	4025C	3912
MAKURA(1st)		12SecDr	2-5/16	2410E	4367
MAKURA(2nd)	N	12SecDr	2-5/16	2410E	4367
MALABAR		16Sec	5-6/13	2412B	4150
MALACCA		12SecDr	12/20-21	2410F	4611
MALADE		10SecLngObs	11/12	2521	4049
MALATHA		10SecLngObs	12/11-12	2521	3944
MALAY		10SecDr2Cpt	11/17	2585D	4527
MALDIVE		12SecDr	12/20-21	2410F	4611
MALFADA		12SecDr	11-12/13	2410A	4234
MALINCHE		10SecDr2Cpt	1-2/13	2585	4091
MALINCHE	N	8Sec5Dbr	10-12/15	4036I	4338
MALLORY		12SecDr	1-4/12	2410	3949
MALLOW		12SecDr	12/20-21	2410F	4611
MALMO		12SecDr	12/20-21	2410F	4611
MALONE		12SecDr	3-4/11	2410	3881
MALOTTE		10SecDr2Cpt	11/17	2585D	4527
MALTA		12SecDr	12/20-21	2410F	4611
MALVERN HILL		BaggBSmkBarber	12/26-27	3951C	6022
MALZAR		12SecDr	2-5/16	2410E	4367
MAMARONECK		36Ch SBS	1/13	2691	4056
MAMBRINO		12SecDr	12/20-21	2410F	4611
MAMMOTH CAVE		BaggBSmkBarber	12/26-27	3951C	6022
MANACOR		12SecDr	6-10/21	2410F	4614
MANASQUAN		26ChDr	5/12	2416	3986
MANASSAS GAP		8SecDr2Cpt	12/29-30	3979A	6334
MANATUCK	N	10SecDr2Cpt	1-2/13	2585	4091
MANAYUNK		12SecDr	2-5/16	2410E	4367
MANCELONA		12SecDr	9-10/11	2410	3924
MANCHESTER	N	12SecDr	2-5/16	2410E	4367

CAR NAME	ST	CAR TYPE	BUILT	PLAN	LOT
MANCHURIA		10SecDr2Cpt	11-12/11	2585	3942
MANDALAY		12SecDr	12/20-21	2410F	4611
MANEKO		10SecLngObs	12/11-12	2521	3944
MANHASSETT	N	12SecDr	1-4/12	2410	3949
MANHATTAN		Private (GS)	11/16	2502B	4422
MANICE		12SecDr	10-12/16	2410F	4431
MANIPUR		12SecDr	12/20-21	2410F	4611
MANITOBA CLUB	RN	8SecRestObsLng	10/13	4025D	4202
MANITOU		12SecDr	2-3/27	3410A	6055
MANITOWOC	N	12SecDr	7-8/28	3410B	6184
MANIVET		12SecDr	7-8/13	2410A	4192
MANLEY		10SecDr2Cpt	11/17	2585D	4527
MANLIUS		16Sec	6-7/20	2412F	4568
MANNHEIM		12SecDr	1-4/12	2410	3949
MANNING		10SecDr2Cpt	7-8/16	2585D	4424
MANNSVILLE		12SecDr	3-4/11	2410	3881
MANOR		12SecDrCpt	5/17	2411C	4494
MANORVILLE		12SecDr	5/14/10	2410	3773
MANOVAR		12SecDr	6-10/21	2410F	4614
MANOWN		12SecDr	10-11/11	2410	3936
MANSFIELD		12SecDr	12/20/10	2410	3864
MANSON		12SecDr	7-8/15	2410D	4322
MANTEO		12SecDr	2-3/27	3410A	6055
MANTOLOKING		12SecDr	6-9/26	3410A	4969
MANUEL					
ACUNA FCP-219	SN	10SecDr2Cpt	12/25-26	3585A	4933
DOBLADO NdeM	N	14Sec	3/27	3958	6054
JOSE OTHON NdeM-185	N	8SecDr3Dbr	1/25	4090	4843
MANUFACTURERS' CLUB		8SecBLngSunRm	8-9/29	3989A	6274
MANVER		12SecDr	5-6/17	2410F	4497
MANYASKA		16Sec	4/17	2412F	4484
MANZANO PEAK		10Sec2Dr	9-10/29	3584B	6284
MANZANOLA		10SecDr2Cpt	8-9/13	2585A	4201
MAPLE GROVE		12SecDr	12/16/10	2410	3864
MAPLE PARK		12SecDr	7-8/28	3410B	6184
MAPLE SHADE		BaggBClubSmkBarber	7/23	2951B	4698
MAPLE VIEW		10Sec2Dr	1/12	2584	3945
MAPLEHURST		BaggBClubSmkBarber	2-3/12	2602	3948
MAPLETON		10SecDr2Cpt	7-8/16	2585D	4424
MAQUON		12SecDr	12/10-11	2410	3866
MARASH		12SecDr	2-5/16	2410E	4367
MARATHON		12SecDr	10-12/15	2410D	4338
MARBLEHEAD		12SecDr	2-3/27	3410A	6055
MARBRIERS		12SecDr	4-5/13	2410	4106
MARCELO		12SecDr	3-6/20	2410F	4565
MARCHAND		12SecDr	1-3/14	2410A	4249
MARCHMONT		12SecDr	2-3/13	2410	4105
MARCO		12SecDr	12/20-21	2410F	4611
MARCO POLO		Private	6/27	3972	6037
MARCUS A. HANNA		14Sec	7/30	3958A	6376

CAR NAME	ST	CAR TYPE	BUILT	PLAN	LOT
MARCUS DALY		8SecDr2Cpt	5-6/29	3979A	6261
MARCUS LOEW	N	12SecDr	4/26	3410A	4943
MARDENIS		12SecDr	1-3/17	2410F	4450
MARDONIA		12SecDr	2-3/13	2410	4105
MARENGO		12SecDr	12/10-11	2410	3866
MARGELAIN		26ChDr	3-4/14	2416A	4264
MARGESON		12SecDr	11-12/13	2410A	4234
MARGRAVE		12SecDr	12/10-11	2410	3866
MARGUERITE		28ChDr	5-7/26	3416	4958
MARIANA		12SecDr	2-3/27	3410A	6055
MARIANO					
AZUELA NdeM-186	N	10Sec3Dbr	5/17	3411	4494
ESCOBEDO NdeM-187	N	12SecDr	4-6/24	3410	4763
MARICOURT		10SecDr2Cpt	7-8/16	2585D	4424
MARIE		28ChDr	6-7/27	3416A	6087
MARIETTA		10SecLngObs	11/12	2521	4049
MARIGOLD		28ChDr	5-6/24	3416	4761
MARIN		LngBDorm(5Sec)	7-8/27	3981	6061
MARINER	RN	17StsRestLng	4-5/13	4019A	4136
MARINETTE		16Sec	9-10/12	2412	4037
MARINO ABASOLO NdeM	N	12SecDr	6-8/24	3410	4764
MARIO					
TALAVERA FCP-223	SN	10SecDr2Cpt	9-10/26	3585A	4996
MARIPOSA		10SecLngObs	11/12	2521	4049
MARK BEAUBIEN		3Cpt2DrLngObs	11-12/26	3959	6015
MARK TWAIN		12SecDr	1-2/29	3410B	6220
MARKEEN		12SecDr	1-2/15	2410B	4311
MARKEL		28ChDr	5-7/26	3416	4958
MARKHAM		12SecDr	10-12/16	2410F	4431
MARLEY		16Sec	10/13	2412B	4216
MARLINTON		12SecDr	11-12/13	2410A	4234
MARLOWE		10SecDr2Cpt	7-8/16	2585D	4424
MARMARTH		12SecDr	1-3/14	2410A	4249
MARMITON		12SecDr	4-5/13	2410	4106
MARMOT		12SecDr	7/16	2410F	4412
MARNE		12SecDr	3-6/20	2410F	4565
MAROMA		12SecDr	12/10-11	2410	3866
MAROON PEAK		10Sec2Dr	12/26-27	3584A	6031
MARQUAND		10SecDr2Cpt	6-7/15	2585B	4328
MARQUESANT		12SecDr	2-3/13	2410	4105
MARQUIS LAFAYETTE	N	12SecDr	10-12/23	3410	4724
MARSH RUN		BaggBClubSmkBarber	7/23	2951B	4698
MARSH RUN	R	6DbrBLng	7/23	4015	4698
MARSHALLTOWN		12SecDr	7-8/28	3410B	6184
MARSHALLVILLE		12SecDr	6-9/26	3410A	4969
MARSHFIELD		12SecDr	2-3/27	3410A	6055
MARSHLAND		16Sec	9-10/12	2412	4037
MARSOVAN		12SecDr	2-5/16	2410E	4367
MARSTON		10SecDr2Cpt	11/17	2585D	4527
MARTEL		10SecDr2Cpt	5/13	2585A	4141

CAR NAME	ST	CAR TYPE	BUILT	PLAN	LOT
MARTELLO TOWER	RN	8SecDr3Dbr	1/25	4090	4843
MARTENE		26ChDr	4-5/13	2416A	4136
MARTHA		32Ch	6/30	3916C	6363
MARTHA	R	18Ch16StsLng	6/30	4075A	6363
MARTHA WASHINGTON		12SecDr	8-12/25	3410A	4894
MARTIN VAN BUREN	N	30ChBLibObs	4/16	2901A	4365
MARTINDALE		12SecDr	3-4/11	2410	3881
MARTINEZ		10SecDr2Cpt	10-11/12	2585	4067
MARTINIQUE		12SecDr	1-2/15	2410B	4311
MARTINSBURG		12SecDr	7-8/12	2410	4014
MARTINSVILLE		12SecDr	10/18/10	2410	3814
MARTISCO		12SecDr	3-4/11	2410	3881
MARTYNIA		24ChDrB	5/17	2417D	4493
MARVEL PASS		8SecDr2Cpt	7-8/30	3979A	6377
MARVIN		12SecDr	5-8/18	2410F	4540
MARWOOD		12SecDr	11/25/10	2410	3864
MARY		28ChDr	6-7/27	3416A	6087
MARY WASHINGTON	N	12SecDr	2-4/24	3410	4762
MARYDEL		16Sec	2/13	2412	4101
MARYLAND	N	32ChObs	8/19/10	2420	3805
MARYLAND		28ChObs	8/24/10	2421	3806
MARYLAND		3CptDrLngBSunRm	7/29	3975C	6275
MARYLAND	N	32ChObs	9-10/24	3420	4785
MARYLAND CLUB	RN	8SecRestObsLng	9-10/14	4025B	4292
MASADA		10SecLngObs	12/11-12	2521	3944
MASARDIS		12SecDr	4-5/13	2410	4106
MASCOMA		12SecDr	7-8/14	2410A	4277
MASK AND WIG CLUB		13ChRestLng	6/30	3992A	6361
MASKELL		10SecDr2Cpt	7-8/16	2585D	4424
MASONVILLE		12SecDr	5-6/17	2410F	4497
MASPERO		12SecDr	11-12/13	2410A	4234
MASSABESIC		12SecDr	7-8/14	2410A	4277
MASSACHUSETTS	N	DrSbrB16StsLng	4/29	3988D	6221
MASSACHUSETTS BAY		32ChDr	12/29-30	3917B	6318
MASSACK		12SecDr	10-12/16	2410F	4431
MASSENET		6Cpt3Dr	11-12/24	3523A	4833
MASSILLON		12SecDr	10/31/10	2410	3860
MATABANICK		12SecDr	2-5/16	2410E	4367
MATACHIN		12SecDr	2-3/13	2410	4105
MATAGORDA		10SecDr2Cpt	8-9/13	2585A	4201
MATAMOROS NdeM	SN	12SecDr	2-5/25	3410	4845
MATANE		10SecDr2Cpt	7-8/16	2585D	4424
MATANZAS		23StsBLng	11/26	3967	4979
MATAQUA		12SecDr	2-3/13	2410	4105
MATCHLESS	RN	10ChRestLng	7/27/10	4019A	3804
MATHER		10SecDr2Cpt	11/17	2585D	4527
MATHESON		10SecDr2Cpt	11/17	2585D	4527
MATHEWS		12SecDr	12/20-21	2410F	4611
MATILDA		24ChDrB	4/13	2417	4108
MATINA		12SecDr	1-2/15	2410B	4311

CAR NAME	ST	CAR TYPE	BUILT	PLAN	LOT
MATINICUS		12SecDr	10/22-23	2410H	4647
MATLOCK		12SecDr	12/10-11	2410	3866
MATOAKA	N	12SecDr	10-12/26	3410A	6023
MATOGUE		12SecDr	7-8/14	2410A	4277
MATTHIAS W. BALDWIN	N	30ChDr	8-9/30	4000A	6384
MATTITUCK		16Sec	5-6/16	2412F	4386
MATTOON		12SecDr	9-10/11	2410	3924
MAUDE		28ChDr	4-6/25	3416	4864
MAUGANSVILLE		12SecDr	6-9/26	3410A	4969
MAUMEE		10SecDr2Cpt	7-8/16	2585D	4424
MAUMEE VALLEY		DrSbrBLngObs	4/29	3988	6221
MAUPIN		12SecDr	1-3/14	2410A	4249
MAUREPAS		12SecDr	9-10/13	2410A	4215
MAURITIUS		12SecDr	10-12/15	2410D	4338
MAUSTON		12SecDr	1-3/14	2410A	4249
MAVOURNEEN		12SecDr	3-6/15	2410B	4319
MAXIMILIAN		12SecDr	2-5/25	3410	4845
MAXIMUS		12SecDr	6-10/21	2410F	4614
MAXWELL	N	12SecDr	4-5/13	2410	4106
MAYENCE		12SecDr	1-3/17	2410F	4450
MAYENNE		12SecDr	12/20-21	2410F	4611
MAYETTA		12SecDr	7-8/15	2410D	4322
MAYFIELD CLUB		8SecBObsLng	2-3/29	3989	6229
MAYITA		16Sec	12/14-15	2412C	4304
MAYMONT		12SecDr	1-3/14	2410A	4249
MAYNARD		12SecDr	1-3/14	2410A	4249
MAYO		12SecDr	12/20-21	2410F	4611
MAYPORT		12SecDr	5/25/10	2410	3794
MAYRAN NdeM	SN	14Sec	9-10/25	3958	4869
MAYTHORNE		12SecDr	1-3/14	2410A	4249
MAYVIEW		12SecDr	4-6/14	2410A	4271
MAYWOOD		36Ch	6/16	2916	4389
MAZATLAN		10SecDr2Cpt	1-2/13	2585	4091
MAZATLAN NdeM-188	SN	8SecDr2Cpt	4-5/30	3979A	6359
McADAM		12SecDr	1-3/25	3410	4844
McADENVILLE		12SecDr	2-5/25	3410	4845
McAFEE		12SecDr	2-5/25	3410	4845
McALEER		12SecDr	2-5/25	3410	4845
McALESTER		12SecDr	10/22-23	2410H	4647
McALISTERVILLE		12SecDr	2-5/25	3410	4845
McALLEN		12SecDr	4-6/24	3410	4763
McALMONT		12SecDr	1-3/25	3410	4844
McALPIN		12SecDr	4-6/24	3410	4763
McANDREWS		12SecDr	1-3/25	3410	4844
McANNA		12SecDr	2-5/25	3410	4845
McANULTY		12SecDr	2-5/25	3410	4845
McARA		12SecDr	2-5/25	3410	4845
McARDLES		12SecDr	2-5/25	3410	4845
McARTHUR		12SecDr	4-6/24	3410	4763
McAULIFFE		12SecDr	1-3/25	3410	4844

CAR NAME	ST	CAR TYPE	BUILT	PLAN	LOT
McAVAN		12SecDr	2-5/25	3410	4845
McAVOY		12SecDr	4-6/24	3410	4763
McBAIN		12SecDr	4-6/24	3410	4763
McBEE		12SecDr	2-5/25	3410	4845
McBETH		12SecDr	4-6/24	3410	4763
McBRAYER		12SecDr	1-3/25	3410	4844
McBRIDE		12SecDr	4-6/24	3410	4763
McBRIDESVILLE		12SecDr	2-5/25	3410	4845
McBURG		12SecDr	2-5/25	3410	4845
McBURNEY		12SecDr	2-5/25	3410	4845
McCABE		12SecDr	4-6/24	3410	4763
McCAFFREY		12SecDr	2-5/25	3410	4845
McCAGG		12SecDr	2-5/25	3410	4845
McCALEB		12SecDr	2-5/25	3410	4845
McCALL		12SecDr	4-6/24	3410	4763
McCALLSBURG		12SecDr	2-5/25	3410	4845
McCALLUM		12SecDr	1-3/25	3410	4844
McCALMONT		12SecDr	2-5/25	3410	4845
McCAMMON		12SecDr	2-5/25	3410	4845
McCAMPBELL		12SecDr	1-3/25	3410	4844
McCANDLESS		12SecDr	2-5/25	3410	4845
McCANN		12SecDr	4-6/24	3410	4763
McCANTS		12SecDr	2-5/25	3410	4845
McCARLEY		12SecDr	1-3/25	3410	4844
McCARR		12SecDr	1-3/25	3410	4844
McCARROLL		12SecDr	2-5/25	3410	4845
McCARRON LAKE		10SecDrCpt	3-5/30	3973A	6338
McCARTERS		12SecDr	2-5/25	3410	4845
McCARTHY		12SecDr	4-6/24	3410	4763
McCARTNEY		12SecDr	2-5/25	3410	4845
McCARTYVILLE		12SecDr	2-5/25	3410	4845
McCASKILL		12SecDr	1-3/25	3410	4844
McCASLIN		12SecDr	2-5/25	3410	4845
McCAUL		12SecDr	2-5/25	3410	4845
McCAULLEY		12SecDr	1-3/25	3410	4844
McCAUSLAND		12SecDr	1-3/25	3410	4844
McCAW		12SecDr	2-5/25	3410	4845
McCAYSVILLE		12SecDr	2-5/25	3410	4845
McCHESNEY		12SecDr	2-5/25	3410	4845
McCHEYNE		12SecDr	2-5/25	3410	4845
McCLAINVILLE		12SecDr	1-3/25	3410	4844
McCLAMMY		12SecDr	2-5/25	3410	4845
McCLANAHAN		12SecDr	2-5/25	3410	4845
McCLAUGHREYS		12SecDr	2-5/25	3410	4845
McCLAVE		12SecDr	2-5/25	3410	4845
McCLEARY		12SecDr	4-6/24	3410	4763
McCLELLAN		12SecDr	4-6/24	3410	4763
McCLELLANVILLE		12SecDr	2-5/25	3410	4845
McCLENACHAN		12SecDr	2-5/25	3410	4845
McCLENDON		12SecDr	2-5/25	3410	4845

CAR NAME	ST	CAR TYPE	BUILT	PLAN	LOT	CAR NAME	ST	CAR TYPE	BUILT	PLAN	LOT
McCLENNY		12SecDr	2-5/25	3410	4845	McCOSH		12SecDr	2-5/25	3410	4845
McCLERNAND		12SecDr	2-5/25	3410	4845	McCOSKRY		12SecDr	2-5/25	3410	4845
McCLINCHY		12SecDr	2-5/25	3410	4845	McCOWN		12SecDr	2-5/25	3410	4845
McCLINTOCK		12SecDr	1-3/25	3410	4844	McCOY		12SecDr	4-6/24	3410	4763
McCLINTOCKBURG		12SecDr	2-5/25	3410	4845	McCOYSBERG		12SecDr	1-3/25	3410	4844
McCLISH		12SecDr	2-5/25	3410	4845	McCRACKEN		12SecDr	1-3/25	3410	4844
McCLOSKEY		12SecDr	1-3/25	3410	4844	McCRANEY		12SecDr	1-3/25	3410	4844
McCLUNEY		12SecDr	1-3/25	3410	4844	McCRAY		12SecDr	4-6/24	3410	4763
McCLUNG		12SecDr	2-5/25	3410	4845	McCREADYVILLE		12SecDr	2-5/25	3410	4845
McCLURE		12SecDr	4-6/24	3410	4763	McCREANOR		12SecDr	2-5/25	3410	4845
McCLURG		12SecDr	2-5/25	3410	4845	McCREDIE		12SecDr	1-3/25	3410	4844
McCOLLESTER		12SecDr	2-5/25	3410	4845	McCREERY		12SecDr	2-5/25	3410	4845
McCOLLOM		12SecDr	1-3/25	3410	4844	McCREIGHT		12SecDr	2-5/25	3410	4845
McCOMAS		12SecDr	1-3/25	3410	4844	McCRIE		12SecDr	2-5/25	3410	4845
McCOMB		12SecDr	10/22-23	2410H	4647	McCRIMMON		12SecDr	2-5/25	3410	4845
McCONAUGHY		12SecDr	2-5/25	3410	4845	McCRORY		12SecDr	1-3/25	3410	4844
McCONDY		12SecDr	2-5/25	3410	4845	McCULLERS		12SecDr	2-5/25	3410	4845
McCONKEY		12SecDr	2-5/25	3410	4845	McCULLOUGH		12SecDr	4-6/24	3410	4763
McCONNELL		12SecDr	4-6/24	3410	4763	McCULLY		12SecDr	2-5/25	3410	4845
McCONNELLSBURG		12SecDr	2-5/25	3410	4845	McCUMBER		12SecDr	2-5/25	3410	4845
McCONNELLSVILLE		12SecDr	1-3/25	3410	4844	McCUNE		12SecDr	4-6/24	3410	4763
McCONNICO		12SecDr	2-5/25	3410	4845	McCUNEVILLE		12SecDr	2-5/25	3410	4845
McCOOK		12SecDr	4-6/24	3410	4763	McCURDY		12SecDr	1-3/25	3410	4844
McCOOL		12SecDr	1-3/25	3410	4844	McCURTAIN		12SecDr	2-5/25	3410	4845
McCORD		12SecDr	4-6/24	3410	4763	McCUTCHENVILLE		12SecDr	1-3/25	3410	4844
McCORDSVILLE		12SecDr	1-3/25	3410	4844	McCUTCHEON		12SecDr	4-6/24	3410	4763
McCORKLE		12SecDr	1-3/25	3410	4844	McDADE		12SecDr	6-8/24	3410	4764
McCORMICK		12SecDr	4-6/24	3410	4763	McDANIEL		12SecDr	4-6/24	3410	4763

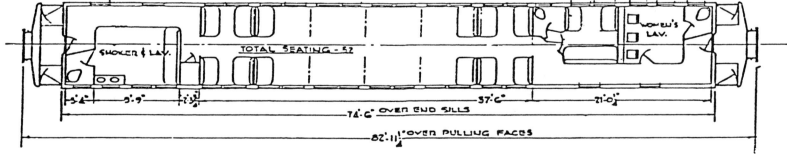

NOTE:
THESE CARS PURCHASED FROM PULLMAN 1-19-49
"McCAULLEY" CONV. TO COACH 2202, 12-C-55
"EAST MONROE" " " " 2203, 8-8-56
"McCALL" " " " 2204, 8-7-56

Figure I-61   McCAULLEY, 1 of 5 Plan 3410 12-section drawing room sleepers sold to GM & O in 1948, was withdrawn from
Pullman lease in May 1950 after only two years' active service with that road. It carried Coach number 2202 from December
1955 until it was retired in 1962. The Author's Collection.

CAR NAME	ST	CAR TYPE	BUILT	PLAN	LOT
McDAVID		12SecDr	6-8/24	3410	4764
McDEARMON		12SecDr	2-5/25	3410	4845
McDERMOTT		12SecDr	6-8/24	3410	4764
McDILL		12SecDr	6-8/24	3410	4764
McDOEL		12SecDr	2-5/25	3410	4845
McDONALD		12SecDr	6-8/24	3410	4764
McDONALDSVILLE		12SecDr	2-5/25	3410	4845
McDONOGHVILLE		12SecDr	2-5/25	3410	4845
McDONOUGH		12SecDr	6-8/24	3410	4764
McDOUGAL		12SecDr	6-8/24	3410	4764
McDOW		12SecDr	6-8/24	3410	4764
McDOWELL		12SecDr	6-8/24	3410	4764
McDUFF		12SecDr	2-5/25	3410	4845
McDUFFIE		12SecDr	2-5/25	3410	4845
McEACHERN		12SecDr	2-5/25	3410	4845
McELDERRY		12SecDr	1-3/25	3410	4844
McELHANY		12SecDr	2-5/25	3410	4845
McELHATTAN		12SecDr	2-5/25	3410	4845
McELHERAN		12SecDr	2-5/25	3410	4845
McELLIGOTT		12SecDr	2-5/25	3410	4845
McELRATH		12SecDr	2-5/25	3410	4845
McELROY		12SecDr	6-8/24	3410	4764
McELWAIN		12SecDr	2-5/25	3410	4845
McENTEE		12SecDr	2-5/25	3410	4845
McEWEN		12SecDr	6-8/24	3410	4764
McEWENSVILLE		12SecDr	1-3/25	3410	4844
McFADDIN		12SecDr	6-8/24	3410	4764
McFALL		12SecDr	6-8/24	3410	4764
McFANN		12SecDr	2-5/25	3410	4845
McFARLAND		12SecDr	6-8/24	3410	4764
McFERRIN		12SecDr	1-3/25	3410	4844
McFORD		12SecDr	2-5/25	3410	4845
McGAHEYSVILLE		12SecDr	2-5/25	3410	4845
McGARY		12SecDr	1-3/25	3410	4844
McGAVICK		12SecDr	2-5/25	3410	4845
McGEE		12SecDr	6-8/24	3410	4764
McGEORGE		12SecDr	2-5/25	3410	4845
McGERVEY		12SecDr	1-3/25	3410	4844
McGIBSON		12SecDr	2-5/25	3410	4845
McGIFFERT		12SecDr	2-5/25	3410	4845
McGILL		12SecDr	6-8/24	3410	4764
McGILL UNIVERSITY	RN	12Sec2Dbr	5-6/17	4046	4497
McGILLICUDDY	N	12SecDr	2-5/25	3410	4845
McGILLIVRAY		12SecDr	2-5/25	3410	4845
McGINTY		12SecDr	2-5/25	3410	4845
McGIRK'S		12SecDr	2-5/25	3410	4845
McGIRR		12SecDr	1-3/25	3410	4844
McGIRTH		12SecDr	2-5/25	3410	4845
McGIVNEY		12SecDr	2-5/25	3410	4845
McGLASHENS		12SecDr	2-5/25	3410	4845

CAR NAME	ST	CAR TYPE	BUILT	PLAN	LOT
McGLASSON		12SecDr	2-5/25	3410	4845
McGLINN		12SecDr	1-3/25	3410	4844
McGLONE		12SecDr	2-5/25	3410	4845
McGONIGLE		12SecDr	1-3/25	3410	4844
McGOVERN		12SecDr	2-5/25	3410	4845
McGOWAN		12SecDr	6-8/24	3410	4764
McGRADY		12SecDr	2-5/25	3410	4845
McGRATH		12SecDr	6-8/24	3410	4764
McGRATTANS		12SecDr	1-3/25	3410	4844
McGRAVE		12SecDr	2-5/25	3410	4845
McGRAW		12SecDr	6-8/24	3410	4764
McGRAWSVILLE		12SecDr	1-3/25	3410	4844
McGREGOR		12SecDr	6-8/24	3410	4764
McGREGORVILLE		12SecDr	2-5/25	3410	4845
McGRIFF		12SecDr	1-3/25	3410	4844
McGROARTY		12SecDr	2-5/25	3410	4845
McGRUDER		12SecDr	1-3/25	3410	4844
McGUFFEY		12SecDr	1-3/25	3410	4844
McGUGGIN		12SecDr	2-5/25	3410	4845
McGUIRE		12SecDr	6-8/24	3410	4764
McHANEY		12SecDr	2-5/25	3410	4845
McHARG		12SecDr	1-3/25	3410	4844
McHATTON		12SecDr	2-5/25	3410	4845
McHENRY		12SecDr	6-8/24	3410	4764
McHESSOR		12SecDr	2-5/25	3410	4845
McHINCH		12SecDr	2-5/25	3410	4845
McHUGH		12SecDr	6-8/24	3410	4764
McILHENNY		12SecDr	2-5/25	3410	4845
McILVAINE		12SecDr	2-5/25	3410	4845
McINDOES		12SecDr	2-5/25	3410	4845
McINNIS		12SecDr	1-3/25	3410	4844
McINTOSH		12SecDr	6-8/24	3410	4764
McINTYRE		12SecDr	6-8/24	3410	4764
McIVER		12SecDr	1-3/25	3410	4844
McJUNKIN		12SecDr	2-5/25	3410	4845
McKAIN		12SecDr	2-5/25	3410	4845
McKAMIE		12SecDr	2-5/25	3410	4845
McKASTOE		12SecDr	2-5/25	3410	4845
McKAY		12SecDr	1-3/25	3410	4844
McKEAGES		12SecDr	2-5/25	3410	4845
McKEANSBURG		12SecDr	1-3/25	3410	4844
McKEEFRY		12SecDr	2-5/25	3410	4845
McKEEN		12SecDr	2-4/24	3410	4762
McKEESPORT		12SecDr	6-8/24	3410	4764
McKEEVER		12SecDr	6-8/24	3410	4764
McKELDER		12SecDr	1-3/25	3410	4844
McKELL		12SecDr	2-5/25	3410	4845
McKELLAR		12SecDr	2-5/25	3410	4845
McKELVIE		12SecDr	1-3/25	3410	4844
McKENDREE		12SecDr	1-3/25	3410	4844

CAR NAME	ST	CAR TYPE	BUILT	PLAN	LOT
McKENNA		12SecDr	6-8/24	3410	4764
McKEON		12SecDr	2-5/25	3410	4845
McKERROW		12SecDr	2-5/25	3410	4845
McKEWAN		12SecDr	2-5/25	3410	4845
McKIBBIN		12SecDr	2-5/25	3410	4845
McKIMM		12SecDr	1-3/25	3410	4844
McKINLEY		12SecDr	6-8/24	3410	4764
McKINLEYVILLE		12SecDr	2-5/25	3410	4845
McKINNELL		12SecDr	2-5/25	3410	4845
McKINNEYSBURG		12SecDr	2-5/25	3410	4845
McKINNON		12SecDr	1-3/25	3410	4844
McKINSTRY		12SecDr	2-5/25	3410	4845
McKIRDY		12SecDr	1-3/25	3410	4844
McKISSOCK		12SecDr	2-5/25	3410	4845
McKITTRICK		12SecDr	6-8/24	3410	4764
McKNIGHT		12SecDr	6-8/24	3410	4764
McKNIGHTSTOWN		12SecDr	1-3/25	3410	4844
McKONE		12SecDr	2-5/25	3410	4845
McKOWEN		12SecDr	2-5/25	3410	4845
McKOWNVILLE		12SecDr	2-5/25	3410	4845
McKULLO		12SecDr	2-5/25	3410	4845
McLAGGAN		12SecDr	2-5/25	3410	4845
McLANAHAN		12SecDr	2-5/25	3410	4845
McLANDBURGH		12SecDr	2-5/25	3410	4845
McLAREN		12SecDr	6-8/24	3410	4764
McLARTY		12SecDr	2-5/25	3410	4845
McLAUGHLIN		12SecDr	1-3/25	3410	4844
McLAWS		12SecDr	2-5/25	3410	4845
McLEAN		12SecDr	6-8/24	3410	4764
McLEANSBORO		12SecDr	1-3/25	3410	4844
McLEANSVILLE		12SecDr	1-3/25	3410	4844
McLEISH		12SecDr	6-8/24	3410	4764
McLEMORE		12SecDr	2-5/25	3410	4845
McLEMORESVILLE		12SecDr	2-5/25	3410	4845
McLENNAN		12SecDr	6-8/24	3410	4764
McLEOD		12SecDr	6-8/24	3410	4764
McLEROY		12SecDr	2-5/25	3410	4845
McLOON		12SecDr	2-5/25	3410	4845
McLOUTH		12SecDr	2-5/25	3410	4845
McMADA		12SecDr	2-5/25	3410	4845
McMAHAN		12SecDr	6-8/24	3410	4764
McMANN		12SecDr	6-8/24	3410	4764
McMANUS		12SecDr	1-3/25	3410	4844
McMARTINVILLE		12SecDr	2-5/25	3410	4845
McMASTER		12SecDr	2-5/25	3410	4845
McMASTER UNIVERSITY	RN	12Sec2Dbr	5-6/17	4046	4497
McMASTERVILLE		12SecDr	2-5/25	3410	4845
McMEEKEN		12SecDr	6-8/24	3410	4764
McMICHAELS		12SecDr	2-5/25	3410	4845
McMICKEN		12SecDr	2-5/25	3410	4845
McMILLAN		12SecDr	6-8/24	3410	4764
McMINNVILLE		12SecDr	1-3/25	3410	4844
McMORRAN		12SecDr	1-3/25	3410	4844
McMULLIN		12SecDr	1-3/25	3410	4844
McMURDO		12SecDr	1-3/25	3410	4844
McMURPHY		12SecDr	1-3/25	3410	4844
McMURRAY		12SecDr	6-8/24	3410	4764
McMURRICH		12SecDr	2-5/25	3410	4845
McNAB		12SecDr	6-8/24	3410	4764
McNAIR		12SecDr	6-8/24	3410	4764
McNALLY		12SecDr	6-8/24	3410	4764
McNAMARA		12SecDr	1-3/25	3410	4844
McNAMEE		12SecDr	2-5/25	3410	4845
McNARRONS		12SecDr	2-5/25	3410	4845
McNATT		12SecDr	2-5/25	3410	4845
McNAUGHTON		12SecDr	6-8/24	3410	4764
McNEAL		12SecDr	6-8/24	3410	4764
McNEARS		12SecDr	2-5/25	3410	4845
McNEELY		12SecDr	2-5/25	3410	4845
McNEICE		12SecDr	2-5/25	3410	4845
McNIERNEY		12SecDr	2-5/25	3410	4845
McNISH		12SecDr	2-5/25	3410	4845
McNULTA		12SecDr	2-5/25	3410	4845
McPAUL		12SecDr	6-8/24	3410	4764
McPEEK		12SecDr	1-3/25	3410	4844
McPHAIL		12SecDr	1-3/25	3410	4844
McPHEE		12SecDr	2-5/25	3410	4845
McPHERSON		12SecDr	6-8/24	3410	4764
McPHERSONVILLE		12SecDr	2-5/25	3410	4845
McPHETRES		12SecDr	2-5/25	3410	4845
McQUADE		12SecDr	6-8/24	3410	4764
McQUAIG		12SecDr	2-5/25	3410	4845
McQUARRIE		12SecDr	1-3/25	3410	4844
McQUEEN		12SecDr	6-8/24	3410	4764
McQUESTEN		12SecDr	2-5/25	3410	4845
McQUILLEN		12SecDr	2-5/25	3410	4845
McRAE		12SecDr	6-8/24	3410	4764
McRANEY		12SecDr	1-3/25	3410	4844
McRAVEN		12SecDr	6-8/24	3410	4764
McREYNOLDS		12SecDr	6-8/24	3410	4764
McROBERTS		12SecDr	6-8/24	3410	4764
McROSS		12SecDr	6-8/24	3410	4764
McRUER		12SecDr	2-5/25	3410	4845
McSHAN		12SecDr	2-5/25	3410	4845
McSHEA		12SecDr	6-8/24	3410	4764
McSHERRY		12SecDr	6-8/24	3410	4764
McSHERRYSTOWN		12SecDr	2-5/25	3410	4845
McSPADDEN		12SecDr	6-8/24	3410	4764
McSWEEN		12SecDr	2-5/25	3410	4845
McSWEENEY		12SecDr	2-5/25	3410	4845

CAR NAME	ST	CAR TYPE	BUILT	PLAN	LOT
McSWEYN		12SecDr	2-5/25	3410	4845
McSWYNES		12SecDr	2-5/25	3410	4845
McTAGGART		12SecDr	6-8/24	3410	4764
McTAVISH		12SecDr	6-8/24	3410	4764
McTWIGGAN		12SecDr	2-5/25	3410	4845
McTYERS		12SecDr	2-5/25	3410	4845
McVEAN		12SecDr	2-5/25	3410	4845
McVEIGH		12SecDr	6-8/24	3410	4764
McVEYTOWN		12SecDr	1-3/25	3410	4844
McVICAR		12SecDr	2-5/25	3410	4845
McVILLE		12SecDr	1-3/25	3410	4844
McVITTIES		12SecDr	2-5/25	3410	4845
McWADE		12SecDr	2-5/25	3410	4845
McWHORTER		12SecDr	6-8/24	3410	4764
McWILLIAMS		12SecDr	6-8/24	3410	4764
McWILLIE		12SecDr	2-5/25	3410	4845
McZENA		12SecDr	2-5/25	3410	4845
MEADE		LngBDorm(5Sec)	7-8/27	3981	6061
MEADENHAM		12SecDr	2-3/13	2410	4105
MEADOW LARK		3Sbr2CptDrBLngSunR	3/30	3991A	6352
MEADOW LILY		34Ch	4/26	3419	4956
MEADOW VIOLET		30ChDr	8-9/30	4000A	6384
MEADOWSIDE		12SecDr	7/13	2410A	4179
MEADOWSWEET		28ChDr	5-7/26	3416	4958
MEATH		12SecDr	12/20-21	2410F	4611
MECCA		12SecDr	5-8/18	2410F	4540
MECHANICSBURG		12SecDr	6-9/26	3410A	4969
MECHLIN		12SecDr	7/13	2410A	4179
MECONAH		10SecLngObs	12/11-12	2521	3944
MEDARY		12SecDr	10-12/16	2410F	4431
MEDBURY		10SecDr2Cpt	7-8/16	2585D	4424
MEDELLIN		12SecDr	7/16	2410F	4412
MEDFIELD		36Ch	6/16	2916	4389
MEDFORD		26ChDr	2-3/14	2416A	4248
MEDIAN		10SecDr2Cpt	11/17	2585D	4527
MEDICINE LAKE		10SecDr2Cpt	5-6/24	3585	4770
MEDILL		10SecDr2Cpt	12/16-17	2585D	4443
MEDONTE		12SecDr	12/10-11	2410	3866
MEDUSA	N	26ChDr	4/12	2416	3950
MEDWAY		12SecDr	1-2/16	2410E	4356
MEEKER		12SecDr	2-5/25	3410	4845
MEEKERS		12SecDr	10/22/10	2410	3814
MEGARA		12SecDr	2-5/16	2410E	4367
MEGEATH		10SecDr2Cpt	7-8/16	2585D	4424
MEIROL		12SecDr	3-6/15	2410B	4319
MELANIE		26ChDr	4/12	2416	3950
MELCHOR OCAMPO NdeM	SN	12SecDr	5-8/18	2410K	4540
MELDRIM		12SecDr	7-8/15	2410D	4322
MELFORT		12SecDr	11-12/13	2410A	4234
MELINDA		26ChDr	4/12	2416	3950

CAR NAME	ST	CAR TYPE	BUILT	PLAN	LOT
MELLBRAKE		12SecDr	9-10/15	2410D	4327
MELMORE	N	12SecDr	10-12/15	2410D	4338
MELROSE		12SecDr	3-4/11	2410	3881
MELROSE PARK		12SecDrCpt	8-9/11	2411	3922
MELSTONE		12SecDr	1-3/14	2410A	4249
MELUNA		12SecDr	2-5/16	2410E	4367
MELVALE		12SecDrCpt	6/13	2411A	4159
MELVINA		12SecDr	12/10-11	2410	3866
MEMLING		16Sec	6-7/13	2412B	4160
MEMNON		12SecDr	7-8/15	2410D	4322
MEMORIAL HALL		4CptLngObs	4/26	3960	4944
MENA	RN	10Sec3Dbr	2-5/25	4087B	4845
MENADO		12SecDr	12/10-11	2410	3866
MENDELSSOHN		6Cpt3Dr	11-12/24	3523A	4833
MENDENHALL		12SecDr	5-6/17	2410F	4497
MENDON		10SecLngObs	11/12	2521	4049
MENES		12SecDr	12/20-21	2410F	4611
MENHADEN		12SecDr	11/12/10	2410	3860
MENLO PARK		10Sec2Dr	2/13	2584	4092
MENOKEN		16Sec SBS	4/15	2412B	4310
MENTANA		12SecDr	10-12/15	2410D	4338
MENTASTA		16Sec	10-11/14	2412C	4296
MENTONE		10SecDr2Cpt	10-11/12	2585	4067
MENTOR		12SecDr	10-11/11	2410	3936
MERAZO		12SecDr	10-12/16	2410F	4431
MERCARA		12SecDr	12/10-11	2410	3866
MERCEDES		26ChDr	3-4/14	2416A	4264
MERCER		12SecDr	10/31/10	2410	3814
MERCHANTS CLUB	RN	8SecRestObsLng	9-10/14	4025B	4292
MERCHANTVILLE		12SecDr	5-6/23	2410H	4699
MERCURY		20StsCoachBLng	4/30	3996	6312
MERCURY	R	16ChB15StsLng	4/30	3996C	6312
MEREDITH		BaggBClubSmk	7/13	2415	4158
MEREDITH COLLEGE	RN	10Sec2DbrCpt	6/13	4042A	4159
MERIDA NdeM-212	SN	8SecDr2Cpt	10-12/28	3979A	6205
MERION		12SecDr	10/27/10	2410	3860
MERISE		26ChDr	6/11	2416	3902
MERIVALE		12SecDr	1-3/14	2410A	4249
MERIWETHER LEWIS	RN	10SecDr2Dbr	1/25	4074C	4843
MERIWETHER LEWIS	N	10SecDrCpt	4-6/27	3973C	6043
MERLERA		12SecDr	12/10-11	2410	3866
MERLIN		12SecDr	1-2/16	2410E	4356
MERMENTAU		12SecDrCpt	6/13	2411A	4159
MERMENTAU	R	24ChLng	6/13	4009	4159
MERMILL		12SecDr	7/11	2410	3913
MERRIAM PARK		14Sec	4/27	3958	6041
MERRICK		LngBDorm(5Sec)	7-8/27	3981	6061
MERRILLAN		14Sec	8-10/28	3958A	6181
MERRIMAC		12SecDr	4-5/13	2410	4106
MERRITON		12SecDr	7-8/15	2410D	4322

CAR NAME	ST	CAR TYPE	BUILT	PLAN	LOT	CAR NAME	ST	CAR TYPE	BUILT	PLAN	LOT
MERSHON		16Sec	7-8/11	2412	3912	MIACOMET		12SecDr	11-12/13	2410A	4234
MERTA		26ChDr	6/11	2416	3902	MIAMI-BILTMORE	N	Recreation	2-5/16	3966	4367
MERTENSIA		12SecDr	7/11	2410	3913	MIANUS		12SecDr	11-12/13	2410A	4234
MERTISEN		12SecDr	11-12/13	2410A	4234	MICAWBER		12SecDr	3-6/20	2410F	4565
MERTOLA		12SecDr	12/10-11	2410	3866	MICHAEL FARADAY		14Sec	3/27	3958	6054
MERVYN		24ChDrB	4/13	2417	4108	MICHEAUD		12SecDr	9-10/13	2410A	4215
MESA PEAK		10Sec2Dr	12/28-29	3584B	6213	MICHELANGELO		6Cpt3Dr	7/25	3523A	4887
MESCALERO		7Cpt2Dr SBS	11/11	2522	3941	MICHICAGO		12SecDr	3-6/15	2410B	4319
MESQUITE		10SecDr2Cpt	8-9/13	2585A	4201	MICHIGAMME		8SecDr2Cpt	7-8/27	3979	6052
METAMORA		12SecDr	3-4/17	2410F	4485	MICKLETON		12SecDr	5-6/17	2410F	4497
METAPAN		12SecDr	7-8/15	2410D	4322	MIDAS		12SecDr	12/20-21	2410F	4611
METCALF		12SecDr	5-6/17	2410F	4497	MIDDLEBURY		12SecDr	9-10/11	2410	3924
METHUEN		12SecDr	9-10/15	2410D	4327	MIDDLEBUSH		12SecDr	7-9/17	2410F	4515
METROPOLITAN CLUB	N	8SecBLngSunRm	8-9/29	3989A	6274	MIDDLEFIELD		12SecDr	3-4/11	2410	3881
METTLER		12SecDr	6-9/26	3410A	4969	MIDDLEPOINT		12SecDr	10/24/10	2410	3814
METUCHEN		12SecDr	2-5/16	2410E	4367	MIDDLESEX		12SecDr	2-4/24	3410	4762
MEUSE	N	7Dr	12/20	2583B	4621	MIDDLETOWN	N	12SecDr	3/15/07	1963D	3402
MEXICALPA FC-DS	SN	8Sec5Dbr	8-9/4	4036B	3922	MIDDLETOWN	N	12SecDr	1-2/29	3410B	6220
MEXICALI NdeM-213	SN	14Sec	10/26	3958	4997	MIDLAKE		12SecDr	1-2/16	2410E	4356
MEXICO		12SecDr	12/20-21	2410F	4611	MIDLAND		12SecDr	10/17/10	2410	3814
MEZCALAPA FC-DS	SN	8Sec5Dbr	8-9/11	4036B	3922	MIDLAND HILLS CLUB	RN	8SecRestLngObs	11/12	4025D	4049
MGRS-55 NdeM	SN	8sec4Dbr	10-11/12	4022B	4067	MIDLAND TRAIL	RN	8SecDrLngObs	1-5/18	4024	4531
MGRS-57 NdeM	SN	8sec4Dbr	7-9/17	4022A	4515	MIDLOTHIAN		12SecDr	10-12/26	3410A	6023
MGRS-58 NdeM	SN	14Sec	9-10/25	3958	4869	MIDMONT		12SecDr	11-12/13	2410A	4234
MGRS-60 NdeM	SN	10Sec3Dbr	8-9/11	3411	3922	MIDVALE		10SecDr2Cpt	7-8/16	2585D	4424
MGRS-61 NdeM	SN	8SecDr2Cpt	10-12/28	3979A	6205	MIDWAY		12SecDr	11/26/10	2410	3864
MGRS-63 NdeM	SN	8SecDr2Cpt	10-12/28	3979A	6205	MIFFLIN		24ChDrB	4/13	2417	4108
MGRS-64 NdeM	SN	14Sec	9-10/25	3958	4869	MIKADO		12SecDr	2-3/15	2410B	4318
MGRS-65 NdeM	SN	12SecDr	4-6/24	3410	4763	MIL CUMBRES NdeM	SN	10SecDrCpt	4-6/27	3973C	6043
MGRS-66 NdeM	SN	12SecDr	5-8/25	3410A	4868	MILANO		10SecDr2Cpt	12/16-17	2585D	4443
MGRS-67 NdeM	SN	8SecDr2Cpt	7-8/30	3979A	6377	MILBANK		12SecDr	1-3/14	2410A	4249
MGRS-68 NdeM	SN	14Sec	3/27	3958	6054	MILBELL		12SecDr	5-6/17	2410F	4497
MGRS-69 NdeM	SN	12SecDr	10-12/23	3410	4724	MILDRED		26ChDr	3-4/14	2416A	4264
MGRS-70 NdeM	SN	8sec4Dbr	7-9/17	4022A	4515	MILES CITY		10SecDrCpt	4-6/27	3973	6043
MGRS-71 NdeM	SN	12SecDr	2-5/25	3410	4845	MILESBURG		12SecDr	5-6/17	2410F	4497
MGRS-73 NdeM	SN	12SecDr	5-8/25	3410A	4868	MILFANWY		12SecDr	6-10/21	2410F	4614
MGRS-74 NdeM	SN	14Sec	9-10/25	3958	4869	MILFORD		12SecDr	5-6/17	2410F	4497
MGRS-77 NdeM	SN	8SecDr2Cpt	2-3/30	3979A	6353	MILL CREEK		28ChDr	6-7/27	3416A	6087
MGRS-78 NdeM	SN	8SecDr2Cpt	10-12/28	3979A	6205	MILLARD		10SecDr2Cpt	11-12/11	2585	3942
MGRS-79 NdeM	SN	10Sec3Dbr	5/17	3411	4494	MILLAUDON		10SecDr2Cpt	7-8/16	2585D	4424
MGRS-80 NdeM	SN	12SecDr	1-3/25	3410	4844	MILLBORO		12SecDr	11-12/13	2410A	4234
MGRS-82 NdeM	SN	8SecDr3Dbr	5-6/25	4090	4846	MILLBROOK		16Sec	9-10/12	2412	4037
MGRS-84 NdeM	SN	8SecDr2Cpt	10-12/28	3979A	6205	MILLBURY		36Ch SBS	1/13	2691	4056
MGRS-85 NdeM	SN	8SecDr2Cpt	10-12/28	3979A	6205	MILLDALE		36Ch	6/16	2916	4389
MGRS-88 NdeM	SN	12SecDr	2-5/25	3410	4845	MILLDYKE		16Sec	5-6/16	2412F	4386
MGRS-89 NdeM	SN	8SecDr3Dbr	1/25	4090	4843	MILLER		12SecDr	7-9/17	2410F	4515
MGRS-94 NdeM	SN	12SecDr	6-8/24	3410	4764	MILLERSBURG		12SecDr	5/4/10	2410	3772
MGRS-95 NdeM	SN	8Sec5Dbr	7/16/10	4036	3800	MILLERSBURG		26ChDr	9-11/22	2416D	4649
MGRS-96 NdeM	SN	8SecDr2Cpt	7-8/30	3979A	6377	MILLERTON		12SecDr	7/11	2410	3913
Mi 31	LN	Lounge (Rexall)	6/16	2916D	4389	MILLET		6Cpt3Dr	7/25	3523A	4887

CAR NAME	ST	CAR TYPE	BUILT	PLAN	LOT	CAR NAME	ST	CAR TYPE	BUILT	PLAN	LOT
MILLINGTON		12SecDr	7/11	2410	3913	MIQUELON		10SecDr2Cpt	7-8/16	2585D	4424
MILLMONT		12SecDr	11/25/10	2410	3864	MIRABILIS		12SecDr	3-6/20	2410F	4565
MILLVILLE		12SecDrCpt	6/30/10	2411	3800	MIRADERO		16Sec	6-7/20	2412F	4568
MILLWOOD		16Sec	7-8/11	2412	3912	MIRAFLORES		10SecLngObs	11/12	2521	4049
MILMAY		12SecDr	11-12/13	2410A	4234	MIRAGE		10SecDr2Cpt	7-8/16	2585D	4424
MILNER PASS		8SecDr2Cpt	12/29-30	3979A	6334	MIRAMAR		10SecDr2Cpt	1-2/13	2585	4091
MILNOR	N	12SecDr	9/28/10	2410	3813	MIRASOL	N	12SecDr	3-6/15	2410B	4319
MILO		12SecDr	12/20-21	2410F	4611	MIRELLA		26ChDr	4/12	2416	3950
MILOMA		16Sec	4/17	2412F	4484	MIRIAM		28ChDr	6-7/27	3416A	6087
MILROY		12SecDr	12/13/10	2410	3864	MIRZA		12SecDr	12/20-21	2410F	4611
MILTON		12SecDr	12/12-13	2410	4076	MISANTLA NdeM-215	SN	10SecDrCpt	5/30	3973A	6358
MILTON H. SMITH	N	12SecDr	7/16	2410F	4412	MISHAWAKA		12SecDr	9-10/11	2410	3924
MILTONVALE		10SecDr2Cpt	10-11/13	2585A	4218	MISLANCO		12SecDr	1-3/14	2410A	4249
MILWAUKEE		6SbrLngSunRm	8/28	3974A	6183	MISSION DOLORES	R	6Sbr2DbrLngSunRm	8/28	3974F	6183
MIMOSA		28ChDr	5-7/26	3416	4958	MISSION DOLORES	N	6SbrLngSunRm	8/28	3974A	6183
MINARD		10SecLngObs	11/12	2521	4049	MISSION PEAK		10Sec2Dr	12/26-27	3584A	6031
MINATITLAN NdeM-214	SN	14Sec	9-10/25	3958	4869	MISSION SANTA YNEZ	R	6Sbr2DbrLngSunRm	8/28	3974F	6183
MINDEN		10SecDr2Cpt	1-2/13	2585	4091	MISSION SANTA YNEZ	N	6SbrLngSunRm	8/28	3974A	6183
MINEOLA		12SecDr	3-6/15	2410B	4319	MISSISQUOI	N	5CptBLng	1/12	2505B	3969
MINERAL SPRING		28ChDr	6-7/27	3416A	6087	MISSISSIPPI		10SecDrCpt	4-6/27	3973	6043
MINERIC		12SecDr	7/13	2410A	4179	MISSOURI TOWER	RN	8SecDr4Dbr	4-5/26	4090E	4961
MINERVA		26ChDr	6/11	2416	3902	MITCHELL		10SecDr2Cpt	11/17	2585D	4527
MINIKAHDA CLUB	RN	2Sbr2Dbr2CptDrLng	12/29	3995B	6301	MITCHELL TOWER	RN	8SecDr3Dbr	12/25-26	4090F	4932
MINIOTA		12SecDr	9-10/13	2410A	4215	MITHRAS		12SecDr	3-6/20	2410F	4565
MINITONAS		12SecDr	11-12/13	2410A	4234	MITLA NdeM	SN	8SecDr2Cpt	3/29	3979A	6237
MINNA		24ChDrB	6-7/14	2417B	4265	MITYLENE		10SecDr2Cpt	7-8/16	2585D	4424
MINNAMERE	N	12SecDr	7-8/15	2410D	4322	MIZPAH		10SecLngObs	12/11-12	2521	3944
MINNEAPOLIS	N	10Sec3Dbr	5/17	3411C	4494	MOBERLY	N	12SecDr	12/15/10	2410	3864
MINNEAPOLIS		6SbrLngSunRm	8/28	3974A	6183	MOBILE	N	12SecDr	12/15/10	2410	3864
MINNEAPOLIS CLUB	RN	8SecRestLngObs	11/12	4025A	4049	MOBRAY		12SecDr	3-6/20	2410F	4565
MINNEHAHA		6SbrLngObs	5/27	3974	6045	MOBRIDGE	N	12SecDr	9-10/15	2410D	4327
MINNEHAHA	R	6Sbr2DbrObsLng	5/27	3974E	6045	MOCANAQUA		12SecDr	7-9/17	2410F	4515
MINNEISKA		10SecDrCpt	4-6/27	3973	6043	MODART		12SecDr	7/13	2410A	4179
MINNEKAHDA CLUB		4Sbr2CptDrLng	12/29	3995	6301	MOFFETT		10SecDr2Cpt	11/17	2585D	4527
MINNEQUA		12SecDr	10/18/10	2410	3814	MOGADORE		12SecDr	6-10/21	2410F	4614
MINNESOTA	N	6Sbr2DbrObsLng	5/27	3974E	6045	MOGOLLON		7Cpt2Dr	9/13	2522A	4208
MINNESOTA CLUB		4Sbr2CptDrLng	12/29	3995	6301	MOHAWK		10SecDr2Cpt	11/17	2585D	4527
MINNESOTA CLUB	R	2Sbr2Dbr2CptDrLng	12/29	3995B	6301	MOHAWK RIVER	N	26ChDrObs	5/25	3957	4862
MINNETONKA	R	6Sbr2DbrObsLng	5/27	3974E	6045	MOHAWK VALLEY		DrSbrBLngObs	4/29	3988	6221
MINNETONKA		6SbrLngObs	5/27	3974	6045	MOHAWK VIEW		12SecDrCpt	5/17	2411C	4494
MINNEWASKA		12SecDr	5-7/13	2410	4149	MOIRA		12SecDr	1-4/12	2410	3949
MINOA		16Sec	2-3/11	2412	3878	MOIZANT		12SecDr	10-12/16	2410F	4431
MINOCQUA		10SecDrCpt	4-6/27	3973	6043	MOKANNA		12SecDr	6-10/21	2410F	4614
MINORCA		16Sec	1/16	2412E	4352	MOLASSE		12SecDr	2-5/16	2410E	4367
MINOT		12SecDr	12/20-21	2410F	4611	MOLIERE	N	12SecDr	1-2/16	2410E	4356
MINOT LIGHT		36Ch	12/29-30	3916B	6319	MOLINEUX		10SecDr2Cpt	11/17	2585D	4527
MINOT LIGHT	R	28Ch10StsLng	12/29-30	3916D	6319	MOLLIERE	N	12SecDr	1-2/16	2410E	4356
MINOTOLA		12SecDr	10/18/10	2410	3814	MOLLY PITCHER		28ChDr	1-2/27	3416A	6032
MINTAKA		10SecLngObs	12/11-12	2521	3944	MOMBRUN		12SecDr	7-8/13	2410A	4192
MINUTE MEN	N	12ChB14StsLng	6/30	3992F	6361	MOMUS		10SecDr2Cpt	11-12/11	2585	3942

CAR NAME	ST	CAR TYPE	BUILT	PLAN	LOT
MONA		26ChDr	6/11	2416	3902
MONACO		10SecDr2Cpt	1-2/13	2585	4091
MONCKTON		10SecDr2Cpt	11/17	2585D	4527
MONCURE		10SecLngObs	2/24/11	2521	3865
MONDELL		10SecDr2Cpt	7-8/16	2585D	4424
MONDOVI		16Sec	4/17	2412F	4484
MONESSEN		12SecDr	7/11	2410	3913
MONETA		12SecDr	1-3/14	2410A	4249
MONHEGAN		12SecDr	10/22-23	2410H	4647
MONICO		16Sec	4/17	2412F	4484
MONITEAU		12SecDr	4-5/13	2410	4106
MONMOUTH		16Sec	9-10/12	2412	4037
MONMOUTH	N	12SecDr	2-5/25	3410	4845
MONOCACY		12SecDr	5-6/17	2410F	4497
MONSON		10SecDr2Cpt	7-8/16	2585D	4424
MONTAGUE		12SecDr	2-5/16	2410E	4367
MONTANDON		12SecDr	9/8/10	2410	3794
MONTANESCA		BaggBClubSmkBarber	2-3/12	2602	3948
MONTARA		LngBDorm(5Sec)	7-8/27	3981	6061
MONTAUBAN		10SecDr2Cpt	7-8/16	2585D	4424
MONTAUK		12SecDr	10-12/26	3410A	6023
MONTAYNE		16Sec	5-6/16	2412F	4386
MONTCALM		10SecLngObs	12/11-12	2521	3944
MONTCHANIN		12SecDr	7-8/13	2410A	4192
MONTCLAIR		BaggBClubSmkBarber	2-3/12	2602	3948
MONTDIDIER		7Dr	12/20	2583B	4621
MONTE ALBAN NdeM	SN	10SecDrCpt	9-10/17	4031	4525
MONTE BALDO		10SecLngObs	11-12/25	3521A	4923
MONTE BELLO		10Sec2Dr	2/13	2584	4092
MONTE BLANCO		10SecLngObs	11-12/25	3521A	4923
MONTE CARLO		10SecLngObs	11-12/25	3521A	4923
MONTE CAVO		10SecLngObs	11-12/25	3521A	4923
MONTE CRISTO		10SecLngObs	11-12/25	3521A	4923
MONTE GRANDE		10SecLngObs	11-12/25	3521A	4923
MONTE LAGUNA		10SecLngObs	11-12/25	3521A	4923
MONTE LEONE		10SecLngObs	11-12/25	3521A	4923
MONTE LIRIO		10SecLngObs	11-12/25	3521A	4923
MONTE ROSA		10SecLngObs	11-12/25	3521A	4923
MONTE ROTONDO		10SecLngObs	11-12/25	3521A	4923
MONTE SANTO		10SecLngObs	11-12/25	3521A	4923
MONTE TABOR		10SecLngObs	11-12/25	3521A	4923
MONTE VISO		10SecLngObs	11-12/25	3521A	4923
MONTE VISTA		10SecLngObs	11-12/25	3521A	4923
MONTEAGLE		12SecDr	7/11	2410	3913
MONTEAU		12SecDr	12/10-11	2410	3866
MONTECITO		16Sec	10/13	2412B	4217
MONTEITH		16Sec	7-8/11	2412	3912
MONTELLO		10SecDr2Cpt	10-11/12	2585	4067
MONTEMORA		12SecDr	12/10-11	2410	3866
MONTERO		12SecDr	10-12/16	2410F	4431

CAR NAME	ST	CAR TYPE	BUILT	PLAN	LOT
MONTEVALDO		12SecDr	1-3/17	2410F	4450
MONTEVIDEO NdeM-218	SN	6Sec6Dbr	9-10/17	4084	4525
MONTEZUMA		10SecLngObs	11/12	2521	4049
MONTFORT		10SecDr2Cpt	7-8/16	2585D	4424
MONTICELLO	N	12SecDr	10-12/23	3410	4724
MONTIEL		12SecDr	10-12/15	2410D	4338
MONTMOLIN		10SecDr2Cpt	11/17	2585D	4527
MONTMORENCI		12SecDr	1-4/12	2410	3949
MONTOUR		10SecDr2Cpt	10-11/12	2585	4067
MONTOWESE		24ChDrB	4/16	2417C	4364
MONTOYA		10SecDr2Cpt	8-9/13	2585A	4201
MONTREAL		3CptDrLngBSunRm	6/29	3975C	6262
MONTREAL UNIVERSITY	RN	12Sec2Dbr	5-6/17	4046	4497
MONTROSE		16Sec	10/13	2412B	4217
MONTVIEW		12SecDrCpt	3/11	2411	3880
MONTWAIT		36Ch	6/16	2916	4389
MONUMENT SQUARE		6Cpt3Dr	7-8/26	3523A	4970
MOON BROOK CLUB		8SecBLngSunRm	8-9/29	3989A	6274
MOON BROOK CLUB	RN	8SecRestLngObs	1-5/18	4025	4531
MOORE		6Cpt3Dr	11-12/24	3523A	4833
MOORESTOWN		16Sec	6/8/10	2412	3801
MOORFIELD		12SecDr	9-10/11	2410	3924
MOOSE LAKE		10SecDr2Cpt	11-12/27	3585B	6123
MOQUI		10SecLngObs	11/12	2521	4049
MORADO		12SecDr	1/14	2410A	4240
MORANGE		12SecDr	7/13	2410A	4179
MOREA		12SecDr	11/25/10	2410	3864
MOREHEAD	N	12SecDr	1-3/14	2410A	4249
MORENCI		10SecLngObs	11/12	2521	4049
MORGAN		12SecDr	12/26/10	2410	3864
MORGANTOWN		28ChObs	3/14	2421	4251
MORGANZA		12SecDr	5-6/23	2410H	4699
MORISCO		12SecDr	12/10-11	2410	3866
MORNING GLORY		8SecDr2Cpt	4-5/30	3979A	6359
MORNING STAR		12SecDr	10/22-23	2410H	4647
MOROCCO		12SecDr	12/20-21	2410F	4611
MORONTS		12SecDr	4-6/14	2410A	4271
MORPETH		10SecDr2Cpt	11/17	2585D	4527
MORRELL		12SecDr	12/12/10	2410	3864
MORRIS		12SecDrCpt	6/13	2411	4151
MORRIS MANSION		13Dbr	6/30	3997	6314
MORRISANIA		12SecDr	3-4/11	2410	3881
MORRISDALE		12SecDr	7/11	2410	3913
MORRISON		14Sec	6/28	3958A	6174
MORRISTOWN		16Sec	10/13	2412B	4217
MORRISVILLE		12SecDrCpt	6/24/10	2411	3800
MORROW		12SecDr	5/26/10	2410	3794
MORSTEIN		12SecDr	12/12/10	2410	3864
MORTIMER		12SecDr	3-4/11	2410	3881
MORTON		12SecDr	10/20/10	2410	3814

CAR NAME	ST	CAR TYPE	BUILT	PLAN	LOT
MORTON PARK		16Sec	9-10/12	2412	4037
MORVEN		12SecDr	12/10-11	2410	3866
MOSGROVE		12SecDr	7-8/12	2410	4014
MOSHANNON		12SecDr	2-5/16	2410E	4367
MOSS ROSE	N	12SecDr	8-12/25	3410A	4894
MOTOSA		12SecDr	7/13	2410A	4179
MOTT HAVEN		10Sec2Dr	5/13	2584	4140
MOULTON		12SecDr	9-10/11	2410	3924
MOULTRIE		12SecDr	2-3/15	2410B	4318
MOUND CITY		BaggClubBSmk	3/25	3415A	4848
MOUNDS		12SecDr	10/22-23	2410H	4647
MOUNT ADAMS	N	8SecBLngSunRm	8-9/29	3989T	6274
MOUNT AIRY	N	10SecLibLng	1-2/28	3521B	6128
MOUNT AIRY		24ChDrObs	4/14	2819	4263
MOUNT CELO		10SecLibLng	1-2/28	3521B	6128
MOUNT CORY	N	10SecLibLng	1-2/28	3521B	6128
MOUNT GRETNA		10SecDr2Cpt	7-8/23	2585D	4707
MOUNT HOLLY		12SecDr	12/14/10	2410	3864
MOUNT JEFFERSON	N	8SecBLngSunRm	8-9/29	3989T	6274
MOUNT JOY		10SecDr2Cpt	7-8/23	2585D	4707
MOUNT MANSFIELD		10SecLibLng	1-2/28	3521B	6128
MOUNT MANSFIELD	N	10SecLngObs	12/11-12	2521	3944
MOUNT MITCHELL		10SecLibLng	1-2/28	3521B	6128
MOUNT PLEASANT		Bagg24Ch SBS	1/13	2692	4057
MOUNT RIGA		12SecDrCpt	3/11	2411	3880
MOUNT ROYAL		10SecLibLng	1-2/28	3521B	6128
MOUNT ROYAL	N	10SecLngObs	11/13	2521	4221
MOUNT SUMMIT	N	10SecLibLng	1-2/28	3521B	6128
MOUNT UNION		10SecDr2Cpt	7-8/23	2585D	4707
MOUNT VERNON		10SecLibLng	1-2/28	3521B	6128
MOUNT VERNON	N	8SecDr2Cpt	3/29	3979A	6237
MOUNTAIN BURG		10SecLngObs	6-7/23	2521C	4690
MOUNTAIN CAVE		10SecLngObs	6-7/23	2521C	4690
MOUNTAIN CITY		10SecLngObs	6-7/23	2521C	4690
MOUNTAIN CREEK		10SecLngObs	6-7/23	2521C	4690
MOUNTAIN DALE		10SecLngObs	6-7/23	2521C	4690
MOUNTAIN GLEN		10SecLngObs	6-7/23	2521C	4690
MOUNTAIN GROVE		10SecLngObs	6-7/23	2521C	4690
MOUNTAIN HOME		10SecLngObs	6-7/23	2521C	4690
MOUNTAIN KING		10SecLngObs	6-7/23	2521C	4690
MOUNTAIN LAKE		10SecLngObs	6-7/23	2521C	4690
MOUNTAIN MILLS		10SecLngObs	6-7/23	2521C	4690
MOUNTAIN PARK		10SecLngObs	6-7/23	2521C	4690
MOUNTAIN QUEEN		10SecLngObs	6-7/23	2521C	4690
MOUNTAIN ROAD		10SecLngObs	6-7/23	2521C	4690
MOUNTAIN SIDE		10SecLngObs	6-7/23	2521C	4690
MOUNTAIN SPRINGS		10SecLngObs	7/14/23	2521C	4690
MOUNTAIN TOP		10SecLngObs	6-7/23	2521C	4690
MOUNTAIN VALLEY		10SecLngObs	6-7/23	2521C	4690
MOUNTAIN VIEW		10SecLngObs	6-7/23	2521C	4690
MOUNTAIN VILLE		10SecLngObs	6-7/23	2521C	4690
MOUNTAINEER		8SecDr2Cpt	5-6/29	3979A	6261
MOUNTVILLE		16Sec	6/6/10	2412	3801
MOYLAN		12SecDr	12/20/10	2410	3864
MOYSTON		12SecDr	7-9/17	2410F	4515
MOYUNE		12SecDr	1-2/16	2410E	4356
MOZART		6Cpt3Dr	11-12/24	3523A	4833
MT. ADAMS		10SecLngObs	10/24	3521	4816
MT. ALLISON UNIVERSITY	RN	12Sec2Dbr	5-6/17	4046	4497
MT. ANGEL	N	10SecLngObs	10/24	3521	4816
MT. ANGELES		10SecLngObs	10-12/26	3521A	4998
MT. BAKER		10SecLngObs	10/24	3521	4816
MT. BAXTER		10SecLngObs	10/24	3521	4816
MT. BLACKMORE		10SecLngObs	10/24	3521	4816
MT. BOLIVAR		10SecLngObs	10-12/26	3521A	4998
MT. BRECKENRIDGE		10SecLngObs	10-12/26	3521A	4998
MT. BRODERICK		10SecLngObs	10-12/26	3521A	4998
MT. CALLAHAN		10SecLngObs	10-12/26	3521A	4998
MT. CHAUVENET		10SecLngObs	10-12/26	3521A	4998
MT. COWEN		10SecLngObs	10/24	3521	4816
MT. DALL		10SecLngObs	10-12/26	3521A	4998
MT. DARWIN		10SecLngObs	10/24	3521	4816
MT. DELANO		10SecLngObs	10/24	3521	4816
MT. DESERT		10SecLngObs	11-12/23	3521	4742
MT. DOANE		10SecLngObs	10/24	3521	4816
MT. DOUGLAS		10SecLngObs	10/24	3521	4816
MT. DRUM		10SecLngObs	10-12/26	3521A	4998
MT. ETNA		10SecLngObs	11-12/23	3521	4742
MT. EVANS		10SecLngObs	11-12/23	3521	4742
MT. EVERETT		10SecLngObs	11-12/23	3521	4742
MT. FORAKER		10SecLngObs	11-12/23	3521	4742
MT. GIBBS		10SecLngObs	10-12/26	3521A	4998
MT. HARVARD		10SecLngObs	11-12/23	3521	4742
MT. HAYES		10SecLngObs	11-12/23	3521	4742
MT. HILGARD		10SecLngObs	10-12/26	3521A	4998
MT. HILLERS		10SecLngObs	10-12/26	3521A	4998
MT. HOFFMAN		10SecLngObs	10-12/26	3521A	4998
MT. HOLMES		10SecLngObs	11-12/23	3521	4742
MT. HOME		10SecLngObs	10/24	3521	4816
MT. HOOD		10SecLngObs	11-12/23	3521	4742
MT. HULL		10SecLngObs	10-12/26	3521A	4998
MT. HUMPHREY		10SecLngObs	10/24	3521	4816
MT. JARVIS		10SecLngObs	10-12/26	3521A	4998
MT. JOY		12SecDrCpt	6/24/10	2411	3800
MT. KING		10SecLngObs	10-12/26	3521A	4998
MT. LANGFORD		10SecLngObs	10-12/26	3521A	4998
MT. LANGLEY		10SecLngObs	10/24	3521	4816
MT. LEIDY		10SecLngObs	10-12/26	3521A	4998
MT. LOWE		10SecLngObs	11-12/23	3521	4742
MT. LYELL		10SecLngObs	10/24	3521	4816

CAR NAME	ST	CAR TYPE	BUILT	PLAN	LOT		CAR NAME	ST	CAR TYPE	BUILT	PLAN	LOT
MT. MARCY		10SecLngObs	11-12/23	3521	4742		MT. STEVENSON		10SecLngObs	10/24	3521	4816
MT. MORAN		10SecLngObs	10/24	3521	4816		MT. TAYLOR		10SecLngObs	10-12/26	3521A	4998
MT. NEBO		10SecLngObs	10-12/26	3521A	4998		MT. THIELSEN		10SecLngObs	10-12/26	3521A	4998
MT. NORRIS		10SecLngObs	10/24	3521	4816		MT. TOM		10SecLngObs	10-12/26	3521A	4998
MT. OLYMPUS		10SecLngObs	11-12/23	3521	4742		MT. TURNBULL		10SecLngObs	10-12/26	3521A	4998
MT. ORD		10SecLngObs	10-12/26	3521A	4998		MT. WASHBURN		10SecLngObs	10/24	3521	4816
MT. OURAY		10SecLngObs	10-12/26	3521A	4998		MT. WATKINS		10SecLngObs	10-12/26	3521A	4998
MT. PEALE		10SecLngObs	10-12/26	3521A	4998		MT. WHITNEY		10SecLngObs	11-12/23	3521	4742
MT. RAINIER		10SecLngObs	11-12/23	3521	4742		MT. WOOD		10SecLngObs	10-12/26	3521A	4998
MT. RUSHMORE	N	10SecDrCpt	8/28	3973A	6185		MT. WRANGELL		10SecLngObs	10-12/26	3521A	4998
MT. SHASTA		10SecLngObs	11-12/23	3521	4742		MUIRHEAD		12SecDr	2-5/16	2410E	4367
MT. SHERIDAN		10SecLngObs	10/24	3521	4816		MUIRKIRK		12SecDr	4-6/14	2410A	4271
MT. SILLIMAN		10SecLngObs	10/24	3521	4816		MUJER					
MT. SLYEN		10SecLngObs	10-12/26	3521A	4998		DIVINA	N	10SecDrCpt	5/30	3973A	6358
MT. SNEFFELS		10SecLngObs	10/24	3521	4816		DIVINA FCP-203	SN	10SecDr2Cpt	10/24	3585A	4817
MT. SPURR		10SecLngObs	10-12/26	3521A	4998		MULLIKIN		12SecDr	7-9/17	2410F	4515

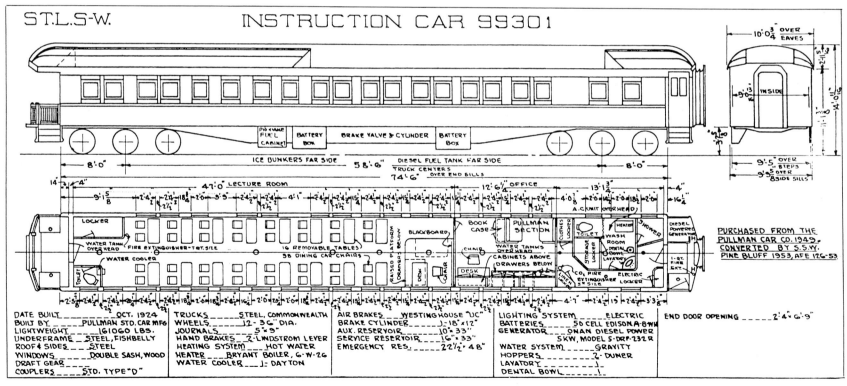

Figure I-62 The 10-section lounge observation MT. SHERIDAN, which ran on the Cotton Belt "Lone Star" between Memphis and Shreveport, was retired in 1952 to become Instruction Car 99301. It is shown here in a StLSW diagram from the Author's Collection.

CAR NAME	ST	CAR TYPE	BUILT	PLAN	LOT
MULTNOMAH		12SecDr	10-12/26	3410A	6023
MUMFORD		16Sec	5-6/16	2412F	4386
MUNCIE		12SecDr	10/13/10	2410	3814
MUNDARE		12SecDr	1-3/14	2410A	4249
MUNGER		12SecDr	1-3/17	2410F	4450
MUNSTER		12SecDr	12/15/10	2410	3864
MUNSTER		12SecDr	12/20-21	2410F	4611
MUPHRID		10SecLngObs	12/11-12	2521	3944
MURCHISON		12SecDr	2-5/16	2410E	4367
MURDOCK	N	12SecDr	1-3/17	2410F	4450
MURIEL		28ChDr	6-7/30	3416B	6372
MURILLO		6Cpt3Dr	7/25	3523A	4887
MURRAY	N	12SecDr	5/9/10	2410	3773
MURRAY BAY		3Sbr2CptDrBLngSunR	11/29	3991	6276
MURRAY HILL		6SbrBLngSunRm	12/29	3994B	6291
MUSCATINE		LngBDorm(5Sec)	7-8/27	3981	6061
MUSINA		10SecLngObs	11/12	2521	4049
MUSSER		28ChDr	5-6/24	3416	4761
MYOLA		12SecDr	12/10-11	2410	3866
MYRA		24ChDrB	5/13	2417A	4137
MYRICA		12SecDr	12/10-11	2410	3866
MYRREL		26ChDr	3-4/14	2416A	4264
MYRTLE		32Ch	6/30	3916C	6363
MYRTUS		12SecDr	12/10-11	2410	3866
NACHITA		12SecDr	1-3/14	2410A	4249
NACORA		16Sec	12/14-15	2412C	4304
NADA		24ChDrB	6-7/14	2417B	4265
NADINE		24ChDrB	4/13	2417	4108
NADURA	N	12SecDr	2-3/13	2410	4105
NAGANOOK		12SecDr	6-7/17	2410F	4503
NAGAWICKA		12SecDr	4-5/27	3410B	6042
NAGINEY		12SecDr	7-9/17	2410F	4515
NAGOYA		12SecDr	12/20-21	2410F	4611
NAHANT		34Ch	8/13	2764A	4195
NAHMA		10SecDrCpt	4-6/27	3973	6043
NAIAD		26ChDr	3-4/14	2416A	4264
NAMOUNA		12SecDr	12/20-21	2410F	4611
NAMOZINE		12SecDr	4-5/13	2410	4106
NAMPA		12SecDr	12/20-21	2410F	4611
NAMUR		12SecDr	2-3/15	2410B	4318
NANCY		28ChDr	4-6/25	3416	4864
NANNETTE		26ChDr	6/11	2416	3902
NANSEMOND	N	10SecLngObs	10-12/26	3521A	4998
NANSEMOND COUNTY	N	10SecLngObs	10-12/26	3521F	4998
NANTES		16Sec	6-7/20	2412F	4568
NANTICOKE		BaggBClubSmk	9/14/10	2415	3834
NANTUCKET LIGHT		36Ch	12/29-30	3916B	6319
NAPANOCH		12SecDr	9-10/15	2410D	4327
NAPIER		12SecDr	7-9/17	2410F	4515
NARAVISA		12SecDr	1-3/14	2410A	4249

CAR NAME	ST	CAR TYPE	BUILT	PLAN	LOT
NARBERTH		12SecDrCpt	6/22/10	2411	3800
NARCISO					
MENDOZA NdeM-219	SN	8SecDr2Cpt	12/29-30	3979A	6334
NARCISSUS		28ChDr	5-6/24	3416	4761
NARDIN		10SecDr2Cpt	8-9/13	2585A	4201
NARINDA		12SecDr	12/10-11	2410	3866
NARKA		12SecDr	12/20-21	2410F	4611
NARLON		16Sec	4/17	2412F	4484
NARRAGANSETT BAY		32ChDr	12/29-30	3917B	6318
NARROWS LIGHT		36Ch	12/29-30	3916B	6319
NASBY		12SecDr	12/20-21	2410F	4611
NASHOTAH		12SecDr	4-5/27	3410B	6042
NASSAU		12SecDr	2-4/24	3410	4762
NASSAU HALL		4CptLngObs	10/25	3960	4889
NATALIE		28ChDr	8/24	3416	4801
NATCHEZ		12SecDr	8-10/26	3410A	4945
NATHAN HALE		12SecDr	5-8/25	3410A	4868
NATHANAEL GREENE		12SecDr	5-8/25	3410A	4868
NATHANIEL BACON		12SecDr	4/26	3410A	4943
NATIONAL		24ChDrB	12/10	2417	3863
NATIONAL		Private (GS)	1/18/10	2492	3812
NATIONAL CITY		3Cpt2DrLngObs	8-10/25	3959	4886
NATIONAL HEIGHTS		3Cpt2DrLngObs	8-10/25	3959	4886
NATIONAL HILL		3Cpt2DrLngObs	8-10/25	3959	4886
NAUSHON		12SecDr	11-12/13	2410A	4234
NAUTLA NdeM-220	SN	12SecDr	6/8/24	3410	4764
NAUVOO		12SecDr	10-12/26	3410A	6023
NAVAJO		7Cpt2Dr (SBS)	11/11	2522	3941
NAVAN		12SecDr	7-8/15	2410D	4322
NAVARINO		12SecDr	10-12/15	2410D	4338
NAVARRE		16Sec	10/13	2412B	4217
NAVASOTA		10SecDr2Cpt	8-9/13	2585A	4201
NAVASSA		10SecLngObs	2/24/11	2521	3865
NAVESINK LIGHT		36Ch	12/29-30	3916B	6319
NAVIGATOR		16Sec	1-5/18	2412F	4531
NEALMONT		12SecDr	2-5/25	3410	4845
NEAPOLITAN		12SecDr	2-3/15	2410B	4318
NECAXA NdeM-282	SN	12SecDr	10-12/23	3410	4724
NEEDLE PEAK	N	10Sec2Dr	9-10/29	3584B	6284
NEEDLES		7Cpt2Dr	9/13	2522A	4208
NEGAUNEE		12SecDr	9-10/15	2410D	4327
NEHALEM		12SecDr	4/15	2410C	4309
NEHASANE		6CptLngObs	8/9/10	2413	3802
NELMOOR		12SecDr	7-9/17	2410F	4515
NEMACK		12SecDr	1-3/14	2410A	4249
NEMACOLIN		12SecDr	5-6/23	2410H	4699
NEMAHA		8SecDr2Cpt	7-8/27	3979	6052
NEMOURS		12SecDr	12/20-21	2410F	4611
NENANA		16Sec	10-11/14	2412C	4296
NEOCOMIAN		12SecDr	6-7/17	2410F	4503

CAR NAME	ST	CAR TYPE	BUILT	PLAN	LOT	CAR NAME	ST	CAR TYPE	BUILT	PLAN	LOT
NATIONAL MINE		BaggBSmkBarber	10-11/25	3951	4885	NEW WINCHESTER		14Sec	9-10/25	3958	4869
NATIONAL PARK		BaggBSmkBarber	10-11/25	3951	4885	NEW YAMA		14Sec	10/26	3958	4997
NATIONAL PRESS CLUB	N	8SecBLngSunRm	8-9/29	3989A	6274	NEW YORK		Private (GS)	1/14	2492A	4210
NATIONAL ROAD		BaggBSmkBarber	10-11/25	3951	4885	NEW YORK		25ChDrObs	12/12	2669A	4053
NATIONAL VIEW		3Cpt2DrLngObs	8-10/25	3959	4886	NEW YORK UNIVERSITY	RN	12Sec2Dbr	7-9/17	4046	4515
NATOMA		12SecDr	3-6/15	2410B	4319	NEWARK		26ChDr	5/12	2416	3986
NATRONA	N	24ChDrB	12/10	2417	3863	NEWBERG		12SecDr	2-3/27	3410A	6055
NATURAL BRIDGE		12SecDr	12/10-11	2410	3866	NEWBURYPORT		12SecDr	10-12/26	3410A	6023
NAUGHTON		12SecDr	10-12/16	2410F	4431	NEWCASTLE		16Sec	1-5/18	2412F	4531
NAUMKEAG	N	10SecDr2Cpt	6-7/15	2585B	4328	NEWFANE		12SecDr	7-8/15	2410D	4322
NEODAK		6CptLngObs	8/12/10	2413	3802	NEWFIELD		12SecDr	9/17/10	2410	3794
NEPESTA		10SecDr2Cpt	8-9/13	2585A	4201	NEWGARD		12SecDr	2-3/13	2410	4105
NEPPERHAN		12SecDr	7/11	2410	3913	NEWGATE		12SecDr	5-7/13	2410	4149
NEPTUNE		12SecDr	12/20-21	2410F	4611	NEWHALL		12SecDrCpt	6/13	2411A	4159
NEREUS		12SecDr	12/20-21	2410F	4611	NEWINGTON		12SecDr	9-10/11	2410	3924
NESCOPECK		12SecDr	5-6/23	2410H	4699	NEWKIRK		12SecDr	7-9/17	2410F	4515
NESHAMINY		12SecDr	11/8/10	2410	3860	NEWMARKET		12SecDr	12/12-13	2410	4076
NESHAMINY		12SecDr	7-9/17	2410F	4515	NEWPORT		Private (GS)	7/5/16	2502B	4422
NESHOBE		12SecDr	11-12/13	2410A	4234	NEWTON FALLS		10SecDr2Cpt	7-8/23	2585D	4707
NESHUGA		12SecDr	3-6/15	2410B	4319	NEWTON HOOK		12SecDrCpt	5/17	2411C	4494
NESTOR		12SecDr	9-10/20	2410F	4590	NEWVILLE		26ChDr	9-11/22	2416D	4649
NETHERBY		16Sec	7-8/11	2412	3912	NEZ PERCE		12SecDr	4/15	2410C	4309
NETHERHILL		12SecDr	11-12/13	2410A	4234	NIAGARA FALLS		10Cpt	3/14	2505A	4252
NEVADO de TOLUCA	N	8Sec5Dbr	10-12/15	4036I	4338	NIAGARA VALLEY		DrSbrBLngObs	4/29	3988	6221
NEW LINCOLN		14Sec	10-11/29	3958A	6285	NEVILLE		12SecDr	12/10-11	2410	3866
NEW LOGAN		14Sec	10-11/29	3958A	6285	NEVIS		12SecDr	5-8/18	2410F	4540
NEW LONDON		BaggBClubSmk	6/16	2919	4392	NEW ACTION		14Sec	9-10/25	3958	4869
NEW LYME		10SecDr2Cpt	7-8/23	2585D	4707	NEW ACTON	N	14Sec	9-10/25	3958	4869
NEW MANITOU		14Sec	10-11/29	3958A	6285	NEW ALBANY		14Sec	9-10/25	3958	4869
NEW MARSHFIELD		14Sec	9-10/25	3958	4869	NEW ALBIN		14Sec	9-10/25	3958	4869
NEW MARTINSVILLE		14Sec	9-10/25	3958	4869	NEW ARCADIE		14Sec	10/26	3958	4997
NEW MEADOWS		14Sec	10/26	3958	4997	NEW ATHENS		14Sec	9-10/25	3958	4869
NEW MIDWAY		14Sec	10/26	3958	4997	NEW AUBURN		14Sec	9-10/25	3958	4869
NEW MILLPORT		14Sec	9-10/25	3958	4869	NEW BLAINE		14Sec	9-10/25	3958	4869
NEW NICOLLET		14Sec	10-11/29	3958A	6285	NEW BOSTON		14Sec	9-10/25	3958	4869
NEW OMAHA		14Sec	10-11/29	3958A	6285	NEW BOULDER		14Sec	10-11/29	3958A	6285
NEW ORLEANS		14Sec	10/26	3958	4997	NEW BRITIAN		40Ch8StsDrBObs	SBS	1/13	2690	4055
NEW ORLEANS CLUB	RN	10SecRestLngObs	5-6/13	4027A	4150	NEW BROCKTON		14Sec	9-10/25	3958	4869
NEW PARIS		14Sec	9-10/25	3958	4869	NEW BUFFALO		14Sec	9-10/25	3958	4869
NEW PHALEN		14Sec	10-11/29	3958A	6285	NEW CANAAN		36Ch	8/27	3916A	6077
NEW POINT		14Sec	9-10/25	3958	4869	NEW CANTON		14Sec	9-10/25	3958	4869
NEW PORTAGE		10SecDr2Cpt	7-8/23	2585D	4707	NEW CAPITOL		14Sec	10-11/29	3958A	6285
NEW PRESTON		BaggBClubSmk	6/16	2919	4392	NEW CARLISLE		14Sec	9-10/25	3958	4869
NEW PROSPECT		14Sec	10/26	3958	4997	NEW CASTLE	N	14Sec	3/27	3958	6054
NEW RIEGEL		14Sec	9-10/25	3958	4869	NEW COLORADO		14Sec	10-11/29	3958A	6285
NEW RIVER GORGE	N	8Rmt3Dbr2Dr	9-10/25	4176	4899	NEW COLUMBIA		14Sec	10/26	3958	4997
NEW ROCHELLE		40Ch8StsdrBObs	SBS	1/13	2690	4055	NEW COMO		14Sec	10-11/29	3958A	6285
NEW SALEM		14Sec	9-10/25	3958	4869	NEW CONCORD		14Sec	9-10/25	3958	4869
NEW TRENTON		14Sec	9-10/25	3958	4869	NEW DALTON		14Sec	9-10/25	3958	4869
NEW WATERFORD		10SecDr2Cpt	7-8/23	2585D	4707	NEW DAYTON		14Sec	9-10/25	3958	4869

CAR NAME	ST	CAR TYPE	BUILT	PLAN	LOT
NEW DENVER		14Sec	9-10/25	3958	4869
NEW DOMINION		14Sec	10/26	3958	4997
NEW DOVER		14Sec	9-10/25	3958	4869
NEW DURHAM		12SecDrCpt	3/14	2411A	4269
NEW ENGLAND		BaggBSmkBarber	12/26-27	3951C	6022
NEW ERA		10Sec2Dr	5/13	2584	4140
NEW ERIN		14Sec	9-10/25	3958	4869
NEW FARNAM		14Sec	10-11/29	3958A	6285
NEW FRANKLIN		14Sec	9-10/25	3958	4869
NEW GENEVA		14Sec	9-10/25	3958	4869
NEW GROVE		14Sec	10/26	3958	4997
NEW HAMBURG		14Sec	9-10/25	3958	4869
NEW HANOVER		14Sec	9-10/25	3958	4869
NEW HARMONY		14Sec	10/26	3958	4997
NEW HAVEN		36Ch	8/27	3916A	6077
NEW HAVEN CLUB		8SecBLngSunRm	10/30	3989C	6393
NEW HENNEPIN		14Sec	10-11/29	3958A	6285
NEW HOLLAND		12SecDrCpt	6/21/10	2411	3800
NEW HUDSON		14Sec	9-10/25	3958	4869
NEW JASPER		14Sec	9-10/25	3958	4869
NEW JERSEY		32ChObs	8/18/10	2420	3805
NEW JERSEY	N	32ChObs	9-10/24	3420	4785
NEW LEBANON		14Sec	9-10/25	3958	4869
NEW LEXINGTON		14Sec	9-10/25	3958	4869
NIANGUA		12SecDr	4-5/13	2410	4106
NIANTIC BAY		32ChDr	12/29-30	3917B	6318
NICANDER		12SecDrCpt	10-12/21	2411C	4624
NICARAGUA NdeM	SN	12SecDr	1-2/16	2410I	4356
NICE		12SecDr	9-10/20	2410F	4590
NICIAS		12SecDr	11-12/13	2410A	4234
NICOBAR		12SecDr	9-10/20	2410F	4590
NICOLA		26ChDr	6/11	2416	3902
NICOLAS					
BRAVO NdeM	SN	12SecDr	2-4/24	3410	4762
URCELAY FCP-222	SN	10SecDr2Cpt	12/23-24	3585C	4728
NICOLET		12SecDr	9-10/20	2410F	4590
NIGHT ARCH		14Sbr	5/28	3980A	6158
NIGHT BAY		14Sbr	12/29	3980B	6304
NIGHT BEAM		14Sbr	12/29	3980D	6304
NIGHT CLOUD		14Sbr	12/29	3980D	6304
NIGHT COVE		14Sbr	12/29	3980B	6304
NIGHT FERN		14Sbr	12/29	3980B	6304
NIGHT FLYER		14Sbr	5/28	3980A	6158
NIGHT GLOW		14Sbr	12/29	3980D	6304
NIGHT HARBOR		14Sbr	5/28	3980A	6158
NIGHT HAVEN		14Sbr	12/29	3980B	6304
NIGHT LAND		14Sbr	5/28	3980A	6158
NIGHT LIGHT		14Sbr	5/28	3980A	6158
NIGHT LINE		14Sbr	5/28	3980A	6158
NIGHT MANTLE		14Sbr	5/28	3980A	6158
NIGHT RIDGE		14Sbr	12/29	3980B	6304
NIGHT ROUTE		14Sbr	12/29	3980D	6304
NIGHT SKY		14Sbr	5/28	3980A	6158
NIGHT TRAIL		14Sbr	12/29	3980B	6304
NIGHT VIEW		14Sbr	5/28	3980A	6158
NIGHT VISTA		14Sbr	5/28	3980A	6158
NIGHTFALL		14Sbr	3-4/27	3980	6056
NIGHTGLEN		14Sbr	3-4/27	3980	6056
NIGHTINGALE		14Sbr	3-4/27	3980	6056
NIGHTLEA		14Sbr	3-4/27	3980	6056
NIGHTMUSE		14Sbr	3-4/27	3980	6056
NIGHTRUN		14Sbr	3-4/27	3980	6056
NIGHTSIDE		14Sbr	3-4/27	3980	6056
NIGHTSTAR		14Sbr	3-4/27	3980	6056
NIGHTVALE		14Sbr	3-4/27	3980	6056
NIGHTWATCH		14Sbr	3-4/27	3980	6056
NIKLAC		12SecDr	2-3/15	2410B	4318
NIMROD		12SecDr	6-10/21	2410F	4614
NINA		28ChDr	4-6/25	3416	4864
NINESHKA		16Sec	12/14-15	2412C	4304
NIOBRARA		12SecDr	10-12/26	3410A	6023
NIPIGON		16Sec	5-6/13	2412B	4150
NIPISSING		12SecDr	12/20-21	2410F	4611
NISBET		26ChDr	4-5/17	2416C	4492
NISCUBA		16Sec	12/14-15	2412C	4304
NISSWA		12SecDr	4/15	2410C	4309
NIVERVILLE		12SecDr	9-10/11	2410	3924
NIXON		12SecDr	10-12/16	2410F	4431
NIZINA		16Sec	10-11/14	2412C	4296
NOANK		32ChDr	10/24	3917	4803
NOANK	R	24ChDr10StsLng	6/30	3917E	4803
NOANK(2nd)	N	12ChDrBLngSunRm	2-3/30	4002	6326
NOBSCOT	N	36Ch	6/16	2916	4389
NOBSCOTT		36Ch	6/16	2916	4389
NOCTURNE	RN	16DuSr	7/23	4029	4698
NOEL		28ChDr	5-7/26	3416	4958
NOKOMIS		12SecDr	3-6/20	2410F	4565
NOLANDO		12SecDr	8-10/26	3410A	4945
NOMAD	RN	22Ch12StsBLng	8/13	4032	4195
NOOKSACK		16Sec SBS	4/15	2412D	4310
NORA		26ChDr	3-4/14	2416A	4264
NORDEN		12SecDr	5-7/13	2410	4149
NORFOLK		12SecDr	7-8/28	3410B	6184
NORFOLK COUNTY	N	10SecLngObs	10-12/26	3521G	4998
NORLINA		10SecLngObs	2/24/11	2521	3865
NORLINA	N	10SecDr2Cpt	7-8/23	2585E	4707
NORMAN TOWER	RN	8SecDr3Dbr	12/26-27	4090D	6031
NORMAN W. KITTSON		LngBBarberSunRm	5-6/29	3990	6249
NORMOYLE		12SecDr	9-10/20	2410F	4590
NORQUAY		12SecDr	11-12/13	2410A	4234

CAR NAME	ST	CAR TYPE	BUILT	PLAN	LOT
NORRISTOWN		12SecDr	5/26/10	2410	3794
NORTH BEND		4Cpt2DrLngObs	3/13	2703	4100
NORTH BERNE		10SecDr2Cpt	7-8/23	2585D	4707
NORTH EASTON		32ChDr	8/27	3917A	6078
NORTH GARDEN		10SecLngObs	2/25/11	2521	3865
NORTH GATE		10SecLngObs	12/15	2521B	4362
NORTH GROVE		10SecDr2Cpt	7-8/23	2585D	4707
NORTH HAVEN		Bagg24Ch SBS	1/13	2692	4057
NORTH ILION		12SecDrCpt	5/17	2411C	4494
NORTH JUDSON		10SecDr2Cpt	7-8/23	2585D	4707
NORTH LAKE		14Sec	8-10/28	3958A	6181
NORTH ROSE		12SecDrCpt	3/11	2411	3880
NORTH STAR		12SecDr	3-6/15	2410B	4319
NORTH VERNON		12SecDrCpt	3/14	2411A	4269
NORTH VINELAND		10SecDr2Cpt	7-8/23	2585D	4707
NORTHAMPTON		36Ch SBS	1/13	2691	4056
NORTHCOTE		12SecDr	2-3/13	2410	4105
NORTHERN LIGHT		30ChDr	2-3/30	4000	6324
NORTHFIELD		16Sec	6-7/20	2412F	4568
NORTHFORD		24ChDrB	4/16	2417C	4364
NORTHLAND	N	12SecDr	2-3/13	2410	4105
NORTHUMBERLAND		12SecDr	5-6/23	2410H	4699
NORVEIL		12SecDr	10-11/11	2410	3936
NORWAY		12SecDr	9-10/20	2410F	4590
NORWICH		36Ch	6/16	2916	4389
NORWICH UNIVERSITY	N	12Sec2Dbr	2-5/25	4046B	4845
NORWOOD		12SecDr	9/9/10	2410	3794
NOTRE DAME		10Sec2Dr	1/12	2584	3947
NOVESTA		12SecDr	2-3/13	2410	4105
NOWATA		12SecDr	3-6/20	2410F	4565
NOYAN		12SecDr	7/11	2410	3913
NOYO		12SecDr	8-10/26	3410A	4945
NUGENT		12SecDr	3-4/17	2410F	4485
NUMA		12SecDr	5-8/18	2410F	4540
NUNDA		26ChDr	4-5/16	2416C	4366
NUREMBURG		12SecDr	2-3/13	2410	4105
NUTWOOD		12SecDr	1/14	2410A	4240
NYANDO		16Sec	7-8/11	2412	3912
OAK BAY	RN	12RmtSbr4Dbr	1-3/25	4172	4844
OAK CITY	RN	12RmtSbr4Dbr	2-5/25	4172	4845
OAK DOME	RN	12RmtSbr4Dbr	6-8/24	4172	4764
OAK GLEN	RN	12RmtSbr4Dbr	2-5/25	4172	4845
OAK GROVE	RN	12RmtSbr4Dbr	1-3/25	4172	4844
OAK HALL	RN	12RmtSbr4Dbr	2-5/25	4172	4845
OAK HILLS	RN	12RmtSbr4Dbr	2-5/25	4172	4845
OAK HOUSE	RN	12RmtSbr4Dbr	2-5/25	4172	4845
OAK LANE	RN	12RmtSbr4Dbr	2-5/25	4172	4845
OAK LAWN	RN	12RmtSbr4Dbr	2-5/25	4172	4845
OAK LEAF	RN	12RmtSbr4Dbr	2-5/25	4172	4845
OAK MILLS	RN	12RmtSbr4Dbr	1-3/25	4172	4844
OAK PARK		12SecDr	7-8/28	3410B	6184
OAK PASS	RN	12RmtSbr4Dbr	2-5/25	4172	4845
OAK POST	RN	12RmtSbr4Dbr	1-3/25	4172	4844
OAK RUN	RN	12RmtSbr4Dbr	2-5/25	4172	4845
OAK SHADE	RN	12RmtSbr4Dbr	2-5/25	4172	4845
OAK TREE	RN	12RmtSbr4Dbr	2-5/25	4172	4845
OAK VALE	RN	12RmtSbr4Dbr	5-8/25	4172	4868
OAK VIEW	RN	12RmtSbr4Dbr	2-5/25	4172	4845
OAK WOOD	RN	12RmtSbr4Dbr	2-5/25	4172	4845
OAKBOURNE		12SecDr	12/16/10	2410	3864
OAKDALE		12SecDr	7-8/28	3410B	6184
OAKENWALD		16Sec	12/14-15	2412C	4304
OAKFIELD		12SecDr	10/26/10	2410	3814
OAKFIELD	N	34ChDr	10/24	3419	4784
OAKHILL	N	12SecDr	5-7/13	2410	4149
OAKLAND		LngBDorm(5Sec)	7-8/27	3981	6061
OAKMONT		16Sec	10/13	2412B	4216
OAKNER		12SecDr	2-5/16	2410E	4367
OAKS		12SecDr	2-5/25	3410	4845
OAKVILLE		12SecDr	2-4/24	3410	4762
OBERLIN COLLEGE	RN	10Sec2DbrCpt	3/11	4042A	3880
OBERON		16Sec	10/13	2412B	4216
OCANA		12SecDr	10-12/15	2410D	4338
OCCIDENTAL		12SecDr	12/12-13	2410	4076
OCEAN BEACH	RN	42StsBLng	3/13	4023	4107
OCEAN CITY		24ChDrB	3/14	2417A	4250
OCEAN PARK		10Sec2Dr	2/13	2584	4092
OCEAN VIEW		10Sec2Dr	2/13	2584	4092
OCEANUS		12SecDr	9-10/20	2410F	4590
OCHELATA		10SecDr2Cpt	8-9/13	2585A	4201
OCMULGEE	P	12SecDr SBS	4/17	2410G	4454
OCONEE	P	12SecDr SBS	4/17	2410G	4454
OCONOMOWOC		6Cpt3Dr	5/27	3523B	6044
OCOSTA		12SecDr	4/15	2410C	4309
OCOTZINGO NdeM-221	SN	8SecDr2Cpt	7-8/30	3979A	6377
OCTAGON HOUSE		13Dbr	6-7/30	3997A	6314
OCTAGON TOWER	RN	8SecDr3Dbr	12/26-27	4090D	6031
OCTAVIA		28ChDr	6-7/27	3416A	6087
OCTORARO		6CptLngObs	8/10/10	2413	3802
ODENTON		6CptLngObs	8/8/10	2413	3802
ODIN		23ChDrSunRm	9/30	4002A	6385
OGDEN CANYON		3CptDrLngBSunRm	6/29	3975C	6262
OGEECHEE	P	12SecDr SBS	4/17	2410G	4454
OGONTZ		12SecDr	3-6/20	2410F	4565
OGRETTA		26ChDr	3-4/14	2416A	4264
OHIO VALLEY	RN	8SecDrBLngObs	12/18	4030	4489
OHLMAN		12SecDr	1-4/12	2410	3949
OIL CITY		12SecDr	11/18/10	2410	3864
OJOS TAPATIOS	N	10SecDrCpt	3-5/30	3973A	6338

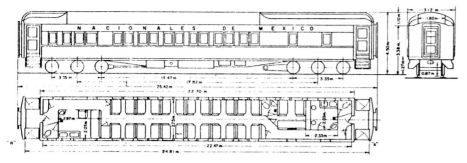

Figure I-63 The former ORANGE ROAD at age 45 was one of 34 Plan 3410 sleepers on the NdeM roster in 1970. Having been sold to the New York Central in 1948 and withdrawn from Pullman lease in October 1956, it was sold to NdeM in 1957 and leased back to Pullman in March 1958 for continued service under the name ISLA DE TIBURON on the Nationales De Mexico. Drawing from NdeM diagram, the Author's Collection.

CAR NAME	ST	CAR TYPE	BUILT	PLAN	LOT
OJOS					
TAPATIOS FCP-206	SN	10SecDr2Cpt	12/25-26	3585A	4933
OKABENA		8SecDr2Cpt	7-8/29	3979A	6272
OKANAGON		12SecDr	12/10-11	2410	3866
OKASAN		12SecDr	2-5/16	2410E	4367
OKAUCHEE		6Cpt3Dr	5/27	3523B	6044
OKOBOJI		8SecDr2Cpt	7-8/27	3979	6052
OLANCHA		10SecDr2Cpt	10-11/12	2585	4067
OLANTA		16Sec	7-8/11	2412	3912
OLCOTT		12SecDr	1-3/17	2410F	4450
OLD COLONY		BaggBSmkBarber	12/26-27	3951C	6022
OLD ELM CLUB	RN	8SecCptRestLngObs	4/17	4026	4484
OLD FERRY	N	24ChDrB	12/10	2417	3863
OLD FORGE		12SecDrCpt	5/17	2411C	4494
OLD IRONSIDES		BaggBSmkBarber	12/26-27	3951C	6022
OLD MANSE		13Dbr	6-7/30	3997A	6314
OLD WHITE	RN	8Rmt3Dbr2Dr	9-10/25	4176	4899
OLINDA		12SecDr	11-12/13	2410A	4234
OLIVER					
ELLSWORTH		30ChDr	2-3/30	4000	6324
HAZARD PERRY		12SecDr	4/26	3410A	4943
WENDELL HOLMES		14Sec	3/27	3958	6054
WOLCOTT		28ChDr	4-6/25	3416	4864
OLIVIA		28ChDr	6-7/30	3416B	6372
OLMSTEAD	N	12SecDr	3/17/10	2410	3794
OLYMPIC		16Sec	5-6/13	2412B	4150
OLYPHANT		16Sec	5-6/16	2412F	4386
OMAHA CLUB		8SecBLngSunRm	8-9/29	3989A	6274
OMAR		12SecDr	9-10/20	2410F	4590
OMENA		12SecDr	9-10/20	2410F	4590
OMESIA		12SecDr	10/14	2410B	4293
ONALASKA		8SecDr2Cpt	7-8/29	3979A	6272

CAR NAME	ST	CAR TYPE	BUILT	PLAN	LOT
ONALUSKA PASS		8SecDr2Cpt	2-3/30	3979A	6353
ONAMIA		16Sec	6-7/20	2412F	4568
ONASCO		12SecDr	3-6/20	2410F	4565
ONAWA		12SecDr	6-10/21	2410F	4614
ONCHIOTA		12SecDr	7/11	2410	3913
ONDAWA		12SecDr	3-6/15	2410B	4319
ONEIDA	N	16Sec	12/14-15	2412C	4304
ONEMHA		12SecDr	3-6/15	2410B	4319
ONEONTA		12SecDr	10-12/26	3410A	6023
ONINTA		16Sec	12/14-15	2412C	4304
ONMOOR		12SecDr	3-6/15	2410B	4319
ONONDAGA		16Sec	4/17	2412F	4484
ONSTED		12SecDr	4-6/14	2410A	4271
ONTARIO		12SecDr	3-4/11	2410	3881
ONTELAUNEE		12SecDr	7-9/17	2410F	4515
ONTONAGON		8SecDr2Cpt	7-8/29	3979A	6272
ONTWOOD		BaggBClubSmkBarber	2-3/12	2602	3948
ONWARD	RN	17StsRestLng	3-4/14	4019B	4264
OOSTANAULA	P	12SecDr	2/21	2410G	4598
OPHELIA		24ChDrB	5/17	2417D	4493
OPORTO		12SecDr	9-10/20	2410F	4590
ORADELL		12SecDr	11-12/13	2410A	4234
ORALIE		26ChDr	4-5/13	2416A	4136
ORAMEL		16Sec	10/13	2412B	4216
ORANGE BAY		12SecDr	8-12/25	3410A	4894
ORANGE BEND		12SecDr	8-12/25	3410A	4894
ORANGE BLOSSOM		12SecDr	8-12/25	3410A	4894
ORANGE CENTER		12SecDr	8-12/25	3410A	4894
ORANGE CITY		12SecDr	8-12/25	3410A	4894
ORANGE COUNTY		12SecDr	8-12/25	3410A	4894
ORANGE COVE		12SecDr	8-12/25	3410A	4894
ORANGE CREEK		12SecDr	8-12/25	3410A	4894
ORANGE DALE		12SecDr	8-12/25	3410A	4894
ORANGE FARM		12SecDr	8-12/25	3410A	4894
ORANGE GROVE		12SecDr	8-12/25	3410A	4894
ORANGE HEIGHTS		12SecDr	8-12/25	3410A	4894
ORANGE HILL		12SecDr	8-12/25	3410A	4894
ORANGE HOME		12SecDr	8-12/25	3410A	4894
ORANGE ISLAND		12SecDr	8-12/25	3410A	4894
ORANGE LAKE		12SecDr	8-12/25	3410A	4894
ORANGE MILLS		12SecDr	8-12/25	3410A	4894
ORANGE PARK		12SecDr	8-12/25	3410A	4894
ORANGE PORT		12SecDr	8-12/25	3410A	4894
ORANGE RIVER		12SecDr	8-12/25	3410A	4894
ORANGE ROAD		12SecDr	8-12/25	3410A	4894
ORANGE SPRINGS		12SecDr	8-12/25	3410A	4894
ORANGE STATE		12SecDr	8-12/25	3410A	4894
ORANGE TOWN		12SecDr	8-12/25	3410A	4894
ORANGE VALLEY		12SecDr	8-12/25	3410A	4894
ORANGE VILLE		12SecDr	8-12/25	3410A	4894

CAR NAME	ST	CAR TYPE	BUILT	PLAN	LOT
ORCHARD DALE		8Sec5Sbr	1/30	3993	6288
ORCHARD FARM		8Sec5Sbr	1/30	3993	6288
ORCHARD GARDENS		8Sec5Sbr	1/30	3993	6288
ORCHARD HILL		10Sec2Dr	5/13	2584	4140
ORCHARD LAKE		8Sec5Sbr	1/30	3993	6288
ORDWAY		12SecDr	6-10/21	2410F	4614
OREGON CLUB		8SecBLngSunRm	3/30	3989B	6349
OREVAL		12SecDr	6-7/17	2410F	4503
ORIANNA		16Sec	6-7/20	2412F	4568
ORIENTAL		12ChDrBLngSunRm	7/30	4002	6364
ORIENTAL(2nd)	N	12ChDrBLngSunRm	2-3/30	4002	6326
ORINOCO		12SecDr	6-7/17	2410F	4503
ORION		12SecDr	11-12/13	2410A	4234
ORIS		24ChDrB	5/13	2417A	4137
ORKNEY		12SecDr	1-3/17	2410F	4450
ORLANTHE		26ChDr	4/12	2416	3950
ORLEANS		12SecDr	10/14/10	2410	3814
ORLEANS		12SecDr	9-10/20	2410F	4590
ORMACINDA		12SecDr	9-10/15	2410D	4327
ORMSBEE		12SecDr	7-9/17	2410F	4515
ORMUS		12SecDr	7-8/15	2410D	4322
ORO BLANCO FCS-BC	L	10SecDr2Cpt	6-7/25	3585A	4888
ORONTES		12SecDr	12/10-11	2410	3866
ORPHA		26ChDr	3-4/14	2416A	4264
ORPHEUS		12SecDr	3-6/20	2410F	4565
ORRMOORE		12SecDr	2-3/15	2410B	4318
ORRTON		12SecDr	12/15/10	2410	3864
ORSTON		12SecDr	7-8/13	2410A	4192
ORTEGA		12SecDr	12/10-11	2410	3866
ORTLEY		26ChDr	4-5/17	2416C	4492
ORTOLAN		12SecDr	12/10-11	2410	3866
ORVISTON		12SecDr	7/11	2410	3913
OSAGE		12SecDr	6-7/17	2410F	4503
OSBORNE	N	12SecDr	4/19/10	2410	3770
OSCAR NEWTON	N	12SecDr	2-5/25	3410	4845
OSCAR UNDERWOOD	N	12SecDr	8-10/26	3410A	4945
OSCAWANA		12SecDr	6-10/21	2410F	4614
OSCEOLA		12SecDr	6-7/17	2410F	4503
OSGOOD		16Sec	5-6/16	2412F	4386
OSIRIS		12SecDr	3-6/20	2410F	4565
OSMOND		12SecDr	2-3/13	2410	4105
OSSINING		12SecDr	1-4/12	2410	3949
OSSIPEE		12SecDr	2-5/16	2410E	4367
OSTEND		12SecDr	2-5/25	3410	4845
OSTERBURG		26ChDr	5/23	2416D	4691
OSTRANDER		12SecDr	1-4/12	2410	3949
OSWALD		12SecDr	11-12/13	2410A	4234
OSWEGO VALLEY		DrSbrBLngObs	4/29	3988	6221
OTERO		LngBDorm(5Sec)	7-8/27	3981	6061
OTHELLO		12SecDr	6-7/17	2410F	4503

CAR NAME	ST	CAR TYPE	BUILT	PLAN	LOT
OTIS		12SecDr	5-8/18	2410F	4540
OTRANTO		12SecDr	6-7/17	2410F	4503
OTSEGO		12SecDr	10-12/26	3410A	6023
OTTAWA		12SecDr	9-10/11	2410	3924
OTTERBURN		16Sec	1/16	2412E	4352
OTTO CARMICHAEL	N	12SSecDr	6-9/26	3410A	4969
OTTUMWA		12SecDr	5-8/18	2410F	4540
OUTCALT		12SecDr	7-9/17	2410F	4515
OVERBERG		14Sec	2/28	3958A	6126
OVERBROOK		12SecDrCpt	3/11	2411	3880
OVERCOT		14Sec	2/28	3958A	6126
OVERDALE		14Sec	2/28	3958A	6126
OVERFIELD		BaggBClubSmkBarber	2-3/12	2602	3948
OVERHILLS		12SecDr	10-12/26	3410A	6023
OVERLAND		14Sec	2/28	3958A	6126
OVERLEA		14Sec	2/28	3958A	6126
OVERTOP		14Sec	2/28	3958A	6126
OVERVIEW		14Sec	2/28	3958A	6126
OVIEDO		12SecDr	3-6/20	2410F	4565
OWANECO		12SecDr	3-6/15	2410B	4319
OWASSA		12SecDr	2-3/15	2410B	4318
OWATONNA		12SecDr	3-6/15	2410B	4319
OWENDALE		12SecDr	9-10/11	2410	3924
OWLSHEAD		12SecDr	10/22-23	2410H	4647
OXFORD		12SecDr	5-8/18	2410F	4540
OXNARD		16Sec	4/17	2412F	4484
OXUS		12SecDr	9-10/20	2410F	4590
OZANA		12SecDr	7-8/15	2410D	4322
OZULUAMA NdeM-222	SN	12SecDr	2-5/25	3410	4845
P. G. T. BEAUREGARD		10Sec2Dr	9-10/25	3584A	4899
P. G. T. BEAUREGARD		10Sec2Dr	9-10/29	3584B	6284
P.G.T. BEAUREGARD	N	8SecDr2Cpt	8-9/29	3979A	6283
PACHAUG		12SecDr	7-8/14	2410A	4277
PACKARD		12SecDr	9-10/17	2410F	4525
PADDLEFORD		12SecDr	9-10/11	2410	3924
PADDOCK		12SecDr	9-10/17	2410F	4525
PADERNO		12SecDr	1-3/17	2410F	4450
PADRE KINO	N	10SecDrCpt	8/29	3973A	6273
PADRE KINO FCP-205	SN	10SecDr2Cpt	9-10/26	3585A	4996
PADUA		12SecDr	9-10/20	2410F	4590
PAGE		26ChDr	9-11/22	2416D	4649
PAGODA		12SecDr	7/16	2410F	4412
PAHASAHPA		8SecDr2Cpt	7-8/27	3979	6052
PAINE WINGATE		30ChDr	2-3/30	4000	6324
PAINESVILLE		8CptLng	4/11	2540	3879
PAINESVILLE	R	7CptBLng	4/11	2540F	3879
PALAMON		12SecDr	2-5/16	2410E	4367
PALANQUIN		12SecDr	2-3/13	2410	4105
PALATINE		12SecDr	9-10/11	2410	3924
PALFREY		12SecDrCpt	10-12/21	2411C	4624

CAR NAME	ST	CAR TYPE	BUILT	PLAN	LOT	CAR NAME	ST	CAR TYPE	BUILT	PLAN	LOT
PALISADE CANYON	N	3CptDrLngBSunRm	6/29	3975U	6262	PALOUSE FALLS		10SecDrCpt	5/30	3973A	6358
PALLADIUM		12SecDr	3-6/20	2410F	4565	PAMELIA		12SecDr	1-4/12	2410	3949
PALLANZA		16Sec	5-6/13	2412B	4150	PAMPA		7Dr	10/13	2583A	4209
PALLING(1st)		12SecDr	2-5/16	2410E	4367	PAMPANGA		12SecDr	6-7/17	2410F	4503
PALLING(2nd)	N	12SecDr	2-5/16	2410E	4367	PAN-AMERICAN		12SecDr	4/26	3410A	4943
PALM BEACH		Private (GS)	11/16	2502B	4422	PANAMA NdeM	SN	12SecDr	1-3/14	2410I	4249
PALM BEACH-BILTMORE	RN	Recreation	4-6/21	3966	4613	PANCOAST		12SecDr	7-8/13	2410A	4192
PALM CITY		3CptDrLngBSunRm	12/28-29	3975C	6217	PANDORA		16Sec	6-7/20	2412F	4568
PALM GROVE		3CptDrLngBSunRm	12/28-29	3975C	6217	PANGBORN		12SecDr	1-3/17	2410F	4450
PALM ISLANDS		3CptDrLngBSunRm	1/30	3975F	6337	PANOVA		12SecDr	3-6/20	2410F	4565
PALM KEY		3CptDrLngBSunRm	1/30	3975F	6337	PANSY		28ChDr	4-6/25	3416	4864
PALM LANE		3CptDrLngBSunRm	12/28-29	3975C	6217	PAOLI		26ChDr	5/23	2416D	4691
PALM POINT		3CptDrLngBSunRm	1/30	3975F	6337	PAQUITA		10SecDr2Cpt	8-9/13	2585A	4201
PALM SPRINGS		10SecLngObs	11/13	2521	4221	PARAMONT		16Sec	6-7/20	2412F	4568
PALM VALLEY		3CptDrLngBSunRm	1/30	3975F	6337	PARICUTIN	N	8Sec5Dbr	9-10/15	4036I	4327
PALM VILLA		3CptDrLngBSunRm	12/28-29	3975C	6217	PARICUTIN NdeM	N	10SecDrCpt	9-10/17	4031	4525
PALMA ALTA	RN	17StsRestLng	12/14-15	4018A	4304	PARIMA		12SecDr	12/10-11	2410	3866
PALMA REAL	N	17StsRestLng	7-8/11	4018A	3912	PARIS		12SecDr	9-10/20	2410	4590
PALMDALE		10SecDr2Cpt	10-11/12	2585	4067	PARIS GIBSON		8SecDr2Cpt	5-6/29	3979A	6261
PALMER		12SecDr	3-4/11	2410	3881	PARISIAN		12SecDr	12/12-13	2410	4076
PALMER LAKE		10SecLngObs	9-10/14	2521A	4292	PARK CITY		14Sec	3-4/30	3958A	6357
PALMER WOODS	N	8SbrObsLng	4/30	4005A	6340	PARK FALLS		14Sec	3-4/30	3958A	6357
PALMERTON		12SecDr	7/11	2410	3913	PARK FIELD		14Sec	7/30	3958A	6376
PALMYRA		BaggBClubSmk	9/8/10	2415	3834	PARK GATE		10Sec2Dr	1/12	2584	3947
PALO		12SecDr	9-10/20	2410F	4590	PARK GLADE		14Sec	3-4/30	3958A	6357
PALO ALTO		10Sec2Dr	2/13	2584	4092	PARK GROVE		14Sec	3-4/30	3958A	6357
PALOMA		16Sec	5-6/13	2412B	4150	PARK HEAD		14Sec	7/30	3958A	6376
PALOMAR		12SecDr	1-3/14	2410A	4249	PARK HEIGHTS		14Sec	3-4/30	3958A	6357

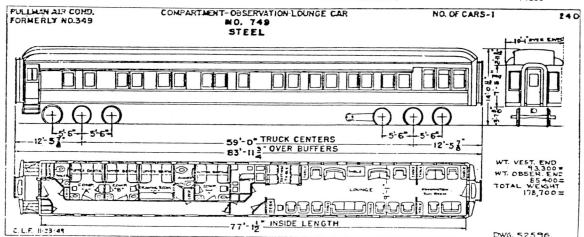

Figure I-64 PALM POINT was sold to the Northen Pacific in 1948 where it became the only standard sleeper lounge car on the railroad and acquired the number 349 in addition to its name. About 1953 the name was removed in favor of the NP No. 749 and it was finally retired in 1965. The Author's Collection.

CAR NAME	ST	CAR TYPE	BUILT	PLAN	LOT
PARK HILL		10Sec2Dr	1/12	2584	3947
PARK LANE		14Sec	3-4/30	3958A	6357
PARK MANOR		12SecDrCpt	8-9/11	2411	3922
PARK PLACE		14Sec	7/30	3958A	6376
PARK POINT		14Sec	7/30	3958A	6376
PARK RAPIDS		14Sec	3-4/30	3958A	6357
PARK RIDGE		14Sec	3-4/30	3958A	6357
PARK RIVER		14Sec	3-4/30	3958A	6357
PARK ROAD	N	14Sec	10-11/29	3958A	6285
PARK SHORE		14Sec	7/30	3958A	6376
PARK SLOPE		14Sec	3-4/30	3958A	6357
PARK SPRINGS		14Sec	3-4/30	3958A	6357
PARK SPUR		14Sec	3-4/30	3958A	6357
PARK SUMMIT	N	14Sec	10-11/29	3958A	6285
PARK TERRACE	N	14Sec	10-11/29	3958A	6285
PARK VALLEY		14Sec	3-4/30	3958A	6357
PARK VISTA	N	14Sec	10-11/29	3958A	6285
PARK WOOD		14Sec	7/30	3958A	6376
PARKDALE		12SecDr	9-10/17	2410F	4525
PARKER		12SecDr	10-12/16	2410F	4431
PARKERTOWN		12SecDr	10/17/10	2410	3814
PARKERTOWN		12SecDr	10-12/16	2410F	4431
PARKESBURG		6CptLngObs	8/11/10	2413	3802
PARKESBURG	N	12SecDr	8-12/25	3410A	4894
PARKINGTON		12SecDr	10-12/20	2410F	4591
PARKSIDE		12SecDr	12/10-11	2410	3866
PARKSLEY		12SecDr	7-8/13	2410A	4192
PARKTON		12SecDr	12/19/10	2410	3864
PARKVIEW		12SecDr	7-9/17	2410F	4515
PARKVILLE		12SecDr	6-9/26	3410A	4969
PARLIN		12SecDr	7/16	2410F	4412
PARMACHENE		12SecDr	11-12/13	2410A	4234
PARMENIO		12SecDr	6-7/17	2410F	4503
PARMENTER	N	12SecDr	9/28/10	2410	3813
PARNELIA		28ChDr	4/12	2418	3951
PARNOB		12SecDr	2-5/25	3410	4845
PAROWAN		12SecDr	7/16	2410F	4412
PARSONS		12SecDr	10/22-23	2410H	4647
PARTHENIA		12SecDr	3-6/15	2410B	4319
PARVIN		12SecDr	1-3/14	2410A	4249
PASADENA		10SecDr2Cpt	10-11/13	2585A	4218
PASCAGOULA	N	12SecDr	3-6/20	2410F	4565
PASHA		12SecDr	9-10/20	2410F	4590
PASSAIC	N	26ChDr	7/25/10	2416	3804
PASSUMPSIC		12SecDr	7-8/14	2410A	4277
PASTORES		12SecDr	1-3/14	2410A	4249
PATANOPA		12SecDr	7-8/14	2410A	4277
PATAPSCO		12SecDr	11/17/10	2410	3864
PATAULA	P	12SecDr	2/21	2410G	4598
PATCHEN		12SecDr	9-10/11	2410	3924

CAR NAME	ST	CAR TYPE	BUILT	PLAN	LOT
PATHFINDER		12SecDr	8-10/26	3410A	4945
PATONSEE		16Sec	12/14-15	2412C	4304
PATRICK HENRY		3Cpt2DrLngObs	8-10/25	3959	4886
PATRIOT		Private (GS)	9/17	2502C	4490
PATRIOT	RN	17StsRestLng	4-5/13	4019A	4136
PATTON		12SecDr	5/20/10	2410	3794
PATUXENT		12SecDr	7-9/17	2410F	4515
PAUL JONES		12SecDr	8-12/25	3410A	4894
PAUL REVERE		12SecDr	8-12/25	3410A	4894
PAULDING		16Sec	7-8/11	2412	3912
PAULSBORO		6CptLngObs	8/4/10	2413	3802
PAUTONGA		12SecDr	10/14	2410B	4293
PAVIA		12SecDr	9-10/20	2410F	4590
PAWLING		12SecDr	3-4/11	2410	3881
PAWNEE		10SecDr2Cpt	1-2/13	2585	4091
PAXICO		12SecDr	7-8/15	2410D	4322
PAYNE		6Cpt3Dr	11-12/24	3523A	4833
PEACE TOWER	RN	8SecDr3Dbr	5-6/25	4090	4846
PEACEDALE		12SecDr	11-12/13	2410A	4234
PEARSALL		16Sec	5-6/16	2412F	4386
PEARSON		12SecDr	10-12/16	2410F	4431
PECK		28ChDr	4-6/25	3416	4864
PECONIC		12SecDr	9-10/13	2410A	4215
PECOS		12SecDr	8-10/26	3410A	4945
PEDERNAL		10SecDr2Cpt	8-9/13	2585A	4201
PEDLEY		12SecDr	2-5/16	2410E	4367
PEDRO					
C. MORALES FCP-224	SN	10SecDr2Cpt	12/23-24	3585	4728
INFANTE FCP-230	SN	10SecDr2Cpt	10-11/13	2585E	4218
PEEKSKILL		16Sec	12/18	2412X	4489
PEERLESS		16Sec	12/14-15	2412C	4304
PEERMONT		12SecDr	11/5/10	2410	3860
PEGASUS		12SecDr	9-10/20	2410F	4590
PEGRAM		12SecDr	9-10/17	2410F	4525
PEKING		12SecDr	9-10/20	2410F	4590
PELADO		12SecDr	1-3/17	2410F	4450
PELEUS		12SecDr	9-10/20	2410F	4590
PELHAM		12SecDr	10-12/26	3410A	6023
PELHAM MANOR		32ChDr	8/27	3917A	6078
PELHAM OAK		8SecDr2Cpt	2-3/30	3979A	6353
PELION		12SecDr	2-5/16	2410E	4367
PELLSTON		12SecDr	2-5/25	3410	4845
PEMBERTON		26ChDr	5/12	2416	3986
PEMBINA		12SecDr	2-3/27	3410A	6055
PEMBRIDGE		12SecDr	12/12-13	2410	4076
PEMBROKE	N	12SecDr	8-10/26	3410A	4945
PEMBROKE	N	10Sec3Dbr	7/7/10	3411	3800
PEMISCOT		12SecDr	4-5/13	2410	4106
PENALOSA		10SecDr2Cpt	6-7/15	2585B	4328
PENANG		12SecDr	12/10-11	2410	3866

CAR NAME	ST	CAR TYPE	BUILT	PLAN	LOT
PENBRYN		12SecDrCpt	3/11	2411	3880
PENCADER		12SecDrCpt	7/16/10	2411	3800
PENDER		12SecDr	10-12/16	2410F	4431
PENDERGAST		12SecDr	3-4/17	2410F	4485
PENELOPE		28ChDr	6-7/27	3416A	6087
PENETANG		12SecDr	2-5/16	2410E	4367
PENLYN		28ChDr	4/12	2418	3951
PENN SQUARE		6Cpt3Dr	7/25	3523A	4887
PENN YAN		10Sec2Dr	1/12	2584	3947
PENNELL		12SecDr	7/16	2410F	4412
PENNHURST		24ChDrB	1/14	2417A	4239
PENNSBORO		12SecDr	7-8/12	2410	4014
PENNSVILLE		12SecDr	7-9/17	2410F	4515
PENNSYLVANIA		32ChObs	8/19/10	2420	3805
PENNSYLVANIA	N	28ChObs	8/24/10	2421	3806
PENOBSCOT		12SecDr	9-10/20	2410F	4590
PENOLA		12SecDr	12/10-11	2410	3866
PENROD		12SecDr	11-12/13	2410A	4234
PENROSE		12SecDr	9-10/17	2410F	4525
PENSADOR					
MEXICANO NdeM-224	N	14Sec	9-10/25	3958	4869
PENSAUKEN		6CptLngObs	8/2/10	2413	3802
PENTHEUS		12SecDr	3-6/20	2410F	4565
PENTON		12SecDr	2-5/25	3410	4845
PENVIR		16Sec	4/17	2412F	4484
PEONY		28ChDr	5-7/26	3416	4958
PEORIA		12SecDrCpt	6/13	2411	4151
PEORIA	N	10SecLngObs	11/12	2521	4049
PEORIA	R	10Sec12ChObs	11/12	2521J	4049
PEOTONE		12SecDr	1-4/12	2410	3949
PEPIN		12SecDr	11-12/29	3410B	6299
PEPPER PIKE					
COUNTRY CLUB	RN	8SecCptRestLngObs	4/17	4026A	4484
PEQUAKET		12SecDr	7-8/14	2410A	4277
PEQUEST		26ChDrObs	12/12	2669	4053
PEQUOT		34Ch	8/13	2764A	4195
PERALTA		12SecDr	11-12/13	2410A	4234
PERCIS		26ChDr	3-4/14	2416A	4264
PERDIX		12SecDrCpt	6/20/10	2411	3800
PERICLES		16Sec	6-7/20	2412F	4568
PERKINS		12SecDr	9-10/17	2410F	4525
PERKIOMEN		BaggBClubSmkBarber	8/15/10	2414A	3803
PERLETTA		26ChDr	6/11	2416	3902
PERROT		14Sec	6/28	3958A	6174
PERRY		12SecDr	3-4/17	2410F	4485
PERRYMAN		12SecDr	7-9/17	2410F	4515
PERRYVILLE		6CptLngObs	8/13/10	2413	3802
PERSEUS		12SecDr	9-10/20	2410F	4590
PERSHING SQUARE		6Cpt3Dr	10-12/29	3523C	6290
PERSICO		12SecDr	7-8/15	2410D	4322
PESHTIGO		16Sec	6-7/20	2412F	4568
PETALUMA		12SecDr	8-10/26	3410A	4945
PETER COOPER		12SecDr	8-12/25	3410A	4894
PETERSBURG		12SecDr	11-12/13	2410A	4234
PETRONA		12SecDr	12/10-11	2410	3866
PETTISVILLE		12SecDr	3-4/17	2410F	4485
PEVERIL		12SecDr	9-10/20	2410F	4590
PEWAUKEE		12SecDr	4-5/27	3410B	6042
PEYRAUD		12SecDr	9-10/17	2410F	4525
PEYTON RANDOLPH		12SecDr	8-12/25	3410A	4894
PHAEDRA		26ChDr	4/12	2416	3950
PHAETON		12SecDr	2-3/15	2410B	4318
PHALANX		12SecDr	9-10/11	2410	3924
PHALERON		12SecDr	2-5/16	2410E	4367
PHANTON		12SecDr	3-6/15	2410B	4319
PHANTON		12SecDr	9-10/20	2410F	4590
PHARSALIA		16Sec	1/16	2412E	4352
PHELPS		12SecDr	3-4/11	2410	3881
PHILADELPHIA		26ChDr	5/12	2416	3986
PHILADELPHIA		Private (GS)	1/14	2502A	4211
PHILEMON		26ChDr	3/13	2416	4107
PHILINDA		26ChDr	3-4/14	2416A	4264
PHILIPPA		26ChDr	6/11	2416	3902
PHILIPPINES		12SecDr	9-10/20	2410F	4590
PHILIPSBURG		12SecDr	7-9/17	2410F	4515
PHILIPSE MANOR		13Dbr	6-7/30	3997A	6314
PHILLIP SCHUYLER		30ChDr	2-3/30	4000	6324
PHILLIPSTON		12SecDr	1-4/12	2410	3949
PHILMONT		12SecDr	3-4/11	2410	3881
PHLOX		28ChDr	5-7/26	3416	4958
PHOENIX		LngBDorm(5Sec)	7-8/27	3981	6061
PHYLLIS		24ChDrB	4/13	2417	4108
PHYLLIS	R	26StsLngBDr	4/13	2417J	4108
PHYLLIS	R	16Ch10StsCafeBDr	4/13	2417H	4108
PICACHO		12SecDr	9-10/15	2410D	4327
PICKERING	N	12SecDr	10/3/10	2410	3813
PICKERT		12SecDr		2410F	4590
PIEDMONT COLLEGE	RN	10Sec2DbrCpt	12/16-17	4042	4443
PIERCE		12SecDr	10/22-23	2410H	4647
PIERCEFIELD		12SecDr	9-10/11	2410	3924
PIERETTE		26ChDr	3-4/14	2416A	4264
PIERRE	N	10SecDrCpt	8/28	3973A	6185
PIERRE CHOUTEAU		14Sec	6/30	3958A	6373
PIERRE CHOUTEAU	R	6Sec6Dbr	6/30	4084B	6373
PIERRE L'ENFANT	N	12SecDr	2-5/25	3410	4845
PIERRE L'ENFANT	R	10SecDr2Dbr	2-5/25	4074E	4845
PIERROT		12SecDr	7-8/15	2410D	4322
PIKE		12SecDr	2-5/25	3410	4845
PIKE'S PEAK		10Sec2Dr	12/26-27	3584A	6031
PIKEVILLE		12SecDr	2-5/25	3410	4845

CAR NAME	ST	CAR TYPE	BUILT	PLAN	LOT
PILGRIM		Private (GS)	9/17	2502C	4490
PILGRIM		12ChDrBLngSunRm	2-3/30	4002	6326
PILGRIM	N	30ChBLibObs	4/16	2901A	4365
PILGRIM	N	12ChDrBLngSunRm	7/30	4002	6364
PILGRIM HALL		4CptLngObs	10/25	3960	4889
PILOT	RN	17StsRestLng	4-5/13	4019A	4136
PILOT PEAK		10Sec2Dr	12/26-27	3584A	6031
PIMA		7Dr (SBS)	12/11	2583	3940
PIMLICO		12SecDr	9-10/20	2410F	4590
PINCONNING		12SecDr	7/11	2410	3913
PINDAR		12SecDr	12/10-11	2410	3866
PINDUS		12SecDr	12/10-11	2410	3866
PINE GLEN		12SecDrCpt	8-9/11	2411	3922
PINE GROVE		10SecLngObs	12/15	2521B	4362
PINEWOLD		16Sec	9-10/12	2412	4037
PINEWOLD	R	12Sec4EnclSec	9-10/12	2412J	4037
PINNACLE PEAK		10Sec2Dr	12/26-27	3584A	6031
PINNACLE TOWER	RN	8SecDr3Dbr	12/26-27	4090D	6031
PINON PINE		10SecDrCpt	5/30	3973A	6358
PINTADO		12SecDr	1-3/14	2410A	4249
PINTARD	N	12SecDr	5/27/10	2410	3794
PINTO		7Dr	11/17	2583B	4516
PINTO PEAK		10Sec2Dr	12/28-29	3584B	6213
PIONEER		24ChDrB	12/10	2417	3863
PIONEER(1st)		Private (GS)	9/17	2502C	4490
PIONEER(2nd)		Private (GS)	8/27	3972	6037
PIPESTONE		12SecDr	10-12/26	3410A	6023
PIPILA FCP-212	SN	10SecDr2Cpt	8-10/23	3585C	4725
PIQUA		12SecDr	10/31/10	2410	3860
PIQUA	N	12SecDr	5-8/25	3410A	4894
PIRANO		12SecDr	12/10-11	2410	3866
PIROL		12SecDr	10-12/15	2410D	4338
PISA		12SecDr	9-10/20	2410F	4590
PITCAIRN		6CptLngObs	8/10/10	2413	3802
PITMAN		12SecDr	2-5/16	2410E	4367
PITTSBURGH		12SecDr	11/19/10	2410	3860
PITTSBURGH CLUB	RN	8SecRestLngObs	2/25/11	4025B	3865
PITTSFORD		12SecDr	7/11	2410	3913
PITTSTON		12SecDr	7-8/12	2410	4014
PIUTE		7Dr (SBS)	12/11	2583	3940
PIXLEY		12SecDr	9-10/17	2410F	4525
PIZARRO		10SecDr2Cpt	1-2/13	2585	4091
PLACER		16Sec	4/17	2412F	4484
PLACERVILLE		12SecDr	12/27-28	3410B	6127
PLAINFIELD		12SecDr	6-7/17	2410F	4503
PLAINS		12SecDr	7-8/15	2410D	4322
PLAINSBORO		24ChDrB	1/14	2417A	4239
PLAINVILLE		12SecDr	10/20/10	2410	3814
PLAN DE					
BARRANCAS FCP-227	SN	10SecDr2Cpt	11-12/11	2585	3942

CAR NAME	ST	CAR TYPE	BUILT	PLAN	LOT
PLANADA		10SecDr2Cpt	8-9/13	2585A	4201
PLANDOME		12SecDr	5-7/13	2410	4149
PLANET		12SecDr	2-5/16	2410E	4367
PLANTER		12SecDr	1-2/15	2410B	4311
PLATANUS		16Sec	9-10/12	2412	4037
PLATEAU		12SecDr	9-10/17	2410F	4525
PLAZA		12SecDr	9-10/17	2410F	4525
PLEASANT HILL		10SecLngObs	9-10/14	2521A	4292
PLEASANTVILLE		12SecDr	5-6/23	2410H	4699
PLEVNA		12SecDr	10-12/15	2410D	4338
PLUMADORE		12SecDr	1-4/12	2410	3949
PLUMAS		12SecDr	8-10/26	3410A	4945
PLUTARCH		12SecDrCpt	10-12/21	2411C	4624
PLYMOUTH		12SecDr	11/9/10	2410	3860
PLYMOUTH		32ChDr	10/24	3917	4803
PLYMOUTH	R	24ChDr10StsLng	6/30	3917E	4803
PLYMOUTH LIGHT		36Ch	12/29-30	3916B	6319
PLYMOUTH ROCK	N	30ChBLibObs	4/16	2901A	4365
PLYMOUTH ROCK		12ChDrBLngSunRm	2-3/30	4002	6326
PLYMOUTH(2nd)	N	24ChDr10StsLng	6/30	3917E	4803
PLYMPTON		36Ch SBS	1/13	2691	4056
POCAHONTAS		12SecDr	8-12/25	3410A	4894
POCATELLO		12SecDr	4/30	3410B	6360
POCOMOKE		12SecDr	7-8/13	2410A	4192
POCOPSON		12SecDr	7-8/13	2410A	4192
POHICK CHURCH	N	12SecDr	1-3/25	3410	4844
POINCAIRE	N	12SecDr	6-10/21	2410F	4614
POINCARE		12SecDr	6-10/21	2410F	4614
POINT ABBAGE		10Sec2Dr	5-6/25	3584	4846
POINT ADAMS		10Sec2Dr	5-6/25	3584	4846
POINT AIRY		10Sec2Dr	12/25-26	3584A	4932
POINT ALDEN		10Sec2Dr	12/25-26	3584A	4932
POINT ALEXANDER		10Sec2Dr	12/24-25	3584	4836
POINT ALLERTON		10Sec2Dr	4-5/26	3584A	4961
POINT AMELIA		10Sec2Dr	5-6/25	3584	4846
POINT ANGELES		10Sec2Dr	12/25-26	3584A	4932
POINT ARENA		10Sec2Dr	5-6/25	3584	4846
POINT ARGUELLO		10Sec2Dr	12/24-25	3584	4836
POINT ARROWSMITH		10Sec2Dr	4-5/26	3584A	4961
POINT BANK		10Sec2Dr	5-6/25	3584	4846
POINT BARROW		10Sec2Dr	12/24-25	3584	4836
POINT BAY		10Sec2Dr	5-6/25	3584	4846
POINT BEDE	N	10Sec2Dr	9-10/25	3584A	4899
POINT BELCHER		10Sec2Dr	5-6/25	3584	4846
POINT BELL		10Sec2Dr	12/24-25	3584	4836
POINT BENNETT		10Sec2Dr	5-6/25	3584	4846
POINT BLACK		10Sec2Dr	5-6/25	3584	4846
POINT BLANCO		10Sec2Dr	12/25-26	3584A	4932
POINT BLUFF		10Sec2Dr	12/24-25	3584	4836
POINT BONITA		10Sec2Dr	12/24-25	3584	4836

CAR NAME	ST	CAR TYPE	BUILT	PLAN	LOT	CAR NAME	ST	CAR TYPE	BUILT	PLAN	LOT
POINT BREAKER		10Sec2Dr	4-5/26	3584A	4961	POINT LOOKOUT		10Sec2Dr	1/25	3584	4843
POINT BROWN		10Sec2Dr	5-6/25	3584	4846	POINT LUZON		10Sec2Dr	12/25-26	3584A	4932
POINT BUCHON		10Sec2Dr	5-6/25	3584	4846	POINT MALCOLM	N	10Sec2Dr	9-10/25	3584A	4899
POINT CARMEL		10Sec2Dr	12/24-25	3584	4836	POINT MANBY		10Sec2Dr	1/25	3584	4843
POINT CASE		10Sec2Dr	4-5/26	3584A	4961	POINT MARION		10Sec2Dr	12/25-26	3584A	4932
POINT CASTILLA		10Sec2Dr	4-5/26	3584A	4961	POINT MARSH		10Sec2Dr	5-6/25	3584	4846
POINT CASWELL		10Sec2Dr	4-5/26	3584A	4961	POINT MURDOCH		10Sec2Dr	1/25	3584	4843
POINT CLAIRE		10Sec2Dr	12/24-25	3584	4836	POINT NIPIGON	N	10Sec2Dr	9-10/25	3584A	4899
POINT CLARK		10Sec2Dr	5-6/25	3584	4846	POINT ORIENT		10Sec2Dr	4-5/26	3584A	4961
POINT COLLIE		10Sec2Dr	5-6/25	3584	4846	POINT PALOWAY	N	10Sec2Dr	9-10/25	3584A	4899
POINT COMFORT		10Sec2Dr	1/25	3584	4843	POINT PEDRO		10Sec2Dr	4-5/26	3584A	4961
POINT CONE		10Sec2Dr	5-6/25	3584	4846	POINT PELEE		10Sec2Dr	1/25	3584	4843
POINT COOPER		10Sec2Dr	12/24-25	3584	4836	POINT PHILLIP	N	10Sec2Dr	9-10/25	3584A	4899
POINT CULVER		10Sec2Dr	12/24-25	3584	4836	POINT PLEASANT		10Sec2Dr	1/25	3584	4843
POINT CUTHBERT		10Sec2Dr	4-5/26	3584A	4961	POINT POLO		10Sec2Dr	5-6/25	3584	4846
POINT DALE		10Sec2Dr	12/24-25	3584	4836	POINT PRIME		10Sec2Dr	4-5/26	3584A	4961
POINT DANGER		10Sec2Dr	5-6/25	3584	4846	POINT RENO		10Sec2Dr	12/25-26	3584A	4932
POINT DELGADA		10Sec2Dr	12/24-25	3584	4836	POINT REYES		10Sec2Dr	12/24-25	3584	4836
POINT DETOUR		10Sec2Dr	5-6/25	3584	4846	POINT RICARDO		10Sec2Dr	1/25	3584	4843
POINT DOUGLAS		10Sec2Dr	1/25	3584	4843	POINT RICH		10Sec2Dr	4-5/26	3584A	4961
POINT DOVER		10Sec2Dr	12/24-25	3584	4836	POINT RICHARDS		10Sec2Dr	4-5/26	3584A	4961
POINT DUME		10Sec2Dr	12/24-25	3584	4836	POINT RICHMOND		10Sec2Dr	4-5/26	3584A	4961
POINT EASTERN		10Sec2Dr	1/25	3584	4843	POINT ROBERTS		10Sec2Dr	5-6/25	3584	4846
POINT EDWARD		10Sec2Dr	4-5/26	3584A	4961	POINT ROCK		10Sec2Dr	12/24-25	3584	4836
POINT EGMONT		10Sec2Dr	12/25-26	3584A	4932	POINT ROMAINE		10Sec2Dr	1/25	3584	4843
POINT ELLIS		10Sec2Dr	4-5/26	3584A	4961	POINT RONALD		10Sec2Dr	1/25	3584	4843
POINT EMMONS		10Sec2Dr	12/25-26	3584A	4932	POINT SAL		10Sec2Dr	5-6/25	3584	4846
POINT ESPADA		10Sec2Dr	12/25-26	3584A	4932	POINT SHIRLEY		10Sec2Dr	1/25	3584	4843
POINT EUGENIA		10Sec2Dr	4-5/26	3584A	4961	POINT SPENCER		10Sec2Dr	1/25	3584	4843
POINT FAYETTE		10Sec2Dr	12/25-26	3584A	4932	POINT ST. JOSEPH	N	10Sec2Dr	9-10/25	3584A	4899
POINT FERMIN		10Sec2Dr	5-6/25	3584	4846	POINT SUR		10Sec2Dr	5-6/25	3584	4846
POINT FINAL		10Sec2Dr	5-6/25	3584	4846	POINT TABLE		10Sec2Dr	5-6/25	3584	4846
POINT FORTUNE		10Sec2Dr	12/25-26	3584A	4932	POINT TARRANT		10Sec2Dr	4-5/26	3584A	4961
POINT FRANKLIN		10Sec2Dr	12/25-26	3584A	4932	POINT TERRACE		10Sec2Dr	12/25-26	3584A	4932
POINT GABRIEL		10Sec2Dr	12/25-26	3584A	4932	POINT TRINIDAD		10Sec2Dr	12/25-26	3584A	4932
POINT GAMMON		10Sec2Dr	5-6/25	3584	4846	POINT TRUTH		10Sec2Dr	5-6/25	3584	4846
POINT GORDA		10Sec2Dr	12/25-26	3584A	4932	POINT TUPPER		10Sec2Dr	4-5/26	3584A	4961
POINT GORE		10Sec2Dr	5-6/25	3584	4846	POINT VICTOR		10Sec2Dr	1/25	3584	4843
POINT GRENVILLE		10Sec2Dr	12/25-26	3584A	4932	POINT VIEW		10Sec2Dr	12/25-26	3584A	4932
POINT HANSON		10Sec2Dr	5-6/25	3584	4846	POINT VINCENTE		10Sec2Dr	12/24-25	3584	4836
POINT HARBOR	N	10Sec2Dr	9-10/25	3584A	4899	POINT VIVIAN		10Sec2Dr	1/25	3584	4843
POINT HOPE		10Sec2Dr	12/24-25	3584	4836	POINT WASHINGTON	N	10Sec2Dr	9-10/25	3584A	4899
POINT ISABEL		10Sec2Dr	1/25	3584	4843	POINT WILLOUGHBY		10Sec2Dr	4-5/26	3584A	4961
POINT JUDITH		10Sec2Dr	1/25	3584	4843	POINT WILSON		10Sec2Dr	5-6/25	3584	4846
POINT LANG		10Sec2Dr	5-6/25	3584	4846	POINT YORKE		10Sec2Dr	1/25	3584	4843
POINT LAY		10Sec2Dr	1/25	3584	4843	POITIERS		12SecDr	9-10/20	2410F	4590
POINT LEVIS	N	10Sec2Dr	9-10/25	3584A	4899	POKAGON		16Sec	7-8/11	2412	3912
POINT LOBOS		10Sec2Dr	12/24-25	3584	4836	POLAND		12SecDr	3-4/11	2410	3881
POINT LOMA		10Sec2Dr	12/24-25	3584	4836	POLARIS		7Cpt2Dr	12/11	2522	3943
POINT LONSDALE		10Sec2Dr	1/25	3584	4843	POLE STAR		12SecDr	10/22-23	2410H	4647

CAR NAME	ST	CAR TYPE	BUILT	PLAN	LOT	CAR NAME	ST	CAR TYPE	BUILT	PLAN	LOT
POLLUX		12SecDr	9-10/20	2410F	4590	POPLAR CLIFF	RN	6Sec6Dbr	10-12/21	4060	4624
POLSON		12SecDr	9-10/17	2410F	4525	POPLAR CORNERS	RN	6Sec6Dbr	12/12-13	4084C	4076
POMEROY		28ChDr	11/25/10	2418	3862	POPLAR COUNTRY	RN	6Sec6Dbr	5-6/23	4084	4699
POMFRET		12SecDr	1-2/15	2410B	4311	POPLAR COURT	RN	6Sec6Dbr	10-12/21	4060	4624
POMONA		24ChDrB	12/10	2417	3863	POPLAR COVE	RN	6Sec6Dbr	10-12/21	4060	4624
POMPTON		12SecDr	2-3/15	2410B	4318	POPLAR CREEK	RN	6Sec6Dbr	5/17	4060	4494
PONCA		7Dr	10/13	2583A	4209	POPLAR CREST	RN	6Sec6Dbr	5/17	4060	4494
PONCE DE LEON		3Cpt2DrLngObs	11-12/26	3959	6015	POPLAR CROSSING	RN	6Sec6Dbr	4-5/13	4084C	4106
PONCELOT		12SecDr	2-3/15	2410B	4318	POPLAR DALE	RN	6Sec6Dbr	10-12/21	4060	4624
PONCHATOULA		12SecDr	4-5/13	2410	4106	POPLAR DELL	RN	6Sec6Dbr	4-5/13	4084C	4106
POND LILY		34Ch	4/26	3419	4956	POPLAR DOME	RN	6Sec6Dbr	7/16	4084	4412
PONDERAY		12SecDr	2-3/13	2410	4105	POPLAR FALLS	RN	6Sec6Dbr	10-12/21	4060	4624
PONTIAC		12SecDr	10-12/26	3410A	6023	POPLAR FARM	RN	6Sec6Dbr	5/17	4060	4494
PONTOTOC		12SecDr	7-8/15	2410D	4322	POPLAR FLAT	RN	6Sec6Dbr	1-3/17	4084	4450
POOR RICHARD CLUB		13ChRestLng	6/30	3992A	6361	POPLAR FOREST	RN	6Sec6Dbr	10-12/21	4060	4624
POPE PIUS XI	LN	Private (Cardinals)	3/11	2503	3847	POPLAR GAP	RN	6Sec6Dbr	3-4/17	4084	4485
POPLAR ACRES	RN	6Sec6Dbr	8-9/20	4084	4574	POPLAR GARDENS	RN	6Sec6Dbr	10-12/16	4084	4431
POPLAR ARBOR	RN	6Sec6Dbr	5-6/23	4084	4699	POPLAR GLADE	RN	6Sec6Dbr	10-12/16	4084	4431
POPLAR ARCH	RN	6Sec6Dbr	12/10-11	4084C	3866	POPLAR GLEN	RN	6Sec6Dbr	10-12/21	4060	4624
POPLAR BAY	RN	6Sec6Dbr	10-12/21	4060	4624	POPLAR GORGE	RN	6Sec6Dbr	5-6/23	4084	4699
POPLAR BLUFF	RN	6Sec6Dbr	10-12/21	4060	4624	POPLAR GROVE	RN	6Sec6Dbr	10-12/21	4060	4624
POPLAR BOROUGH	RN	6Sec6Dbr	5-6/11	4084C	3903	POPLAR HAVEN	RN	6Sec6Dbr	10-12/21	4060	4624
POPLAR BRANCH	RN	6Sec6Dbr	2-5/21	4084	4612	POPLAR HEIGHTS	RN	6Sec6Dbr	10-12/21	4060	4624
POPLAR BROOK	RN	6Sec6Dbr	10-12/21	4060	4624	POPLAR HIGHLANDS	RN	6Sec6Dbr	10-12/21	4060	4624
POPLAR CAMP	RN	6Sec6Dbr	9-10/11	4084C	3924	POPLAR HILL	RN	6Sec6Dbr	10-12/21	4060	4624
POPLAR CASTLE	RN	6Sec6Dbr	8/23	4060	4697	POPLAR HOLLOW	RN	6Sec6Dbr	3-4/17	4084	4485
POPLAR CENTER	RN	6Sec6Dbr	12/12-13	4084C	4076	POPLAR ISLE	RN	6Sec6Dbr	5-6/23	4084	4699
POPLAR CITY	RN	6Sec6Dbr	10-12/21	4060	4624	POPLAR JUNCTION	RN	6Sec6Dbr	6-8/24	4084A	4764

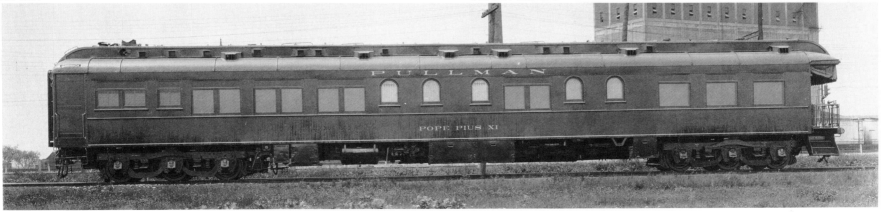

Figure I-65 Pullman private car SUPERB has a long and interesting history. In November 1925 it was renamed LOS ANGELES; in June 1926 it was painted bright red and carried the tail sign on "The Cardinals' Train" as POPE PIUS XI. In July 1922, it returned to SUPERB. In March 1944 it became a supply and porter car at Camp Kilmer, New Jersey; in June 1945 it was sold to the Charleston & Western Carolina RR as 101; in 1958 to ACL 301; in 1967 to SCL 301; and in January 1969 it was retired and placed with the Atlanta Chapter of the NRHS. Smithsonian Institution Photograph P-30180.

CAR NAME	ST	CAR TYPE	BUILT	PLAN	LOT
POPLAR KNOLL	RN	6Sec6Dbr	10-12/21	4060	4624
POPLAR LAKE	RN	6Sec6Dbr	5/17	4060	4494
POPLAR LANE	RN	6Sec6Dbr	10-12/23	4084A	4724
POPLAR LEAF	RN	6Sec6Dbr	5/17	4060	4494
POPLAR LODGE	RN	6Sec6Dbr	10-12/21	4060	4624
POPLAR MANOR	RN	6Sec6Dbr	2-5/25	4084A	4845
POPLAR MEADOW	RN	6Sec6Dbr	10-12/21	4060	4624
POPLAR MINES	RN	6Sec6Dbr	1-3/17	4084	4450
POPLAR PARK	RN	6Sec6Dbr	10-12/21	4060	4624
POPLAR PATH	RN	6Sec6Dbr	1-3/17	4084	4450
POPLAR PIKE	RN	6Sec6Dbr	8/23	4060	4697
POPLAR PLACE	RN	6Sec6Dbr	9-10/17	4084	4525
POPLAR PLAINS	RN	6Sec6Dbr	10-12/21	4060	4624
POPLAR POINT	RN	6Sec6Dbr	8/23	4060	4697
POPLAR PORT	RN	6Sec6Dbr	6-10/21	4084	4614
POPLAR RANGE	RN	6Sec6Dbr	5-6/23	4084	4699
POPLAR REALM	RN	6Sec6Dbr	5-6/23	4084	4699
POPLAR REST	RN	6Sec6Dbr	10-12/21	4060	4624
POPLAR RIDGE	RN	6Sec6Dbr	5/17	4060	4494
POPLAR ROAD	RN	6Sec6Dbr	2-5/25	4084A	4845
POPLAR RUN	RN	6Sec6Dbr	2-5/25	4084A	4845
POPLAR SHORE	RN	6Sec6Dbr	6-10/21	4084	4614
POPLAR SLOPE	RN	6Sec6Dbr	2-5/16	4084	4367
POPLAR SPRINGS	RN	6Sec6Dbr	2-5/25	4084A	4845
POPLAR SQUARE	RN	6Sec6Dbr	7-9/17	4084	4515
POPLAR STREET	RN	6Sec6Dbr	2-5/25	4084A	4845
POPLAR SUMMIT	RN	6Sec6Dbr	2-5/25	4084A	4845
POPLAR TERRACE	RN	6Sec6Dbr	1-3/25	4084A	4844
POPLAR TOWN	RN	6Sec6Dbr	12/20-21	4084	4611
POPLAR TRAIL	RN	6Sec6Dbr	1-3/25	4084A	4844
POPLAR TREE	RN	6Sec6Dbr	9-10/20	4084	4590
POPLAR VALE	RN	6Sec6Dbr	5-6/23	4084	4699
POPLAR VALLEY	RN	6Sec6Dbr	4-6/24	4084A	4763
POPLAR VIEW	RN	6Sec6Dbr	8/23	4060	4697
POPLAR VILLA	RN	6Sec6Dbr	10-12/16	4084	4431
POPLAR VILLAGE	RN	6Sec6Dbr	1-3/25	4084A	4844
POPLAR VISTA	RN	6Sec6Dbr	2-5/25	4084A	4845
POPLAR WOODS	RN	6Sec6Dbr	6-8/24	4084A	4764
POPOCATEPETL	N	8Sec5Dbr	1-3/14	4036E	4249
POQUOSON		12SecDr	4-5/13	2410	4106
PORT BURWELL		14Sec	7/29	3958A	6271
PORT BYRON		12SecDrCpt	7/21/10	2411	3800
PORT CLINTON		10Cpt	3/14	2505A	4252
PORT COLBORNE	N	14Sec	6/29	3958A	6263
PORT COLBURN		14Sec	6/29	3958A	6263
PORT COLUMBUS	N	4CptLngObs	10/25	3960	4889
PORT EWEN		12SecDrCpt	3/11	2411	3880
PORT HOPE		14Sec	6/29	3958A	6263
PORT HURON		14Sec	6/29	3958A	6263
PORT LEWIS		14Sec	6/29	3958A	6263
PORT NELSON		14Sec	7/29	3958A	6271

CAR NAME	ST	CAR TYPE	BUILT	PLAN	LOT
PORT ROWAN		14Sec	7/29	3958A	6271
PORT ROYAL		12SecDr	12/19/10	2410	3864
PORT STANLEY		14Sec	6/29	3958A	6263
PORTAGE		12SecDr	11/23/10	2410	3864
PORTAGE VALLEY		DrSbrBLngObs	4/29	3988	6221
PORTAGEVILLE		12SecDr	2-4/24	3410	4762
PORTALES		10SecDr2Cpt	8-9/13	2585A	4201
PORTCHESTER		36Ch	8/27	3916A	6077
PORTLAND		12SecDr	6-10/21	2410F	4614
PORTMORE		12SecDr	9-10/20	2410F	4590
PORTSLADE		10SecDr2Cpt	12/15	2585C	4351
PORTSMOUTH		12SecDr	2-3/27	3410A	6055
PORTSMOUTH LIGHT		36Ch	12/29-30	3916B	6319
PORTUGAL		12SecDr	9-10/20	2410F	4590
PORTULACA		28ChDr	5-7/26	3416	4958
POSEYVILLE	N	12SecDr	1-3/14	2410A	4249
POTOKA		12SecDr	3-6/20	2410F	4565
POTOMAC	N	8SecDr2Cpt	3/29	3979A	6237
POTRERO		12SecDr	9-10/13	2410A	4215
POTTER PALMER		14Sec	3-4/30	3958A	6357
POTTSTOWN		12SecDr	4/29/10	2410	3772
POTTSVILLE	N	BaggBClubSmk	2-4/09	2136C	3660
POUDRE LAKE		10SecDrCpt	10-12/29	3973A	6300
POULTNEY		12SecDr	7/16	2410F	4412
POWASSAN		12SecDr	2-5/16	2410E	4367
POWELTON		12SecDr	12/19/10	2410	3864
POWHATAN	N	12SecDr	8-12/25	3410A	4894
POWHATAN	N	10Sec3Dbr	5/17	3411	4494
POYSER		12SecDr	9-10/17	2410F	4525
POZA RICA NdeM-225	SN	12SecDr	8-12/25	3410A	4894
PRADO		7Dr	11/17	2583B	4516
PRAIRIE CENTER		14Sec	6/29	3958A	6263
PRAIRIE CITY		14Sec	6/29	3958A	6263
PRAIRIE CREEK		14Sec	7/29	3958A	6271
PRAIRIE DU CHIEN		10SecDrCpt	4-6/27	3973	6043
PRAIRIE GROVE		14Sec	7/29	3958A	6271
PRAIRIE HOME		14Sec	6/29	3958A	6263
PRAIRIE LAWN		14Sec	7/29	3958A	6271
PRAIRIE POINT		14Sec	7/29	3958A	6271
PRAIRIE RIDGE		14Sec	7/29	3958A	6271
PRAIRIE RIVER		14Sec	7/29	3958A	6271
PRAIRIE ROAD		14Sec	7/29	3958A	6271
PRAIRIE STATE		BaggBSmkBarber	10-11/25	3951	4885
PRAIRIE TRAIL		14Sec	7/29	3958A	6271
PRAIRIE VIEW		14Sec	6/29	3958A	6263
PRATT		LngBDorm(5Sec)	7-8/27	3981	6061
PREMIER ROSE	N	12SecDr	8-12/25	3410A	4894
PRENTISS		12SecDr	9-10/17	2410F	4525
PRESCOTT	N	12SecDr	2-3/13	2410	4105
PRESIDENT		12SecDr	4/26	3410A	4943
PRESIDENT ADAMS		7Cpt2Dr	4-5/23	2522C	4689

CAR NAME	ST	CAR TYPE	BUILT	PLAN	LOT
PRESIDENT CLEVELAND		7Cpt2Dr	4-5/23	2522C	4689
PRESIDENT GARFIELD		7Cpt2Dr	4-5/23	2522C	4689
PRESIDENT GRANT		7Cpt2Dr	4-5/23	2522C	4689
PRESIDENT HARRISON		7Cpt2Dr	4-5/23	2522C	4689
PRESIDENT JACKSON		7Cpt2Dr	4-5/23	2522C	4689
PRESIDENT JEFFERSON		7Cpt2Dr	4-5/23	2522C	4689
PRESIDENT LINCOLN		7Cpt2Dr	4-5/23	2522C	4689
PRESIDENT ROOSEVELT		7Cpt2Dr	4-5/23	2522C	4689
PRESIDENT WASHINGTON		7Cpt2Dr	4-5/23	2522C	4689
PRESIDENTE					
LERDO NdeM	SN	12SecDr	6-7/17	2410K	4503
PRESIDIO		12SecDr	8-10/26	3410A	4945
PRESS CLUB		8SecBLngSunRm	8-9/29	3989A	6274
PRESTON		12SecDr	12/12-13	2410	4076
PRIARIE HOME		14Sec	6/29	3958A	6263
PRIARIE VIEW		14Sec	6/29	3958A	6263
PRIMATE		12SecDr	9-10/17	2410F	4525
PRIMOS		12SecDr	12/27/10	2410	3864
PRIMROSE		12SecDr	10-12/16	2410F	4431
PRIMULA		10SecDr2Cpt	12/15	2585C	4351
PRINCESS ADELAIDE		16ChBLngObs	9/25	3961	4910
PRINCESS JOAN		16ChBLngObs	9/25	3961	4910
PRINCESS MARGARET		16ChBLngObs	9/25	3961	4910
PRINCETON		12SecDr	8-9/12	2410	4035
PRINCETON UNIVERSITY	RN	12Sec2Dbr	7-9/17	4046	4515
PRINCEVILLE	N	12SecDr	8-9/12	2410	4035
PRINCIPIO		12SecDr	2-4/24	3410	4762
PRINDLE		12SecDr	10-12/16	2410F	4431
PRINGLE HOUSE		13Dbr	6-7/30	3997A	6314
PRIOR LAKE		10SecDrCpt	10-12/29	3973A	6300
PRITCHARD		12SecDr	3-4/17	2410F	4485
PROCTOR		12SecDr	3-4/17	2410F	4485
PROFESSIONAL CLUB		13ChRestLng	12/29	3992	6289
PROGRESO NdeM-226	SN	12SecDr	5-8/25	3410A	4868
PROGRESS	RN	17StsRestLng	3-4/14	4019B	4264
PROMETHEUS		12SecDr	2-3/13	2410	4105
PROPELLER CLUB	RN	8SecRestLngObs	12/17-18	4025	4528
PROSPECT		12SecDr	12/12-13	2410	4076
PROSPECT PEAK		10Sec2Dr	12/26-27	3584A	6031
PROSSER		12SecDr	9-10/17	2410F	4525
PROVINCETOWN		32ChDr	10/24	3917	4803
PROVINCETOWN	R	24ChDr10StsLng	6/30	3917E	4803
PROVINCETOWN(2nd)	N	24ChDr10StsLng	5/26	3917E	4957
PROVISO		10SecDr2Cpt	5/13	2585A	4141
PRUDENCE		26ChDr	5/23	2416D	4691
PUEBLO		10SecDr2Cpt	12/16-17	2585D	4443
PUERTO		12SecDr	1-3/17	2410F	4450
PUERTO					
JUAREZ NdeM-227	SN	8SecDr2Cpt	12/29-30	3979A	6334
MARQUEZ NdeM-228	SN	14Sec	11-12/28	3958A	6212

CAR NAME	ST	CAR TYPE	BUILT	PLAN	LOT
PUERTO					
RICO	N	10SecDr2Cpt	10/24	3585A	4817
VALLARTA NdeM-229	SN	14Sec	8-10/28	3958A	6181
PUGET SOUND		10SecDrCpt	4-6/27	3973	6043
PULASKI		12SecDr	10/18/10	2410	3814
PUNTA					
DELGADA NdeM-230	SN	14Sec	10-11/26	3958A	6012
PUNTA GORDA		12SecDr	9-10/13	2410A	4215
PURBECK		12SecDr	2-5/16	2410E	4367
PURCELL		12SecDr	9-10/17	2410F	4525
PURCELL RANGE		10SecDrCpt	5/30	3973A	6358
PURETEST	LN	Private (Rexall)	7/5/16	2502B	4422
PURITAN		12SecDr	10-12/26	3410A	6023
PUTNAM		12SecDr	12/12-13	2410	4076
PYBURN		12SecDr	9-10/17	2410F	4525
PYRAMID PEAK		10Sec2Dr	12/26-27	3584A	6031
PYRENEES		12SecDr	10/22-23	2410H	4647
PYRITON		12SecDr	1-3/14	2410A	4249
QUADRANT PEAK	N	10Sec2Dr	9-10/29	3584B	6284
QUAKER VALLEY		28ChDr	6-7/27	3416A	6087
QUARREN		12SecDr	9-10/13	2410A	4215
QUARTER CENTURY CLUB	N	8SecBLngSunRm	8-9/29	3989K	6274
QUEBEC		3CptDrLngBSunRm	6/29	3975C	6262
QUEEN ANNE		26ChDrObs	5/25	3957	4862
QUEEN CITY CLUB	RN	8SecRestLngObs	2/24/11	4025B	3865
QUEEN ELIZABETH		26ChDrObs	5/25	3957	4862
QUEEN ISABELLA		26ChDrObs	5/25	3957	4862
QUEEN MARY		26ChDrObs	5/25	3957	4862
QUEEN'S PARK		8SecDr2Cpt	5-6/29	3979A	6261
QUEENS UNIVERSITY	RN	12Sec2Dbr	5-6/17	4046	4497
QUEENSBORO		12SecDr	7-8/13	2410A	4192
QUEENSTOWN		12SecDr	4-5/13	2410	4106
QUEENSVILLE		12SecDr	6-9/26	3410A	4969
QUEENSVILLE	N	12SecDr	4/26	3410A	4943
QUENEMO		7Cpt2Dr	11/17	2522B	4517
QUEPONCO		12SecDr	7-9/17	2410F	4515
QUERINO		10SecDr2Cpt	8-9/13	2585A	4201
QUINCY		Bagg24Ch	6/16	2915	4388
QUINCY MANSION		13Dbr	6/30	3997	6314
QUINNIMONT		12SecDr	1-3/14	2410A	4249
QUINOLA		12SecDr	3-6/20	2410F	4565
QUINTARD		12SecDr	7-8/15	2410D	4322
QUIVERO		7Cpt2Dr	11/17	2522B	4517
RABIDA		12SecDr	7/13	2410A	4179
RACHITA		12SecDr	9-10/13	2410A	4215
RACINE		16Sec	5-6/13	2412B	4150
RACQUET CLUB		8SecBLngSunRm	8-9/29	3989A	6274
RADCLIFFE		16Sec	1-5/18	2412F	4531
RADCLIFFE COLLEGE	RN	10Sec2DbrCpt	6-13	4042A	4159
RADEBAUGH		12SecDr	7-9/17	2410F	4515

CAR NAME	ST	CAR TYPE	BUILT	PLAN	LOT
RADFORD		12SecDr	1-2/16	2410E	4356
RADIANCE ROSE	N	12SecDr	8-12/25	3410A	4894
RADIANT		12SecDr	12/12-13	2410	4076
RADISSON		12SecDr	1-2/15	2410B	4311
RADIUM		12SecDr	9-10/20	2410F	4590
RADLEY		12SecDr	2-5/25	3410	4845
RADNOR		16Sec	1-5/18	2412F	4531
RAFAEL					
DELGADO NdeM-231	N	12SecDr	1-3/25	3410	4844
RAHWAY		26ChDr	5/12	2416	3986
RAINBOW CANYON		3CptDrLngBSunRm	6/29	3975C	6262
RAINSFORD		34Ch	8/13	2764A	4195
RALEIGH		12SecDr	11-12/13	2410A	4234
RALPH WALDO EMERSON		12SecDr	1-2/29	3410B	6220
RALSTON '		12SecDr	7-9/17	2410F	4515
RAMADAN		12SecDr	1-2/16	2410E	4356
RAMAGE		12SecDr	10-12/16	2410F	4431
RAMAPO		12SecDr	9-10/20	2410F	4590

CAR NAME	ST	CAR TYPE	BUILT	PLAN	LOT
RAMBLER	RN	10ChRestLng	6/11	4019	3902
RAMBLER	RN	14Ch20StsLng	8/27	4075	6077
RAMBLER ROSE	N	12SecDr	8-12/25	3410A	4894
RAMBO		32Ch	11/15/10	2419	3861
RAMON LOPEZ					
VELARDE NdeM	SN	12SecDr	8-10/26	3410A	4945
RAMONA		24ChDrB	4/16	2417C	4364
RAMPUR		12SecDr	9-10/20	2410F	4590
RAMSAY		12SecDr	10-12/16	2410F	4431
RAMSDELL	N	12SecDr	11/15/10	2410	3860
RANCHER		8SecDr2Cpt	5-6/29	3979A	6261
RANDON		12SecDrCpt	6/13	2411A	4159
RANEE		12SecDr	4-5/13	2410	4106
RANELAGH		16Sec	10-11/14	2412C	4296
RANGER		8SecDr2Cpt	5-6/29	3979A	6261
RANGOON		12SecDr	9-10/20	2410F	4590
RANKIN		16Sec	6-7/20	2412F	4568
RANKOKA		12SecDr	11-12/13	2410A	4234

Figure I-66 Many Pullman sleepers lived well beyond their time as Bunk, Kitchen, or other railroad work train tasks. Former 12-section drawing room RANKOKA was withdrawn from Pullman lease in February 1953 and served in temporary coach service until June 1961 when it became B&O X-4381, the bunk and office car for the Cumberland wreck train. The local shops modified one vestibule to the open platform as shown. In June 1982 this car was renumbered Chessie System 911602. Photograph from a color slide by the Author.

CAR NAME	ST	CAR TYPE	BUILT	PLAN	LOT	CAR NAME	ST	CAR TYPE	BUILT	PLAN	LOT
RANSFORD		16Sec	5-6/16	2412F	4386	RED ELM		12SecDr	8-12/25	3410A	4894
RANSOM		10SecDr2Cpt	12/16-17	2585D	4443	RED FEATHER		12SecDr	8-12/25	3410A	4894
RANSTON		12SecDr	1-2/16	2410E	4356	RED FERN		12SecDr	8-12/25	3410A	4894
RAPHAEL		6Cpt3Dr	7/25	3523A	4887	RED FIELD		12SecDr	8-12/25	3410A	4894
RAPHAEL SEMMES	N	12SecDr	8-10/26	3410A	4945	RED FORK		12SecDr	8-12/25	3410A	4894
RAPID CITY		10SecDrCpt	4-6/27	3973	6043	RED GRANITE		12SecDr	8-12/25	3410A	4894
RAPID CITY	N	10SecDrCpt	8/28	3973A	6185	RED GRAVEL		12SecDr	8-12/25	3410A	4894
RARITAN		12SecDr	11-12/13	2410A	4234	RED HILL		12SecDr	8-12/25	3410A	4894
RATHAUS TOWER	RN	8SecDr3Dbr	5-6/25	4090	4846	RED HOOK		12SecDr	8-12/25	3410A	4894
RATON		7Cpt2Dr	9/13	2522A	4208	RED HOUSE		12SecDr	8-12/25	3410A	4894
RAVANEL		12SecDr	1-2/16	2410E	4356	RED JACKET		30ChDr	2-3/30	4000	6324
RAVENDEN		12SecDr	2-5/16	2410E	4367	RED JUNIPER		12SecDr	8-12/25	3410A	4894
RAVENNA		12SecDr	2-4/24	3410	4762	RED KEY		12SecDr	8-12/25	3410A	4894
RAVENSCROFT		16Sec	10-11/14	2412C	4296	RED LAKE		12SecDr	8-12/25	3410A	4894
RAVENSWORTH		12SecDrCpt	3/11	2411	3880	RED LAWN		12SecDr	8-12/25	3410A	4894
RAVINIA		14Sec	8-10/28	3958A	6181	RED LEVEL		12SecDr	8-12/25	3410A	4894
RAWLINS		12SecDr	12/27-28	3410B	6127	RED LINE		12SecDr	8-12/25	3410A	4894
RAYLAND		12SecDr	10/17/10	2410	3814	RED LION		12SecDr	8-12/25	3410A	4894
RAYMILTON		12SecDr	9-10/11	2410	3924	RED LODGE		12SecDr	8-12/25	3410A	4894
RAYMORE		12SecDr	11-12/13	2410A	4234	RED MAN		12SecDr	8-12/25	3410A	4894
RAYNER		12SecDr	10-12/16	2410F	4431	RED MAPLE	N	12SecDr	8-12/25	3410A	4894
RAYVILLE		12SecDr	9-10/11	2410	3924	RED MILL		12SecDr	8-12/25	3410A	4894
RAYWOOD		12SecDrCpt	6/13	2411A	4159	RED OAK		12SecDr	8-12/25	3410A	4894
READVILLE		32ChDr	5/26	3917	4957	RED ORE		12SecDr	8-12/25	3410A	4894
READVILLE	R	24ChDr10StsLng	5/26	3917E	4957	RED PHEASANT		12SecDr	8-12/25	3410A	4894
READVILLE(2nd)	N	24ChDr10StsLng	5/26	3917E	4957	RED PINE		12SecDr	8-12/25	3410A	4894
REAS PASS		8SecDr2Cpt	12/29-30	3979A	6334	RED POINT		12SecDr	8-12/25	3410A	4894
REBA		24ChDrB	6-7/14	2417B	4265	RED RAPIDS		12SecDr	8-12/25	3410A	4894
REBECCA		26ChDr	6/11	2416	3902	RED RIDGE		12SecDr	8-12/25	3410A	4894
RED ARROW	N	5CptLngObs	8/9/10	2413A	3802	RED RIVER		12SecDr	8-12/25	3410A	4894
RED ASH		12SecDr	8-12/25	3410A	4894	RED ROCK PASS		8SecDr2Cpt	12/29-30	3979A	6334
RED BANK		12SecDr	3/26/10	2410	3769	RED RUN		12SecDr	8-12/25	3410A	4894
RED BAY		12SecDr	8-12/25	3410A	4894	RED SPRUCE		12SecDr	8-12/25	3410A	4894
RED BERRY		12SecDr	8-12/25	3410A	4894	RED WING		14Sec	4/27	3958	6041
RED BIRD		12SecDr	8-12/25	3410A	4894	REDDING		12SecDr	10-12/16	2410F	4431
RED BLUFF		12SecDr	8-12/25	3410A	4894	REDINGTON		12SecDr	1-2/15	2410B	4311
RED BRANCH		12SecDr	8-12/25	3410A	4894	REDWOOD		10SecDr2Cpt	1-2/13	2585	4091
RED BRIDGE		12SecDr	8-12/25	3410A	4894	REDWOOD PEAK		10Sec2Dr	12/28-29	3584B	6213
RED BRUSH		12SecDr	8-12/25	3410A	4894	REESE		12SecDr	2-5/25	3410	4845
RED BUD		12SecDr	8-12/25	3410A	4894	REGAL		12SecDr	5-8/18	2410F	4540
RED BUTTES		12SecDr	8-12/25	3410A	4894	REHOBOTH		12SecDr	7-9/17	2410F	4515
RED CANON		12SecDr	8-12/25	3410A	4894	REINBECK		16Sec	1-5/18	2412F	4531
RED CEDAR		12SecDr	8-12/25	3410A	4894	REITA		24ChDrB	4/13	2417	4108
RED CLAY		12SecDr	8-12/25	3410A	4894	RELIANT	RN	22Ch12StsBLng	8/13	4032	4195
RED CLIFF		12SecDr	8-12/25	3410A	4894	RELIUS		12SecDr	3-6/11	2410	3881
RED CLOUD		12SecDr	8-12/25	3410A	4894	REMBRANDT		12SecDr	7-9/17	2410F	4515
RED CREEK		12SecDr	8-12/25	3410A	4894	REMINGTON		16Sec	1/16	2412E	4352
RED CROSS		12SecDr	8-12/25	3410A	4894	REMNOY		12SecDr	9-10/13	2410A	4215
RED DEER		12SecDr	8-12/25	3410A	4894	REMOLA		30ChDr	11/12	2668	4051
RED DESERT		12SecDr	8-12/25	3410A	4894	REMSEN		10Cpt	8/15/10	2505	3833

CAR NAME	ST	CAR TYPE	BUILT	PLAN	LOT
RENAULT		10SecDr2Cpt	12/16-17	2585D	4443
RENDVILLE		12SecDr	7/11	2410	3913
RENOVO		24ChDrB	4/13	2417	4108
RENSSELAER TECH	N	BaggBSmk	12/26-27	3951G	6022
RENWICK		12SecDr	6-7/16	2410F	4385
REPOSO		12SecDr	3-6/15	2410B	4319
REPTON		16Sec	1-5/18	2412F	4531
REPUBLIC	N	3CptDrLngBSunRm	12/28-29	3975C	6217
RESEARCH #2	LN	Exhibit (Rexall)	6/16	2916B	4389
RESERVATION PEAK	N	10Sec2Dr	9-10/29	3584B	6284
RESTOULE		12SecDr	1-2/16	2410E	4356
RETORT		12SecDr	2-5/25	3410	4845
REVILLON		12SecDr	7-8/15	2410D	4322
REXFORD		34Ch	8/13	2764A	4195
REXIS		28ChDr	5-7/26	3416	4958
REYBOLD		12SecDr	7-9/17	2410F	4515
REYBURN		10SecDr2Cpt	12/16-17	2585D	4443
REYNARD		12SecDr	1-2/15	2410B	4311
REYNOLDS		12SecDr	10/20/10	2410	3814
RHEEMS	N	BaggBClubSmk	2-4/09	2136C	3660
RHEIMS		7Dr	12/20	2583B	4621
RHINECLIFF		BaggClubBSmk	11/24	3415	4808
RHODE ISLAND	N	SbrDrB16StsLng	7/30	3988D	6356
RIADA		12SecDr	10-12/15	2410D	4338
RIALTO		12SecDr	11/11/10	2410	3860
RICARDO		12SecDr	6-7/17	2410F	4503
RICE LAKE		10SecDrCpt	10-12/29	3973A	6300
RICHARD BEATTY MELLON	N	CptDrBObsLng	9-10/27	3975V	6076
RICHARD BEATTY MELLON	N	8SecDr2Cpt	7-8/30	3979A	6377
RICHARD DALE		12SecDr	4/26	3410A	4943
RICHARD HENRY LEE		30ChDr	9-10/24	3418	4783
RICHARD HENRY LEE		30ChDr	8-9/30	4000A	6384
RICHARD MONTGOMERY		12SecDr	4/26	3410A	4943
RICHARDSON		10SecDr2Cpt	12/16-17	2585D	4443
RICHELIEU	N	5CptBLng	3/14	2505B	4252
RICHLAND		16Sec	6-7/20	2412F	4568
RICHMOND		12SecDr	2-5/21	2410F	4612
RICHVILLE		12SecDr	3-4/11	2410	3881
RICHWOOD		12SecDr	6/28	3410B	6173
RIDDLETON		12SecDr	7-9/17	2410F	4515
RIDERWOOD		12SecDr	7-9/17	2410F	4515
RIDGE FARM		10Sec2Dr	1/12	2584	3947
RIDGEDALE		BaggBClubSmkBarber	2-3/12	2602	3948
RIDGEMOOR		16Sec	10-11/14	2412C	4296
RIDGETOWN		12SecDr	3-4/17	2410F	4485
RIDGEVILLE		10Cpt	8/26/10	2505	3833
RIDGEWAY		12SecDr	7/11	2410	3913
RIDGEWOOD					
COUNTRY CLUB		8SecBLngSunRm	2/31	3989D	6396
RIDLEY PARK		12SecDr	2-5/16	2410E	4367
RIEGELSVILLE		12SecDr	7-9/17	2410F	4515
RIENZI		16Sec	5-6/13	2412B	4150
RILEY		12SecDrCpt	10-12/21	2411C	4624
RILLITO		12SecDr	6-7/17	2410F	4503
RILLTON		12SecDr	7-9/17	2410F	4515
RIMERSBURG		12SecDr	7-9/17	2410F	4515
RIMERTON		12SecDr	7-9/17	2410F	4515
RINCON		LngBDorm(5Sec)	7-8/27	3981	6061
RINGOES		12SecDr	12/20/10	2410	3864
RIO					
ATOYAC NdeM-232	SN	10SecDr2Dbr	5-6/25	4074C	4843
BALSAS	N	10SecDr2Cpt	8-9/13	2585A	4201
BALSAS NdeM	SN	10SecDr2Dbr	5-6/25	4074C	4843
BLANCO		CptDrLngBSunRm	10/27	3975	6047
BLANCO NdeM	SN	14Sec	9-10/25	3958	4869
BRAVO	N	10SecDr2Cpt	12/15	2585C	4351
BRAVO NdeM-238	SN	10SecDr2Dbr	1-3/25	4074E	4844
COLORADO NdeM	SN	10SecDr2Dbr	2-5/25	4074E	4845
COLORADO FCS-BC	SN	10SecDr2Cpt	9-10/26	3585A	4996
CONCHOS NdeM-234	SN	10SecDr2Dbr	2-5/25	4074E	4845
DE JANEIRO NdeM	SN	6Sec6Dbr	10-12/21	4060	4624
FRIO	N	10SecDr2Cpt	8-9/13	2585A	4201
GILA NdeM-235	SN	10SecDr2Dbr	1-3/25	4074E	4844
GRANDE		10Sec2Dr	2/13	2584	4092
GRIJALVA	N	10SecDr2Cpt	8-9/13	2585A	4201
GUADIANA NdeM-236	SN	10SecDr2Dbr	5-6/25	4074C	4843
GUAYALEJO	N	10SecDr2Cpt	11-12/11	2585	3942
LERMA	N	10SecDr2Cpt	1-2/13	2585	4091
MAZATLAN	RN	6SecRest SBS	4/15	4017	4310
MEZCALA	N	10SecDr2Cpt	6-7/15	2585B	4328
PANUCA NdeM	N	10SecDr2Dbr	5-6/25	4074E	4843
PANUCO	N	10SecDr2Cpt	12/15	2585C	4351
PAPAGAYO	N	10SecDr2Cpt	8-10/23	3585	4725
PAPALOAPAN	N	10SecDr2Cpt	1-2/13	2585	4091
SABINAS NdeM-238	N	10SecDr2Dbr	1-3/25	4074E	4844
SANTIAGO	N	6SecRest SBS	4/15	4017	4310
SINALOA	RN	6SecRest SBS	4/15	4017	4310
SONORA	RN	6SecRest SBS	4/15	4017	4310
SUCHIATE NdeM-239	N	10SecDr2Dbr	5-6/25	4074C	4843
TAMESI NdeM-233	N	10SecDr2Dbr	5-6/25	4074C	4843
USUMACINTA	N	10SecDr2Cpt	8-9/13	2585A	4201
VERDE NdeM-240	SN	10SecDr2Dbr	5-6/25	4074C	4843
YAQUI	RN	6SecRest SBS	4/15	4017	4310
RIOVISTA		12SecDr	9-10/13	2410A	4215
RIPLEY		BaggBClubSmkBarber	9/24	2951C	4805
RIPON		12SecDr	1-2/16	2410E	4356
RIPPLEMEAD		12SecDr	11-12/13	2410A	4234
RIPPLETON		12SecDr	7/11	2410	3913
RISING STAR		12SecDr	10/22-23	2410H	4647
RISLEY		12SecDr	3-4/17	2410F	4485

CAR NAME	ST	CAR TYPE	BUILT	PLAN	LOT	CAR NAME	ST	CAR TYPE	BUILT	PLAN	LOT
RITCHIE		12SecDr	7-9/17	2410F	4515	ROCK LAND		8SecDr2Cpt	7-8/29	3979A	6272
RITTENHOUSE SQUARE		6Cpt3Dr	10-12/29	3523C	6290	ROCK LAWN		8SecDr2Cpt	7-8/29	3979A	6272
RITTER		12SecDr	3-4/17	2410F	4485	ROCK PARK		8SecDr2Cpt	7-8/29	3979A	6272
RIVAL		16Sec	12/14-15	2412C	4304	ROCK PASS		8SecDr2Cpt	7-8/29	3979A	6272
RIVER VIEW		10Cpt	3/14	2505A	4252	ROCK RAPIDS		8SecDr2Cpt	7-8/29	3979A	6272
RIVERDALE		16Sec	1-5/18	2412F	4531	ROCK RIVER		10SecDrCpt	8/28	3973A	6185
RIVERDALE PARK		8SecDr2Cpt	5-6/29	3979A	6261	ROCK RUN		8SecDr2Cpt	7-8/29	3979A	6272
RIVERGATE		12SecDr	7/11	2410	3913	ROCK SPRINGS		8SecDr2Cpt	7-8/29	3979A	6272
RIVERSIDE		12SecDr	3-4/11	2410	3881	ROCK STREAM		8SecDr2Cpt	7-8/29	3979A	6272
RIVERSIDE	N	12SecDr	10/22-23	2410H	4647	ROCK VALLEY		8SecDr2Cpt	7-8/29	3979A	6272
RIVERTON		12SecDr	10-12/26	3410A	6023	ROCK WOOD		8SecDr2Cpt	7-8/29	3979A	6272
ROALD AMUNDSEN		Private	8/29	3972B	6246	ROCKARU		12SecDr	3-6/20	2410F	4565
ROANOKE COUNTY	N	10SecLngObs	10-12/26	3521A	4998	ROCKBURN		26ChDr	5/12	2416	3986
ROARING CAMP	RN	17RmtSec	10-11/25	4068C	4885	ROCKET TOWER	RN	8SecDr3Dbr	12/28-29	4090C	6213
ROBERT BOYLE		14Sec	3/27	3958	6054	ROCKFORD COLLEGE	RN	10Sec2DbrCpt	11/17	4042	4527
ROBERT E. LEE		3Cpt2DrLngObs	8-10/25	3959	4886	ROCKHAM		12SecDr	12/12-13	2410	4076
ROBERT F. HOKE		10Sec2Dr	9-10/25	3584A	4899	ROCKLIN		12SecDr	12/27-28	3410B	6127
ROBERT F. HOKE		10Sec2Dr	9-10/29	3584B	6284	ROCKMART		12SecDrCpt	3/11	2411	3880
ROBERT F. HOKE	N	8SecDr2Cpt	8-9/29	3979A	6283	ROCKMERE		12SecDr	5/3/10	2410	3772
ROBERT FULTON		14Sec	10-11/26	3958	6012	ROCKPORT		12SecDr	4-6/24	3410	4763
ROBERT MORRIS		12SecDr	5-8/25	3410A	4868	ROCKROSE		28ChDr	5-7/26	3416	4958
ROBERT PEARY		Private	7/27	3972	6037	ROCKVILLE		12SecDr	11/17/10	2410	3864
ROBERT R. LIVINGSTON		30ChDr	9-10/24	3418	4783	ROCKWYN		12SecDr	3-6/15	2410B	4319
ROBERT R. LIVINGSTON		30ChDr	8-9/30	4000A	6384	ROCKY HILL		BaggBSmkBarber	12/26-27	3951C	6022
ROBERT TOOMBS		10Sec2Dr	9-10/25	3584A	4899	RODANO		12SecDr	12/10-11	2410	3866
ROBERT TOOMBS		10Sec2Dr	9-10/29	3584B	6284	RODBOURN		12SecDr	3-6/20	2410F	4565
ROBERT Y. HAYNE	N	10Sec2Dr	9-10/25	3584A	4899	RODMAN		16Sec	6-7/20	2412F	4568
ROBERTA		26ChDr	6/11	2416	3902	RODNEY		12SecDr	10-11/11	2410	3936
ROBERTSDALE		12SecDr	7/11	2410	3913	ROEBLING		12SecDr	12/22/10	2410	3864
ROBESON		10SecDr2Cpt	12/16-17	2585D	4443	ROEBLING		14Sec	10-11/26	3958	6012
ROBINA		24ChDrB	6-7/14	2417B	4265	ROEBUCK		12SecDr	3-6/20	2410F	4565
ROBINSON		12SecDr	3-4/17	2410F	4485	ROENTGEN		14Sec	10-11/26	3958	6012
ROBINWOOD		12SecDr	3-4/11	2410	3881	ROGER SHERMAN		30ChDr	9-10/24	3418	4783
ROCHAMBEAU		12SecDr	7/13	2410A	4179	ROGER WILLIAMS		12SecDr	5-8/25	3410A	4868
ROCHELLE		16Sec	5-6/13	2412B	4150	ROGERS		BaggBClubSmkBarber	8/17/10	2414A	3803
ROCHESTER CLUB	RN	8SecCptRestObs	12/17-18	4026	4528	ROLFE		28ChDr	5-7/26	3416	4958
ROCK ACRES		8SecDr2Cpt	7-8/29	3979A	6272	ROLLAND		12SecDr	7/16	2410F	4412
ROCK BEND		8SecDr2Cpt	7-8/29	3979A	6272	ROLLINS		12SecDr	5-8/18	2410F	4540
ROCK CABIN	N	8SecDr2Cpt	10-12/28	3979A	6205	ROLLINS COLLEGE	RN	10Sec2DbrCpt	7-8/16	4042	4424
ROCK CAMP	N	8SecDr2Cpt	10-12/28	3979A	6205	ROLYAT		12SecDr	3-6/20	2410F	4565
ROCK CLIFF		8SecDr2Cpt	7-8/29	3979A	6272	ROME		12SecDr	1-2/30	3410B	6351
ROCK CRAG		8SecDr2Cpt	7-8/29	3979A	6272	ROMELINK		16Sec	1-5/18	2412F	4531
ROCK CREEK	N	23ChDrLibObs	4/14	2819	4263	ROMNEY		6Cpt3Dr	7/25	3523A	4887
ROCK DELL		8SecDr2Cpt	7-8/29	3979A	6272	ROMONT		12SecDr	9-10/20	2410F	4590
ROCK GAP		8SecDr2Cpt	12/29-30	3979A	6334	RONALDSON		16Sec	5-6/16	2412F	4386
ROCK GLEN	N	23ChDrLibObs	4/14	2819	4263	RONDAXE		10Cpt	8/15/10	2505	3833
ROCK GORGE		8SecDr2Cpt	7-8/29	3979A	6272	RONEY PLAZA	N	36Ch	10/24	3916	4802
ROCK HARBOR		8SecDr2Cpt	7-8/29	3979A	6272	RONK		12SecDr	3/16/10	2410	3794
ROCK HAVEN		8SecDr2Cpt	7-8/29	3979A	6272	RONNEBY		12SecDr	9-10/20	2410F	4590
ROCK ISLE		8SecDr2Cpt	7-8/29	3979A	6272	RONVILLE		12SecDr	3-6/15	2410B	4319

CAR NAME	ST	CAR TYPE	BUILT	PLAN	LOT
ROOKWOOD		12SecDr	2-3/15	2410B	4318
ROSA BONHEUR		34Ch	1/27	3419A	6033
ROSALIE		28ChDr	6-7/27	3416A	6087
ROSALIND		28ChDr	8/24	3416	4801
ROSCOMMON		12SecDr	9-10/11	2410	3924
ROSE		24ChDrB	5/13	2417A	4137
ROSEBANK		12SecDr	3-6/15	2410B	4319
ROSEBUD		28ChDr	5-7/26	3416	4958
ROSECLEER		26ChDr	4/12	2416	3950
ROSECRANS		12SecDr	3-6/20	2410F	4565
ROSEISLE		12SecDr	1-3/14	2410A	4249
ROSELAND	N	12SecDr	12/22/10	2410	3864
ROSELLA		26ChDr	6/11	2416	3902
ROSEMONT		24ChDrB	4/13	2417	4108
ROSENEATH		12SecDr	12/12-13	2410	4076
ROSETON		12SecDr	3-4/11	2410	3881
ROSEWOOD		12SecDr	3-6/20	2410F	4565
ROSIERE		16Sec	9/11	2412	3923
ROSINA		26ChDr	4-5/13	2416A	4136
ROSLINDALE		12SecDr	5-7/13	2410	4149
ROSSCLAIR		12SecDr	1-2/15	2410B	4311
ROSSINI		6Cpt3Dr	11-12/24	3523A	4833
ROSSITER		12SecDr	9-10/11	2410	3924
ROSSMOYNE		12SecDr	1/14	2410A	4240
ROSTAND		12SecDr	11-12/13	2410A	4234
ROSTHERN		12SecDr	1-2/15	2410B	4311
ROSTRAVER		12SecDr	1-4/12	2410	3949
ROSWELL		12SecDr	6-9/26	3410A	4969
ROTARY CLUB		13ChRestLng	12/29	3992	6289
ROTHERHAM		10SecDr2Cpt	12/15	2585C	4351
ROTHERWOOD		12SecDr	2-3/15	2410B	4318
ROTHRUCK		12SecDr	7-9/17	2410F	4515
ROTHSAY		12SecDr	9-10/20	2410F	4590
ROUBAIX		12SecDr	6-7/16	2410F	4385
ROUEN		12SecDr	7-8/15	2410D	4322
ROUGEMONT		12SecDr	7-8/13	2410A	4192
ROUMANIA		12SecDr	9-10/20	2410F	4590
ROUND TOWER	RN	8SecDr3Dbr	1/25	4090	4843
ROUNDCROFT		12SecDr	2-5/16	2410E	4367
ROWANTY		12SecDr	4-5/13	2410	4106
ROWENA		26ChDr	3-4/14	2416A	4264
ROWLAND		12SecDr	11/17/10	2410	3860
ROWSLEY		12SecDr	7-8/13	2410A	4192
ROWTON		12SecDr	10-12/15	2410D	4338
ROXANA		28ChDr	4-6/25	3416	4864
ROXBURG		12SecDr	7-9/17	2410F	4515
ROXTON		12SecDr	2-4/24	3410	4762
ROYAL OAK		8SecDr2Cpt	2-3/30	3979A	6353
ROYALL HOUSE		13Dbr	6-7/30	3997A	6314
ROYCROFT		12SecDr	2-3/15	2410B	4318

CAR NAME	ST	CAR TYPE	BUILT	PLAN	LOT
ROYERSFORD		12SecDr	7-9/17	2410F	4515
ROYERTON		12SecDr	9-10/11	2410	3924
RUBENS		6Cpt3Dr	7/25	3523A	4887
RUBIDOUX		12SecDr	8-10/26	3410A	4945
RUBINAT		12SecDr	12/12-13	2410	4076
RUBY		24ChDrB	5/13	2417A	4137
RUDYARD		16Sec	5-6/13	2412B	4150
RUFFSDALE		12SecDr	2-4/24	3410	4762
RUGGLES	N	12SecDr	11/8/10	2410	3860
RUHEMA		24ChDrB	4/16	2417C	4364
RULETON		12SecDr	1-3/14	2410A	4249
RUMLEY		12SecDr	2-4/24	3410	4762
RUMSON	RN	30ChBLibObs	4/16	2901A	4365
RUNYON		12SecDr	7-9/17	2410F	4515
RUPERT		16Sec	1-5/18	2412F	4531
RUREMONT	N	12SecDr	3-4/11	2410	3881
RUSCOMB		12SecDr	9-10/11	2410	3924
RUSHVILLE		12SecDr	9-10/11	2410	3924
RUSKIN		7Cpt2Dr	11/13	2522A	4222
RUSSELL		12SecDr	3-4/11	2410	3881
RUSSELL		12SecDr	10/22-23	2410H	4647
RUSTIC		28ChDr	5-6/24	3416	4761
RUTGERS UNIVERSITY	RN	12Sec2Dbr	7-9/17	4046	4515
RUTGERS UNIVERSITY	N	12Sec2Dbr	5-8/18	4046B	4540
RUTH		26ChDr	5/23	2416D	4691
RUTHERGLEN		12SecDr	12/12-13	2410	4076
RUXTON		24ChDrB	4/13	2417	4108
RYDE		12SecDr	11/18/10	2410	3864
RYEGATE		12SecDr	7-8/14	2410A	4277
RYERSON		12SecDr	7/16	2410F	4412
RYLANDER		12SecDr	11-12/13	2410A	4234
RYNDAM		12SecDr	11-12/13	2410A	4234
SABLON		12SecDr	7-8/13	2410A	4192
SABULA		12SecDr	2-4/24	3410	4762
SACAJAWEA		12SecDr	4/15	2410C	4309
SACANDAGA		10Cpt	8/16/10	2505	3833
SACHEM		12SecDr	9-10/20	2410F	4590
SACILE		16Sec	1/16	2412E	4352
SACKETT'S HARBOR		BaggBSmkBarber	12/26-27	3951C	6022
SACO		12SecDr	2-3/27	3410A	6055
SACRAMENTO		10SecDr2Cpt	10-11/12	2585	4067
SAGAMORE		12SecDr	12/12-13	2410	4076
SAGEHURST		10SecDr2Cpt	1-2/13	2585	4091
SAGINAW		10SecDr2Cpt	5/13	2585A	4141
SAGITA		16Sec	12/14-15	2412C	4304
SAGUENAY		12SecDr	4-5/13	2410	4106
SAKANA		12SecDr	3-6/15	2410B	4319
SALADIN		12SecDr	6-7/17	2410F	4503
SALAMANCA		12SecDr	7-8/12	2410	4014
SALEM		12SecDr	6-10/21	2410F	4614

CAR NAME	ST	CAR TYPE	BUILT	PLAN	LOT
SALEM COLLEGE	RN	10Sec2DbrCpt	12/14-15	4042B	4304
SALEM OAK		8SecDr2Cpt	2-3/30	3979A	6353
SALINAS PEAK		10Sec2Dr	9-10/29	3584B	6284
SALINEVILLE		12SecDr	5-6/23	2410H	4699
SALIX		12SecDr	2-4/24	3410	4762
SALLING		12SecDr	10-12/20	2410F	4591
SALLY ANN FURNACE	N	30ChDr	9-10/24	3418	4783
SALMON P. CHASE		14Sec	3-4/30	3958A	6357
SALO		12SecDr	10-12/20	2410F	4591
SALONIKA		12SecDr	3-6/20	2410F	4565
SALPHRONA		12SecDr	1-2/15	2410B	4311
SALT LAKE CITY	N	10SecLngObs	9-10/14	2521A	4292
SALTSBURG		12SecDr	2-5/25	3410	4845
SALUDA		12SecDr	2-3/15	2410B	4318
SALUSKIN		16Sec SBS	4/15	2412D	4310
SALVADOR DIAZ MIRON NdeM-243	SN	8SecDr3Dbr	1/25	4090	4843
SALVERLEY		12SecDr	1-3/17	2410F	4450
SAM DAVIS	N	12SecDr	11-12/13	2410A	4234
SAM HOUSTON	N	10SecDrCpt	4-6/27	3973	6043
SAM HOUSTON	RN	10SecDr2Dbr	1/25	4074C	4843
SAMAMISH		16Sec SBS	4/15	2412D	4310
SAMANTHA		26ChDr	3-4/14	2416A	4264
SAMITAR		12SecDr	12/10-11	2410	3866
SAMOSET		12SecDr	10-12/26	3410A	6023
SAMUEL ADAMS		12SecDr	4/26	3410A	4943
SAMUEL BARD		14Sec	10-11/26	3958	6012
SAMUEL MORSE		14Sec	10-11/26	3958	6012

CAR NAME	ST	CAR TYPE	BUILT	PLAN	LOT
SAMUEL P. LANGLEY		14Sec	3/27	3958	6054
SAMUEL SLATER	N	14Sec	9-10/25	3958	4869
SAMUEL SPENCER	N	10Sec2Dr	9-10/25	3584A	4899
SAN ANTONIO		12SecDr	10/22-23	2410H	4647
SAN ARDO	N	7CptLngObs	10/12	2635	4041
SAN BENITO		10SecLngObs	12/16-17	2521C	4442
SAN BERNARDINO		10SecLngObs	12/16-17	2521C	4442
SAN DIEGO		10SecLngObs	9-10/14	2521A	4292
SAN FRANCISCO		12SecDr	8-10/26	3410A	4945
SAN ISIDORO	N	10SecDr2Cpt	10-11/13	2585A	4218
SAN JACINTO		10SecLngObs	9-10/14	2521A	4292
SAN JOAQUIN		10Sec2Dr	2/13	2584	4092
SAN JOSE		10Sec2Dr	2/13	2584	4092
SAN JUAN	N	10SecLngObs	12/16-17	2521C	4442
SAN LUCAS	N	7CptLngObs	10/12	2635	4041
SAN MARCIAL		10SecLngObs	12/16-17	2521C	4442
SAN MARCOS		12SecDr	10/22-23	2410H	4647
SAN MATEO		10SecLngObs	12/16-17	2521C	4442
SAN PABLO		10Sec2Dr	2/13	2584	4092
SAN PEDRO	N	10SecLngObs	12/16-17	2521C	4442
SAN RAMON		10SecLngObs	12/16-17	2521C	4442
SAN SALVADOR NdeM-142	N	6Sec6Dbr	1-3/17	4084	4450
SANDBORN		12SecDr	10/19/10	2410	3814
SANDERSON		12SecDr	12/10-11	2410	3866
SANDFIELD		12SecDr	1-2/16	2410E	4356
SANDOWN		10SecDr2Cpt	11-12/11	2585	3942
SANDUSKY VALLEY		DrSbrBLngObs	4/29	3988	6221
SANDY HOOK		BaggBSmkBarber	12/26-27	3951C	6022

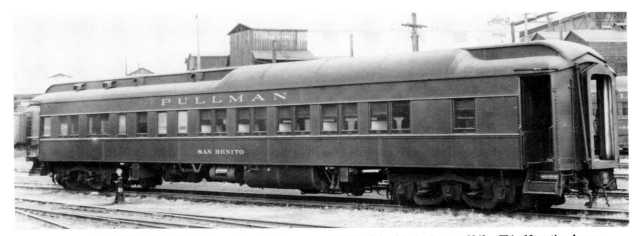

Figure I-67 SAN BENITO is shown here in Vancouver, British Columbia in August 1948. This 10-section lounge observation remained in Pullman ownership and was placed in government storage at Utah General Depot, Ogden, from 1954 to 1961 when it was sold for scrap to Luria Brothers. The Robert Wayner Collection.

CAR NAME	ST	CAR TYPE	BUILT	PLAN	LOT	CAR NAME	ST	CAR TYPE	BUILT	PLAN	LOT
SANDY RIDGE		28ChDr	6-7/27	3416A	6087	SCENIC RAVINE	RN	10SecDr2Dbr	12/26-27	4074	6031
SANFORD		7Cpt2Dr	11/13	2522A	4222	SCENIC RIDGE	RN	10SecDr2Dbr	9-10/29	4074	6284
SANKATY LIGHT		36Ch	12/29-30	3916B	6319	SCENIC RIDGE	R	4Sec6Rmt4Dbr	9-10/29	4182	6284
SANSILLA		16Sec	12/14-15	2412C	4304	SCENIC SLOPE	RN	10SecDr2Dbr	12/28-29	4074	6213
SANTA FE		10SecLngObs	9-10/14	2521A	4292	SCENIC TRAIL	RN	10SecDr2Dbr	12/28-29	4074	6213
SANTA MARIA		10Sec2Dr	2/13	2584	4092	SCENIC VALLEY	RN	10SecDr2Dbr	12/28-29	4074	6213
SANTA RITA		10SecLngObs	9-10/14	2521A	4292	SCHENECTADY		12SecDr	12/12-13	2410	4076
SANTIAGO		12SecDr	9-10/20	2410F	4590	SCHENEVUS		12SecDr	1-3/17	2410F	4450
SANTIAGO		23StsBLng	11/26	3967	4979	SCHENLEY		24ChDrB	4/13	2417	4108
SANTIAGO NdeM	SN	6Sec6Dbr	1-3/17	4084	4450	SCHENLEY	R	18Ch10StsCafeBDr	4/13	2417H	4108
SANTOS						SCHILLER		6Cpt3Dr	11-12/23	3523	4726
DEGOLLADO NdeM-242	N	12SecDr	5-8/25	3410A	4868	SCHODACK		12SecDr	1-4/12	2410	3949
SAOMA		12SecDr	3-6/15	2410B	4319	SCHOFIELD	N	16Sec	5-6/16	2412F	4386
SAPULPA		12SecDr	3-6/20	2410F	4565	SCHOHARIE		16Sec	4/17	2412F	4484
SAQUITO		12SecDr	9-10/13	2410A	4215	SCHOHARIE VALLEY		SbrDrBLngObs	7/30	3988A	6356
SARAH		30ChDr	11/12	2668	4051	SCHUBERT		16Sec	5-6/16	2412F	4386
SARAH BERNHARDT		30ChDr	1/27	3418A	6034	SCHUBERT		6Cpt3Dr	11-12/24	3523A	4833
SARANE		26ChDr	4/12	2416	3950	SCHUMANN		6Cpt3Dr	11-12/24	3523A	4833
SARASOTA		12SecDr	3-6/20	2410F	4565	SCHUYLER MANSION		13Dbr	8/30	3997A	6392
SARATOGA		16Sec	12/18	2412X	4489	SCHUYLKILL		12SecDr	7-9/17	2410F	4515
SARBEN		12SecDr	11-12/13	2410A	4234	SCIOTO COUNTRY CLUB	RN	8SecBLngObs	6/21/10	4025C	3801
SARDINIA		12SecDr	9-10/20	2410F	4590	SCORPIO		12SecDr	6-10/21	2410F	4614
SARDONYX		16Sec	6-7/20	2412F	4568	SCOTLAND		12SecDr	9-10/20	2410F	4590
SARPEDON		12SecDr	3-6/20	2410F	4565	SCOTT		6Cpt3Dr	12/23-24	3523	4741
SARSFIELD		12SecDr	10-12/20	2410F	4591	SCOTT CIRCLE		4CptLngObs	4/26	3960	4944
SARTWELL		12SecDr	7-9/17	2410F	4515	SCOTT HAVEN		10Sec2Dr	1/12	2584	3947
SARVER		12SecDr	7-9/17	2410F	4515	SCOTTDALE		10Cpt	8/22/10	2505	3833
SATTERFIELD	N	12SecDr	10-12/15	2410D	4338	SCOVILLE	N	12SecDr	11/17/10	2410	3864
SATURN		20StsCoachBLng	4/30	3996	6312	SCRANTON		12SecDr	2-3/15	2410B	4318
SATURN	R	16ChB15StsLng	4/30	3996C	6312	SCRIBNER		12SecDr	5-8/18	2410F	4540
SATURN CLUB	RN	8SecRestObsLng	9-10/14	4025A	4292	SCUTARI		12SecDr	9-10/20	2410F	4590
SAUGERTIES		12SecDr	9-10/11	2410	3924	SEA GIRT		12SecDr	4/15/10	2410	3771
SAUGUS		12SecDrCpt	6/13	2411A	4159	SEA ISLAND	N	6SecDrLng	10-11/14	4010	4296
SAUTRELLE		12SecDr	4-5/13	2410	4106	SEABOARD	RN	6SecDrLng	10-11/14	4010	4296
SAVERTON		12SecDr	7-8/15	2410D	4322	SEABREEZE	RN	6SecDrLng	12/14-15	4010	4304
SAXE		6Cpt3Dr	11-12/24	3523A	4833	SEABROOK		12SecDr	12/22/10	2410	3864
SAYBROOK		36Ch SBS	1/13	2691	4056	SEAFARER	RN	17StsRestLng	3-4/14	4019A	4264
SAYVILLE		12SecDr	3-6/20	2410F	4565	SEAFORD		26ChDr	2-3/14	2416A	4248
SCANLON		10SecDr2Cpt	12/16-17	2585D	4443	SEASIDE	RN	6SecDrLng	10/13	4010	4216
SCARBOROUGH		12SecDr	1-4/12	2410	3949	SEASIDE PARK		BaggBClubSmkBarber	7/23	2951B	4698
SCARCLIFF		12SecDr	1-2/15	2410B	4311	SEATTLE		12SecDr	4-6/24	3410	4763
SCARSDALE		12SecDr	4-6/14	2410A	4271	SEAVIEW	RN	6SecDrLng	1/16	4010	4352
SCENIC CASCADE	RN	10SecDr2Dbr	9-10/29	4074	6284	SEBAGO		12SecDr	7-8/15	2410D	4322
SCENIC CASCADE	R	4Sec6Rmt4Dbr	9-10/29	4182	6284	SEBASTIAN		12SecDr	10-12/15	2410D	4338
SCENIC FALLS	RN	10SecDr2Dbr	12/28-29	4074	6213	SEBEKA		12SecDr	10-12/20	2410F	4591
SCENIC FOREST	RN	10SecDr2Dbr	12/26-27	4074	6031	SECANE		BaggBClubSmkBarber	8/19/10	2414A	3803
SCENIC GLADE	RN	10SecDr2Dbr	9-10/29	4074	6284	SECURITAS		12SecDr	2-3/15	2410B	4318
SCENIC HIGHLANDS	RN	10SecDr2Dbr	12/26-27	4074	6031	SEDGEMOOR		12SecDr	10-12/15	2410D	4338
SCENIC ISLAND	RN	10SecDr2Dbr	9-10/25	4074	4899	SEDGWICK		10SecDr2Cpt	11-12/11	2585	3942
SCENIC RAPIDS	RN	10SecDr2Dbr	9-10/25	4074A	4899	SEINE		12SecDr	9-10/20	2410F	4590

CAR NAME	ST	CAR TYPE	BUILT	PLAN	LOT
SELBYVILLE		12SecDr	7-9/17	2410F	4515
SELMA		12SecDr	10/19/10	2410	3814
SEMLOH		12SecDr	1-2/15	2410B	4311
SEMLOH		12SecDr	6-10/21	2410F	4614
SEMPACH		12SecDr	10-12/15	2410D	4338
SEMPLE		12SecDr	3-4/17	2410F	4485
SENACHWINE		16Sec	10-11/14	2412C	4296
SENATE		12SecDr	4/26	3410A	4943
SENATOR					
HENRY CABOT LODGE	N	BaggBSmkBarber	12/26-27	3951C	6022
LEWIS	N	12SecDr	2-5/25	3410	4845
MORGAN G. BULKELEY	N	BaggBSmkBarber	12/26-27	3951C	6022
SENECA FALLS		10Sec2Dr	5/13	2584	4140
SENECA VALLEY		SbrDrBLngObs	7/30	3988A	6356
SENNETT		12SecDr	3-4/17	2410F	4485
SENORITA		26ChDr	3-4/14	2416A	4264
SENTA		24ChDrB	5/13	2417A	4137
SENTINEL PASS		8SecDr2Cpt	7-8/30	3979A	6377
SENTINEL PEAK		10Sec2Dr	12/26-27	3584A	6031
SENTINEL RANGE		10SecDrCpt	5/30	3973A	6358
SENTINUM		16Sec	1/16	2412E	4352
SEQUOYAH		7Cpt2Dr	9/13	2522A	4208
SERAPIS		16Sec	5-6/13	2412B	4150
SERGIUS		12SecDr	6-7/17	2410F	4503
SERRA		12SecDr	8-10/26	3410A	4945
SERVIAN		12SecDr	12/12-13	2410	4076
SETH WARNER		12SecDr	8-12/25	3410A	4894
SETUCKIT		12SecDr	6-10/21	2410F	4614
SEVENA		26ChDr	3-4/14	2416A	4264
SEWANEE	N	14Sec	9-10/25	3958	4869
SEWANEE	RN	10Sec3Dbr	2-5/25	4087B	4845
SEWAREN		12SecDr	7-8/15	2410D	4322
SEWELL	N	12SecDr	5/25/10	2410	3794
SHABBONA		12SecDr	8-9/12	2410	4035
SHADE GAP		8SecDr2Cpt	12/29-30	3979A	6334
SHADWELL		12SecDr	1-3/14	2410A	4249
SHAFFER		12SecDr	7-9/17	2410F	4515
SHAFTON		12SecDr	7-9/17	2410F	4515
SHAKESPEARE		6Cpt3Dr	12/23-24	3523	4741
SHAKOPEE		10SecDrCpt	4-6/27	3973	6043
SHALETON		12SecDr	7-9/17	2410F	4515
SHALIMAR		12SecDr	6-10/21	2410F	4614
SHALOTT		12SecDr	4-5/13	2410	4106
SHAMOKIN		12SecDr	11/19/10	2410	3864
SHANNOCK		12SecDr	11-12/13	2410A	4234
SHANNON		12SecDr	10-12/20	2410F	4591
SHAPLEIGH		12SecDr	4-5/13	2410	4106
SHAPONACK		12SecDr	1-2/15	2410B	4311
SHARDLOW		12SecDr	1-2/15	2410B	4311
SHARON		12SecDr	1-2/16	2410E	4356
SHARONVILLE		12SecDr	9-10/11	2410	3924
SHARPSBURG		12SecDr	7-9/17	2410F	4515
SHATTUCK		12SecDr	3-6/20	2410F	4565
SHAWMONT		12SecDr	11/8/10	2410	3860
SHAWMUT		12SecDr	10-12/26	3410A	6023
SHAWNEE		12SecDr	5-6/23	2410H	4699
SHEDDEN		12SecDr	10-12/20	2410F	4591
SHEFFIELD		32ChDr	8/27	3917A	6078
SHELDON		12SecDr	9-10/11	2410	3924
SHELLEY		6Cpt3Dr	12/23-24	3523	4741
SHELOCTA		12SecDr	6-10/21	2410F	4614
SHENANGO		12SecDr	10-12/20	2410F	4591
SHENLEY		12SecDr	10-12/20	2410F	4591
SHENSTONE		12SecDr	12/12-13	2410	4076
SHEOMET		12SecDr	6-10/21	2410F	4614
SHEPAUG		34Ch	8/13	2764A	4195
SHERATON		12SecDr	2-5/16	2410E	4367
SHERIDAN		12SecDr	5-8/18	2410F	4540
SHETLAND		12SecDr	9-10/20	2410F	4590
SHEVLIN		12SecDr	1-3/14	2410A	4249
SHINDLE		12SecDr	7-9/17	2410F	4515
SHINUMO		7Cpt2Dr SBS	11/11	2522	3941
SHIP ROAD		12SecDr	4/30/10	2410	3772
SHIPPENSBURG		12SecDr	5-6/23	2410H	4699
SHIREMANSTOWN		12SecDr	5-6/23	2410H	4699
SHIRVAN		16Sec	5-6/16	2412F	4386
SHORE LARK		3Sbr2CptDrBLngSunR	3/30	3991A	6352
SHOREHAM		12SecDr	9-10/11	2410	3924
SHORELINE		12SecDr	5-8/18	2410F	4540
SHOREWOOD		14Sec	8-10/28	3958A	6181
SHOSHONE TRIBE	RN	6SecDr4Dbr	4-5/26	4092	4961
SHOW ME	N	8SecDr3Dbr	1/25	4090B	4843
SHOWELL		12SecDr	7-9/17	2410F	4515
SHRADERS		12SecDr	6-9/26	3410A	4969
SHREVEPORT CLUB	RN	10SecRestLngObs	6-7/13	4027A	4160
SHREWSBURY		10SecDr2Cpt	12/15	2585C	4351
SHRIEVER		12SecDr	6-7/17	2410F	4503
SHRINE TOWER	RN	8SecDr3Dbr	5-6/25	4090E	4846
SHUBRICK		12SecDr	2-5/16	2410E	4367
SHUMLA		12SecDr	6-7/17	2410F	4503
SIAM		12SecDr	9-10/20	2410F	4590
SIBLEY		14Sec	6/28	3958A	6174
SIBYL		24ChDrB	4/13	2417	4108
SIDMAR		12SecDr	1-2/16	2410E	4356
SIDNEY		12SecDr	12/12-13	2410	4076
SIDNEY LANIER	RN	10SecDr2Dbr	9-10/25	4074B	4899
SIEBERS TOWER	RN	8SecDr3Dbr	5-6/25	4090	4846
SIERRA					
DE JUAREZ NdeM	SN	10SecDrCpt	5-7/18	4031	4540
MADRE		10Sec2Dr	2/13	2584	4092

CAR NAME	ST	CAR TYPE	BUILT	PLAN	LOT
SIERRA					
MADRE	N	8Sec5Dbr	6-7/16	4036I	4385
VISTA		10Sec2Dr	2/13	2584	4092
SIESTA		12SecDr	3-6/20	2410F	4565
SIGNAL PEAK		10Sec2Dr	12/28-29	3584B	6213
SIGRID		26ChDr	6/11	2416	3902
SIKESTON		10SecDr2Cpt	6-7/15	2585B	4328
SILHOUETTE TOWER	RN	8SecDr3Dbr	8-12/25	4090F	4894
SILLISTRIA		12SecDr	10-12/15	2410D	4338
SILOAM SPRINGS	RN	10Sec3Dbr	2-5/25	4087B	4845
SILSILI		12SecDr	6-10/21	2410F	4614
SILVER BEACH		3Cpt2DrLngLibObs	12/24	3950D	4835
SILVER BOW		3Cpt2DrLngLibObs	12/24	3950D	4835
SILVER BROOK		3Cpt2DrLngLibObs	12/24	3950D	4835
SILVER CITY		3Cpt2DrLngLibObs	12/24	3950D	4835
SILVER CREEK		12SecDr	3-6/20	2410F	4565
SILVER LAKE		12SecDrCpt	8-9/11	2411	3922
SILVER LEAF		3Cpt2DrLngLibObs	12/24	3950D	4835
SILVER PEAK		3Cpt2DrLngLibObs	12/24	3950D	4835
SILVER PLUME		3Cpt2DrLngLibObs	12/24	3950D	4835
SILVER SPRINGS		3Cpt2DrLngLibObs	12/24	3950D	4835
SILVER SPRUCE		10SecDrCpt	5/30	3973A	6358
SILVER STAR		12SecDr	10/22-23	2410H	4647
SILVER VALLEY		3Cpt2DrLngLibObs	12/24	3950D	4835
SILVERTON		12SecDrCpt	3/11	2411	3880
SIMODA		10SecDr2Cpt	1-2/13	2585	4091
SIMON KENTON	N	10SecDrCpt	4-6/27	3973	6043
SIMON KENTON	RN	10SecDr2Dbr	1/25	4074C	4843
SIMONIDES		12SecDrCpt	10-12/21	2411C	4624
SIMPLON		12SecDr	7-8/13	2410A	4192
SIMPLON PASS		8SecDr2Cpt	7-8/30	3979A	6377
SIMSBURY		32ChDr	6/16	2917	4390
SINBAD		12SecDr	6-10/21	2410F	4614
SINGARA		12SecDr	10-12/15	2410D	4338
SINGING TOWER	RN	8SecDr3Dbr	10-12/26	4090F	6023
SINOPE		12SecDr	10-12/15	2410D	4338
SIOUX CITY		10SecDrCpt	4-6/27	3973	6043
SIOUX FALLS		10SecDrCpt	4-6/27	3973	6043
SIR HENRY					
W. THORNTON	N	10SecDr2Cpt	10-12/20	2585D	4592
SIRDAR		12SecDr	11-12/14	2410B	4297
SIRIUS		12SecDr	10-12/20	2410F	4591
SITKA		12SecDr	10-12/26	3410A	6023
SIVERLY		12SecDr	7-9/17	2410F	4515
SIWANOY CLUB		8SecBObsLng	2-3/29	3989	6229
SKIPANON		12SecDr	4/15	2410C	4309
SKOKIE CLUB		8SecBObsLng	2-3/29	3989	6229
SKYLAND		7Cpt2Dr	11/13	2522A	4222
SLATINGTON		12SecDr	7-8/12	2410	4014
SLEEPY HOLLOW		BaggBSmkBarber	12/26-27	3951C	6022

CAR NAME	ST	CAR TYPE	BUILT	PLAN	LOT
SLIDELL		12SecDr	6-7/17	2410F	4503
SLIGO		12SecDr	10-12/20	2410F	4591
SMALLEY		12SecDr	7-9/17	2410F	4515
SMALLWOOD		12SecDr	11-12/13	2410A	4234
SMITH COLLEGE	RN	10Sec2DbrCpt	12/14-15	4042B	4304
SMITHBORO		12SecDr	4/28/10	2410	3772
SMITHDALE		12SecDr	4-6/14	2410A	4271
SMITHFIELD		12SecDr	3-6/20	2410F	4565
SMITHSONIAN		12SecDr	4/26	3410A	4943
SMITHTON		12SecDr	1-4/12	2410	3949
SMITHVILLE		28ChDr	5-6/24	3416	4761
SMOCK		28ChDr	5-7/26	3416	4958
SMYRNA		BaggBClubSmkBarber	8/22/10	2414A	3803
SMYSER		12SecDr	7-9/17	2410F	4515
SNEDIKER		12SecDr	2-5/16	2410E	4367
SNOQUALMIE		12SecDr	4/15	2410C	4309
SOBRADES		12SecDr	6-10/21	2410F	4614
SOBRANTE		12SecDr	2-5/16	2410E	4367
SOBRE					
LAS OLAS	N	10SecDrCpt	3-5/30	3973A	6338
LAS OLAS FCP-207	SN	10SecDr2Cpt	12/25-26	3585A	4933
SOCRATES		12SecDrCpt	10-12/21	2411C	4624
SOHO		12SecDr	10-12/20	2410F	4591
SOISSONS		7Dr	12/20	2583B	4621
SOLANO		12SecDr	8-10/26	3410A	4945
SOLDANI		12SecDr	3-6/20	2410F	4565
SOLEDAD		12SecDr	11-12/13	2410A	4234
SOLFATARA		12SecDr	9-10/13	2410A	4215
SOLFERINO		10SecDr2Cpt	12/15	2585C	4351
SOLON		12SecDr	6-10/21	2410F	4614
SOLWAY		12SecDr	12-10-11	2410	3866
SOMENA		10SecDr2Cpt	11-12/11	2585	3942
SOMERFIELD		12SecDr	7-8/12	2410	4014
SOMERSET	N	12SecDr	7-8/12	2410	4014
SOMERSET CLUB	RN	10SecRestLngObs	1-5/18	4027	4531
SOMERVILLE		12SecDr	4/21/10	2410	3770
SOMME		7Dr	12/20	2583B	4621
SOMNUS		12SecDr	7-8/15	2410D	4322
SOMONAUK		16Sec	5-6/13	2412B	4150
SONOGEE		12SecDr	1-3/17	2410F	4450
SONOMA		10SecDr2Cpt	1-2/13	2585	4091
SONORA PEAK		10Sec2Dr	12/28-29	3584B	6213
SOPHOCLES		12SecDrCpt	10-12/21	2411C	4624
SOPHRONIA		26ChDr	6/11	2416	3902
SOUTH GAP		8SecDr2Cpt	12/29-30	3979A	6334
SOUTH NORWALK		36Ch	8/27	3916A	6077
SOUTH PASS		8SecDr2Cpt	12/29-30	3979A	6334
SOUTH SHORE		12SecDr	3-6/15	2410B	4319
SOUTH WILTON		BaggBClubSmk	6/16	2919	4392
SOUTHERN COLLEGE	RN	10Sec2DbrCpt	11/17	4042	4527

Figure I-68 This Seaboard work train car was built in 1912, Lot 4014, Plan 2410, as SPRINGVILLE. In December 1941 it was converted to Tourist Car 1316 and sold to the Seaboard Airline in December 1947 where it was converted to work train service. This photograph was taken March 23, 1967 in Creedmore, North Carolina by R. S. Short.

CAR NAME	ST	CAR TYPE	BUILT	PLAN	LOT
SOUTHEY		6Cpt3Dr	12/23-24	3523	4741
SOUTHPORT		32ChDr	6/16	2917	4390
SOUTHWICK		12SecDr	9-10/11	2410	3924
SOVEREIGN					
OF THE SEAS		34Ch	2/30	4001	6325
SPAIN		12SecDr	9-10/20	2410F	4590
SPALDING		10SecDr2Cpt	12/16-17	2585D	4443
SPANGLER		12SecDr	1-4/12	2410	3949
SPANISH CREST	N	12SecDr	6-8/24	3410F	4764
SPANISH OAK		8SecDr2Cpt	7-8/30	3979A	6377
SPANISH PEAKS		10Sec2Dr	12/26-27	3584B	6031
SPANISH RANGE	N	12SecDr	6-8/24	3410F	4764
SPARKS		10SecDr2Cpt	11-12/11	2585	3942
SPARTA		14Sec	4/27	3958	6041
SPARTA	R	6Sec6Dbr	4/27	4084B	6041
SPARTACUS		12SecDr	1-3/17	2410F	4450
SPEAR LAKE		10SecDr2Cpt	11-12/27	3585B	6123
SPEECEVILLE		12SecDr	6-9/26	3410A	4969
SPENSER		6Cpt3Dr	12/23-24	3523	4741
SPICELAND		12SecDr	1-4/12	2410	3949
SPIRE PEAK		10Sec2Dr	12/28-29	3584B	6213
SPIRIT OF '76	N	12SecDr	10-12/26	3410A	6023
SPIRIT OF ST. LOUIS	N	3Cpt2DrLngLibObs	12/29-30	3950E	4744
SPOFFORD		12SecDrCpt	6/13	2411A	4159
SPOKANE		12SecDr	4/15	2410C	4309
SPRAGUE	N	12SecDr	6-7/17	2410F	4503
SPRAKERS		12SecDr	9-10/11	2410	3924
SPRING ARBOR		10Sec2Dr	1/12	2584	3947

CAR NAME	ST	CAR TYPE	BUILT	PLAN	LOT
SPRING BROOK CLUB		8SecBLngSunRm	8-9/29	3989A	6274
SPRING BROOK CLUB	RN	8SecRestLngObs	4/17	4025	4484
SPRING CITY		28ChDr	6-7/27	3416A	6087
SPRING GAP		8SecDr2Cpt	12/29-30	3979A	6334
SPRING MEADOW		28ChDr	6-7/27	3416A	6087
SPRINGBORO		12SecDr	1/14	2410A	4240
SPRINGBORO	N	12SecDr	5-8/25	3410A	4868
SPRINGDALE		24ChDrB	4/13	2417	4108
SPRINGFIELD		10Cpt	8/18/10	2505	3833
SPRINGFIELD	R	5CptBLng	8/18/10	2505C	3833
SPRINGHAVEN		16Sec	6-7/20	2412F	4568
SPRINGVILLE		12SecDr	7-8/12	2410	4014
SPRITE		10SecDr2Cpt	12/15	2585C	4351
SPROUL		12SecDr	2-5/25	3410	4845
SPRUCE CREEK		BaggBClubSmkBarber	7/23	2951B	4698
SPUYTEN DUYVIL		10Cpt	3/14	2505A	4252
SPUYTEN DUYVIL	R	5CptBLng	3/14	2505B	4252
ST. ADELE		12SecDr	2-4/24	3410	4762
ST. ALBANS		12SecDr	10-12/23	3410	4724
ST. ALEXIS		12SecDr	10-12/23	3410	4724
ST. ANGELE		12SecDr	2-4/24	3410	4762
ST. ANSELME		12SecDr	2-4/24	3410	4762
ST. ANSGAR	N	12SecDr	2-4/24	3410	4762
ST. ANSGER		12SecDr	2-4/24	3410	4762
ST. ANTHONY		12SecDr	10-12/23	3410	4724
ST. ARMAND		12SecDr	10-12/23	3410	4724
ST. ARSENE		12SecDr	2-4/24	3410	4762
ST. AUBERT		12SecDr	10-12/23	3410	4724

CAR NAME	ST	CAR TYPE	BUILT	PLAN	LOT
ST. AUBERT		12SecDr	2-4/24	3410	4762
ST. AUGUSTINE		12SecDr	10/22-23	2410H	4647
ST. BENEDICT		12SecDr	2-4/24	3410	4762
ST. BERNARD		12SecDr	10-12/23	3410	4724
ST. BERNICE		12SecDr	2-4/24	3410	4762
ST. BETHLEHEM		12SecDr	2-4/24	3410	4762
ST. BRIDES		12SecDr	2-4/24	3410	4762
ST. CARVAN		12SecDr	2-4/24	3410	4762
ST. CATHERINE		12SecDr	2-4/24	3410	4762
ST. CHARLES		12SecDr	10-12/23	3410	4724
ST. CLAIR		12SecDr	10-12/23	3410	4724
ST. CLAUDE		12SecDr	10-12/23	3410	4724
ST. CLOUD		12SecDr	10-12/23	3410	4724
ST. COLLINS		12SecDr	2-4/24	3410	4762
ST. CROIX		12SecDr	10-12/23	3410	4724
ST. DAVIDS		12SecDr	4/23/10	2410	3771
ST. DELPHINE		12SecDr	2-4/24	3410	4762
ST. DENIS		12SecDr	10-12/23	3410	4724
ST. EDWARD		12SecDr	2-4/24	3410	4762
ST. ELMO		12SecDr	10-12/23	3410	4724
ST. FRANCIS		12SecDr	10-12/23	3410	4724
ST. GABRIEL		12SecDr	2-4/24	3410	4762
ST. GENEVIEVE		12SecDr	2-4/24	3410	4762
ST. GEORGE		12SecDr	10-12/23	3410	4724
ST. GERMAIN		12SecDr	10-12/23	3410	4724
ST. GOTHARD PASS		8SecDr2Cpt	7-8/30	3979A	6377
ST. HELENA		12SecDr	10-12/23	3410	4724
ST. HELIER		12SecDr	10-12/23	3410	4724
ST. HILAIRE		12SecDr	2-4/24	3410	4762
ST. HUBERTS	N	12SecDr	10-12/23	3410	4724
ST. IGNACE		12SecDr	10-12/23	3410	4724
ST. IVES		12SecDr	10-12/23	3410	4724
ST. JAMES		12SecDr	10-12/23	3410	4724
ST. JOHNS		12SecDr	10-12/23	3410	4724
ST. JOSEPH		12SecDr	10-12/23	3410	4724
ST. JULIEN		12SecDr	10-12/23	3410	4724
ST. LAMBERT		12SecDr	2-4/24	3410	4762
ST. LAWRENCE		12SecDr	10-12/23	3410	4724
ST. LEON		12SecDr	2-4/24	3410	4762
ST. LEONARDS		12SecDr	10-12/23	3410	4724
ST. LOUIS		12SecDr	10-12/23	3410	4724
ST. LOUIS CLUB	RN	8SecRestLngObs	12/11-12	4025A	3944
ST. LUCIEN		12SecDr	10-12/23	3410	4724
ST. MALO		12SecDr	10-12/23	3410	4724
ST. MARIE		12SecDr	10-12/23	3410	4724
ST. MARTINS		12SecDr	10-12/23	3410	4724
ST. MARY OF THE LAKE	LN	Dining Car(Rexall)	11-12/25	3952C	4916
ST. MIHIEL		7Dr	12/20	2583B	4621
ST. NICHOLAS		12SecDr	10-12/23	3410	4724
ST. NORBERT		12SecDr	10-12/23	3410	4724

CAR NAME	ST	CAR TYPE	BUILT	PLAN	LOT
ST. PANCRAS		12SecDr	10-12/23	3410	4724
ST. PAUL	N	12SecDr	10-12/23	3410	4724
ST. PAUL CLUB	RN	8SecRestLngObs	11/12	4025A	4049
ST. PETER		12SecDr	10-12/23	3410	4724
ST. PETERSBURG		12SecDr	9-10/13	2410A	4215
ST. PETERSBURG		12SecDr	10/22-23	2410H	4647
ST. PIERRE		12SecDr	10-12/23	3410	4724
ST. QUENTIN		7Dr	12/20	2583B	4621
ST. REGIS		12SecDr	10-12/23	3410	4724
ST. ROSE		12SecDr	2-4/24	3410	4762
ST. SERVAN		12SecDr	10-12/23	3410	4724
ST. THOMAS		12SecDr	10-12/23	3410	4724
ST. VICTOR		12SecDr	10-12/23	3410	4724
STAATSBURG		BaggClubBSmk	11/24	3415	4808
STAATSBURGH		12SecDr	3-4/11	2410	3881
STACIA		12SecDr	2-3/15	2410B	4318
STAG HOUND		20ChBLng	2/30	3999	6327
STALWART	RN	22Ch12StsBLng	8/13	4032	4195
STAMFORD		40Ch8StsBDrObs	8/13	2766	4197
STANDISH		12SecDr	3-6/20	2410F	4565
STANFIELD		26ChDr	5/12	2416	3986
STANHOPE	N	12SecDr	10/6/10	2410	3813
STANWICK		12SecDr	11/12/10	2410	3860
STAR BAY		14Sec	8-10/28	3958A	6181
STAR BEACH		14Sec	11-12/28	3958A	6212
STAR BEAM		14Sec	8-10/28	3958A	6181
STAR BROOK		14Sec	8-10/28	3958A	6181
STAR CITY		28ChDr	6-7/27	3416A	6087
STAR CLUSTER		14Sec	10-11/29	3958A	6285
STAR CRAG		14Sec	8-10/28	3958A	6181
STAR CREST		14Sec	8-10/28	3958A	6181
STAR DELL		14Sec	8-10/28	3958A	6181
STAR DRIFT		14Sec	11-12/28	3958A	6212
STAR FINCH		14Sec	8-10/28	3958A	6181
STAR FLOWER		14Sec	11-12/28	3958A	6212
STAR GLEN		14Sec	8-10/28	3958A	6181
STAR GRASS		14Sec	8-10/28	3958A	6181
STAR ISLAND		14Sec	11-12/28	3958A	6212
STAR LAKE		10SecDr2Cpt	11-12/27	3585B	6123
STAR LAND		14Sec	8-10/28	3958A	6181
STAR LIGHT		14Sec	10-11/29	3958A	6285
STAR LILY		34Ch	4/26	3419	4956
STAR MANOR		14Sec	8-10/28	3958A	6181
STAR MOUNT		14Sec	10-11/29	3958A	6285
STAR PEAK		14Sec	8-10/28	3958A	6181
STAR POINT		14Sec	8-10/28	3958A	6181
STAR RAPIDS		14Sec	11-12/28	3958A	6212
STAR SPUR		14Sec	11-12/28	3958A	6212
STAR STONE		14Sec	8-10/28	3958A	6181
STAR SUMMIT		14Sec	11-12/28	3958A	6212

CAR NAME	ST	CAR TYPE	BUILT	PLAN	LOT
STAR TRAIL		14Sec	8-10/28	3958A	6181
STAR VALLEY		14Sec	11-12/28	3958A	6212
STAR VIEW		14Sec	11-12/28	3958A	6212
STAR VISTA		14Sec	11-12/28	3958A	6212
STARKEY		12SecDr	2-5/25	3410	4845
STARVED ROCK		BaggBSmkBarber	12/26-27	3951C	6022
STATE CAPITAL	RN	10Sec3Dbr	2-5/25	4087B	4845
STATE LEGION	RN	32StsBSunRm	6/26	3964E	4965
STATESMAN		12SecDr	4/26	3410A	4943
STAUNTON		12SecDr	11-12/13	2410A	4234
STEDMAN		12SecDrCpt	10-12/21	2411C	4624
STEELTON		12SecDr	5/6/10	2410	3773
STEILACOOM		12SecDr	4/15	2410C	4309
STEINERS		12SecDr	2-4/24	3410	4762
STELLA		28ChDr	4-6/25	3416	4864
STELVIO PASS		8SecDr2Cpt	7-8/30	3979A	6377
STEPHEN					
COLLINS FOSTER	N	12SecDr	9-10/15	2410D	4327
DECATUR		12SecDr	8-12/25	3410A	4894
GIRARD		12SecDr	5-8/25	3410A	4868
HOPKINS		12SecDr	4/26	3410A	4943
STEPHIA		12SecDr	3-6/15	2410B	4319
STERLING		12SecDr	7-8/28	3410B	6184
STEUBENVILLE		12SecDr	4/8/10	2410	3770
STEUBENVILLE		12SecDrCpt	5/17	2411C	4494
STEVENS		12SecDr	2-5/25	3410	4845
STICKNEY		12SecDr	9-10/11	2410	3924
STILLWATER		12SecDr	7-8/28	3410B	6184
STILLWELL		12SecDr	9-10/11	2410	3924
STOCKDALE		12SecDr	2-4/24	3410	4762
STOCKLEY		12SecDr	2-4/24	3410	4762
STOCKPORT		12SecDr	10/24/10	2410	3814
STOCKPORT		BaggClubBSmk	11/24	3415	4808
STOCKWELL		12SecDr	9-10/11	2410	3924
STODDARD		12SecDr	4-5/13	2410	4106
STONE HARBOR		26ChDr	9-11/22	2416D	4649
STONE HAVEN		BaggBClubBSmk	6/16	2919	4392
STONE MOUNTAIN		BaggBSmkBarber	12/26-27	3951C	6022
STONEBORO		12SecDr	10-11/11	2410	3936
STONEHAM		12SecDr	12/12-13	2410	4076
'STONEWALL' JACKSON		10Sec2Dr	9-10/25	3584A	4899
'STONEWALL' JACKSON		10Sec2Dr	9-10/29	3584B	6284
'STONEWALL' JACKSON	N	8SecDr2Cpt	8-9/29	3979A	6283
STONY POINT		10Sec2Dr	5/13	2584	4140
STORM KING		12SecDrCpt	7/14/10	2411	3800
STORM PEAK		10Sec2Dr	12/26-27	3584A	6031
STORY		12SecDrCpt	10-12/21	2411C	4624
STOUTSVILLE		28ChDr	5-6/24	3416	4761
STOVER		12SecDr	2-4/24	3410	4762
STOVERTON		12SecDr	10-11/11	2410	3936

CAR NAME	ST	CAR TYPE	BUILT	PLAN	LOT
STOWE		6Cpt3Dr	11-12/24	3523A	4833
STRABANE		12SecDr	10-12/20	2410F	4591
STRABO		12SecDrCpt	10-12/21	2411C	4624
STRADELLA PASS		8SecDr2Cpt	7-8/30	3979A	6377
STRAFFORD	N	BaggBClubSmk	2-4/09	2136C	3660
STRALSUND		12SecDr	10-12/15	2410D	4338
STRASBOURG TOWER	RN	8SecDr3Dbr	5-6/25	4090	4846
STRATFIELD		12SecDr	1-2/15	2410B	4311
STRATHEARN		12SecDr	9-10/13	2410A	4215
STRATHMORE		12SecDr	12/10-11	2410	3866
STRATTON		10SecDr2Cpt	12/15	2585C	4351
STREATOR		10SecDr2Cpt	12/16-17	2585D	4443
STRELITSO		12SecDr	4-5/13	2410	4106
STRELNA		16Sec	10-11/14	2412C	4296
STRELSA		30ChDr	11/12	2668	4051
STRICKLAND	N	12SecDr	3-4/11	2410	3881
STRONACH		12SecDr	2-4/24	3410	4762
STRUTHERS		12SecDr	10/19/10	2410	3814
STRYKER		BaggBClubSmkBarber	9/24	2951C	4805
STURGIS		12SecDr	9-10/11	2410	3924
STYMPHALE		12SecDr	2-5/16	2410E	4367
SUBLETTE		10SecDr2Cpt	8-9/13	2585A	4201
SUDBURY		12SecDr	7-8/14	2410A	4277
SUENO		12SecDr	10-12/20	2410F	4591
SUEZ		12SecDr	10-12/20	2410F	4591
SUFFERN		12SecDr	3-6/20	2410F	4565
SUGAR LOAF MOUNTAIN	N	14Sec	3-4/30	3958A	6357
SULLINS		12SecDr	6-10/21	2410F	4614
SULPHUR SPRINGS	RN	10Sec3Dbr	2-5/25	4087B	4845
SUMMERDALE		24ChDrB	1/14	2417A	4239
SUMMERFIELD		12SecDr	9-10/13	2410A	4215
SUMMERLAND		12SecDrCpt	3/11	2411	3880
SUMMIT GROVE		6SbrBLngSunRm	12/29	3994B	6291
SUMMIT LAKE		6SbrBLngSunRm	12/29	3994B	6291
SUMMIT PEAK		10Sec2Dr	12/26-27	3584A	6031
SUMNER		16Sec	2/13	2412	4101
SUMTER		7Cpt2Dr	11/13	2522A	4222
SUN-DAWN		2CptDrLngBSunRm	9-10/27	3975B	6076
SUN-GLOW		2CptDrLngBSunRm	9-10/27	3975B	6076
SUN-GOLD		2CptDrLngBSunRm	9-10/27	3975B	6076
SUNAPEE		12SecDr	7-8/14	2410A	4277
SUNBEAM		2CptDrLngBSunRm	9-10/27	3975B	6076
SUNBROOK	N	26ChDr	5/12	2416	3986
SUNBROOK		12SecDr	2-5/25	3410	4845
SUNBURST		2CptDrLngBSunRm	9-10/27	3975B	6076
SUNBURST ROSE	N	12SecDr	8-12/25	3410A	4894
SUNBURY		12SecDr	2-5/16	2410E	4367
SUNDALE		12SecDr	1-3/14	2410A	4249
SUNDERLAND		10Cpt	8/16/10	2505	3833
SUNDRIDGE		12SecDr	1-2/16	2410E	4356

CAR NAME	ST	CAR TYPE	BUILT	PLAN	LOT
SUNGIR		12SecDr	10-12/20	2410F	4591
SUNLIGHT		2CptDrLngBSunRm	9-10/27	3975B	6076
SUNNYMEAD	N	12SecDr	9-10/13	2410A	4215
SUNNYSIDE		12SecDr	10-12/26	3410A	6023
SUNRISE		2CptDrLngBSunRm	9-10/27	3975B	6076
SUNSET BEACH		4Cpt2DrLngObs	4-5/24	3950A	4760
SUNSET BEACH	R	3Cpt2DrLngLibObs	4-5/24	3950C	4760
SUNSET CAPE		4Cpt2DrLngObs	4-5/24	3950A	4760
SUNSET CAPE	R	3Cpt2DrLngLibObs	4-5/24	3950C	4760
SUNSET HEIGHTS		4Cpt2DrLngObs	4-5/24	3950A	4760
SUNSET HEIGHTS	R	3Cpt2DrLngLibObs	4-5/24	3950C	4760
SUNSET LAKE		4Cpt2DrLngObs	4-5/24	3950A	4760
SUNSET LAKE	R	3Cpt2DrLngLibObs	4-5/24	3950C	4760
SUNSET PARK		4Cpt2DrLngObs	4-5/24	3950A	4760

CAR NAME	ST	CAR TYPE	BUILT	PLAN	LOT
SUNSET PARK	R	3Cpt2DrLngLibObs	4-5/24	3950C	4760
SUNSET PEAK		4Cpt2DrLngObs	4-5/24	3950A	4760
SUNSET PEAK	R	3Cpt2DrLngLibObs	4-5/24	3950C	4760
SUNSET ROCK		4Cpt2DrLngObs	4-5/24	3950A	4760
SUNSET ROCK	R	3Cpt2DrLngLibObs	4-5/24	3950C	4760
SUNSET TRAIL		4Cpt2DrLngObs	4-5/24	3950A	4760
SUNSET TRAIL	R	3Cpt2DrLngLibObs	4-5/24	3950C	4760
SUNSET VIEW		4Cpt2DrLngObs	4-5/24	3950A	4760
SUNSET VIEW	R	3Cpt2DrLngLibObs	4-5/24	3950C	4760
SUNSHINE		2CptDrLngBSunRm	9-10/27	3975B	6076
SUNSTAR ROSE	N	12SecDr	8-12/25	3410A	4894
SUPERB		Private (GS)	3/11	2503	3847
SUPERIOR		12SecDr	4-6/24	3410	4763

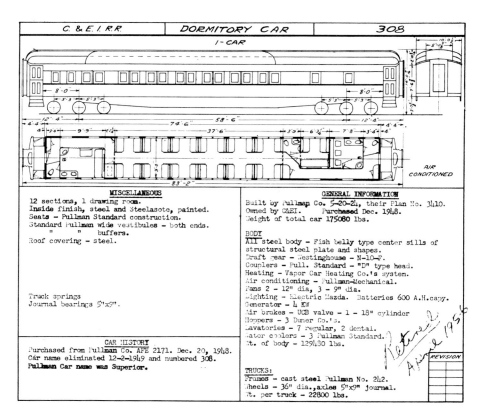

Figure I-69 This diagram for C & EI Dormitory 308, the former Plan 3410, Lot 4763 SUPERIOR, has a cartoon-like side elevation. It is wise to consult photographs for the general appearance of cars; however, diagrams are usually an excellent source for dimensions, car history, and general data about the car. The Author's Collection.

CAR NAME	ST	CAR TYPE	BUILT	PLAN	LOT	CAR NAME	ST	CAR TYPE	BUILT	PLAN	LOT
SUPERVIA		24ChDrB	5/17	2417D	4493	SWANNANOA		10SecLngObs	10/11	2521	3926
SUPLEE		12SecDr	2-5/25	3410	4845	SWARTHMORE COLLEGE	RN	10Sec2DbrCpt	11/17	4042	4527
SUPREME		16Sec	6-7/20	2412F	4568	SWEDEN		12SecDr	9-10/20	2410F	4590
SURFSIDE		40Ch8StsBDrObs	8/13	2766	4197	SWEET BRIAR		10SecLngObs	2/24/11	2521	3865
SURINAM		12SecDr	10-12/15	2410D	4338	SWEET BRIAR COLLEGE	RN	10Sec2DbrCpt	11/17	4042	4527
SURPRISE		30ChDr	2-3/30	4000	6324	SWEETWATER		12SecDr	2-3/15	2410B	4318
SUSAN B. ANTHONY	N	28ChDr	4-6/25	3416	4864	SWINBURNE		6Cpt3Dr	12/23-24	3523	4741
SUSANNE		26ChDr	3-4/14	2416A	4264	SWISSVALE		24ChDrB	1/14	2417A	4239
SUSQUEHANNA		12SecDr	10-12/26	3410A	6023	SWITZERLAND		16Sec	6-7/13	2412B	4160
SUSQUEHANNA CLUB	N	8SecBlngSunRm	10/30	3989F	6393	SYBERTON		12SecDr	2-4/24	3410	4762
SUSSEX		12SecDr	3-6/20	2410F	4565	SYCAMORE		12SecDr	10-12/20	2410F	4591
SUSSEX TOWER	RN	8SecDr3Dbr	12/26-27	4090D	6031	SYLMAR		12SecDr	7-9/17	2410F	4515
SUTTON		12SecDr	2-5/25	3410	4845	SYLPH		12SecDr	3-6/15	2410B	4319
SWAN LAKE		10SecDr2Cpt	11-12/27	3585B	6123	SYLVAN GLEN		10SecLngObs	12/15	2521B	4362
SWANINGTON		12SecDr	12/10-11	2410	3866	SYLVAN PASS		8SecDr2Cpt	12/29-30	3979A	6334

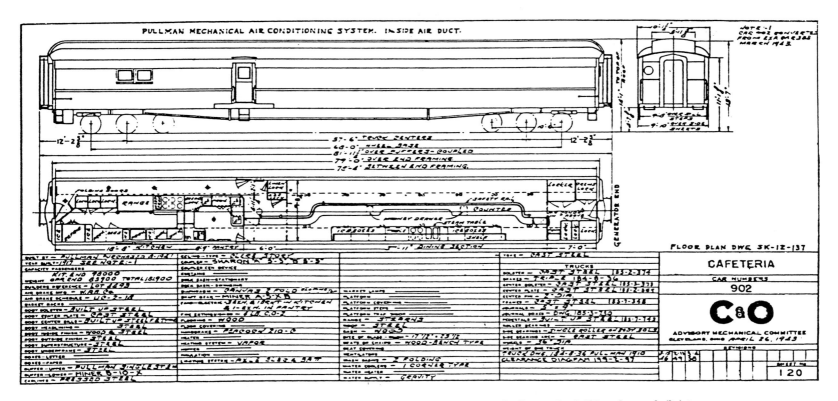

Figure I-70 SUPERVIA, 1 of the 325 parlor cars sold in 1941-1942, went to the Chesapeake & Ohio and was rebuilt into Express Car 383. In March and April 1943, this car became C&O 902, along with three sisters, when they were again converted to the so-called "Blackout Diners." The above diagram represents the layout of Diner 902 after another modification in 1948 when it became a cafeteria. The Author's Collection.

CAR NAME	ST	CAR TYPE	BUILT	PLAN	LOT
SYMPHONY (Rexall)	LN	Dining Counter	10-11/14	4004A	4296
SYRACUSE		12SecDr	1-2/30	3410B	6351
SYSONBY		12SecDr	7-8/15	2410D	4322
TABITHA		26ChDr	6/11	2416	3902
TABLE ROCK		4Cpt2DrLngObs	3/13	2703	4100
TABOR		12SecDr	2-4/24	3410	4762
TABRIZ		16Sec	1-5/18	2412F	4531
TACNA		12SecDr	10-12/15	2410D	4338
TACOMA		12SecDr	4-6/24	3410	4763
TACONY		12SecDr	11/15/10	2410	3860
TACONY		26ChDr	5/23	2416D	4691
TADOUSAC		12SecDr	2-5/16	2410E	4367
TAIT'S TOWER	RN	8SecDr3Dbr	5-6/25	4090	4846
TALANA		12SecDr	10-12/15	2410D	4338
TALARIA		24ChDrB	5/17	2417D	4493
TALBOTT	N	12SecDr	11/15/10	2410	3860
TALEVILLE		12SecDr	6-10/21	2410F	4614
TALIPOT		12SecDr	6-10/21	2410F	4614
TALISMAN	N	12SecDr	3/15/07	1963D	3402
TALISMAN		12SecDr	3-6/20	2410F	4565
TALLADEGA		10SecLngObs	10/11	2521	3926
TALLAHASSEE		12SecDr	9-10/13	2410A	4215
TALMADGE		12SecDr	2-5/16	2410E	4367
TALMO		16Sec	2/13	2412	4101
TAMALPAIS		7Cpt2Dr	9/13	2522A	4208
TAMAQUA		12SecDr	7-8/13	2410A	4192
TAMARACK		12SecDr	10-12/20	2410F	4591
TAMARISK		28ChDr	5-7/26	3416	4958
TAMBINE		24ChDrB	1/14	2417A	4239
TAMUS		12SecDr	3-6/15	2410B	4319
TAMWORTH		12SecDr	7-8/14	2410A	4277
TANAYA		12SecDr	9-10/13	2410A	4215
TANCRED		16Sec	1-5/18	2412F	4531
TANGIER		16Sec	1-5/18	2412F	4531
TANGLEWOOD		12SecDr	3-6/20	2410F	4565
TANTALLON		7Cpt2Dr	1/4/11	2522	3867
TAOS PEAK		10Sec2Dr	12/28-29	3584B	6213
TAPPAN		12SecDr	3-4/11L	2410	3881
TARENTUM		24ChDrB	4/13	2417	4108
TARENTUM	R	26StsLngBDr	4/13	2417J	4108
TARLETON		12SecDr	7-8/14	2410A	4277
TARNESDA		16Sec	6-7/20	2412F	4568
TARRANT		16Sec	1-5/18	2412F	4531
TARRYTOWN		16Sec	12/18	2412X	4489
TARTARUS		12SecDr	3-6/20	2410F	4565
TARTARY		12SecDr	6-10/21	2410F	4614
TASCOTT		12SecDr	7-8/13	2410A	4192
TASMANIA		12SecDr	6-7/17	2410F	4503
TATA NACHO FCP-232	SN	10SecDr2Cpt	12/23-24	3585C	4728
TATLOW		12SecDr	2-5/16	2410E	4367
TATNALL		16Sec	1-5/18	2412F	4531
TAUNTON		36Ch	8/27	3916A	6077
TAURUS		12SecDr	6-10/21	2410F	4614
TAVARES		12SecDr	6-7/17	2410F	4503
TAXCO NdeM	SN	8SecDr2Cpt	3/29	3979A	6237
TAZEWELL		28ChDr	5-6/24	3416	4761
TEAPA FC-DS	SN	8Sec5Dbr	7/14/10	4036G	3800
TECATE FCS-BC	SN	10SecDr2Cpt	1-2/24	3585	4743
TEGUCIGALPA NdeM-276	SN	6SecDr6Dbr	5-6/11	4084C	3903
TEHACHAPI		7Cpt2Dr	9/13	2522A	4208
TEHERAN		12SecDr	6-10/21	2410F	4614
TEHUANTEPEC NdeM-264	SN	8SecDr3Dbr	5-6/25	4090A	4843
TELOS		12SecDr	5-8/18	2410F	4540
TEMECULA		12SecDr	6-7/17	2410F	4503
TEMESVAR		12SecDr	10-12/15	2410D	4338
TEMPERANCE		12SecDr	3-6/20	2410F	4565
TEMPLAR		12SecDr	2-5/16	2410E	4367
TEMPLE		12SecDr	10/22-23	2410H	4647
TEMPLE TOWER	RN	8SecDr3Dbr	5-6/25	4090E	4846
TEMPLE UNIVERSITY	RN	12Sec2Dbr	7-9/17	4046	4515
TEMPLETON		16Sec	6/4/10	2412	3801
TEMPUS		12SecDr	3-6/15	2410B	4319
TENADORES		12SecDr	1-3/14	2410A	4249
TENAYUCA NdeM-265	SN	7Cpt2Dr	12/11	2522E	3943
TENDA PASS		8SecDr2Cpt	7-8/30	3979A	6377
TENEDOS		16Sec	1-5/18	2412F	4531
TENINO		12SecDr	4/15	2410C	4309
TENNECOTT		12SecDr	6-7/17	2410F	4503
TENNESSEE	N	10SecLngObs	10-12/26	3521J	4998
TENNESSEE	N	3CptDrLngBSunRm	6/29	3975C	6262
TENNIELL		12SecDr	1-3/17	2410F	4450
TENNYSON		6Cpt3Dr	12/23-24	3523	4741
TENOCHTITLAN NdeM	SN	8secDr2Cpt	8-9/29	3979A	6283
TENSAS		12SecDr	8-10/26	3410A	4945
TEOCELO NdeM-266	N	14Sec	9-10/25	3958	4869
TEODORO					
LARREY FCP-213	SN	10SecDr2Cpt	5-6/24	3585F	4770
TEPIC NdeM-268	SN	14Sec	10/26	3958	4997
TERMINAL TOWER	RN	8SecDr3Dbr	1/25	4090	4843
TERRACE		12SecDr	3-4/11	2410	3881
TERRE HAUTE		12SecDr	10/21/10	2410	3814
TERREBONNE		12SecDr	8-10/26	3410A	4945
TERRELL		12SecDr	6-7/17	2410F	4503
TESTOUT ROSE	N	12SecDr	8-12/25	3410A	4894
TETON		12SecDr	7-8/28	3410B	6184
TETON PASS		8SecDr2Cpt	12/29-30	3979A	6334
TETON PEAK		10Sec2Dr	12/28-29	3584B	6213
TEVIOT		7Cpt2Dr	1/4/11	2522	3867
TEXAS CITY		2CptDrLngBSunRm	9-10/27	3975B	6076
TEXAS CITY	R	CptDrLngBSunRm	9-10/27	3975G	6076

CAR NAME	ST	CAR TYPE	BUILT	PLAN	LOT
TEXAS PLAINS		2CptDrLngBSunRm	9-10/27	3975B	6076
TEXAS PLAINS	R	CptDrLngBSunRm	9-10/27	3975G	6076
TEXAS RANGER		2CptDrLngBSunRm	9-10/27	3975B	6076
TEXAS ROUTE		2CptDrLngBSunRm	9-10/27	3975B	6076
TEXAS ROUTE	R	CptDrLngBSunRm	9-10/27	3975G	6076
THACKERAY		6Cpt3Dr	12/23-24	3523	4741
THADDEUS STEVENS		12SecDr	5-8/25	3410A	4868
THAIS		30ChDr	11/12	2668	4051
THAMES		12SecDr	6-10/21	2410F	4614
THASOS		16Sec	1-5/18	2412F	4531
THASOS	R	12Sec4EnclSec	1-5/18	2412K	4531
THAYER	N	12SecDr	10/17/10	2410	3814
THE AIRWAY	N	BaggBSmkMail	10-11/25	3951	4885
THE BROADWAY	N	CptDrLngBSunRm	9-10/27	3975G	6076
THE CITADEL	N	10SecDr2Cpt	12/15	2585C	4351
THE MALL		12SecDr	4/26	3410A	4943
THEBES		12SecDr	6-10/21	2410F	4614
THELMA		26ChDr	5/23	2416D	4691
THENARD		10SecDr2Cpt	1-2/13	2585	4091
THENDARA		12SecDr	10-12/26	3410A	6023
THEODORA		26ChDr	3-4/14	2416A	4264
THEODORE O'HARA	N	12SecDr	11-12/13	2410A	4234
THEODOSIA		26ChDr	6/11	2416	3902
THEOLINDA		26ChDr	4/12	2416	3950
THERESA		28ChDr	4-6/25	3416	4864
THESEUM		12SecDr	2-5/16	2410E	4367
THESPIS		12SecDrCpt	10-12/21	2411C	4624
THETFORD		12SecDr	7-8/14	2410A	4277
THETIS		12SecDr	6-10/21	2410F	4614
THIBET		12SecDr	6-10/21	2410F	4614
THISTLE		16Sec	1-5/18	2412F	4531
THOMAR		12SecDr	7-8/15	2410D	4322
THOMAS A. HENDRICKS		14Sec	3-4/30	3958A	6357
THOMAS BURKE		8SecDr2Cpt	5-6/29	3979A	6261
THOMAS COOK	N	12SecDr	4/26	3410A	4943
THOMAS H. CALLOWAY	N	10Sec2Dr	9-10/25	3584A	4899
THOMAS JEFFERSON		32ChObs	9-10/24	3420	4785
THOMAS JEFFERSON		12ChDrBLngSunRm	9/30	4002C	6385
THOMAS MACDONOUGH		12SecDr	1-2/29	3410B	6220
THOMAS PAINE		12SecDr	8-12/25	3410A	4894
THOMAS RUFFIN		10Sec2Dr	9-10/25	3584A	4899
THOMAS RUFFIN		10Sec2Dr	9-10/29	3584B	6284
THOMAS RUFFIN	R	10SecDr2Dbr	9-10/29	4074D	6284
THOMASVILLE		12SecDr	9-10/13	2410A	4215
THOMOND		12SecDr	2-3/15	2410B	4318
THOMPSON		12SecDr	2-5/25	3410	4845
THOMSON PEAK		10Sec2Dr	9-10/29	3584B	6284
THORA		26ChDr	6/11	2416	3902
THORBURN	N	12SecDr	10/22/10	2410	3814
THOREAU		12SecDrCpt	10-12/21	2411C	4624
THORNDALE		12SecDr	11/14/10	2410	3860
THORNDALE		26ChDr	9-11/22	2416D	4649
THORNHOPE		12SecDr	6-9/26	3410A	4969
THORNTON		16Sec	1-5/18	2412F	4531
THOROLD		12SecDr	1-3/14	2410A	4249
THRACIA		12SecDr	6-7/17	2410F	4503

Figure I-71 The silver B&O work train paint is wearing thin and revealing the blue and gray of the former B&O coach 4723, a 1956 B&O rebuild from 12-1 sleeper THENDARA. Chessie System renumbered this car to 911559. This photograph by the author was taken in Somerset, Pennsylvania in September 1977.

231

CAR NAME	ST	CAR TYPE	BUILT	PLAN	LOT
THREE FORKS		10SecDrCpt	4-6/27	3973	6043
THREE OAKS		10Sec2Dr	1/12	2584	3947
THRIFT-T-SLEEPER #1	N	8SecDr3Dbr	5-6/25	4090E	4846
THRIFT-T-SLEEPER #2	N	8SecDr3Dbr	5-6/25	4090E	4846
THRIFT-T-SLEEPER #3	N	8SecDr3Dbr	5-6/25	4090E	4846
THURIO		12SecDr	10-12/20	2410F	4591
THURLOW		BaggBClubSmkBarber	8/17/10	2414A	3803
THURSTON		16Sec	7-8/11	2412	3912
TIANA		12SecDr	3-6/20	2410F	4565
TIBERIUS		12SecDr	6-10/21	2410F	4614
TICKNOR		10SecDr2Cpt	12/16-17	2585D	4443
TICONDEROGA		10SecDr2Cpt	5/13	2585A	4141
TIE BINDERS	N	3Sbr2CptDrBLngSunR	11/29	3991F	6276
TIFFIN		12SecDr	10/25/10	2410	3814
TIFFIN		26ChDr	4-5/17	2416C	4492
TIGER LILY		34Ch	4/26	3419	4956
TIJUANA FCS-BC	SN	10SecDr2Cpt	9-10/26	3585A	4996
TILBURY		12SecDr	4-6/14	2410A	4271
TILDEN		16Sec	1-5/18	2412F	4531
TIMES SQUARE		6Cpt3Dr	7/25	3523A	4887
TIMKEN		12SecDr	6-7/17	2410F	4503
TIMOTHY PICKERING		12SecDr	8-12/25	3410A	4894
TIMPAS		10SecDr2Cpt	8-9/13	2585A	4201
TINEMAN		7Cpt2Dr	1/5/11	2522	3867
TINTIC		10SecDr2Cpt	10-11/13	2585A	4218
TIOGA VALLEY		SbrDrBLngObs	7/30	3988A	6356
TIONESTA		24ChDrB	1/14	2417A	4239
TIPPERARY		12SecDr	6-10/21	2410F	4614
TIPTON		26ChDr	2-3/14	2416A	4248
TIPTON	R	30StsBLng	2-3/14	2416J	4248
TISONIA		12SecDr	3-6/20	2410F	4565
TITAN		12SecDr	10-12/20	2410F	4591
TITUSVILLE		BaggBClubSmkBarber	8/31/10	2414A	3803
TIVOLI		BaggClubBSmk	11/24	3415	4808
TIZATLAN NdeM-269	SN	7Cpt2Dr	9/13	2522F	4208
TIZIMIN FCS-BC	N	8Sec5Dbr	5-8/18	4036F	4540
TLACOTALPAN NdeM-270	N	14Sec	7/30	3958A	6376
TOANO	N	12SecDr	1-3/14	2410A	4249
TOANO		12SecDr	10-12/20	2410F	4591
TOBIAS LEAR	RN	10SecDr2Cpt	9-10/17	4031	4525
TOBYHANNA	N	26ChDr	4-5/13	2416A	4136
TOCALOMA		12SecDr	9-10/13	2410A	4215
TOFIELD		12SecDr	2-5/16	2410E	4367
TOKAY		12SecDr	11-12/13	2410A	4234
TOLEDO		16Sec	12/18	2412X	4489
TOLEDO COUNTRY CLUB	RN	8SecBLngObs	6/6/10	4025C	3801
TOLENAS		12SecDr	2-5/16	2410E	4367
TOLENTINO		10SecDr2Cpt	12/15	2585C	4351
TOLMINO		12SecDr	2-5/16	2410E	4367
TOLSTOI		6Cpt3Dr	12/23-24	3523	4741
TOLTEC		12SecDr	1-3/17	2410F	4450
TOMAH		10SecDrCpt	4-6/27	3973	6043
TOMAHAWK		10SecDrCpt	4-6/27	3973	6043
TOMHICKEN		26ChDr	4-5/17	2416C	4492
TOMOKA		12SecDr	7/13	2410A	4179
TOMPKINS		12SecDr	4-6/11	2410A	4271
TOMS RIVER		BaggBClubSmkBarber	7/23	2951B	4698
TONAWANDA VALLEY		DrSbrBLngObs	4/29	3988	6221
TONDEE		16Sec	1-5/18	2412F	4531
TONOOR		12SecDr	6-10/21	2410F	4614
TONQUIN		12SecDr	1-3/17	2410F	4450
TONTOS		12SecDr	8-10/26	3410A	4945
TOPAZ		32ChDr	8/27	3917A	6078
TOPEKA		10SecDr2Cpt	12/16-17	2585D	4443
TOPINABEE		12SecDr	1-4/12	2410	3949
TOPOCK		12SecDr	6-7/17	2410F	4503
TOPOLOBAMPO NdeM-271	SN	12SecDr	5-8/25	3410A	4868
TOPSTONE		36Ch SBS	1/13	2691	4056
TORAZZO TOWER	RN	8SecDr3Dbr	1/25	4090	4843
TORBANK	N	12SecDr	11-12/13	2410A	4234
TORBERT		12SecDr	10-11/11	2410	3936
TOREADOR	RN	12ChRestLng	10/13	4033	4217
TORONTO		3CptDrLngBSunRm	6/29	3975C	6262
TORONTO UNIVERSITY	N	12Sec2Dbr	5-6/17	4046	4497
TORRANCE		12SecDr	6-9/26	3410A	4969
TORRESDALE		BaggBClubSmkBarber	8/31/10	2414A	3803
TOTTEN		10SecDr2Cpt	12/16-17	2585D	4443
TOWANTIC		12SecDr	6-7/17	2410F	4503
TOWER GROVE		10SecLngObs	9-10/14	2521A	4292
TOWN CLUB	RN	8SecRestLngObs	12/17-18	4025	4528
TOWNSEND		12SecDr	12/21/10	2410	3864
TOXAWAY		10SecLngObs	10/11	2521	3926
TRACEMARIE		16Sec	6-7/20	2412F	4568
TRACY		12SecDr	6-7/16	2410F	4385
TRAFFIC CLUB		8SecBLngSunRm	8-9/29	3989A	6274
TRAFFORD		12SecDr	4/19/10	2410	3770
TRAFFORD		12SecDr	10-11/16	2410F	4431
TRANIO		12SecDr	10-12/20	2410F	4591
TRAPPER		8SecDr2Cpt	5-6/29	3979A	6261
TRASKWOOD		10SecDr2Cpt	6-7/15	2585B	4328
TRAUTMAN		12SecDr	4-6/14	2410A	4271
TRAVEL CLUB	RN	8SecRestLngObs	4/17	4025	4484
TRAWDEN		16Sec	1-5/18	2412F	4531
TRAYMORE		12SecDr	1-2/15	2410B	4311
TREATY OAK		8SecDr2Cpt	4-5/30	3979A	6359
TREDEGAR		12SecDr	11-12/13	2410A	4234
TREMAINE		16Sec	5-6/16	2412F	4386
TREMONT		32ChDr	8/27	3917A	6078
TREMPALEAU		12SecDr	10/22-23	2410H	4647
TRENTON		12SecDr	5-6/23	2410H	4699

CAR NAME	ST	CAR TYPE	BUILT	PLAN	LOT
TRENTON	N	12SecDr	5-8/25	3410A	4894
TREONTA		7Cpt2Dr	12/11	2522	3943
TREVELYAN		12SecDr	7-8/13	2410A	4192
TREVILIAN		12SecDr	11-12/13	2410A	4234
TRIANDA		12SecDr	12/10-11	2410	3866
TRIBUNE		12SecDr	6-10/21	2410F	4614
TRIGO		12SecDr	6-7/17	2410F	4503
TRILIUM		12SecDr	12/10-11	2410	3866
TRIMBLE		12SecDr	9-10/11	2410	3924
TRIMOUNT	N	12ChBLngSunRm	6/26	3964	4965
TRINDLE SPRING		28ChDr	6-7/27	3416A	6087
TRINGA		12SecDr	3-6/15	2410B	4319
TRINIDAD		12SecDr	6-10/21	2410F	4614
TRINITY		12SecDr	8-10/26	3410A	4945
TRINITY COLLEGE	RN	10Sec2DbrCpt	11/17	4042	4527
TRINWAY		12SecDr	3/2/10	2410	3769
TRIPOLI		12SecDr	6-10/21	2410F	4614
TRISTAM DALTON		30ChDr	2-3/30	4000	6324
TRISTRAM DALTON	N	30ChDr	2-3/30	4000	6324
TRITONIA		12SecDr	5-7/13	2410	4149
TRIUMPH	R	10ChRestLng	4/12	4019A	3950
TRIVIA		26ChDr	6/11	2416	3902
TROCADERO	RN	Recreation	2-5/16	3966	4367
TROLLOPE		16Sec	1-5/18	2412F	4531
TROMBLEY		16Sec	7-8/11	2412	3912
TROPICO		10SecDr2Cpt	1-2/13	2585	4091
TROSACH		7Cpt2Dr	1/5/11	2522	3867
TROTWOOD		12SecDr	9-10/11	2410	3924
TROUBADOR	RN	22Ch12StsBLng	8/13	4032	4195
TROUTDALE		12SecDr	11-12/29	3410B	6299
TROWBRIDGE		16Sec	7-8/11	2412	3912
TROY		12SecDr	1-2/30	3410B	6351
TRUESDALE		16Sec	5-6/16	2412F	4386
TRUMAN		10SecDr2Cpt	12/16-17	2585D	4443
TRURO		12SecDr	10-22-23	2410H	4647
TRUXTON		12SecDr	11-12/13	2410A	4234
TUALATIN		12SecDr	2-5/16	2410E	4367
TUBEROSE		28ChDr	5-7/26	3416	4958
TUCKAHOE	N	6SecDr2Cpt2Dbr	2/16	2547C	4363
TUCKERMAN		12SecDrCpt	10-12/21	2411C	4624
TUCUMCARI		LngBDorm(5Sec)	7-8/27	3981	6061
TUFTS COLLEGE	RN	10Sec2DbrCpt	1-5/18	4042B	4531
TULIP		26ChDr	5/23	2416D	4691
TULLYTOWN	N	BaggBClubSmk	2-4/09	2136C	3660
TULPEHOCKEN		26ChDr	5/23	2416D	4691
TULSA		12SecDr	1-3/17	2410F	4450
TUOLUMNE		12SecDr	8-10/26	3410A	4945
TURIN		12SecDr	7-8/13	2410A	4192
TURLOCK		12SecDr	10-12/26	3410A	6023
TURNER		16Sec	1-5/18	2412F	4531

CAR NAME	ST	CAR TYPE	BUILT	PLAN	LOT
TURNER'S GAP		8SecDr2Cpt	12/29-30	3979A	6334
TURPIN		28ChDr	5-6/24	3416	4761
TURTLE CREEK		12SecDr	10/27/10	2410	3814
TUSCALOOSA		12SecDr	10-12/26	3410A	6023
TUSCARORA		12SecDr	11/16/10	2410	3860
TUXPAN NdeM-273	SN	14Sec	3/27	3958	6054
TUXTLA					
GUTIERREZ NdeM-274	SN	14Sec	7/30	3958A	6376
TWILIGHT	N	32ChObs	8/18/10	2420H	3805
TWIN CITIES		14Sec	8-10/28	3958A	6181
TWIN STAR		12SecDr	10/22-23	2410H	4647
TYBEE		16Sec	1-5/18	2412F	4531
TYBURN		12SecDr	3-6/20	2410F	4565
TYCONDA		12SecDr	2-5/16	2410E	4367
TYLER		12SecDr	6-7/16	2410F	4385
TYNEMOUTH		16Sec	1-5/18	2412F	4531
TYPHON		12SecDr	6-10/21	2410F	4614
TYRE		12SecDr	5-8/18	2410F	4540
TYROL		16Sec	6-7/13	2412B	4160
TYRONE		12SecDrCpt	6/16/10	2411	3800
TYSON	N	12SecDr	10/25/10	2410	3814
TZARARAACUA NdeM	N	8SecRestLngObs	7-8/11	4025C.3912	
UANDI		12SecDr	3-6/15	2410B	4319
UINTAH		10SecDr2Cpt	10-11/13	2585A	4218
UKRAINE		12SecDr	10-12/20	2410F	4591
ULUNDI		10SecDr2Cpt	12/15	2585C	4351
UMATILLA		12SecDr	4/30	3410B	6360
UMBRIA		12SecDr	10-12/20	2410F	4591
UNCAS		12SecDr	2-3/27	3410A	6055
UNCAS ROAD		12SecDrCpt	5/17	2411C	4494
UNDERCLIFF		12SecDr	9-10/15	2410D	4327
UNDINA		24ChDrB	5/17	2417D	4493
UNION LEAGUE CLUB		8SecBLngSunRm	8-9/29	3989A	6274
UNIONPORT		12SecDr	10/25/10	2410	3814
UNIONTOWN	N	BaggBClubSmk	2-4/09	2136C	3660
UNIONVILLE		12SecDr	1-2/30	3410B	6351
UNIVERSITY					
CITY		8SbrObsLng	4/30	4005	6340
CLUB		8SecBLngSunRm	8-9/29	3989A	6274
OF PITTSBURGH	N	4CptLngObs	10/25	3960A	4889
UPLAND		BaggBClubSmkBarber	9/2/10	2414A	3803
UPPREST	N	12SecDr	6-8/24	3410	4764
UPPVIEW	N	12SecDr	5-8/25	3410A	4868
UPPWIN	N	12SecDr	6-8/24	3410	4764
UPPWORTH	N	12SecDr	5-8/25	3410A	4868
UPSAL		12SecDr	11/4/10	2410	3860
UPSAL	N	30ChDr	8-9/30	4000A	6384
UPTON		12SecDr	2-4/24	3410	4762
URANUS		12SecDr	6-7/17	2410F	4503
URBANA		12SecDr	2-5/25	3410	4845

CAR NAME	ST	CAR TYPE	BUILT	PLAN	LOT
URUGUAY		12SecDr	10-12/20	2410F	4591
USELDA		28ChDr	4/12	2418	3951
USEPPA		12SecDr	10-12/26	3410A	6023
USUMACINTA FC-DS	SN	8Sec5Dbr	6/13	4036B	4159
UTAH		12SecDr	6-10/21	2410F	4614
UTAHVILLE		12SecDr	6-9/26	3410A	4969
UTICA		12SecDr	1-2/30	3410B	6351
UTOWANA		12SecDr	3-6/15	2410B	4319
UTOWANA		12SecDr	2-5/21	2410F	4612
UVONIA		12SecDr	6-7/17	2410F	4503
V:XI G.B.C.	N	24ChDr10StsLng	6/30	3917E	4803
VACA		7Cpt2Dr SBS	11/11	2522	3941
VACHERIE		12SecDr	11-12/13	2410A	4234
VAIL		26ChDr	9-11/22	2416D	4649
VALCARTIER		12SecDr	2-3/15	2410B	4318
VALCOUR		12SecDr	7/16	2410F	4412
VALDEZ		12SecDr	1-2/16	2410E	4356
VALDONA		12SecDr	4-5/13	2410	4106
VALENCIA		14Sec	7/29	3958A	6271
VALENTIN					
GOMEZ FARIAS NdeM	N	12SecDr	5-8/25	3410A	4868
VALENTINE		12SecDr	7-8/28	3410B	6184
VALERIAN		12SecDr	2-5/21	2410F	4612
VALIANT	RN	10ChRestLng	6/11	4019B	3902
VALLEJO		12SecDr	10-12/20	2410F	4591
VALLEY FORGE		BaggBSmkBarber	12/26-27	3951C	6022
VALLEY FORGE	N	8SecDr2Cpt	3/29	3979A	6237
VALLEY FORGE	RN	8SecDr3Dbr	1/25	4090A	4843
VALLEY PARK		10SecLngObs	9-10/14	2521A	4292
VALLEY PASS		8SecDr2Cpt	12/29-30	3979A	6334
VALLEYFIELD		12SecDr	9-10/11	2410	3924
VALLICITO		12SecDr	9-10/13	2410A	4215
VALLON		12SecDr	5-7/13	2410	4149
VALLONIA		12SecDr	3-6/20	2410F	4565
VALMA		32ChObs	11/12	2420	4050
VALOIS		12SecDr	7-8/15	2410D	4322
VALPA		12SecDr	2-5/21	2410F	4612
VALPARAISO		10SecDr2Cpt	12/15	2585C	4351
VALPARAISO					
UNIVERSITY	RN	12Sec2Dbr	7-9/17	4046	4515
VALRICO		10SecLngObs	2/25/11	2521	3865
VAN CORTLANDT		BaggBSmkBarber	8/30	3987	6371
VAN DORP		12SecDr	10-12/20	2410F	4591
VAN RENSSELAER		BaggBSmkBarber	8/30	3987	6371
VAN TWILLER		BaggBSmkBarber	8/30	3987	6371
VAN WERT		12SecDr	10/22/10	2410	3814
VAN WERT		12SecDrCpt	5/17	2411C	4494
VAN WINKLE		BaggBSmkBarber	8/30	3987	6371
VANCOUVER		12SecDr	4-6/24	3410	4763
VANCURA		12SecDr	2-5/16	2410E	4367

CAR NAME	ST	CAR TYPE	BUILT	PLAN	LOT
VANDALIA		26ChDr	2-3/14	2416A	4248
VANDEN		12SecDr	9-10/13	2410A	4215
VANDERBILT	N	12SecDr	7/13	2410A	4179
VANDERGRIFT		BaggBClubSmkBarber	8/25/10	2414A	3803
VANDERLYN		12SecDr	2-3/15	2410B	4318
VANDOR		12SecDr	4-5/13	2410	4106
VANDYKE		6Cpt3Dr	7/25	3523A	4887
VANESSA		12SecDr	3-6/20	2410F	4565
VANKIRK		12SecDr	2-4/24	3410	4762
VANOMI		12SecDr	2-5/16	2410E	4367
VANORA		12SecDr	1-3/14	2410A	4249
VANSCOY		12SecDr	11-12/13	2410A	4234
VARICK		12SecDr	7/16	2410F	4412
VARNADO		12SecDr	7-8/13	2410A	4192
VARNUM		12SecDr	1-2/15	2410B	4311
VARSITY		12SecDr	12/12-13	2410	4076
VARUNA		12SecDr	3-6/15	2410B	4319
VASHON		12SecDr	1-3/14	2410A	4249
VASSAR COLLEGE	RN	10Sec2DbrCpt	11/17	4042	4527
VAUBAN		12SecDr	1-2/16	2410E	4356
VAUXHALL		12SecDr	11-12/14	2410B	4297
VELESKA		12SecDr	10-12/20	2410F	4591
VELINCO		12SecDr	10-12/20	2410F	4591
VELINO M. PREZA NdeM	SN	12SecDr	6-8/24	3410	4764
VENDOLA		12SecDr	10-12/20	2410F	4591
VENDOME		12SecDr	4-5/13	2410	4106
VENEZUELA NdeM	SN	12SecDr	9-10/13	2410I	4215
VENICE		12SecDr	6-7/17	2410F	4503
VENTA		12SecDr	2-5/21	2410F	4612
VENTNOR		12SecDr	10/24/10	2410	3814
VENTOSE		12SecDr	2-5/16	2410E	4367
VENTURA		16Sec	4/17	2412F	4484
VENUS		32ChObs	11/12	2420	4050
VERA		24ChDrB	5/13	2417A	4137
VERA	R	26StsLngBDr	5/13	2417J	4137
VERADA		12SecDr	2-5/16	2410E	4367
VERBANK	N	12SecDr	11-12/13	2410A	4234
VERDI		6Cpt3Dr	11-12/24	3523A	4833
VERDUN		16Sec	6-7/20	2412F	4568
VERENDRYE		12SecDr	2-5/25	3410	4845
VERGANA		12SecDr	3-6/15	2410B	4319
VERGENNES		12SecDr	7-8/14	2410A	4277
VERMILION VALLEY		DrSbrBLngObs	4/29	3988	6221
VERMILLION		12SecDr	1-4/12	2410	3949
VERMONT	N	SbrDrB16StsLng	7/30	3988D	6356
VERNALIS		12SecDr	9-10/13	2410A	4215
VERNETTE		26ChDr	4-5/16	2416C	4366
VERNON		12SecDr	2-5/16	2410E	4367
VEROCHIO		12SecDr	3-6/20	2410F	4565
VERONA		12SecDr	6-7/17	2410F	4503

CAR NAME	ST	CAR TYPE	BUILT	PLAN	LOT
VERONESE		6Cpt3Dr	7/25	3523A	4887
VERPLANCK MANSION		13Dbr	8/30	3997A	6392
VERSHIRE		12SecDr	7-8/14	2410A	4277
VESPER		12SecDr	6-7/17	2410F	4503
VESTALIA		24ChDrB	5/17	2417D	4493
VESUVIUS	N	12SecDr	8-12/25	3410A	4894
VETA PASS		8SecDr2Cpt	12/29-30	3979A	6334
VEYTIA		12SecDr	6-7/17	2410F	4503
VICAM					
PUEBLO NdeM-228	SN	12SecDr	2-5/25	3410	4845
VICENTE					
GUERRERO NdeM	SN	12SecDr	2-4/24	3410	4762
SUAREZ NdeM-279	SN	14Sec	3/27	3958	6054
VICKER	N	12SecDr	8-12/25	3410A	4894
VICKER	N	10Sec3Dbr	3/11	3411	3880
VICONTOUR		12SecDr	2-5/16	2410E	4367
VICTORIA		26ChDr	3/13	2416	4107
VICTORIA TOWER	RN	8SecDr3Dbr	4-5/26	4090	4961
VICTORIA UNIVERSITY	RN	12Sec2Dbr	5-6/17	4046	4497
VICTORY	N	12SecDr	1-3/25	3410	4844
VIDETTE		12SecDr	3-6/15	2410B	4319

CAR NAME	ST	CAR TYPE	BUILT	PLAN	LOT
VIDONIA		12SecDr	12/10-11	2410	3866
VIENTO		12SecDr	2-5/16	2410E	4367
VIEUX CARRE	RN	10ChRestLng	8/13	4056	4195
VIKING		23ChDrSunRm	9/30	4002A	6385
VILAS		12SecDr	10-12/20	2410F	4591
VILLA ADA	RN	10Sec3Dbr	10-11/12	3411A	4067
VILLA ADRIANA	RN	10Sec3Dbr	10-11/13	3411A	4218
VILLA ALBANI	RN	10Sec3Dbr	12/15	3411A	4351
VILLA ALTA	RN	10Sec3Dbr	8-9/11	3411	3922
VILLA ANITA	RN	10Sec3Dbr	3/11	3411	3880
VILLA ARTISTIC	RN	10Sec3Dbr	6/17/10	3411	3800
VILLA BEAUTIFUL	RN	10Sec3Dbr	6-7/15	3411A	4328
VILLA CHARMING	RN	10Sec3Dbr	5/17	3411	4494
VILLA CHEER	RN	10Sec3Dbr	5/17	3411	4494
VILLA CLARA	RN	10Sec3Dbr	11-12/11	3411A	3942
VILLA COLONIAL	RN	10Sec3Dbr	12/15	3411A	4351
VILLA COMFORT	RN	10Sec3Dbr	3/14	3411	4269
VILLA EASE	RN	10Sec3Dbr	3/11	3411	3880
VILLA ENCHANTING	RN	10Sec3Dbr	6/13	3411	4151
VILLA FALLS	RN	10Sec3Dbr	6/13	3411	4159
VILLA FLORA	RN	10Sec3Dbr	11-12/11	3411A	3942

Figure I-72 VILLA FALLS, a 10-section, 3-double-bedroom, rebuilt in August 1930 from DELCAMBRE, is shown here as B&O work train car X-4567 in December 1965 at Willard, Ohio. The Author's Collection.

CAR NAME	ST	CAR TYPE	BUILT	PLAN	LOT
VILLA GARDEN	RN	10Sec3Dbr	6/13	3411	4159
VILLA GLORIA	RN	10Sec3Dbr	11-12/11	3411A	3942
VILLA GRANDE	RN	10Sec3Dbr	3/11	3411	3880
VILLA GROVE	RN	10Sec3Dbr	3/14	3411	4269
VILLA HEIGHTS	RM	10Sec3Dbr	6/13	3411	4151
VILLA HERMOSA	RN	10Sec3Dbr	7/18/10	3411	3800
VILLA IDEAL	RN	10Sec3Dbr	12/15	3411A	4351
VILLA IMPERIAL	RN	10Sec3Dbr	10-11/13	3411A	4218
VILLA JUAREZ NdeM-280	SN	12SecDr	2-5/25	3410	4845
VILLA MAJESTIC	RN	10Sec3Dbr	10-11/13	3411A	4218
VILLA MARTHA	RN	10Sec3Dbr	11-12/11	3411A	3942
VILLA MEDICI	RN	10Sec3Dbr	1-2/13	3411A	4091
VILLA NOVA		12SecDrCpt	8-9/11	2411	3922
VILLA NOVA	R	10Sec3Dbr	8-9/11	3411	3922
VILLA NUEVA	RN	10Sec3Dbr	5/17	3411	4494
VILLA PALATIAL	RN	10Sec3Dbr	6-7/15	3411A	4328
VILLA PARK	RN	10Sec3Dbr	8-9/11	3411	3922
VILLA PEERLESS	RN	10Sec3Dbr	3/11	3411	3880
VILLA REAL	RN	10Sec3Dbr	3/11	3411	3880
VILLA REGAL	RN	10Sec3Dbr	10-11/12	3411A	4067
VILLA REST	RN	10Sec3Dbr	7-8/10	3411	3800
VILLA RICA	RN	10Sec3Dbr	6/13	3411	4159
VILLA RIDGE	RN	10Sec3Dbr	3/14	3411	4269
VILLA ROAD	RN	10Sec3Dbr	6/13	3411	4159
VILLA ROSA	RN	10Sec3Dbr	11-12/11	3411A	3942
VILLA ROYAL	RN	10Sec3Dbr	11-12/11	3411A	3942
VILLA SERENA	RN	10Sec3Dbr	5/17	3411	4494
VILLA SUPERB	RN	10Sec3Dbr	1-2/13	3411A	4091
VILLA TRAIL	RN	10Sec3Dbr	6/13	3411	4151
VILLA VERDE	RN	10Sec3Dbr	3/11	3411	3880
VILLAHERMOSA NdeM	SN	8SecDr2Cpt	4-5/30	3979A	6359
VILLIERS		12SecDr	1-3/17	2410F	4450
VIMY RIDGE		7Dr	12/20	2583B	4621
VINDEX		12SecDr	2-3/15	2410B	4318
VINEYARD		12SecDrCpt	5/17	2411C	4494
VINORA		26ChDr	3-4/14	2416A	4264
VINTON		12SecDr	2-5/16	2410E	4367
VIOLA		24ChDrB	4/13	2417	4108
VIOLET		28ChDr	5-7/26	3416	4958
VIRGINIA	RN	10StsRestLng	3-4/14	4019	4264
VIRGINIA DARE		28ChDr	1-2/27	3416A	6032
VIRGINIA MANOR		12SecDr	12/10-11	2410	3866
VIROQUA		14Sec	7/29	3958A	6271
VISALIA		12SecDr	12/10-11	2410	3866
VIVIAN		26ChDr	4-5/16	2416C	4366
VIVITA		26ChDr	6/11	2416	3902
VOLARE		12SecDr	2-5/21	2410F	4612
VOLCAN de COLIMA	N	8Sec5Dbr	2-5/21	4036I	4612
VOLGARA		12SecDr	10-12/20	2410F	4591
VOLNEY		12SecDr	3-6/20	2410F	4565

CAR NAME	ST	CAR TYPE	BUILT	PLAN	LOT
VOLUSIA		12SecDr	2-5/16	2410E	4367
VOORHEES		12SecDr	1-3/17	2410F	4450
VORANT		12SecDr	3-6/15	2410B	4319
VOYAGER	RN	10Sec4DuSr	7-8/11	2412L	3912
VULCAN	N	12SecDr	8-12/25	3410A	4894
W.C.C. CLAIBORNE		BaggBSmkBarber	10-11/25	3951	4885
W.L. MAPOTHER	N	12SecDr	7/11	2410	3913
W.W. FINLEY	N	10Sec2Dr	9-10/25	3584A	4899
WABASH COLLEGE	RN	10Sec2DbrCpt	6/13	4042A	4151
WABASHA		14Sec	4/27	3958	6041
WACHUSETT		8SecDr2Cpt	7-8/27	3979	6052
WACO		12SecDr	10/22-23	2410H	4647
WACOUTA		12SecDr	4-5/27	3140B	6042
WACOUTA	R	6Sec6Dbr	4-5/27	4084A	6042
WADE HAMPTON		10Sec2Dr	9-10/25	3584A	4899
WADE HAMPTON		10Sec2Dr	9-10/29	3584B	6284
WADENA		12SecDr	4-6/24	3410	4763
WADSWORTH		12SecDr	11-12/13	2410A	4234
WADSWORTH OAK		8SecDr2Cpt	2-3/30	3979A	6353
WADSWORTH OAK	N	8SecDr2Cpt	7-8/30	3979A	6377
WAHMEDA		12SecDr	3-6/20	2410F	4565
WAHPETON		12SecDr	4-6/24	3410	4763
WAINWRIGHT		12SecDr	4-6/14	2410A	4271
WAKAMBA		12SecDr	11-12/14	2410B	4297
WAKEFIELD		12SecDr	12/10-11	2410	3866
WAKEFIELD	N	12SecDr	2-5/25	3410	4845
WAKELAND	N	12SecDr	12/10-11	2410	3866
WAKPALA		8SecDr2Cpt	7-8/27	3979	6052
WALBRIDGE		12SecDr	10/21/10	2410	3814
WALDAMEER		8CptLng	4/11	2540	3879
WALDAMEER	R	6CptBLng	4/11	2540E	3879
WALDECK		12SecDr	1-2/16	2410E	4356
WALDRON		12SecDr	1-4/12	2410	3949
WALES		12SecDr	5-8/18	2410F	4540
WALKERTON		12SecDr	1-4/12	2410	3949
WALL STREET	RN	14ChDrBLng	6/13	4007	4159
WALL STREET	N	24ChDr10StsLng	6/30	3917E	4803
WALLACE		12SecDr	1-2/16	2410E	4356
WALLACETON		26ChDr	5/23	2416D	4691
WALLINGFORD		36Ch	8/27	3916A	6077
WALLINGTON		12SecDr	11/12/10	2410	3860
WALLKILL		12SecDr	1-4/12	2410	3949
WALLOON		12SecDr	10-12/20	2410F	4591
WALPI		7Cpt2Dr SBS	11/11	2522	3941
WALSALL		12SecDr	10-12/20	2410F	4591
WALTER REED		14Sec	3/27	3958	6054
WALTERSBURG		12SecDr	2-4/24	3410	4762
WALTON		12SecDr	2-5/25	3410	4845
WALWORTH		12SecDr	3-4/11	2410	3881
WAMPUM		12SecDr	10/24/10	2410	3814

CAR NAME	ST	CAR TYPE	BUILT	PLAN	LOT
WAMSUTTA		32ChDr	10/24	3917	4803
WAMSUTTA	R	24ChDr10StsLng	6/30	3917E	4803
WAMSUTTA	N	24ChDr10StsLng	6/30	3917E	4803
WANAKENA		16Sec	10-11/14	2412C	4296
WANATAH		12SecDr	11/10/10	2410	3860
WANDERER	RN	10Sec4DuSr	2/13	2412L	4101
WANDIN		16Sec	9/11	2412	3923
WANNASKA		16Sec	6-7/20	2412F	4568
WAPAKONETA		12SecDr	1-4/12	2410	3949
WARM SPRINGS		10SecLngObs	2/25/11	2521	3865
WARNERS		12SecDr	3-4/11	2410	3881
WARREN		BaggBClubSmkBarber	8/26/10	2414A	3803
WARRENVILLE		12SecDrCpt	3/11	2411	3880
WARSAW		10SecDr2Cpt	12/15	2585C	4351
WARWICK LIGHT		36Ch	12/29-30	3916B	6319
WASEDA		12SecDr	11-12/14	2410B	4297
WASEPI		12SecDr	1-4/12	2410	3949
WASHBURNE		16Sec	12/14-15	2412C	4304
WASHINGTON		26ChDr	5/12	2416	3986
WASHINGTON		Private (GS)	1/14	2502A	4211
WASHINGTON CIRCLE		4CptLngObs	4/26	3960	4944
WASHINGTON CLUB		8SecBLngSunRm	3/30	3989B	6349
WASHINGTON COLLEGE	RN	10Sec2DbrCpt	12/16-17	4042	4443
WASHINGTON ELM	RN	8SecDrLngObs	12/17-18	4024	4528
WASHINGTON HALL		4CptLngObs	4/26	3960	4944
WASHINGTON MONUMENT		34Ch	8/30	4001A	6386
WASHINGTON SQUARE		6Cpt3Dr	7/25	3523A	4887
WASHITA		12SecDr	3-6/15	2410B	4319
WASHITA		12SecDr	8-10/26	3410A	4945
WASKADA		12SecDr	3-6/15	2410B	4319
WASSAIC		12SecDr	3-4/11	2410	3881
WASSONIA		16Sec	10-11/14	2412C	4296
WASTREL		12SecDr	12/12-13	2410	4076
WATANGA		12SecDr	3-6/15	2410B	4319
WATCHING TOWER	RN	8SecDr3Dbr	9-10/25	4090D	4899
WATER LILY		34Ch	4/26	3419	4956
WATER OAK		8SecDr2Cpt	7-8/30	3979A	6377
WATERBURY		36Ch	8/27	3916A	6077
WATERFORD		12SecDr	12/28/10	2410	3864
WATERPORT		12SecDr	3-4/11	2410	3881
WATERTOWN		8CptLng	4/11	2540	3879
WATERTOWN		12SecDr	4-5/27	3410B	6042
WATERVIEW	N	8CptLng	4/11	2540	3879
WATERVIEW	R	6CptBLng	4/11	2540E	3879
WATERVILLE		32ChDr	6/16	2917	4390
WATERVLIET	N	12SecDr	7-8/15	2410D	4322
WATKINS		BaggBClubSmkBarber	8/23/10	2414A	3803
WATSONTOWN		BaggBClubSmkBarber	9/2/10	2414A	3803
WATTS		28ChDr	4-6/25	3416	4864
WAUBUN		12SecDr	9-10/15	2410D	4327
WAUKEGAN		12SecDr	5-8/18	2410F	4540
WAUPONSEE		12SecDr	1-4/12	2410	3949
WAUSAU		10SecDrCpt	4-6/27	3973	6043
WAUSEON		12SecDr	4-6/14	2410A	4271
WAUWATOSA		12SecDr	6/28	3410B	6173
WAVERLY	N	12SecDr	10/22-23	2410H	4647
WAVERLY		10SecDr2Cpt	12/16-17	2585D	4443
WAWA		12SecDr	4/16/10	2410	3771
WAXAHACHIE		12SecDr	10/22-23	2410H	4647
WAYCROSS		7Cpt2Dr	11/13	2522A	4222
WAYFARER	R	10ChRestLng	6/11	4019	3902
WAYFARER	RN	14Ch20StsLng	8/27	4075	6077
WAYNEPORT		8CptLng	4/11	2540	3879
WAYNEPORT	R	6CptBLng	4/11	2540E	3879
WAYNESBURG		12SecDr	10/25/10	2410	3814
WAYNESVILLE		12SecDr	1/14	2410A	4240
WAYNETOWN		12SecDr	3-4/17	2410F	4485
WAYNOKA		10SecDr2Cpt	8-9/13	2585A	4201
WAYSIDE		12SecDr	4-5/13	2410	4106
WAYZATA		12SecDr	4-6/24	3410	4763
WEALDEN		12SecDr	2-5/16	2410E	4367
WEBB C. BALL	N	2CptDrBLngSunRm	9-10/27	3975V	6076
WEDGEWOOD		12SecDr	3-4/17	2410F	4485
WEEDSPORT		12SecDr	3-4/17	2410F	4485
WEEPER'S TOWER	RN	8SecDr3Dbr	1/25	4090	4843
WEETON	N	12SecDr	9-10/20	2410F	4590
WEIDLER		12SecDr	3-4/17	2410F	4485
WEIMAR		12SecDr	7-8/15	2410D	4322
WEIRTON		12SecDr	2-5/25	3410	4845
WELAKA		12SecDr	2-3/15	2410B	4318
WELBY		12SecDr	1-3/14	2410A	4249
WELLESLEY		12SecDr	2-5/16	2410E	4367
WELLESLEY COLLEGE	N	10Sec2DbrCpt	7/20/10	4042A	3800
WELLFORD		12SecDrCpt	3/11	2411	3880
WELLGATE		16Sec	5-6/16	2412F	4386
WELLINGTON		12SecDr	7-8/13	2410A	4192
WELLSBURG		12SecDr	10/26/10	2410	3814
WELLSVILLE		12SecDr	3/15/10	2410	3769
WENACHUS		12SecDr	9-10/15	2410D	4327
WENATCHEE		12SecDr	4-6/24	3410	4763
WENLOCK		12SecDr	1-2/16	2410E	4356
WENONAH		12SecDr	10/22/10	2410	3814
WESCOTT	N	12SecDr	1-2/16	2410E	4356
WESLEY		12SecDr	1-2/16	2410E	4356
WESLEYAN COLLEGE	N	8SecDr2Cpt	4-5/30	3979A	6359
WEST END		12SecDrCpt	3/14	2411A	4269
WEST TOWER	RN	8SecDr3Dbr	9-10/25	4090D	4899
WEST VIEW		10SecLngObs	12/15	2521A	4362
WEST VIRGINIA	RN	10StsRestLng	6/11	4019	3902
WEST WILLOW		BaggBClubSmkBarber	7/23	2951B	4698

CAR NAME	ST	CAR TYPE	BUILT	PLAN	LOT
WEST WILLOW	R	6DbrBLng	7/23	4015	4698
WESTBORO		8CptLng	4/11	2540	3879
WESTBORO	R	6CptBLng	4/11	2540E	3879
WESTDALE		32ChDr	6/16	2917	4390
WESTERN					
ONTARIO UNIVERSITY	N	12Sec2Dbr	5-6/17	4046	4497
WESTERVILLE		12SecDr	6-9/26	3410A	4969
WESTFIELD		12SecDr	1-2/30	3410B	6351
WESTGATE		10SecDr2Cpt	1-2/13	2585	4091
WESTLAND		12SecDr	10/24/10	2410	3814
WESTMORELAND		12SecDr	10-12/26	3410A	6023
WESTOVER	N	8SecDr2Cpt	3/29	3979A	6237
WESTPORT		32ChDr	8/27	3917A	6078
WESTRAY		12SecDr	1-3/14	2410A	4249
WESTWARD HO		12ChDrBLngSunRm	7/30	4002	6364
WESTWARD HO(2nd)	N	14ChDrBLng	6/13	4007	4159
WETAMOO		10SecDr2Cpt	12/15	2585C	4351
WETMORE		BaggBClubSmkBarber	8/30/10	2414A	3803
WEVERTON		12SecDr	4/30	3410B	6360
WEWONDA		16Sec	10-11/14	2412C	4296
WEXFORD		10SecDr2Cpt	12/15	2585C	4351
WEYBRIDGE		12SecDr	3-6/20	2410F	4565
WHARNCLIFFE	N	12SecDr	8-12/25	3410A	4894
WHARTON		12SecDr	1-4/12	2410	3949
WHEATON		12SecDr	7-8/28	3410B	6184
WHEATSTONE		14Sec	10-11/26	3958	6012

CAR NAME	ST	CAR TYPE	BUILT	PLAN	LOT
WHEELER PEAK		10Sec2Dr	12/26-27	3584A	6031
WHETHAM		24ChDrB	1/14	2417A	4239
WHIGVILLE		12SecDr	2-5/25	3410	4845
WHIPPLE		12SecDr	10/26/10	2410	3814
WHISTLER		6Cpt3Dr	7/25	3523A	4887
WHITAKER		12SecDr	2-5/25	3410	4845
WHITE BEAR LAKE		10SecDrCpt	3-5/30	3973A	6338
WHITE CASTLE	RN	12Sec2Dbr	8-9/20	4046A	4574
WHITE CLOUD	RN	12Sec2Dbr	8-9/20	4046A	4574
WHITE COTTAGE	RN	12Sec2Dbr	10-12/26	4046B	6023
WHITE CREEK	RN	12Sec2Dbr	8-9/20	4046A	4574
WHITE DIAMOND		24ChLngSunRm	11/27	3984	6092
WHITE EAGLE	RN	12Sec2Dbr	8-9/20	4046A	4574
WHITE HARBOR	RN	12Sec2Dbr	8-9/20	4046A	4574
WHITE HAVEN	RN	12Sec2Dbr	8-9/20	4046A	4574
WHITE KNOB	RN	12Sec2Dbr	2-5/25	4046B	4845
WHITE MOUNTAINS	RN	12Sec2Dbr	5-6/17	4046A	4497
WHITE OAK		8SecDr2Cpt	7-8/30	3979A	6377
WHITE PLAINS		12SecDrCpt	3/11	2411	3880
WHITE PLAINS	RN	12Sec2Dbr	5-8/18	4046A	4540
WHITE RIVER	RN	12Sec2Dbr	5-6/17	4046A	4497
WHITE STAR	RN	12Sec2Dbr	5-6/17	4046A	4497
WHITE SULPHUR		12SecDr	12/10-11	2410	3866
WHITE SWAN	RN	12Sec2Dbr	5-6/17	4046A	4497
WHITEHALL		12SecDr	12/10-11	2410	3866
WHITEHEAD		12SecDr	10/22-23	2410H	4647

Figure I-73 DAFX-13 containing a KC 135 Training Simulator was originally built as an 8-section drawing room, 2-compartment sleeper named WILD ROSE. It was sold to the government in January 1962 for rebuilding and is shown here at Fort Eustis, Virginia, after retirement, in a February 1987 photograph by the Author.

Figure I-74   WILLOW TRAIL, 1 of 25 cars to carry the prefix "Willow," had 7 drawing rooms and remained the property of Pullman until it was sold to the Terminal Railroad Association of St. Louis in December 1953 where it was assigned to work train service as number 492.   Smithsonian Institution Photograph P-33004.

CAR NAME	ST	CAR TYPE	BUILT	PLAN	LOT	CAR NAME	ST	CAR TYPE	BUILT	PLAN	LOT
WHITEPORT		12SecDr	4-6/14	2410A	4271	BEAUMONT		14Sec	3/27	3958	6054
WHITESBORO		12SecDr	3-4/11	2410	3881	BLOUNT	N	12SecDr	2-3/15	2410B	4318
WHITFORD		12SecDrCpt	6/18/10	2411	3800	CLARK	N	10SecDrCpt	4-6/27	3973	6043
WHITNEY		BaggBClubSmk	9/10/10	2415	3834	CLARK	RN	10SecDr2Dbr	1/25	4074C	4843
WHITSETT		12SecDr	1-4/12	2410	3949	DAVIDSON	RN	10Sec3Dbr	3/11	3411	3880
WHITSON		12SecDr	1-2/15	2410B	4311	DAVIDSON		10Sec2Dr	9-10/25	3584A	4899
WIAKA		12SecDr	7-8/15	2410D	4322	DAVIDSON		8SecDr2Cpt	8-9/29	3979A	6283
WICKLIFFE		12SecDr	1-4/12	2410	3949	DAVIDSON	RN	8Sec5Dbr	11/16	4036F	4431
WICKLOW		12SecDr	10-12/20	2410F	4591	ELLERY		12SecDr	4/26	3410A	4943
WIERWOOD		26ChDr	9-11/22	2416D	4649	GILBERT		14Sec	3/27	3958	6054
WIGWAM		12SecDr	10-12/26	3410A	6023	HARVEY		14Sec	3/27	3958	6054
WILBUR WRIGHT		14Sec	10-11/26	3958	6012	LEWIS SHARKEY		10Sec2Dr	9-10/25	3584A	4899
WILBURTHA		24ChDrB	4/13	2417	4108	LEWIS SHARKEY		8SecDr2Cpt	8-9/29	3979A	6283
WILBURTHA	R	18Ch10StsCafeBDr	4/13	2417H	4108	LEWIS SHARKEY	RN	8Sec5Dbr	2-5/16	4036F	4367
WILCOX		26ChDr	4-5/17	2416C	4492	LEWIS SHARKEY	RN	10Sec3Dbr	3/11	3411	3880
WILD ROSE		8SecDr2Cpt	4-5/30	3979A	6359	MOULTRIE		BaggBSmkBarber	10-11/25	3951	4885
WILDOMAR		10SecDr2Cpt	8-9/13	2585A	4201	OSLER		14Sec	3/27	3958	6054
WILDWOOD		12SecDr	3-6/20	2410F	4565	PENN		12SecDr	8-12/25	3410A	4894
WILKESBARRE		12SecDr	7-8/12	2410	4014	PENN	N	12SecDr	1-2/29	3410B	6220
WILKINSBURG		BaggBClubSmkBarber	8/23/10	2414A	3803	RUFUS KING	RN	10Sec3Dbr	7/7/10	3411	3800
WILKINSON		12SecDr	1-4/12	2410	3949	RUFUS KING		10Sec2Dr	9-10/25	3584A	4899
WILLAPA		16Sec SBS	4/15	2412D	4310	RUFUS KING		8SecDr2Cpt	8-9/29	3979A	6283
WILLIAM						RUFUS KING	RN	8Sec5Dbr	11/16	4036F	4431
AIKEN	N	10Sec2Dr	9-10/25	3584A	4899	WATTS FOLWELL		8SecDr2Cpt	12/29-30	3979A	6334
BAINBRIDGE		12SecDr	8-12/25	3410A	4894	WYATT BIBB	RN	10Sec3Dbr	5/17	3411	4494

CAR NAME	ST	CAR TYPE	BUILT	PLAN	LOT	CAR NAME	ST	CAR TYPE	BUILT	PLAN	LOT
WYATT BIBB		10Sec2Dr	9-10/25	3584A	4899	WILLOW PARK		7Dr	2/29	3583A	6214
WYATT BIBB	RN	8Sec5Dbr	6-7/17	4036F	4503	WILLOW POINT		7Dr	10/26	3583	4973
WYATT RIBB		8SecDr2Cpt	8-9/29	3979A	6283	WILLOW RANGE		7Dr	12/29	3583A	6317
WILLIAMS GROVE		28ChDr	6-7/27	3416A	6087	WILLOW RIDGE		7Dr	12/29	3583A	6317
WILLIAMSBURG	N	26ChDr	5/12	2416	3986	WILLOW RIVER		7Dr	2/29	3583A	6214
WILLIAMSBURG	N	8SecDr2Cpt	8-9/29	3979A	6283	WILLOW ROAD		7Dr	12/29	3583A	6317
WILLIAMSBURG	RN	8SecDr3Dbr	1/25	4090A	4843	WILLOW SHORE		7Dr	12/29	3583A	6317
WILLIAMSON		12SecDr	3-4/11	2410	3881	WILLOW SLOPE		7Dr	12/29	3583A	6317
WILLIS		12SecDrCpt	10-12/21	2411C	4624	WILLOW SPRING		7Dr	12/24	3583	4834
WILLISTON		12SecDr	4-6/24	3410	4763	WILLOW TRAIL		7Dr	2/29	3583A	6214
WILLMAR		12SecDr	4-6/24	3410	4763	WILLOW VALLEY		7Dr	2/29	3583A	6214
WILLOUGHBY		12SecDr	4-6/14	2410A	4271	WILLOWS		10SecDr2Cpt	12/16-17	2585D	4443
WILLOW BANK		7Dr	12/24	3583	4834	WILMA		26ChDr	6/11	2416	3902
WILLOW BAY		7Dr	2/29	3583A	6214	WILMERDING		16Sec	6/9/10	2412	3801
WILLOW BLUFF		7Dr	12/29	3583A	6317	WILMINGTON		26ChDr	5/12	2416	3986
WILLOW BROOK		7Dr	12/24	3583	4834	WILMORE		12SecDrCpt	6/27/10	2411	3800
WILLOW CLIFF		7Dr	12/29	3583A	6317	WILMOT		12SecDr	11-12/13	2410A	4234
WILLOW CREEK		7Dr	12/24	3583	4834	WILSON		BaggBClubSmkBarber	9/3/10	2414A	3803
WILLOW DELL		7Dr	10/26	3583	4973	WILTON		12SecDr	12/10-11	2410	3866
WILLOW FALLS		7Dr	2/29	3583A	6214	WIMBLEY TOWER	RN	8SecDr3Dbr	12/26-27	4090	6031
WILLOW GLADE		7Dr	12/29	3583A	6317	WINANS	N	12SecDrCpt	6/28/10	2411	3800
WILLOW GLEN		7Dr	12/24	3583	4834	WINBURNE		12SecDr	9-10/11	2410	3924
WILLOW GROVE		7Dr	12/24	3583	4834	WINCHESTER		12SecDr	7-8/12	2410	4014
WILLOW HILL		7Dr	10/26	3583	4973	WINDIGO		12SecDr	10-12/20	2410F	4591
WILLOW ISLAND		7Dr	10/26	3583	4973	WINDOM		10SecDr2Cpt	8-9/13	2585A	4201
WILLOW LAKE		7Dr	10/26	3583	4973	WINDSOR		12SecDr	10/31/10	2410	3860
WILLOW OAK		8SecDr2Cpt	7-8/30	3979A	6377	WINDSOR BEACH	RN	10SecRestLngObs	4/17	4027	4484

Figure I-75 This builder photograph of **WILLIAM RUFUS KING**, an 8-section drawing room, 2-compartment built in 1929 under Plan 3979A for service on Southern's "Crescent Limited" is one of four cars that carried this name. In June 1932 it was renamed "STONEWALL" JACKSON and became the property of the C&O in 1948 where it was sold to NdeM in September 1950 and renamed CACAHUAMILPA. Smithsonian Institution Photograph P-33838.

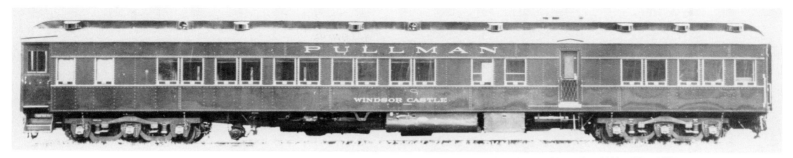

Figure I-76   WINDSOR CASTLE was a 10-section restaurant / observation lounge, rebuilt in November 1932 to Plan 4027 from PENVIR and was sold to Southern Railway in 1948. It was withdrawn from Pullman lease in October 1953 and by 1954 it had been converted to Southern Coach 1041.   Pullman Photograph P-37409.   The Arthur D. Dubin Collection.

CAR NAME	ST	CAR TYPE	BUILT	PLAN	LOT	CAR NAME	ST	CAR TYPE	BUILT	PLAN	LOT
WINDSOR CASTLE	RN	10SecRestLngObs	4/17	4027	4484	WOODBURY		26ChDr	5/12	2416	3986
WINDSOR FOREST	RN	10SecRestLngObs	12/17-18	4027	4528	WOODCREST		26ChDr	9-11/22	2416D	4649
WINDSOR LOCKS	RN	10SecRestLngObs	1-5/18	4027	4531	WOODHILL CLUB	RN	10SecRestLngObs	4/17	4027	4484
WINDSOR PARK	RN	10SecRestLngObs	4/17	4027	4484	WOODINGTON		12SecDr	10/31/10	2410	3814
WINDSOR SPRINGS	RN	10SecRestLngObs	1-5/18	4027	4531	WOODLAND		12SecDr	4/30	3410B	6360
WINGATE	N	12SecDrCpt	6/13	2411A	4159	WOODMERE		16Sec	7-8/11	2412	3912
WINGDALE		12SecDr	1-4/12	2410	3949	WOODPORT		BaggBClubSmkBarber	2-3/12	2602	3948
WINNETKA		16Sec	5-6/13	2412B	4150	WOODRUFF		12SecDr	1-4/12	2410	3949
WINNETT		10SecDrCpt	4-6/27	3973	6043	WOODS HOLE		36Ch	8/27	3916A	6077
WINNIPEG		12SecDr	4-6/24	3410	4763	WOODS LAKE		12SecDrCpt	3/11	2411	3880
WINONA		12SecDr	10/22-23	2410H	4647	WOODSLEE		12SecDr	9-10/11	2410	3924
WINOOSKI		12SecDr	7-8/14	2410A	4277	WOODSTOCK		12SecDr	6-9/26	3410A	4969
WINTER		12SecDrCpt	10-12/21	2411C	4624	WOODSTOWN		BaggBClubSmkBarber	9/10/10	2414A	3803
WINTERBURN		12SecDr	11-12/13	2410A	4234	WOODWORTH		12SecDrCpt	10-12/21	2411C	4624
WINTHROP		12SecDrCpt	6/22/10	2411	3800	WOOSTER		12SecDr	5-6/23	2410H	4699
WINTHROP COLLEGE	RN	10Sec2DbrCpt	3/11	4042A	3880	WORDSWORTH		6Cpt3Dr	12/23-24	3523	4741
WIOBELL		12SecDr	10-12/20	2410F	4591	WORLAND		10SecDr2Cpt	6-7/15	2585B	4328
WIROCK		10SecDrCpt	4-6/27	3973	6043	WORTHAM		12SecDr	12-10-11	2410	3866
WISEBURN		10SecDr2Cpt	8-9/13	2585A	4201	WORTHINGTON		12SecDr	10/21/10	2410	3814
WISSINOMING		BaggBClubSmkBarber	9/1/10	2414A	3803	WREXHAM TOWER	RN	8SecDr3Dbr	12/28-29	4090	6213
WISTERIA		28ChDr	5-6/24	3416	4761	WRIGHTSVILLE		BaggBClubSmk	9/14/10	2415	3834
WITMER		26ChDr	2-3/14	2416A	4248	WYACONDA		12SecDr	4-5/13	2410	4106
WITTENBERG COLLEGE	RN	10Sec2DbrCpt	6/20/10	4042A	3800	WYANDOTTE		12SecDr	4/20/10	2410	3771
WIXOM		12SecDr	11-12/13	2410A	4234	WYEBROOKE		BaggBClubSmk	9/15/10	2415	3834
WOFFORD COLLEGE	N	10Sec2DbrCpt	6/13	4042A	4159	WYMORE		12SecDr	9-10/15	2410D	4327
WOLCOTT		12SecDr	10/29/10	2410	3814	WYNKOOP		26ChDr	9-11/22	2416D	4649
WOLFEBORO		12SecDr	3-6/20	2410F	4565	WYNNEWOOD		12SecDr	10/27/10	2410	3814
WOLSEY		12SecDr	5-8/18	2410F	4540	WYNOOCHE		16Sec SBS	4/15	2412D	4310
WONALANCET		10SecDr2Cpt	12/15	2585C	4351	WYOCENA		12SecDr	4-5/27	3410B	6042
WONDER PASS		8SecDr2Cpt	7-8/30	3979A	6377	WYOCENA	R	6Sec6Dbr	4-5/27	4084A	6042
WOOD LILY		34Ch	4/26	3419	4956	WYTHEVILLE	N	12SecDr	8-12/25	3410A	4894
WOOD VIOLET		30ChDr	8-9/30	4000A	6384	XOCHIMILCO NdeM	N	8SecDr2Cpt	3/29	3979A	6237
WOODBINE		BaggBClubSmkBarber	9/10/10	2414A	3803	YACHT CLUB	N	8SecBLngSunRm	8-9/29	3989A	6274
WOODBRIDGE		12SecDr	11/4/10	2410	3860	YADKIN		12SecDr	10-12/26	3410A	6023

CAR NAME	ST	CAR TYPE	BUILT	PLAN	LOT
YAKIMA		12SecDr	10-12/26	3410A	6023
YALE CLUB	RN	8SecRest6StsObsLng	9-10/14	4025D	4292
YALE UNIVERSITY	RN	12Sec2Dbr	7-9/17	4046	4515
YAQUI		12SecDr	2-5/21	2410F	4612
YARDVILLE		12SecDr	4/27/10	2410	3772
YARMOUTH		32ChDr	8/27	3917A	6078
YELLOW OAK		8SecDr2Cpt	7-8/30	3979A	6377
YELLOWSTONE PARK		2Cpt3Dr2PvtRm	6/26	3962	4895
YELLOWSTONE PARK	R	4Cpt4Dr	6/26	3962A	4895
YELLOWTAIL	N	12SecDr	1-3/14	2410A	4249
YELVERTON		12SecDr	4-6/14	2410A	4271
YEMASSEE		7Cpt2Dr	11/13	2522A	4222
YOKOHAMA	N	12SecDr	10-12/20	2410F	4591
YOKOHOMA		12SecDr	10-12/20	2410F	4591

CAR NAME	ST	CAR TYPE	BUILT	PLAN	LOT
YONDOTEGA CLUB		8SecBLngSunRm	10/30	3989C	6393
YORK		26ChDr	4-5/17	2416C	4492
YORK HALL	N	8SecDr3Dbr	1/25	4090A	4843
YORK HAVEN		12SecDr	4/15/10	2410	3771
YORKTOWN	N	8SecDr2Cpt	3/29	3979A	6237
YORKTOWN	RN	8SecDr3Dbr	1/25	4090A	4843
YORKVILLE		12SecDr	1-4/12	2410	3949
YOSEMITE PARK		2Cpt3Dr2PvtRm	6/26	3962	4895
YOSEMITE PARK	R	4Cpt4Dr	6/26	3962A	4895
YOUNGDALE		12SecDr	1-4/12	2410	3949
YOUNGSTOWN					
COUNTRY CLUB		8SecBLngSunRm	2/31	3989D	6396
YOUNGWOOD		12SecDr	5-6/23	2410H	4699
YOUNT PEAK		10Sec2Dr	12/28-29	3584B	6213

Figure I-77 YALE CLUB, an 8-section restaurant / observation lounge, was rebuilt from CASABLANCA in September 1936. It was sold to the Union Pacific in 1948, withdrawn from Pullman lease in October 1958 and scrapped. Smithsonian Institution Photograph P-40042.

Figure I-78 ZEPHYR TOWER was rebuilt and stream-styled to an eight-section drawing room, 3-double bedroom, Plan 4090C, in July 1941. It became the property of CB & Q in 1948, and in December 1966 it was withdrawn from Pullman lease and became Dormitory 662. Photographed in September 1965 in Kansas City, Missouri. The Robert Wayner Collection.

CAR NAME	ST	CAR TYPE	BUILT	PLAN	LOT	CAR NAME	ST	CAR TYPE	BUILT	PLAN	LOT
YPSILANTI		12SecDr	1-4/12	2410	3949	ZEBULON B. VANCE		10Sec2Dr	9-10/29	3584B	6284
YUCCA		12SecDr	2-5/21	2410F	4612	ZELLA		24ChDrB	5/13	2417A	4137
YUKON		12SecDr	2-5/21	2410F	4612	ZELYA		26ChDr	3-4/14	2416A	4264
YVETTE		30ChDr	11/12	2668	4051	ZENITH		12SecDr	10-12/20	2410F	4591
YVONNE		26ChDr	3-4/14	2416A	4264	ZENO		12SecDrCpt	10-12/21	2411C	4624
ZACHARY TAYLOR		12SecDr	4/26	3410A	4943	ZEPHON		12SecDr	10-12/20	2410F	4591
ZADORA		24ChDrB	6-7/14	2417B	4265	ZEPHYR TOWER	RN	8SecDr3Dbr	12/28-29	4090C	6213
ZAMBRA		12SecDr	1-3/14	2410A	4249	ZINNIA		28ChDr	5-7/26	3416	4958
ZANA		26ChDr	4-5/16	2416C	4366	ZION CANYON		3CptDrLngBSunRm	6/29	3975C	6262
ZANMORE		12SecDr	1-4/12	2410	3949	ZIPRA		26ChDr	4-5/16	2416C	4366
ZANZIBAR		12SecDr	10-12/20	2410F	4591	ZIRAHUEN NdeM	SN	14Sec	9-10/25	3958	4869
ZEARING		12SecDr	1-4/12	2410	3949	ZUNI		12SecDr	10-12/20	2410F	4591
ZEBULON B. VANCE		10Sec2Dr	9-10/25	3584A	4899	ZURICH		26ChDr	4-5/17	2416C	4492

THE LIGHTWEIGHT CARS

Competition, Speed, Shine And Fancy Paint, 1934-1956

Figure I-79   Illinois Central BLOOMINGTON, one of the numerous cars retired when Amtrak was born, represents the end of an era for most of the Pullman-built fleet of lightweight cars. It is shown parked in Chicago on 29 August 1971, three months after Amtrak began operations.   Railroad Avenue Entriprises Photograph PN-4409.

CAR NAME OR NUMBER		ST	CAR TYPE	DATE BUILT	CAR BLDR	PLAN	LOT	CAR NAME OR NUMBER		ST	CAR TYPE	DATE BUILT	CAR BLDR	PLAN	LOT
100	SP	S	10Rmt5Dbr	2-3/41	PS	4072E	6641	301	SP	S	13Dbr	3/41	PS	4071D	6643
101	SP	S	10Rmt5Dbr	2-3/41	PS	4072E	6641	302	SP	S	13Dbr	3/41	PS	4071D	6643
102	SP	S	10Rmt5Dbr	2-3/41	PS	4072E	6641	303	SP	S	13Dbr	3/41	PS	4071D	6643
103	SP	S	10Rmt5Dbr	2-3/41	PS	4072E	6641	304	CB&Q	S	4Rmt4DbrDome4DuRmt	6-11/54	Budd	9535	9669
104	SP	S	10Rmt5Dbr	2-3/41	PS	4072E	6641	304	SP	S	13Dbr	3/41	PS	4071D	6643
105	SP	S	10Rmt5Dbr	2-3/41	PS	4072E	6641	305	(2nd)CB&Q	S	4Rmt4DbrDome4DuRmt	6-11/54	Budd	9535	9669
106	SP	S	10Rmt5Dbr	2-3/41	PS	4072E	6641	305	SP	S	13Dbr	3/41	PS	4071D	6643
107	SP	S	10Rmt5Dbr	2-3/41	PS	4072E	6641	306	SP	S	13Dbr	3/41	PS	4071D	6643
108	SP	S	10Rmt5Dbr	2-3/41	PS	4072E	6641	306	SP&S	S	4Rmt4DbrDome4DuRmt	6-11/54	Budd	9535	9669
109	SP	S	10Rmt5Dbr	2-3/41	PS	4072E	6641	307	NP	S	4Rmt4DbrDome4DuRmt	6-11/54	Budd	9535	9669
200	SP	S	4Dbr4Cpt2Dr	3/41	PS	4069G	6642	307	NP	S	13Dbr	3/41	PS	4071D	6643
201	SP	S	4Dbr4Cpt2Dr	3/41	PS	4069G	6642	308	NP	S	4Rmt4DbrDome4DuRmt	6-11/54	Budd	9535	9669
202	SP	S	4Dbr4Cpt2Dr	3/41	PS	4069G	6642	309	NP	S	4Rmt4DbrDome4DuRmt	6-11/54	Budd	9535	9669
203	SP	S	4Dbr4Cpt2Dr	3/41	PS	4069G	6642	310	NP	S	4Rmt4DbrDome4DuRmt	6-11/54	Budd	9535	9669
204	SP	S	4Dbr4Cpt2Dr	3/41	PS	4069G	6642	311	NP	S	4Rmt4DbrDome4DuRmt	6-11/54	Budd	9535	9669
205	SP	S	4Dbr4Cpt2Dr	3/41	PS	4069G	6642	312	NP	S	4Rmt4DbrDome4DuRmt	6-11/54	Budd	9535	9669
206	SP/CRI&P	S	4Dbr4Cpt2Dr	3-4/42	PS	4069H	6668	313	NP	S	4Rmt4DbrDome4DuRmt	6-11/54	Budd	9535	9669
207	SP/CRI&P	S	4Dbr4Cpt2Dr	3-4/42	PS	4069H	6668	314	NP	S	4Rmt4DbrDome4DuRmt	12/57	Budd	9535	9669
208	SP/CRI&P	S	4Dbr4Cpt2Dr	3-4/42	PS	4069H	6668	350	NP	N	8DuRmt6Rmt3DbrCpt	7-9/48	PS	4119	6781
209	SP/CRI&P	S	4Dbr4Cpt2Dr	3-4/42	PS	4069H	6668	351	NP	N	8DuRmt6Rmt3DbrCpt	7-9/48	PS	4119	6781
210	SP/CRI&P	S	4Dbr4Cpt2Dr	3-4/42	PS	4069H	6668	352	NP	N	8DuRmt6Rmt3DbrCpt	7-9/48	PS	4119	6781
211	SP/CRI&P	S	4Dbr4Cpt2Dr	3-4/42	PS	4069H	6668	353	NP	N	8DuRmt6Rmt3DbrCpt	7-9/48	PS	4119	6781
212	SP/CRI&P	S	4Dbr4Cpt2Dr	3-4/42	PS	4069H	6668	354	NP	N	8DuRmt6Rmt3DbrCpt	7-9/48	PS	4119	6781
213	SP/CRI&P	S	4Dbr4Cpt2Dr	3-4/42	PS	4069H	6668	355	NP	N	8DuRmt6Rmt3DbrCpt	7-9/48	PS	4119	6781
214	SP/CRI&P	S	4Dbr4Cpt2Dr	3-4/42	PS	4069H	6668	356	NP	N	8DuRmt6Rmt3DbrCpt	7-9/48	PS	4119	6781
215	SP/CRI&P	S	4Dbr4Cpt2Dr	3-4/42	PS	4069H	6668	357	NP	N	8DuRmt6Rmt3DbrCpt	7-9/48	PS	4119	6781
216	SP/CRI&P	S	4Dbr4Cpt2Dr	3-4/42	PS	4069H	6668	358	NP	N	8DuRmt6Rmt3DbrCpt	7-9/48	PS	4119	6781
300	SP	S	13Dbr	3/41	PS	4071D	6643	359	NP	N	8DuRmt6Rmt3DbrCpt	7-9/48	PS	4119	6781

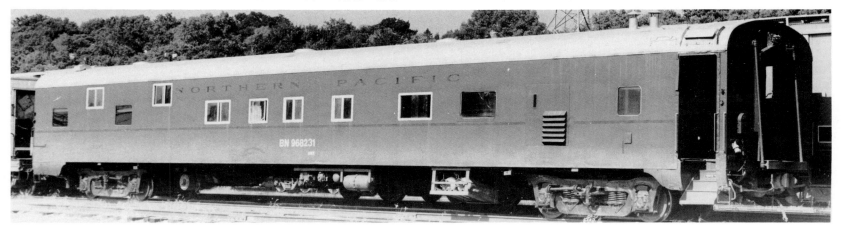

Figure I-80 Former Northern Pacific NP-356 was built under Pullman Plan 4119, Lot 6781 in 1948. It is shown in this George Cockle photograph taken 11 September 1982 at Omaha, Nebraska as Burlington Northern Bunk Car 968231. Changes typical to Pullman cars in W/T service include removal of diaphragms, traps, and dutch doors; substitution of sliding windows for sealed sash; air brakes changed from D-22 to AB freight car-type; addition of rooftop vents and smoke jacks, gas bottles under car for heat and hot water, and louvered panel in car for mechanical room.

CAR NAME OR NUMBER		ST	CAR TYPE	DATE BUILT	CAR BLDR	PLAN	LOT		CAR NAME OR NUMBER		ST	CAR TYPE	DATE BUILT	CAR BLDR	PLAN	LOT
360	NP	N	8DuRmt6Rmt3DbrCpt	7-9/48	PS	4119	6781		500	CRI&P/SP	S	6Sec6Rmt4Dbr	4-5/42	PS	4099	6669
361	NP	N	8DuRmt6Rmt3DbrCpt	7-9/48	PS	4119	6781		501	CRI&P/SP	S	6Sec6Rmt4Dbr	4-5/42	PS	4099	6669
362	NP	N	8DuRmt6Rmt3DbrCpt	7-9/48	PS	4119	6781		502	CRI&P/SP	S	6Sec6Rmt4Dbr	4-5/42	PS	4099	6669
363	NP	N	8DuRmt6Rmt3DbrCpt	7-9/48	PS	4119	6781		503	CRI&P/SP	S	6Sec6Rmt4Dbr	4-5/42	PS	4099	6669
364	NP	S	10Rmt6Dbr	5-7/50	PS	4140A	6874		504	CRI&P/SP	S	6Sec6Rmt4Dbr	4-5/42	PS	4099	6669
365	NP	S	10Rmt6Dbr	5-7/50	PS	4140A	6874		505	CRI&P/SP	S	6Sec6Rmt4Dbr	4-5/42	PS	4099	6669
366	SP&S	N	6Rmt8DuRmt3DbrCpt	7-9/48	PS	4119	6781		506	CRI&P/SP	S	6Sec6Rmt4Dbr	4-5/42	PS	4099	6669
367	NP	S	6Rmt8DuRmt3DbrCpt	10-11/54	PS	4192	6939		507	CRI&P/SP	S	6Sec6Rmt4Dbr	4-5/42	PS	4099	6669
368	NP	S	6Rmt8DuRmt3DbrCpt	10-11/54	PS	4192	6939		508	CRI&P/SP	S	6Sec6Rmt4Dbr	4-5/42	PS	4099	6669
369	NP	S	6Rmt8DuRmt3DbrCpt	10-11/54	PS	4192	6939		509	CRI&P/SP	S	6Sec6Rmt4Dbr	4-5/42	PS	4099	6669
370	NP	S	6Rmt8DuRmt3DbrCpt	10-11/54	PS	4192	6939		510	CRI&P/SP	S	6Sec6Rmt4Dbr	4-5/42	PS	4099	6669
371	NP	S	6Rmt8DuRmt3DbrCpt	10-11/54	PS	4192	6939		511	CRI&P/SP	S	6Sec6Rmt4Dbr	4-5/42	PS	4099	6669
372	NP	S	6Rmt8DuRmt3DbrCpt	10-11/54	PS	4192	6939		512	CRI&P/SP	S	6Sec6Rmt4Dbr	4-5/42	PS	4099	6669
375	NP	RN	4Rmt3DbrDome2DuRmt	6-11/54	Budd	9535A	9669		513	SP	N	6Sec6Rmt4Dbr	5-6/42	PS	4099	6669
376	NP	RN	4Rmt3DbrDome2DuRmt	6-11/54	Budd	9535A	9669		514	SP	N	6Sec6Rmt4Dbr	5-6/42	PS	4099	6669
377	NP	RN	4Rmt3DbrDome2DuRmt	6-11/54	Budd	9535A	9669		515	SP	N	6Sec6Rmt4Dbr	5-6/42	PS	4099	6669
378	NP	RN	4Rmt3DbrDome2DuRmt	6-11/54	Budd	9535A	9669		516	SP	N	6Sec6Rmt4Dbr	5-6/42	PS	4099	6669
379	NP	RN	4Rmt4DbrDome2DuRmt	12/57	Budd	9535A	9669		1600	WAB	S	24ChDrLngREObs	10/47	AC&F	9001	2769
380	CB&Q	RN	4Rmt4DbrDome2DuRmt	6-11/54	Budd	9535A	9669		1601	WAB	S	23ChDrLngDomeREObs	1/50	Budd	9525	9652
390	NP	RN	4DbrCptBLngREObs	6-7/48	PS	4120	6781		1602	WAB	S	21ChDrDome	7/52	PS	7551B	6904
391	NP	RN	4DbrCptBLngREObs	6-7/48	PS	4120	6781		5740	Milw (MT.SPOKANE)	L	14Sec(Touralux)	6/28/47	Milw	4504A	
392	NP	RN	4DbrCptBLngREObs	6-7/48	PS	4120	6781		5741	Milw						
393	NP	RN	4DbrCptBLngREObs	6-7/48	PS	4120	6781			(MT.WASHINGTON)	L	14Sec(Touralux)	6/28/47	Milw	4504A	
394	NP	RN	4DbrCptBLngREObs	6-7/48	PS	4120	6781		5742	Milw (MT.McKINLEY)	L	14Sec(Touralux)	6/23/47	Milw	4504A	
400	SP(1st)	S	2DbrCptDrObsBLng	4/41	PS	4082A	6644		5743	Milw (MT.BOSLEY)	L	14Sec(Touralux)	6/27/47	Milw	4504A	
400	SP(2nd)	N	2DbrCptDrBLngREObs	7/39	PS	4082	6567		5744	Milw (MT.RAINIER)	L	14Sec(Touralux)	6/19/47	Milw	4504A	
401	SP(1st)	S	2DbrCptDrObsBLng	4/41	PS	4082A	6644		5745	Milw (MT.RUSHMORE)	L	14Sec(Touralux)	6/19/47	Milw	4504A	
401	SP(2nd)	N	2DbrCptDrBLngREObs	6/40	PS	4082	6608		5746	Milw (MT.ST.HELENS)	L	14Sec(Touralux)	6/19/47	Milw	4504A	
480	CB&Q	N	8DuRmt6Rmt4Dbr	7-9/48	PS	4119	6781		5747	Milw (MT.WILSON)	L	14Sec(Touralux)	6/19/47	Milw	4504A	
481	CB&Q	N	8DuRmt6Rmt4Dbr	7-9/48	PS	4119	6781		5748	Milw (MT.HOPE)	L	14Sec(Touralux)	6/19/47	Milw	4504A	
482	CB&Q	N	8DuRmt6Rmt4Dbr	7-9/48	PS	4119	6781		5749	Milw (MT.STUART)	L	14Sec(Touralux)	6/23/47	Milw	4504A	
483	CB&Q	RN	4DbrCptBLngREObs	6-7/48	PS	4120	6781		5750	Milw (MT.HAROLD)	L	14Sec(Touralux)	6/27/47	Milw	4504A	

Figure I-81 NP-375 was built by Budd in 1954 for service on the Northern Pacific's "North Coast Limited" as NP-307, a 4-roomette, 4-double-bedroom, 4-duplex roomette car with dome. This car was rebuilt and renumbered in 1967 and became Amtrak's 9220 in 1971. NP was one of a few roads that specified flat stainless panels below the windows instead of the standard fluting; it also painted its stainless cars. Railroad Avenue Enterprises Photograph PN-2672.

CAR NAME OR NUMBER		ST	CAR TYPE	DATE BUILT	CAR BLDR	PLAN	LOT	CAR NAME OR NUMBER		ST	CAR TYPE	DATE BUILT	CAR BLDR	PLAN	LOT
5751	Milw (MT.ANGELES)	L	14Sec(Touralux)	6/19/47	Milw	4504A		9025	SP	S	10Rmt6DbrBE	5-6/50	Budd	9522A	9678
5752	Milw (MT.CHITTENDEN)	L	14Sec(Touralux)	6/9/47	Milw	4504		9026	SP	S	10Rmt6DbrBE	5-6/50	Budd	9522A	9678
								9027	SP	S	10Rmt6DbrBE	5-6/50	Budd	9522A	9678
5753	Milw (MT.JUPITER)	L	14Sec(Touralux)	6/23/47	Milw	4504		9028	SP	S	10Rmt6DbrBE	5-6/50	Budd	9522A	9678
5754	Milw (MT.TACOMA)	L	14Sec(Touralux)	6/19/47	Milw	4504		9029	SP	S	10Rmt6DbrBE	5-6/50	Budd	9522A	9678
9000	SP	S	10Rmt6Dbr	3-6/50	Budd	9522	9660	9030	SP	S	10Rmt6Dbr	5-7/50	PS	4140C	6874
9001	SP	S	10Rmt6Dbr	3-6/50	Budd	9522	9660	9031	SP	S	10Rmt6Dbr	5-7/50	PS	4140C	6874
9002	SP	S	10Rmt6Dbr	3-6/50	Budd	9522	9660	9032	SP	S	10Rmt6Dbr	5-7/50	PS	4140C	6874
9003	SP	S	10Rmt6Dbr	3-6/50	Budd	9522	9660	9033	SP	S	10Rmt6Dbr	5-7/50	PS	4140C	6874
9004	SP	S	10Rmt6Dbr	3-6/50	Budd	9522	9660	9034	SP	S	10Rmt6Dbr	5-7/50	PS	4140C	6874
9005	SP	S	10Rmt6Dbr	3-6/50	Budd	9522	9660	9035	SP	S	10Rmt6Dbr	5-7/50	PS	4140C	6874
9006	SP	S	10Rmt6Dbr	3-6/50	Budd	9522	9660	9036	SP	S	10Rmt6Dbr	5-7/50	PS	4140C	6874
9007	SP	S	10Rmt6Dbr	3-6/50	Budd	9522	9660	9037	SP	S	10Rmt6Dbr	5-7/50	PS	4140C	6874
9008	SP	S	10Rmt6Dbr	3-6/50	Budd	9522	9660	9038	SP	S	10Rmt6Dbr	5-7/50	PS	4140C	6874
9009	SP	S	10Rmt6Dbr	3-6/50	Budd	9522	9660	9039	SP	S	10Rmt6Dbr	5-7/50	PS	4140C	6874
9010	SP	S	10Rmt6Dbr	3-6/50	Budd	9522	9660	9040	SP	S	10Rmt6Dbr	5-7/50	PS	4140C	6874
9011	SP	S	10Rmt6Dbr	3-6/50	Budd	9522	9660	9041	SP	S	10Rmt6Dbr	5-7/50	PS	4140C	6874
9012	SP	S	10Rmt6Dbr	3-6/50	Budd	9522	9660	9042	SP	S	10Rmt6Dbr	5-7/50	PS	4140C	6874
9013	SP	S	10Rmt6Dbr	3-6/50	Budd	9522	9660	9043	SP	S	10Rmt6Dbr	5-7/50	PS	4140C	6874
9014	SP	S	10Rmt6Dbr	3-6/50	Budd	9522	9660	9044	SP	S	10Rmt6Dbr	5-7/50	PS	4140C	6874
9015	SP	S	10Rmt6Dbr	3-6/50	Budd	9522	9660	9045	SP	S	10Rmt6Dbr	5-7/50	PS	4140C	6874
9016	SP	S	10Rmt6Dbr	3-6/50	Budd	9522	9660	9046	SP	N	10Rmt6Dbr	5-7/50	PS	4140C	6874
9017	SP	S	10Rmt6Dbr	3-6/50	Budd	9522	9660	9047	SP	N	10Rmt6Dbr	5-7/50	PS	4140C	6874
9018	SP	S	10Rmt6Dbr	3-6/50	Budd	9522	9660	9048	SP	N	10Rmt6Dbr	5-7/50	PS	4140C	6874
9019	SP	S	10Rmt6Dbr	3-6/50	Budd	9522	9660	9049	SP	S	10Rmt6Dbr	5-7/50	PS	4140C	6874
9020	SP	S	10Rmt6Dbr	3-6/50	Budd	9522	9660	9050	SP	S	10Rmt6Dbr	5-7/50	PS	4140C	6874
9021	SP	S	10Rmt6Dbr	3-6/50	Budd	9522	9660	9051	SP	S	10Rmt6Dbr	5-7/50	PS	4140C	6874
9022	SP	S	10Rmt6Dbr	3-6/50	Budd	9522	9660	9052	SP	S	10Rmt6Dbr	5-7/50	PS	4140C	6874
9023	SP	S	10Rmt6Dbr	3-6/50	Budd	9522	9660	9053	SP	S	10Rmt6DbrBE	7/50	PS	4140D	6874
9024	SP	S	10Rmt6Dbr	3-6/50	Budd	9522	9660	9054	SP	S	10Rmt6DbrBE	7/50	PS	4140D	6874

Figure I-82 SP-9118 was a 4-compartment, 4-double bedroom, 2-drawing room car, built by Pullman in 1950 for Southern Pacific's "Cascade." This Wilber Whittaker photograph taken in Oakland, California in May 1952 shows the car in two-tone gray paint for the "Cascade." It received the silver with red stripe color scheme later for general service. This photograph shows the visual effect produced by the full width diaphragms; the canvas sides were painted to enhance the notion of the train as a unit.

CAR NAME OR NUMBER	ST	CAR TYPE	DATE BUILT	CAR BLDR	PLAN	LOT
9055 SP	N	10Rmt6DbrBE	7/50	PS	4140D	6874
9056 SP	N	10Rmt6DbrBE	7/50	PS	4140D	6874
9100 SP	SN	4Dbr4Cpt2Dr	3-4/42	PS	4069H	6668
9101 SP	SN	4Dbr4Cpt2Dr	3-4/42	PS	4069H	6668
9102 SP	N	4Dbr4Cpt2Dr	11/40	PS	4069F	6636
9103 SP	N	4Dbr4Cpt2Dr	11/40	PS	4069F	6636
9104 SP	N	4Dbr4Cpt2Dr	3/41	PS	4069G	6642
9105 SP	N	4Dbr4Cpt2Dr	3/41	PS	4069G	6642
9106 SP	N	4Dbr4Cpt2Dr	3/41	PS	4069G	6642
9107 SP	N	4Dbr4Cpt2Dr	3/41	PS	4069G	6642
9108 SP	N	4Dbr4Cpt2Dr	3/41	PS	4069G	6642
9109 SP	N	4Dbr4Cpt2Dr	3/41	PS	4069G	6642
9110 SP	N	4Dbr4Cpt2Dr	3-4/42	PS	4069H	6668
9111 SP	N	4Dbr4Cpt2Dr	3-4/42	PS	4069H	6668
9112 SP	N	4Dbr4Cpt2Dr	3-4/42	PS	4069H	6668
9113 SP	N	4Dbr4Cpt2Dr	3-4/42	PS	4069H	6668
9114 SP	N	4Dbr4Cpt2Dr	3-4/42	PS	4069H	6668
9115 SP	N	4Dbr4Cpt2Dr	3-4/42	PS	4069H	6668
9116 SP	N	4Dbr4Cpt2Dr	4-5/40	PS	4069D	6605
9117 SP	N	4Dbr4Cpt2Dr	4-5/40	PS	4069D	6605
9118 SP	S	4Dbr4Cpt2Dr	7/50	PS	4069M	6871
9119 SP	S	4Dbr4Cpt2Dr	7/50	PS	4069M	6871
9120 SP	S	4Dbr4Cpt2Dr	7/50	PS	4069M	6871
9150 SP	N	6Sec6Rmt4Dbr	5-6/42	PS	4099	6669
9151 SP	N	6Sec6Rmt4Dbr	5-6/42	PS	4099	6669
9152 SP	N	6Sec6Rmt4Dbr	5-6/42	PS	4099	6669
9153 SP	N	6Sec6Rmt4Dbr	5-6/42	PS	4099	6669
9154 SP	N	6Sec6Rmt4Dbr	5-6/42	PS	4099	6669
9155 SP	N	6Sec6Rmt4Dbr	5-6/42	PS	4099	6669
9156 SP	N	6Sec6Rmt4Dbr	5-6/42	PS	4099	6669
9157 SP	N	6Sec6Rmt4Dbr	4-5/42	PS	4099	6669
9158 SP	N	6Sec6Rmt4Dbr	4-5/42	PS	4099	6669
9159 SP	N	6Sec6Rmt4Dbr	4-5/42	PS	4099	6669
9160 SP	N	6Sec6Rmt4Dbr	4-5/42	PS	4099	6669
9161 SP	N	6Sec6Rmt4Dbr	4-5/42	PS	4099	6669
9162 SP	N	6Sec6Rmt4Dbr	5-6/42	PS	4099	6669
9163 SP	N	6Sec6Rmt4Dbr	5-6/42	PS	4099	6669
9164 SP	N	6Sec6Rmt4Dbr	5-6/42	PS	4099	6669
9165 SP	N	6Sec6Rmt4Dbr	5-6/42	PS	4099	6669
9200 SP	N	10Rmt5Dbr	11/40	PS	4072D	6636
9201 SP	N	10Rmt5Dbr	11/40	PS	4072D	6636
9202 SP	N	10Rmt5Dbr	2-3/41	PS	4072E	6641
9203 SP	N	10Rmt5Dbr	2-3/41	PS	4072E	6641
9204 SP	N	10Rmt5Dbr	2-3/41	PS	4072E	6641
9205 SP	N	10Rmt5Dbr	2-3/41	PS	4072E	6641
9206 SP	N	10Rmt5Dbr	2-3/41	PS	4072E	6641
9207 SP	N	10Rmt5Dbr	2-3/41	PS	4072E	6641
9208 SP	N	10Rmt5Dbr	2-3/41	PS	4072E	6641
9209 SP	N	10Rmt5Dbr	2-3/41	PS	4072E	6641
9210 SP	N	10Rmt5Dbr	2-3/41	PS	4072E	6641
9211 SP	N	10Rmt5Dbr	2-3/41	PS	4072E	6641
9250 SP	N	12DuRmt5Dbr	12/37	PS	4066A	6525
9251 SP	N	12DuRmt5Dbr	12/37	PS	4066A	6525
9300 SP	S	22Rmt	7/50	PS	4122B	6872

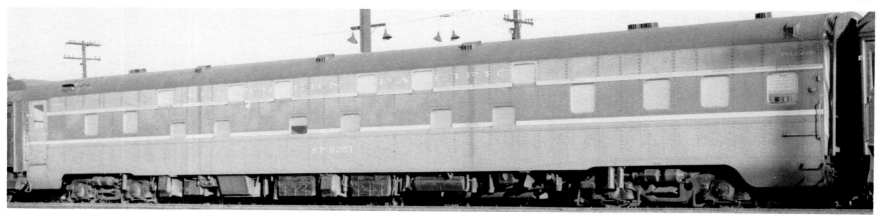

Figure I-83 Pullman built the first lightweight cars with duplex accommodations in 1937 for the seventh and eighth "City of - - -" trains. SP-9251 was built as ROSE BOWL, renamed TELEGRAPH HILL in 1941, and the name was removed and replaced with a number in 1947. The riveted aluminum construction with belt rail marks this as a traditional car, using standard car techniques with weight saving materials. The underbody is a complete mass of air brakes, Waukesha ice engines, propane bottles, battery boxes, steam pipes, and 32V DC generating equipment. Photograph by Wilber C. Whittaker.

Figure I-84 FERRY BUILDING was delivered in November 1940 by Pullman in Plan 4072D, Lot 6636, along with RINCON HILL, for service in the revised eighth and ninth trains respectively. Both cars were sold to SP in 1946. FERRY BUILDING became SP-9200 in 1950 and RINCON HILL became SP-9201 in 1949. They ride on distinctive floating journal trucks. In a nod to standard cars, the "City of" trains carried ersatz belt and letterboard rails. Photograph by Wilber C. Whittaker.

CAR NAME OR NUMBER	ST	CAR TYPE	DATE BUILT	CAR BLDR	PLAN	LOT
9301 SP	S	22Rmt	7/50	PS	4122B	6872
9302 SP	S	22Rmt	7/50	PS	4122B	6872
9303 SP	S	22Rmt	7/50	PS	4122B	6872
9304 SP	S	22Rmt	7/50	PS	4122B	6872
9305 SP	S	22Rmt	7/50	PS	4122B	6872
9306 SP	S	22Rmt	7/50	PS	4122B	6872
9350 SP	N	13Dbr	3/41	PS	4071D	6643
9351 SP	N	13Dbr	3/41	PS	4071D	6643
9352 SP	N	13Dbr	3/41	PS	4071D	6643
9353 SP	N	13Dbr	3/41	PS	4071D	6643
9354 SP	N	13Dbr	3/41	PS	4071D	6643
9355 SP	N	13Dbr	3/41	PS	4071D	6643
9356 SP	N	13Dbr	3/41	PS	4071D	6643
9357 SP	N	13Dbr	3/41	PS	4071D	6643
9400 SP	S	12Dbr	7/50	PS	4139A	6873
9401 SP	S	12Dbr	7/50	PS	4139A	6873
9402 SP	N	12Dbr	7/50	PS	4139A	6873
9403 SP	N	12Dbr	7/50	PS	4139A	6873
9500 SP	N	2DbrCptDrBLngREObs	7/39	PS	4082	6567
9501 SP	N	2DbrCptDrBLngREObs	6/40	PS	4083	6608
ABBOTT MANOR CP-10301	S	5DbrCpt4Sec4Rmt	11/54-55	Budd		9658
ABERDEEN NP	S	8DuRmt6Rmt4Dbr	7-9/48	PS	4119	6781
ABRAHAM LINCOLN (DS)UP		10SecCptDbrArtic	10/34	PS	4035A	6428

CAR NAME OR NUMBER	ST	CAR TYPE	DATE BUILT	CAR BLDR	PLAN	LOT
ACUARIO NdeM-597	SN	4Sec8DuRmt4Dbr		PS	4107?	
ADVANCE Pool		14DuSrArtic	8/36	PS	4050	6478
ADVENTURER NYC	N	DrMbrObsBLng	5/38	PS	4079A	6547
AFGANISTAN NdeM-526	SN	10Rmt6Dbr	9/48-/49	PS	4123	6790
AGASSIZ GLACIER GN-1182	S	16DuRmt4Dbr	12/50	PS	4108A	6890
AGAWAM RIVER NYC	S	10Rmt6Dbr	9/48-/49	PS	4123	6790
AHERN GLACIER GN-1171	S	16DuRmt4Dbr	1-2/47	PS	4108	6751
AIR FORCE ACADEMY CRI&PS		8Rmt6Dbr	11-12/54	PS	4195	6944
AKAMINA PASS GN-1372		6Rmt5Dbr2Cpt	10-11/50	PS	4180	6877
ALABAMA PINE L&N-3450	S	6Sec6Rmt4Dbr	3-5/53	PS	4183	6909
ALABAMA RIVER WofA	S	10Rmt6Dbr	9-10/49	PS	4140	6814
ALACHUA COUNTY ACL	S	10Rmt6Dbr	9-10/49	PS	4140B	6809
ALAN HARDMAN JonesProp	SN	4Dbr4Cpt2Dr	9-10/40	PS	4069E	6617
ALAPAHA RIVER SOU	S	10Rmt6Dbr	9-10/49	PS	4140	6814
ALBANIA NdeM-340	SN	10Rmt6Dbr	9/48-/49	PS	4123	6790
ALBANY HARBOR NYC	S	22Rmt	4-6/49	Budd	9501	9661
ALDER CREEK Milw-12	S	8DbrSKYTOPLng	11/48-49	PS	4138	6775
ALDER FALLS PRR	S	6DbrBLng	3-5/49	PS	4131	6792
ALEMENIA NdeM-341	SN	10Rmt6Dbr	9/48-/49	PS	4123	6790
ALEXANDRA FALLS CN-2105 LW	SN	14Rmt4Dbr	1-6/48	PS	4153	6769
JOHNSON CASSATT PRR	S	2DrCptDbrBEObsLng	3-4/49	PS	4134	6792
M. BYERS PRR	N	10Rmt6Dbr	11/48-49	PS	4140	6792

Figure I-85   SP-9353, a 13-double-bedroom product of Pullman, was built in 1941 as SP-303. Pullman had by then settled on the smooth side, carbon steel rectangular truss car body that characterizes most of its lightweight cars. It was renumbered in May 1951 and is shown here in "Lark" service at San Francisco on March 9, 1952. Photographed by Wilber C. Whittaker.

CAR NAME OR NUMBER	ST	CAR TYPE	DATE BUILT	CAR BLDR	PLAN	LOT
ALFRED E.HUNT PRR	N	12DuSr4Dbr	1-4/49	PS	4130	6792
ALFRED HUNT PRR	N	12DuSr4Dbr	1-4/49	PS	4130	6792
ALLAN MANOR CP-10302	S	5DbrCpt4Sec4Rmt	11/54-55	Budd		9658
ALLEGHENY B&O	SN	10Rmt6Dbr	2-7/50	PS	4167	6864
CLUB C&O	S	5DbrBEObsBLng	8/50	PS	4165	6863
COUNTY PRR	S	13Dbr	4/38	PS	4071A	6541
RAPIDS PRR-8444	S	10Rmt6Dbr	9-10/50	AC&F	9008	3212
RIVER NYC	S	10Rmt6Dbr	9/48-/49	PS	4123	6790
ALLIANCE INN PRR-8241	S	21Rmt	1-6/49	Budd	9513	9667
ALLYNS POINT NH	S	14Rmt4Dbr	12/49/50	PS	4159	6822
ALPINE						
CAMP UP	S	14Sec	1-4/54	AC&F	9017	3816
CREST UP	S	14Sec	1-4/54	AC&F	9017	3816
GROVE UP	S	14Sec	1-4/54	AC&F	9017	3816
LAKE UP	S	14Sec	1-4/54	AC&F	9017	3816
LODGE UP	S	14Sec	1-4/54	AC&F	9017	3816
MEADOW UP	S	14Sec	1-4/54	AC&F	9017	3816
PARK UP	S	14Sec	1-4/54	AC&F	9017	3816
PASS UP	S	14Sec	1-4/54	AC&F	9017	3816
PEAK UP	S	14Sec	1-4/54	AC&F	9017	3816
RIVER UP	S	14Sec	1-4/54	AC&F	9017	3816
ROAD UP	S	14Sec	1-4/54	AC&F	9017	3816
SCENE UP	S	14Sec	1-4/54	AC&F	9017	3816
STREAM UP	S	14Sec	1-4/54	AC&F	9017	3816
VIEW UP	S	14Sec	1-4/54	AC&F	9017	3816
ALTAMAHA RIVER SOU	S	10Rmt6Dbr	9-10/49	PS	4140	6814
AMACUZAC NdeM	SN	4DbrBLngObs	5/38	PS	4079A	6547

CAR NAME OR NUMBER	ST	CAR TYPE	DATE BUILT	CAR BLDR	PLAN	LOT
AMAGANSETT LIRR-2000	SN	5DbrBEObsBLng	8/50	PS	4165	6863
AMAZON RIVER NYC	S	10Rmt6Dbr	9/48-/49	PS	4123	6790
AMERICAN						
ACE JT	S	6Sec6Rmt4Dbr	5-6/42	PS	4099	6669
ADVENTURE JT	S	6Sec6Rmt4Dbr	5-6/42	PS	4099	6669
AMBASSADOR JT	S	6Sec6Rmt4Dbr	5-6/42	PS	4099	6669
ARMY JT	S	6Sec6Rmt4Dbr	5-6/42	PS	4099	6669
BEACON JT	S	6Sec6Rmt4Dbr	5-6/42	PS	4099	6669
BEAUTY JT	S	6Sec6Rmt4Dbr	5-6/42	PS	4099	6669
BORDER UP	S	6Sec6Rmt4Dbr	3-4/50	AC&F	9005	3070
BRIGADE JT	S	6Sec6Rmt4Dbr	5-6/42	PS	4099	6669
BUFFALO JT	S	6Sec6Rmt4Dbr	5-6/42	PS	4099	6669
CANYON JT	S	6Sec6Rmt4Dbr	5-6/42	PS	4099	6669
CAPTAIN JT	S	6Sec6Rmt4Dbr	5-6/42	PS	4099	6669
CHARM JT	S	6Sec6Rmt4Dbr	5-6/42	PS	4099	6669
CLASSIC JT	S	6Sec6Rmt4Dbr	5-6/42	PS	4099	6669
CLIPPER JT	S	6Sec6Rmt4Dbr	5-6/42	PS	4099	6669
COMMAND JT	S	6Sec6Rmt4Dbr	5-6/42	PS	4099	6669
CONSULATE UP	S	6Sec6Rmt4Dbr	3-4/50	AC&F	9005	3070
COURIER JT	S	6Sec6Rmt4Dbr	5-6/42	PS	4099	6669
CRUISER JT	S	6Sec6Rmt4Dbr	5-6/42	PS	4099	6669
DAIRYLAND JT	N	6Sec6Rmt4Dbr	5-6/42	PS	4099	6669
EAGLE JT	S	6Sec6Rmt4Dbr	5-6/42	PS	4099	6669
ELM JT	S	6Sec6Rmt4Dbr	5-6/42	PS	4099	6669
EMBASSY C&NW	S	6Sec6Rmt4Dbr	3-4/50	AC&F	9005	3073
EMBLEM JT	S	6Sec6Rmt4Dbr	5-6/42	PS	4099	6669
ENSIGN JT	S	6Sec6Rmt4Dbr	5-6/42	PS	4099	6669

CAR NAME OR NUMBER	ST	CAR TYPE	DATE BUILT	CAR BLDR	PLAN	LOT	CAR NAME OR NUMBER	ST	CAR TYPE	DATE BUILT	CAR BLDR	PLAN	LOT
AMERICAN							AMERICAN						
ESCORT JT	S	6Sec6Rmt4Dbr	5-6/42	PS	4099	6669	PARK JT	S	6Sec6Rmt4Dbr	5-6/42	PS	4099	6669
FALLS JT	S	6Sec6Rmt4Dbr	5-6/42	PS	4099	6669	PATROL JT	S	6Sec6Rmt4Dbr	5-6/42	PS	4099	6669
FLYER JT	S	6Sec6Rmt4Dbr	5-6/42	PS	4099	6669	PLAINS JT	S	6Sec6Rmt4Dbr	5-6/42	PS	4099	6669
FORTRESS JT	S	6Sec6Rmt4Dbr	5-6/42	PS	4099	6669	PROGRESS JT	S	6Sec6Rmt4Dbr	5-6/42	PS	4099	6669
FORUM JT	N	6Sec6Rmt4Dbr	5-6/42	PS	4099	6669	RAMPART JT	N	6Sec6Rmt4Dbr	5-6/42	PS	4099	6669
GENERAL UP	S	6Sec6Rmt4Dbr	3-4/50	AC&F	9005	3070	RANGER JT	N	6Sec6Rmt4Dbr	5-6/42	PS	4099	6669
GUARD JT	N	6Sec6Rmt4Dbr	5-6/42	PS	4099	6669	RAPIDS JT	S	6Sec6Rmt4Dbr	5-6/42	PS	4099	6669
HAVEN JT	S	6Sec6Rmt4Dbr	5-6/42	PS	4099	6669	RIVER JT	S	6Sec6Rmt4Dbr	5-6/42	PS	4099	6669
HEIGHTS JT	S	6Sec6Rmt4Dbr	5-6/42	PS	4099	6669	ROSE JT	S	6Sec6Rmt4Dbr	5-6/42	PS	4099	6669
HILLS JT	S	6Sec6Rmt4Dbr	5-6/42	PS	4099	6669	ROYAL JT	S	6Sec6Rmt4Dbr	5-6/42	PS	4099	6669
HOME JT	S	6Sec6Rmt4Dbr	5-6/42	PS	4099	6669	SAILOR JT	S	6Sec6Rmt4Dbr	5-6/42	PS	4099	6669
INDIAN JT	S	6Sec6Rmt4Dbr	5-6/42	PS	4099	6669	SCENE JT	S	6Sec6Rmt4Dbr	5-6/42	PS	4099	6669
LAKE JT	S	6Sec6Rmt4Dbr	5-6/42	PS	4099	6669	SENTRY JT	S	6Sec6Rmt4Dbr	5-6/42	PS	4099	6669
LIBERTY Erie	S	6Sec6Rmt4Dbr	6/42	PS	4099	6669	SHORES JT	S	6Sec6Rmt4Dbr	5-6/42	PS	4099	6669
LIFE Erie	S	6Sec6Rmt4Dbr	6/42	PS	4099	6669	SKIES JT	S	6Sec6Rmt4Dbr	5-6/42	PS	4099	6669
LIGHT JT	S	6Sec6Rmt4Dbr	5-6/42	PS	4099	6669	SKIPPER JT	S	6Sec6Rmt4Dbr	5-6/42	PS	4099	6669
MANOR JT	S	6Sec6Rmt4Dbr	5-6/42	PS	4099	6669	SOLDIER JT	S	6Sec6Rmt4Dbr	5-6/42	PS	4099	6669
MARINE JT	S	6Sec6Rmt4Dbr	5-6/42	PS	4099	6669	STAR JT	S	6Sec6Rmt4Dbr	5-6/42	PS	4099	6669
MERCHANT JT	N	6Sec6Rmt4Dbr	5-6/42	PS	4099	6669	SUNSET C&NW	S	6Sec6Rmt4Dbr	3-4/50	AC&F	9005	3073
MILEMASTER Pool	S	2DbrCptDrObsBLng	7/39	PS	4082	6567	TRAILS JT	S	6Sec6Rmt4Dbr	5-6/42	PS	4099	6669
MONITOR JT	S	6Sec6Rmt4Dbr	5-6/42	PS	4099	6669	TROOPER JT	S	6Sec6Rmt4Dbr	5-6/42	PS	4099	6669
NAVY JT	S	6Sec6Rmt4Dbr	5-6/42	PS	4099	6669	UNITY Erie	S	6Sec6Rmt4Dbr	6/42	PS	4099	6669

**Figure I-86   AMERICAN SHORES is 1 of 60 6-section, 6-roomette, 4-double-bedroom cars in Pullman Standard Lot 6669
built in 1942 for joint CNW-UP-SP "Overland" service. The pairs of upper berth "loopholes" were on many prewar
(pre-1942) sleepers. Wilber Whittaker photographed this Union Pacific car in September 1954 in San Francisco. It was
withdrawn from Pullman lease October 10, 1967 and converted to UP bunk car 906088.**

CAR NAME OR NUMBER	ST	CAR TYPE	DATE BUILT	CAR BLDR	PLAN	LOT
AMERICAN						
VIEW UP	S	6Sec6Rmt4Dbr	3-4/50	AC&F	9005	3070
WAY Erie	S	6Sec6Rmt4Dbr	6/42	PS	4099	6669
WOODLAND JT	S	6Sec6Rmt4Dbr	5-6/42	PS	4099	6669
AMHERST MANOR CP-10303	S	5DbrCpt4Sec4Rmt	11/54-55	Budd		9658
AMON G.CARTER PRR-8441	N	10Rmt6Dbr	9-10/50	AC&F	9008	3200
ANACOSTIA						
RIVER PRR-8359	S	14Rmt2Dr	2-3/50	AC&F	9007	3098
ANDERSON INN PRR-8242	S	21Rmt	1-6/49	Budd	9513	9667
ANDREW						
W. MELLON PRR-8387	N	4Dbr4Cpt2Dr	9-12/48	AC&F	9009	2982
ANDROMEDA NdeM-588	SN	4Sec8DuRmt4Dbr	12/46-47	PS	4107A	6751
ANGEL ISLAND JT	S	4Dbr4Cpt2Dr	11/40	PS	4069F	6636
ANGELA ANN JonesProp	SN	4Dbr4Cpt2Dr	9-10/40	PS	4069E	6617
ANTELOPE VALLEY ATSF	S	6Sec6Rmt4Dbr	6/42	PS	4099	6669
APPLE GROVE CP-14101		10Sec5Dbr	/49-/50	CPRR		
AQUARIUS FCP	SN	10Rmt6Dbr	9/48-/49	PS	4123	6790
ARABIA NdeM-524	SN	10Rmt6Dbr	9/48-/49	PS	4123	6790
ARCADIA JT	S	4Dbr4Cpt2Dr	11/40	PS	4069F	6636
ARGELIA NdeM-532	SN	10Rmt6Dbr	9/48-/49	PS	4123	6790
ARGENTINA FEC	S	10Rmt6Dbr	9-10/49	PS	4140	6814
ARGENTINA NdeM-563	SN	10Rmt6Dbr	12/48-49	Budd	9510	9660
ARIES FCP	SN	10Rmt6Dbr	9/48-/49	PS	4123	6790
ARKANSAS RIVER MP	S	6Sec6Rmt4Dbr	5/42	PS	4099	6669
ARLINGTON CLUB NP	S	5DbrREObsBLng	6-7/48	PS	4120	6781
ARMON B. KING MKT-1506	S	14Rmt4Dbr	1-6/48	PS	4153	6769
AROOSTOOK RIVER NYC	S	10Rmt6Dbr	9/48-/49	PS	4123	6790
ARROW CREEK Milw-14	S	8DbrSKYTOPLng	11/48-49	PS	4138	6775
ARROYO SECO JT	S	11DbrArtic	12/37	PS	4067A	6525
ARTHUR STILLWELL KCS	S	14Rmt4Dbr	5/48	PS	4153	6795
ASH GROVE CP-14102		10Sec5Dbr	/49-/50	CPRR		
ASHLEY RIVER ACL	S	14Rmt2Dr	2-3/50	AC&F	9007	3091
ASHTABULA COUNTY NYC	S	13Dbr	10/40	PS	4071C	6618
ASHTABULA HARBOR NYC	S	22Rmt	4-6/49	Budd	9501	9661
ASHTABULA INN PRR-8243	S	21Rmt	1-6/49	Budd	9513	9667
ASPEN FALLS PRR	S	6DbrBLng	3-5/49	PS	4131	6792
ATHENS PRR-8333	S	10Rmt6Dbr	6-8/49	Budd	9503	9662
ATLANTA SAL	S	10Rmt6Dbr	5-6/49	PS	4140A	6796
AUGLAISE B&O	S	10Rmt6Dbr	3/50	PS	4167	6864
AUGUST A. BUSCH PRR	N	12DuSr4Dbr	1-4/49	PS	4130	6792
AUGUSTA COUNTY N&W	S	10Rmt6Dbr	11/49-50	Budd	9523	9660
AUGUSTE						
CHOTEAU SLSF-1454	S	14Rmt4Dbr	1-6/48	PS	4153	6769
AULD COVE CN-1182	N	4Dbr4Cpt2Dr	3-4/38	PS	4069B	6540
AUSABLE RIVER NYC	S	10Rmt6Dbr	9/48-/49	PS	4123	6790
AUSTRIA NdeM-342	SN	10Rmt6Dbr	9/48-/49	PS	4123	6790
AVON PARK SAL-74	S	11Dbr	12/55-56	PS	4198A	6968
AYLMER MANOR CP-10304	S	5DbrCpt4Sec4Rmt	11/54-55	Budd		9658
AZALEA FEC	S	5DbrREObsBLng	2-3/50	PS	4162	6814
B.F. JONES PRR-8388	N	4Dbr4Cpt2Dr	9-12/48	AC&F	9009	2982
B.F. JONES PRR-8287	N	21Rmt	1-6/49	Budd	9513	9667

Figure I-87 ASPEN FALLS, a former PRR 6-double-bedroom/buffet lounge is shown in Penn Central green livery at Chicago in 1969. In 1974 it went to Amtrak 3207 with the same name, was retired in 1975 and sold to a private owner. The stripe follows the belt rail, a long horizonal fishplate found on most standard cars, but used by only a few railroads on lightweight cars. Railroad Avenue Enterprises Photograph PN-2439.

CAR NAME OR NUMBER	ST	CAR TYPE	DATE BUILT	CAR BLDR	PLAN	LOT
BABBLING BROOK NYC	S	5DbrBLngObs	5-6/49	Budd	9506	9664
BACOBI ATSF	S	4Dbr4Cpt2Dr	12/39	PS	4069C	6597
BAD AXLE RIVER GN-1264	R	8DuRmt6DbrCpt	11-12/50	PS	4181A	6889
BAD AXLE RIVER GN-1264	S	8DuRmt4Sec3DbrCpt	11-12/50	PS	4181	6889
BADDECK CN-1904	SN	8DbrSKYTOPLng	11/48-49	PS	4138	6775
BAHAMIAN FEC	S	10Rmt6Dbr	9-10/49	PS	4140B	6809
BAILEY'S BEACH NH-527	S	6Sec6Rmt4Dbr	11/54-55	PS	4194	W6942
BAKER UP	S	5DbrBLng	4/56	PS	4199	6959
BALDY MOUNTAIN JT	S	4DbrBEObsBLng	11/40	PS	4096A	6636
BALSAM FALLS PRR	S	6DbrBLng	3-5/49	PS	4131	6792
BALSAM GROVE CP-14103		10Sec5Dbr	/49-/50	CPRR		
BALTIMORE COUNTY PRR	S	10Rmt6Dbr	9-10/49	PS	4140B	6809
BANANA ROAD IC	S	6Sec6Rmt4Dbr	4/42	PS	4099	6669
BAR HARBOR NYC	S	22Rmt	4-6/49	Budd	9501	9661
BARCLAY PC-4216	SN	6Sec6Rmt4Dbr	11/54-55	PS	4194	W6942
BARREN RIVER L&N	S	10Rmt6Dbr	4/50	PS	4140	6814
BATON ROUGE IC	S	11Dbr	7/53	PS	4168A	6913
BATTLE RIVER CN-2147	SN	10Rmt6Dbr	5-6/49	PS	4129A	6797
BAY BRIDGE NYC	N	4Dbr4Cpt2Dr	3-4/38	PS	4069B	6540
BAY CITY C&O	S	10Rmt6Dbr	2-7/50	PS	4167	6864
BAY PINES SAL-50	S	5DbrCpt4Sec4Rmt	11/55	Budd	9537	9658
BAY STATE NH-550	S	6DbrBLng	1/55	PS	4193	W6941
BAYFIELD MANOR CP-10305	S	5DbrCpt4Sec4Rmt	11/54-55	Budd		9658
BEAR						
FLAG (DS)UP	N	14DuSrArtic	8/36	PS	4050	6478
LAKE UP	RN	4CptDr4DbrArtic	12/37	PS	4063C	6525
BEAR						
MOUNTAIN BRIDGE NYC	N	4Dbr4Cpt2Dr	3-4/38	PS	4069B	6540
BEAUFORT COUNTY ACL	S	10Rmt6Dbr	9-10/49	PS	4140B	6809
BEAVER						
COVE CN-1186	SN	4Dbr4Cpt2Dr	9-10/40	PS	4069E	6617
FALLS INN PRR-8244	S	21Rmt	1-6/49	Budd	9513	9667
RIVER NYC	S	10Rmt6Dbr	9/48-/49	PS	4123	6790
TAIL POINT NH	S	14Rmt4Dbr	12/49/50	PS	4159	6822
BEDFORD CN-1099	S	7CptBLngObs	3/54	PS	4190	6935
BEDFORD INN PRR-8245	S	21Rmt	1-6/49	Budd	9513	9667
BEDLOE'S ISLAND NYC	S	DrMbrObsBLng	5/38	PS	4079	6547
BEECH FALLS PRR	S	6DbrBLng	3-5/49	PS	4131	6792
BEECH GROVE CP-14104		10Sec5Dbr	/49-/50	CPRR		
BELGICA NdeM-343	SN	10Rmt6Dbr	9/48-/49	PS	4123	6790
BELL MANOR CP-10306	S	5DbrCpt4Sec4Rmt	11/54-55	Budd		9658
BELLE						
ISLE DOME C&O-1850	S	Dome5RmtDbr3Dr	9/48	Budd	9524	9669
RIVER CN-2138	SN	10Rmt6Dbr	9-10/49	PS	4140B	6809
BELLEVILLE IC	S	11Dbr	7/53	PS	4168A	6913
BENJAMIN						
BAKEWELL PRR	N	12DuSr4Dbr	1-4/49	PS	4130	6792
HILL FS-BC	SN	10Rmt6Dbr	2-7/50	PS	4167	6864
LODER Erie	S	10Rmt6Dbr	5-6/49	PS	4129A	6797
R. MILAM MKT-1501	S	14Rmt4Dbr	1-6/48	PS	4153	6769
BENNINGTON FALLS NYC	S	6DbrBLng	8-9/40	PS	4086B	6612
BENTON IC	S	11Dbr	7/53	PS	4168A	6913
BENTON HARBOR NYC	S	22Rmt	4-6/49	Budd	9501	9661

Figure I-88   BENTON is an 11-double-bedroom car built by Pullman Standard for the Illinois Central Railroad in 1953. Note that the name Pullman enjoys the center of the letterboard. This car was used in the consists of the "Panama Limited" and "City of Miami." The conspicuously different third sash from left is an emergency exit window. In the summer of 1966 it was painted for service on the "North Coast Limited." From the Collection of Howard Ameling.

CAR NAME OR NUMBER	ST	CAR TYPE	DATE BUILT	CAR BLDR	PLAN	LOT
BERKELEY COUNTY ACL	S	10Rmt6Dbr	9-10/49	PS	4140B	6809
BETAHTAKIN ATSF	S	4DrDbrREOBSLng	1-2/38	PS	4070	6532
BEVERLY HILLS JT	S	12SecArtic	12/37	PS	4064A	6525
BIG HORN PASS GN-1379	S	6Rmt5Dbr2Cpt	10-11/50	PS	4180	6877
BIG MOOSE LAKE NYC	S	6DbrBLng	12/48-49	PS	4124	6790
BIG PINEY (DS)UP		12SecArtic	6/36	PS	4053	6486
BILBAO NdeM-	SN	6Dbr5Cpt	7-8/52	Budd	9534	9641
BILLINGS NP	S	8DuRmt6Rmt4Dbr	7-9/48	PS	4119	6781
BILTABITO ATSF	S	4DrDbrREOBSLng	1-2/38	PS	4070	6532
BILTMORE PC-4212	SN	6Sec6Rmt4Dbr	11/54-55	PS	4194	W6942
BIRCH BAY NYC	S	22Rmt	10/48	PS	4122	6790
BIRCH FALLS PRR	S	6DbrBLng	9/40	PS	4086A	6612
BIRCH GROVE CP-14105		10Sec5Dbr	/49-/50	CPRR		
BIRCH RIVER PRR	S	10Rmt6Dbr	9-10/49	PS	4140	6814
BIRMANIA NdeM-518	SN	10Rmt6Dbr	9/48-/49	PS	4123	6790
BIRMINGHAM SAL	S	10Rmt6Dbr	5-6/49	PS	4140A	6796
BISMARK NP	S	8DuRmt6Rmt4Dbr	7-9/48	PS	4119	6781
BLACK POINT NH	S	14Rmt4Dbr	12/49/50	PS	4159	6822
BLACKFOOT						
GLACIER GN-1170	S	16DuRmt4Dbr	1-2/47	PS	4108	6751
BLAIR MANOR CP-10307	S	5DbrCpt4Sec4Rmt	11/54-55	Budd		9658
BLANCHARD RIVER NYC	S	10Rmt6Dbr	9/48-/49	PS	4123	6790
BLEWETTE PASS GN-1382	S	6Rmt5Dbr2Cpt	10-11/50	PS	4180	6877
BLISS MANOR CP-10308	S	5DbrCpt4Sec4Rmt	11/54-55	Budd		9658
BLOOMINGTON IC	S	11Dbr	7/53	PS	4168A	6913
BLUE BAY ATSF-204	S	10Rmt3Dbr2Cpt	11/47-48	PS	4145	6757
BLUE BELL ATSF-205	S	10Rmt3Dbr2Cpt	11/47-48	PS	4145	6757
BLUE BIRD ACL	RN	7Dbr2Dr	2-3/50	AC&F	9018	3091
BLUE BOY WAB	S	12Rmt4Dbr	2/50	AC&F	9004	3074
BLUE CLOUD WAB	S	12Rmt4Dbr	2/50	AC&F	9004	3074
BLUE FLAG ATSF-206	S	10Rmt3Dbr2Cpt	11/47-48	PS	4145	6757
BLUE GAZELLE WAB	S	12Rmt4Dbr	2/50	AC&F	9004	3074
BLUE GEM ATSF-207	S	10Rmt3Dbr2Cpt	11/47-48	PS	4145	6757
BLUE GRASS ATSF-208	S	10Rmt3Dbr2Cpt	11/47-48	PS	4145	6757
BLUE GRASS STATE IC	S	6Sec6Rmt4Dbr	4/42	PS	4099	6669
BLUE GROVE ATSF-209	S	10Rmt3Dbr2Cpt	11/47-48	PS	4145	6757

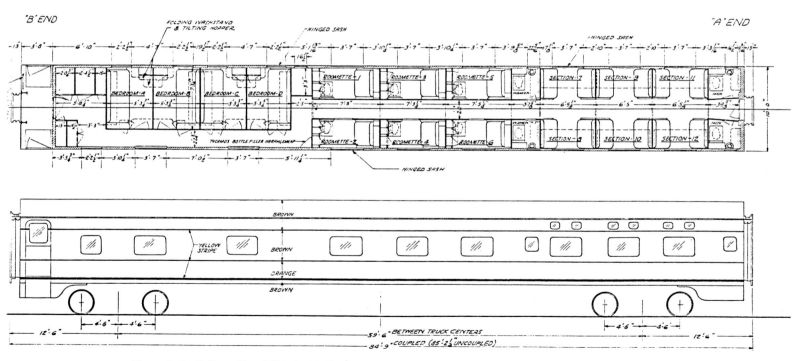

Figure I-89 Pullman Plan 4099 included 12 "6-6-4" sleepers assigned to the Illinois Central "Panama Limited." Shown above is the IC diagram for those cars which included the BLUE GRASS STATE. Most railroads prepared detailed diagram sheets of their cars for use by the various mechanical and operating departments. The Author's Collection.

CAR NAME OR NUMBER	ST	CAR TYPE	DATE BUILT	CAR BLDR	PLAN	LOT
BLUE HEART ATSF-210	S	10Rmt3Dbr2Cpt	11/47-48	PS	4145	6757
BLUE HERON ATSF-211	S	10Rmt3Dbr2Cpt	11/47-48	PS	4145	6757
BLUE HILL ATSF-212	S	10Rmt3Dbr2Cpt	11/47-48	PS	4145	6757
BLUE HORIZON WAB	S	12Rmt4Dbr	2/50	AC&F	9004	3074
BLUE ISLAND ATSF-213	S	10Rmt3Dbr2Cpt	11/47-48	PS	4145	6757
BLUE KNIGHT WAB	S	12Rmt4Dbr	2/50	AC&F	9004	3074
BLUE LAKE ATSF-214	S	10Rmt3Dbr2Cpt	11/47-48	PS	4145	6757
BLUE MOON ATSF-215	S	10Rmt3Dbr2Cpt	11/47-48	PS	4145	6757
BLUE MOTT ATSF-216	S	10Rmt3Dbr2Cpt	11/47-48	PS	4145	6757
BLUE MOUND ATSF-217	S	10Rmt3Dbr2Cpt	11/47-48	PS	4145	6757
BLUE POINT ATSF-218	S	10Rmt3Dbr2Cpt	11/47-48	PS	4145	6757
BLUE POND ATSF-219	S	10Rmt3Dbr2Cpt	11/47-48	PS	4145	6757
BLUE RAPIDS PRR-8442	S	10Rmt6Dbr	9-10/50	AC&F	9008	3200
BLUE RIDGE ATSF-220	S	10Rmt3Dbr2Cpt	11/47-48	PS	4145	6757
BLUE RIDGE CLUB C&O	S	5DbrBEObsBLng	8/50	PS	4165	6863
BLUE SKIES WAB	S	12Rmt4Dbr	2/50	AC&F	9004	3074
BLUE SPRINGS ATSF-221	S	10Rmt3Dbr2Cpt	11/47-48	PS	4145	6757
BLUE VALLEY ATSF	S	6Sec6Rmt4Dbr	6/42	PS	4099	6669
BLUE WATER ATSF-223	S	10Rmt3Dbr2Cpt	11/47-48	PS	4145	6757

CAR NAME OR NUMBER	ST	CAR TYPE	DATE BUILT	CAR BLDR	PLAN	LOT
BLUEGRASS CLUB C&O	S	5DbrBEObsBLng	8/50	PS	4165	6863
BOBOLINK B&O	S	16DuRmt4Dbr	4-6/54	Budd	9536	9658
BOCA GRANDE SAL-60	S	5Dbr2Cpt2Dr	1/56	PS	4201A	6969
BOIS de						
SIOUX RIVER GN-1274	R	8DuRmt6DbrCpt	11-12/50	PS	4181A	6889
SIOUX RIVER GN-1274	S	8DuRmt4Sec3DbrCpt	11-12/50	PS	4181	6889
BOISE UP	S	5DbrBLng	4/56	PS	4199	6959
BOLIVIA NdeM-564	SN	10Rmt6Dbr	12/48-49	Budd	9510	9660
BONAMPAK NdeM-108	L	8Sec3Cpt	/52	Schi	100570	
BONNE FEMME RIVER NYC	S	10Rmt6Dbr	9/48-/49	PS	4123	6790
BONNIE BROOK NYC	S	5DbrBLngObs	5-6/49	Budd	9506	9664
BOOTHBAY HARBOR NYC	S	22Rmt	4-6/49	Budd	9501	9661
BOSTON HARBOR NYC	S	22Rmt	4-6/49	Budd	9501	9661
BOULDER UP	N	12SecArtic	12/37	PS	4064	6525
BOULDER CANYON (DS)UP		11SecArtic	5/36	PS	4038B	6434
BOULDER STREAM NYC	S	6DbrBLng	3-5/49	Budd	9505	9663
BRADDOCK INN PRR-8246	S	21Rmt	1-6/49	Budd	9513	9667
BRADENTON PRR-8334	S	10Rmt6Dbr	6-8/49	Budd	9503	9662
BRADLEY IC	S	11Dbr	7/53	PS	4168A	6913

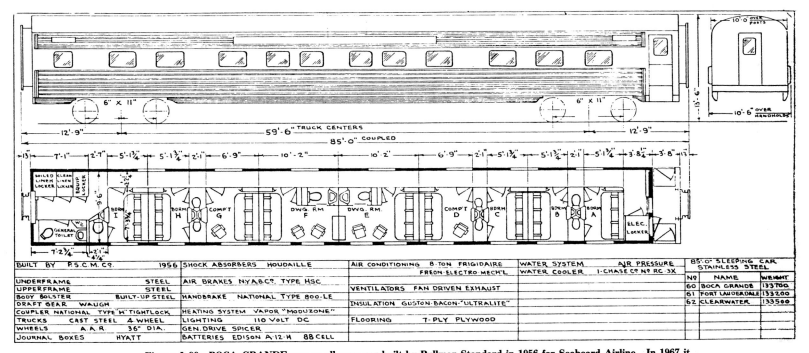

Figure I-90 BOCA GRANDE was an all-room car built by Pullman Standard in 1956 for Seaboard Airline. In 1967 it passed to Seaboard Coast Line, in 1971 to Amtrak, and was retired and sold to the Pacific South West Railway Museum. Drawings like these were also known as "Clearance Diagrams," as they dimensioned the official size of the cars. The Author's Collection.

CAR NAME OR NUMBER	ST	CAR TYPE	DATE BUILT	CAR BLDR	PLAN	LOT
BRAINERD NP	S	8DuRmt6Rmt4Dbr	7-9/48	PS	4119	6781
BRANDYWINE						
RAPIDS PRR-8438	N	10Rmt6Dbr	8-10/50	AC&F	9008	3200
BRANT MANOR CP-10309	S	5DbrCpt4Sec4Rmt	11/54-55	Budd		9658
BRAZIL FEC	S	10Rmt6Dbr	9-10/49	PS	4140	6814
BRAZIL NdeM-565	SN	10Rmt6Dbr	1-3/49	Budd	9502	9662
BRIGHAM YOUNG D&RG-1272	S	10Rmt6Dbr	7/50	PS	4167	6864
BRILLIANT						
COVE CN (1187)	P	4Dbr4Cpt2Dr	7-8/39	PS	4069D	6571
BROCK MANOR CP-10310	S	5DbrCpt4Sec4Rmt	11/54-55	Budd		9658
BRONX RIVER NYC	S	10Rmt6Dbr	9/48-/49	PS	4123	6790
BROOKHAVEN IC	S	11Dbr	7/53	PS	4168A	6913
BROOKLYN						
BRIDGE NYC	N	4Dbr4Cpt2Dr	3-4/38	PS	4069B	6540
BRIDGE AMT-2552	SN	8Rmt6Dbr	11-12/54	PS	4195	6944
BRYAN COUNTY ACL	SN	10Rmt6Dbr	2-7/50	PS	4167	6864
BUCHANAN COUNTY N&W	S	10Rmt6Dbr	3/49	PS	4140	6792
BUCKLEY BAY CN-2022	S	10Rmt5Dbr	6/54	PS	4186A	6925

CAR NAME OR NUMBER	ST	CAR TYPE	DATE BUILT	CAR BLDR	PLAN	LOT
BUCKS COUNTY PRR	S	10Rmt6Dbr	9-10/49	PS	4140B	6809
BUCYRUS INN PRR-8247	S	21Rmt	1-6/49	Budd	9513	9667
BUFFALO						
BAYOU CRI&P	S	8Rmt6Dbr	11-12/54	PS	4195	6944
HARBOR NYC	S	22Rmt	4-6/49	Budd	9501	9661
RAPIDS PRR-8443	S	10Rmt6Dbr	9-10/50	AC&F	9008	3200
BULGARIA NdeM	SN	10Rmt6Dbr	9/48-/49	PS	4123	6790
BURBOIS RIVER NYC	S	10Rmt6Dbr	9/48-/49	PS	4123	6790
BURRARD CN-1098	S	7CptBLngObs	3/54	PS	4190	6935
BURTON MANOR CP-10311	S	5DbrCpt4Sec4Rmt	11/54-55	Budd		9658
BUSH RIVER PRR	S	10Rmt6Dbr	9-10/49	PS	4140	6814
BUTLER INN PRR-8248	S	21Rmt	1-6/49	Budd	9513	9667
BUTLER MANOR CP-10312	S	5DbrCpt4Sec4Rmt	11/54-55	Budd		9658
BUTTE NP	S	8DuRmt6Rmt4Dbr	7-9/48	PS	4119	6781
BUTTERMILK FALLS NYC	S	6DbrBLng	9/39	PS	4086	6573
BUTTERNUT FALLS PRR	N	6DbrBLng	9/40	PS	4086A	6612
BYRD ISLAND RF&	S	21Rmt	9-10/49	PS	4156B	6809
CABIN CREEK PRR	S	12DuSr4Dbr	1-4/49	PS	4130	6792

Figure I-91 **CAPARRA** was built for joint PRR-RF&P-ACL-FEC Florida trains in 1949. Stainless steel sheathing was standard for all cars in this service even though Pullman, Budd, and AC&F stainless cars were visually very different. **CAPARRA** was withdrawn from Pullman lease in December 1966 and sold to the Canadian National in May 1967 where it became CN-2136 **RIVIERE CLOCHE**. Railroad Avenue Enterprises Photograph 1359.

CAR NAME OR NUMBER	ST	CAR TYPE	DATE BUILT	CAR BLDR	PLAN	LOT
CABOT MANOR CP-10313	S	5DbrCpt4Sec4Rmt	11/54-55	Budd		9658
CABRILLO JT	S	4Dbr4Cpt2Dr	11/40	PS	4069F	6636
CACAPON B&O	S	14Rmt4Dbr	4-6/48	PS	4153B	6776
CACHE LA POUDRE (DS)UP		12SecArtic	6/36	PS	4053	6486
CACHEMIRA NdeM-527	SN	10Rmt6Dbr	9/48-/49	PS	4123	6790
CAIRO IC	S	10Rmt6Dbr	4/50	PS	4167	6864
CALIFORNIA						
REPUBLIC (DS)UP	N	3DbrCptREObsBLng	8/36	PS	4051A	6478
CALUMET IC	SN	10Rmt6Dbr	1-2/50	PS	4167A	6866
CALUMET RIVER NYC	S	10Rmt6Dbr	9/48-/49	PS	4123	6790
CALVERT IC	SN	10Rmt6Dbr	2-7/50	PS	4167	6864
CAMBOYA NdeM-516	SN	10Rmt6Dbr	9/48-/49	PS	4123	6790
CAMBRIA COUNTY PRR	S	13Dbr	6-7/39	PS	4071B	6572
CAMBRIDGE INN PRR-8249	S	21Rmt	1-6/49	Budd	9513	9667
CAMDEN SAL-53	S	5DbrCpt4Sec4Rmt	11/55	Budd	9537	9658
CAMELLIA ACL	S	4Rmt5DbrCpt4Sec	12/54	PS	4196	6949
CAMERON MANOR CP-10314	S	5DbrCpt4Sec4Rmt	11/54-55	Budd		9658
CAMINADA BAY NYC	S	22Rmt	10/48	PS	4122	6790
CAMPBELL COUNTY N&W	S	10Rmt5Dbr	11/49-50	Budd	9523	9660
CANADA NdeM-	SN	10Rmt6Dbr	1-3/49	Budd	9502	9662
CANADIAN RIVER SLSF-1463	S	14Rmt4Dbr	1-6/48	PS	4153	6769
CANARY VALLEY NYC	S	10Rmt6Dbr	1-3/49	Budd	9502	9662
CANNON FALLS Milw-5775		8SecCoachTouralux	6/28/47	Milw		
CANTON IC	SN	10Rmt6Dbr	1-2/50	PS	4167A	6866
CANYON RIVER MP	S	10Rmt6Dbr	8-9/48	Budd	9504	9660
CAPARRA FEC	S	10Rmt6Dbr	9-10/49	PS	4140B	6809
CAPE						
BRETON CN-1087	S	2Cpt2DbrBLng	5/54	PS	4189A	6928
BRULE CN-1083	S	2Cpt2DbrBLng	5/54	PS	4189A	6928
CANSO CN-1086	S	2Cpt2DbrBLng	5/54	PS	4189A	6928

CAR NAME OR NUMBER	ST	CAR TYPE	DATE BUILT	CAR BLDR	PLAN	LOT
CAPE						
CHICNECTO CN-1088	S	2Cpt2DbrBLng	5/54	PS	4189A	6928
COD BAY NYC	S	22Rmt	10/48	PS	4122	6790
FEAR RIVER ACL	RN	4Cpt4Dr	3-4/38	PS	6013	6540
FEAR RIVER ACL	S	14Rmt2Dr	2-3/50	AC&F	9007	3091
PORCUPINE CN-1084	S	2Cpt2DbrBLng	5/54	PS	4189A	6928
RACE CN-1085	S	2Cpt2DbrBLng	5/54	PS	4189A	6928
ROSIER CN-1082	S	2Cpt2DbrBLng	5/54	PS	4189A	6928
TORMENTINE CN-1089	S	2Cpt2DbrBLng	5/54	PS	4189A	6928
VINCENT HARBOR NYC	S	22Rmt	4-6/49	Budd	9501	9661
CAPRICORNUS FCP	SN	10Rmt6Dbr	9/48-/49	PS	4123	6790
CARBONDALE IC	S	10Rmt6Dbr	4/50	PS	4167	6864
CARDINAL B&O	S	16DuRmt4Dbr	4-6/54	Budd	9536	9658
CARIBOU VALLEY NYC	S	10Rmt6Dbr	1-3/49	Budd	9502	9662
CARLETON MANOR CP-10315	S	5DbrCpt4Sec4Rmt	11/54-55	Budd		9658
CARNEGIE INN PRR-8250	S	21Rmt	1-6/49	Budd	9513	9667
CAROLINE COUNTY RF&P	S	10Rmt6Dbr	9-10/49	PS	4140B	6809
CASCADE						
BANKS Pool	S	10Rmt5Dbr	3-4/40	PS	4072B	6606
BASIN Pool	S	10Rmt5Dbr	3-4/40	PS	4072B	6606
BAY PRR	S	10Rmt5Dbr	7-8/40	PS	4072C	6610
BEND PRR	S	10Rmt5Dbr	7-8/40	PS	4072C	6610
BLUFF Pool	S	10Rmt5Dbr	5/40	PS	4072B	6606
BOULDERS Pool	S	10Rmt5Dbr	5/40	PS	4072B	6606
BOWER PRR	S	10Rmt5Dbr	7-8/40	PS	4072C	6610
BRIM PRR	S	10Rmt5Dbr	7-8/40	PS	4072C	6610
BRINK PRR	S	10Rmt5Dbr	7-8/40	PS	4072C	6610
CANYON PRR	S	10Rmt5Dbr	7-8/40	PS	4072C	6610
CHANNEL NYC	S	10Rmt5Dbr	6-7/40	PS	4072C	6610
CHASM PRR	S	10Rmt5Dbr	7-8/40	PS	4072C	6610

Figure I-92 This 10-roomette, 5-double-bedroom sleeper was 1 of 8 cars in the second lot of this configuration built in 1940 for Pullman Pool service. CASCADE BLUFF was painted B&O blue and gray in 1942. Cascade sleepers introduced lightweight accommodations to that railroad's patrons in trains 1 & 2, The "National Limited," and 5 & 6, The "Capitol Limited." According to B&O records, this car was leased to that Railroad in December 1945 and sold to the B&O in December 1948. B&O Railroad Historical Society Photograph.

CAR NAME OR NUMBER	ST	CAR TYPE	DATE BUILT	CAR BLDR	PLAN	LOT	CAR NAME OR NUMBER	ST	CAR TYPE	DATE BUILT	CAR BLDR	PLAN	LOT
CASCADE							CASCADE						
CLIFF PRR	S	10Rmt5Dbr	7-8/40	PS	4072C	6610	ELF NYC	S	10Rmt5Dbr	6/39	PS	4072B	6565
COVE PRR	S	10Rmt5Dbr	7-8/40	PS	4072C	6610	FALLS PRR	S	10Rmt5Dbr	4/38	PS	4072A	6542
CRAG PRR	S	10Rmt5Dbr	7-8/40	PS	4072C	6610	FAUN NYC	S	10Rmt5Dbr	6/39	PS	4072B	6565
CRYSTALS NYC	S	10Rmt5Dbr	6-7/40	PS	4072C	6610	FOREST PRR	S	10Rmt5Dbr	6/39	PS	4072B	6565
DAWN NYC	S	10Rmt5Dbr	4/38	PS	4072A	6542	GARDENS PRR	S	10Rmt5Dbr	6/39	PS	4072B	6565
DEN PRR	S	10Rmt5Dbr	7-8/40	PS	4072C	6610	GLADE NYC	S	10Rmt5Dbr	4/38	PS	4072A	6542
DRIVE B&O	S	10Rmt5Dbr	10/43	PS	4072F	6679	GLEN NYC	S	10Rmt5Dbr	4/38	PS	4072A	6542
ECHO PRR	S	10Rmt5Dbr	7-8/40	PS	4072C	6610	GLORY NYC	S	10Rmt5Dbr	4/38	PS	4072A	6542

Figure I-93 CASCADE CLIFF was built in the third lot of 10-roomette, 5-double-bedroom sleepers in August 1940 for service on the Pennsylvania Railroad. It was sold to the United States Army in 1967 to be utilized as an ambulance personnel car, USA 89575. Amtrak acquired it in 1973 and in turn sold it two years later to Kirk Potter without applying Amtrak paint. Photograph by Wilber C. Whittaker.

Figure I-94 CASCADE FALLS, in the initial lot of cars built in 1938 as a 10-roomette, 5-double-bedroom, was assigned to the PRR blue ribbon fleet. Aside from the early Pullman 43R trucks, this early lightweight carries a modern paint scheme that epitomizes the late art deco movement in railroading. Pullman Photograph 41254. The Arthur Dubin Collection.

CAR NAME OR NUMBER	ST	CAR TYPE	DATE BUILT	CAR BLDR	PLAN	LOT	CAR NAME OR NUMBER	ST	CAR TYPE	DATE BUILT	CAR BLDR	PLAN	LOT
CASCADE							CASCADE						
GORGE PRR	S	10Rmt5Dbr	7-8/40	PS	4072C	6610	PASS PRR	S	10Rmt5Dbr	6/39	PS	4072B	6565
GREEN NYC	S	10Rmt5Dbr	6/39	PS	4072B	6565	PEAK PRR	S	10Rmt5Dbr	7-8/40	PS	4072C	6610
GROTTO NYC	S	10Rmt5Dbr	4/38	PS	4072A	6542	PINNACLE PRR	S	10Rmt5Dbr	7-8/40	PS	4072C	6610
GULLY Pool	S	10Rmt5Dbr	6/40	PS	4072B	6606	PLATEAU PRR	S	10Rmt5Dbr	7-8/40	PS	4072C	6610
HEIGHTS PRR	S	10Rmt5Dbr	7-8/40	PS	4072C	6610	POND PRR	S	10Rmt5Dbr	7-8/40	PS	4072C	6610
HILLS PRR	S	10Rmt5Dbr	6/39	PS	4072B	6565	POOL PRR	S	10Rmt5Dbr	7-8/40	PS	4072C	6610
HOLLOW PRR	S	10Rmt5Dbr	7-8/40	PS	4072C	6610	RAINBOW NYC	S	10Rmt5Dbr	6/39	PS	4072B	6565
KNOLL PRR	S	10Rmt5Dbr	7-8/40	PS	4072C	6610	RANGE PRR	S	10Rmt5Dbr	7-8/40	PS	4072C	6610
LAKE PRR	S	10Rmt5Dbr	7-8/40	PS	4072C	6610	RAPIDS PRR	S	10Rmt5Dbr	7-8/40	PS	4072C	6610
LANE NYC	S	10Rmt5Dbr	6/39	PS	4072B	6565	RAVINE PRR	S	10Rmt5Dbr	7-8/40	PS	4072C	6610
LEDGE PRR	S	10Rmt5Dbr	7-8/40	PS	4072C	6610	RIPPLE NYC	S	10Rmt5Dbr	4/38	PS	4072A	6542
LOCKS Pool	S	10Rmt5Dbr	5/40	PS	4072B	6606	ROAR PRR	S	10Rmt5Dbr	7-8/40	PS	4072C	6610
MANTLE PRR	S	10Rmt5Dbr	7-8/40	PS	4072C	6610	ROCKS NYC	S	10Rmt5Dbr	4/38	PS	4072A	6542
MARBLE NYC	S	10Rmt5Dbr	6/39	PS	4072B	6565	RUN NYC	S	10Rmt5Dbr	6-7/40	PS	4072C	6610
MEADOW PRR	S	10Rmt5Dbr	7-8/40	PS	4072C	6610	SHOALS Pool	S	10Rmt5Dbr	3-4/40	PS	4072B	6606
MELODY PRR	S	10Rmt5Dbr	7-8/40	PS	4072C	6610	SLOPE PRR	S	10Rmt5Dbr	7-8/40	PS	4072C	6610
MILLS Pool	S	10Rmt5Dbr	3-4/40	PS	4072B	6606	SOUND B&O	S	10Rmt5Dbr	10/43	PS	4072F	6679
MIRAGE PRR	S	10Rmt5Dbr	7-8/40	PS	4072C	6610	SPIRIT NYC	S	10Rmt5Dbr	4/38	PS	4072A	6542
MIST NYC	S	10Rmt5Dbr	6/39	PS	4072B	6565	SPRAY NYC	S	10Rmt5Dbr	6-7/40	PS	4072C	6610
MOON NYC	S	10Rmt5Dbr	6/39	PS	4072B	6565	SPRINGS NYC	S	10Rmt5Dbr	6-7/40	PS	4072C	6610
MUSIC B&O	S	10Rmt5Dbr	10/43	PS	4072F	6679	STREAM NYC	S	10Rmt5Dbr	6-7/40	PS	4072C	6610
PARK PRR	S	10Rmt5Dbr	4/38	PS	4072A	6542	SUMMIT NYC	S	10Rmt5Dbr	6-7/40	PS	4072C	6610

Figure I-95 Howard Ameling photographed George Payne's restored CATALPA FALLS on June 26, 1983 at Bellevue, Ohio while it was in the consist of a rail-fan excursion. This car is one of the least altered and best preserved 6-double bedroom/buffet lounge cars in the United States.

CAR NAME OR NUMBER	ST	CAR TYPE	DATE BUILT	CAR BLDR	PLAN	LOT
CASCADE						
TERRACE PRR	S	10Rmt5Dbr	6/39	PS	4072B	6565
TIMBER PRR	S	10Rmt5Dbr	7-8/40	PS	4072C	6610
TORRENT NYC	S	10Rmt5Dbr	6-7/40	PS	4072C	6610
TRAILS PRR	S	10Rmt5Dbr	6/39	PS	4072B	6565
VALE NYC	S	10Rmt5Dbr	4/38	PS	4072A	6542
VALLEY NYC	S	10Rmt5Dbr	4/38	PS	4072A	6542
WATERS NYC	S	10Rmt5Dbr	6-7/40	PS	4072C	6610
WAVES NYC	S	10Rmt5Dbr	6-7/40	PS	4072C	6610
WHIRL NYC	S	10Rmt5Dbr	6-7/40	PS	4072C	6610
WHISPER NYC	S	10Rmt5Dbr	6-7/40	PS	4072C	6610
WONDER NYC	S	10Rmt5Dbr	6-7/40	PS	4072C	6610
WOODS NYC	S	10Rmt5Dbr	4/38	PS	4072A	6542
CASCO BAY NYC	S	22Rmt	10/48	PS	4122	6790
CASHIERS VALLEY SOU	S	14Rmt4Dbr	10-11/49	PS	4153C	6814
CASS RIVER NYC	S	10Rmt6Dbr	9/48-/49	PS	4123	6790
CASSIOPEDIA FCP	SN	10Rmt6Dbr	9/48-/49	PS	4123	6790
CASTLE VALLEY NYC	S	10Rmt6Dbr	1-3/49	Budd	9502	9662
CASTLETON BRIDGE NYC	N	4Dbr4Cpt2Dr	3-4/38	PS	4069B	6540
CATALPA FALLS PRR	S	6DbrBLng	3-5/49	PS	4131	6792

CAR NAME OR NUMBER	ST	CAR TYPE	DATE BUILT	CAR BLDR	PLAN	LOT
CATAWBA RIVER SOU	S	10Rmt6Dbr	9-10/49	PS	4140	6814
CATAWISSA RAPIDS PRR	S	10Rmt6Dbr	3-4/49	PS	4129	6792
CATOCTIN B&O	S	10Rmt6Dbr	3/50	PS	4167	6864
CAYUGA LAKE NYC	S	6DbrBLng	12/48-49	PS	4124	6790
CEDAR						
CITY UP	S	5DbrBLng	4/56	PS	4199	6959
CREEK PRR	S	12DuSr4Dbr	1-4/49	PS	4130	6792
FALLS PRR	S	6DbrBLng	9/40	PS	4086A	6612
GROVE CP-14106		10Sec5Dbr	/49-/50	CPRR		
VALLEY NYC	S	10Rmt6Dbr	1-3/49	Budd	9502	9662
CEDARTOWN SAL-55	S	5DbrCpt4Sec4Rmt	11/55	Budd	9537	9658
CENTER CREEK PRR	S	12DuSr4Dbr	1-4/49	PS	4130	6792
CENTRAL CITY IC	SN	10Rmt6Dbr	1-2/50	PS	4167A	6866
CENTRAL PARK AMT-2553	SN	8Rmt6Dbr	11-12/54	PS	4195	6944
CENTRAL VALLEY NYC	S	10Rmt6Dbr	1-3/49	Budd	9502	9662
CENTRALIA IC	S	10Rmt6Dbr	4/50	PS	4167	6864
CEYLAN NdeM-521	SN	10Rmt6Dbr	9/48-/49	PS	4123	6790
CHACO ATSF	S	17RmtSec	1-2/38	PS	4068B	6532
CHAGRIN VALLEY NYC	S	10Rmt6Dbr	1-3/49	Budd	9502	9662
CHAISTLA ATSF	S	4DrDbrREOBSLng	1-2/38	PS	4070	6532

Figure I-96   Illinois Central Railroad acquired the 10-roomette, 6-double-bedroom CENTRAL CITY from N&W in June 1965. It was formerly NKP's CITY OF ST. LOUIS and was sold to Butterworth Tours in 1971. IC retained its art deco lettering and orange and brown paint scheme until Amtrak. This photograph was taken in Chicago in July 1968. Railroad Avenue Enterprises Photograph PN-2852.

CAR NAME OR NUMBER	ST	CAR TYPE	DATE BUILT	CAR BLDR	PLAN	LOT
CHALEUR BAY CN-2024	S	10Rmt5Dbr	6/54	PS	4186A	6925
CHAMA VALLEY ATSF	S	6Sec6Rmt4Dbr	6/42	PS	4099	6669
CHAMBERLIN						
DOME C&O-1852	S	Dome5RmtDbr3Dr	9/48	Budd	9524	9669
CHAMBERSBURG						
INN PRR-8251	S	21Rmt	1-6/49	Budd	9513	9667
CHAMPAIGN IC	S	10Rmt6Dbr	4/50	PS	4167	6864
CHAMPLAIN VALLEY NYC	S	10Rmt6Dbr	1-3/49	Budd	9502	9662
CHANDEYSSON IC	SN	10Rmt6Dbr	1-2/50	PS	4167A	6866
CHANEY GLACIER GN-1184	S	16DuRmt4Dbr	12/50	PS	4108A	6890
CHARLES						
A. WICKERSHAM WofA	N	5DbrREObsBLng	2-3/50	PS	4162	6814
CITY IC	SN	10Rmt6Dbr	1-2/50	PS	4167A	6866
LOCKHART PRR	N	12DuSr4Dbr	1-4/49	PS	4130	6792
MICHAEL SCHWAB PRR	N	6DbrBLng	3-5/49	PS	4131	6792
MINOT Erie	S	10Rmt6Dbr	5-6/49	PS	4129A	6797
CHARLOTTE SAL	S	10Rmt6Dbr	5-6/49	PS	4140A	6796
CHARLOTTE HARBOR NYC	S	22Rmt	4-6/49	Budd	9501	9661
CHARTIERS CREEK PRR	S	12DuSr4Dbr	1-4/49	PS	4130	6792
CHATEAU						
ARGENSON CP-14201	S	8DuRmtDr3Dbr4Sec	7-11/54	Budd		9690
BIENVILLE CP-14202	S	8DuRmtDr3Dbr4Sec	7-11/54	Budd		9690
BRULE CP-14203	S	8DuRmtDr3Dbr4Sec	7-11/54	Budd		9690
CADILLAC CP-14204	S	8DuRmtDr3Dbr4Sec	7-11/54	Budd		9690
CLOSSE CP-14205	S	8DuRmtDr3Dbr4Sec	7-11/54	Budd		9690
DENONVILLE CP-14206	S	8DuRmtDr3Dbr4Sec	7-11/54	Budd		9690
DOLLARD CP-14207	S	8DuRmtDr3Dbr4Sec	7-11/54	Budd		9690
DOLLIER CP-14208	S	8DuRmtDr3Dbr4Sec	7-11/54	Budd		9690
IBERVILLE CP-14209	S	8DuRmtDr3Dbr4Sec	7-11/54	Budd		9690
JOLLIET CP-14210	S	8DuRmtDr3Dbr4Sec	7-11/54	Budd		9690
LASALLE CP-14211	S	8DuRmtDr3Dbr4Sec	7-11/54	Budd		9690
LATOUR CP-14212	S	8DuRmtDr3Dbr4Sec	7-11/54	Budd		9690
LAUSON CP-14213	S	8DuRmtDr3Dbr4Sec	7-11/54	Budd		9690
LAVAL CP-14214	S	8DuRmtDr3Dbr4Sec	7-11/54	Budd		9690
LEMOYNE CP-14215	S	8DuRmtDr3Dbr4Sec	7-11/54	Budd		9690
LEVIS CP-14216	S	8DuRmtDr3Dbr4Sec	7-11/54	Budd		9690
MAISONNEUVE CP-14217	S	8DuRmtDr3Dbr4Sec	7-11/54	Budd		9690
MARQUETTE CP-14218	S	8DuRmtDr3Dbr4Sec	7-11/54	Budd		9690
MONTCALM CP-14219	S	8DuRmtDr3Dbr4Sec	7-11/54	Budd		9690
PAPINEAU CP-14220	S	8DuRmtDr3Dbr4Sec	7-11/54	Budd		9690
RADISSON CP-14221	S	8DuRmtDr3Dbr4Sec	7-11/54	Budd		9690
RICHELIEU CP-14222	S	8DuRmtDr3Dbr4Sec	7-11/54	Budd		9690
RIGAUD CP-14223	S	8DuRmtDr3Dbr4Sec	7-11/54	Budd		9690
ROBERVAL CP-14224	S	8DuRmtDr3Dbr4Sec	7-11/54	Budd		9690
ROUVILLE CP-14225	S	8DuRmtDr3Dbr4Sec	7-11/54	Budd		9690
SALABERRY CP-14226	S	8DuRmtDr3Dbr4Sec	7-11/54	Budd		9690
VERCHERES CP-14228	S	8DuRmtDr3Dbr4Sec	7-11/54	Budd		9690
VERENNES CP-14227	S	8DuRmtDr3Dbr4Sec	7-11/54	Budd		9690
VIGER CP-14229	S	8DuRmtDr3Dbr4Sec	7-11/54	Budd		9690
CHATEAUGAY RIVER NYC	S	10Rmt6Dbr	9/48-/49	PS	4123	6790
CHATHAM COUNTY ACL	S	10Rmt6Dbr	9-10/49	PS	4140B	6809
CHATHAM CREEK PRR	S	12DuSr4Dbr	1-4/49	PS	4130	6792
CHATTAHOOCHEE						
RIVER A&WP	S	10Rmt6Dbr	9-10/49	PS	4140	6814
CHAUMONT BAY NYC	S	22Rmt	10/48	PS	4122	6790
CHEBANSE IC	SN	10Rmt6Dbr	2-7/50	PS	4167	6864
CHEBOYGAN HARBOR NYC	S	22Rmt	4-6/49	Budd	9501	9661
CHECOESLO-						
VAQUIA NdeM-345	SN	10Rmt6Dbr	9/48-/49	PS	4123	6790
CHENANGO DL&W	S	10Rmt 6Dbr	6-7/49	AC&F	9008A	3089
CHERRY BLOSSOM AMT-2557	SN	8Rmt6Dbr	11-12/54	PS	4195	6944
CHERRY CREEK PRR	S	12DuSr4Dbr	1-4/49	PS	4130	6792
CHERRY VALLEY NYC	S	10Rmt6Dbr	1-3/49	Budd	9502	9662
CHESAPEAKE BAY NYC	S	22Rmt	10/48	PS	4122	6790
CHESTER PRR-8335	S	10Rmt6Dbr	6-8/49	Budd	9503	9662
CHESTER COUNTY PRR	S	10Rmt6Dbr	9-10/49	PS	4140B	6809
CHESTER INN PRR-8252	S	21Rmt	1-6/49	Budd	9513	9667
CHESTERFIELD RF&P	S	10Rmt6Dbr	5-6/49	PS	4140A	6796
CHEYENNE UP	S	5DbrBLng	4/56	PS	4199	6959
CHICAGO CB&Q	S	8DuRmt6Rmt4Dbr	7-9/48	PS	4119	6781
CHICAGO RIVER NYC	S	10Rmt6Dbr	9/48-/49	PS	4123	6790
CHICAGOLAND IC-3540	S	4Dbr4Cpt2Dr	4/42	PS	4069H	6668
CHICHEN-ITZA NdeM-117	L	8Sec3Cpt	/52	Schi	100570	
CHICKAMAUGA						
PINE NC&SL-200	S	6Sec6Rmt4Dbr	3-5/53	PS	4183	6909
CHICKIES CREEK PRR	S	12DuSr4Dbr	1-4/49	PS	4130	6792
CHICOPEE FALLS NYC	S	6DbrBLng	8-9/40	PS	4086B	6612
CHICOPEE RIVER NYC	S	10Rmt6Dbr	9/48-/49	PS	4123	6790
CHILCOTIN RIVER CN-2150	SN	10Rmt6Dbr	5-6/49	PS	4129A	6797
CHILE FEC	S	10Rmt6Dbr	9-10/49	PS	4140	6814
CHIMAVO ATSF	S	17RmtSec	7/38	PS	4068F	6553
CHIMNEY CREEK PRR	S	12DuSr4Dbr	1-4/49	PS	4130	6792
CHINATOWN (1st) JT		12SecArtic	12/37	PS	4064	6525
CHINATOWN (2nd) JT	S	4Dbr4Cpt2Dr	11/40	PS	4069F	6636
CHINLE ATSF	S	17RmtSec	1-2/38	PS	4068B	6532
CHINOOK COVE CN-1184	N	4Dbr4Cpt2Dr	3-4/38	PS	4069B	6540
CHIPPEWA						
BAY NYC	S	22Rmt	10/48	PS	4122	6790
CREEK PRR	S	12DuSr4Dbr	1-4/49	PS	4130	6792
FALLS Milw-5771		8SecCoachTouralux	6/23/47	Milw		
RIVER Milw-25	S	8DuRmt6Rmt4Dbr	10-12/48	PS	4135	6775
CHIPRE NdeM-536	SN	10Rmt6Dbr	9/48-/49	PS	4123	6790
CHITTENANGO FALLS NYC	S	6DbrBLng	9/39	PS	4086	6573
CHOTEAU COULEE GN-1190	RN	4DbrCpt6RmtObsLng	1-2/47	PS	4109K	6751
CHRISTIE MANOR CP-10316	S	5DbrCpt4Sec4Rmt	11/54-55	Budd		9658
CHUMSTICK RIVER GN-1265	S	8DuRmt4Sec3DbrCpt	11-12/50	PS	4181	6889
CHURCHILL FALLS CN-2095	SN	14Rmt4Dbr	1-6/48	PS	4153	6769
CHUSKA ATSF	S	4DrDbrREOBSLng	1-2/38	PS	4070	6532
CIMARRON						
RIVER SLSF-1466	S	14Rmt4Dbr	1-6/48	PS	4153	6769

Figure I-97 American Car & Foundry built 9 10-roomette, 6-double-bedroom cars for the Delaware Lackawanna & Western for joint service with the Nickel Plate in Chicago-Hoboken operation. CHENANGO, shown here after 1960 when it became part of the Erie-Lackawanna fleet, was withdrawn from Pullman lease and scrapped in 1969. Bob Pennisi Photograph No. 2590.

CAR NAME OR NUMBER	ST	CAR TYPE	DATE BUILT	CAR BLDR	PLAN	LOT
CIMARRON						
VALLEY ATSF	S	6Sec6Rmt4Dbr	6/42	PS	4099	6669
CINCINNATI INN PRR-8264	N	21Rmt	1-6/49	Budd	9513	9667
CINEMA (DS)UP		11 SecArtic	5/36	PS	4037B	6434
CITRUS VALLEY ATSF	N	6Sec6Rmt4Dbr	6/42	PS	4099	6669
CITY OF						
AKRON PRR	S	18Rmt	3/38	PS	4068D	6539
ALBANY NYC	S	17RmtSec	3-4/38	PS	4068E	6539
ALDERSON C&O	S	10Rmt6Dbr	2-7/50	PS	4167	6864
ALDERSON(2nd) C&O	N	10Rmt6Dbr	2-7/50	PS	4167	6864
ALEXANDRIA C&O	S	10Rmt6Dbr	2-7/50	PS	4167	6864
ALTOONA PRR	S	18Rmt	12/38	PS	4068H	6563
ANN ARBOR NYC	S	18Rmt	6-7/39	PS	4068H	6570
ARLINGTON C&O	S	10Rmt6Dbr	2-7/50	PS	4167	6864
ASHLAND C&O	S	10Rmt6Dbr	2-7/50	PS	4167	6864
ATHENS C&O	S	10Rmt6Dbr	2-7/50	PS	4167	6864
BALTIMORE PRR	S	18Rmt	3/38	PS	4068D	6539

CAR NAME OR NUMBER	ST	CAR TYPE	DATE BUILT	CAR BLDR	PLAN	LOT
City Of						
BECKLEY(1st) C&O	S	10Rmt6Dbr	2-7/50	PS	4167	6864
BECKLEY(2nd) C&O	N	10Rmt6Dbr	2-7/50	PS	4167	6864
BENTON HARBOR C&O	S	10Rmt6Dbr	2-7/50	PS	4167	6864
BOSTON NYC	S	17RmtSec	3-4/38	PS	4068E	6539
BUFFALO NYC	S	17RmtSec	3-4/38	PS	4068E	6539
BUFFALO NKP-200	S	10Rmt6Dbr	1-2/50	PS	4167A	6866
CAMDEN PRR	S	18Rmt	12/38	PS	4068H	6563
CANTON PRR	S	18Rmt	12/38	PS	4068H	6563
CHARLESTON(1st) C&O	S	10Rmt6Dbr	2-7/50	PS	4167	6864
CHARLESTON(2nd) C&O	N	10Rmt6Dbr	2-7/50	PS	4167	6864
CHARLOTTESVILLE C&O	S	10Rmt6Dbr	2-7/50	PS	4167	6864
CHICAGO NYC	S	17RmtSec	3-4/38	PS	4068E	6539
CHICAGO NKP-151	S	5DbrBLng	5/50	PS	4169	6867
CINCINNATI PRR	S	18Rmt	3/38	PS	4068D	6539
CLEVELAND NYC	S	17RmtSec	3-4/38	PS	4068E	6539
CLEVELAND NKP-150	S	5DbrBLng	5/50	PS	4169	6867

Figure I-98   The CITY OF ALDERSON was built in 1950 as a part of Robert Young's Chessie fleet and carried the name
CITY OF PLYMOUTH. In 1951, the sale of a number of cars in this lot to other railroads necessitated a reshuffle of car
names. The half stainless fluting was unique to the large C & O-PM-NKP joint order of 1946, and proved a water trap and
subsequent maintenance headache.   Railroad Avenue Enterprises Photograph PN-4402.

Figure I-99   CITY OF CLEVELAND represents one of the few instances where more than one car in the Pullman fleet was
assigned the same name. In this case, NKP insisted on naming the cars for cities on its line despite the fact that NYC had a
car built in 1938 which carried the same name. Pullman solved this problem by referring to the car as NKP-150 (CITY OF
CLEVELAND). By 1969, when this photograph was taken at Chicago, NKP was part of N & W and its name obliterated from
all equipment.   Railroad Avenue Enterprises Photograph PN-2682.

CAR NAME OR NUMBER	ST	CAR TYPE	DATE BUILT	CAR BLDR	PLAN	LOT
CITY OF						
CLIFTON FORGE C&O	S	10Rmt6Dbr	2-7/50	PS	4167	6864
COLUMBUS PRR	S	18Rmt	3/38	PS	4068D	6539
COVINGTON C&O	S	10Rmt6Dbr	2-7/50	PS	4167	6864
DAYTON NYC	S	17RmtSec	3-4/38	PS	4068E	6539
DETROIT NYC	S	17RmtSec	3-4/38	PS	4068E	6539
ELIZABETH PRR	S	18Rmt	12/38	PS	4068H	6563
ERIE PRR	S	18Rmt	12/38	PS	4068H	6563
ERIE NKP-205	S	10Rmt6Dbr	1-2/50	PS	4167A	6866
FINDLEY NKP-210	S	10Rmt6Dbr	1-2/50	PS	4167A	6866
FLINT C&O	S	10Rmt6Dbr	2-7/50	PS	4167	6864
FOSTORIA(1st) C&O	S	10Rmt6Dbr	2-7/50	PS	4167	6864
FOSTORIA(2nd) C&O	N	10Rmt6Dbr	2-7/50	PS	4167	6864
FRANKFORT C&O	S	10Rmt6Dbr	2-7/50	PS	4167	6864
FT.WAYNE PRR	S	18Rmt	12/38	PS	4068H	6563
FT.WAYNE NKP-212	S	10Rmt6Dbr	1-2/50	PS	4167A	6866
GARY PRR	S	18Rmt	12/38	PS	4068H	6563
GRAND RAPIDS C&O	S	10Rmt6Dbr	2-7/50	PS	4167	6864
HAMPTON C&O	S	10Rmt6Dbr	2-7/50	PS	4167	6864
HARRISBURG PRR	S	18Rmt	12/38	PS	4068H	6563
HINTON C&O	S	10Rmt6Dbr	2-7/50	PS	4167	6864
HOLLAND C&O	S	10Rmt6Dbr	2-7/50	PS	4167	6864
HUNTINGTON C&O	S	10Rmt6Dbr	2-7/50	PS	4167	6864
INDIANAPOLIS NYC	S	17RmtSec	3-4/38	PS	4068E	6539
INDIANAPOLIS NKP-203	S	10Rmt6Dbr	1-2/50	PS	4167A	6866
JACKSON IC	S	18Rmt	4/42	PS	4068K	6670
JOHNSTOWN PRR	S	18Rmt	12/38	PS	4068H	6563
KALAMAZOO NYC	S	18Rmt	9/40	PS	4068J	6616
KENOVA C&O	S	10Rmt6Dbr	2-7/50	PS	4167	6864
KOKOMO NKP-208	S	10Rmt6Dbr	1-2/50	PS	4167A	6866
LANCASTER PRR	S	18Rmt	12/38	PS	4068H	6563
LANSING NYC	S	18Rmt	6-7/39	PS	4068H	6570
LEXINGTON C&O	S	10Rmt6Dbr	2-7/50	PS	4167	6864
LIMA NKP-211	S	10Rmt6Dbr	1-2/50	PS	4167A	6866
LOGAN(1st) C&O	S	10Rmt6Dbr	2-7/50	PS	4167	6864
LOGAN(2nd) C&O	N	10Rmt6Dbr	2-7/50	PS	4167	6864
LORAIN NKP-202	S	10Rmt6Dbr	1-2/50	PS	4167A	6866
LOUISVILLE PRR	S	18Rmt	3/38	PS	4068D	6539
LUDINGTON C&O	S	10Rmt6Dbr	2-7/50	PS	4167	6864
LYNCHBURG C&O	S	10Rmt6Dbr	2-7/50	PS	4167	6864
MARION(1st) C&O	S	10Rmt6Dbr	2-7/50	PS	4167	6864
MARION(2nd) C&O	N	10Rmt6Dbr	2-7/50	PS	4167	6864
MAYSVILLE C&O	S	10Rmt6Dbr	2-7/50	PS	4167	6864
MIDLAND C&O	S	10Rmt6Dbr	2-7/50	PS	4167	6864
MONTGOMERY(1st) C&O	S	10Rmt6Dbr	2-7/50	PS	4167	6864
MONTGOMERY(2nd) C&O	N	10Rmt6Dbr	2-7/50	PS	4167	6864
MT.HOPE C&O	S	10Rmt6Dbr	2-7/50	PS	4167	6864
MUNCIE NKP-209	S	10Rmt6Dbr	1-2/50	PS	4167A	6866
MUSKEGON C&O	S	10Rmt6Dbr	2-7/50	PS	4167	6864
NEW BRUNSWICK PRR	S	18Rmt	12/38	PS	4068H	6563
NEW ORLEANS IC	S	18Rmt	4/42	PS	4068K	6670
CITY OF						
NEW YORK PRR	S	18Rmt	3/38	PS	4068D	6539
NEWARK PRR	S	18Rmt	3/38	PS	4068D	6539
NEWPORT C&O	S	10Rmt6Dbr	2-7/50	PS	4167	6864
NEWPORT NEWS C&O	S	10Rmt6Dbr	2-7/50	PS	4167	6864
NORFOLK PRR	S	18Rmt	12/38	PS	4068H	6563
PAINESVILLE NKP-204	S	10Rmt6Dbr	1-2/50	PS	4167A	6866
PEORIA NYC	S	18Rmt	9/40	PS	4068J	6616
PEORIA NKP-207	S	10Rmt6Dbr	1-2/50	PS	4167A	6866
PETOSKEY(1st) C&O	S	10Rmt6Dbr	2-7/50	PS	4167	6864
PETOSKEY(2nd) C&O	N	10Rmt6Dbr	2-7/50	PS	4167	6864
PHEOBUS C&O	S	10Rmt6Dbr	2-7/50	PS	4167	6864
PHILADELPHIA PRR	S	18Rmt	3/38	PS	4068D	6539
PIKEVILLE C&O	S	10Rmt6Dbr	2-7/50	PS	4167	6864
PITTSBURGH PRR	S	18Rmt	3/38	PS	4068D	6539
PITTSFIELD NYC	S	18Rmt	6-7/39	PS	4068H	6570
PLYMOUTH C&O	S	10Rmt6Dbr	2-7/50	PS	4167	6864
PORTSMOUTH C&O	S	10Rmt6Dbr	2-7/50	PS	4167	6864
POUGHKEEPSIE NYC	S	18Rmt	9/40	PS	4068J	6616
READING PRR	S	18Rmt	12/38	PS	4068H	6563
RICHMOND C&O	S	10Rmt6Dbr	2-7/50	PS	4167	6864
ROCHESTER NYC	S	17RmtSec	3-4/38	PS	4068E	6539
RONCEVERTE(1st) C&O	S	10Rmt6Dbr	2-7/50	PS	4167	6864
RONCEVERTE(2nd) C&O	N	10Rmt6Dbr	2-7/50	PS	4167	6864
SAGINAW C&O	S	10Rmt6Dbr	2-7/50	PS	4167	6864
SCHENECTADY NYC	S	18Rmt	9/40	PS	4068J	6616
SOUTH BEND NYC	S	18Rmt	6-7/39	PS	4068H	6570
SPRINGFIELD NYC	S	18Rmt	6-7/39	PS	4068H	6570
ST.ALBANS C&O	S	10Rmt6Dbr	2-7/50	PS	4167	6864
ST.JOSEPH C&O	S	10Rmt6Dbr	2-7/50	PS	4167	6864
ST.LOUIS PRR	S	18Rmt	3/38	PS	4068D	6539
ST.LOUIS NKP-201	S	10Rmt6Dbr	1-2/50	PS	4167A	6866
ST.MARYS NKP 215	N	18Rmt	8/37	PS	4068	6526
STAUNTON C&O	S	10Rmt6Dbr	2-7/50	PS	4167	6864
SYRACUSE NYC	S	18Rmt	6-7/39	PS	4068H	6570
TERRE HAUTE PRR	S	18Rmt	12/38	PS	4068H	6563
TOLEDO NYC	S	17RmtSec	3-4/38	PS	4068E	6539
TOLEDO NKP-206	S	10Rmt6Dbr	1-2/50	PS	4167A	6866
TRENTON PRR	S	18Rmt	12/38	PS	4068H	6563
TROY NYC	S	18Rmt	9/40	PS	4068J	6616
UTICA NYC	S	18Rmt	6-7/39	PS	4068H	6570
VIRGINIA BEACH C&O	S	10Rmt6Dbr	2-7/50	PS	4167	6864
WASHINGTON PRR	S	18Rmt	3/38	PS	4068D	6539
WAYNESBORO C&O	S	10Rmt6Dbr	2-7/50	PS	4167	6864
WHEELING PRR	S	18Rmt	12/38	PS	4068H	6563
WILLIAMSBURG C&O	S	10Rmt6Dbr	2-7/50	PS	4167	6864
WILLIAMSPORT PRR	S	18Rmt	12/38	PS	4068H	6563
WILMINGTON PRR	S	18Rmt	3/38	PS	4068D	6539
WINCHESTER C&O	S	10Rmt6Dbr	2-7/50	PS	4167	6864
WORCESTER NYC	S	18Rmt	6-7/39	PS	4068H	6570
YONKERS NYC	S	18Rmt	9/40	PS	4068J	6616

CAR NAME OR NUMBER	ST	CAR TYPE	DATE BUILT	CAR BLDR	PLAN	LOT
CITY OF						
YORK PRR	S	18Rmt	12/38	PS	4068H	6563
YOUNGSTOWN PRR	S	18Rmt	3/38	PS	4068D	6539
CITY POINT NH	S	14Rmt4Dbr	12/49/50	PS	4159	6822
CIVIC CENTER JT	S	4Dbr4Cpt2Dr	11/40	PS	4069F	6636
CLARENDON COUNTY ACL	S	10Rmt6Dbr	9-10/49	PS	4140B	6809
CLARION RAPIDS PRR	S	10Rmt6Dbr	3-4/49	PS	4129	6792
CLARKSDALE IC	SN	10Rmt6Dbr	2-7/50	PS	4167	6864
CLEAR CREEK PRR	S	12DuSr4Dbr	1-4/49	PS	4130	6792
CLEARFIELD RAPIDS PRR	S	10Rmt6Dbr	3-4/49	PS	4129	6792
CLEARWATER SAL-62	S	5Dbr2Cpt2Dr	1/56	PS	4201A	6969
CLEARWATER RIVER CN-2149	SN	10Rmt6Dbr	5-6/49	PS	4129A	6797
CLEVELAND HARBOR NYC	S	22Rmt	4-6/49	Budd	9501	9661
CLIFF CREEK PRR	S	12DuSr4Dbr	1-4/49	PS	4130	6792
CLIFF HOUSE JT	S	4Dbr4Cpt2Dr	11/40	PS	4069F	6636
CLIFTON IC	SN	10Rmt6Dbr	2-7/50	PS	4167	6864
CLINTON PRR-8336	S	10Rmt6Dbr	6-8/49	Budd	9503	9662
CLINTON RIVER NYC	S	10Rmt6Dbr	9/48-/49	PS	4123	6790
CLOUD CREEK PRR	S	12DuSr4Dbr	1-4/49	PS	4130	6792
CLUB BOCA						
DEL MONTE NdeM-372	SN	6DbrBLng	8-9/40	PS	4086B	6612
CLUB						
CREEK PRR	S	12DuSr4Dbr	1-4/49	PS	4130	6792
ESMERALDA NdeM-373	SN	6DbrBLng	9/39	PS	4086	6573
FORTIN NdeM-371	SN	6DbrBLng	9/39	PS	4086	6573
MALTRATA NdeM	P	4DbrREObsBLng	5/38	PS	4079	6547
METLAC NdeM	N	4DbrREObsBLng	5/38	PS	4079A	6547
CLYDE CREEK PRR	S	12DuSr4Dbr	1-4/49	PS	4130	6792
COATSVILLE INN PRR-8253	S	21Rmt	1-6/49	Budd	9513	9667
COBDEN IC	SN	10Rmt6Dbr	2-7/50	PS	4167	6864
COCONINO ATSF		4DrDbrREOBSLng	1-2/38	PS	4070	6532
COFFEE CREEK Milw-15	S	8DbrSKYTOPLng	11/48-49	PS	4138	6775
COHOCTON DL&W	S	10Rmt 6Dbr	6-7/49	AC&F	9008A	3089
COLDWATER CREEK PRR	S	12DuSr4Dbr	1-4/49	PS	4130	6792
COLES COUNTY IC	SN	10Rmt6Dbr	2-7/50	PS	4167	6864
COLLEGE CREEK PRR	S	12DuSr4Dbr	1-4/49	PS	4130	6792
COLLETON COUNTY ACL	S	10Rmt6Dbr	9-10/49	PS	4140B	6809
COLLINSVILLE INN PRR-8250	S	21Rmt	1-6/49	Budd	9513	9667
COLOMBIA FEC	S	10Rmt6Dbr	9-10/49	PS	4140B	6809
COLOMBIA(2nd) NdeM-567	SN	10Rmt6Dbr	1-3-49	Budd	9502	9662
COLONEL FORDYCE KCS	S	14Rmt4Dbr	5/48	PS	4153	6795
COLONIAL						
ARMS PRR	S	3DbrDrBLng	5-6/49	PS	4132	6792
BEACH RF&P	S	6DbrBLng	11-12/49	AC&F	9002	3093
CABINS PRR	S	3DbrDrBLng	5-6/49	PS	4132	6792
CONGRESS PRR	S	3DbrDrBarLng	5/38	PS	4078	6551
CRAFTS PRR	S	3DbrDrBLng	5-6/49	PS	4132	6792
DAMES PRR	S	3DbrDrBarLng	5/38	PS	4078	6551
DOORWAYS PRR	S	3DbrDrBLng	5-6/49	PS	4132	6792
FATHERS PRR	S	3DbrDrBarLng	5/38	PS	4078	6551
FLAGS PRR	S	3DbrDrBLng	5-6/49	PS	4132	6792
GOVERNORS PRR	S	3DbrDrBarLng	5/38	PS	4078	6551
HOUSES PRR	S	3DbrDrBLng	5-6/49	PS	4132	6792
INNS PRR	S	3DbrDrBarLng	5/38	PS	4078	6551
LANTERNS PRR	S	3DbrDrBLng	5-6/49	PS	4132	6792
MANSIONS PRR	S	3DbrDrBarLng	5/38	PS	4078	6551
SCOUTS PRR	S	3DbrDrBLng	5-6/49	PS	4132	6792
STAGES PRR	S	3DbrDrBarLng	5/38	PS	4078	6551
STATESMAN PRR	S	3DbrDrBarLng	5/38	PS	4078	6551
TRAILS PRR	S	3DbrDrBarLng	5/38	PS	4078	6551
COLORADO RIVER MP	S	6Sec6Rmt4Dbr	5/42	PS	4099	6669
COLORES (DS)UP	S	5DbrCptREObsLng	6/36	PS	4055	6486
COLUMBIA SAL	S	10Rmt6Dbr	5-6/49	PS	4140A	6796
COMMODORE PC-4215	SN	6Sec6Rmt4Dbr	11/54-55	PS	4194	W6942
CONEMAUGH RAPIDS PRR	S	10Rmt6Dbr	3-4/49	PS	4129	6792
CONESTOGA RAPIDS PRR	S	10Rmt6Dbr	3-4/49	PS	4129	6792

Figure I-100 CLUB CREEK, 1 of 24 cars in Plan 4130, Lot 6792, delivered in 1949 to the Pennsylvania Railroad, represents an update of the innovative design Pullman first delivered in 1937. This car was withdrawn from Pullman lease in July 1968 and by May 1969 when this photograph was taken in Chicago, its Keystone heritage had been masked by the Penn Central logo. Railroad Avenue Enterprises Photograph PN-2859.

CAR NAME OR NUMBER	ST	CAR TYPE	DATE BUILT	CAR BLDR	PLAN	LOT
CONEWAGO CREEK PRR	S	12DuSr4Dbr	1-4/49	PS	4130	6792
CONEWAGO RAPIDS PRR	S	10Rmt6Dbr	3-4/49	PS	4129	6792
CONNEAUT HARBOR NYC	S	22Rmt	4-6/49	Budd	9501	9661
CONNECTICUT RIVER NYC	S	10Rmt6Dbr	9/48-/49	PS	4123	6790
CONNOQUENESSING						
CREEK PRR	S	12DuSr4Dbr	1-4/49	PS	4130	6792
CONODOGUINET CREEK PRR	S	12DuSr4Dbr	1-4/49	PS	4130	6792
COOK COUNTY IC	SN	10Rmt6Dbr	1-2/50	PS	4167A	6866
COOK COUNTY NYC	S	13Dbr	4/38	PS	4071A	6541
COOKS FALLS NYC	S	6DbrBLng	8-9/40	PS	4086B	6612
COOPER RIVER ACL	S	14Rmt2Dr	2-3/50	AC&F	9007	3091
COOSA RIVER CNO&TP	S	10Rmt6Dbr	9-10/49	PS	4140	6814
COPAKE FALLS NYC	S	6DbrBLng	9/39	PS	4086	6573
COREA NdeM-529	SN	10Rmt6Dbr	9/48-/49	PS	4123	6790
CORINTH IC	SN	10Rmt6Dbr	2-7/50	PS	4167	6864
CORNFIELD POINT NH	S	14Rmt4Dbr	12/49/50	PS	4159	6822
CORNWALL MANOR CP-10317	S	5DbrCpt4Sec4Rmt	11/54-55	Budd		9658
CORRAL COULEE GN-1192	RN	4DbrCpt6RmtObsLng	1-2/47	PS	4109K	6751
COSHOCTON INN PRR-8255	S	21Rmt	1-6/49	Budd	9513	9667
COSTA RICA NdeM-568	SN	10Rmt6Dbr	1-3/49	Budd	9502	9662
COTTONWOOD VALLEY ATSF	S	6Sec6Rmt4Dbr	6/42	PS	4099	6669
COUNCIL BLUFFS IC	SN	10Rmt6Dbr	1-2/50	PS	4167A	6866
COUNTRY CREEK PRR	S	12DuSr4Dbr	1-4/49	PS	4130	6792
COVINGTON IC	S	10Rmt6Dbr	4/50	PS	4167	6864
CRAIG MANOR CP-10318	S	5DbrCpt4Sec4Rmt	11/54-55	Budd		9658
CRANBERRY LAKE NYC	S	6DbrBLng	12/48-49	PS	4124	6790
CRANE CREEK PRR	S	12DuSr4Dbr	1-4/49	PS	4130	6792
CRESCENT BEACH NH-528	S	6Sec6Rmt4Dbr	11/54-55	PS	4194	W6942
CRESCENT CITY SOU	S	2DrMbrBLng	12/49	PS	4160	6814
CRESCENT HARBOR SOU	S	2DrMbrBLng	12/49	PS	4160	6814
CRESCENT MOON SOU	S	2DrMbrBLng	12/49	PS	4160	6814
CRESCENT SHORES SOU	S	2DrMbrBLng	12/49	PS	4160	6814
CROSS CREEK PRR	S	12DuSr4Dbr	1-4/49	PS	4130	6792
CROTON FALLS NYC	S	6DbrBLng	9/39	PS	4086	6573
CROTON RIVER NYC	S	10Rmt6Dbr	9/48-/49	PS	4123	6790
CRUZ DEL SUR NdeM-594	SN	8DuRmt6Rmt4Dbr	12/46-47	PS	4107	6751
CRYSTAL CREEK PRR	S	12DuSr4Dbr	1-4/49	PS	4130	6792
CRYSTAL FALLS Milw-5773		8SecCoachTouralux	6/19/47	Milw		
CRYSTAL RIVER MP	S	10Rmt6Dbr	8-9/48	Budd	9504	9660
CRYSTAL SPRINGS IC	SN	10Rmt6Dbr	2-7/50	PS	4167	6864
CRYSTAL STREAM NYC	S	6DbrBLng	3-5/49	Budd	9505	9663
CUAUHTEMOC FCS-BC	SN	6DbrBLng	12/48-49	PS	4124	6790
CUBA FEC	S	10Rmt6Dbr	9-10/49	PS	4140B	6809
CUBA NdeM-569	SN	10Rmt6Dbr	1-3/49	Budd	9502	9662
CUITZEO NdeM	P	4DbrREObsBLng	5/38	PS	4079A	6547
CULVER INN PRR-8256	S	21Rmt	1-6/49	Budd	9513	9667
CULVER WHITE GM&O	S	8RmtCpt3Dbr4Sec	7/50	AC&F	9012	3208
CUMBERLAND COUNTY ACL	S	10Rmt6Dbr	9-10/49	PS	4140B	6809
CUMBERLAND VALLEY NYC	S	10Rmt6Dbr	1-3/49	Budd	9502	9662
CURLY PINE L&N-3451	S	6Sec6Rmt4Dbr	3-5/53	PS	4183	6909
CURRENT RIVER NYC	S	10Rmt6Dbr	9/48-/49	PS	4123	6790
CUT BANK PASS GN-1167	S	8DuRmt4Dbr4Sec	12/46-47	PS	4107	6751
CUYAHOGA B&O	SN	10Rmt6Dbr	2-7/50	PS	4167	6864
CUYAHOGA COUNTY NYC	S	13Dbr	4/38	PS	4071A	6541
CYNTHIA IC	SN	10Rmt6Dbr	1-2/50	PS	4167A	6866
CYPRESS CREEK PRR	S	12DuSr4Dbr	1-4/49	PS	4130	6792
CYPRESS FALLS PRR	S	6DbrBLng	9/40	PS	4086A	6612
CYRUS H.K. CURTIS PRR	N	10Rmt6Dbr	3-4/49	PS	4129	6792
DAN RIVER SOU	S	10Rmt6Dbr	9-10/49	PS	4140	6814
DANA B&O	SN	5DbrBEObsBLng	8/50	PS	4165	6863
DANIEL						
CRAIG McCALLUM ERIE	S	10Rmt6Dbr	5-6/49	PS	4129A	6797
DARLINGTON COUNTY ACL	S	10Rmt6Dbr	9-10/49	PS	4140B	6809
DARTMOUTH						
COLLEGE I B&M-31	N	6Sec6Rmt4Dbr	11/54-55	PS	4194	W6942
COLLEGE II B&M-32	N	6Sec6Rmt4Dbr	11/54-55	PS	4194	W6942
DAVID CROCKETT MKT-1503	S	14Rmt4Dbr	1-6/48	PS	4153	6769
DAVID MOFFAT D&RGW-M1	S	8Sec2SrLngObs	11/17/41	Budd		97102
DAVID MOFFAT D&RG-1271	S	10Rmt6Dbr	7/50	PS	4167	6864
DAWSON MANOR CP-10319	S	5DbrCpt4Sec4Rmt	11/54-55	Budd		9658
DAWSON PASS GN-1162	S	8DuRmt4Dbr4Sec	12/46-47	PS	4107	6751
DAYTON RAPIDS PRR-8435	N	10Rmt6Dbr	8-10/50	AC&F	9008	3200
DECATUR IC	N	10Rmt5Dbr	6-7/40	PS	4072C	6610
DEEP						
RIVER CN-2139	SN	10Rmt6Dbr	9-10/49	PS	4140B	6809
WOODS PINE L&N-3452	S	6Sec6Rmt4Dbr	3-5/53	PS	4183	6909
DEER RIVER NYC	S	10Rmt6Dbr	9/48-/49	PS	4123	6790
DEKALB COUNTY NYC	S	13Dbr	7/39	PS	4071B	6572
DELAWARE						
BAY NYC	S	22Rmt	10/48	PS	4122	6790
COUNTY PRR	S	10Rmt6Dbr	9-10/49	PS	4140B	6809
RAPIDS PRR-8437	N	10Rmt6Dbr	8-10/50	AC&F	9008	3200
RIVER PRR	S	10Rmt6Dbr	9-10/49	PS	4140	6814
DENARGO JT	S	4Rmt3CptDr4Dbr	5/39	PS	4083	6568
DENEHOTSO ATSF	S	4DrDbrREOBSLng	1-2/38	PS	4070	6532
DES MOINES RIVER NYC	S	10Rmt6Dbr	9/48-/49	PS	4123	6790
DES PLAINES RIVER NYC	S	10Rmt6Dbr	9/48-/49	PS	4123	6790
DETROIT LAKES NP	S	8DuRmt6Rmt4Dbr	7-9/48	PS	4119	6781
DETROIT RIVER NYC	S	10Rmt6Dbr	9/48-/49	PS	4123	6790
DICKERSON NP	S	8DuRmt6Rmt4Dbr	7-9/48	PS	4119	6781
DILLON COUNTY ACL	SN	10Rmt6Dbr	2-7/50	PS	4167	6864
DINAMARCA NdeM-346	SN	10Rmt6Dbr	9/48-/49	PS	4123	6790
DINNEBITO ATSF	S	14Sec	12/37-38	PS	4065	6532
DISTRICT						
OF COLUMBIA C&O	S	10Rmt6Dbr	2-7/50	PS	4167	6864
DIXIE PINE L&N-3465	N	6Sec6Rmt4Dbr	3-5/53	PS	4183	6909
DOMINICANA NdeM-570	SN	10Rmt6Dbr	1-3/49	Budd	9502	9662
DORCAS BAY NYC	S	22Rmt	10/48	PS	4122	6790
DORCHESTER BAY NYC	S	22Rmt	10/48	PS	4122	6790
DOROTHY ANN JonesProp	SN	12Dbr	4-5/49	PS	4125	6790

Figure I-101 Seventeen 6-section, 6-roomette, 4-double-bedroom sleepers were delivered in Worcester Lot 6942 in 1955, 11 for New Haven, 2 for BAR, and 4 to B&M. DARTMOUTH COLLEGE I (B&M-31), built as HAMPTON BEACH for B&M, was withdrawn from Pullman lease in June 1966 and sold to the Canadian National where it became GREENDALE. Most Pullman cars for the New England roads were assembled at the old Osgood-Bradley plant in Worcester, Massachusetts, of parts supplied by Pullman at Chicago. Railroad Avenue Enterprises Photograph PN-1321 taken August 1960 at White River Junction, Vermont.

CAR NAME OR NUMBER	ST	CAR TYPE	DATE BUILT	CAR BLDR	PLAN	LOT
DOUGLAS MANOR CP-10320	S	5DbrCpt4Sec4Rmt	11/54-55	Budd		9658
DOUPHIN RIVER CN-2145	SN	10Rmt6Dbr	10-11/48	PS	4137	6775
DR.						
ARCE NdeM-315	SN	18Rmt	12/38	PS	4068H	6563
BALMIS NdeM-313	SN	18Rmt	12/38	PS	4068H	6563
DURAN NdeM-316	SN	18Rmt	12/38	PS	4068H	6563
ERAZO NdeM-309	SN	18Rmt	3/38	PS	4068D	6539
GARCIA DIEGO NdeM-311	SN	18Rmt	12/38	PS	4068H	6563
ICAZA NdeM-322	SN	18Rmt	3/38	PS	4068D	6539
JIMENEZ NdeM-320	SN	18Rmt	3/38	PS	4068D	6539
LAVISTA NdeM-305	SN	18Rmt	12/38	PS	4068H	6563
LICEAGA NdeM-306	SN	18Rmt	12/38	PS	4068H	6563
LUCIO NdeM	SN	18Rmt	3/38	PS	4068D	6539
MARQUEZ NdeM-317	SN	18Rmt	3/38	PS	4068D	6539
MARTINEZ DEL RIO NdeM-310	SN	18Rmt	12/38	PS	4068H	6563
NAVARRO NdeM-307	SN	18Rmt	3/38	PS	4068D	6539

CAR NAME OR NUMBER	ST	CAR TYPE	DATE BUILT	CAR BLDR	PLAN	LOT
DR.						
NORMA NdeM-323	SN	18Rmt	12/38	PS	4068H	6563
OLVERA NdeM-312	SN	18Rmt	12/38	PS	4068H	6563
RIO DE LA LOZA NdeM-304	SN	18Rmt	12/38	PS	4068H	6563
TERRES NdeM-321	SN	18Rmt	12/38	PS	4068H	6563
UGARTE NdeM-319	SN	18Rmt	3/38	PS	4068D	6539
VELASCO NdeM-308	SN	18Rmt	12/38	PS	4068H	6563
VERTIZ NdeM-314	SN	18Rmt	3/38	PS	4068D	6539
DRAPER MANOR CP-10321	S	5DbrCpt4Sec4Rmt	11/54-55	Budd		9658
DREAM CLOUD GM	S	8DuRmt3Cpt2Dr	5/47	PS	4128	6780
DREAM LAKE UP	RN	4CptDr4DbrArtic	12/37	PS	4063C	6525
DREAMLAND B&O	S	24Sr8Dbr	2/58	Budd	9540	9691
DRUMMOND MANOR CP-10322	S	5DbrCpt4Sec4Rmt	11/54-55	Budd		9658
DUBUQUE CB&Q	S	8DuRmt6Rmt4Dbr	7-9/48	PS	4119	6781
DUFFERIN MANOR CP-10323	S	5DbrCpt4Sec4Rmt	11/54-55	Budd		9658
DUKE UNIVERSITY N&W	S	10Rmt6Dbr	11/49-50	Budd	9523	9660
DUNKIRK HARBOR NYC	S	22Rmt	4-6/49	Budd	9501	9661

CAR NAME OR NUMBER	ST	CAR TYPE	DATE BUILT	CAR BLDR	PLAN	LOT	CAR NAME OR NUMBER	ST	CAR TYPE	DATE BUILT	CAR BLDR	PLAN	LOT
DUNSMUIR MANOR CP-10324	S	5DbrCpt4Sec4Rmt	11/54-55	Budd		9658	EAGLE CHARM PRR-8432	S	10Rmt6Dbr	8-10/50	AC&F	9008	3200
DUPLEX ROOMETTE I Pool		24DuRmt	4/42	PS	4100	6673	EAGLE CHASM MP	S	10Rmt6Dbr	8-9/48	Budd	9504	9660
DURANT IC	SN	10Rmt5Dbr	6-7/40	PS	4072C	6610	EAGLE CHIEF PRR-8433	S	10Rmt6Dbr	8-10/50	AC&F	9008	3200
DUTCHESS COUNTY NYC	S	13Dbr	4/38	PS	4071A	6541	EAGLE CIRCLE MP-631	S	14Rmt2DbrDr	7-8/48	PS	4154	6758
DUVAL COUNTY ACL	S	10Rmt6Dbr	9-10/49	PS	4140B	6809	EAGLE CITY T&P-16	S	14Rmt4Dbr	2-7/48	PS	4153A	6758
DYERSBURG IC	SN	10Rmt5Dbr	6-7/40	PS	4072C	6610	EAGLE CLIFF MP	S	5DbrBLng	7/48	PS	4110	6758
E.H.HARRIMAN (DS)UP		10SecCptDbrArtic	10/34	PS	4035A	6428	EAGLE COUNTRY MP	S	14Rmt2DbrDr	7-8/48	PS	4154	6758
E.T.WEIR PRR-8259	N	21Rmt	1-6/49	Budd	9513	9667	EAGLE COUNTRY T&P-14	S	14Rmt2DbrDr	2-7/48	PS	4153A	6758
EADS BRIDGE NYC	N	4Dbr4Cpt2Dr	3-4/38	PS	4069B	6540	EAGLE COVE PRR-8434	S	10Rmt6Dbr	8-10/50	AC&F	9008	3200
EAGLE BAY T&P-1	S	14Rmt4Dbr	2-7/48	PS	4153A	6758	EAGLE CREEK MP	S	14Rmt4Dbr	2-7/48	PS	4153A	6758
EAGLE BEACH T&P-2	S	14Rmt4Dbr	2-7/48	PS	4153A	6758	EAGLE CREST MP	S	14Rmt4Dbr	2-7/48	PS	4153A	6758
EAGLE BEAM PRR-8430	S	10Rmt6Dbr	8-10/50	AC&F	9008	3200	EAGLE DAM MP	S	14Rmt4Dbr	2-7/48	PS	4153A	6758
EAGLE BLUFF PRR-8431	S	10Rmt6Dbr	8-10/50	AC&F	9008	3200	EAGLE DIVIDE MP	S	14Rmt4Dbr	2-7/48	PS	4153A	6758
EAGLE BRIDGE MP	S	14Rmt2DbrDr	7-8/48	PS	4154	6758	EAGLE DOMAIN T&P-9	S	14Rmt4Dbr	2-7/48	PS	4153A	6758
EAGLE BROOK T&P-7	S	14Rmt4Dbr	2-7/48	PS	4153A	6758	EAGLE EYE PRR-8435	S	10Rmt6Dbr	8-10/50	AC&F	9008	3200
EAGLE BUTTE MP	S	10Rmt6Dbr	8-9/48	Budd	9504	9660	EAGLE FLIGHT T&P-17	S	14Rmt2DbrDr	7-8/48	PS	4154	6758
EAGLE CALL T&P-8	S	14Rmt4Dbr	2-7/48	PS	4153A	6758	EAGLE FOREST IGN	S	14Rmt4Dbr	2-7/48	PS	4153A	6758
EAGLE CANYON T&P-18	S	5DbrBLng	7/48	PS	4110	6758	EAGLE GLIDE MP	S	14Rmt2DbrDr	7-8/48	PS	4154	6758
EAGLE CHAIN MP	S	14Rmt4Dbr	2-7/48	PS	4153A	6758	EAGLE GRAND PRR-8436	S	10Rmt6Dbr	8-10/50	AC&F	9008	3200

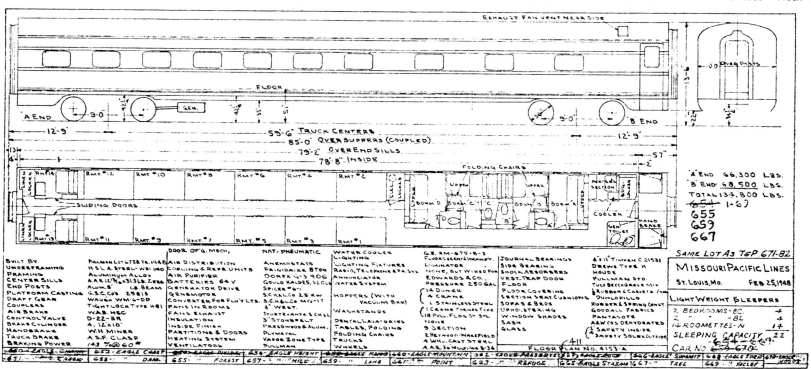

Figure I-102 Thirty-eight "EAGLE" cars were built for MP, T&P, and IGN in Pullman Lot 6758, Plan 4253A during February-July 1948. Here we see the Circa 1969 Missouri Pacific diagram for some of these cars. The Author's Collection.

CAR NAME OR NUMBER	ST	CAR TYPE	DATE BUILT	CAR BLDR	PLAN	LOT
EAGLE HAVEN MP	S	10Rmt6Dbr	8/56	Budd	9538	9660
EAGLE HEAD PRR-8437	S	10Rmt6Dbr	8-10/50	AC&F	9008	3200
EAGLE HEIGHT IGN	S	14Rmt4Dbr	2-7/48	PS	4153A	6758
EAGLE HILL MP	S	14Rmt4Dbr	2-7/48	PS	4153A	6758
EAGLE HOLLOW MP	S	10Rmt6Dbr	8/56	Budd	9538	9660
EAGLE ISLAND T&P-3	S	14Rmt4Dbr	2-7/48	PS	4153A	6758
EAGLE KNOB MP	S	14Rmt4Dbr	2-7/48	PS	4153A	6758
EAGLE LAKE MP	S	14Rmt4Dbr	2-7/48	PS	4153A	6758
EAGLE LAND T&P-13	S	14Rmt4Dbr	2-7/48	PS	4153A	6758
EAGLE LIGHT T&P-12	S	14Rmt4Dbr	2-7/48	PS	4153A	6758
EAGLE LODGE MP	S	10Rmt6Dbr	8/56	Budd	9538	9660
EAGLE MARSH MP	S	14Rmt4Dbr	2-7/48	PS	4153A	6758
EAGLE MEADOW MP	S	10Rmt6Dbr	8/56	Budd	9538	9660
EAGLE MOUNTAIN MP	S	14Rmt4Dbr	2-7/48	PS	4153A	6758
EAGLE NEST VALLEY ATSF	S	6Sec6Rmt4Dbr	6/42	PS	4099	6669
EAGLE OAK PRR-8438	S	10Rmt6Dbr	8-10/50	AC&F	9008	3200
EAGLE PARK PRR-8439	S	10Rmt6Dbr	8-10/50	AC&F	9008	3200
EAGLE PASS PRR-8440	S	10Rmt6Dbr	8-10/50	AC&F	9008	3200
EAGLE PATH T&P-5	S	14Rmt4Dbr	2-7/48	PS	4153A	6758
EAGLE POINT MP	S	14Rmt4Dbr	2-7/48	PS	4153A	6758
EAGLE PRESERVE IGN	S	14Rmt4Dbr	2-7/48	PS	4153A	6758
EAGLE RAPIDS MP	S	10Rmt6Dbr	8/56	Budd	9538	9660
EAGLE REFUGE IGN	S	14Rmt4Dbr	2-7/48	PS	4153A	6758
EAGLE REST T&P-15	S	14Rmt4Dbr	2-7/48	PS	4153A	6758
EAGLE RIDGE T&P-19	S	5DbrBLng	7/48	PS	4110	6758
EAGLE RIVER MP	S	6Sec6Rmt4Dbr	5/42	PS	4099	6669
EAGLE ROAD T&P-4	S	14Rmt4Dbr	2-7/48	PS	4153A	6758
EAGLE ROCK MP	S	14Rmt4Dbr	2-7/48	PS	4153A	6758
EAGLE SPIRIT T&P-11	S	14Rmt4Dbr	2-7/48	PS	4153A	6758
EAGLE STREAM MP	S	14Rmt4Dbr	2-7/48	PS	4153A	6758
EAGLE SUMMIT MP	S	14Rmt4Dbr	2-7/48	PS	4153A	6758
EAGLE TRAIL T&P-6	S	14Rmt4Dbr	2-7/48	PS	4153A	6758
EAGLE TREE IGN	S	14Rmt4Dbr	2-7/48	PS	4153A	6758
EAGLE TURN MP	S	14Rmt4Dbr	2-7/48	PS	4153A	6758
EAGLE VALLEY MP	S	14Rmt4Dbr	2-7/48	PS	4153A	6758
EAGLE VIEW MP	S	10Rmt6Dbr	8/56	Budd	9538	9660
EAGLE VILLAGE MP	S	14Rmt2DbrDr	7-8/48	PS	4154	6758
EAGLE WATCH T&P-10	S	14Rmt4Dbr	2-7/48	PS	4153A	6758
EAGLE WOODS IGN	S	14Rmt4Dbr	2-7/48	PS	4153A	6758
EAST POINT NH	S	14Rmt4Dbr	12/49/50	PS	4159	6822
EAST RIVER NYC	S	10Rmt6Dbr	9/48-/49	PS	4123	6790
EASTPORT CN-1110	S	8DuRmt4Sec4Dbr	1-3/54	PS	4124A	6922
EASTVIEW CP	SN	5DbrBLngObs	5-6/49	Budd	9506	9664
EASTVIEW CN-1111	S	8DuRmt4Sec4Dbr	1-3/54	PS	4124A	6922
ECUADOR NdeM-571	SN	10Rmt6Dbr	1-3-49	Budd	9502	9662
ECUM SECUM RIVER CN-2133	SN	10Rmt6Dbr	9-10/49	PS	4140	6814
EDEN VALLEY NYC	S	10Rmt6Dbr	1-3/49	Budd	9502	9662
EDENWOLD CN-1112	S	8DuRmt4Sec4Dbr	1-3/54	PS	4124A	6922
EDGECOMBE COUNTY ACL	S	10Rmt6Dbr	9-10/49	PS	4140B	6809
EDGELEY CN-1113	S	8DuRmt4Sec4Dbr	1-3/54	PS	4124A	6922
EDGERTON CN-1118	S	8DuRmt4Sec4Dbr	1-3/54	PS	4124A	6922
EDISTO ISLAND ACL	S	21Rmt	9-10/49	PS	4156B	6809
EDMONTON CN-1114	S	8DuRmt4Sec4Dbr	1-3/54	PS	4124A	6922
EDMUNDSON CN-1115	S	8DuRmt4Sec4Dbr	1-3/54	PS	4124A	6922
EDNA JonesProp	SN	5DbrBLngObs	5-6/49	Budd	9506	9664
EDSON CN-1116	S	8DuRmt4Sec4Dbr	1-3/54	PS	4124A	6922
EDWARDSVILLE CN-1117	S	8DuRmt4Sec4Dbr	1-3/54	PS	4124A	6922
EGIPTO NdeM-531	SN	10Rmt6Dbr	9/48-/49	PS	4123	6790
EKHART CN-1119	S	8DuRmt4Sec4Dbr	1-3/54	PS	4124A	6922
EL SALVADOR NdeM-573	SN	10Rmt6Dbr	1-3-49	Budd	9502	9662
ELBERTON PRR-8337	S	10Rmt6Dbr	6-8/49	Budd	9503	9662
ELBOW RIVER CN-2148	SN	10Rmt6Dbr	5-6/49	PS	4129A	6797
ELCOTT CN-1120	S	8DuRmt4Sec4Dbr	1-3/54	PS	4124A	6922
ELDERBANK CN-1121	S	8DuRmt4Sec4Dbr	1-3/54	PS	4124A	6922
ELDORADO CN-1159	S	8DuRmt4Sec4Dbr	1-3/54	PS	4124A	6922
ELEAZAR LORD Erie	S	10Rmt6Dbr	5-6/49	PS	4129A	6797
ELGIN CN-1123	S	8DuRmt4Sec4Dbr	1-3/54	PS	4124A	6922
ELGIN MANOR CP-10325	S	5DbrCpt4Sec4Rmt	11/54-55	Budd		9658
ELIZABETH CN-1124	S	8DuRmt4Sec4Dbr	1-3/54	PS	4124A	6922
ELK RIVER MP	S	10Rmt6Dbr	8-9/48	Budd	9504	9660
ELKHART COUNTY NYC	S	13Dbr	7/39	PS	4071B	6572
ELKHART RIVER NYC	S	10Rmt6Dbr	9/48-/49	PS	4123	6790
ELLERSLIE CN-1125	S	8DuRmt4Sec4Dbr	1-3/54	PS	4124A	6922
ELLISTONE CN-1126	S	8DuRmt4Sec4Dbr	1-3/54	PS	4124A	6922
ELM FALLS PRR	S	6DbrBLng	9/40	PS	4086A	6612
ELM GROVE CP-14107		10Sec5Dbr	/49-/50	CPRR		
ELMIRA CN-1127	S	8DuRmt4Sec4Dbr	1-3/54	PS	4124A	6922
ELMIRA INN PRR-8257	S	21Rmt	1-6/49	Budd	9513	9667
ELMSDALE CN-1128	S	8DuRmt4Sec4Dbr	1-3/54	PS	4124A	6922
ELNORA CN-1129	S	8DuRmt4Sec4Dbr	1-3/54	PS	4124A	6922
ELROSE CN-1130	S	8DuRmt4Sec4Dbr	1-3/54	PS	4124A	6922
EMERALD CN-1131	S	8DuRmt4Sec4Dbr	1-3/54	PS	4124A	6922
EMERSON CN-1132	S	8DuRmt4Sec4Dbr	1-3/54	PS	4124A	6922
EMORY & HENRY COLLEGE N&W	S	10Rmt6Dbr	11/49-50	Budd	9523	9660
EMORY RIVER CNO&TP	S	10Rmt6Dbr	9-10/49	PS	4140	6814
EMPEROR CN-1133	S	8DuRmt4Sec4Dbr	1-3/54	PS	4124A	6922
ENDAKO CN-1134	S	8DuRmt4Sec4Dbr	1-3/54	PS	4124A	6922
ENDCLIFFE CN-1135	S	8DuRmt4Sec4Dbr	1-3/54	PS	4124A	6922
ENDEAVOUR CN-1136	S	8DuRmt4Sec4Dbr	1-3/54	PS	4124A	6922
ENFIELD CN-1137	S	8DuRmt4Sec4Dbr	1-3/54	PS	4124A	6922
ENGLEE CN-1138	S	8DuRmt4Sec4Dbr	1-3/54	PS	4124A	6922
ENNISHORE CN-1139	S	8DuRmt4Sec4Dbr	1-3/54	PS	4124A	6922
ENOREE RIVER SOU	S	10Rmt6Dbr	9-10/49	PS	4140	6814
ENTERPRISE CN-1140	S	8DuRmt4Sec4Dbr	1-3/54	PS	4124A	6922
ENTRANCE CN-1141	S	8DuRmt4Sec4Dbr	1-3/54	PS	4124A	6922
ENTWISTLE CN-1142	S	8DuRmt4Sec4Dbr	1-3/54	PS	4124A	6922
EQUITY CN-1143	S	8DuRmt4Sec4Dbr	1-3/54	PS	4124A	6922
ERICKSON CN-1144	S	8DuRmt4Sec4Dbr	1-3/54	PS	4124A	6922

CAR NAME OR NUMBER	ST	CAR TYPE	DATE BUILT	CAR BLDR	PLAN	LOT
ERIE COUNTY NYC	S	13Dbr	7/39	PS	4071B	6572
ERIE HARBOR NYC	S	22Rmt	4-6/49	Budd	9501	9661
ERINVIEW CN-1145	S	8DuRmt4Sec4Dbr	1-3/54	PS	4124A	6922
ERNESTOWN CN-1146	S	8DuRmt4Sec4Dbr	1-3/54	PS	4124A	6922
ERWOOD CN-1147	S	8DuRmt4Sec4Dbr	1-3/54	PS	4124A	6922
ESCOCIA NdeM-572	SN	10Rmt6Dbr	12/48-49	Budd	9510	9660
ESCUMINAC CN-1148	S	8DuRmt4Sec4Dbr	1-3/54	PS	4124A	6922
ESPANA NdeM-347	SN	10Rmt6Dbr	9/48-/49	PS	4123	6790
ESSEX RF&P	S	10Rmt6Dbr	5-6/49	PS	4140A	6796
ESSEX CN-1149	S	8DuRmt4Sec4Dbr	1-3/54	PS	4124A	6922
ESSEX FALLS NYC	S	6DbrBLng	8-9/40	PS	4086B	6612
ESTANCIA VALLEY ATSF	N	6Sec6Rmt4Dbr	6/42	PS	4099	6669
ESTCOURT CN-1150	S	8DuRmt4Sec4Dbr	1-3/54	PS	4124A	6922
ETHELBERT CN-1151	S	8DuRmt4Sec4Dbr	1-3/54	PS	4124A	6922
ETIOPIA NdeM-544	SN	10Rmt6Dbr	9/48-/49	PS	4123	6790
ETOWAH RIVER CNO&TP	S	10Rmt6Dbr	9-10/49	PS	4140	6814
EUCLID CN-1152	S	8DuRmt4Sec4Dbr	1-3/54	PS	4124A	6922
EUGENE FIELD SLSF-1456	S	14Rmt4Dbr	1-6/48	PS	4153	6769
EUREKA CN-1153	S	8DuRmt4Sec4Dbr	1-3/54	PS	4124A	6922
EVANDALE CN-1154	S	8DuRmt4Sec4Dbr	1-3/54	PS	4124A	6922
EVANGELINE CN-1155	S	8DuRmt4Sec4Dbr	1-3/54	PS	4124A	6922
EVANSTON CN-1156	S	8DuRmt4Sec4Dbr	1-3/54	PS	4124A	6922
EVELYN CN-1157	S	8DuRmt4Sec4Dbr	1-3/54	PS	4124A	6922
EVENTIDE VALLEY NYC	S	10Rmt6Dbr	1-3/49	Budd	9502	9662
EVERETT CN-1158	S	8DuRmt4Sec4Dbr	1-3/54	PS	4124A	6922
EVERGLADES SCL-6400	SN	4Rmt5DbrCpt4Sec	12/54	PS	4196	6949
EXCELSIOR CN-1122	S	8DuRmt4Sec4Dbr	1-3/54	PS	4124A	6922
EXETER CN-1160	S	8DuRmt4Sec4Dbr	1-3/54	PS	4124A	6922
EXPLOITS RIVER CN-2075	SN	10Rmt6Dbr	9/48-/49	PS	4123	6790
EXTEW CN-1161	S	8DuRmt4Sec4Dbr	1-3/54	PS	4124A	6922
EXTREMADURA NdeM	SN	10Rmt6Dbr	4/50	PS	4167	6864
F.FAY JR. JonesProp	SN	10Rmt6Dbr	9-10/49	PS	4140B	6809
FAIRFAX RIVER RF&P	S	14Rmt2Dr	2-3/50	AC&F	9007	3094
FAIRLESS HILLS PRR-8442	N	10Rmt6Dbr	9-10/50	AC&F	9008	3200
FAIRPORT HARBOR NYC	S	22Rmt	4-6/49	Budd	9501	9661
FALL BROOK NYC	S	5DbrBLngObs	5-6/49	Budd	9506	9664
FARGO NP	S	8DuRmt6Rmt4Dbr	7-9/48	PS	4119	6781
FARMINGTON RIVER NYC	S	10Rmt6Dbr	9/48-/49	PS	4123	6790
FEATHER RIVER NYC	S	10Rmt6Dbr	9/48-/49	PS	4123	6790
FEDERAL VIEW PRR	S	2MbrDbrREObsBLng	5/38	PS	4080	6548
FERNWOOD IC	SN	22Rmt	10/48	PS	4122	6790
FERRY BUILDING JT	S	10Rmt5Dbr	11/40	PS	4072D	6636
FIGUEROA JT	S	4Dbr4Cpt2Dr	11/40	PS	4069F	6636
FINLANDIA NdeM-348	SN	10Rmt6Dbr	9/48-/49	PS	4123	6790
FIR FALLS PRR	S	6DbrBLng	3-5/49	PS	4131	6792
FIR GROVE CP-14108		10Sec5Dbr	/49-/50	CPRR		
FIREBRAND PASS GN-1378	S	6Rmt5Dbr2Cpt	10-11/50	PS	4180	6877
FISHERMENS WHARF JT		4Cpt3DrArtic	12/37	PS	4063B	6525
FISHING RAPIDS PRR	S	10Rmt6Dbr	3-4/49	PS	4129	6792
FLATHEAD RIVER GN-1192	S	2DbrDrObsBLng	1-2/47	PS	4109	6751
FLINT RIVER SOU	S	10Rmt6Dbr	9-10/49	PS	4140	6814

CAR NAME OR NUMBER	ST	CAR TYPE	DATE BUILT	CAR BLDR	PLAN	LOT
FLORDIA						
FLOWERS C&EI-904	S	4Rmt5DbrCpt4Sec	12/54	PS	4196	6949
LAKES C&EI-905	S	4Rmt5DbrCpt4Sec	12/54	PS	4196	6949
SUNSET NC&StL-250	S	4Rmt5DbrCpt4Sec	12/54	PS	4196	6949
SURF NC&StL-251	S	4Rmt5DbrCpt4Sec	12/54	PS	4196	6949
TRAVELER L&N-3449	S	4Rmt5DbrCpt4Sec	12/54	PS	4196	6949
FLORENCIA NdeM-652	SN	10Rmt6Dbr	4/49	PS	4140	6792
FLOSSMOOR IC	SN	22Rmt	10/48	PS	4122	6790
FOREST CANYON CRI&P	S	8Sec5Dbr	11/40	PS	4095	6631
FOREST STREAM NYC	S	6DbrBLng	3-5/49	Budd	9505	9663
FORT DODGE IC	SN	22Rmt	10/48	PS	4122	6790
FORT LAUDERDALE SAL-61	S	5Dbr2Cpt2Dr	1/56	PS	4201A	6969
FORTUNE BAY CN-2027	S	10Rmt5Dbr	6/54	PS	4186A	6925
FORWARD Pool	S	8Sec2Dbr2Cpt	11/36	PS	4057	6494
FRANCIA NdeM-349	SN	10Rmt6Dbr	9/48-/49	PS	4123	6790
FRANCIS P.						
BLAIR SLSF-1453	S	14Rmt4Dbr	1-6/48	PS	4153	6769
FRANK THOMPSON PRR	S	2DrCptDbrLngBEObs	br3-4/49	PS	4134	6792
FRANKLIN						
COUNTY N&W	S	10Rmt6Dbr	11/49-50	Budd	9523	9660
INN PRR-8258	S	21Rmt	1-6/49	Budd	9513	9667
MANOR CP-10326	S	5DbrCpt4Sec4Rmt	11/54-55	Budd		9658
FRASER MANOR CP-10327	S	5DbrCpt4Sec4Rmt	11/54-55	Budd		9658
FRASER RIVER GN-1271	S	8DuRmt4Sec3DbrCpt	11-12/50	PS	4181	6889
FRENCH RAPIDS PRR-8339	S	10Rmt 6Dbr	3/49	AC&F	9008	3079
FRENCH RIVER CNO&TP	S	10Rmt6Dbr	9-10/49	PS	4140	6814
FULTON IC	SN	22Rmt	10/48	PS	4122	6790
FUNDY CN-1902	SN	8DbrSKYTOPLng	11/48-49	PS	4138	6775
GALENA IC	N	4Dbr4Cpt2Dr	3-4/38	PS	4069B	6540
GALLATIN RIVER Milw-22	S	8DuRmt6Rmt4Dbr	10-12/48	PS	4135	6775
GANADO ATSF	S	14Sec	12/37-38	PS	4065	6532
GARDEN						
OF THE GODS CRI&P	S	8Sec2Dbr2Cpt	10/39	PS	4058C	6585
GARDENIA ACL	S	4Rmt5DbrCpt4Sec	12/54	PS	4196	6949
GARDINERS BAY NYC	S	22Rmt	10/48	PS	4122	6790
GARY HARBOR NYC	S	22Rmt	4-6/49	Budd	9501	9661
GASCONADE						
RIVER SLSF-1459	S	14Rmt4Dbr	1-6/48	PS	4153	6769
GASPE CN-1905	SN	8DbrSKYTOPLng	11/48-49	PS	4138	6775
GEMINI FCP	SN	10Rmt6Dbr	9/48-/49	PS	4123	6790
GENERAL BEAUREGARD IC	S	3DbrCptDrBarLng	4/42	PS	4101	6671
GENERAL JACKSON IC	S	3DbrCptDrBarLng	4/42	PS	4101	6671
GENESEE						
FALLS NYC	S	6DbrBLng	9/39	PS	4086	6573
RAPIDS PRR-8340	S	10Rmt 6Dbr	3/49	AC&F	9008	3079
RIVER B&O	P	2DbrCptDrObsBLng	7/39	PS	4082	6567
RIVER NYC	S	2DbrCptDrObsBLng	7/39	PS	4082	6567
VALLEY NYC	S	10Rmt6Dbr	1-3/49	Budd	9502	9662
GEORGE						
BROOKE ROBERTS PRR	S	2DrCptDbrBEObsLng	3-4/49	PS	4134	6792
G. VEST SLSF-1455	S	14Rmt4Dbr	1-6/48	PS	4153	6769

Figure I-103 FOREST STREAM, a 6-double-bedroom / buffet lounge, was built by Budd in 1949 for the New York Central. It is shown here in Chicago during August 1971 with a black Penn Central letterboard. In 1971, it became Amtrak FOREST STREAM 3200. The fluted stainless steel roof is characteristic of Budd-built cars, contrasting with the jig-welded stainless roof used on Pullman-built cars. Railroad Avenue Enterprises Photograph PN-2545.

CAR NAME OR NUMBER	ST	CAR TYPE	DATE BUILT	CAR BLDR	PLAN	LOT
GEORGE						
WASHINGTON BRIDGE NYCN		4Dbr4Cpt2Dr	3-4/38	PS	4069B	6540
WESTINGHOUSE PRR	N	10Rmt5Dbr	7-8/40	PS	4072C	6610
GEORGIA PINE NC&SL-201	S	6Sec6Rmt4Dbr	3-5/53	PS	4183	6909
GILA RIVER NYC	S	10Rmt6Dbr	9/48-/49	PS	4123	6790
GILMAN IC	N	4Dbr4Cpt2Dr	3-4/38	PS	4069B	6540
GLYNN COUNTY ACL	S	10Rmt6Dbr	9-10/49	PS	4140B	6809
GOLD CREEK Milw-16	S	8DbrSKYTOPLng	11/48-49	PS	4138	6775
GOLDEN						
BANNER CRI&P	SN	6Sec6Rmt4Dbr	4-5/42	PS	4099	6669
CANYON SP	SN	6Sec6Rmt4Dbr	4-5/42	PS	4099	6669
CAVERN SP(9116)	N	4Dbr4Cpt2Dr	4-5/40	PS	4069D	6605
CHARIOT CRI&P	SN	6Sec6Rmt4Dbr	4-5/42	PS	4099	6669

CAR NAME OR NUMBER	ST	CAR TYPE	DATE BUILT	CAR BLDR	PLAN	LOT
GOLDEN						
CLOUD SP	SN	6Sec6Rmt4Dbr	4-5/42	PS	4099	6669
CRAG SP	SN	4Dbr4Cpt2Dr	3-4/42	PS	4069H	6668
CREEK SP	SN	6Sec6Rmt4Dbr	4-5/42	PS	4099	6669
CREST SP-9047	S	10Rmt6Dbr	5-7/50	PS	4140C	6874
DAWN SP-9055	S	10Rmt6DbrBE	7/50	PS	4140D	6874
DESERT CRI&P	SN	4Dbr4Cpt2Dr	3-4/42	PS	4069H	6668
DIAL CRI&P	N	4Dbr4Cpt2Dr	4-5/40	PS	4069D	6605
DIVAN CRI&P	S	2DbrDrREObsBLng	8/48	PS	4127	6761
DOME CRI&P	SN	6Sec6Rmt4Dbr	4-5/42	PS	4099	6669
DREAM CRI&P	SN	4Dbr4Cpt2Dr	3-4/42	PS	4069H	6668
FLEECE CRI&P	N	4Dbr4Cpt2Dr	4-5/40	PS	4069D	6605
GATE PARK JT		12SecArtic	12/37	PS	4064	6525

CAR NAME OR NUMBER	ST	CAR TYPE	DATE BUILT	CAR BLDR	PLAN	LOT
GOLDEN						
HORIZON CRI&P	S	8Rmt6Dbr	11-12/54	PS	4195	6944
HOUR CRI&P	SN	4Dbr4Cpt2Dr	3-4/42	PS	4069H	6668
JOURNEY CRI&P	S	8Rmt6Dbr	11-12/54	PS	4195	6944
LOCKET CRI&P	SN	6Sec6Rmt4Dbr	4-5/42	PS	4099	6669
MEADOW CRI&P	S	8Rmt6Dbr	11-12/54	PS	4195	6944
MESA CRI&P	SN	6Sec6Rmt4Dbr	4-5/42	PS	4099	6669
MISSION SP	SN	4Dbr4Cpt2Dr	3-4/42	PS	4069H	6668
MOON SP	SN	4Dbr4Cpt2Dr	3-4/42	PS	4069H	6668
ORANGE SP	S	12dbr	7/50	PS	4139A	6873
PLAIN SP	SN	6Sec6Rmt4Dbr	4-5/42	PS	4099	6669
PLAZA CRI&P	SN	6Sec6Rmt4Dbr	4-5/42	PS	4099	6669
POPPY SP	S	12dbr	7/50	PS	4139A	6873
RIVER SP	SN	4Dbr4Cpt2Dr	3-4/42	PS	4069H	6668
SEA SP	SN	6Sec6Rmt4Dbr	4-5/42	PS	4099	6669
SPIRE CRI&P	S	8Rmt6Dbr	11-12/54	PS	4195	6944
STAR SP-9048	S	10Rmt6Dbr	5-7/50	PS	4140C	6874
STRAND SP	SN	4Dbr4Cpt2Dr	3-4/42	PS	4069H	6668
SUN SP-9046	S	10Rmt6Dbr	5-7/50	PS	4140C	6874
SUNSET CRI&P	SN	4Dbr4Cpt2Dr	3-4/42	PS	4069H	6668
TOWER CRI&P	S	8Rmt6Dbr	11-12/54	PS	4195	6944
TRAIL SP(9117)	N	4Dbr4Cpt2Dr	4-5/42	PS	4069D	6605
TRIANGLE AMT-2554	SN	8Rmt6Dbr	11-12/54	PS	4195	6944
VALLEY SP	SN	6Sec6Rmt4Dbr	4-5/42	PS	4099	6669
VISTA CRI&P	S	2DbrDrREObsBLng	8/48	PS	4127	6761
WAVE SP-9056	S	10Rmt6DbrBE	7/50	PS	4140D	6874
WEST CRI&P	SN	4Dbr4Cpt2Dr	3-4/42	PS	4069H	6668
GOLFO DE						
CALIFORNIA NdeM	SN	4Dbr4Cpt2Dr	7-8/39	PS	4069D	6571
CORTES NdeM-291	SN	4Dbr4Cpt2Dr	9-10/40	PS	4069E	6617
DARIEN NdeM-148	SN	4Dbr4Cpt2Dr	9-10/40	PS	4069E	6617
MEXICO NdeM-149	SN	4Dbr4Cpt2Dr	3-4/38	PS	4069B	6540
GOSHEN POINT NH	S	14Rmt4Dbr	12/49/50	PS	4159	6822
GOVERNORS ISLAND PRR	S	21Rmt	9-10/49	PS	4156B	6809
GRACE BAY CN-2025	S	10Rmt5Dbr	6/54	PS	4186A	6925
GRAN TEOCALLI NdeM-150	L	8Sec3Cpt	/52	Schi	100570	
GRAND						
CODROY RIVER CN-2131	SN	10Rmt6Dbr	9-10/49	PS	4140	6814
COULEE GN-1197	RN	4DbrDr6RmtLngObs	12/50	PS	4109K	6878
ISLAND UP	S	5DbrBLng	4/56	PS	4199	6959
MESA CRI&P	S	8Sec2Dbr2Cpt	10/39	PS	4058C	6585
RAPIDS INN PRR-8259	S	21Rmt	1-6/49	Budd	9513	9667
RIVER SLSF-1462	S	14Rmt4Dbr	1-6/48	PS	4153	6769
GRANDE RIVIERE CN-2135	SN	10Rmt6Dbr	9-10/49	PS	4140B	6809
GRANGER CRI&P	S	8DuRmt6Rmt4Dbr	8-9/48	PS	4116	6761
GRANITE FALLS Milw-5772		8SecCoachTouralux	6/9/47	Milw		
GRANT MANOR CP-10328	S	5DbrCpt4Sec4Rmt	11/54-55	Budd		9658
GRASS RIVER NYC	S	10Rmt6Dbr	9/48-/49	PS	4123	6790
GREAT PECONIC BAY NYC	S	22Rmt	10/48	PS	4122	6790
GREAT SOUTH BAY NYC	S	22Rmt	10/48	PS	4122	6790
GRECIA NdeM-500	SN	10Rmt6Dbr	9/48-/49	PS	4123	6790
GREELEY UP	N	12SecArtic	12/37	PS	4064	6525
GREEN						
BANK CN-1177	S	6Rmt4Dbr6Sec	4/54	PS	4183A	6923
BROOK CN-1164	S	6Rmt4Dbr6Sec	4/54	PS	4183A	6923
BUSH CN-1169	S	6Rmt4Dbr6Sec	4/54	PS	4183A	6923
CABIN CN-1167	S	6Rmt4Dbr6Sec	4/54	PS	4183A	6923
COURT CN-1165	S	6Rmt4Dbr6Sec	4/54	PS	4183A	6923
GABLES CN-1190	SN	6Sec6Rmt4Dbr	11/54-55	PS	4194	W6942
HARBOUR CN-1171	S	6Rmt4Dbr6Sec	4/54	PS	4183A	6923
HARBOUR CN-1193	SN	6Sec6Rmt4Dbr	11/54-55	PS	4194	W6942
HILL CN-1172	S	6Rmt4Dbr6Sec	4/54	PS	4183A	6923
LANE CN-1173	S	6Rmt4Dbr6Sec	4/54	PS	4183A	6923
PINE L&N-3453	S	6Sec6Rmt4Dbr	3-5/53	PS	4183	6909
POINT CN-1162	S	6Rmt4Dbr6Sec	4/54	PS	4183A	6923
RAPIDS PRR-8341	S	10Rmt 6Dbr	3/49	AC&F	9008	3079

Figure I-104 The Rivarossi (AHM) HO gauge smooth side 10-roomette, 6-double-bedroom car was modeled after this American Car & Foundry 1949 prototype. GREEN RAPIDS was withdrawn from Pullman lease in July 1968 and scrapped in January 1971. Railroad Avenue Enterprises Photograph PN-2542.

CAR NAME OR NUMBER	ST	CAR TYPE	DATE BUILT	CAR BLDR	PLAN	LOT
GREEN RIDGE CN-1179	N	6Rmt4Dbr6Sec	4/54	PS	4183A	6923
GREEN RIVER L&N	S	10Rmt6Dbr	4/50	PS	4140	6814
GREEN RIVER CN-1179	S	6Rmt4Dbr6Sec	4/54	PS	4183A	6923
GREENBRIER C&O	S	11Dbr	6/50	PS	4168	6865
GREENBRIER CN-1178	S	6Rmt4Dbr6Sec	4/54	PS	4183A	6923
GREENDALE CN-1192	SN	6Sec6Rmt4Dbr	11/54-55	PS	4194	W6942
GREENFIELD CN-1170	S	6Rmt4Dbr6Sec	4/54	PS	4183A	6923
GREENHURST CN-1194	SN	6Sec6Rmt4Dbr	11/54-55	PS	4194	W6942
GREENING CN-1166	S	6Rmt4Dbr6Sec	4/54	PS	4183A	6923
GREENMOUNT CN-1163	S	6Rmt4Dbr6Sec	4/54	PS	4183A	6923
GREENOCK CN-1191	SN	6Sec6Rmt4Dbr	11/54-55	PS	4194	W6942
GREENSBURG INN PRR-8260	S	21Rmt	1-6/49	Budd	9513	9667
GREENSHIELDS CN-1168	S	6Rmt4Dbr6Sec	4/54	PS	4183A	6923
GREENVALE CN-1175	S	6Rmt4Dbr6Sec	4/54	PS	4183A	6923
GREENVIEW CN-1174	S	6Rmt4Dbr6Sec	4/54	PS	4183A	6923
GREENVILLE IC	SN	4Dbr4Cpt2Dr	9-10/40	PS	4069E	6617
GREENWALD CN-1195	SN	6Sec6Rmt4Dbr	11/54-55	PS	4194	W6942
GREENWAY CN-1176	S	6Rmt4Dbr6Sec	4/54	PS	4183A	6923
GREENWICH CN-1181	S	6Rmt4Dbr6Sec	4/54	PS	4183A	6923
GREENWOOD CN-1180	S	6Rmt4Dbr6Sec	4/54	PS	4183A	6923
GREENWOOD PRR-8338	S	10Rmt6Dbr	6-8/49	Budd	9503	9662
GRENADA IC	SN	4Dbr4Cpt2Dr	9-10/40	PS	4069E	6617
GRINNEL						
GLACIER GN-1172	S	16DuRmt4Dbr	1-2/47	PS	4108	6751
GROVE BEACH NH-529	S	6Sec6Rmt4Dbr	11/54-55	PS	4194	W6942
GUATEMALA FEC	S	10Rmt6Dbr	9-10/49	PS	4140	6814
GUATEMALA NdeM-574	SN	10Rmt6Dbr	1-3-49	Budd	9502	9662
GULF COAST CRI&P	S	8DuRmt6Rmt4Dbr	8-9/48	PS	4116	6761
GULF STREAM NYC	S	6DbrBLng	3-5/49	Budd	9505	9663
GULFPORT IC	S	2Dbr2CptDrObsLng	4/42	PS	4102	6672
GULL B&O	S	16DuRmt4Dbr	4-6/54	Budd	9536	9658
GULL POINT NH	S	14Rmt4Dbr	12/49/50	PS	4159	6822
GUNNING COVE CN-1183	N	4Dbr4Cpt2Dr	3-4/38	PS	4069B	6540
GUNNISON RIVER MP	S	6Sec6Rmt4Dbr	5/42	PS	4099	6669
GUNSIGHT PASS GN-1160	S	8DuRmt4Dbr4Sec	12/46-47	PS	4107	6751
GUYANDOTTE B&O	S	10Rmt6Dbr	3/50	PS	4167	6864
H.J. HEINZ PRR	N	10Rmt6Dbr	11/48-49	PS	4140	6792
H.J. HEINZ PRR-8387	N	4Dbr4Cpt2Dr	9-12/48	AC&F	9009	2982
HACKENSACK						
RIVER PRR-8360	S	14Rmt2Dr	2-3/50	AC&F	9007	3098
HAGERSTOWN INN PRR-8261	S	21Rmt	1-6/49	Budd	9513	9667
HAINES PASS GN-1381	S	6Rmt5Dbr2Cpt	10-11/50	PS	4180	6877
HAKATAI ATSF	S	14Sec	12/37-38	PS	4065	6532
HALEYVILLE IC	P	4Dbr4Cpt2Dr	7-8/39	PS	4069D	6571
HALIFAX COUNTY ACL	SN	10Rmt6Dbr	2-7/50	PS	4167	6864
HAMILTON COUNTY PRR	S	13Dbr	6-7/39	PS	4071B	6572
HAMILTON INN PRR-8262	S	21Rmt	1-6/49	Budd	9513	9667
HAMMOND IC	SN	4Dbr4Cpt2Dr	9-10/40	PS	4069E	6617
HAMPDEN-SIDNEY						
COLLEGE N&W	S	10Rmt6Dbr	11/49-50	Budd	9523	9660
HAMPTON						
BEACH B&M-31	S	6Sec6Rmt4Dbr	11/54-55	PS	4194	W6942
ROADS DOME C&O-1851	S	Dome5RmtDbr3Dr	9/48	Budd	9524	9669
HANGING GLACIER GN-1173	S	16DuRmt4Dbr	1-2/47	PS	4108	6751
HANOVER COUNTY RF&P	S	10Rmt6Dbr	9-10/49	PS	4140B	6809
HAPPY VALLEY NYC	S	10Rmt6Dbr	1-3/49	Budd	9502	9662
HARBOR COVE PRR	S	3DbrBLng	12/48	PS	4141	6792
HARBOR POINT PRR	S	2DbrBarLng	5/38	PS	4077	6550
HARBOR REST PRR	S	3DbrBLng	12/48	PS	4141	6792
HARBOR SPRINGS PRR	S	2DbrBarLng	5/38	PS	4077	6550
HARLEM RIVER NYC	SN	10Rmt6Dbr	9/48-/49	PS	4123	6790
HARNETT COUNTY ACL	S	10Rmt6Dbr	9-10/49	PS	4140B	6809
HARRISON						
GLACIER GN-1177	S	16DuRmt4Dbr	1-2/47	PS	4108	6751
HART PASS GN-1376	S	6Rmt5Dbr2Cpt	10-11/50	PS	4180	6877
HARVEY IC	SN	4Dbr4Cpt2Dr	9-10/40	PS	4069E	6617
HARVEY COUCH KCS	S	14Rmt4Dbr	5/48	PS	4153	6795
HASTA ATSF	S	4Dbr4Cpt2Dr	12/39	PS	4069C	6597
HATTIESBURG IC	P	4Dbr4Cpt2Dr	7-8/39	PS	4069D	6571
HAVANA FEC	S	10Rmt6Dbr	9-10/49	PS	4140B	6809
HAVASU ATSF	S	14Sec	12/37-38	PS	4065	6532
HAVERSTRAW BAY NYC	S	22Rmt	10/48	PS	4122	6790
HAWAII (DS)UP		11 SecArtic	5/36	PS	4037B	6434
HAY RIVER CN-2092	SN	10Rmt6Dbr	9/48-/49	PS	4123	6790
HAZELHURST IC	P	4Dbr4Cpt2Dr	7-8/39	PS	4069D	6571
HEARNE MANOR CP-10329	S	5DbrCpt4Sec4Rmt	11/54-55	Budd		9658
HEBER						
C. KIMBALL D&RGW-M2	S	8Sec2SrLngObs	11/17/41	Budd		97102
C. KIMBALL D&RG-1273	S	10Rmt6Dbr	7/50	PS	4167	6864
HELENA NP	S	8DuRmt6Rmt4Dbr	7-9/48	PS	4119	6781
HEMLOCK FALLS PRR	S	6DbrBLng	9/40	PS	4086A	6612
HENDERSON SAL-54	S	5DbrCpt4Sec4Rmt	11/55	Budd	9537	9658
HENDERSON HARBOR NYC	S	22Rmt	4-6/49	Budd	9501	9661
HENRY						
CLAY FRICK PRR	N	12DuSr4Dbr	1-4/49	PS	4130	6792
HUDSON BRIDGE NYC	N	4Dbr4Cpt2Dr	3-4/38	PS	4069B	6540
PHIPPS PRR	N	6DbrBLng	3-5/49	PS	4131	6792
S. SPANG PRR-8247	N	21Rmt	1-6/49	Budd	9513	9667
SHAW SLSF-1452	S	14Rmt4Dbr	1-6/48	PS	4153	6769
W.OLIVER PRR	N	10Rmt6Dbr	11/48-49	PS	4140	6792
HERKIMER COUNTY NYC	S	13Dbr	7/39	PS	4071B	6572
HIALEAH SAL-72	S	11Dbr	12/55-56	PS	4198A	6968
HICKORY CREEK NYC	S	5DbrREObsBLng	8-9/48	PS	4126	6790
HICKORY FALLS PRR	S	6DbrBLng	9/40	PS	4086A	6612
HIDDEN VALLEY ATSF	S	6Sec6Rmt4Dbr	6/42	PS	4099	6540
HIGH BRIDGE NYC	N	4Dbr4Cpt2Dr	3-4/38	PS	4069B	6540
HILLSBOROUGH COUNTY ACLS		10Rmt6Dbr	9-10/49	PS	4140B	6809
HIWASSEE VALLEY SOU	S	14Rmt4Dbr	10-11/49	PS	4153C	6814
HOCKING RIVER NYC	S	10Rmt6Dbr	9/48-/49	PS	4123	6790
HOLANDA NdeM-575	SN	10Rmt6Dbr	12/48-49	Budd	9510	9660

Figure I-105   Ringling Bros., Barnum & Bailey Circus made no attempt to disguise the heritage of HICKORY CREEK,
which traveled the water-level route on The Twentieth Century Limited for 20 years.   The large observation room (solarium)
windows made the cars distinctive, though the idea was borrowed from standard solarium cars of the 1920s.   In 1968 it was
sold, along with a number of NYC all-bedroom cars, to RBBB who fielded a second show for the 1969 season, known as
"The Blue Unit."   HICKORY CREEK carried the tail sign for this train.   Wilber C. Whittaker Photograph circa 1970.

CAR NAME OR NUMBER	ST	CAR TYPE	DATE BUILT	CAR BLDR	PLAN	LOT	CAR NAME OR NUMBER	ST	CAR TYPE	DATE BUILT	CAR BLDR	PLAN	LOT
HOLIDAY PINE L&N-3454	S	6Sec6Rmt4Dbr	3-5/53	PS	4183	6909	HORSESHOE FALLS CN-2100	SN	14Rmt4Dbr	1-6/48	PS	4153	6769
HOLLINS COLLEGE N&W	S	10Rmt6Dbr	11/49-50	Budd	9523	9660	HORSESHOE FALLS NYC	S	6DbrBLng	8-9/40	PS	4086B	6612
HOLLYWOOD							HOSKINNINI ATSF	S	14Sec	12/37-38	PS	4065	6532
BEACH(1st) SAL-20	S	5DbrBLng	1/56	PS	4202	6970	HOTAUTA ATSF	S	14Sec	12/37-38	PS	4065	6532
BEACH(2nd) SCL-6200	SN	6DbrBLng	8/49	AC&F	9003	3045	HOTEVILLA ATSF	S	4Dbr4Cpt2Dr	1/38	PS	4069A	6532
HOLSTON RIVER CNO&TP	S	10Rmt6Dbr	9-10/49	PS	4140	6814	HOUSATONIC RIVER NYC	S	10Rmt6Dbr	9/48-/49	PS	4123	6790
HOME HARBOR NYC	S	22Rmt	4-6/49	Budd	9501	9661	HUALPAI ATSF	S	4Dbr4Cpt2Dr	1/38	PS	4069A	6532
HOMESTEAD C&O	S	11Dbr	6/50	PS	4168	6865	HUDSON BAY CN-2023	S	10Rmt5Dbr	6/54	PS	4186A	6925
HOMEWOOD IC	P	4Dbr4Cpt2Dr	7-8/39	PS	4069D	6571	HUDSON COUNTY PRR	S	10Rmt6Dbr	9-10/49	PS	4140B	6809
HONDURAS FEC	S	10Rmt6Dbr	9-10/49	PS	4140B	6809	HUDSON GLACIER GN-1183	S	16DuRmt4Dbr	12/50	PS	4108A	6890
HONDURAS NdeM-576	SN	10Rmt6Dbr	12/48-49	Budd	9510	9660	HUDSON RAPIDS PRR-8441	S	10Rmt6Dbr	9-10/50	AC&F	9008	3200
HONEOYE FALLS NYC	S	6DbrBLng	8-9/40	PS	4086B	6612	HUDSON RIVER NYC	S	10Rmt6Dbr	9/48-/49	PS	4123	6790
HONEY-BIRD ACL	RN	7Dbr2Dr	2-3/50	AC&F	9018	3091	HUGH HENRY						
HONOLULU (DS)UP		11 SecArtic	5/36	PS	4037B	6434	BRACKENRIDGE PRR	N	12DuSr4Dbr	1-4/49	PS	4130	6792
HOOVER DAM CNW	N	4DbrBEObsBLng	11/40	PS	4096A	6636	HUMBER BAY NYC	S	22Rmt	10/48	PS	4122	6790
HOPE FALLS NYC	S	6DbrBLng	8-9/40	PS	4086B	6612	HUMMING BIRD ACL	RN	7Dbr2Dr	2-3/50	AC&F	9018	3091

Figure I-106 HONEY-BIRD, a 7-drawing room, 2-bedroom car, was rebuilt in June 1961 from American Car & Foundry 14-roomette, 2-drawing room ASHLEY RIVER. Nee' Atlantic Coast Line, it carries the letterboard of Seaboard Coast Line in September 1972 despite the fact that it had been sold to Hamburg Industries in 1967 and to Amtrak as No. 2301 in 1971. In 1976 it became Auto-Train 252. Railroad Avenue Enterprises Photograph PN-4636.

CAR NAME OR NUMBER	ST	CAR TYPE	DATE BUILT	CAR BLDR	PLAN	LOT	CAR NAME OR NUMBER	ST	CAR TYPE	DATE BUILT	CAR BLDR	PLAN	LOT
HUNGRIA NdeM-501	SN	10Rmt6Dbr	9/48-/49	PS	4123	6790	IMPERIAL						
HUNTER MANOR CP-10330	S	5DbrCpt4Sec4Rmt	11/54-55	Budd		9658	BEACH JT	S	4Dbr4Cpt2Dr	3-4/42	PS	4069H	6668
HUNTERS POINT JT	S	4Dbr4Cpt2Dr	11/40	PS	4069F	6636	BENCH PRR-8385	S	4Dbr4Cpt2Dr	9-12/48	AC&F	9009	2982
HUNTING VALLEY NYC	S	10Rmt6Dbr	1-3/49	Budd	9502	9662	BIRD JT	S	4Dbr4Cpt2Dr	3-4/42	PS	4069H	6668
HUNTINGTON							BOWER PRR-8386	S	4Dbr4Cpt2Dr	9-12/48	AC&F	9009	2982
BAY NYC	S	22Rmt	10/48	PS	4122	6790	BRINK PRR-8387	S	4Dbr4Cpt2Dr	9-12/48	AC&F	9009	2982
RAPIDS PRR-8439	N	10Rmt6Dbr	8-10/50	AC&F	9008	3200	CANYON NYC	S	4Dbr4Cpt2Dr	3-4/38	PS	4069B	6540
HURON							CAPE JT	S	4Dbr4Cpt2Dr	3-4/42	PS	4069H	6668
RAPIDS PRR-8342	S	10Rmt 6Dbr	3/49	AC&F	9008	3079	CARRIAGE NYC	S	4Dbr4Cpt2Dr	9-10/40	PS	4069E	6617
RIVER NYC	S	10Rmt6Dbr	9/48-/49	PS	4123	6790	CASTLE NYC	S	4Dbr4Cpt2Dr	3-4/38	PS	4069B	6540
HYDE PARK IC	P	4Dbr4Cpt2Dr	7-8/39	PS	4069D	6571	CHAMBER NYC	S	4Dbr4Cpt2Dr	7-8/39	PS	4069D	6571
ILLINOIS RAPIDS PRR	S	10Rmt6Dbr	3-4/49	PS	4129	6792	CHARIOT NYC	S	4Dbr4Cpt2Dr	9-10/40	PS	4069E	6617
ILLINOIS RIVER NYC	S	10Rmt6Dbr	9/48-/49	PS	4123	6790	CHATEAU NYC	S	4Dbr4Cpt2Dr	3-4/38	PS	4069B	6540
IMPERIAL							CITY NYC	S	4Dbr4Cpt2Dr	9-10/40	PS	4069E	6617
ARCH NYC	S	4Dbr4Cpt2Dr	3-4/38	PS	4069B	6540	CLIFF PRR-8388	S	4Dbr4Cpt2Dr	9-12/48	AC&F	9009	2982
BAND JT	S	4Dbr4Cpt2Dr	3-4/42	PS	4069H	6668	CLIPPER JT	S	4Dbr4Cpt2Dr	4-5/40	PS	4069D	6605
BANNER JT	S	4Dbr4Cpt2Dr	4-5/40	PS	4069D	6605	COURT NYC	S	4Dbr4Cpt2Dr	3-4/38	PS	4069B	6540
BAY NYC	S	4Dbr4Cpt2Dr	3-4/38	PS	4069B	6540	CREST PRR	S	4Dbr4Cpt2Dr	7-8/39	PS	4069D	6571

CAR NAME OR NUMBER	ST	CAR TYPE	DATE BUILT	CAR BLDR	PLAN	LOT	CAR NAME OR NUMBER	ST	CAR TYPE	DATE BUILT	CAR BLDR	PLAN	LOT
IMPERIAL							IMPERIAL						
CROWN NYC	S	4Dbr4Cpt2Dr	9-10/40	PS	4069E	6617	FLOWER JT	S	4Dbr4Cpt2Dr	3-4/42	PS	4069H	6668
DOMAIN NYC	S	4Dbr4Cpt2Dr	9-10/40	PS	4069E	6617	FOREST NYC	S	4Dbr4Cpt2Dr	3-4/38	PS	4069B	6540
DOME NYC	S	4Dbr4Cpt2Dr	9-10/40	PS	4069E	6617	FOUNTAIN NYC	S	4Dbr4Cpt2Dr	3-4/38	PS	4069B	6540
DRIVE JT	S	4Dbr4Cpt2Dr	3-4/42	PS	4069H	6668	GARDEN NYC	S	4Dbr4Cpt2Dr	3-4/38	PS	4069B	6540
EMBLEM NYC	S	4Dbr4Cpt2Dr	9-10/40	PS	4069E	6617	GATE JT	S	4Dbr4Cpt2Dr	3-4/42	PS	4069H	6668
EMPIRE NYC	S	4Dbr4Cpt2Dr	9-10/40	PS	4069E	6617	GUARD JT	S	4Dbr4Cpt2Dr	4-5/40	PS	4069D	6605
ESTATE NYC	S	4Dbr4Cpt2Dr	7-8/39	PS	4069D	6571	HARBOR JT	S	4Dbr4Cpt2Dr	3-4/42	PS	4069H	6668
FALLS NYC	S	4Dbr4Cpt2Dr	3-4/38	PS	4069B	6540	HIGHLANDS NYC	S	4Dbr4Cpt2Dr	3-4/38	PS	4069B	6540
FIELDS PRR-8389	S	4Dbr4Cpt2Dr	9-12/48	AC&F	9009	2982	HILLS PRR-8390	S	4Dbr4Cpt2Dr	9-12/48	AC&F	9009	2982
FLEET NYC	S	4Dbr4Cpt2Dr	9-10/40	PS	4069E	6617	HORN JT	S	4Dbr4Cpt2Dr	3-4/42	PS	4069H	6668

Figure I-107 IMPERIAL HOUR, built by Pullman under wartime allocation in 1942 for joint overland service, was sold to Union Pacific in 1947 and remained in service until October 1968. In this March 9, 1952 photograph at Oakland, California, the car is barely a decade old and in its prime. It can be found today in UP work train service as 906210. Wilber C. Whittaker Photograph.

Figure I-108 IMPERIAL PALM at the age of 44 has been "white lined" and is nearing its final hour as UP 906211 MP&M Derrick Bunk Car in this George Cockle photograph taken in Omaha during 1986. Though rebuilt in 1972, it has been "Condemned," a railroad term for authorization for removal from service, sale, or scrapping. Note the corrosion of the roof and lower side sheets.

CAR NAME OR NUMBER	ST	CAR TYPE	DATE BUILT	CAR BLDR	PLAN	LOT
IMPERIAL						
HOUR JT	S	4Dbr4Cpt2Dr	3-4/42	PS	4069H	6668
HOUSE NYC	S	4Dbr4Cpt2Dr	3-4/38	PS	4069B	6540
JEWEL NYC	S	4Dbr4Cpt2Dr	7-8/39	PS	4069D	6571
LAWN PRR	S	4Dbr4Cpt2Dr	7-8/39	PS	4069D	6571
LEA PRR-8391	S	4Dbr4Cpt2Dr	9-12/48	AC&F	9009	2982
LEAF JT	S	4Dbr4Cpt2Dr	3-4/42	PS	4069H	6668
LETTER JT	S	4Dbr4Cpt2Dr	3-4/42	PS	4069H	6668
LOCH PRR-8392	S	4Dbr4Cpt2Dr	9-12/48	AC&F	9009	2982
MAJESTY NYC	S	4Dbr4Cpt2Dr	9-10/40	PS	4069E	6617
MANOR NYC	N	4Dbr4Cpt2Dr	7-8/39	PS	4069D	6571
MANSION NYC	S	4Dbr4Cpt2Dr	3-4/38	PS	4069B	6540
MANTLE PRR	S	4Dbr4Cpt2Dr	7-8/39	PS	4069D	6571
MARK JT	S	4Dbr4Cpt2Dr	3-4/42	PS	4069H	6668
MEADOWS PRR-8393	S	4Dbr4Cpt2Dr	9-12/48	AC&F	9009	2982
MOUNTAIN NYC	S	4Dbr4Cpt2Dr	7-8/39	PS	4069D	6571
PALACE NYC	S	4Dbr4Cpt2Dr	3-4/38	PS	4069B	6540
PALM JT	S	4Dbr4Cpt2Dr	3-4/42	PS	4069H	6668
PARK PRR	S	4Dbr4Cpt2Dr	3/38	PS	4069B	6540
PASS PRR	S	4Dbr4Cpt2Dr	3/38	PS	4069B	6540
PATH PRR	S	4Dbr4Cpt2Dr	7-8/39	PS	4069D	6571
PEAK PRR-8394	S	4Dbr4Cpt2Dr	9-12/48	AC&F	9009	2982
PLATEAU PRR	S	4Dbr4Cpt2Dr	3/38	PS	4069B	6540
POINT PRR	S	4Dbr4Cpt2Dr	3/38	PS	4069B	6540
QUEEN NYC	S	4Dbr4Cpt2Dr	7-8/39	PS	4069D	6571
RANCH JT	S	4Dbr4Cpt2Dr	3-4/42	PS	4069H	6668
RANGE PRR	S	4Dbr4Cpt2Dr	3/38	PS	4069B	6540
RIDGE PRR-8395	S	4Dbr4Cpt2Dr	9-12/48	AC&F	9009	2982
ROAD PRR-8396	S	4Dbr4Cpt2Dr	9-12/48	AC&F	9009	2982
ROBE JT	S	4Dbr4Cpt2Dr	3-4/42	PS	4069H	6668

CAR NAME OR NUMBER	ST	CAR TYPE	DATE BUILT	CAR BLDR	PLAN	LOT
IMPERIAL						
ROCK JT	S	4Dbr4Cpt2Dr	3-4/42	PS	4069H	6668
SANDS JT	S	4Dbr4Cpt2Dr	3-4/42	PS	4069H	6668
SCEPTRE NYC	S	4Dbr4Cpt2Dr	7-8/39	PS	4069D	6571
SOVERIGN NYC	S	4Dbr4Cpt2Dr	7-8/39	PS	4069D	6571
STATE NYC	S	4Dbr4Cpt2Dr	7-8/39	PS	4069D	6571
TEMPLE NYC	S	4Dbr4Cpt2Dr	9-10/40	PS	4069E	6617
TERRACE PRR	S	4Dbr4Cpt2Dr	3/38	PS	4069B	6540
TERRAIN PRR-8397	S	4Dbr4Cpt2Dr	9-12/48	AC&F	9009	2982
THRONE JT	S	4Dbr4Cpt2Dr	4-5/40	PS	4069D	6605
TRAIL PRR	S	4Dbr4Cpt2Dr	7-8/39	PS	4069D	6571
TREES PRR-8398	S	4Dbr4Cpt2Dr	9-12/48	AC&F	9009	2982
VALE PRR-8399	S	4Dbr4Cpt2Dr	9-12/48	AC&F	9009	2982
VALLEY NYC	S	4Dbr4Cpt2Dr	7-8/39	PS	4069D	6571
VIEW PRR	S	4Dbr4Cpt2Dr	7-8/39	PS	4069D	6571
INDIA NdeM-519	SN	10Rmt6Dbr	9/48-/49	PS	4123	6790
INDIA POINT NH	S	14Rmt4Dbr	12/49/50	PS	4159	6822
INDIAN ARROW(1st) ATSF	S	24DuRmt	6-8/47	PS	4100B	6757
INDIAN ARROW(2nd) ATSF	RN	11Dbr	6-8/47	PS	6007A	6757
INDIAN CANOE ATSF	S	24DuRmt	6-8/47	PS	4100B	6757
INDIAN DRUM ATSF	S	24DuRmt	6-8/47	PS	4100B	6757
INDIAN FALLS ATSF	S	24DuRmt	6-8/47	PS	4100B	6757
INDIAN FLUTE ATSF	S	24DuRmt	6-8/47	PS	4100B	6757
INDIAN LAKE ATSF	S	24DuRmt	6-8/47	PS	4100B	6757
INDIAN LAKE(2nd) ATSF	RN	11Dbr	6-8/47	PS	6007A	6757
INDIAN MAID ATSF	S	24DuRmt	6-8/47	PS	4100B	6757
INDIAN MESA ATSF	S	24DuRmt	6-8/47	PS	4100B	6757
INDIAN PASS SP&S	S	8DuRmt4Dbr4Sec	2/50	PS	4107	6828
INDIAN PONY ATSF	S	24DuRmt	6-8/47	PS	4100B	6757
INDIAN RIVER NYC	S	10Rmt6Dbr	9/48-/49	PS	4123	6790

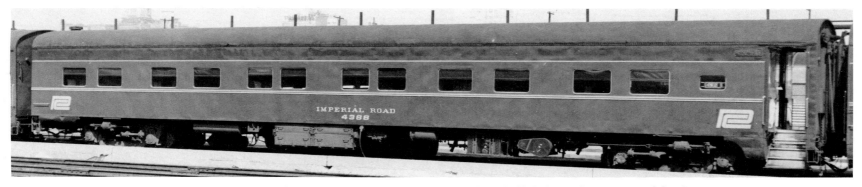

Figure I-109 IMPERIAL ROAD was a postwar American Car & Foundry 4-double-bedroom, 4-compartment, 2-drawing room sister to IMPERIAL HOUR delivered to the Pennsylvania Railroad in 1948. It is shown here in tuscan red with a clean letterboard and the new Penn Central number and logo in May 1969. Though of the same configuration, notice how different its window arrangement is from the 1942 cars and the absence of the upper berth windows. The undercar mechanical layout also differs considerably. Railroad Avenue Enterprises Photograph PN-2440.

CAR NAME OR NUMBER	ST	CAR TYPE	DATE BUILT	CAR BLDR	PLAN	LOT
INDIAN SCOUT ATSF	S	24DuRmt	6-8/47	PS	4100B	6757
INDIAN SONG ATSF	S	24DuRmt	6-8/47	PS	4100B	6757
INDIAN SQUAW ATSF	S	24DuRmt	6-8/47	PS	4100B	6757
INDIANA HARBOR NYC	S	22Rmt	4-6/49	Budd	9501	9661
INDIGO CN-2000		24DuRmt	10/49-50	CC&F	4100C	
INDOCHINA NdeM-577	SN	10Rmt6Dbr	12/48-49	Budd	9510	9660
INGELOW CN-2001	L	24DuRmt	10/49-50	CC&F	4100C	
INGERSOLL CN-2003	L	24DuRmt	10/49-50	CC&F	4100C	
INGLATERRA NdeM-502	SN	10Rmt6Dbr	9/48-/49	PS	4123	6790
INGONISH CN-2009		24DuRmt	10/49-50	CC&F	4100C	
INGRAMPORT CN-2002		24DuRmt	10/49-50	CC&F	4100C	
INKERMAN CN-2004	L	24DuRmt	10/49-50	CC&F	4100C	
INNES CN-2005	L	24DuRmt	10/49-50	CC&F	4100C	
INTERNATIONAL						
BRIDGE NYC	N	4Dbr4Cpt2Dr	3-4/38	PS	4069B	6540
INTERVALE CN-2011		24DuRmt	10/49-50	CC&F	4100C	
INUYA PASS GN-1383	S	6Rmt5Dbr2Cpt	10-11/50	PS	4180	6877
INVERMAY CN-2010		24DuRmt	10/49-50	CC&F	4100C	
INVERNESS CN-2006		24DuRmt	10/49-50	CC&F	4100C	
INWOOD CN-2007		24DuRmt	10/49-50	CC&F	4100C	
IONA CN-2008		24DuRmt	10/49-50	CC&F	4100C	
IRAK NdeM-523	SN	10Rmt6Dbr	9/48-/49	PS	4123	6790
IRAN NdeM-540	SN	10Rmt6Dbr	9/48-/49	PS	4123	6790
IRIS CN-2012		24DuRmt	10/49-50	CC&F	4100C	
IRLANDA NdeM-503	SN	10Rmt6Dbr	9/48-/49	PS	4123	6790
IRMA CN-2013		24DuRmt	10/49-50	CC&F	4100C	
IRONDALE CN-2014		24DuRmt	10/49-50	CC&F	4100C	
IROQUOIS CN-2015		24DuRmt	10/49-50	CC&F	4100C	
IROQUOIS RAPIDS PRR	S	10Rmt6Dbr	3-4/49	PS	4129	6792
IROQUOIS RIVER NYC	S	10Rmt6Dbr	9/48-/49	PS	4123	6790
IRVINE CN-2016	L	24DuRmt	10/49-50	CC&F	4100C	
ISABELLA CN-2017	L	24DuRmt	10/49-50	CC&F	4100C	
ISLANDIA NdeM-505	SN	10Rmt6Dbr	9/48-/49	PS	4123	6790
ISLETA ATSF	S	8SecDr2Cpt	4/37	Budd	9517	967
ISLEVIEW CN-2018		24DuRmt	10/49-50	CC&F	4100C	
ISRAEL NdeM-538	SN	10Rmt6Dbr	9/48-/49	PS	4123	6790
ITALIA NdeM-504	SN	10Rmt6Dbr	9/48-/49	PS	4123	6790
ITUNA CN-2019		24DuRmt	10/49-50	CC&F	4100C	
JACK'S NARROWS PRR	S	2DrCptDbrObsBLng	5/38	PS	4081	6549
JACKSONVILLE SAL	S	10Rmt6Dbr	5-6/49	PS	4140A	6796
JACOB J.						
VANDERGRIFT PRR-8241	N	21Rmt	1-6/49	Budd	9513	9667
JADITO ATSF	S	4Dbr4Cpt2Dr	1/38	PS	4069A	6532
JAMAICA FEC	S	4Rmt5DbrCpt4Sec	12/54	PS	4196	6949
JAMAICA BAY NYC	S	22Rmt	10/48	PS	4122	6790
JAMES						
B. BONHAM MKT-1505	S	14Rmt4Dbr	1-6/48	PS	4153	6769
BAY NYC	S	22Rmt	10/48	PS	4122	6790
BOWIE MKT-1504	S	14Rmt4Dbr	1-6/48	PS	4153	6769
FRANCIS JonesProp	SN	12Dbr	4-5/49	PS	4125	6790

CAR NAME OR NUMBER	ST	CAR TYPE	DATE BUILT	CAR BLDR	PLAN	LOT
JAMES						
GORE KING Erie	S	10Rmt6Dbr	5-6/49	PS	4129A	6797
HAY REED PRR	N	12DuSr5Dbr	6/39	PS	4066B	6566
HAY REED PRR-8287	N	21Rmt	1-6/49	Budd	9513	9667
Mc CREA PRR	S	2DrCptDbrBEObsLng	3-4/49	PS	4134	6792
O'HARA PRR	N	10Rmt6Dbr	11/48-49	PS	4140	6792
PARK PRR	N	10Rmt6Dbr	11/48-49	PS	4140	6792
PARK JR. PRR	N	10Rmt6Dbr	11/48-49	PS	4140	6792
RIVER SLSF-1461	S	14Rmt4Dbr	1-6/49	PS	4153	6769
RIVER VALLEY SOU	S	14Rmt4Dbr	10-11/49	PS	4153C	6814
W. FANNIN MKT-1500	S	14Rmt4Dbr	1-6/48	PS	4153	6769
JAMESTOWN NP	S	8DuRmt6Rmt4Dbr	7-9/48	PS	4119	6781
JANICE WINIFRED JonesProp	SN	12Dbr	4-5/49	PS	4125	6790
JARVIS MANOR CP-10331	S	5DbrCpt4Sec4Rmt	11/54-55	Budd		9658
JAY BIRD ACL	RN	7Dbr2Dr	2-3/50	AC&F	9018	3091
JEANETTE INN PRR-8263	S	21Rmt	1-6/49	Budd	9513	9667
JEFFERSON						
COUNTY PRR	S	13Dbr	6-7/39	PS	4071B	6572
PASS GN-1375	S	6Rmt5Dbr2Cpt	10-11/50	PS	4180	6877
RIVER Milw-24	S	8DuRmt6Rmt4Dbr	10-12/48	PS	4135	6775
JEFFERY SCAIFE PRR-8278	N	21Rmt	1-6/49	Budd	9513	9667
JOB EDSON KCS	S	14Rmt4Dbr	5/48	PS	4153	6795
JOHN						
EDGAR THOMSON PRR	N	2DrCptDbrObsBLng	5/38	PS	4081	6549
EVANS D&RG-1270	S	10Rmt6Dbr	7/50	PS	4167	6864
PITCAIRN PRR	N	10Rmt6Dbr	11/48-49	PS	4140	6792
PITCAIRN PRR-8268	N	21Rmt	1-6/49	Budd	9513	9667
JORDANIA NdeM-525	SN	10Rmt6Dbr	9/48-/49	PS	4123	6790
JOSEPH						
HORNE PRR	N	12DuSr4Dr	1-4/49	PS	4130	6792
PULITZER SLSF-1350	S	2DbrDrREObsLng	5/48	PS	4121	6792
JUAREZ FSC-BC	SN	4Dbr4Cpt2Dr	7-8/39	PS	4069D	6571
JUDGE						
MILTON BROWN GM&O	S	8RmtCpt3Dbr4Sec	7/50	AC&F	9012	3208
JUNIATA NARROWS PRR	S	2DrCptDbrObsBLng	5/38	PS	4081	6549
JUNIATA RAPIDS PRR	S	10Rmt6Dbr	11/48-49	PS	4140	6792
JUNIPER FALLS PRR	S	6DbrBLng	9/40	PS	4086A	6612
JUPITER NdeM-589	SN	4Sec8DuRmt4Dbr	12/46-47	PS	4107	6751
KABAH NdeM-174	L	8Sec3Cpt	/52	Schi	100570	
KAIBITO ATSF	S	4Dbr4Cpt2Dr	1/38	PS	4069A	6532
KAKABEKA FALLS CN-2101	SN	14Rmt4Dbr	1-6/48	PS	4153	6769
KALAMAZOO RAPIDS PRR	S	10Rmt6Dbr	3-4/49	PS	4129	6792
KALAMAZOO RIVER NYC	S	10Rmt6Dbr	9/48-/49	PS	4123	6790
KANE INN PRR-8264	S	21Rmt	1-6/49	Budd	9513	9667
KANKAKEE RAPIDS PRR	S	10Rmt6Dbr	3-4/49	PS	4129	6792
KANKAKEE RIVER NYC	S	10Rmt6Dbr	9/48-/49	PS	4123	6790
KASKASKIA RAPIDS PRR	S	10Rmt6Dbr	3-4/49	PS	4129	6792
KAW VALLEY ATSF	S	6Sec6Rmt4Dbr	6/42	PS	4099	6669
KAYENTA ATSF	S	4Dbr4Cpt2Dr	1/38	PS	4069A	6532
KEENE VALLEY NYC	S	10Rmt6Dbr	1-3/49	Budd	9502	9662

Figure I-110 Sublettered for subsidiary Louisiana & Arkansas, KCS 14-roomette, 4-double-bedroom Pullman JOB EDSON illustrates Pullman's version of a postwar riveted aluminum car body. Thousands of rivet heads somewhat spoiled the desired smooth side effect and the weight savings of aluminum were not significant. Unlike AC&F, Pullman encouraged car buyers to stick with its standard carbon steel car body. John P. Hankey Collection.

Figure I-111 L. S. HUNGERFORD was renamed from the Pullman car DUPLEX ROOMETTE I in 1949. It was built by Pullman in April 1942 as a sample proposed standard floor plan and placed in Pool service. It appears that no additional orders were received. Still in Pullman ownership, it was leased to the B&O from 1952 until 1962. In September 1967 it was sold to NdeM and became their PARICUTIN 562. NdeM diagram from the Author's Collection.

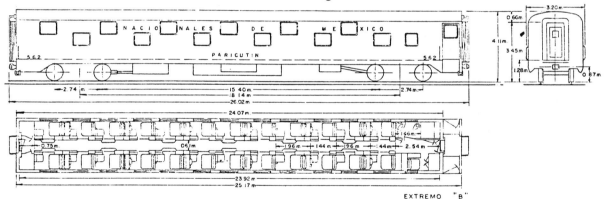

CAR NAME OR NUMBER	ST	CAR TYPE	DATE BUILT	CAR BLDR	PLAN	LOT
KENIA NdeM-561	N	10Rmt6Dbr	12/48-49	Budd	9510	9660
KENNESAW MOUNTAIN SAL-17	S	6DbrBLng	8/49	AC&F	9003	3045
KENTUCKY PINE L&N-3455	S	6Sec6Rmt4Dbr	3-5/53	PS	4183	6909
KENTUCKY RIVER L&N	S	10Rmt6Dbr	4/50	PS	4140	6814
KEUKA VALLEY NYC	S	10Rmt6Dbr	1-3/49	Budd	9502	9662
KEYSTONE STATE NH-551	S	6DbrBLng	1/55	PS	4193	W6941
KEZAR FALLS NYC	S	6DbrBLng	8-9/40	PS	4086B	6612
KIETSIEL ATSF	S	4Dbr4Cpt2Dr	1/38	PS	4069A	6532
KING AND QUEEN RF&P	S	10Rmt6Dbr	4/49	PS	4140	6792

CAR NAME OR NUMBER	ST	CAR TYPE	DATE BUILT	CAR BLDR	PLAN	LOT
KING COAL IC	S	6Sec6Rmt4Dbr	4/42	PS	4099	6669
KING COTTON IC	S	6Sec6Rmt4Dbr	4/42	PS	4099	6669
KING GEORGE RF&P	S	10Rmt6Dbr	4/49	PS	4140	6792
KING WILLIAM RF&P	S	10Rmt6Dbr	4/49	PS	4140	6792
KING'S COVE CN-118	SN	4Dbr4Cpt2Dr	9-10/40	PS	4069E	6617
KINGFISHER B&O	S	16DuRmt4Dbr	4-6/54	Budd	9536	9658
KINGS RIVER NYC	S	10Rmt6Dbr	9/48-/49	PS	4123	6790
KINTLA GLACIER GN-1181	S	16DuRmt4Dbr	12/50	PS	4108A	6890
KISSIMEE RIVER ACL	RN	4Cpt4Dr	3-4/38	PS	6013	6540
KITTATINNY DL&W	S	10Rmt 6Dbr	6-7/49	AC&F	9008A	3089

CAR NAME OR NUMBER	ST	CAR TYPE	DATE BUILT	CAR BLDR	PLAN	LOT
KLAMATH VALLEY NYC	S	10Rmt6Dbr	1-3/49	Budd	9502	9662
KLETHLA ATSF	S	4Dbr4Cpt2Dr	1/38	PS	4069A	6532
KOKOMO INN PRR-8265	S	21Rmt	1-6/49	Budd	9513	9667
KOKOSING B&O	S	10Rmt6Dbr	3/50	PS	4167	6864
KOOTENAI RIVER GN-1193	S	2DbrDrObsBLng	1-2/47	PS	4109	6751
L.S. HUNGERFORD Pool	N	24DuRmt	4/42	PS	4100	6673
LA						
COSTA CRI&P	S	22Rmt	8/48	PS	4122A	6761
JOLLA CRI&P	S	12Dbr	8/48	PS	4139	6761
MIRADA CRI&P	S	2DbrDrREObsBLng	8/48	PS	4127A	6761
PALMA CRI&P	S	12Dbr	8/48	PS	4139	6761
PORTE COUNTY NYC	S	13Dbr	10/40	PS	4071C	6618
QUINTA CRI&P	S	4Dbr4Cpt2Dr	8/48	PS	4069L	6761
LACKAWANNA DL&W	S	10Rmt 6Dbr	6-7/49	AC&F	9008A	3089
LAGO						
ARAL NdeM-325	SN	16DuRmtCpt3Dbr	10-11/48	PS	4146	6774
BAIKAL NdeM-330	SN	16DuRmtCpt3Dbr	10-11/48	PS	4146	6774
CHAPALA FCP	SN	10Rmt6Dbr	9-11/48	Budd	9509	9660
CHAPULTPEC FCP	SN	6Dbr5Cpt	7-8/52	Budd	9534	9641

CAR NAME OR NUMBER	ST	CAR TYPE	DATE BUILT	CAR BLDR	PLAN	LOT
LAGO						
CUITZERO FCP	SN	10Rmt6Dbr	2-7/50	PS	4167	6864
DE LOS OSOS NdeM-331	SN	16DuRmtCpt3Dbr	10-11/48	PS	4146	6774
ERIE NdeM-332	SN	16DuRmtCpt3Dbr	10-11/48	PS	4146	6774
HURON NdeM-327	SN	16DuRmtCpt3Dbr	10-11/48	PS	4146	6774
LAGODA NdeM-335	SN	16DuRmtCpt3Dbr	10-11/48	PS	4146	6774
MICHIGAN NdeM-328	SN	16DuRmtCpt3Dbr	10-11/48	PS	4146	6774
ONTARIO NdeM-334	SN	16DuRmtCpt3Dbr	10-11/48	PS	4146	6774
PATZCUARO FCP	SN	10Rmt6Dbr	9-11/48	Budd	9509	9660
SUPERIOR NdeM-324	SN	16DuRmtCpt3Dbr	10-11/48	PS	4146	6774
TANGANYIKA NdeM-32	SN	16DuRmtCpt3Dbr	10-11/48	PS	4146	6774
TEXCOCO FCP	SN	6Dbr5Cpt	7-8/52	Budd	9534	9641
VICTORIA NdeM-326	SN	16DuRmtCpt3Dbr	10-11/48	PS	4146	6774
WINNIPEG NdeM-333	SN	16DuRmtCpt3Dbr	10-11/48	PS	4146	6774
LAGUNA ATSF	S	8SecDr2Cpt	4/37	Budd	9517	967
LAIRD MANOR CP-10332	S	5DbrCpt4Sec4Rmt	11/54-55	Budd		9658
LAKE						
CHATCOLET Milw-7	S	10Rmt6Dbr	10-11/48	PS	4137	6775
COEUR d'ALENE Milw-2	S	10Rmt6Dbr	10-11/48	PS	4137	6775

Figure I-112 The Milwaukee Road's Pullman-built LAKE NASHOTAH, painted in the UP style, is shown here in Chicago in August 1968, two years after it was withdrawn from Pullman lease. The round lavatory window is an innovation credited to The Milwaukee Road; the stretcher window amidships and the oval windows in the end bedrooms add up to an odd appearance. Railroad Avenue Enterprises Photograph PN-3779.

CAR NAME OR NUMBER	ST	CAR TYPE	DATE BUILT	CAR BLDR	PLAN	LOT
LAKE						
COUNTY PRR	S	13Dbr	6-7/39	PS	4071B	6572
CRESCENT Milw-11	S	10Rmt6Dbr	10-11/48	PS	4137	6775
KAPOWSIN Milw-10	S	10Rmt6Dbr	10-11/48	PS	4137	6775
KEECHELUS Milw-3	S	10Rmt6Dbr	10-11/48	PS	4137	6775
NASHOTAH Milw-9	S	10Rmt6Dbr	10-11/48	PS	4137	6775
NOKOMIS CRI&P	S	8Rmt6Dbr	11-12/54	PS	4195	6944
OCONOMOWOC Milw-5	S	10Rmt6Dbr	10-11/48	PS	4137	6775
PEND OREILLE Milw-6	S	10Rmt6Dbr	10-11/48	PS	4137	6775
PEPIN Milw-4	S	10Rmt6Dbr	10-11/48	PS	4137	6775
PEWAUKEE Milw-8	S	10Rmt6Dbr	10-11/48	PS	4137	6775
WALES SAL-41	S	10Rmt6Dbr	6-8/49	Budd	9503	9662
LAKESIDE JT	S	4Dbr4Cpt2Dr	11/40	PS	4069F	6636
LANAI (DS)UP		7Dbr2CptArtic	5/36	PS	4039D	6434
LANCASTER RF&P	S	10Rmt6Dbr	5-6/49	PS	4140A	6796
LANCASTER COUNTY PRR	S	13Dbr	4/38	PS	4071A	6541
LAND O'STRAWBERRIES IC	S	6Sec6Rmt4Dbr	4/42	PS	4099	6669
LANE WOODY JonesProp	SN	4Dbr4Cpt2Dr	7-8/39	PS	4069D	6571
LAOS NdeM-517	SN	10Rmt6Dbr	9/48-/49	PS	4123	6790
LARCH FALLS PRR	S	6DbrBLng	3-5/49	PS	4131	6792
LATROBE INN PRR-8266	S	21Rmt	1-6/49	Budd	9513	9667
LAUREL STREAM NYC	S	6DbrBLng	3-5/49	Budd	9505	9663
LAWN LAKE CRI&P	S	10Sec4Rmt	10/39	PS	4088	6586
LEBANON VALLEY NYC	S	10Rmt6Dbr	1-3/49	Budd	9502	9662
LEHIGH RAPIDS PRR	S	10Rmt6Dbr	3-4/49	PS	4129	6792
LEO FCP	SN	10Rmt6Dbr	9/48-/49	PS	4123	6790
LEONOR LEE KCS	S	14Rmt4Dbr	5/48	PS	4153	6795
LEWIS & CLARK PASS GN-1384	S	6Rmt5Dbr2Cpt	10-11/50	PS	4180	6877

CAR NAME OR NUMBER	ST	CAR TYPE	DATE BUILT	CAR BLDR	PLAN	LOT
LEWISTOWN INN PRR-8267	S	21Rmt	1-6/49	Budd	9513	9667
LEWISTOWN NARROWS PRR	S	2DrCptDbrObsBLng	5/38	PS	4081	6549
LIBANO NdeM-539	SN	10Rmt6Dbr	9/48-/49	PS	4123	6790
LIBERIA NdeM-578	SN	10Rmt6Dbr	12/48-49	Budd	9510	9660
LIBIA NdeM-579	SN	10Rmt6Dbr	8-9/48	Budd	9504	9660
LIBRA FCP	SN	10Rmt6Dbr	9/48-/49	PS	4123	6790
LICKING RIVER NYC	S	10Rmt6Dbr	9/48-/49	PS	4123	6790
LIGHT PINE L&N-3456	S	6Sec6Rmt4Dbr	3-5/53	PS	4183	6909
LIGONIER						
RAPIDS PRR-8430	N	10Rmt6Dbr	8-10/50	AC&F	9008	3200
LIMA INN PRR-8268	S	21Rmt	1-6/49	Budd	9513	9667
LINCOLN PARK AMT-2561	SN	8Rmt6Dbr	11-12/54	PS	4195	6944
LINCOLN PASS GN-1166	S	8DuRmt4Dbr4Sec	12/46-47	PS	4107	6751
LINDEN FALLS PRR	S	6DbrBLng	3-5/49	PS	4131	6792
LITTLE						
FOX RIVER NYC	S	10Rmt6Dbr	9/48-/49	PS	4123	6790
MIAMI RAPIDS PRR	S	10Rmt6Dbr	3-4/49	PS	4129	6792
MIAMI RIVER NYC	S	10Rmt6Dbr	9/48-/49	PS	4123	6790
NECK BAY NYC	S	22Rmt	10/48	PS	4122	6790
OSAGE RIVER NYC	S	10Rmt6Dbr	9/48-/49	PS	4123	6790
LOBLOLLY PINE C&EI-902	S	6Sec6Rmt4Dbr	3-5/53	PS	4183	6909
LOCH						
ARKAIG NP-335	SN	24Sr8Dbr	12/59	Budd	9540	9691
AWE NP-336	SN	24Sr8Dbr	12/59	Budd	9540	9691
KATRINE NP-330	SN	24SR8Dbr	12/59	Budd	9540	9691
LEVEN NP-326	L	24Sr8Dbr	12/59	Budd	9540	9691
LOCHY NP-332	SN	24Sr8Dbr	12/59	Budd	9540	9691
LOMOND NP-327	L	24Sr8Dbr	12/59	Budd	9540	9691

Figure I-113   LOCH LEVEN, Northern Pacific 326, was built by Budd in 1959. The Budd Company had the final word over The Pullman Company when it constructed this last lot of lightweight sleeping cars. Ironically, these last sleepers utilized Pullman's duplex room ideas of the mid-1930s. LOCH LEVEN was leased to the Northern Pacific until 1964 when it was purchased by the railroad. This photograph from the Collection of Howard Ameling was taken in Chicago in June 1970 prior to its sale to Amtrak.

CAR NAME OR NUMBER	ST	CAR TYPE	DATE BUILT	CAR BLDR	PLAN	LOT
LOCH						
LONG NP-331	SN	24Sr8Dbr	12/59	Budd	9540	9691
NESS NP-328	L	24Sr8Dbr	12/59	Budd	9540	9691
RANNOCH NP-334	SN	24Sr8Dbr	12/59	Budd	9540	9691
SLOY NP-325	L	24Sr8Dbr	12/59	Budd	9540	9691
TARBET NP-329	SN	24Sr8Dbr	12/59	Budd	9540	9691
TAY NP-333	SN	24Sr8Dbr	12/59	Budd	9540	9691
LOCK HAVEN INN PRR-8269	S	21Rmt	1-6/49	Budd	9513	9667
LOCUST FALLS PRR	S	6DbrBLng	9/40	PS	4086A	6612
LOGAN PASS GN-1164	S	8DuRmt4Dbr4Sec	12/46-47	PS	4107	6751
LOGANSPORT INN PRR-8270	S	21Rmt	1-6/49	Budd	9513	9667
LONESOME PINE L&N-3457	S	6Sec6Rmt4Dbr	3-5/53	PS	4183	6909
LONG LEAF PINE L&N-3458	S	6Sec6Rmt4Dbr	3-5/53	PS	4183	6909
LONG POINT NH	S	14Rmt4Dbr	12/49/50	PS	4159	6822
LOOKOUT POINT NH	S	14Rmt4Dbr	12/49/50	PS	4159	6822
LOPEZ COLLADA FCS-BC	SN	14Rmt4Dbr	2-7/48	PS	4153A	6758
LORAIN COUNTY NYC	S	13Dbr	10/40	PS	4071C	6618
LORDS POINT NH	S	14Rmt4Dbr	12/49/50	PS	4159	6822
LORNE MANOR CP-10333	S	5DbrCpt4Sec4Rmt	11/54-55	Budd		9658
LOS FELIZ JT	S	4Dbr4Cpt2Dr	11/40	PS	4069F	6636

CAR NAME OR NUMBER	ST	CAR TYPE	DATE BUILT	CAR BLDR	PLAN	LOT
LOUISIANA PINE L&N-3459	S	6Sec6Rmt4Dbr	3-5/53	PS	4183	6909
LOUVETTI GREEN JonesProp	SN	10Rmt6Dbr	12/48-49	Budd	9510	9660
LOYALHANNA						
RAPIDS PRR-843	N	10Rmt6Dbr	8-10/50	AC&F	9008	3200
LOYALSOCK						
RAPIDS PRR-8434	N	10Rmt6Dbr	8-10/50	AC&F	9008	3200
LUCAS COUNTY NYC	S	13Dbr	4/38	PS	4071A	6541
LUTHER						
CALVIN NORRIS SOU	N	5DbrREObsBLng	2-3/50	PS	4162	6814
LUXEMBURGO NdeM-580	SN	10Rmt6Dbr	8-9/48	Budd	9504	9660
LYCOMING						
RAPIDS PRR-8440	N	10Rmt6Dbr	8-10/50	AC&F	9008	3200
LaGRANDE UP	S	5DbrBLng	4/56	PS	4199	6959
MABOU RIVER CN-2077	SN	10Rmt6Dbr	9/48-/49	PS	4123	6790
MacKENZIE						
MANOR CP-10335	S	5DbrCpt4Sec4Rmt	11/54-55	Budd		9658
MACKINAC HARBOR NYC	S	22Rmt	4-6/49	Budd	9501	9661
MACKINAW RAPIDS PRR	S	10Rmt6Dbr	11/48-49	PS	4140	6792
MADISON RIVER Milw-19	S	8DuRmt6Rmt4Dbr	10-12/48	PS	4135	6775
MADRID NdeM-639	SN	10Rmt6Dbr	6-8/49	Budd	9503	9662

Figure I-114 This is 1 of 29 6-section, 6-roomette, 4-double-bedroom cars built by Pullman in 1953 for the L&N system. The incongruous presence of open sections in these late lightweight sleepers was mandated by United States Government travel allowances, which would only reimburse for sleeper section space. It was withdrawn from Pullman lease in May 1969, sold to Edwards International in 1970, and it is assumed that this car was then sold to the NdeM. The Howard Ameling Collection.

CAR NAME OR NUMBER	ST	CAR TYPE	DATE BUILT	CAR BLDR	PLAN	LOT
MAGIC BROOK PRR	S	12DuSr5Dbr	6/39	PS	4066B	6566
MAGNOLIA FEC	S	6DbrBLng	11-12/49	AC&F	9002	3095
MAGNOLIA GARDENS ACL	RN	11Dbr	9-10/49	PS	6007	6809
MAGNOLIA STATE IC	S	6Sec6Rmt4Dbr	4/42	PS	4099	6669
MAHONE CN-1900	SN	8DbrSKYTOPLng	11/48-49	PS	4138	6775
MAHONING B&O	S	14Rmt4Dbr	4-6/48	PS	4153B	6776
MAHONING						
RAPIDS PRR-8448	S	10Rmt6Dbr	9-10/50	AC&F	9008	3212
RIVER NYC	S	10Rmt6Dbr	9/48-/49	PS	4123	6790
MAHOPAC FALLS NYC	S	6DbrBLng	8-9/40	PS	4086B	6612
MAIDEN BROOK PRR	S	12DuSr5Dbr	4-5/38	PS	4066B	6538
MAITO ATSF	S	17RmtSec	1-2/38	PS	4068B	6532
MAJOR BROOK PRR	S	12DuSr5Dbr	4-5/38	PS	4066B	6538
MALI(1st) NdeM-558	SN	12Dbr	4-5/49	PS	4125	6790
MALI(2nd) NdeM-560	SN	10Rmt6Dbr	12/48-49	Budd	9510	9660
MALPEQUE CN-1901	SN	8DbrSKYTOPLng	11/48-49	PS	4138	6775
MAMMONASSET						
BEACH NH-530	S	6Sec6Rmt4Dbr	11/54-55	PS	4194	W6942
MANATEE RIVER ACL	RN	4Cpt4Dr	3-4/38	PS	6013	6540
MANATEE RIVER ACL	S	14Rmt2Dr	2-3/50	AC&F	9007	3091
MANHASSET BAY NYC	S	22Rmt	10/48	PS	4122	6790
MANHATTAN ISLAND NYC	S	DrMbrObsBLng	5/38	PS	4079	6547
MANISTEE RAPIDS PRR	S	10Rmt6Dbr	3-4/49	PS	4129	6792
MANISTEE RIVER NYC	S	10Rmt6Dbr	9/48-/49	PS	4123	6790
MANITOBA CLUB GN-1198	RN	8uRmt2DbrBLng	1-2/47	PS	4108B	6751
MANITOU CN-1701	SN	4Rmt5DbrCpt4Sec	12/54	PS	4196	6949
MANOMET POINT NH	S	14Rmt4Dbr	12/49/50	PS	4159	6822
MANOR BROOK PRR	S	12DuSr5Dbr	6/39	PS	4066B	6566
MANSFIELD INN PRR-8271	S	21Rmt	1-6/49	Budd	9513	9667
MANY GLACIER GN-1174	S	16DuRmt4Dbr	1-2/47	PS	4108	6751
MAPLE BROOK PRR	S	12DuSr5Dbr	4-5/38	PS	4066B	6538
MAPLE FALLS PRR	S	6DbrBLng	9/40	PS	4086A	6612
MAPLE GROVE CP-14109		10Sec5Dbr	/49-/50	CPRR		
MAPLE VALLEY NYC	S	10Rmt6Dbr	1-3/49	Budd	9502	9662
MAR						
ADRIATICO NdeM	P	10Rmt5Bdr	4/38	PS	4072A	6542
AMARILLO NdeM-200	P	10Rmt5Dbr	6/39	PS	4072B	6565
BALTICO NdeM-190	P	10Rmt5Bdr	4/38	PS	4072A	6542
BLANCO NdeM	P	10Rmt5Bdr	4/38	PS	4072A	6542
BROOK PRR	S	12DuSr5Dbr	4-5/38	PS	4066B	6538
CASPIO NdeM-191	SN	10Rmt5Bdr	4/38	PS	4072A	6542
JONICO NdeM	SN	10Rmt5Bdr	4/38	PS	4072A	6542
MEDITTERANEO NdeM	SN	10Rmt5Bdr	4/38	PS	4072A	6542
NEGRO NdeM	SN	10Rmt5Dbr	6-7/40	PS	4072C	6610
ROJO NdeM	SN	10Rmt5Bdr	4/38	PS	4072A	6542
TIRRENO NdeM	SN	10Rmt5Bdr	4/38	PS	4072A	6542
MAR DE						
MARMARA NdeM-206	SN	10Rmt5Dbr	6-7/40	PS	4072C	6610
NORUEGA NdeM	SN	10Rmt5Dbr	6-7/40	PS	4072C	6610
TIMOR NdeM-202	SN	10Rmt5Dbr	6-7/40	PS	4072C	6610

CAR NAME OR NUMBER	ST	CAR TYPE	DATE BUILT	CAR BLDR	PLAN	LOT
MAR DE						
BERING NdeM-208	SN	10Rmt5Bdr	4/38	PS	4072A	6542
CHINA NdeM-199	SN	10Rmt5Dbr	6/39	PS	4072B	6565
CRETA NdeM	SN	10Rmt5Bdr	4/38	PS	4072A	6542
FILIPINAS NdeM-201	SN	10Rmt5Dbr	6-7/40	PS	4072C	6610
GROENLANDIA NdeM-211	SN	10Rmt5Dbr	6-7/40	PS	4072C	6610
IRLANDA NdeM	SN	10Rmt5Bdr	4/38	PS	4072A	6542
JAVA NdeM-210	SN	10Rmt5Dbr	6-7/40	PS	4072C	6610
LIGURIA NdeM-203	SN	10Rmt5Dbr	6-7/40	PS	4072C	6610
MAR DEL						
CORAL NdeM-192	SN	10Rmt5Bdr	4/38	PS	4072A	6542
JAPON NdeM-205	SN	10Rmt5Dbr	6/39	PS	4072B	6565
NORTE NdeM-197	SN	10Rmt5Dbr	6/39	PS	4072B	6565
PLATA NdeM-204	SN	10Rmt5Dbr	6-7/40	PS	4072C	6610
MARBLE CREEK Milw-17	S	8DbrSKYTOPLng	11/48-49	PS	4138	6775
MARGAREE RIVER CN-2076	SN	10Rmt6Dbr	9/48-/49	PS	4123	6790
MARIAS RIVER CB&Q-1194	S	2DbrDrObsBLng	1-2/47	PS	4109	6751
MARIETTA INN PRR-8272	S	21Rmt	1-6/49	Budd	9513	9667
MARION COUNTY ACL	S	10Rmt6Dbr	9-10/49	PS	4140B	6809
MARSH BROOK PRR	S	12DuSr5Dbr	9/40	PS	4066C	6611
MARTE NdeM-590	SN	4Sec8DuRmt4Dbr	6/51	PS	4107	6895
MARTIN BROOK PRR	S	12DuSr5Dbr	6/39	PS	4066B	6566
MARVIN KENT Erie	S	10Rmt6Dbr	5-6/49	PS	4129A	6797
MARY						
ELIZABETH JonesProp	SN	4Dbr4Cpt2Dr	7-8/39	PS	4069D	6571
MASSILLON INN PRR-8273	S	21Rmt	1-6/49	Budd	9513	9667
MATTINICOCK LIRR-2065	SN	Parlor/Coach	12/49/50	PS	4159	6822
MATTITUCK LIRR-2066	SN	Parlor/Coach	12/49/50	PS	4159	6822
MATUNUCK BEACH NH-531	S	6Sec6Rmt4Dbr	11/54-55	PS	4194	W6942
MAUMEE RIVER B&O	P	2DbrCptDrObsBLng	7/39	PS	4082	6567
MAUMEE RIVER NYC	S	2DbrCptDrObsBLng	7/39	PS	4082	6567
MAUMEE VALLEY NYC	S	10Rmt6Dbr	1-3/49	Budd	9502	9662
MAY BROOK PRR	S	12DuSr5Dbr	6/39	PS	4066B	6566
Mc DOWELL COUNTY N&W	S	10Rmt6Dbr	3/49	PS	4140	6792
MEADOW BROOK PRR	S	12DuSr5Dbr	4-5/38	PS	4066B	6538
MEADOW VALLEY NYC	S	10Rmt6Dbr	1-3/49	Budd	9502	9662
MECOX LIRR-2067	SN	Parlor/Coach	12/49/50	PS	4159	6822
MEMORY BROOK PRR	S	12DuSr5Dbr	9/40	PS	4066C	6611
MEMPHIS IC	S	2Dbr2CptDrObsLng	4/42	PS	4102	6672
MERCER COUNTY N&W	S	10Rmt6Dbr	11/49-50	Budd	9523	9660
MERCURIO NdeM-591	SN	4Sec8DuRmt4Dbr		PS	4107?	
MEREMEC RIVER SLSF-1457	S	14Rmt4Dbr	1-6/48	PS	4153	6769
MERRICK LIRR-2068	SN	Parlor/Coach	12/49/50	PS	4159	6822
MERRIMACK RIVER NYC	S	10Rmt6Dbr	9/48-/49	PS	4123	6790
MESCALERO VALLEY ATSF	S	6Sec6Rmt4Dbr	6/42	PS	4099	6669
METALINE						
FALLS Milw-5774		8SecCoachTouralux	6/27/47	Milw		
METCALF B&O	SN	5DbrBEObsBLng	8/50	PS	4165	6863
METROPOLITAN VIEW PRR	S	2MbrDbrREObsBLng	5/38	PS	4080	6548
MEXICALI FS-BC	SN	10Rmt6Dbr	2-7/50	PS	4167	6864

Figure I-115 MARSH BROOK, in tuscan red, represents 1 of the 4 cars built in the third lot of this design delivered to the Pennsylvania Railroad in 1940. Note that the double-bedroom windows at the ends of the car are smaller than the later version CLUB CREEK (See page 265) and the vestibule is on the opposite end. It has depressed-center sides in common with the 1937 riveted aluminum car SP-9251 (See page 248) and does not have the PRR belt rail as does CLUB CREEK. The Wilber C. Whittaker Collection.

CAR NAME OR NUMBER	ST	CAR TYPE	DATE BUILT	CAR BLDR	PLAN	LOT
MEXICO BAY NYC	S	22Rmt	10/48	PS	4122	6790
MGRS-76 NdeM	SN	4DbrBLngObs	5/38	PS	4079A	6547
MIAMI SAL-43	S	10Rmt6Dbr	6-8/49	Budd	9503	9662
MIAMI BEACH SAL-18	S	5DbrBLng	1/56	PS	4202	6970
MIAMI BEACH SCL-6601	SN	6DbrBLng	8/49	AC&F	9003	3045
MIAMI RIVER NYC	S	10Rmt6Dbr	9/48-/49	PS	4123	6790
MICHIGAN						
CITY HARBOR NYC	S	22Rmt	4-6/49	Budd	9501	9661
MIDDLE BROOK PRR	S	12DuSr5Dbr	6/39	PS	4066B	6566
MIDDLE RIVER PRR	S	10Rmt6Dbr	9-10/49	PS	4140	6814
MILK RIVER GN-1263	S	8DuRmt4Sec3DbrCpt	11-12/50	PS	4181	6889
MILK RIVER GN-1263	R	8DuRmt6DbrCpt	11-12/50	PS	4181A	6889
MILL BROOK PRR	S	12DuSr5Dbr	4-5/38	PS	4066B	6538
MINEOLA LIRR-2069	SN	Parlor/Coach	12/49/50	PS	4159	6822
MINERAL BROOK PRR	S	12DuSr5Dbr	4-5/38	PS	4066B	6538
MINGO COUNTY N&W	S	10Rmt6Dbr	11/49-50	Budd	9523	9660
MINNEAPOLIS CLUB CB&Q	S	5DbrREObsBLng	6-7/48	PS	4120	6781
MINNESOTA RIVER Milw-31	S	8DuRmt6Rmt4Dbr	11/54	PS	4192A	6945
MINNESOTA VALLEY NYC	S	10Rmt6Dbr	12/48-49	Budd	9510	9660
MIRROR BROOK PRR	S	12DuSr5Dbr	4-5/38	PS	4066B	6538
MIRROR LAKE NYC	S	6DbrBLng	12/48-49	PS	4124	6790
MISHAUM POINT NH	S	14Rmt4Dbr	12/49/50	PS	4159	6822
MISSISSIPPI						
PINE L&N-3460	S	6Sec6Rmt4Dbr	3-5/53	PS	4183	6909
RAPIDS PRR-8445	S	10Rmt6Dbr	9-10/50	AC&F	9008	3212
RIVER GN-1190	S	2DbrDrObsBLng	1-2/47	PS	4109	6751
VALLEY SOU	S	14Rmt4Dbr	10-11/49	PS	4153C	6814
MISSOULA NP	S	8DuRmt6Rmt4Dbr	7-9/48	PS	4119	6781
MISSOURI RIVER GN-1191	S	2DbrDrObsBLng	1-2/47	PS	4109	6751
MISSOURI VALLEY NYC	S	10Rmt6Dbr	12/48-49	Budd	9510	9660
MITLA NdeM-216	L	8Sec3Cpt	/52	Schi	100570	
MOBILE BAY NYC	S	22Rmt	10/48	PS	4122	6790
MOBILE RIVER L&N	S	10Rmt6Dbr	9-10/49	PS	4140	6814
MOCKINGBIRD B&O	S	16DuRmt4Dbr	4-6/54	Budd	9536	9658
MOENCOPI ATSF	S	4Dbr4Cpt2Dr	1/38	PS	4069A	6532
MOHAVE ATSF	S	4Dbr4Cpt2Dr	1/38	PS	4069A	6532
MOHAWK VALLEY NYC	S	10Rmt6Dbr	12/48-49	Budd	9510	9660
MONCK MANOR CP-10336	S	5DbrCpt4Sec4Rmt	11/54-55	Budd		9658
MONOCACY B&O	S	14Rmt4Dbr	4-6/48	PS	4153B	6776
MONOGAHELA RIVER NYC	S	10Rmt6Dbr	9/48-/49	PS	4123	6790
MONOMOY POINT NH	S	14Rmt4Dbr	12/49/50	PS	4159	6822
MONONGAHELA						
RAPIDS PRR	S	10Rmt6Dbr	3-4/49	PS	4129	6792
MONROE COUNTY NYC	S	13Dbr	4/38	PS	4071A	6541
MONROE HARBOR NYC	S	22Rmt	4-6/49	Budd	9501	9661
MONTANA CLUB NP	S	5DbrREObsBLng	6-7/48	PS	4120	6781
MONTAUK LIRR-2070	SN	Parlor/Coach	12/49/50	PS	4159	6822
MONTE						
ACONOGUA NdeM	P	6Sec6Rmt4Dbr	3-5/53	PS	4183	6909
ALBAN NdeM-217	L	8Sec3Cpt	/52	Schi	100570	
ANNATURNA NdeM-609	P	6Sec6Rmt4Dbr	3-5/53	PS	4183	6909
AZUL NdeM-336	SN	6Sec6Rmt4Dbr	3-4/50	AC&F	9005	3073
BLANCO NdeM-284	SN	6Sec6Rmt4Dbr	5-6/42	PS	4099	6669
CAUCASO NdeM-285	SN	6Sec6Rmt4Dbr	5-6/42	PS	4099	6669
IRAZU NdeM-286	SN	6Sec6Rmt4Dbr	5-6/42	PS	4099	6669
KILIMANJARO NdeM-613	P	6Sec6Rmt4Dbr	3-5/53	PS	4183	6909
LIBANO NdeM-287	SN	6Sec6Rmt4Dbr	5-6/42	PS	4099	6669
McKINLEY NdeM-615	P	6Sec6Rmt4Dbr	3-5/53	PS	4183	6909

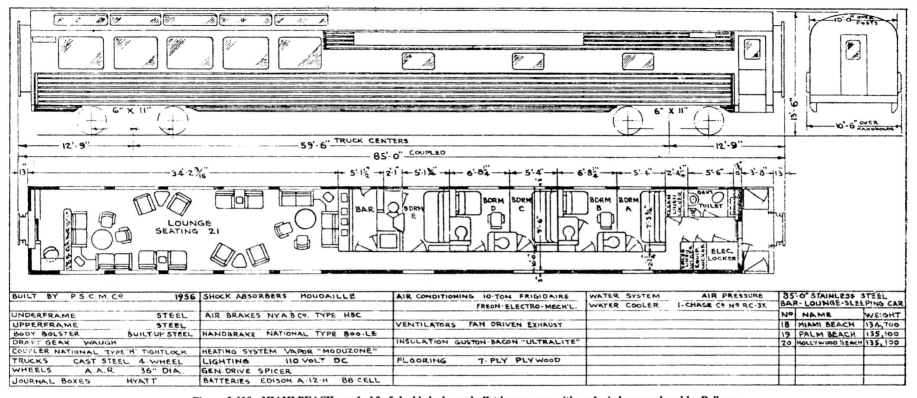

BUILT BY P.S.C.M.Cº		1956	SHOCK ABSORBERS HOUDAILLE	AIR CONDITIONING 10-TON FRIGIDAIRE	WATER SYSTEM	AIR PRESSURE	85'-0" STAINLESS STEEL BAR-LOUNGE-SLEEPING CAR		
				FREON·ELECTRO-MECH'L.	WATER COOLER	1-CHASE Cº Nº RC-3X			
UNDERFRAME	STEEL		AIR BRAKES N.Y.A.B Cº. TYPE HSC				Nº	NAME	WEIGHT
UPPERFRAME	STEEL			VENTILATORS FAN DRIVEN EXHAUST			18	MIAMI BEACH	134,700
BODY BOLSTER	BUILT-UP STEEL		HANDBRAKE NATIONAL TYPE 800-LE				19	PALM BEACH	135,100
DRAFT GEAR WAUGH				INSULATION GUSTON-BACON "ULTRALITE"			20	HOLLYWOOD BEACH	135,100
COUPLER NATIONAL TYPE "H" TIGHTLOCK			HEATING SYSTEM VAPOR "MODUZONE"						
TRUCKS CAST STEEL 4-WHEEL			LIGHTING 110 VOLT DC	FLOORING 7-PLY PLYWOOD					
WHEELS A.A.R. 36" DIA.			GEN. DRIVE SPICER						
JOURNAL BOXES HYATT			BATTERIES EDISON A-12-H 88-CELL						

Figure I-116 **MIAMI BEACH was 1 of 3, 5-double-bedroom buffet lounge cars with roof windows produced by Pullman Standard in 1956 becoming Seaboard Coast Line SUN VIEW in 1967. These unique, much talked about, all-stainless cars maximized sunlight, as befitted the Florida trains. The sunroofs also created a greenhouse effect that the car's A/C system could not rectify. Amtrak purchased it in 1971 and it was retired in 1971, thence sold to Mike Kenelea. The Author's Collection.**

CAR NAME OR NUMBER	ST	CAR TYPE	DATE BUILT	CAR BLDR	PLAN	LOT
MONTE						
OLIMPO NdeM-288	SN	6Sec6Rmt4Dbr	5-6/42	PS	4099	6669
PARNASO NdeM-289	SN	6Sec6Rmt4Dbr	5-6/42	PS	4099	6669
SINAI NdeM-290	SN	6Sec6Rmt4Dbr	5-6/42	PS	4099	6669
VERDE NdeM-337	SN	6Sec6Rmt4Dbr	3-4/50	AC&F	9005	3073
MONTEREY BAY NYC	S	22Rmt	10/48	PS	4122	6790
MONTGOMERY COUNTY NYC	S	13Dbr	10/40	PS	4071C	6618
MONTICELLO C&O		11Dbr	6/50	PS	4168	6865
MONUMENT BEACH NH-536	S	6Sec6Rmt4Dbr	11/54-55	PS	4194	W6942
MONUMENT VALLEY ATSF	S	6Sec6Rmt4Dbr	6/42	PS	4099	6669
MOONLIGHT DOME B&O(7600)	SN	Dome5RmtDbr3Dr	9/48	Budd	9524	9669
MOOR BROOK PRR	S	12DuSr5Dbr	6/39	PS	4066B	6566
MOOSE RIVER CN-2132	SN	10Rmt6Dbr	9-10/49	PS	4140	6814
MORGAN POINT NH	S	14Rmt4Dbr	12/49/50	PS	4159	6822
MORICHES LIRR-2071	SN	Parlor/Coach	12/49/50	PS	4159	6822

CAR NAME OR NUMBER	ST	CAR TYPE	DATE BUILT	CAR BLDR	PLAN	LOT
MORMON TRAIL (DS)UP		7Dbr2CptArtic	5/36	PS	4039D	6434
MORNING BROOK PRR	S	12DuSr5Dbr	4-5/38	PS	4066B	6538
MORNING VALLEY NYC	S	10Rmt6Dbr	12/48-49	Budd	9510	9660
MORROW BROOK PRR	S	12DuSr5Dbr	9/40	PS	4066C	6611
MORRUECOS NdeM-533	SN	10Rmt6Dbr	9/48-/49	PS	4123	6790
MOSES CLEVELAND NKP	N	18Rmt	8/37	PS	4068	6526
MOUNT						
ALBREDA CN-1102	S	5Cpt3Dr	6/54	PS	4187A	6926
EDITH CAVELL CN-1100	S	5Cpt3Dr	6/54	PS	4187A	6926
FITZWILLIAM CN-1103	S	5Cpt3Dr	6/54	PS	4187A	6926
HOOD SP&S	S	6Rmt3DbrBLng	2/50	PS	4163	6829
RESPLENDENT CN-1104	S	5Cpt3Dr	6/54	PS	4187A	6926
ROBSON CN-1101	S	5Cpt3Dr	6/54	PS	4187A	6926
ST. HELENS SP&S	S	6Rmt3DbrBLng	2/50	PS	4163	6829
TEKARRA CN-1105	S	5Cpt3Dr	6/54	PS	4187A	6926

CAR NAME OR NUMBER	ST	CAR TYPE	DATE BUILT	CAR BLDR	PLAN	LOT
MOUNTAIN PINE L&N-3461	S	6SecRmt4Dbr	3-5/53	PS	4183	6909
MOUNTAIN STREAM NYC	S	6DbrBLng	3-5/49	Budd	9505	9663
MOUNTAIN VIEW CP	SN	5DbrBLngObs	6-7/49	Budd	9508	9636
MOUNTAIN VIEW PRR	S	2MbrDbrREObsBLng	1/49	PS	4133	6792
MOUSE RIVER GN-1268	S	8DuRmt4Sec3DbrCpt	11-12/50	PS	4181	6889
MT.ANGELES Milw-5751		14SecTourist	6/19/47	Milw		
MT.BOSLEY Milw-5743		14SecTourist	6/27/47	Milw		
MT.CHITTENDEN Milw-5752	RN	14SecTourist	6/9/47	Milw		
MT.HAROLD Milw-5750		14SecTourist	6/27/47	Milw		
MT.HOPE Milw-5748		14SecTourist	6/19/47	Milw		
MT.JUPITER Milw-5753	RN	14SecTourist	6/23/47	Milw		
MT.McKINLEY Milw-5742		14SecTourist	6/23/47	Milw		
MT.RAINIER Milw-5744		14SecTourist	6/19/47	Milw		
MT.RUSHMORE Milw-5745		14SecTourist	6/19/47	Milw		
MT.SPOKANE Milw-5740		14SecTourist	6/28/47	Milw		
MT.ST.HELENS Milw-5746		14SecTourist	6/19/47	Milw		
MT.STUART Milw-5749		14SecTourist	6/23/47	Milw		
MT.TACOMA Milw-5754	RN	14SecTourist	6/19/47	Milw		
MT.VERNON C&O	S	11Dbr	6/50	PS	4168	6865
MT.WASHINGTON Milw-5741		14SecTourist	6/28/47	Milw		
MT.WILSON Milw-5747		14SecTourist	6/19/47	Milw		
MUSCATATUCK B&O	S	10Rmt6Dbr	3/50	PS	4167	6864
MUSKEGON RAPIDS PRR	S	10Rmt6Dbr	3-4/49	PS	4129	6792
MUSKINGUM B&O	S	14Rmt4Dbr	4-6/48	PS	4153B	6776
MUSKINGUM RIVER NYC	S	10Rmt6Dbr	9/48-/49	PS	4123	6790
MUSKINGUM RIVER Pool	S	2DbrCptDrREObsLng	6/40	PS	4082	6608
MYRTLE BEACH ACL	S	6DbrBLng	11-12/49	AC&F	9002	3090
MYSTIC BROOK PRR	S	12DuSr5Dbr	9/40	PS	4066C	6611
MYSTIC RIVER NYC	S	10Rmt6Dbr	9/48-/49	PS	4123	6790

CAR NAME OR NUMBER	ST	CAR TYPE	DATE BUILT	CAR BLDR	PLAN	LOT
MacDONALD MANOR CP-10334	S	5DbrCpt4Sec4Rmt	11/54-55	Budd		9658
NABOR FLORES FCS-BC	SN	14Rmt4Dbr	2-7/48	PS	4153A	6758
NAHANT BAY NYC	S	22Rmt	10/48	PS	4122	6790
NAISCOOT RIVER CN-2141	SN	10Rmt6Dbr	9-10/49	PS	4140B	6809
NANTASKET BEACH NH-532	S	6Sec6Rmt4Dbr	11/54-55	PS	4194	W6942
NAPATREE POINT NH	S	14Rmt4Dbr	12/49/50	PS	4159	6822
NAPEAGUE LI 2059	P	Parlor/Coach	6-7/39	PS	4071B	6572
NAPPANEE B&O	SN	5DbrBEObsBLng	8/50	PS	4165	6863
NASH COUNTY ACL	S	10Rmt6Dbr	9-10/49	PS	4140B	6809
NASHWEAK RIVER CN-2134	SN	10Rmt6Dbr	9-10/49	PS	4140	6814
NASLINI ATSF	S	4Dbr4Cpt2Dr	1/38	PS	4069A	6532
NASSAU FEC	S	4Rmt5DbrCpt4Sec	12/54	PS	4196	6949
NASSAU COUNTY ACL	S	10Rmt6Dbr	9-10/49	PS	4140B	6809
NATIONAL BORDER UP	S	6Sec6Rmt4Dbr	11/55-56	PS	4197	6957
COLORS WAB	S	6Sec4Dbr6Rmt	12/55	PS	4197	6966
COMMAND UP	S	6Sec6Rmt4Dbr	11/55-56	PS	4197	6957
CONSOLATE UP	S	6Sec6Rmt4Dbr	11/55-56	PS	4197	6957
DOMAIN UP	S	6Sec6Rmt4Dbr	11/55-56	PS	4197	6957
EMBASSY UP	S	6Sec6Rmt4Dbr	11/55-56	PS	4197	6957
EMBLEM UP	S	6Sec6Rmt4Dbr	11/55-56	PS	4197	6957
FORUM UP	S	6Sec6Rmt4Dbr	11/55-56	PS	4197	6957
FRONTIER UP	S	6Sec6Rmt4Dbr	11/55-56	PS	4197	6957
HOMES WAB	S	6Sec4Dbr6Rmt	12/55	PS	4197	6966
PROGRESS UP	S	6Sec6Rmt4Dbr	11/55-56	PS	4197	6957
SCENE UP	S	6Sec6Rmt4Dbr	11/55-56	PS	4197	6957
SHORES UP	S	6Sec6Rmt4Dbr	11/55-56	PS	4197	6957
UNITY WAB	S	6Sec4Dbr6Rmt	12/55	PS	4197	6966

Figure I-117 The Baltimore & Ohio Railroad received from The Pullman Company in 1950 10 10-roomette, 6-double-bedroom sleepers named for rivers in B&O territory. These cars were originally ordered by the C&O in 1946, obvious by the window configuration indicating roomettes at either end with bedrooms in the center. MUSCATATUCK was sold to Ringling Bros., Barnum & Bailey Combined Shows, Inc., in October 1970, and became RBBB 140 in 1972 on The Blue Unit. The B&O Railroad Historical Society Collection.

Figure I-118 NATIONAL BORDER, a 6-section, 6-roomette, 4-double-bedroom was built in the last year of Pullman sleeper production. It is somewhat surprising that section accommodations were included in this lot of cars as late as 1956, and even more so that there would be upper berth windows. UP, like Pennsy, used belt rails on lightweight equipment, and ordered riveted roof sheets rather than the standard welded roofs. This photograph, taken in August 1976 by Wilber C. Whittaker at Sonora, California shows the markings of its new owner, Great Western Tours.

CAR NAME OR NUMBER	ST	CAR TYPE	DATE BUILT	CAR BLDR	PLAN	LOT
NATIONAL VIEW UP	S	6Sec6Rmt4Dbr	11/55-56	PS	4197	6957
NATURAL BRIDGE C&O	S	11Dbr	6/50	PS	4168	6865
NAVA ATSF	S	4Dbr4Cpt2Dr	12/39	PS	4069C	6597
NAVAJO ATSF	S	3Cpt2DrDbrLngObs	4/37	Budd	9518	967
NAVAJO VALLEY NYC	S	10Rmt6Dbr	12/48-49	Budd	9510	9660
NECHAKO RIVER CN-2151	SN	10Rmt6Dbr	5-6/49	PS	4129A	6797
NEOSHO RIVER SLSF-1464	S	14Rmt4Dbr	1-6/48	PS	4153	6769
NEPAL NdeM-528	SN	10Rmt6Dbr	9/48-/49	PS	4123	6790
NEPONSET RIVER NYC	S	10Rmt6Dbr	9/48-/49	PS	4123	6790
NEPTUNO NdeM-592	SN	8DuRmt6Rmt4Dbr	12/46-47	PS	4107	6751
NESCONSET LI-2060	P	Parlor/Coach	6-7/39	PS	4071B	6572
NESHANNOCK RIVER NYC	S	10Rmt6Dbr	9/48-/49	PS	4123	6790
NETZAHUALCOYOTL FCS-BC	SN	6DbrBLng	3-5/49	Budd	9505	9663
NEW						
CASTLE INN PRR-8274	S	21Rmt	1-6/49	Budd	9513	9667
LISBON Milw-28	S	16DuRmt4Dbr	1-2/49	PS	4147D	6775
RIVER CLUB C&O	S	5DbrBEObsBLng	8/50	PS	4165	6863
YORK BAY NYC	S	22Rmt	10/48	PS	4122	6790
YORK COUNTY PRR	S	13Dbr	4/38	PS	4071A	6541
NIAGARA COUNTY NYC	S	13Dbr	4/38	PS	4071A	6541
NIAGARA RIVER NYC	S	10Rmt6Dbr	9/48-/49	PS	4123	6790
NIANGUA RIVER SLSF-1460	S	14Rmt4Dbr	1-6/48	PS	4153	6769
NIANTIC RIVER NYC	S	10Rmt6Dbr	9/48-/49	PS	4123	6790
NICARAGUA NdeM-581	SN	10Rmt6Dbr	8-9/48	Budd	9504	9660
NICHOLAS FIRESTONE PRR	N	3DbrDrBLng	5-6/49	PS	4132	6792
NIPIGON RIVER CN-2085	SN	10Rmt6Dbr	9/48-/49	PS	4123	6790
NOBLE COUNTY NYC	S	13Dbr	10/40	PS	4071C	6618
NONKOWEAP ATSF	S	4Dbr4Cpt2Dr	1/38	PS	4069A	6532
NORFOLK SAL	S	10Rmt6Dbr	5-6/49	PS	4140A	6796

CAR NAME OR NUMBER	ST	CAR TYPE	DATE BUILT	CAR BLDR	PLAN	LOT
NORRISTOWN INN PRR-8275	S	21Rmt	1-6/49	Budd	9513	9667
NORTH						
BEACH JT	S	4Dbr4Cpt2Dr	11/40	PS	4069F	6636
PLATTE UP	S	5DbrBLng	4/56	PS	4199	6959
POINT NH	S	14Rmt4Dbr	12/49/50	PS	4159	6822
RIVER NYC	S	10Rmt6Dbr	9/48-/49	PS	4123	6790
STAR CN-1095	SN	6DbrBLng	11-12/49	AC&F	9002	3095
TWIN LAKES BAR-80	S	6Sec6Rmt4Dbr	11/54-55	PS	4194	W6942
WIND CN-1096	SN	6DbrBLng	11-12/49	AC&F	9002	3095
WOODS CRI&P	S	8DuRmt6Rmt4Dbr	8-9/48	PS	4116	6761
NORTHAMPTON COUNTY ACL	S	10Rmt6Dbr	9-10/49	PS	4140B	6809
NORTHERN						
DAIRYLAND CNW	S	16DuRmtCpt3Dbr	10-11/48	PS	4146	6774
FORESTS CNW	S	16DuRmtCpt3Dbr	10-11/48	PS	4146	6774
LAKES CRI&P	S	8DuRmt6Rmt4Dbr	8-9/48	PS	4116	6761
PINES CSPM&O	S	16DuRmtCpt3Dbr	10-11/48	PS	4146	6774
PROGRESS CNW	S	16DuRmtCpt3Dbr	10-11/48	PS	4146	6774
STAR CSPM&O	S	16DuRmtCpt3Dbr	10-11/48	PS	4146	6774
STATES CSPM&O	S	16DuRmtCpt3Dbr	10-11/48	PS	4146	6774
STREAMS CNW	S	16DuRmtCpt3Dbr	10-11/48	PS	4146	6774
SUMMER CNW	S	16DuRmtCpt3Dbr	10-11/48	PS	4146	6774
TRAILS CNW	S	16DuRmtCpt3Dbr	10-11/48	PS	4146	6774
VACATIONS CNW	S	16DuRmtCpt3Dbr	10-11/48	PS	4146	6774
WILD LIFE CSPM&O	S	16DuRmtCpt3Dbr	10-11/48	PS	4146	6774
WONDERLAND CSPM&O	S	16DuRmtCpt3Dbr	10-11/48	PS	4146	6774
NORUEGA NdeM-506	SN	10Rmt6Dbr	9/48-/49	PS	4123	6790
NOYACK LI 2061	P	Parlor/Coach	6-7/39	PS	4071B	6572
NUTMEG STATE NH-552	S	6DbrBLng	1/55	PS	4193	W6941
NYC-10800	L	24Sr8Dbr	10/59	Budd	9540	9691

CAR NAME OR NUMBER	ST	CAR TYPE	DATE BUILT	CAR BLDR	PLAN	LOT
NYC-10801	L	24Sr3Dbr	10/59	Budd	9540	9691
NYC-10802	L	24Sr8Dbr	10/59	Budd	9540	9691
NYC-10803	L	24Sr8Dbr	10/59	Budd	9540	9691
OAHU (DS)UP		11SecArtic	5/36	PS	4038B	6434
OAK FALLS PRR	S	6DbrBLng	9/40	PS	4086A	6612
OAK GROVE CP-14110		10Sec5Dbr	/49-/50	CPRR		
OAK HARBOR NYC	S	22Rmt	4-6/49	Budd	9501	9661
OBERLIN GLACIER GN-1175	S	16DuRmt4Dbr	1-2/47	PS	4108	6751
OCALA SAL-70	S	11Dbr	12/55-56	PS	4198A	6968
OCEAN BEACH NH-533	S	6Sec6Rmt4Dbr	11/54-55	PS	4194	W6942
OCEAN MIST UP	S	5Dbr2Cpt2Dr	5/56	PS	4200	6960
OCEAN SANDS UP	S	5Dbr2Cpt2Dr	12/54	AC&F	9016	3815
OCEAN SCENE UP	S	5Dbr2Cpt2Dr	5/56	PS	4200	6960
OCEAN SUNSET UP	S	5Dbr2Cpt2Dr	5/56	PS	4200	6960
OCEAN VIEW UP	S	5Dbr2Cpt2Dr	12/54	AC&F	9016	3815
OCMULGEE RIVER SOU	S	10Rmt6Dbr	9-10/49	PS	4140	6814
OCTORARO RAPIDS PRR	S	10Rmt6Dbr	11/48-49	PS	4140	6792
ODESSA LAKE CRI&P	S	10Sec4Rmt	10/39	PS	4088	6586
ODGENSBURG HARBOR NYC	S	22Rmt	4-6/49	Budd	9501	9661
OGALLALA (DS)UP	S	5DbrCptREObsLng	6/36	PS	4055	6486
OGDEN UP	S	5DbrBLng	4/56	PS	4199	6959
OGEECHEE RIVER ACL	S	14Rmt2Dr	2-3/50	AC&F	9007	3091
OHIO						
RAPIDS PRR-8446	S	10Rmt6Dbr	9-10/50	AC&F	9008	3212
RIVER CLUB C&O	S	5DbrBEObsBLng	8/50	PS	4165	6863
STATE UNIVERSITY N&W	S	10Rmt6Dbr	11/49-50	Budd	9523	9660
OKEFENOKEE SCL-6401	SN	4Rmt5DbrCpt4Sec	12/54	PS	4196	6949
OLD						
ORCHARD BEACH B&M-32	S	6Sec6Rmt4Dbr	11/54-55	PS	4194	W6942
POINT COMFORT C&O	S	10Rmt6Dbr	2-7/50	PS	4167	6864

CAR NAME OR NUMBER	ST	CAR TYPE	DATE BUILT	CAR BLDR	PLAN	LOT
OLEAN INN PRR-8276	S	21Rmt	1-6/49	Budd	9513	9667
OLEANDER FEC	S	6DbrBLng	11-12/49	AC&F	9002	3095
OLIVE GROVE CP-14111		10Sec5Dbr	/49-/50	CPRR		
OLVERA JT	S	4Dbr4Cpt2Dr	11/40	PS	4069F	6636
OMAHA UP	S	5DbrBLng	4/56	PS	4199	6959
OMAN NdeM-541	SN	10Rmt6Dbr	9/48-/49	PS	4123	6790
ONEIDA COUNTY NYC	S	13Dbr	7/39	PS	4071B	6572
ONEIDA LAKE NYC	S	6DbrBLng	12/48-49	PS	4124	6790
ONONDAGA COUNTY NYC	S	13Dbr	4/38	PS	4071A	6541
ONONDAGA VALLEY NYC	S	10Rmt6Dbr	12/48-49	Budd	9510	9660
OPALE NdeM-626	SN	4Rmt5DbrCpt4Sec	3/54	PS	4196	6949
OPEQUON B&O	S	10Rmt6Dbr	3/50	PS	4167	6864
ORAIBI ATSF	S	6Dbr2Cpt2Dr	4/37	Budd	9519	967
ORANGE GROVE CP-14112		10Sec5Dbr	/49-/50	CPRR		
ORCHARD VALLEY NYC	S	10Rmt6Dbr	12/48-49	Budd	9510	9660
OREGON TRAIL (DS)UP		8SecCptDbrArtic	10/34	PS	4034A	6428
ORIENTE FEC	S	10Rmt6Dbr	9-10/49	PS	4140B	6809
ORIOLE B&O	S	16DuRmt4Dbr	4-6/54	Budd	9536	9658
ORION NdeM-593	SN	8DuRmt4Dbr4Sec	12/46-47	PS	4107	6751
ORLANDO SAL	S	10Rmt6Dbr	5-6/49	PS	4140A	6796
ORTEGA RIVER ACL	RN	4Cpt4Dr	3-4/38	PS	6013	6540
OSAGE RIVER SLSF-1458	S	14Rmt4Dbr	1-6/48	PS	4153	6769
OSCEOLA COUNTY ACL	S	10Rmt6Dbr	9-10/49	PS	4140B	6809
OSLER MANOR CP-10337	S	5DbrCpt4Sec4Rmt	11/54-55	Budd		9658
OSWEGATCHIE RIVER NYC	S	10Rmt6Dbr	9/48-/49	PS	4123	6790
OSWEGO RIVER NYC	S	10Rmt6Dbr	9/48-/49	PS	4123	6790
OTISCO LAKE NYC	S	6DbrBLng	12/48-49	PS	4124	6790
OTOWI ATSF	S	17RmtSec	1-2/38	PS	4068B	6532
OTSEGO LAKE NYC	S	6DbrBLng	12/48-49	PS	4124	6790
OTTAWA RIVER NYC	S	10Rmt6Dbr	9/48-/49	PS	4123	6790

Figure I-119 Silver trucks and yellow paint, perhaps the most difficult colors to keep clean in the railroad business, grace the OCEAN VIEW, a Union Pacific 5-double-bedroom, 2-compartment, 2-drawing room car built by American Car & Foundry in December 1954. Union Pacific obviously spent considerable time and effort to maintain these beautifully decorated cars. This car continued its colorful history when it donned white paint and purple trucks in 1971 as Auto-Train 302. Railroad Avenue Enterprises Photograph PN-2811.

CAR NAME OR NUMBER	ST	CAR TYPE	DATE BUILT	CAR BLDR	PLAN	LOT	CAR NAME OR NUMBER	ST	CAR TYPE	DATE BUILT	CAR BLDR	PLAN	LOT
OTTER							PACIFIC						
LAKE NYC	S	6DbrBLng	12/48-49	PS	4124	6790	LIGHT UP	S	10Rmt6Dbr	12/49-50	Budd	9522	9660
RIVER SOU	S	10Rmt6Dbr	9-10/49	PS	4140	6814	LODGE UP	S	10Rmt6Dbr	12/49-50	Budd	9522	9660
OVERLAND TRAIL (DS)UP		8SecCptDbrArtic	10/34	PS	4034A	6428	MEADOW UP	S	10Rmt6Dbr	12/49-50	Budd	9522	9660
PACIFIC							MIST UP	S	10Rmt6Dbr	12/49-50	Budd	9522	9660
BAY UP	S	10Rmt6Dbr	12/49-50	Budd	9522	9660	NORTHWEST UP	S	10Rmt6Dbr	12/49-50	Budd	9522	9660
BEACH UP	S	10Rmt6Dbr	12/49-50	Budd	9522	9660	OCEAN UP	S	10Rmt6Dbr	12/49-50	Budd	9522	9660
BEAUTY UP	S	10Rmt6Dbr	12/49-50	Budd	9522	9660	PARK UP	S	10Rmt6Dbr	12/49-50	Budd	9522	9660
BEND UP	S	10Rmt6Dbr	12/49-50	Budd	9522	9660	PATROL UP	S	10Rmt6Dbr	12/49-50	Budd	9522	9660
BRIDGE UP	S	10Rmt6Dbr	12/49-50	Budd	9522	9660	PEAK UP	S	10Rmt6Dbr	12/49-50	Budd	9522	9660
CAPE UP	S	10Rmt6Dbr	12/49-50	Budd	9522	9660	PLATEAU UP	S	10Rmt6Dbr	12/49-50	Budd	9522	9660
CASTLE UP	S	10Rmt6Dbr	12/49-50	Budd	9522	9660	RANGE UP	S	10Rmt6Dbr	12/49-50	Budd	9522	9660
COMMAND UP	S	10Rmt6Dbr	12/49-50	Budd	9522	9660	REST UP	S	10Rmt6Dbr	12/49-50	Budd	9522	9660
COVE UP	S	10Rmt6Dbr	12/49-50	Budd	9522	9660	RIDGE UP	S	10Rmt6Dbr	12/49-50	Budd	9522	9660
CREST UP	S	10Rmt6Dbr	12/49-50	Budd	9522	9660	SANDS UP	S	10Rmt6Dbr	12/49-50	Budd	9522	9660
CRUISER UP	S	10Rmt6Dbr	12/49-50	Budd	9522	9660	SCENE UP	S	10Rmt6Dbr	12/49-50	Budd	9522	9660
DOMAIN UP	S	10Rmt6Dbr	12/49-50	Budd	9522	9660	SHORE UP	S	10Rmt6Dbr	12/49-50	Budd	9522	9660
EMBLEM UP	S	10Rmt6Dbr	12/49-50	Budd	9522	9660	SKIES UP	S	10Rmt6Dbr	12/49-50	Budd	9522	9660
EMPIRE UP	S	10Rmt6Dbr	12/49-50	Budd	9522	9660	SLOPE UP	S	10Rmt6Dbr	12/49-50	Budd	9522	9660
FALLS UP	S	10Rmt6Dbr	12/49-50	Budd	9522	9660	SPRAY UP	S	10Rmt6Dbr	12/49-50	Budd	9522	9660
FOREST UP	S	10Rmt6Dbr	12/49-50	Budd	9522	9660	SUNSET UP	S	10Rmt6Dbr	12/49-50	Budd	9522	9660
FORUM UP	S	10Rmt6Dbr	12/49-50	Budd	9522	9660	TERRACE UP	S	10Rmt6Dbr	12/49-50	Budd	9522	9660
GARDENS UP	S	10Rmt6Dbr	12/49-50	Budd	9522	9660	TRAIL UP	S	10Rmt6Dbr	12/49-50	Budd	9522	9660
GUARD UP	S	10Rmt6Dbr	12/49-50	Budd	9522	9660	UNION UP	S	10Rmt6Dbr	12/49-50	Budd	9522	9660
HARBOR UP	S	10Rmt6Dbr	12/49-50	Budd	9522	9660	VIEW UP	S	10Rmt6Dbr	12/49-50	Budd	9522	9660
HEIGHTS UP	S	10Rmt6Dbr	12/49-50	Budd	9522	9660	WATERS UP	S	10Rmt6Dbr	12/49-50	Budd	9522	9660
HILLS UP	S	10Rmt6Dbr	12/49-50	Budd	9522	9660	WAVES UP	S	10Rmt6Dbr	12/49-50	Budd	9522	9660
HOME UP	S	10Rmt6Dbr	12/49-50	Budd	9522	9660	PACOLET RIVER SOU	S	10Rmt6Dbr	9-10/49	PS	4140	6814
ISLAND UP	S	10Rmt6Dbr	12/49-50	Budd	9522	9660	PAINT ROCK VALLEY SOU	S	14Rmt4Dbr	10-11/49	PS	4153C	6814

Figure I-120 Speed, shine, and fancy paint were emphasized in all respects when the Union Pacific ordered this 1950 stainless steel car PACIFIC CASTLE from Budd in the standard UP livery. The letterboard was flat, rather than the standard fluted letterboard with nameplate. In 1971, Amtrak applied the number 2605, later changed to 2438. Photograph by Wilber C. Whittaker.

CAR NAME OR NUMBER	ST	CAR TYPE	DATE BUILT	CAR BLDR	PLAN	LOT
PAISES BAJOS NdeM-507	SN	10Rmt6Dbr	9/48-/49	PS	4123	6790
PAKISTAN NdeM-520	SN	10Rmt6Dbr	9/48-/49	PS	4123	6790
PALENQUE NdeM-223	L	8Sec3Cpt	/52	Schi	100570	
PALM ARCH ATSF	S	10Rmt6Dbr	6-9/51	AC&F	9014	3359
PALM BEACH SAL-19	S	5DbrBLng	1/56	PS	4202	6970
PALM BEACH SCL-6602	SN	6DbrBLng	8/49	AC&F	9003	3045
PALM DOME ATSF	S	10Rmt6Dbr	6-9/51	AC&F	9014	3359
PALM FALLS PRR	S	6DbrBLng	3-5/49	PS	4131	6792
PALM GROVE CP-14113		10Sec5Dbr	/49-/50	CPRR		
PALM HAVEN ATSF	S	10Rmt6Dbr	6-9/51	AC&F	9014	3359
PALM LEAF ATSF	S	10Rmt6Dbr	6-9/51	AC&F	9014	3359
PALM LOCH ATSF	S	10Rmt6Dbr	6-9/51	AC&F	9014	3359
PALM LORE ATSF	S	10Rmt6Dbr	6-9/51	AC&F	9014	3359
PALM PATH ATSF	S	10Rmt6Dbr	6-9/51	AC&F	9014	3359
PALM STAR ATSF	S	10Rmt6Dbr	6-9/51	AC&F	9014	3359
PALM STREAM ATSF	S	10Rmt6Dbr	6-9/51	AC&F	9014	3359
PALM SUMMIT ATSF	S	10Rmt6Dbr	6-9/51	AC&F	9014	3359
PALM TOP ATSF	S	10Rmt6Dbr	6-9/51	AC&F	9014	3359
PALM TOWER ATSF	S	10Rmt6Dbr	6-9/51	AC&F	9014	3359
PALM VIEW ATSF	S	10Rmt6Dbr	6-9/51	AC&F	9014	3359
PALOS VERDES JT	S	4Dbr4Cpt2Dr	11/40	PS	4069F	6636
PANAMA FEC	S	14Rmt2Dr	2-3/50	AC&F	9007	3096
PANAMA NdeM-582	SN	10Rmt6Dbr	8/56	Budd	9538	9660
PANTIGO LIRR 2051	SN	Parlor/Coach	9/40	PS	4086A	6612
PAQUOTT LIRR 2063	SN	Parlor/Coach	9/40	PS	4086A	6612
PARADISE GLACIER GN-1185	S	16DuRmt4Dbr	12/50	PS	4108A	6890
PARADISE VALLEY ATSF	S	6Sec6Rmt4Dbr	6/42	PS	4099	6669
PARAGUAY NdeM-583	SN	10Rmt6Dbr	8/56	Budd	9538	9660
PARIA ATSF	S	17RmtSec	1-2/38	PS	4068B	6532
PARICUTIN NdeM-562	SN	24DuRmt	4/42	PS	4100	6673
PARK CREEK PASS GN-1374	S	6Rmt5Dbr2Cpt	10-11/50	PS	4180	6877
RAPIDS PRR-8433	N	10Rmt6Dbr	8-10/50	AC&F	9008	3200
PARRIS ISLAND ACL	S	21Rmt	9-10/49	PS	4156B	6809
PASCO NP	S	8DuRmt6Rmt4Dbr	7-9/48	PS	4119	6781
PASS-a-GRILLE BEACH ACL	S	6DbrBLng	11-12/49	AC&F	9002	3090
PASSAIC RIVER NYC	S	10Rmt6Dbr	9/48-/49	PS	4123	6790
PATAPSCO RIVER PRR	S	10Rmt6Dbr	9-10/49	PS	4140	6814
PATCHOQUE LIRR 2052	SN	Parlor/Coach	9/40	PS	4086A	6612
PATUXENT B&O	S	14Rmt4Dbr	4-6/48	PS	4153B	6776
PAW PAW B&O	S	14Rmt4Dbr	4-6/48	PS	4153B	6776
PEACE RIVER CN-2089	SN	10Rmt6Dbr	9/48-/49	PS	4123	6790
PEACEFUL VALLEY NYC	S	10Rmt6Dbr	12/48-49	Budd	9510	9660
PEACH GROVE CP-14114		10Sec5Dbr	/49-/50	CPRR		
PEACH VALLEY NYC	S	10Rmt6Dbr	12/48-49	Budd	9510	9660
PEARL RIVER L&N	S	10Rmt6Dbr	9-10/49	PS	4140	6814
PECONIC LIRR 2053	SN	Parlor/Coach	9/40	PS	4086A	6612
PECOS RIVER NYC	S	10Rmt6Dbr	9/48-/49	PS	4123	6790
PECOS VALLEY ATSF	S	6Sec6Rmt4Dbr	6/42	PS	4099	6669
PEEKSKILL BAY NYC	S	22Rmt	10/48	PS	4122	6790
PELEE ISLAND NYC	S	DrMbrObsBLng	5/38	PS	4079	6547
PELICAN STATE IC	S	6Sec6Rmt4Dbr	4/42	PS	4099	6669
PEMBINA RIVER CN-2086	SN	10Rmt6Dbr	9/48-/49	PS	4123	6790

Figure I-121   PALM FALLS has seen better days as it sits in the storage yard in Chicago in 1972. It would have a date with the scrapper, Bethlehem Steel Corporation, within months; sister CATALPA FALLS was saved and runs today. The period 1970 to 1980 was one of wholesale car disposition and "recycling." Railroad Avenue Enterprises Photograph PN-2438.

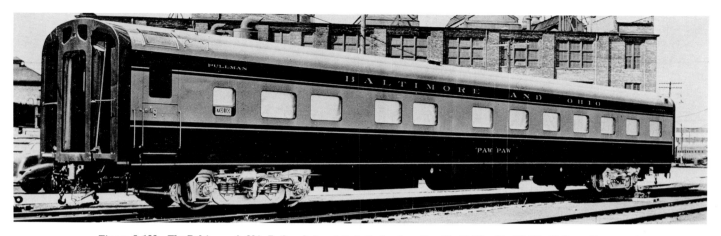

Figure I-122 The Baltimore & Ohio Railroad placed their first order after World War II with The Pullman Company in October 1945 for 8 14-roomette, 4-double-bedroom sleepers like the PAW PAW. Note the wide light gray stripe above the letterboard, a design adopted from the streamlined standard cars. This paint scheme would change in the next two years. Pullman Photograph 58000. The Author's Collection.

CAR NAME OR NUMBER	ST	CAR TYPE	DATE BUILT	CAR BLDR	PLAN	LOT
PEND O'REILLE RIVER GN-1273	S	8DuRmt4Sec3DbrCpt	11-12/50	PS	4181	6889
PENDLETON UP	S	5DbrBLng	4/56	PS	4199	6959
PENNS RAPIDS PRR-8447	S	10Rmt6Dbr	9-10/50	AC&F	9008	3212
PENOBSCOT RIVER NYC	S	10Rmt6Dbr	9/48-/49	PS	4123	6790
PEQUEST DL&W	S	10Rmt6Dbr	6-7/49	AC&F	9008A	3089
PERCH RIVER NYC	S	10Rmt6Dbr	9/48-/49	PS	4123	6790
PERKIOMEN RAPIDS PRR-8432	N	10Rmt6Dbr	8-10/50	AC&F	9008	3200
PERSEO NdeM-595	SN	8DuRmt6Rmt4Dbr	12/46-47	PS	4107	6751

CAR NAME OR NUMBER	ST	CAR TYPE	DATE BUILT	CAR BLDR	PLAN	LOT
PERU NdeM-584	SN	10Rmt6Dbr	8/56	Budd	9538	9660
PETAWAWA RIVER CN-2140	SN	10Rmt6Dbr	9-10/49	PS	4140B	6809
PETER SHOENBERGER PRR-8246	N	21Rmt	1-6/49	Budd	9513	9667
PETERSBURG SAL	S	10Rmt6Dbr	5-6/49	PS	4140A	6796
PETITCODIAC RIVER CN-2080	SN	10Rmt6Dbr	9/48-/49	PS	4123	6790
PETOSKEY INN PRR-8277	S	21Rmt	1-6/49	Budd	9513	9667
PETROLEUM IC	S	6Sec6Rmt4Dbr	4/42	PS	4099	6669
PHANTOM VALLEY CRI&P	S	5DdrBLngREObs	10/39	PS	4089	6587

Figure I-123 PENNS RAPIDS was one of the few cars that was actually painted Penn Central green with a white stripe. It is shown here in April 1970 during the brief interval that Penn Central operated their own sleeping cars. This 1950 AC&F riveted aluminum 10-6 sleeper did not survive to Amtrak ownership. Railroad Avenue Enterprises Photograph PN-2860.

CAR NAME OR NUMBER	ST	CAR TYPE	DATE BUILT	CAR BLDR	PLAN	LOT
PHILADELPHIA COUNTY PRR	S	13Dbr	4/38	PS	4071A	6541
PIEDMONT VALLEY SOU	S	14Rmt4Dbr	10-11/49	PS	4153C	6814
PIEGAN PASS GN-1163	S	8DuRmt4Dbr4Sec	12/46-47	PS	4107	6751
PIERRE						
LACLEDE SLSF-1450	S	14Rmt4Dbr	1-6/48	PS	4153	6769
PIGEON BAY NYC	S	22Rmt	10/48	PS	4122	6790
PIKE COUNTY N&W	S	10Rmt6Dbr	11/49-50	Budd	9523	9660
PINE						
ARROYO ATSF-1617	S	10Rmt6Dbr	12/49-50	Budd	9521	9660
BEACH ATSF-1618	S	10Rmt6Dbr	12/49-50	Budd	9521	9660
BELL ATSF-1619	S	10Rmt6Dbr	12/49-50	Budd	9521	9660
BLUFF ATSF-1620	S	10Rmt6Dbr	12/49-50	Budd	9521	9660
BROOK ATSF-1621	S	10Rmt6Dbr	12/49-50	Budd	9521	9660
CAVERN ATSF-1622	S	10Rmt6Dbr	12/49-50	Budd	9521	9660
COVE ATSF-1623	S	10Rmt6Dbr	12/49-50	Budd	9521	9660
CREEK ATSF-1624	S	10Rmt6Dbr	12/49-50	Budd	9521	9660
CREST ATSF-1625	S	10Rmt6Dbr	12/49-50	Budd	9521	9660
DALE ATSF-1626	S	10Rmt6Dbr	12/49-50	Budd	9521	9660
DAWN ATSF-1628	S	10Rmt6Dbr	12/49-50	Budd	9521	9660
FALLS CN-2099	SN	14Rmt4Dbr	1-6/48	PS	4153	6769
FALLS PRR	S	6DbrBLng	9/40	PS	4086A	6612
FERN ATSF-1627	S	10Rmt6Dbr	12/49-50	Budd	9521	9660
GEM ATSF-1629	S	10Rmt6Dbr	12/49-50	Budd	9521	9660
GORGE ATSF-1630	S	10Rmt6Dbr	12/49-50	Budd	9521	9660
GROVE CP-14115		10Sec5Dbr	/49-/50	CPRR		

CAR NAME OR NUMBER	ST	CAR TYPE	DATE BUILT	CAR BLDR	PLAN	LOT
PINE						
GROVE ATSF-1631	S	10Rmt6Dbr	12/49-50	Budd	9521	9660
HILL ATSF-1632	S	10Rmt6Dbr	12/49-50	Budd	9521	9660
ISLAND ATSF-1633	S	10Rmt6Dbr	12/49-50	Budd	9521	9660
KING ATSF-1634	S	10Rmt6Dbr	12/49-50	Budd	9521	9660
LEAF ATSF-1635	S	10Rmt6Dbr	12/49-50	Budd	9521	9660
LODGE ATSF-1636	S	10Rmt6Dbr	12/49-50	Budd	9521	9660
MESA ATSF-1637	S	10Rmt6Dbr	12/49-50	Budd	9521	9660
PASS ATSF-1638	S	10Rmt6Dbr	12/49-50	Budd	9521	9660
PEAK ATSF-1639	S	10Rmt6Dbr	12/49-50	Budd	9521	9660
RANGE ATSF-1641	S	10Rmt6Dbr	12/49-50	Budd	9521	9660
RAPIDS ATSF-1640	S	10Rmt6Dbr	12/49-50	Budd	9521	9660
RING ATSF-1642	S	10Rmt6Dbr	12/49-50	Budd	9521	9660
SHORE ATSF-1643	S	10Rmt6Dbr	12/49-50	Budd	9521	9660
TREE STATE NH-553	S	6DbrBLng	1/55	PS	4193	W6941
VALLEY NYC	S	10Rmt6Dbr	12/48-49	Budd	9510	
PINEHURST SAL-52	S	5DbrCpt4Sec4Rmt	11/55	Budd	9537	9658
PINELLAS COUNTY ACL	S	10Rmt6Dbr	9-10/49	PS	4140B	6809
PIQUA INN PRR-8278	S	21Rmt	1-6/49	Budd	9513	9667
PISCIS FCP	SN	10Rmt6Dbr	9/48-/49	PS	4123	6790
PITAMAKIN PASS GN-1371	S	6Rmt5Dbr2Cpt	10-11/50	PS	4180	6877
PITCH PINE C&EI-901	S	6Sec6Rmt4Dbr	3-5/53	PS	4183	6909
PLACID BAY UP	S	11Dbr	2-3/56	PS	4198	6958
PLACID HARBOR UP	S	11Dbr	2-3/56	PS	4198	6958
PLACID HAVEN UP	S	11Dbr	2-3/56	PS	4198	6958

Figure I-124 Eleven-double-bedrooms were the accommodations in Union Pacific's PLACID series. All ten of the Pullman cars in this lot were sold to Amtrak in 1971 and acquired the numbers 2260 through 2269. The eleven bedroom cars proved very popular in railroad, Amtrak, and private car service. The plate just ahead of the far window carried a machine readable bar code for the abortive "Automatic Car Identification" system. Railroad Avenue Enterprises Photograph PN-2805.

CAR NAME OR NUMBER	ST	CAR TYPE	DATE BUILT	CAR BLDR	PLAN	LOT	CAR NAME OR NUMBER	ST	CAR TYPE	DATE BUILT	CAR BLDR	PLAN	LOT
PLACID LAKE UP	S	11Dbr	2-3/56	PS	4198	6958	POLONIA NdeM-508	SN	10Rmt6Dbr	9/48-/49	PS	4123	6790
PLACID MEADOW UP	S	11Dbr	2-3/56	PS	4198	6958	POND PINE C&EI-903	S	6Sec6Rmt4Dbr	3-5/53	PS	4183	6909
PLACID SCENE UP	S	11Dbr	2-3/56	PS	4198	6958	POND POINT NH	S	14Rmt4Dbr	12/49/50	PS	4159	6822
PLACID SEA UP	S	11Dbr	2-3/56	PS	4198	6958	PONQUOGUE LIRR 2054	SN	Parlor/Coach	9/40	PS	4086A	6612
PLACID VALE UP	S	11Dbr	2-3/56	PS	4198	6958	PONTE VEDRA BEACH ACL	S	6DbrBLng	11-12/49	AC&F	9002	3090
PLACID VALLEY UP	S	11Dbr	2-3/56	PS	4198	6958	POPLAR GROVE CP-14116		10Sec5Dbr	/49-/50	CPRR		
PLACID WATERS UP	S	11Dbr	2-3/56	PS	4198	6958	POPLAR RIVER GN-1269	S	8DuRmt4Sec3DbrCpt	11-12/50	PS	4181	6889
PLAINSMAN CRI&P	S	8DuRmt6Rmt4Dbr	8-9/48	PS	4116	6761	POPPONESSET						
PLANTATION PINE L&N-3462	S	6Sec6Rmt4Dbr	3-5/53	PS	4183	6909	BEACH NH-534	S	6Sec6Rmt4Dbr	11/54-55	PS	4194	W6942
PLATTE RIVER NYC	S	10Rmt6Dbr	9/48-/49	PS	4123	6790	PORT						
PLAZA DEL REY JT	S	4Dbr4Cpt2Dr	11/40	PS	4069F	6636	BYRON NYC	S	12Dbr	4-5/49	PS	4125	6790
PLEASANT VALLEY ATSF	S	6Sec6Rmt4Dbr	6/42	PS	4099	6669	CHESTER NYC	S	12Dbr	4-5/49	PS	4125	6790
PLUM BROOK NYC	S	5DbrBLngObs	5-6/49	Budd	9506	9664	CLINTON NYC	S	12Dbr	4-5/49	PS	4125	6790
POCONO DL&W	S	10Rmt6Dbr	6-7/49	AC&F	9008A	3089	LAWRENCE NYC	S	12Dbr	4-5/49	PS	4125	6790
POLACCA ATSF	S	4Dbr4Cpt2Dr	1/38	PS	4069A	6532	OF ALBANY NYC	S	12Dbr	4-5/49	PS	4125	6790
POLARIS FCP	SN	10Rmt6Dbr	9/48-/49	PS	4123	6790	OF BOSTON NYC	S	12Dbr	4-5/49	PS	4125	6790
POLK COUNTY ACL	S	10Rmt6Dbr	9-10/49	PS	4140B	6809	OF BUFFALO NYC	S	12Dbr	4-5/49	PS	4125	6790
							OF CHICAGO NYC	S	12Dbr	4-5/49	PS	4125	6790

Figure I-125 This is 1 of 9 10-roomette, 6-double-bedroom sleepers built by American Car & Foundry in 1949 for the Delaware Lackawanna & Western Railroad, which would become Erie-Lackawanna property in 1960. During Hurricane Agnes in 1972, Straits Shows lost most of their train in the flood at Wilkes-Barre, Pennsylvania. Within days, Straits acquired five of the cars from this lot and continued its tour. POCONO became Straits Shows' SYRACUSE 4. Photograph by Robert Pennisi No. 1935.

CAR NAME OR NUMBER	ST	CAR TYPE	DATE BUILT	CAR BLDR	PLAN	LOT
PORT						
OF DETROIT NYC	S	12Dbr	4-5/49	PS	4125	6790
OF LEWISTON NYC	S	12Dbr	4-5/49	PS	4125	6790
OF NEW YORK NYC	S	12Dbr	4-5/49	PS	4125	6790
OF OSWEGO NYC	S	12Dbr	4-5/49	PS	4125	6790
OF WINDSOR NYC	S	12Dbr	4-5/49	PS	4125	6790
ORANGE NYC	S	12Dbr	4-5/49	PS	4125	6790
PORTAGE RIVER NYC	S	10Rmt6Dbr	9/48-/49	PS	4123	6790
PORTLAND SP&S	S	8DuRmt6Rmt4Dbr	7-9/48	PS	4119	6781
PORTSMOUTH SAL	S	10Rmt6Dbr	5-6/49	PS	4140A	6796
PORTSMOUTH SQUARE JT	S	12DuRmt5Dbr	12/37	PS	4066A	6525
PORTUGAL NdeM-509	SN	10Rmt6Dbr	9/48-/49	PS	4123	6790
POTOMAC RIVER SOU	S	10Rmt6Dbr	9-10/49	PS	4140	6814
POWDER RIVER NYC	S	10Rmt6Dbr	9/48-/49	PS	4123	6790
PRAIRIE RIVER CN-2088	SN	10Rmt6Dbr	9/48-/49	PS	4123	6790
PRAIRIE STATE IC	S	6Sec6Rmt4Dbr	4/42	PS	4099	6669
PRESA ABELARDO						
L.RODRIGUEZ FCP	SN	3DbrDrBarLng	5/38	PS	4078	6551
PRESA ADOLFO						
LOPEZ MATEOS NdeM-605	SN	22Rmt	4-6/49	Budd	9501	9661
PRESA						
AZUCAR NdeM-545	N	22Rmt	4-6/49	Budd	9501	9661
BENITO JUAREZ NdeM-59	SN	22Rmt	4-6/49	Budd	9501	9661
CHIQUE NdeM-552	SN	22Rmt	4-6/49	Budd	9501	9661
COINTZIO NdeM-551	SN	22Rmt	4-6/49	Budd	9501	9661
CUAUHTEMOC FCP	SN	3DbrDrBarLng	5/38	PS	4078	6551
DANZHO NdeM-546	SN	22Rmt	4-6/49	Budd	9501	9661
DEL HUMAYA FCP	SN	3DbrDrBarLng	5/38	PS	4078	6551
EL AZUCAR NdeM	N	3DbrDrbARLng	5/38	PS	4078	6551
EL NOVILLA FCP	SN	3DbrDrBarLng	5/38	PS	4078	6551
FALCON NdeM-548	SN	22Rmt	4-6/49	Budd	9501	9661
PRESA FRANCISCO						
VILLA NdeM-600	SN	22Rmt	4-6/49	Budd	9501	9661
PRESA						
GUADALUPE NdeM-599	SN	22Rmt	4-6/49	Budd	9501	9661
INFIERNILLO NdeM-557	SN	22Rmt	4-6/49	Budd	9501	9661
LA AMISTAD NdeM-547	SN	22Rmt	4-6/49	Budd	9501	9661
LA ANGOSTURA NdeM	SN	22Rmt	4-6/49	Budd	9501	9661
LA ANGOSTURA FCP	SN	3DbrDrBarLng	5/38	PS	4078	6551
LAS LAJAS NdeM-601	SN	22Rmt	4-6/49	Budd	9501	9661
PREAS LAZARO						
CARDENAS NdeM-602	SN	22Rmt	4-6/49	Budd	9501	9661
PRESA						
MAL PASO NdeM-550	SN	22Rmt	4-6/49	Budd	9501	9661
PRESA MIGUEL						
ALEMAN NdeM-604	N	22Rmt	4-6/49	Budd	9501	9661
HIDALGO FCP-301	SN	2DRCptDbrObsBLng	5/38	PS	4081	6549
PRESA						
MOCUZARI FCP-302	SN	2MbrDbrREObsBLng	5/38	PS	4080	6548
MORELOS FCP-303	SN	3DbrDrBarLng	5/38	PS	4078	6551
OVIACHIC FCP-300	SN	2DrCptDbrObsBLng	5/38	PS	4081	6549

CAR NAME OR NUMBER	ST	CAR TYPE	DATE BUILT	CAR BLDR	PLAN	LOT
PRESA						
PABELLON NdeM-603	SN	22Rmt	4-6/49	Budd	9501	9661
PRESA PRESIDENTE						
ALEMAN NdeM-604	SN	22Rmt	4-6/49	Budd	9501	9661
PRESA						
SAN LORENZO FCP	SN	3DbrDrBarLng	5/38	PS	4078	6551
SANALONA FCP	SN	3DbrDrBarLng	5/38	PS	4078	6551
SANTA ROSA FCP	SN	3DbrDrBarLng	5/38	PS	4078	6551
SANTA TERESA NdeM-553	SN	22Rmt	4-6/49	Budd	9501	9661
SOLIS NdeM-549	SN	22Rmt	4-6/49	Budd	9501	9661
TECAMACHALCO NdeM-554	SN	22Rmt	4-6/49	Budd	9501	9661
TESOYO NdeM-555	SN	22Rmt	4-6/49	Budd	9501	9661
VALSEQUILLO NdeM-556	SN	22Rmt	4-6/49	Budd	9501	9661
DEL HUMAYA FCP	SN	3DbrDrBarLng	5/38	PS	4078	6551
LA ANGOSTURA FCP	SN	3DbrDrBarLng	5/38	PS	4078	6551
PRIDE						
OF YOUNGSTOWN Erie	S	10Rmt6Dbr	6/54	PS	4186B	6946
PRIEST RIVER GN-1197	S	2DbrDrBLngObs	12/50	PS	4109A	6878
PRINCE						
GEORGE COUNTY ACL	S	10Rmt6Dbr	9-10/49	PS	4140B	6809
PRINCETON INN PRR-8279	S	21Rmt	1-6/49	Budd	9513	9667
PROGRESS Pool		3DbrCptREObsBLng	8/36	PS	4051A	6478
PTARMIGAN PASS GN-1161	S	8DuRmt4Dbr4Sec	12/46-47	PS	4107	6751
PUERTO RICO NdeM-585	SN	10Rmt6Dbr	8/56	Budd	9538	9660
PULASKI COUNTY N&W	S	10Rmt6Dbr	11/49-50	Budd	9523	9660
PUMPELLEY						
GLACIER GN-1186	S	16DuRmt4Dbr	12/50	PS	4108A	6890
PUTNAM COUNTY ACL	S	10Rmt6Dbr	9-10/49	PS	4140B	6809
PUTNAM VALLEY NYC	S	10Rmt6Dbr	12/48-49	Budd	9510	9660
PUYE ATSF	S	4DrDbrREObsLng	7/38	PS	4070A	6553
PYRAMID FALLS CN-2102	SN	14Rmt4Dbr	1-6/48	PS	4153	6769
QUAIL B&O-7106	S	16DuRmt4Dbr	4-6/54	Budd	9536	9658
QUEENSBORO BRIDGE NYC	N	4Dbr4Cpt2Dr	3-4/38	PS	4069B	6540
QUINCY BAY NYC	S	22Rmt	10/48	PS	4122	6790
QUONSET POINT NH	S	14Rmt4Dbr	12/49/50	PS	4159	6822
RACCOON RAPIDS PRR	S	10Rmt6Dbr	11/48-49	PS	4140	6792
RACE POINT NH	S	14Rmt4Dbr	12/49/50	PS	4159	6822
RAINBOW FALLS CN-2105	SN	14Rmt4Dbr	1-6/48	PS	4153	6769
RAINBOW STREAM NYC	S	6DbrBLng	3-5/49	Budd	9505	9663
RAINIER CLUB NP	S	5DbrREObsBLng	6-7/48	PS	4120	6781
RALEIGH SAL	S	10Rmt6Dbr	5-6/49	PS	4140A	6796
RAMPART RANGE CRI&P	S	8Rmt6Dbr	11-12/54	PS	4195	6944
RANDOLPH						
MACON COLLEGE N&W	S	10Rmt6Dbr	11/49-50	Budd	9523	9660
RAPID STREAM NYC	S	6DbrBLng	3-5/49	Budd	9505	9663
RAPIDAN RIVER SOU	S	10Rmt6Dbr	9-10/49	PS	4140	6814
RAPPAHANNOCK RIVER SOU	S	10Rmt6Dbr	9-10/49	PS	4140	6814
RAQUETTE FALLS NYC	S	6DbrBLng	9/39	PS	4086	6573
RAQUETTE LAKE NYC	S	6DbrBLng	12/48-49	PS	4124	6790
RARITAN BAY NYC	S	22Rmt	10/48	PS	4122	6790
RARITAN RAPIDS PRR	S	10Rmt6Dbr	11/48-49	PS	4140	6792

CAR NAME OR NUMBER	ST	CAR TYPE	DATE BUILT	CAR BLDR	PLAN	LOT
RARITAN RIVER PRR	S	10Rmt6Dbr	9-10/49	PS	4140	6814
RASIN RIVER NYC	S	10Rmt6Dbr	9/48-/49	PS	4123	6790
RAVENNA INN PRR-8280	S	21Rmt	1-6/49	Budd	9513	9667
RAYMOND Milw-27	S	16DuRmt4Dbr	1-2/49	PS	4147D	6775
RECONDO VALLEY ATSF	S	6Sec6Rmt4Dbr	6/42	PS	4099	6669
RED						
GAP PASS CB&Q-1168	S	8DuRmt4Dbr4Sec	12/46-47	PS	4107	6751
MOUNTAIN SAL-15	S	6DbrBLng	8/49	AC&F	9003	3045
RIVER VALLEY ATSF	S	6Sec6Rmt4Dbr	6/42	PS	4099	6669
ROCK VALLEY ATSF	S	6Sec6Rmt4Dbr	6/42	PS	4099	6669
REGAL						
ARMS ATSF	S	4Dbr4Cpt2Dr	9-11/50	AC&F	9011	3358
CENTER ATSF-1805	S	4Dbr4Cpt2Dr	12/47-48	PS	4144	6757
CITY ATSF-1806	S	4Dbr4Cpt2Dr	12/47-48	PS	4144	6757
CORPS ATSF	S	4Dbr4Cpt2Dr	9-11/50	AC&F	9011	3358
COURT ATSF	S	4Dbr4Cpt2Dr	9-11/50	AC&F	9011	3358
CREEK ATSF-1807	S	4Dbr4Cpt2Dr	12/47-48	PS	4144	6757
CREST ATSF	S	4Dbr4Cpt2Dr	9-11/50	AC&F	9011	3358
CROSS ATSF-1810	S	4Dbr4Cpt2Dr	12/47-48	PS	4144	6757
CROWN ATSF-1812	S	4Dbr4Cpt2Dr	12/47-48	PS	4144	6757

CAR NAME OR NUMBER	ST	CAR TYPE	DATE BUILT	CAR BLDR	PLAN	LOT
REGAL						
DOME ATSF	S	4Dbr4Cpt2Dr	9-11/50	AC&F	9011	3358
ELM ATSF	S	4Dbr4Cpt2Dr	9-11/50	AC&F	9011	3358
GATE ATSF	S	4Dbr4Cpt2Dr	9-11/50	AC&F	9011	3358
GORGE ATSF-1816	S	4Dbr4Cpt2Dr	12/47-48	PS	4144	6757
GULF ATSF	S	4Dbr4Cpt2Dr	9-11/50	AC&F	9011	3358
HILL ATSF-1818	S	4Dbr4Cpt2Dr	12/47-48	PS	4144	6757
HOUSE ATSF	S	4Dbr4Cpt2Dr	3/48	PS	4144	6833
HUNT ATSF	S	4Dbr4Cpt2Dr	9-11/50	AC&F	9011	3358
INN ATSF	S	4Dbr4Cpt2Dr	9-11/50	AC&F	9011	3358
ISLE ATSF	S	4Dbr4Cpt2Dr	9-11/50	AC&F	9011	3358
LANE ATSF	S	4Dbr4Cpt2Dr	9-11/50	AC&F	9011	3358
LARK ATSF	S	4Dbr4Cpt2Dr	9-11/50	AC&F	9011	3358
MANOR ATSF	S	4Dbr4Cpt2Dr	9-11/50	AC&F	9011	3358
OAK ATSF-1826	S	4Dbr4Cpt2Dr	12/47-48	PS	4144	6757
PASS ATSF-1827	S	4Dbr4Cpt2Dr	12/47-48	PS	4144	6757
RING ATSF-1828	S	4Dbr4Cpt2Dr	12/47-48	PS	4144	6757
RIVER ATSF-1829	S	4Dbr4Cpt2Dr	12/47-48	PS	4144	6757
RUBY ATSF-1830	S	4Dbr4Cpt2Dr	12/47-48	PS	4144	6757
SPA ATSF	S	4Dbr4Cpt2Dr	9-11/50	AC&F	9011	3358

Figure I-126   PUTNAM COUNTY was 1 of 42 cars in Pullman Plan 4140B, Lot 6809, which was shared by RF&P, FEC, PRR, and ACL. Similar cars were purchased by Southern, and all of these immediate postwar stainless sheathed cars had severe, continuing internal corrosion. The stainless skin accelerated galvanic corrosion of the carbon steel framing; Budd-built all-stainless cars did not suffer that problem. These cars were placed in service between New York and Florida. Railroad Avenue Enterprises Photograph PN-2393.

Figure I-127   The standard prewar accommodation of the 4-bedroom, 4-compartment, 2-drawing room car was ordered by the Santa Fe Railroad in 1948. Santa Fe was a good customer for Pullman's cheaper, lighter, less durable stainless sheathed cars. It was a early user of the outside swing hanger truck shown here, and of AAR alternate standard-type CS tightlock couplers. Most of these all-stainless steel Pullman cars in Lot 6757 were sold to Amtrak in 1971. Railroad Avenue Enterprises Photograph PN-2775

CAR NAME OR NUMBER	ST	CAR TYPE	DATE BUILT	CAR BLDR	PLAN	LOT	CAR NAME OR NUMBER	ST	CAR TYPE	DATE BUILT	CAR BLDR	PLAN	LOT
REGAL							RIO						
STREAM ATSF-1832	S	4Dbr4Cpt2Dr	12/47-48	PS	4144	6757	GRANDE VALLEY ATSF	S	6Sec6Rmt4Dbr	6/42	PS	4099	6669
TEMPLE ATSF-1833	S	4Dbr4Cpt2Dr	12/47-48	PS	4144	6757	LAS CANAS FCP	SN	14Rmt4Dbr	2-7/48	PS	4153A	6758
TOWN ATSF-1834	S	4Dbr4Cpt2Dr	12/47-48	PS	4144	6757	LERMA FCP	SN	14Rmt4Dbr	2-7/48	PS	4153A	6758
VALE ATSF-1835	S	4Dbr4Cpt2Dr	12/47-48	PS	4144	6757	MAGDALENA FCP	SN	14Rmt4Dbr	2-7/48	PS	4153A	6758
RESTIGOUCHE RIVER CN-2079	SN	10Rmt6Dbr	9/48-/49	PS	4123	6790	MAYO FCP	SN	14Rmt4Dbr	2-7/48	PS	4153A	6758
RESTLAND B&O	L	24Sr8Dbr	5/59	Budd	9540	9691	MOCORITO FCP	SN	14Rmt4Dbr	2-7/48	PS	4153A	6758
REVERSING FALLS CN-2098	SN	14Rmt4Dbr	1-6/48	PS	4153	6769	MOCTEZUMA FCP	SN	14Rmt4Dbr	2-7/48	PS	4153A	6758
RICE BIRD ACL	RN	7Dbr2Dr	2-3/50	AC&F	9018	3091	MOLOLOA FCP	SN	14Rmt4Dbr	2-7/48	PS	4153A	6758
RICHARD BEATTY							OCORONI FCP	SN	10Rmt6Dbr	2-7/48	PS	4153D	6758
MELLON PRR-8388	N	4Dbr4Cpt2Dr	9-12/48	AC&F	9009	2982	PIAXTLA FCP	SN	14Rmt4Dbr	2-7/48	PS	4153A	6758
RICHARD							PRESIDIO FCP	SN	14Rmt4Dbr	2-7/48	PS	4153A	6758
DALE JonesProp	SN	6DbrBLng	12/48-49	PS	4124	6790	QUELITE FCP	SN	14Rmt4Dbr	2-7/48	PS	4153A	6758
LANE JonesProp	SN	10Rmt6Dbr	12/48-49	Budd	9510	9660	SAN LORENZO FCP	SN	14Rmt4Dbr	2-7/48	PS	4153A	6758
WARNER NYC	SN	6Dbr3Mbr	1-3/49	Budd	9502	9662	SAN MIGUEL FCP	SN	14Rmt4Dbr	2-7/48	PS	4153A	6758
RICHMOND SAL	S	10Rmt6Dbr	5-6/49	PS	4140A	6796	SAN PEDRO FCP	SN	14Rmt4Dbr	2-7/48	PS	4153A	6758
RIDEAU RIVER CN-2094	SN	10Rmt6Dbr	9/48-/49	PS	4123	6790	SANTA CRUZ FCP	SN	14Rmt4Dbr	2-7/48	PS	4153A	6758
RINCON HILL JT	S	10Rmt5Dbr	11/40	PS	4072D	6636	SONORA FCP	SN	14Rmt4Dbr	2-7/48	PS	4153A	6758
RIO							TAMAZULA FCP	SN	10Rmt6Dbr	2-7/48	PS	4153D	6758
ACAPONETA FCP	SN	14Rmt4Dbr	2-7/48	PS	4153A	6758	YAQUI FCP	SN	10Rmt6Dbr	2-7/48	PS	4153D	6758
ALAMOS FCP	SN	14Rmt4Dbr	2-7/48	PS	4153A	6758	RIP VAN						
BALUARTE FCP	SN	14Rmt4Dbr	2-7/48	PS	4153A	6758	WINKLE BRIDGE NYC	N	4Dbr4Cpt2Dr	3-4/38	PS	4069B	6540
CHIMPAS FCP	SN	14Rmt4Dbr	2-7/48	PS	4153A	6758	RIPPLING STREAM NYC	S	6DbrBLng	3-5/49	Budd	9505	9663
COLORADO FCP	SN	14Rmt4Dbr	2-7/48	PS	4153A	6758	RIVANNA RIVER SOU	S	10Rmt6Dbr	9-10/49	PS	4140	6814
CULIACAN FCP	SN	14Rmt4Dbr	2-7/48	PS	4153A	6758	RIVERDALE CP	SN	10Rmt5Dbr	6-7/40	PS	4072C	6610
ELOTA FCP	SN	14Rmt4Dbr	2-7/48	PS	4153A	6758	RIVERDALE CN-2125	SN	10Rmt6Dbr	1-2/50	PS	4167A	6866
FUERTE FCP	SN	14Rmt4Dbr	2-7/48	PS	4153A	6758	RIVERFIELD CN-2128	SN	10Rmt6Dbr	1-2/50	PS	4167A	6866

CAR NAME OR NUMBER	ST	CAR TYPE	DATE BUILT	CAR BLDR	PLAN	LOT
RIVERLEA CN-2126	SN	10Rmt6Dbr	1-2/50	PS	4167A	6866
RIVERSIDE CN-2127	SN	10Rmt6Dbr	1-2/50	PS	4167A	6866
RIVERVIEW CP	SN	5DbrBLngObs	5-6/49	Budd	9506	9664
RIVERVIEW CN-2129	SN	10Rmt6Dbr	1-2/50	PS	4167A	6866
RIVIERE						
CLOCHE CN-2136	SN	10Rmt6Dbr	9-10/49	PS	4140B	6809
RAQUETTE CN-2084	SN	10Rmt6Dbr	9/48-/49	PS	4123	6790
ROUGE CN-2083	SN	10Rmt6Dbr	9/48-/49	PS	4123	6790
ST. FRANCOIS CN-2137	SN	10Rmt6Dbr	9-10/49	PS	4140B	6809
au RENARD CN-2082	SN	10Rmt6Dbr	9/48-/49	PS	4123	6790
de LOUP CN-2081	SN	10Rmt6Dbr	9/48-/49	PS	4123	6790
ROANOKE COLLEGE N&W	S	10Rmt6Dbr	11/49-50	Budd	9523	9660
ROANOKE ISLAND ACL	S	21Rmt	9-10/49	PS	4156B	6809
ROANOKE RIVER NYC	S	10Rmt6Dbr	9/48-/49	PS	4123	6790
ROANOKE VALLEY SOU	S	14Rmt4Dbr	10-11/49	PS	4153C	6814
ROARING RIVER MP	S	10Rmt6Dbr	8-9/48	Budd	9504	9660
ROBIN B&O	S	16DuRmt4Dbr	4-6/54	Budd	9536	9658
ROCK CREEK AMT-2558	SN	8Rmt6Dbr	11-12/54	PS	4195	6944
ROCKY						
COULEE GN-1193	RN	4DbrCpt6RmtObsLng	1-2/47	PS	4109K	6751
CREEK PINE L&N-3463	S	6Sec6Rmt4Dbr	3-5/53	PS	4183	6909
NECK BEACH NH-535	S	6Sec6Rmt4Dbr	11/54-55	PS	4194	W6942
POINT NH	S	14Rmt4Dbr	12/49/50	PS	4159	6822
RIVER NYC	S	10Rmt6Dbr	9/48-/49	PS	4123	6790
RODNEY Milw-30	S	16DuRmt4Dbr	1-2/49	PS	4147D	6775
ROGERS MANOR CP-10338	S	5DbrCpt4Sec4Rmt	11/54-55	Budd		9658
ROGERS PASS GN-1370	S	6Rmt5Dbr2Cpt	10-11/50	PS	4180	6877
ROOMETTE I (NKP)Pool		18Rmt	8/37	PS	4068	6526
ROOMETTE II Pool		18Rmt	4/39	PS	4068G	6556
ROSE BOWL(1st) JT		12DuSr5Dbr	12/37	PS	4066A	6525
ROSE BOWL(2nd) JT	N	18Rmt	12/37	PS	4068A	6525
ROTON POINT NH	S	14Rmt4Dbr	12/49/50	PS	4159	6822
ROUGE RIVER NYC	S	10Rmt6Dbr	9/48-/49	PS	4123	6790
ROUND VALLEY NYC	S	10Rmt6Dbr	12/48-49	Budd	9510	9660
ROUNDOUT RIVER NYC	S	10Rmt6Dbr	9/48-/49	PS	4123	6790
ROYAL ARCH SOU	RN	11Dbr	9-10/49	PS	4140A	6814
ROYAL ARCH SOU	S	5DbrREObsBLng	2-3/50	PS	4162	6814
ROYAL CANAL L&N	S	5DbrREObsBLng	2-3/50	PS	4162	6814
ROYAL COURT SOU	S	5DbrREObsBLng	2-3/50	PS	4162	6814
ROYAL CREST NYC	S	5DbrREObsBLng	2-3/50	PS	4162	6814
ROYAL PALACE WofA	S	5DbrREObsBLng	2-3/50	PS	4162	6814
ROYAL PALM CNO&TP	S	5DbrREObsBLng	2-3/50	PS	4162	6814
ROYAL STREET L&N	S	5DbrREObsBLng	2-3/50	PS	4162	6814
RUMANIA NdeM-510	SN	10Rmt6Dbr	9/48-/49	PS	4123	6790
RUSIA NdeM-511	SN	10Rmt6Dbr	9/48-/49	PS	4123	6790
RUSSIAN HILL JT	S	4DbrBEObsBLng	11/40	PS	4096A	6636
RYE BEACH B&M-33	S	6Sec6Rmt4Dbr	11/54-55	PS	4194	W6942
SABINE RIVER NYC	S	10Rmt6Dbr	9/48-/49	PS	4123	6790
SABLE RIVER CN-2078	SN	10Rmt6Dbr	9/48-/49	PS	4123	6790
SACANDAGA RIVER NYC	S	10Rmt6Dbr	9/48-/49	PS	4123	6790

CAR NAME OR NUMBER	ST	CAR TYPE	DATE BUILT	CAR BLDR	PLAN	LOT
SACKETS HARBOR NYC	S	22Rmt	4-6/49	Budd	9501	9661
SAGGITARIOUS FCP	SN	10Rmt6Dbr	9/48-/49	PS	4123	6790
SAGINAW BAY NYC	S	22Rmt	10/48	PS	4122	6790
SAGTIKOS LIRR-2075	SN	Parlor Coach	11/54-55	PS	4194	W6942
SALAHKAI ATSF	S	8Sec2Dbr2Cpt	12/37-38	PS	4058A	6532
SALEM INN PRR-8281	S	21Rmt	1-6/49	Budd	9513	9667
SALMON RIVER NYC	S	10Rmt6Dbr	9/48-/49	PS	4123	6790
SALONGA LIRR-2076	SN	Parlor Coach	11/54-55	PS	4194	W6942
SALSBURY BEACH B&M-34	S	6Sec6Rmt4Dbr	11/54-55	PS	4194	W6942
SALT RIVER						
VALLEY ATSF	S	6Sec6Rmt4Dbr	6/42	PS	4099	6669
SALUDA RIVER SOU	S	10Rmt6Dbr	9-10/49	PS	4140	6814
SALVADOR FEC	S	21Rmt	9-10/49	PS	4156B	6809
SAMUEL						
KING TIGRETT GM&O	S	8RmtCpt3Dbr4Sec	7/50	AC&F	9012	3208
M. KIER PRR	N	10Rmt6Dbr	11/48-49	PS	4140	6792
REA PRR	S	2DrCptDbrBEObsLng	3-4/49	PS	4134	6792
VAUGHAN MERRICK PRR	N	2DrCptDbrObsBLng	5/38	PS	4081	6549
SAN						
DOMINGUEZ(1st) JT		4Cpt3DrArtic	12/37	PS	4063B	6525
DOMINGUEZ(2nd) JT	N	11DbrArtic	12/37	PS	4067A	6525
FERNANDO JT	S	13RmtSecArtic	12/37	PS	4073	6525
FRANCISCO BAY NYC	S	22Rmt	10/48	PS	4122	6790
GABRIEL JT	S	4Dbr4Cpt2Dr	11/40	PS	4069F	6636
JACINTO CRI&P	S	8Rmt6Dbr	11-12/54	PS	4195	6944
JOAQUIN VALLEY NYC	S	10Rmt6Dbr	12/48-49	Budd	9510	9660
MIGUEL VALLEY ATSF	S	6Sec6Rmt4Dbr	6/42	PS	4099	6669
SANDUSKY BAY NYC	S	22Rmt	10/48	PS	4122	6790
SANDUSKY COUNTY NYC	S	13Dbr	7/39	PS	4071B	6572
SANDY CREEK NYC	S	5DbrREObsBLng	8-9/48	PS	4126	6790
SANDY HOOK BAY NYC	S	22Rmt	10/48	PS	4122	6790
SANDY POINT NH	S	14Rmt4Dbr	12/49/50	PS	4159	6822
SANGAMON B&O	S	14Rmt4Dbr	4-6/48	PS	4153B	6776
SANGAMON RAPIDS PRR	S	10Rmt6Dbr	11/48-49	PS	4140	6792
SANGAMON RIVER NYC	S	10Rmt6Dbr	9/48-/49	PS	4123	6790
SANTA ANITA (DS)UP		11 SecArtic	5/36	PS	4037B	6434
SANTA MONICA JT	S	4Cpt3DrArtic	12/37	PS	4063B	6525
SANTIAM PASS GN-1373	S	6Rmt5Dbr2Cpt	10-11/50	PS	4180	6877
SAPAPONACK LIRR-2074	SN	Parlor Coach	11/54-55	PS	4194	W6942
SARANAC LAKE NYC	S	6DbrBLng	12/48-49	PS	4124	6790
SARASOTA SAL-39	S	10Rmt6Dbr	6-8/49	Budd	9503	9662
SASKATCHEWAN						
RIVER CN-2087	SN	10Rmt6Dbr	9/48-/49	PS	4123	6790
SASSAFRAS FALLS PRR	S	6DbrBLng	3-5/49	PS	4131	6792
SATURNO NdeM-339	SN	12Rmt4Dbr	12/49-50	AC&F	9004	3072
SAUGUS RIVER NYC	S	10Rmt6Dbr	9/48-/49	PS	4123	6790
SAVANNA CB&Q	S	8DuRmt6Rmt4Dbr	7-9/48	PS	4119	6781
SAVANNAH SAL	S	10Rmt6Dbr	5-6/49	PS	4140A	6796
SAVANNAH RIVER ACL	RN	4Cpt4Dr	3-4/38	PS	6013	6540
SAYDATCH ATSF	S	4Dbr4Cpt2Dr	12/39	PS	4069C	6597

CAR NAME OR NUMBER	ST	CAR TYPE	DATE BUILT	CAR BLDR	PLAN	LOT
SCHOHARIE VALLEY NYC	S	10Rmt6Dbr	12/48-49	Budd	9510	9660
SCHUYLKILL B&O	SN	10Rmt6Dbr	2-7/50	PS	4167	6864
SCHUYLKILL RAPIDS PRR	S	10Rmt6Dbr	11/48-49	PS	4140	6792
SCHUYLKILL RIVER PRR	S	10Rmt6Dbr	9-10/49	PS	4140	6814
SCIOTO B&O	S	14Rmt4Dbr	4-6/48	PS	4153B	6776
SCIOTO						
COUNTY N&W	S	10Rmt6Dbr	11/49-50	Budd	9523	9660
RAPIDS PRR-8451	S	10Rmt6Dbr	8/49	Budd	9503	9662
RIVER NYC	S	10Rmt6Dbr	9/48-/49	PS	4123	6790

CAR NAME OR NUMBER	ST	CAR TYPE	DATE BUILT	CAR BLDR	PLAN	LOT
SCOTT M. LOFTIN FEC	N	4Rmt5DbrCpt4Sec	12/54	PS	4196	6949
SEA ISLAND BEACH ACL	S	6DbrBLng	11-12/49	AC&F	9002	3090
SEAL COVE CN (1189)	P	4Dbr4Cpt2Dr	7-8/39	PS	4069D	6571
SEAL ROCKS(1st) JT		11DbrArtic	12/37	PS	4067A	6525
SEAL ROCKS(2nd) JT	N	4Cpt3DrArtic	12/37	PS	4063B	6525
SEAVIEW CP	SN	5DbrBLngObs	5-6/49	Budd	9506	9664
SEBONAC LIRR-2077	SN	Parlor Coach	11/54-55	PS	4194	W6942
SEBOYETA ATSF	S	4Dbr4Cpt2Dr	12/39	PS	4069C	6597
SEBRING SAL-73	S	11Dbr	12/55-56	PS	4198A	6968

Figure I-128   Gulf Mobile & Ohio ordered four cars of Plan 9012, American Car & Foundry Lot 3208 in December 1946. SAMUEL KING TIGRETT, an 8-roomette, compartment, 3-double-bedroom, 4-section car was delivered in July 1950. It was withdrawn from Pullman lease in February 1969 and shown here in June 1969 prior to its sale to Great Western Enterprises where it became their GOLDEN GATE 205. This was an unusual configuration, as befitted a car built by an innovative but hungry car builder for a somewhat eccentric regional carrier. Railroad Avenue Enterprises Photograph PN-3034.

Figure I-129   SAMUEL M. KIER was renamed in July 1956 from SHENANGO RAPIDS, a 10-section, 6-double-bedroom from Pullman Plan 4140, Lot 6792 built in 1949. This photograph, taken in April 1971, does not hint the fact that this car is owned and operated by the Penn Central Railroad, and has been for three years. Many similar cars operated in PRR - eastern road pool service until Amtrak. Railroad Avenue Enterprises Photograph PN-2436.

CAR NAME OR NUMBER	ST	CAR TYPE	DATE BUILT	CAR BLDR	PLAN	LOT
SEGATOA ATSF	S	8Sec2Dbr2Cpt	12/37-38	PS	4058A	6532
SEMINOLE CRI&P	S	8Sec5Dbr	11/40	PS	4095	6631
SEMINOLE COUNTY ACL	S	10Rmt6Dbr	9-10/49	PS	4140B	6809
SENECA LAKE NYC	S	6DbrBLng	12/48-49	PS	4124	6790
SENECA RIVER SOU	S	10Rmt6Dbr	9-10/49	PS	4140	6814
SETAUKET LIRR-2078	SN	Parlor Coach	11/54-55	PS	4194	W6942
SEWICKLEY INN PRR-8282	S	21Rmt	1-6/49	Budd	9513	9667
SEXTON GLACIER GN-1176	S	16DuRmt4Dbr	1-2/47	PS	4108	6751
SHANTO ATSF	S	8Sec2Dbr2Cpt	12/37-38	PS	4058A	6532
SHARON INN PRR-8283	S	21Rmt	1-6/49	Budd	9513	9667
SHEEPSHEAD BAY NYC	S	22Rmt	10/48	PS	4122	6790
SHELBURNE FALLS NYC	S	6DbrBLng	8-9/40	PS	4086B	6612
SHENANDOAH B&O	SN	10Rmt6Dbr	2-7/50	PS	4167	6864
SHENANDOAH CLUB C&O	S	5DbrBEObsBLng	8/50	PS	4165	6863
SHENANDOAH RIVER SOU	S	10Rmt6Dbr	9-10/49	PS	4140	6814
SHENANDOAH VALLEY SOU	S	14Rmt4Dbr	10-11/49	PS	4153C	6814
SHENANGO B&O	S	10Rmt6Dbr	3/50	PS	4167	6864
SHENANGO RAPIDS PRR	S	10Rmt6Dbr	11/48-49	PS	4140	6792
SHENANGO RIVER NYC	S	10Rmt6Dbr	9/48-/49	PS	4123	6790
SHERMAN RAPIDS PRR	S	10Rmt6Dbr	11/48-49	PS	4140	6792

CAR NAME OR NUMBER	ST	CAR TYPE	DATE BUILT	CAR BLDR	PLAN	LOT
SHERWOOD MANOR CP-10339S	S	5DbrCpt4Sec4Rmt	11/54-55	Budd		9658
SHEYENNE RIVER GN-1266	S	8DuRmt4Sec3DbrCpt	11-12/50	PS	4181	6889
SHINNECOCK LIRR-2079	SN	Parlor Coach	11/54-55	PS	4194	W6942
SHIP COVE CN-1185	SN	4Dbr4Cpt2Dr	9-10/40	PS	4069E	6617
SHIPPAN POINT NH	S	14Rmt4Dbr	12/49/50	PS	4159	6822
SHORT						
LEAF PINE L&N-3464	S	6Sec6Rmt4Dbr	3-5/53	PS	4183	6909
SIERRA						
AMATEPEC NdeM-254	SN	13Dbr	10/40	PS	4071C	6618
AZUL NdeM-253	SN	13Dbr	7/39	PS	4071B	6572
CASTIZA NdeM-248	SN	13Dbr	10/40	PS	4071C	6618
SIERRA DE						
GUADARROMA NdeM-262	SN	13Dbr	10/40	PS	4071C	6618
IXTLAN NdeM	SN	13Bdr	4/38	PS	4071A	6541
OAXACA NdeM-256	SN	13Bdr	4/38	PS	4071A	6541
ORO NdeM-251	SN	13Dbr	10/40	PS	4071C	6618
SIERRA						
FRIA NdeM	SN	13Dbr	7/39	PS	4071B	6572
GORDA NdeM-261	SN	13Dbr	7/39	PS	4071B	6572
GRANDE NdeM	SN	13Bdr	4/38	PS	4071A	6541

Figure I-130 The B&O's SHENANDOAH was built by Pullman in 1950 as a part of the large C&O order as the C&O CITY OF ATHENS, later CITY OF PETOSKEY. B&O purchased this 10-roomette, 6-double-bedroom sleeper in February 1957, operating it with the almost identical 10-6 sleepers B&O purchased from Pullman in 1950 when C&O canceled the balance of the order. It continued in service until the advent of Amtrak when it was sold to George Pins who named it KEYSTONE RAPIDS. Now it belongs to the C&O Historical Society. The Author's Collection.

CAR NAME OR NUMBER	ST	CAR TYPE	DATE BUILT	CAR BLDR	PLAN	LOT
SIERRA						
LEONA NdeM-249	SN	13Dbr	7/39	PS	4071B	6572
MADRE NdeM	SN	13Bdr	4/38	PS	4071A	6541
MOJADA NdeM-255	SN	13Bdr	4/38	PS	4071A	6541
MORENA NdeM-246	SN	13Bdr	4/38	PS	4071A	6541
NEVADA NdeM-258	SN	13Bdr	4/38	PS	4071A	6541
TARAHUMARA NdeM-245	SN	13Bdr	4/38	PS	4071A	6541
VENTANA NdeM-260	SN	13Dbr	10/40	PS	4071C	6618
VERDE NdeM-252	SN	13Dbr	7/39	PS	4071B	6572
SILVER						
ARROW CB&Q-412	S	12SecArtic	10/36	Budd	9512	965
ARROYO WP-861	S	10Rmt6Dbr	9-11/48	Budd	9509	9660
ASPEN D&RGW-1120	S	16Sec	10-11/48	Budd	9507	9639
BASIN CB&Q-492	S	10Rmt6Dbr	6-7/56	Budd	9538	9660
BAY WP-866	S	10Rmt6Dbr	9-11/48	Budd	9509	9660
BOULDER CB&Q-488	S	10Rmt6Dbr	6-7/56	Budd	9538	9660
BUTTE CB&Q-425	S	10Rmt6Dbr	9-11/48	Budd	9509	9660
CANYON WP-862	S	10Rmt6Dbr	9-11/48	Budd	9509	9660
CEDAR CB&Q-402	S	16Sec	9/52	Budd	9532	9659
CHANNEL CB&Q-489	S	10Rmt6Dbr	6-7/56	Budd	9538	9660
CHASM CB&Q-430	S	10Rmt6Dbr	6/52	Budd	9531	9660
CLIFF CB&Q-426	S	10Rmt6Dbr	9-11/48	Budd	9509	9660
CRAG CB&Q-429	S	10Rmt6Dbr	6/52	Budd	9531	9660
CRANE WP-851	S	6Dbr5Cpt	7-8/52	Budd	9534	9641
CREEK D&RGW-1133	S	10Rmt6Dbr	9-11/48	Budd	9509	9660
CRESCENT WP-881	S	3DbrDrDomeLngObs	12/48-49	Budd	9511	9659
DALE CB&Q-431	S	10Rmt6Dbr	6/52	Budd	9531	9660
DOLLAR (DS)UP		12SecArtic	6/36	PS	4053	6486
DOVE CB&Q-450	S	6Dbr5Cpt	7-8/52	Budd	9534	9641
FALLS CB&Q-427	S	10Rmt6Dbr	4/48	Budd	9509	9660
FLOWER CB&Q-460	S	6Sec6Rmt4Dbr	8-9/52	Budd	9530	9658
GLACIER D&RGW-1134	S	10Rmt6Dbr	9-11/48	Budd	9509	9660
GLADIOLA CB&Q-461	S	6Sec6Rmt4Dbr	8-9/52	Budd	9530	9658
GORGE D&RGW-1132	S	10Rmt6Dbr	9-11/48	Budd	9509	9660

CAR NAME OR NUMBER	ST	CAR TYPE	DATE BUILT	CAR BLDR	PLAN	LOT
SILVER						
GULL D&RGW-1135	S	6Dbr5Cpt	7-8/52	Budd	9534	9641
HOLLOW CB&Q-487	S	10Rmt6Dbr	6-7/56	Budd	9538	9660
HORIZON CB&Q-375	S	3DbrDrDomeLngObs	12/48-49	Budd	9511	9659
HYACINTH CB&Q-462	S	6Sec6Rmt4Dbr	8-9/52	Budd	9530	9658
IRIS CB&Q-463	S	6Sec6Rmt4Dbr	8-9/52	Budd	9530	9658
ISLE CB&Q-432	S	10Rmt6Dbr	6/52	Budd	9531	9660
LAKE NYC	S	6DbrBLng	12/48-49	PS	4124	6790
LARCH CB&Q-401	S	16Sec	10-11/48	Budd	9507	9639
LOOKOUT CB&Q-378	S	3DbrDrDomeLngObs	12/52	Budd	9534	9659
MAPLE CB&Q-400	S	16Sec	10-11/48	Budd	9507	9639
MEADOW CB&Q-433	S	10Rmt6Dbr	6/52	Budd	9531	9660
MOON CB&Q-446	S	4Rmt4SrDrCpt4Dbr	3/39	Budd	9516	96105
MOUNTAIN WP-863	S	10Rmt6Dbr	9-11/48	Budd	9509	9660
ORCHID CB&Q-464	S	6Sec6Rmt4Dbr	8-9/52	Budd	9530	9658
PALISADE WP-864	S	10Rmt6Dbr	9-11/48	Budd	9509	9660
PALM WP-871	S	16Sec	10-11/48	Budd	9507	9639
PASS D&RGW-1130	S	10Rmt6Dbr	9-11/48	Budd	9509	9660
PELICAN CB&Q-454	S	6Dbr6Cpt	8-9/56	Budd	9539	9641
PENTHOUSE CB&Q-376	S	3DbrDrDomeLngObs	12/48-49	Budd	9511	9659
PINE D&RGW-1121	S	16Sec	10-11/48	Budd	9507	9639
PLAIN CB&Q-434	S	10Rmt6Dbr	6/52	Budd	9531	9660
PLANET WP-882	S	3DbrDrDomeLngObs	12/48-49	Budd	9511	9659
PLATEAU CB&Q-486	S	10Rmt6Dbr	6-7/56	Budd	9538	9660
POINT CB&Q-423	S	10Rmt6Dbr	9-11/48	Budd	9509	9660
POPLAR WP-872	S	16Sec	10-11/48	Budd	9507	9639
PRAIRIE CB&Q-435	S	10Rmt6Dbr	6/52	Budd	9531	9660
QUAIL CB&Q-451	S	6Dbr5Cpt	7-8/52	Budd	9534	9641
RANGE WP-865	S	10Rmt6Dbr	9-11/48	Budd	9509	9660
RAPIDS PRR-8449	S	10Rmt6Dbr	11/48	Budd	9520	9662
RAVINE CB&Q-491	S	10Rmt6Dbr	6-7/56	Budd	9538	9660
REPOSE CB&Q-4903	S	24Sr8Dbr	10-11/56	Budd	9540	9658
REST CB&Q-4902	S	24Sr8Dbr	10-11/56	Budd	9540	9658
RIDGE CB&Q-493	S	10Rmt6Dbr	6-7/56	Budd	9538	9660

Figure I-131   Nine Budd-built 10-roomette, 6-double-bedroom cars were added to the "Denver Zephyr" in October 1956. The aisle side of SILVER RIDGE is shown in this June 1970 photograph. This car was arranged with 6 roomettes on the left side, 6 double-bedrooms in the center and 4 roomettes on the vestibule end, similar to the C&O center-bedroom cars of 1950. At this late date, CB&Q continued to order the old GSC 41-NP truck. Railroad Avenue Enterprises Photograph PN-2450.

CAR NAME OR NUMBER	ST	CAR TYPE	DATE BUILT	CAR BLDR	PLAN	LOT
SILVER						
SCREEN CB&Q-414	S	12SecArtic	10/36	Budd	9514	965
SHORE CB&Q-424	S	10Rmt6Dbr	9-11/48	Budd	9509	9660
SIDES CB&Q-440	S	3Cpt6DbrDrArtic	10/36	Budd	9515	965
SIESTA CB&Q-4900	S	24Sr8Dbr	10-11/56	Budd	9540	9658
SKATES CB&Q-413	S	12SecArtic	10/36	Budd	9512	965
SKY D&RGW-1145	S	3DbrDrDomeLngObs	12/48-49	Budd	9511	9659
SLIPPER CB&Q-445	S	4Rmt4SrDrCpt4Dbr	3/39	Budd	9516	96105
SLOPE CB&Q-436	S	10Rmt6Dbr	6/52	Budd	9531	9660
SLUMBER CB&Q-4901	S	24Sr8Dbr	10-11/56	Budd	9540	9658
SOLARIUM CB&Q-377	S	3DbrDrDomeLngObs	12/48-49	Budd	9511	9659
STATE CB&Q-410	S	12SecArtic	10/36	Budd	9512	965
SUMMIT D&RGW-1131	S	10Rmt6Dbr	9-11/48	Budd	9509	9660
SURF WP-867	S	10Rmt6Dbr	9-11/48	Budd	9509	9660
SWALLOW WP-852	S	6Dbr5Cpt	7-8/52	Budd	9534	9641
SWAN CB&Q-453	S	6Dbr6Cpt	8-9/56	Budd	9539	9641
TERRAIN CB&Q-485	S	10Rmt6Dbr	6-7/56	Budd	9538	9660
THREADS CB&Q-441	S	3Cpt6DbrDrArtic	10/36	Budd	9515	965
THRUSH CB&Q-452	S	6Dbr5Cpt	7-8/52	Budd	9534	9641
TIP CB&Q-411	S	12SecArtic	10/36	Budd	9514	965
TONE CB&Q-415	S	12SecArtic	10/36	Budd	9512	965
TULIP CB&Q-465	S	6Sec6Rmt4Dbr	8-9/52	Budd	9530	9658
VALE CB&Q-490	S	10Rmt6Dbr	6-7/56	Budd	9538	9660
VALLEY CB&Q-428	S	10Rmt6Dbr	9-11/48	Budd	9509	9660
SINGING BROOK NYC	S	5DbrBLngObs	6-7/49	Budd	9508	9636
SINYALA ATSF	S	8Sec2Dbr2Cpt	12/37-38	PS	4058A	6532
SIOUX FALLS Milw-5770		8SecCoachTouralux	6/9/47	Milw		
SIRIA NdeM-522	SN	10Rmt6Dbr	9/48-/49	PS	4123	6790
SIRIUS FCP	SN	10Rmt6Dbr	9/48-/49	PS	4123	6790
SISIBOO FALLS CN-2097	SN	14Rmt4Dbr	1-6/48	PS	4153	6769
SIYEH GLACIER CB&Q-1179	S	16DuRmt4Dbr	1-2/47	PS	4108	6751
SKAGIT RIVER GN-1267	S	8DuRmt4Sec3DbrCpt	11-12/50	PS	4181	6889
SKANEATELES LAKE NYC	S	6DbrBLng	12/48-49	PS	4124	6790
SKEENA RIVER CN-2091	SN	10Rmt6Dbr	9/48-/49	PS	4123	6790
SKYKOMISH RIVER GN-1260	S	8DuRmt4Sec3DbrCpt	11-12/50	PS	4181	6889
SKYKOMISH RIVER GN-1260	R	8DuRmt6DbrCpt	11-12/50	PS	4181A	6889
SKYLAND VALLEY NYC	S	10Rmt6Dbr	12/48-49	Budd	9510	9660
SKYLINE VIEW PRR	S	2MbrDbrREObsBLng	5/38	PS	4080	6548
SLEEPLAND B&O	L	24Sr8Dbr	5/59	Budd	9540	9691
SLUMBERLAND B&O		24Sr8Dbr	2/58	Budd	9540	9691
SMITHTOWN BAY NYC	S	22Rmt	10/48	PS	4122	6790
SMOKY RIVER CN-2090	SN	10Rmt6Dbr	9/48-/49	PS	4123	6790
SNAKE RIVER SP&S-702	S	8DuRmt4Sec3DbrCpt	11-12/50	PS	4181	6889
SNOHOMISH RIVER GN-1262	R	8DuRmt6DbrCpt	11-12/50	PS	4181A	6889
SNOHOMISH RIVER GN-1262	S	8DuRmt4Sec3DbrCpt	11-12/50	PS	4181	6889
SNOWY RANGE (DS)UP		8Sec2DbrCptArtic	6/36	PS	4054	6486
SODUS BAY NYC	S	22Rmt	10/48	PS	4122	6790
SONOMA VALLEY NYC	S	10Rmt6Dbr	12/48-49	Budd	9510	9660
SOUND BEACH NH-537	S	6Sec6Rmt4Dbr	11/54-55	PS	4194	W6942
SOUTH						
BEND CN (2038)	P	18Rmt	6-7/39	PS	4068H	6570
SOUTH						
BRANCH CN-2045	SN	18Rmt	9/40	PS	4068J	6616
BROOK CN (2039)	P	18Rmt	6-7/39	PS	4068H	6570
DURHAM CN (2046)	P	18Rmt	6-7/39	PS	4068H	6570
ELBOW CN-2040	SN	18Rmt	9/40	PS	4068J	6616
FIELD CN (2050)	N	18Rmt	6-7/39	PS	4068H	6570
HAVEN HARBOR NYC	S	22Rmt	4-6/49	Budd	9501	9661
MARCH CN-2047	SN	18Rmt	9/40	PS	4068J	6616
MEGNETAWAN CN (2044)	P	18Rmt	6-7/39	PS	4068H	6570
NELSON CN (2041)	P	18Rmt	6-7/39	PS	4068H	6570
PARIS CN-2048	SN	18Rmt	9/40	PS	4068J	6616
PARRY CN (2042)	P	18Rmt	6-7/39	PS	4068H	6570
PLATTE JT	S	4Rmt3CptDr4Dbr	5/39	PS	4083	6568
PORCUPINE CN (2049)	P	18Rmt	6-7/39	PS	4068H	6570
RIVER CN (2050)	P	18Rmt	6-7/39	PS	4068H	6570
TWIN LAKES BAR-81	S	6Sec6Rmt4Dbr	11/54-55	PS	4194	W6942
SOUTHERN PINE L&N-3465	S	6Sec6Rmt4Dbr	3-5/53	PS	4183	6909
SOUTHERN PINES SAL-51	S	5DbrCpt4Sec4Rmt	11/55	Budd	9537	9658
SOUTHLAND MP	L	24Sr8Dbr	9/59	Budd	9540	9691
SOUTHPORT CN-2043	SN	18Rmt	9/40	PS	4068J	6616
SOUTHWOOD CN-2051	SN	18Rmt	9/40	PS	4068J	6616
SPANISH CREEK Milw-18	S	8DbrSKYTOPLng	11/48-49	PS	4138	6775
SPARKLING RIVER NYC	S	10Rmt6Dbr	9/48-/49	PS	4123	6790
SPEONK LIRR-2080	SN	Parlor Coach	11/54-55	PS	4194	W6942
SPERRY						
GLACIER CB&Q-1178	S	16DuRmt4Dbr	1-2/47	PS	4108	6751
SPIRIT						
OF YOUNGSTOWN Erie	S	10Rmt6Dbr	6/54	PS	4186B	6946
SPLASH PINE C&EI-900	S	6Sec6Rmt4Dbr	3-5/53	PS	4183	6909
SPOKANE CLUB NP	S	5DbrREObsBLng	6-7/48	PS	4120	6781
SPOKANE RIVER GN-1272	S	8DuRmt4Sec3DbrCpt	11-12/50	PS	4181	6889
SPOTSYLVANIA						
COUNTY RF&P	S	10Rmt6Dbr	9-10/49	PS	4140B	6809
SPRING RIVER SLSF-1465	S	14Rmt4Dbr	1-6/48	PS	4153	6769
SPRING VALLEY SOU	S	14Rmt4Dbr	10-11/49	PS	4153C	6814
SPRUCE FALLS PRR	S	6DbrBLng	9/40	PS	4086A	6612
SPRUCE GROVE CP-14117		10Sec5Dbr	/49-/50	CPRR		
SQUAW BONNET (DS)UP		8Sec2DbrCptArtic	6/36	PS	4054	6486
ST.						
CLAIR RIVER NYC	S	10Rmt6Dbr	9/48-/49	PS	4123	6790
CROIX RIVER NYC	S	10Rmt6Dbr	9/48-/49	PS	4123	6790
FRANCIS RIVER NYC	S	10Rmt6Dbr	9/48-/49	PS	4123	6790
JOE RIVER Milw-20	S	8DuRmt6Rmt4Dbr	10-12/48	PS	4135	6775
JOHNS RIVER SOU	S	10Rmt6Dbr	9-10/49	PS	4140	6814
JOSEPH RIVER NYC	S	10Rmt6Dbr	9/48-/49	PS	4123	6790
LAWRENCE RIVER NYC	S	10Rmt6Dbr	9/48-/49	PS	4123	6790
LOUISAN IC	S	4Dbr4Cpt2Dr	4/42	PS	4069H	6668
MARYS RIVER NYC	S	10Rmt6Dbr	9/48-/49	PS	4123	6790
PETERSBURG SAL-38	S	10Rmt6Dbr	6-8/49	PS	9503	9662
REGIS RIVER NYC	S	10Rmt6Dbr	9/48-/49	PS	4123	6790
SIMON ISLAND ACL	S	21Rmt	9-10/49	PS	4156B	6809

CAR NAME OR NUMBER	ST	CAR TYPE	DATE BUILT	CAR BLDR	PLAN	LOT
ST. VRAINS (DS)UP		12SecArtic	6/36	PS	4053	6486
STAMPEDE PASS BN	N	10Rmt6Dbr	5-7/50	PS	4140C	6874
STAR BAY UP	RN	11Dbr	4/56	PS	6008	6959
STAR CREST UP	RN	11Dbr	4/56	PS	6008	6959
STAR LEAF UP	RN	11Dbr	4/56	PS	6008	6959
STAR RANGE UP	RN	11Dbr	4/56	PS	6008	6959
STAR SCENE UP	RN	11Dbr	4/56	PS	6008	6959
STAR VALE UP	RN	11Dbr	4/56	PS	6008	6959
STAR VIEW UP	RN	11Dbr	4/56	PS	6008	6959
STARLIGHT DOME B&O(7601)	SN	Dome5RmtDbr3Dr	9/48	Budd	9524	9669

CAR NAME OR NUMBER	ST	CAR TYPE	DATE BUILT	CAR BLDR	PLAN	LOT
STATE PASS GN-1377	S	6Rmt5Dbr2Cpt	10-11/50	PS	4180	6877
STATEN ISLAND PRR	S	21Rmt	9-10/49	PS	4156B	6809
STEEL CITY AMT-2555	SN	8Rmt6Dbr	11-12/54	PS	4195	6944
STEPHEN F. AUSTIN MKT-1400	S	2DbrDrREObsLng	5/48	PS	4121	6769
STEUBENVILLE INN PRR-8284	S	21Rmt	1-6/49	Budd	9513	9667
STEVENS PASS GN-1180	S	8DuRmt4Dbr4Sec	1/50	PS	4107	6883
STILLWATER RIVER NYC	S	10Rmt6Dbr	9/48-/49	PS	4123	6790
STONE MOUNTAIN SAL-16	S	6DbrBLng	8/49	AC&F	9003	3045
STONEY RAPIDS PRR	S	10Rmt6Dbr	11/48-49	PS	4140	6792

Figure I-132 This B&O Slumbercoach SLUMBERLAND delivered by Budd in February 1958 for service on "The Columbian" had 24 single and 8 double rooms. This car and its sister, DREAMLAND, were sold to the High Iron Company for use in the Golden Spike Centennial Train in 1969. Surprisingly these cars, out of active railroad service for 13 years, were purchased by Amtrak in 1982 and placed back in service under the numbers 2096 and 2097 respectively. The durability of Budd stainless steel cars is one of the few immutable truths left in the railroad business today. The Author's Collection.

CAR NAME OR NUMBER	ST	CAR TYPE	DATE BUILT	CAR BLDR	PLAN	LOT
STRATFORD COUNTY RF&P	S	10Rmt6Dbr	9-10/49	PS	4140B	6809
STRATFORD POINT NH	S	14Rmt4Dbr	12/49/50	PS	4159	6822
STUART KNOTT KCS	S	14Rmt4Dbr	5/48	PS	4153	6795
STUART MANOR CP-10340	S	5DbrCpt4Sec4Rmt	11/54-55	Budd		9658
STURGEON						
BAY NYC	S	22Rmt	10/48	PS	4122	6790
RAPIDS PRR-8452	S	10Rmt6Dbr	8/49	Budd	9503	9662
STUYVESANT FALLS NYC	S	6DbrBLng	8-9/40	PS	4086B	6612
SUDAN NdeM-543	SN	10Rmt6Dbr	9/48-/49	PS	4123	6790
SUECIA NdeM-512	SN	10Rmt6Dbr	9/48-/49	PS	4123	6790
SUGARLAND IC	S	6Sec6Rmt4Dbr	4/42	PS	4099	6669
SUIATTLE PASS GN-1380	S	6Rmt5Dbr2Cpt	10-11/50	PS	4180	6877
SUIZA NdeM-513	SN	10Rmt6Dbr	9/48-/49	PS	4123	6790
SULLIVANS ISLAND ACL	S	21Rmt	9-10/49	PS	4156B	6809
SUMAC FALLS PRR	S	6DbrBLng	3-5/49	PS	4131	6792
SUMATRA NdeM-535	SN	10Rmt6Dbr	9/48-/49	PS	4123	6790
SUMTER COUNTY ACL	S	10Rmt6Dbr	9-10/49	PS	4140B	6809
SUN BEAM SCL-6500	N	5DbrBLng	1/56	PS	4202	6970
SUN CAPE UP	RN	11Dbr	12/49-50	AC&F	6009	3069
SUN ISLE UP	RN	11Dbr	12/49-50	AC&F	6009	3069
SUN LAKE UP	RN	11Dbr	12/49-50	AC&F	6009	3069
SUN LANE UP	RN	11Dbr	12/49-50	AC&F	6009	3069
SUN MANOR UP	RN	11Dbr	12/49-50	AC&F	6009	3069
SUN PARK UP	RN	11Dbr	12/49-50	AC&F	6009	3069
SUN POINT UP	RN	11Dbr	12/49-50	AC&F	6009	3069
SUN RAY SCL-6501	N	5DbrBLng	1/56	PS	4202	6970
SUN REST UP	RN	11Dbr	12/49-50	AC&F	6009	3069
SUN RIDGE UP	RN	11Dbr	12/49-50	AC&F	6009	3069
SUN RIVER GN-1261	S	8DuRmt4Sec3DbrCpt	11-12/50	PS	4181	6889
SUN SKIES UP	RN	11Dbr	12/49-50	AC&F	6009	3069
SUN SLOPE UP	RN	11Dbr	12/49-50	AC&F	6009	3069
SUN VIEW SCL-6502	N	5DbrBLng	1/56	PS	4202	6970
SUN VILLA UP	SN	11Dbr	12/49-50	AC&F	6009	3069
SUNBURY INN PRR-8285	S	21Rmt	1-6/49	Budd	9513	9667
SUNLIGHT DOME B&O(7602)	SN	Dome5RmtDbr3Dr	9/48	Budd	9524	9669
SUNRISE BROOK NYC	S	5DbrBLngObs	6-7/49	Budd	9508	9636
SUNSHINE VALLEY ATSF	S	6Sec6Rmt4Dbr	6/42	PS	4099	6669
SURF BIRD ACL	RN	7Dbr2Dr	2-3/50	AC&F	9018	3091
SURPRISE VALLEY ATSF	S	6Sec6Rmt4Dbr	6/42	PS	4099	6669
SUSPENSION BRIDGE NYC	N	4Dbr4Cpt2Dr	3-4/38	PS	4069B	6540
SUSQUEHANNA RIVER PRR	S	10Rmt6Dbr	9-10/49	PS	4140	6814
SUSSEX COUNTY N&W	S	10Rmt6Dbr	3/49	PS	4140	6792
SUTRO HEIGHTS JT	S	4Dbr4Cpt2Dr	11/40	PS	4069F	6636
SUWANEE RIVER ACL	RN	4Cpt4Dr	3-4/38	PS	6013	6540
SUWANEE RIVER ACL	S	14Rmt2Dr	2-3/50	AC&F	9007	3091
SWAN B&O	S	16DuRmt4Dbr	4-6/54	Budd	9536	9658
SWAN RIVER NYC	S	10Rmt6Dbr	9/48-/49	PS	4123	6790
SWATARA RAPIDS PRR	S	10Rmt6Dbr	11/48-49	PS	4140	6792

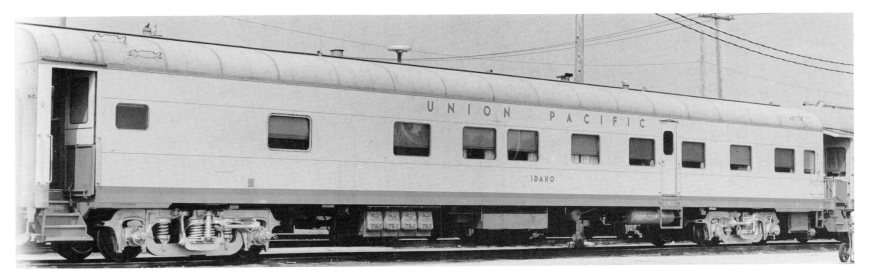

Figure I-133 Union Pacific staff car IDAHO was modified for company use from SUN LANE, an 11-double-bedroom sleeper that had been rebuilt from the AC&F-built 12-roomette, 4-double-bedroom WESTERN MOUNTAIN in August 1965. Lightweight cars rarely undergo that many rebuildings, common with standard cars. Photograph by George Cockle.

CAR NAME OR NUMBER	ST	CAR TYPE	DATE BUILT	CAR BLDR	PLAN	LOT
SWEETWATER VALLEY ATSF	S	6Sec6Rmt4Dbr	6/42	PS	4099	6669
SWIFT						
CURRENT PASS GN-1169	S	8DuRmt4Dbr4Sec	6/51	PS	4107	6895
CURRENT PASS CB&Q-1169	S	8DuRmt4Dbr4Sec	12/46-47	PS	4107	6751
STREAM NYC	S	6DbrBLng	3-5/49	Budd	9505	9663
SYCAMORE FALLS PRR	S	6DbrBLng	9/40	PS	4086A	6612
TABACCO RIVER GN-1270	S	8DuRmt4Sec3DbrCpt	11-12/50	PS	4181	6889
TACOMA CLUB NP	S	5DbrREObsBLng	6-7/48	PS	4120	6781
TAHOMA GLACIER GN-1187	S	16DuRmt4Dbr	12/50	PS	4108A	6890
TAILANDIA NdeM-515	SN	10Rmt6Dbr	9/48-/49	PS	4123	6790
TAJIN NdeM-263	L	7Cpt2Dr	/52	Schi	100571	
TALL PINE L&N-3466	S	6Sec6Rmt4Dbr	3-5/53	PS	4183	6909
TALLAHASSEE SAL-75	S	11Dbr	12/55-56	PS	4198A	6968
TALWIWI ATSF	S	8Sec2Dbr2Cpt	6-7/38	PS	4058B	6553
TAMPA SAL	S	10Rmt6Dbr	5-6/49	PS	4140A	6796
TAMPA BAY NYC	S	22Rmt	10/48	PS	4122	6790
TAOS ATSF	S	6Dbr2Cpt2Dr	4/37	Budd	9519	967
TAOS VALLEY ATSF	S	6Sec6Rmt4Dbr	6/42	PS	4099	6669
TAPACIPA ATSF	S	4Dbr4Cpt2Dr	12/39	PS	4069C	6597
TAURUS FCP	SN	10Rmt6Dbr	9/48-/49	PS	4123	6790
TAWAS BAY NYC	S	22Rmt	10/48	PS	4122	6790
TCHIREGE ATSF	S	4Dbr4Cpt2Dr	7/38	PS	4069C	6553
TELEGRAPH HILL(1st) JT	N	18Rmt	12/37	PS	4068A	6525
TELEGRAPH HILL(2nd) JT	N	12DuSr5Dbr	12/37	PS	4066A	6525

CAR NAME OR NUMBER	ST	CAR TYPE	DATE BUILT	CAR BLDR	PLAN	LOT
TENNESSEE						
PINE NC&SL-202	S	6Sec6Rmt4Dbr	3-5/53	PS	4183	6909
VALLEY SOU	S	14Rmt4Dbr	10-11/49	PS	4153C	6814
TEOTIHUACAN NdeM-267	L	7Cpt2Dr	/52	Schi	100571	
TERRA						
NOVA RIVER CN-2130	SN	10Rmt6Dbr	9-10/49	PS	4140	6814
TERRYTOWN HARBOR NYC	S	22Rmt	4-6/49	Budd	9501	9661
TESUQUE VALLEY ATSF	S	6Sec6Rmt4Dbr	6/42	PS	4099	6669
THAMES RIVER NYC	S	10Rmt6Dbr	9/48-/49	PS	4123	6790
THE BROADMOOR CRI&P	S	8Rmt6Dbr	11-12/54	PS	4195	6944
THE KOKOSING AMT-3270	SN	4CptDbrLng	11/50	AC&F	9013A	3360
THE POTOMAC AMT-2556	SN	8Rmt6Dbr	11-12/54	PS	4195	6944
THE ROOSEVELT PC-4219	SN	6Sec6Rmt4Dbr	11/54-55	PS	4194	W6942
THOMAS						
ALEXANDER SCOTT PRR	S	2DrCptDbrBEObsLng	3-4/49	PS	4134	6792
HART BENTON SLSF-1451	S	14Rmt4Dbr	1-6/48	PS	4153	6769
THOMPSON CANYON CRI&P	S	8Sec5Dbr	11/40	PS	4095	6631
THOMPSON MANOR CP-10341	S	5DbrCpt4Sec4Rmt	11/54-55	Budd		9658
THORNAPPLE RIVER NYC	S	10Rmt6Dbr	9/48-/49	PS	4123	6790
THOUSAND						
ISLANDS NYC	S	DrMbrObsBLng	5/38	PS	4079	6547
ISLANDS BRIDGE NYC	N	4Dbr4Cpt2Dr	3-4/38	PS	4069B	6540
THREE MILE BAY NYC	S	22Rmt	10/48	PS	4122	6790
THRIFTLAND B&O	L	24Sr8Dbr	5/59	Budd	9540	9691

Figure I-134 Union Pacific SUN SLOPE, an all-bedroom car, was rebuilt in September 1965 from WESTERN WONDERLAND, a 12-roomette, 4-double-bedroom car built by American Car & Foundry in 1950. SUN REST was not selected for purchase by Amtrak in 1971. In 1973, it was converted to UP excursion train use. Railroad Avenue Enterprises Photograph PN-2813.

CAR NAME OR NUMBER	ST	CAR TYPE	DATE BUILT	CAR BLDR	PLAN	LOT
THRUSH B&O	S	16DuRmt4Dbr	4-6/54	Budd	9536	9658
THUNDER BAY NYC	S	22Rmt	10/48	PS	4122	6790
THUNDER BAY CN-2026	S	10Rmt5Dbr	6/54	PS	4186A	6925
THUNDER MOUNTAIN CRI&P	S	5DBRREObsBLng	10/39	PS	4089	6587
TIBET NdeM-530	SN	10Rmt6Dbr	9/48-/49	PS	4123	6790
TIDEWATER CLUB C&O	S	5DbrBEObsBLng	8/50	PS	4165	6863
TIFFIN INN PRR-8286	S	21Rmt	1-6/49	Budd	9513	9667
TIFFIN RIVER NYC	S	10Rmt6Dbr	9/48-/49	PS	4123	6790
TIGER RIVER SOU	S	10Rmt6Dbr	9-10/49	PS	4140	6814
TIMBERLAND IC	S	6Sec6Rmt4Dbr	4/42	PS	4099	6669
TIMES SQUARE AMT-2551	SN	8Rmt6Dbr	11-12/54	PS	4195	6944
TIMOTHY						
B.BLACKSTONE GM&O	S	8RmtCpt3Dbr4Sec	7/50	AC&F	9012	3208
TIOGA RAPIDS PRR	S	10Rmt6Dbr	11/48-49	PS	4140	6792
TIOUGHNIOGA DL&W	S	10Rmt 6Dbr	6-7/49	AC&F	9008A	3089
TIPPECANOE RAPIDS PRR	S	10Rmt6Dbr	11/48-49	PS	4140	6792
TIPPECANOE RIVER NYC	S	10Rmt6Dbr	9/48-/49	PS	4123	6790
TOADLENA ATSF	S	8Sec2Dbr2Cpt	12/37-38	PS	4058A	6532
TOBYHANNA DL&W	S	10Rmt 6Dbr	6-7/49	AC&F	9008A	3089
TOHATCHI ATSF	S	8Sec2Dbr2Cpt	12/37-38	PS	4058A	6532
TOLANI ATSF	S	8Sec2Dbr2Cpt	12/37-38	PS	4058A	6532
TOLCHICO ATSF	S	8Sec2Dbr2Cpt	12/37-38	PS	4058A	6532
TOLEDO HARBOR NYC	S	22Rmt	4-6/49	Budd	9501	9661
TOMBIGBEE RIVER SOU	S	10Rmt6Dbr	9-10/49	PS	4140	6814

CAR NAME OR NUMBER	ST	CAR TYPE	DATE BUILT	CAR BLDR	PLAN	LOT
TONALEA ATSF	S	8Sec2Dbr2Cpt	12/37-38	PS	4058A	6532
TONAWANDA HARBOR NYC	S	22Rmt	4-6/49	Budd	9501	9661
TONTO ATSF	S	17RmtSec	1-2/38	PS	4068B	6532
TOPSOIL FALLS CN-2095	SN	14Rmt4Dbr	1-6/48	PS	4153	6769
TORCH RIVER CN-2146	SN	10Rmt6Dbr	10-11/48	PS	4137	6775
TOREVA ATSF	S	8Sec2Dbr2Cpt	12/37-38	PS	4058A	6532
TORONTO HARBOUR PC-4335	SN	10Rmt6Dbr	8/49	Budd	9503	9662
TORONTO ISLANDS PC-4334	SN	10Rmt6Dbr	8/49	Budd	9503	9662
TOROWEAP ATSF	S	8Sec2Dbr2Cpt	12/37-38	PS	4058A	6532
TOWANDA RAPIDS PRR	S	10Rmt6Dbr	11/48-49	PS	4140	6792
TOWER VIEW PRR	S	2MbrDbrREObsBLng	1/49	PS	4133	6792
TOWERING PINE L&N-3467	S	6Sec6Rmt4Dbr	3-5/53	PS	4183	6909
TOWN OF PRINCE C&O	S	10Rmt6Dbr	2-7/50	PS	4167	6864
TOWN OF THURMOND C&O	S	10Rmt6Dbr	2-7/50	PS	4167	6864
TRAIL COULEE CB&Q-1194	RN	4DbrCpt6RmtObsLng	1-2/47	PS	4109K	6751
TRAVERSE BAY NYC	S	22Rmt	10/48	PS	4122	6790
TRAVERSE CITY C&O	S	10Rmt6Dbr	2-7/50	PS	4167	6864
TRI-BORO BRIDGE NYC	N	4Dbr4Cpt2Dr	3-4/38	PS	4069B	6540
TRINITY CN-1903	SN	8DbrSKYTOPLng	11/48-49	PS	4138	6775
TRINITY RIVER NYC	S	10Rmt6Dbr	9/48-/49	PS	4123	6790
TRIPLE						
DIVIDE PASS GN-1165	S	8DuRmt4Dbr4Sec	12/46-47	PS	4107	6751
TROUT LAKE NYC	S	6DbrBLng	12/48-49	PS	4124	6790
TSANKAWI ATSF	S	4Dbr4Cpt2Dr	7/38	PS	4069C	6553

Figure I-135 This car was built by Pullman for service on Southern Railway's "Crescent Limited." TIGER RIVER remained in the Southern-named train fleet for 30 years before it was sold to Amtrak in 1979 with the withdrawal of Southern from the passenger train business. The River cars, of the ubiquitous Plan 4140, fulfilled their projected three-decade service life virtually unchanged. Railroad Avenue Enterprises Photograph PN-4638.

CAR NAME OR NUMBER	ST	CAR TYPE	DATE BUILT	CAR BLDR	PLAN	LOT
TUBA ATSF	S	17RmtSec	1-2/38	PS	4068B	6532
TUGALO RIVER SOU	S	10Rmt6Dbr	9-10/49	PS	4140	6814
TULLY VALLEY NYC	S	10Rmt6Dbr	12/48-49	Budd	9510	9660
TULUM NdeM-272	L	7Cpt2Dr	/52	Schi	100571	
TUNEZ NdeM-534	SN	10Rmt6Dbr	9/48-/49	PS	4123	6790
TUNKHANNOCK DL&W	S	10Rmt 6Dbr	6-7/49	AC&F	9008A	3089
TUPPER LAKE NYC	S	6DbrBLng	12/48-49	PS	4124	6790
TURQUESA NdeM-627	SN	4Rmt5DbrCpt4Sec	3/54	PS	4196	6949
TURQUIA NdeM-537	SN	10Rmt6Dbr	9/48-/49	PS	4123	6790
TURQUOISE SKY CRI&P	S	8Rmt6Dbr	11-12/54	PS	4195	6944
TURTLE BAY NYC	S	22Rmt	10/48	PS	4122	6790
TURTLE RAPIDS PRR	S	10Rmt6Dbr	11/48-49	PS	4140	6792

CAR NAME OR NUMBER	ST	CAR TYPE	DATE BUILT	CAR BLDR	PLAN	LOT
TUSCARAWAS B&O	S	10Rmt6Dbr	3/50	PS	4167	6864
TUSCARORA RAPIDS PRR	S	10Rmt6Dbr	11/48-49	PS	4140	6792
TWELVE MILE COULEE GN-1191	RN	4DbrCpt6RmtObsLng	1-2/47	PS	4109K	6751
TWIN PEAKS(1st) JT		4Cpt3DrArtic	12/37	PS	4063B	6525
TWIN PEAKS(2nd) JT	S	4Dbr4Cpt2Dr	11/40	PS	4069F	6636
TWO OCEAN GLACIER GN-1188	S	16DuRmt4Dbr	12/50	PS	4108A	6890
TYE RIVER SOU	S	10Rmt6Dbr	9-10/49	PS	4140	6814
TYENDE ATSF	S	8Sec2Dbr2Cpt	12/37-38	PS	4058A	6532
TYGART B&O	S	10Rmt6Dbr	3/50	PS	4167	6864
TYRONE INN PRR-8287	S	21Rmt	1-6/49	Budd	9513	9667

Figure I-136 TUNKHANNOCK is shown at Dover, New Jersey in the consist of an Erie-Lackawanna train. The 10-roomette, 6-double-bedroom was sold to the Straits Shows in 1970 who renamed it WINSTON SALEM, North Carolina 3. The car still carries the Lackawanna Route pressed glass prism window in the lavatory by the vestibule in this September, 1966 photograph by Robert Pennisi, 2556.

CAR NAME OR NUMBER	ST	CAR TYPE	DATE BUILT	CAR BLDR	PLAN	LOT
TYUONYI ATSF	S	8Sec2Dbr2Cpt	6-7/38	PS	4058B	6553
UNION						
COUNTY PRR	S	10Rmt6Dbr	9-10/49	PS	4140B	6809
SQUARE(1st) JT		12SecArtic	12/37	PS	4064	6525
SQUARE(2nd) JT	N	12SecArtic	12/37	PS	4064	6525
UNIONTOWN INN PRR-8288	S	21Rmt	1-6/49	Budd	9513	9667
UNIVERSITY						
OF CINCINNATI N&W	S	10Rmt6Dbr	11/49-50	Budd	9523	9660
URANO NdeM-596	SN	8DuRmt6Rmt4Dbr	12/46-47	PS	4107	6751
URBANA INN PRR-8289	S	21Rmt	1-6/49	Budd	9513	9667
URSA MAJOR FCP	SN	10Rmt6Dbr	9/48-/49	PS	4123	6790
URSA MINOR FCP	SN	10Rmt6Dbr	9/48-/49	PS	4123	6790
URUGUAY FEC	S	21Rmt	9-10/49	PS	4156B	6809
URUGUAY NdeM-586	SN	10Rmt6Dbr	8/56	Budd	9538	9660
UXMAL NdeM-276	L	8Sec3Cpt	/52	Schi	100570	
VAL						
ALAIN CN-2052	SN	22Rmt	10/48	PS	4122	6790
BRILLANT CN-2053	SN	22Rmt	10/48	PS	4122	6790
CARTIER CN-2054	SN	22Rmt	10/48	PS	4122	6790
COTE CN-2055	SN	22Rmt	10/48	PS	4122	6790

CAR NAME OR NUMBER	ST	CAR TYPE	DATE BUILT	CAR BLDR	PLAN	LOT
VAL						
DOUCET CN-2060	SN	22Rmt	10/48	PS	4122	6790
GAGNE CN-2071	SN	22Rmt	10/48	PS	4122	6790
JALBERT CN-2062	SN	22Rmt	10/48	PS	4122	6790
MARIE CN-2064	SN	22Rmt	10/48	PS	4122	6790
ROSA CN-2073	SN	22Rmt	10/48	PS	4122	6790
ROYAL CN-2072	SN	22Rmt	10/48	PS	4122	6790
ST. MICHEL CN-2074	SN	22Rmt	10/48	PS	4122	6790
ST. PATRICE CN-2070	SN	22Rmt	10/48	PS	4122	6790
d'AMOUR CN-2057	SN	22Rmt	10/48	PS	4122	6790
d'ESPOIR CN-2058	SN	22Rmt	10/48	PS	4122	6790
d'OR CN-2059	SN	22Rmt	10/48	PS	4122	6790
VALCOURT CN-2056	SN	22Rmt	10/48	PS	4122	6790
VALHALLA CN-2061	SN	22Rmt	10/48	PS	4122	6790
VALJEAN CN-2063	SN	22Rmt	10/48	PS	4122	6790
VALLE DE						
BRAVO NdeM-558	N	12Dbr	4-5/49	PS	4125	6790
MEXICO(2nd) NdeM-559	N	12Dbr	4-5/49	PS	4125	6790
MEXICO(1st) NdeM	SN	10Rmt6Dbr	12/48-49	Budd	9510	9660
VALLEY CITY NP	S	8DuRmt6Rmt4Dbr	7-9/48	PS	4119	6781

Figure I-137   Of the ten cars built in Pullman Plan 4167, Lot 6864, delivered in March 1950 for The B&O Railroad, seven were sold to Ringling Bros., Barnum & Bailey Combined Shows, Inc. One was wrecked on the railroad in 1968, one was sold to a private party, and TYGART, shown here, has been on display at The B&O Railroad Museum, Baltimore, Maryland since 1979. This is the last car painted in the famous Blue and Gray scheme, which it carried through A-Day (May 1, 1971). The Howard Ameling Collection.

CAR NAME OR NUMBER	ST	CAR TYPE	DATE BUILT	CAR BLDR	PLAN	LOT
VALLEY MILLS CN-1015	S	10SecDbrB	7/54	PS	4191A	6934
VALLEY PARK CN-1016	S	10SecDbrB	7/54	PS	4191A	6934
VALLEY RIVER CN-1017	S	10SecDbrB	7/54	PS	4191A	6934
VALLEY ROAD CN-1018	S	10SecDbrB	7/54	PS	4191A	6934
VALLEYFIELD CN-1014	S	10SecDbrB	7/54	PS	4191A	6934
VALLEYVIEW CN-1019	S	10SecDbrB	7/54	PS	4191A	6934
VALLOIS CN-2066	SN	22Rmt	10/48	PS	4122	6790
VALMONT CN-2065	SN	22Rmt	10/48	PS	4122	6790
VALPARAISO CN-2067	SN	22Rmt	10/48	PS	4122	6790
VALPOY CN-2068	SN	22Rmt	10/48	PS	4122	6790

CAR NAME OR NUMBER	ST	CAR TYPE	DATE BUILT	CAR BLDR	PLAN	LOT
VALRITA CN-2069	SN	22Rmt	10/48	PS	4122	6790
VAN WERT INN PRR-8290	S	21Rmt	1-6/49	Budd	9513	9667
VENEZUELA FEC	S	10Rmt6Dbr	9-10/49	PS	4140	6814
VENEZUELA NdeM-587	SN	10Rmt6Dbr	2-7/48	PS	4153D	6758
VENICE SAL-71	S	11Dbr	12/55-56	PS	4198A	6968
VENUS NdeM-338	SN	12Rmt4Dbr	12/49-50	AC&F	9004	3072
VERDE VALLEY ATSF	S	6Sec6Rmt4Dbr	6/42	PS	4099	6669
VERDUGO JT	S	4Dbr4Cpt2Dr	11/40	PS	4069F	6636
VERMILLION						
HARBOR NYC	S	22Rmt	4-6/49	Budd	9501	9661

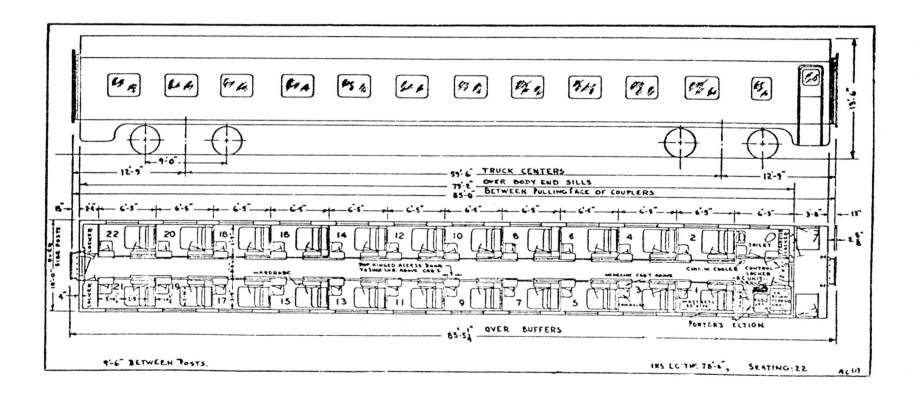

Figure I-138 This Canadian National passenger diagram depicts VALRITA, 1 of 22 22-roomette cars built by Pullman Standard under Plan 4122 Lot 6790 for New York Central that were sold to Canadian National in 1958. This car was the former NYC HAVERSTRAW. In May 1974 it was converted to a baggage dormitory and numbered CN-9486 and is now the property of the Bluewater Chapter of the National Railway Historical Society. The Author's Collection.

CAR NAME OR NUMBER	ST	CAR TYPE	DATE BUILT	CAR BLDR	PLAN	LOT
VERMILLION						
RIVER CN-2144	SN	10Rmt6Dbr	10-11/48	PS	4137	6775
RIVER Milw-23	S	8DuRmt6Rmt4Dbr	10-12/48	PS	4135	6775
VIENTO NEGRO FCS-BC	SN	14Rmt4Dbr	2-7/48	PS	4153A	6758
VINCENNES INN PRR-8291	S	21Rmt	4-5/48	AC&F	9010	2981
VIRGINIA						
BEACH ACL	S	6DbrBLng	11-12/49	AC&F	9002	3090
HOT SPRINGS C&O	S	10Rmt6Dbr	2-7/50	PS	4167	6864
VIRGINIA MILITARY						
INSTITUTE N&W	S	10Rmt6Dbr	11/49-50	Budd	9523	9660
VIRGINIA POLYTECHNIC						
INSTITUTE N&W	S	10Rmt6Dbr	11/49-50	Budd	9523	9660
VISTA CANYON ATSF	S	4DrDbrREObsLng	11/47	PS	4115	6757
VISTA CAVERN ATSF	S	4DrDbrREObsLng	11/47	PS	4115	6757
VISTA CLUB ATSF	S	4DrDbrLngObs	11/50	AC&F	9013	3360
VISTA HEIGHTS ATSF	S	4DrDbrREObsLng	11/47	PS	4115	6757
VISTA PLAINS ATSF	N	4DrDbrREOBSLng	1-2/38	PS	4070E	6532
VISTA VALLEY ATSF	S	4DrDbrREObsLng	11/47	PS	4115	6757
VOLUNTEER STATE IC	S	6Sec6Rmt4Dbr	4/42	PS	4099	6669
VOLUSIA COUNTY ACL	SN	10Rmt6Dbr	2-7/50	PS	4167	6864
WABASH RIVER B&O	P	2DbrCptDrObsBLng	7/39	PS	4082	6567
WABASH RIVER NYC	S	2DbrCptDrObsBLng	7/39	PS	4082	6567
WADE COUNTY ACL	S	10Rmt6Dbr	9-10/49	PS	4140B	6809
WAKE ISLAND PRR	N	2DrCptDbrObsBLng	5/38	PS	4081	6549
WALLA WALLA NP	S	8DuRmt6Rmt4Dbr	7-9/48	PS	4119	6781
WALNUT GROVE CP-14118		10Sec5Dbr	/49-/50	CPRR		
WALTON LAKE NYC	S	6DbrBLng	12/48-49	PS	4124	6790
WANAPITEI RIVER CN-2142	SN	10Rmt6Dbr	10-11/48	PS	4137	6775
WANTAGH LI-(2055)	P	Parlor/Coach	7-8/39	PS	4069D	6571
WAPINITIA						
PASS SP&S-701	S	6Rmt5Dbr2Cpt	10-11/50	PS	4180	6877
WAPPINGERS FALLS NYC	S	6DbrBLng	8-9/40	PS	4086B	6612
WARPATH RIVER CN-2143	SN	10Rmt6Dbr	10-11/48	PS	4137	6775
WARRIER RIVER SOU	S	10Rmt6Dbr	9-10/49	PS	4140	6814
WARSAW INN PRR-8292	S	21Rmt	4-5/48	AC&F	9010	2981
WASHINGTON &						
LEE UNIVERSITY N&W	S	10Rmt6Dbr	11/49-50	Budd	9523	9660
WASHINGTON VIEW PRR	S	2MbrDbrREObsBLng	5/38	PS	4080	6548
WASHTENAW COUNTY NYC	S	13Dbr	10/40	PS	4071C	6618
WAUHATCHIE VALLEY SOU	S	14Rmt4Dbr	10-11/49	PS	4153C	6814
WAUWEPEX LI-(2056)	P	Parlor/Coach	7-8/39	PS	4069D	6571
WAWASEE B&O	SN	5DbrBEObsBLng	8/50	PS	4165	6863
WAYNE COUNTY NYC	S	13Dbr	10/40	PS	4071C	6618
WEST END CN-2029	P	17RmtSec	3-4/38	PS	4068E	6539
WEST PALM BEACH SAL-42	S	10Rmt6Dbr	6-8/49	Budd	9503	9662
WEST RIVER CN-2031	P	17RmtSec	3-4/38	PS	4068E	6539
WEST SHEFFORD CN-2036	P	17RmtSec	3-4/38	PS	4068E	6539
WESTBANK CN-2028	P	17RmtSec	3-4/38	PS	4068E	6539
WESTCHESTER CN-2033	P	17RmtSec	3-4/38	PS	4068E	6539
WESTCHESTER COUNTY NYC	S	13Dbr	4/38	PS	4071A	6541
WESTERN						
ADVENTURE UP	S	12Rmt4Dbr	12/49-50	AC&F	9004	3069
FRONTIER C&NW	S	12Rmt4Dbr	12/49-50	AC&F	9004	3072
HILLS UP	S	12Rmt4Dbr	12/49-50	AC&F	9004	3069
LAKE WAB	S	12Rmt4Dbr	12/49-50	AC&F	9004	3074
LODGE UP	S	12Rmt4Dbr	12/49-50	AC&F	9004	3069
MOUNTAIN UP	S	12Rmt4Dbr	12/49-50	AC&F	9004	3069
PEAK C&NW	S	12Rmt4Dbr	12/49-50	AC&F	9004	3072
PLAINS UP	S	12Rmt4Dbr	12/49-50	AC&F	9004	3069
SCENE WAB	S	12Rmt4Dbr	12/49-50	AC&F	9004	3074
SEA UP	S	12Rmt4Dbr	12/49-50	AC&F	9004	3069
SLOPE UP	S	12Rmt4Dbr	12/49-50	AC&F	9004	3069
STAR UP	S	12Rmt4Dbr	12/49-50	AC&F	9004	3069
SUNSET WAB	S	12Rmt4Dbr	12/49-50	AC&F	9004	3074
TRAIL UP	S	12Rmt4Dbr	12/49-50	AC&F	9004	3069
VALLEY UP	S	12Rmt4Dbr	12/49-50	AC&F	9004	3069
VIEW WAB	S	12Rmt4Dbr	12/49-50	AC&F	9004	3074
WONDERLAND UP	S	12Rmt4Dbr	12/49-50	AC&F	9004	3069
WESTFIELD RIVER NYC	S	10Rmt6Dbr	9/48-/49	PS	4123	6790
WESTGATE CN-2034	P	17RmtSec	3-4/38	PS	4068E	6539
WESTLOCK CN-2030	P	17RmtSec	3-4/38	PS	4068E	6539
WESTPORT CN-2035	P	17RmtSec	3-4/38	PS	4068E	6539
WESTVILLE CN-2032	P	17RmtSec	3-4/38	PS	4068E	6539
WESTWOLD CN-2037	P	17RmtSec	3-4/38	PS	4068E	6539
WESTWOOD CN-2031	N	17RmtSec	3-4/38	PS	4068E	6539
WEWOKA CRI&P	S	8Sec5Dbr	11/40	PS	4095	6631
WHEELING RAPIDS PRR	S	10Rmt6Dbr	11/48-49	PS	4140	6792
WHISPERING						
PINE L&N-3468	S	6Sec6Rmt4Dbr	3-5/53	PS	4183	6909
WHITE						
OAK CN-1012	S	8SecDbrDiner	6/54	PS	4188	6927
PINE L&N-3469	S	6Sec6Rmt4Dbr	3-5/53	PS	4183	6909
RAPIDS CN-1011	S	8SecDbrDiner	6/54	PS	4188	6927
ROCK CN-1010	S	8SecDbrDiner	6/54	PS	4188	6927
SANDS CN-1013	S	8SecDbrDiner	6/54	PS	4188	6927
WHITE SULPHUR						
SPRINGS(1st) C&O	S	10Rmt6Dbr	2-7/50	PS	4167	6864
SPRINGS(2nd) C&O	N	10Rmt6Dbr	2-7/50	PS	4167	6864
WHITEWATER RIVER NYC	S	10Rmt6Dbr	9/48-/49	PS	4123	6790
WHITEWATER VALLEY ATSF	S	6Sec6Rmt4Dbr	6/42	PS	4099	6669
WICKOPOGUE LI (2057)	P	Parlor/Coach	7-8/39	PS	4069D	6571
WILD FOWL BAY NYC	S	22Rmt	10/48	PS	4122	6790
WILD PINE L&N-3470	S	6Sec6Rmt4Dbr	3-5/53	PS	4183	6909
WILKES						
BARRE INN PRR-8293	S	21Rmt	4-5/48	AC&F	9010	2981
WILKINSBURG						
INN PRR-8294	S	21Rmt	4-5/48	AC&F	9010	2981
WILLIAM						
A. GRIFFIN AMT-3228	SN	6DbrBLng	11-12/49	AC&F	9002	3093
B. TRAVIS MKT-1502	S	14Rmt4Dbr	1-6/48	PS	4153	6769

CAR NAME OR NUMBER	ST	CAR TYPE	DATE BUILT	CAR BLDR	PLAN	LOT
WILLIAM BUCHANAN KCS	S	14Rmt4Dbr	5/48	PS	4153	6795
WILLIAM CHAMBERLAIN						
PATTERSON PRR	N	2DrCptDbrObsBLng	5/38	PS	4081	6549
WILLIAM						
EDENBORN KCS	S	14Rmt4Dbr	5/48	PS	4153	6795
REYNOLDS Erie	S	10Rmt6Dbr	5-6/49	PS	4129A	6797
THAW PRR-8252	N	21Rmt	1-6/49	Budd	9513	9667
WALLACE ATTERBURY PRRS		2DrCptDbrBEObsLng	3-4/49	PS	4134	6792
WILLIAMETTE RIVER NYC	S	10Rmt6Dbr	9/48-/49	PS	4123	6790
WILLOUGHBY BAY NYC	S	22Rmt	10/48	PS	4122	6790

CAR NAME OR NUMBER	ST	CAR TYPE	DATE BUILT	CAR BLDR	PLAN	LOT
WILLOW GROVE CP-14119		10Sec5Dbr	/49-/50	CPRR		
WILLS RAPIDS PRR	S	10Rmt6Dbr	11/48-49	PS	4140	6792
WILSHIRE(1st) JT		12SecArtic	12/37	PS	4064	6525
WILSHIRE(2nd) JT	N	12SecArtic	12/37	PS	4064	6525
WILSON POINT NH	S	14Rmt4Dbr	12/49/50	PS	4159	6822
WINDIGO CN-1700	SN	4Rmt5DbrCpt4Sec	12/54	PS	4196	6949
WINDING RIVER NYC	S	10Rmt6Dbr	9/48-/49	PS	4123	6790
WINDY CITY AMT-2559	SN	8Rmt6Dbr	11-12/54	PS	4195	6944
WINGATE BROOK NYC	S	5DbrBLngObs	6-7/49	Budd	9508	9636
WINNIPEG CLUB GN-1199	RN	8uRmt2DbrBLng	1-2/47	PS	4108B	6751

Figure I-139 The Wabash WESTERN LAKE, shown here in 1967 in blue livery, was built by American Car & Foundry in 1950 for joint CNW-WAB-UP service. All cars in this lot were initially painted the UP colors. This car was repainted Wabash blue in January 1957, withdrawn from Pullman lease in May 1967, and subsequently sold to Butterworth Tours. Railroad Avenue Enterprises Photograph PN-2681.

CAR NAME OR NUMBER	ST	CAR TYPE	DATE BUILT	CAR BLDR	PLAN	LOT
WINTER HAVEN SAL-40	S	10Rmt6Dbr	6-8/49	Budd	9503	9662
WISCONSIN						
DELLS Milw-29	S	16DuRmt4Dbr	1-2/49	PS	4147D	6775
RIVER Milw-25	S	8DuRmt6Rmt4Dbr	10-12/48	PS	4135	6775
WISSAHICKON						
RAPIDS PRR-8436	N	10Rmt6Dbr	8-10/50	AC&F	9008	3200
WOLFE MANOR CP-10342	S	5DbrCpt4Sec4Rmt	11/54-55	Budd		9658
WOLVERINE CLUB C&O	S	5DbrBEObsBLng	8/50	PS	4165	6863
WOODLAND STREAM NYC	S	6DbrBLng	3-5/49	Budd	9505	9663
WOOSTER INN PRR-8295	S	21Rmt	4-5/48	AC&F	9010	2981
WREN B&O	S	16DuRmt4Dbr	4-6/54	Budd	9536	9658
WRIGHTSVILLE BEACH ACL	S	6DbrBLng	11-12/49	AC&F	9002	3090
WRIGLEY FIELD AMT-2560	SN	8Rmt6Dbr	11-12/54	PS	4195	6944
WUNNEWETA LI-(2058)	P	Parlor/Coach	7-8/39	PS	4069D	6571
WUPATKI ATSF	S	8Sec2Dbr2Cpt	12/37-38	PS	4058A	6532
WYANDANCH LI-(2062)	P	Parlor/Coach	7-8/39	PS	4069D	6571
WYTHE COUNTY N&W	S	10Rmt6Dbr	11/49-50	Budd	9523	9660
XENIA INN PRR-8296	S	21Rmt	4-5/48	AC&F	9010	2981
XICOTENCATL FCS-BC	SN	4Dbr4Cpt2Dr	9-10/40	PS	4069E	6617
YADKIN RIVER SOU	S	10Rmt6Dbr	9-10/49	PS	4140	6814
YAMPAI ATSF	S	8Sec2Dbr2Cpt	12/37-38	PS	4058A	6532
YELLOW PINE L&N-3471	S	6Sec6Rmt4Dbr	3-5/53	PS	4183	6909
YELLOWSTONE						
RIVER Milw-21	S	8DuRmt6Rmt4Dbr	10-12/48	PS	4135	6775
YEMEN NdeM-542	SN	10Rmt6Dbr	9/48-/49	PS	4123	6790
YERBA BUENA JT	S	4Dbr4Cpt2Dr	11/40	PS	4069F	6636
YORK HARBOR NYC	S	22Rmt	4-6/49	Budd	9501	9661
YORK RIVER SOU	S	10Rmt6Dbr	9-10/49	PS	4140	6814
YORKTOWNE C&O	S	10Rmt6Dbr	2-7/50	PS	4167	6864
YOSEMITE VALLEY NYC	S	10Rmt6Dbr	12/48-49	Budd	9510	9660
YOUGHIOGHENY B&O	S	10Rmt6Dbr	3/50	PS	4167	6864
YUGOSLAVIA NdeM-514	SN	10Rmt6Dbr	9/48-/49	PS	4123	6790
YUKON RIVER CN-2093	SN	10Rmt6Dbr	9/48-/49	PS	4123	6790
ZANESVILLE INN PRR-8297	S	21Rmt	4-5/48	AC&F	9010	2981
ZOAR VALLEY NYC	S	10Rmt6Dbr	12/48-49	Budd	9510	9660
ZUMBRO RIVER Milw-32	S	8DuRmt6Rmt4Dbr	11/54	PS	4192A	6945

EVOLUTION OF INNOVATIONS
IN THE PULLMAN FLEET
1858 - 1969

1858 The remodeling of a Chicago & Alton coach into the first Pullman sleeping car was begun.

1859 No. 9, the first Pullman sleeper, made its first trip from Bloomington, Illinois to Chicago on September 1. A wood-burning stove provided heat. Candles furnished light.

1865 PIONEER, the first sleeper constructed by modern methods of the day, was put into service. Six-wheel trucks were applied to the PIONEER. Wooden panels were used on the exterior of the car. A hand pump maintained the water supply.

1867 The first Pullman "hotel" car was built.

1868 The first Pullman dining car was constructed. Hot water heat replaced coal stoves.

1871 Straight air brakes were applied.

1872 Overhead tanks, with or without a pump, were added to the water supply system.

1873 Oil lamps replaced candles.

1875 Plain automatic brakes were adopted in place of straight air brakes. Parlor cars were introduced into Pullman service.

1879 Paper-center, steel-rim wheels replaced iron wheels.

1886 Electric call bells were added.

1887 The narrow vestibule was invented and applied. Electric lights were used for the first time. Steam, in conjunction with hot water circulation, was applied to the heating apparatus instead of hot water alone.

1889 Quick acting automatic brakes were installed in lieu of plain automatic.

1890 The swinging berth curtain rod was developed. Wooden tongue and groove sheathing was used on the exterior of cars instead of wooden panels.

1891 Pintsch gas burners replaced oil lamps after previous experiments. Anti-telescoping construction was developed.

1893 An air pressure water system replaced overhead tanks. The wide vestibule was developed.

1895 The high speed brake was developed and first applied to six-wheel trucks.

1899 Axle generators and storage batteries for electric lighting supplanted the head-end system. Electric fans were installed. The use of paper-center wheels was abandoned.

1900 Cast-iron hub, steel-rim wheels replaced the paper-center, steel-rim wheels. Roller curtains over vestibule diaphragm connections between cars were added.

1901 Reading lights were placed in berths.

1905 Open plumbing was installed.

1906 The automatic exhaust-type ventilator was applied to the deck of the car. The fumigation of cars became standard practice.

1907 The first steel Pullman, JAMESTOWN, was displayed at the Jamestown Exposition in Virginia.

1908 The vacuum steam heat system supplanted the hot water system. The dental lavatory was first applied.

1909 The first steel sleeper was placed in service. Sliding window screens replaced folding screens. "First Aid" kits were added.

1910 Improved air brakes were adopted. Steel sheathing was applied to the exterior of cars. Improved anti-telescoping construction was adopted.

1913 Electric exhaust fans were added to sleeping cars and to club cars and rooms where smoking was permitted. Solid steel wheels were tried out in place of steel-tired wheels. Water and ice were separated, a tank for each being used.

1914 The vacuum cleaning of cars was begun.

1915 Mirrors were placed in upper berths.

1916 Divided berth curtains replaced old type. The safety ladder replaced the step ladder. Universal control brakes were adopted. Steel plates were used on the exterior of cars. Rods replaced the safety cord used to support the upper berth.

1917 Floor lights were installed. The clasp brake was adopted (two brake shoes being applied on each wheel).

1919 Draft gears were improved. A buffer mechanism was added. Window sash ventilators were added. Safety locking center pins were added to hold the truck to the body. Split front toilet seats replaced solid seats.

1920 The use of heavier couplers was begun.

1921 Various improvements in electric lighting, generator suspensions and regulators were adopted. A new style truck was placed in service.

1922 Luminous berth numbers and "quiet" signs were installed.

1923 Permanent headboards were added. A larger step box supplanted a smaller one. Rubber battery jars replaced lead-lined tanks. Passenger agents and platform men were uniformed. Carpet padding was applied.

1924 A safety razor blade receptacle was added. An anti-pinch device for doors was installed. End door car numbers became standard. A dressing shelf was installed in the mens' washroom. "Watch Your Step" signs were applied to steps and vestibules.

Improved valves and fittings were added to the water system. A new three-position headrest for sleeping-car sections was developed. New toilet indicator locks were added.

1925 Shoe receptacles were added for rooms and compartments. An improved and more attractive design of seat end was adopted. Porcelain washbowls and stands were installed. Womens' lounge rooms with baths were added. Water coolers were arranged so they could be iced from hallways were placed in womens' dressing rooms. The four-tread step was adopted, thereby shortening the distance between the lower step and the step box. Larger fans were provided in smoking rooms. The general application of variable interior designs with plain colors instead of graining, new decorative schemes and improved lighting fixtures were adopted. Observation rooms were provided with lounges and tables. Semi-enclosed section partitions with narrow sliding headboards were installed. The amount of insulating material in car frames was increased and improved. The lighting of toilet rooms was improved for shaving purposes by placing bracket lamps over mirrors. Single Room Cars were introduced.

1926 Gas-filled lamp bulbs were installed throughout the car to improve the general illumination. Berth curtains were changed in color schemes. Reading light fixtures in berths were improved by changing the bulb from 15 to 25 watts. Adjustable reading lamps were added for the drawing room sofa. Lower-berth reading lamps were changed from the recess to the bracket-and-shade type. Additional racks were provided in drawing rooms and compartments.

1927 Brass window sash was employed. A new truck was adopted to accommodate the various types of roller bearing applications. The two-part vestibule door became standard. Car flooring was improved by a change in the materials employed. The electric generator drive was changed and new driving pulleys with belt guards were provided.

1928 Movable revolving chairs were provided for the womens' lavatory. Two-inch metal steam connections were provided between cars in place of rubber hose. The curtain arrangement in the womens' lavatory was improved to prevent entangling when entering womens' room with baggage. The smoking room sofa was improved. An automatic locking device for the train connector was applied. The design of journal boxes was changed, the single guide being adopted. Larger steam pipes were applied. A wardrobe was provided for compartments. Spring mattresses for lower berths were developed and generally applied.

1929 Water coolers were redesigned to overcome overflow. A new washstand faucet with an overhanging spout was adopted in place of the concealed spout. Electric exhaust fans were added generally to improve ventilation when cars were standing. A section seat with a slanting top and with a second stop provided on the seat slide to permit pulling the seat cushion out, to make a wider and more comfortable seat, was adopted. The journal-box design was improved to prevent the ingress of water. A special upper-berth spring mattress was adopted. A non-slip shoe was provided for the upper-berth ladder. A larger rack was provided for the drawing room. The vitrolite shelf was adopted for the smoking room in place of enameled steel.

1930 Mechanically refrigerated air-conditioned car placed in test service. Improvements in brake cylinder protectors and Universal valves adopted. Test application made of individual heat control for rooms, concealed heat and thermostatic heat control applied to parlor and lounge cars. Spring cushions for chairs, settees and couches adopted.

Illuminated display lockers added to buffet cars. Folding upper-berth step ladders developed for bedrooms. Extended supplementary coat hangers for single-occupancy sections. Toilet cabinets applied to drawing rooms and compartments. Extensive use made of chromium plating on observation railings and gates. More legible "Men" and "Women" bulkhead signs adopted. Bottle holders installed in drawing room annexes and compartments. Mattress roll and shelf installed in cars having spring mattresses. Experiments made with rubber buffer mechanism and draft gears. New design sun room end, known as the "Clipper"-type, developed and applied. Automobile-type windows and regulators applied to some observation sun rooms. New window curtain fixture adopted. Upper berths developed and adopted for bedrooms having convertible sofas. Adjustable mirror lights applied as test to washstand mirrors in bedrooms. Convertible sofa bed devised for bedrooms. Double pan hoppers applied to prevent spraying. Improved drawing room sofa adopted.

1931 Adjustable four-position section seat developed and placed in service. Travel accessories sale cabinets installed. New inner spring construction developed for upper berths. Types of positive generator drives applied for test purposes. Cars with private sections with adjoining toilets placed in service. Cars with enclosed section placed in service. Test application of permanent ladders for upper berths. New ice-cooled air-conditioned car perfected. Experimental air and water-cooled air-conditioned car placed in service. One private and five sleeping cars equipped for air conditioning by means of mechanical refrigeration. Chaise lounge section cars placed in service. Two cars equipped with double-deck bedrooms. Air conditioning applied to nineteen cars.

1932 Cars equipped with dressing platforms in upper berths. Improved spring construction for mattress support for convertible sofas in bedrooms. Improved vestibule trap-door lock. Improved buttoning arrangement devised for berth curtains. Improvement in three-position headrest locking plates to prevent dropping. Use of sound-deadening metal and improved platform mechanism. Improved shades applied to reading lamps and bracket fixtures. Improved fastening arrangement for hopper casings in bedrooms to prevent rattling. Design of coat hangers changed to provide more space for hanging trousers and hooks provided at each end so trousers can be suspended from the belt straps. New hinge construction applied to drop tables in bedrooms. Improvement in water tank filling valve by using a weighted cover. Catch developed for the upper part of vestibule doors in order to hold them in partly open position to increase air circulation. Air conditioning applied to 100 cars.

1933 Cars equipped with eight upper and eight lower rooms, air conditioned, placed in service. First lightweight (all-aluminum) sleeping car, GEORGE M. PULLMAN, constructed and put on exhibition at A Century of Progress. New type rigid ladder devised for drawing rooms and compartments. New type adjustable seat arrangement adopted. Foot-operated hopper mechanism adopted. Several types of wearing plates applied to truck center plates to reduce noise. Air conditioning applied to 155 cars.

1934 Installation of double pan hoppers. New type automatic coupler applied. Standard use of metal steam hose. Light excluder for single-occupancy section used in open section cars. Sponge rubber carpet padding adopted as standard in place of hair felt previously used. Air conditioning applied to 1018 cars. First streamlined train, **City of Salina**, designed and built at Pullman. Second streamlined train with Pullman sleepers, **City of Portland**.

1935 Soldered joint fittings adopted as standard for air-conditioned cars. Synthetic varnish (quick four-hour dry) adopted as standard. Safety guards for drawing room sofas for protection of children. Brake shoe keys of improved design. Manganese steel liners for center plates. Magazine and eye-glass case pocket for bedrooms. Use of enamel for exterior of wood sash. New design anti-rattle diaphragm suspension arrangement. Various modern developments in cushions and mattresses, such as rubberized hair of various kinds and sponge rubber cushions. Development of "streamline" car. Individual control of air conditioning in sections. Air conditioning applied to 1,947 cars.

1936 Safety guards adopted for single-occupancy sections. Sponge rubber mattresses adopted for bedrooms. Hoppers (enclosed) for bedrooms, lowered to present a better appearance as a daytime chair. Facial tissues and paper towels adopted as auxiliary service for mens' and womens' rooms. Armrest for convertible sofa backs in bedrooms, adopted as standard. Air conditioning applied to 875 cars. Third streamline train, **City of Los Angeles.** Fourth streamline train, **City of San Francisco.** Sixth and seventh streamline trains, **City of Denver.** Two-car unit, lightweight, streamlined cars, ADVANCE and PROGRESS. Alloy steel truss frame sleeping car, FORWARD.

1937 Developed new standard for Pullman equipment for lightweight cars for high speed service, including:

New and improved type of trucks for four and six wheels.

New truck brakes.

New air brakes for high speed service.

Entire new car body structures:

   (a) Girder type of welded high tensile steel;

   (b) Welded truss frame type with stainless steel siding.

New draft gears and coupler attachments.

New platform and end construction with complete enclosure between cars for single-unit cars.

New materials for inside finish and bright and striking color schemes for interior paint and upholstery.

Greatly increased illumination.

Improvement in air conditioning with full thermostatic control for each individual room and space in cars.

Introduction of porter's call connections between cars.

Development of converter transformers for converting 32 volt direct current to 110 volt alternating current for electric shavers, etc.

Provided new Pullman accommodations as follows:

   (a) New type drawing rooms with folding bed, sofa, upper berth and lounge chairs.

   (b) New type compartment with sofa, upper berth and lounging chairs.

   (c) Improved type of connected bedrooms, fully open folding partition between rooms.

   (d) Roomette accommodation, single room with folding bed.

Disappearing type of swivel and folding hoppers for bedrooms.

Telephone communication between cars.

Latex rubber cushions for all seats and chairs and for all mattresses.

New and modern design for all hardware trimmings and lighting fixtures.

New color scheme for car exteriors.

Duplex car developed and put into service.

1938 Developed Deluxe Drawing Room accommodation. Special lightweight stainless steel exterior, truss frame cars developed for service in Santa Fe **Super Chief.** Guard ropes for Single Occupancy Sections. Lightweight full vitreous washstands. Synthetic interior enamels adopted standard for lightweight cars. Developed new four-wheel truck with helical bolster springs and one-way shock absorber (extensively applied in 1939.) New streamline train, **City of Los Angeles.** New streamline train, **City of San Francisco.** New streamline train for New York Central, **Twentieth Century Limited.** New streamline train for Pennsylvania R.R., **Broadway Limited.** New streamline Pullman cars for Santa Fe, "Chief." New streamline Pullman cars for Santa Fe, "Super Chief."

1939 Propane gas engines for driving air-conditioning compressors and electric generators. Non-breakable rubber buttons versus hard rubber buttons for berth curtains. Metal end door name plates, lightweight cars. Stainless steel hoppers. New underframe construction developed incorporating increased anti-telescoping. Forty additional streamline Pullman cars for important New York Central trains. Forty additional streamline Pullman cars for important Pennsylvania R.R. trains. Exhibition observation car for New York World's Fair.

1940 Development of fluorescent lighting. New shaving mirrors developed for washstand cabinets for Roomettes, Bedrooms, Drawing Room Annexes. New "Zone" heating system adopted. New air brake with Decelostat brake pressure control on each car provided. Streamline Pullman cars for Rock Island **Rocky Mountain Rocket.** Thirteen Streamline Pullman cars for Pullman pool. Fifty additional streamline Pullman cars for important Pennsylvania R.R. trains. Fifty additional streamline Pullman cars for important New York Central R.R. trains. Constructed four Coach-Sleepers for operation in important coach trains to develop a new type, low cost, sleeping accommodation.

1941 Additional streamliner, **City of Los Angeles.** Additional streamliner, **City of San Francisco.** New streamliner LARK for Southern Pacific Company.

1942 Seventy-eight new streamline Pullman cars for "Overland Route." New **Panama Limited** for the Illinois Central. Four Streamline Pullman cars for Erie R.R. Four Streamline Pullman cars for Missouri Pacific R.R. Twenty-four Pullman cars for **Golden State Route.** Twenty-six Streamline Pullman cars for various important Santa Fe trains. Developed Duplex-Roomette, a car containing 24 small room accommodations with all the facilities of the larger rooms including complete privacy, toilet and wash facilities, individual control of heating and air conditioning, modern lighting and a comfortable seat and full-sized bed.

1945 The Sleepy Hollow car seat was developed. The first Vista Dome car began service on Burlington R.R. Roller bearings standard on new passenger cars.

1945-1946 Orders for 3,000 new cars received from railroads.

1947 Pullman Sleeping Car Division was sold to the railroads.

1949 Illinois Central R.R. introduced the all-electric dining car.

1956 Budd introduces the Slumber Coach, an economy sleeper.

1959 Budd built the last sleeping car.

1963 Railway Post Office service was greatly reduced.

1969 The Pullman Company ceased sleeping car operations on January 1.

GLOSSARY OF TERMS AND ABBREVIATIONS

STATUS COLUMN ABBREVIATIONS

SYMBOL	STATUS
N	Name change
SN	Sold, name changed
S	Sold (normally new)
P	Purchased
R	Rebuilt
RN	Rebuilt, name changed
L	Leased from or to

DISPOSITION ABBREVIATIONS

ABBREVIATION	DISPOSITION
( )	Assigned Number not on car
(10)	in service in 1910
(10X)	not in service in 1910
(DS)	Designated Service
÷	with same name
÷/00	after 1900
-/00	before 1900
conv	Converted
dest	Destroyed
dism	Dismantled
fr	From
(GS)	General Service
L-	Pullman Lot Number
LFP	Leased from Pullman
Lto	Leased to
LTP	Leased to Pullman
M/W	Maintenance of way
P-	Pullman Plan number
rblt	Rebuilt
Ret	Retired
RPL	Returned to Pullman lease
rtn	Return
S&P Car	Pullman Supply & Porter Car
S/L	Streamlined
SFS	Sold for scrap
TC	Tourist car
TGS	To Government Storage
W/D	Withdrawn
W/T	Work Train Service
WGS	Withdrawn from Government Storage
WPL	Withdrawn from Pullman lease

CAR TYPE ABBREVIATIONS

ABBREVIATION	DESCRIPTION
(NG)	Narrow gauge
B	Buffet
Bagg	Baggage
Barber	Barber Shop
BE	Blunt End
Boud	Boudoir or Compartment
Ch	Parlor Chair
Club	Club
Comm	Commissary Car
Cpt	Compartment
D	Dining
Dbr	Double Bedroom
DDR	Double Drawing Room
Dorm	Dormitory
Dr	Drawing Room
DuSr	Duplex Single Room
Emigrant	Emigrant car
EnclSec	Enclosed Section
GS	General Service
Lib	Library
Lng	Lounge
Mail	Mail Room
O	Open Section
Obs	Observation
OS	Open Section
Private	Private Car
PTN	Partition
PvtRm	Private room
PvtSec	Private Section
R-End	Round End (Rotunda)
RE	Round End
RecCh	Reclining Chair
Rest	Restaurant
Rmt	Roomette
Sbr	Single Bedroom
SBS	Wood Interior
SDR	Single Drawing Room
Sec	Section
Smk	Smoking
Sr	Stateroom-smoking or single room
Sts	Lounge or coach type seats
SU	Steel Underframe
SunRm	Sun Room
SUV	Steel Underframe & Vestibule
Tourist	Tourist Car

ABBREVIATIONS RAILROADS-ORGANIZATIONS-FACILITIES

ABBREVIATION	NAME
A&WP	Atlanta & West Point Railroad (ACL System)
A-T	Auto-Train
AC&F	American Car & Foundry
ACL	Atlantic Coast Line Railroad
AEC	Atomic Energy Commission
AMT	Amtrak
Assn	Agreement between Pullman & Railroads (Maintenance or Mileage)
ATSF	Atchison, Topeka & Santa Fe Railway
ATSF Assn	Pullman Atchison Topeka & Santa Fe Association
B&A	Boston & Albany Railroad (New York Central System)
B&M	Boston & Maine Railroad
B&O	Baltimore & Ohio Railroad
B&S	The Barney & Smith Manufacturing Company, Dayton, OH
B,D	Bowers, Dure & Company, Wilmington, DL (1871-1886)
B,P&Co	Barney, Parker & Company
BAR	Bangor & Aroostook Railroad
Big-4	Cleveland, Cincinnati, Chicago & St. Louis RR (CCC&StL)
Budd	The Budd Manufacturing Company, Philadelphia, PA
C&A	Chicago & Alton Railroad
C&EI	Chicago & Eastern Illinois Railroad
C&NW	Chicago & North Western Railway
C&NW Ry	The Chicago & Northwestern Railway (Fond Du Lac Shops)
C&O	Chesapeake & Ohio Railway
C&O(PM)	The Chesapeake & Ohio Railway, Pere Marquette District
C&S	Colorado & Southern Railway
CB&Q	Chicago, Burlington & Quincy Railroad
CCC&StL	Cleveland, Cincinnati, Chicago & St. Louis RR (NYC System)
CentralCarCo	Central Car Company (Operated cars for Wisconsin Central RR)
Chem	Chemical Branch of U.S. Army
Chessie	The Chessie System
CIWL	Compagnie Internationale des Wagons-Lits (Int'l Sleeping Car Co.)
CK&N Assn	Chicago, Kansas & Nebraska Association (CRI&P RR)
CM	Colorado Midland Railroad
CMStP&P	Chicago, Milwaukee, St. Paul & Pacific Railroad
CN	Canadian National Railways
CNJ	Central Railroad of New Jersey
CNO&TP	Cincinnati, New Orleans & Texas Pacific Railway (Southern)
CochranCCo.	Cochran Car Company
CofG	Central of Georgia Railway
CP	Canadian Pacific Railroad
CP	Central Pacific Railroad
CP Assn	Pullman Central Pacific Association
CRI&P	Chicago, Rock Island & Pacific R.R. (Rock Island)
CRI&P Assn	Chicago Rock Island & Pacific Association
CStPM&O	Chicago, St. Paul, Minneapolis & Omaha Railway

ABBREVIATION	NAME	ABBREVIATION	NAME
CTC	Central Transportation Company	IC Assn	Pullman Illinois Central Association
Ctr	Center	IC RR	Illinois Central RR
D&H	Delaware & Hudson Railroad	ICC	Interstate Commerce Commission, U.S. Government
D&MSCC	Detroit & Milwaukee Sleeping Car Company	IGN	International Great Northern (Mopac)
D&RG	Denver & Rio Grande Railroad	J&S	Jackson & Sharp Company, Wilmington, DE (1863-1901)
D&RGW	Denver & Rio Grande Western Railroad	John Ringling	Used equipment dealer (One of the Ringling Brothers)
D&SP	Denver South Park & Pacific Railroad	Jones C&MC	Jones Car & Manufacturing Company
Dep	Depot (U.S. Military Supply Installation)	JT	Joint operation
DetC&MCo	Detroit Car & Manufacturing Company, Detroit, MI	KCS	Kansas City Southern Railway
DL&W	Delaware Lackawanna & Western Railroad	KP	Kansas Pacific RR (Union Pacific Eastern Div 1/80 to UP)
DL&W Assn	Pullman Delaware Lackawanna & Western Association	L&N	Louisville & Nashville Railroad
DODX	Department of Defense	LBSCR	London Brighton and South Coast Railway, England
E&A Assn	Pullman Erie & Atlantic Association (Erie Railroad)	LCDR	London Chatham & Dover Railway, England
E&ASCCo	Erie & Atlantic Sleeping Coach Company	LH&StL	Louisville Henderson & St. Louis (L&N System)
E,G	Eaton, Gilbert (Car manufacturers)	LIRR	Long Island Rail Road
E-L	Erie Lackawanna Railroad	LS&MS	Lake Shore & Michigan Southern (New York Central System)
ElmCW	Elmira Car Works, Elmira, NY	LSWR	London & South Western Railway, England
EMD	Electro-Motive Division, General Motors	LV	Lehigh Valley Railroad
Eng	Engineer Branch of U.S. Army	LW	Louisiana Western Railroad (SP System)
Erie	Erie Railroad	M&LT	Morgan's Louisiana & Texas Railroad (SP System)
F.M.Hicks & Co	Rebuilt used equipment and sales	MannBCC	Mann Boudoir Car Company
FC-SUR	Ferrocarril Del Sureste of Mexico	MC RR	Michigan Central Railroad (NYC System)
FCP	Pacific Railroad of Mexico	McComb	McComb Lines (Sleeping Cars)
FCS-BC	Ferrocarril Sonora-Baja California of Mexico	Med	Medical Branch of U.S. Army
FEC	Florida East Coast Railway	Med Dept	Medical Department, US Army
Fitz-Hugh & Co	Used equipment dealer (Circus Equipment)	Mem & Char RR	Memphis & Charleston Railroad (Southern System)
FW&D	Fort Worth & Denver Railway	Mex Cent	Mexican Central Railroad
FW&DC	Fort Worth & Denver City Railway	MGRS	Mexican Government Railroad System
G&B	Gilbert & Bush Company, Troy, NY	Mid Ry	Midland Railway, England
Gen	General (normally General Depot)	Milw	Chicago Milwaukee St. Paul & Pacific Railroad
Georgia	Georgia Railroad & Banking Company	MKT	Missouri-Kansas-Texas Railroad
GH&SA	Galveston, Houston & San Antonio Railroad (SP System)	MP	Missouri Pacific Railroad
GM	General Motors Corp	MP Assn	Pullman Missouri Pacific Association
GM&O	Gulf Mobile & Ohio Railroad	N&W	Norfolk & Western Railway
GMP	George M. Pullman	N-OdeM	Chihuahua Pacific Railway of Mexico
GN	Great Northern Railway	Nash&Chat RR	Nashville & Chattanooga Railroad (1853 to NC&StL)
GNR	Great Northern Railway Company of England	NC&StL	Nashville, Chattanooga & St. Louis Railway (L&N System)
Great Western	The Great Western Railway of Canada, Hamilton, Ontario	NdeM	National Railways of Mexico
GS&F	Georgia Southern & Flordia Railway (Southern System)	New Haven	New Haven Car Company
GSA	General Services Administration	NH-	New York New Haven & Hartford car number
GT RR	The Grand Trunk Railway System, Canada. Montreal Shops	NKP	The New York, Chicago & St. Louis R.R. (Nickel Plate)
GTW	Grand Trunk Western Railroad	NP	Northern Pacific Railway
H&H	Harlan & Hollingsworth Company, Wilmington, DE	NYC	The New York Central System
H&NE	Hillsboro & Northeastern Railroad	NYC&StL	The New York, Chicago, & St. Louis R.R. (Nickel Plate)
H&StJ RR	The Hanibal & St. Joseph Railroad (CB&Q System)	NYLE&W	New York, Lake Erie & Western RR (11/95 to Erie RR)
H/B&Co	Hotchkiss Blue & Co: Used Equipment dealer, Chicago	NYLE&W Assn	Pullman New York Lake Erie & Western Association
HR	Highland Railway, England	NYNH&H	New York, New Haven & Hartford Railroad
I&StLSCCo	Indianapolis & St. Louis Sleeping Car Company	NYWS&B	New York, West Shore & Buffalo RR (12/85 to West Shore RR)

ABBREVIATION	NAME	ABBREVIATION	NAME
O&C	Oregon & California Railroad (SP System)	Sou	Southern Railway System
O&M RR	Ohio & Mississippi Railroad (B&O System)	SouTransCo	Southern Transportation Company
Ohio Falls	Ohio Falls, Jeffersonville, IN (car builder)	SP	Southern Pacific Lines
OPLines	"Old Paine Lines"	SP Assn	Pullman Southern Pacific Association
Ord	Ordnance Branch of U.S. Army	SP&D	Saint Paul & Duluth Railroad (Northern Pacific)
P&LE	Pittsburgh & Lake Erie Railroad	SP&S	Spokane, Portland & Seattle Railway
PC&StL	Pittsburgh, Cincinnati & St. Louis Railroad (PRR System)	SPA	Southern Pacific of Arizona
PCM&StP	Pullman, Chicago, Milwaukee & St. Paul Association	SPC	Southern Pacific of California
PCW	The Pullman Car Works, Calumet Shops, IL	SPdeM	Southern Pacific of Mexico
PCW Det	The Pullman Car Works, Detroit, MI (Former Detroit C&M Co.)	SPNM	Southern Pacific of New Mexico
PFtW&C	Pittsburgh Fort Wayne & Chicago Railroad (PRR System)	StL-SF	St. Louis-San Francisco Railway (FRISCO)
PFtW&StL	Pittsburgh Fort Wayne & St. Louis Railroad	StL-SW	St. Louis Southwestern Railway (Cotton Belt-SP System)
PGE	Pacific Great Eastern Railroad	StLK&NW	St. Louis Keokuk & Northwestern Railroad (CB&Q)
Phil&Erie	Philadelphia & Erie RR (PRR System)	StLSCCo.	St. Louis Sleeping Car Company
PK&RSCC	Pullman, Kimble & Ramsey Sleeping Car Company	T&NO	Texas & New Orleans Railroad (SP System)
PM	Pere Marquette Railway (C&O System)	T&P	Texas & Pacific Railway
PPCC	Pullman's Palace Car Company, 12/99 renamed The Pullman Co.	Taunton C.C.	Taunton Car Company
Pre Pullman	Name in use prior to Pullman ownership	TC	Tennessee Central Railway
PRR	The Pennsylvania Railroad	TC Repair Shop	Transportation Corps Railroad Repair Shop, Ogden, Utah
PS	Pullman Standard	TH&B	Toronto, Hamilton & Buffalo Railroad
Pull	The Pullman Company	Trans	Transportation Corps (Branch of U.S. Army)
Pull Buf	The Pullman Car Shops, Buffalo, NY (Former Wagner Shops)	TW&W Ry	Toledo, Wabash & Western Railway, Toledo Shops (Wabash)
Pull Cal	The Pullman Car Shops, Calumet, IL	UP	Union Pacific Railroad
Pull Ltd	The Pullman Company Limited in England est: Circa 1906	UP Assn	Union Pacific Association
Pull Rich	The Pullman Car Shops, Richmond, CA	UPCC	Union Palace Car Company
Pull StL	The Pullman Car Shops, St. Louis, MO	US Govt	The United States Government
Pull Wilm	The Pullman Car Shops, Wilmington, DL	US Med	The United States Medical Department
Pull/Sou	Pullman Southern Car Company	USAF	United States Air Force
Pullman	The Pullman Company, Chicago, IL	VC RR	Vermont Central RR
PW&B	Philadelphia, Washington & Baltimore RR (PRR system)	Vermont Cent	Vermont Central RR
QM	Quartermaster Corps (Branch of U.S. Army)	WAB	Wabash Railroad
RBBB	Ringling Bros., Barnum & Bailey Combined Shows, Inc.	Wagner	The Wagner Palace Car Company, Buffalo, NY
RDG	The Reading Company	Wagner/Jones	Wagner Palace Car Company and (J.M. Jones & Co. Schnectady, NY?)
RF&P	Richmond Fredericksburg & Potomac R.R.	Wason	Wason Manufacturing Company, Springfield, MA
SAL	Seaboard Air Line R.R.	WC	Wisconsin Central Railroad (90-94 NP)
SBC	Sonora Baja California Railroad of Mexico	WC Assn	Pullman Wisconsin Central Association
Schi	Schindler Wagons SA (Swiss Built)	Western Union	Western Union Telegraph Company
SCL	Seaboard Coast Line (consolidation of SAL & ACL)	WM	Western Maryland Railway
SF&W	Savannah, Florida & Western RR (7/02 to ACL System)	WofA	Western Railway of Alabama
Sig	Signal Corps (Branch of U.S.Army)	Woodruff	Woodruff Sleeping & Parlor Coach Company
SilverPC	The Silver Palace Car Company	WP	Western Pacific Railroad
Soo	Minneapolis, St. Paul & Sault Ste. Marie R.R. (Soo Line)	WS&PCC	Woodruff Sleeping & Palace Coach Company
		WU	Western Union Telegraph Company

INDEX OF ILLUSTRATIONS